The Geology of North America
Volume O-1

Surface Water Hydrology

Edited by

M. G. Wolman
Department of Geography and Environmental Engineering
The Johns Hopkins University
Baltimore, Maryland 21218

H. C. Riggs
U.S. Geological Survey
415 National Center
Reston, Virginia 22092

1990

Acknowledgment

Publication of this volume, one of the synthesis volumes of *The Decade of North American Geology Project* series, has been made possible by members and friends of the Geological Society of America, corporations, and government agencies through contributions to the Decade of North American Geology fund of the Geological Society of America Foundation.

Following is a list of individuals, corporations, and government agencies giving and/or pledging more than $50,000 in support of the DNAG Project:

Amoco Production Company
ARCO Exploration Company
Chevron Corporation
Cities Service Oil and Gas Company
Diamond Shamrock Exploration Corporation
Exxon Production Research Company
Getty Oil Company
Gulf Oil Exploration and Production Company
Paul V. Hoovler
Kennecott Minerals Company
Kerr McGee Corporation
Marathon Oil Company
Maxus Energy Corporation
McMoRan Oil and Gas Company
Mobil Oil Corporation
Occidental Petroleum Corporation
Pennzoil Exploration and Production Company
Phillips Petroleum Company
Shell Oil Company
Caswell Silver
Standard Oil Production Company
Oryx Energy Company (formerly Sun Exploration and Production Company)
Superior Oil Company
Tenneco Oil Company
Texaco, Inc.
Union Oil Company of California
Union Pacific Corporation and its operating companies:
Union Pacific Resources Company
Union Pacific Railroad Company
Upland Industries Corporation
U.S. Department of Energy

Published by The Geological Society of America, Inc.
3300 Penrose Place, P.O. Box 9140, Boulder, Colorado 80301

Printed in U.S.A.

Library of Congress Cataloging-in-Publication Data

Surface water hydrology / edited by M. G. Wolman, H. C. Riggs.
p. cm.—(The Geology of North America ; v. O-1)
"One of the synthesis volumes of The Decade of North American Geology Project series"—T.p. verso.
Includes Bibliographical references.
ISBN 0-8137-5210-8
1. Hydrology—North America. I. Wolman, M. Gordon (Markley Gordon), 1924– . II. Riggs, H. C. III. Geological Society of North America. IV. Decade of North American Geology Project. V. Series.
QE71.G48 1986 vol. O-1
[GB701]
55 s—dc20
[551.47'097] 90-2905
CIP

Front Cover: Popo Agie River, Wind River Range, west-central Wyoming. Photo by H. C. Riggs.

10 9 8 7 6 5 4 3 2

Contents

Preface v

Foreword vii

1. *Introduction* 1
   H. C. Riggs and M. G. Wolman

2. *Influence of the atmosphere* 11
   Harry F. Lins, F. Kenneth Hare, and Krishan P. Singh

3. *Surface waters of North America; Influence of land and vegetation on streamflow* 55
   K. E. Saxton and S. Y. Shiau

4. *Temporal and spatial variability of streamflow* 81
   H. C. Riggs and K. D. Harvey

5. *Floods* 97
   Howard F. Matthai

6. *Low flows and hydrologic droughts* 121
   J. D. Rogers and J. T. Armbruster

7. *Snow and ice* 131
   Mark F. Meier

8. *Hydrology of lakes and wetlands* 159
   Thomas C. Winter and Ming-Ko Woo

9. *Hydrogeochemistry of rivers and lakes* 189
   John D. Hem, Adrian Demayo, and Richard A. Smith

10. *Aquatic biota in North America* 233
    R. Patrick and D. D. Williams

11. Movement and storage of sediment in rivers of the United States and Canada 255
Robert H. Meade, Ted R. Yuzyk, and Terry J. Day

12. The riverscape 281
M. Gordon Wolman, Michael Church, Robert Newbury, Michel Lapointe, Marcel Frenette, E. D. Andrews, Thomas E. Lisle, John P. Buchanan, Stanley A. Schumm, and Brien R. Winkley

13. The influence of man on hydrologic systems 329
Robert M. Hirsch, John F. Walker, J. C. Day, and Raimo Kallio

Index 361

Plates
(in pocket inside back cover)

Plate 1. Chemical composition of the surface waters of North America.
J. D. Hem

Plate 2. Distribution of the difference between precipitation and open-water evaporation in North America.
T. C. Winter

Plate 3. A. Ranges of mean annual runoff.
B. Variation of flow regimes.
H. C. Riggs and K. D. Harvey

Preface

The Geology of North America series has been prepared to mark the Centennial of The Geological Society of America. It represents the cooperative efforts of more than 1,000 individuals from academia, state and federal agencies of many countries, and industry to prepare syntheses that are as current and authoritative as possible about the geology of the North American continent and adjacent oceanic regions.

This series is part of the Decade of North American Geology (DNAG) Project, which also includes eight wall maps at a scale of 1:5,000,000 that summarize the geology, tectonics, magnetic and gravity anomaly patterns, regional stress fields, thermal aspects, seismicity, and neotectonics of North America and its surroundings. Together, the synthesis volumes and maps are the first coordinated effort to integrate all available knowledge about the geology and geophysics of a crustal plate on a regional scale.

The products of the DNAG Project present the state of knowledge of the geology and geophysics of North America in the 1980s, and they point the way toward work to be done in the decades ahead.

A. R. Palmer
General Editor for the volumes
published by the Geological
Society of America

J. O. Wheeler
General Editor for the volumes
published by the Geological
Survey of Canada

Foreword

This book, and a companion volume on Hydrogeology, describe the occurrence and movement of water on the continent and explain the hydrologic effects of geology, physiography, and climate. Chapter 1 provides a very general picture of the geographic distribution of surface water (streamflow) in North America, the effects of land and climate on the amount and variability of streamflow and its impurities, the procedures used for measuring surface-water quality and quantity, and the common descriptors of surface water.

Successive chapters treat the various elements of the land phase of the hydrologic cycle, beginning with the source of water, precipitation. Streamflow is the residual of precipitation after abstractions by evaporation and transpiration. These abstractions are shown to be related to the character of the land as well as to the climate. The resulting streamflows throughout North America are described in three ways: (1) the distribution of monthly and annual averages in time and space, (2) floods, and (3) low flows and hydrologic droughts. Considerable surface water also occurs in lakes and wetlands in some regions. The extent of these waters and their relation to streamflow is the next subject.

Hydrology encompasses water quality as well as water quantity. Three types of water constituents—dissolved solids, biota, and sediment—are described in separate chapters. An attempt is made to relate each of these constituents to geologic, climatic, and land characteristics that influence their distribution and behavior. Riverscapes from different climatic and physiographic regions are described briefly to illustrate the way in which geologic structure and history, climate, and hydrology influence the look of a given river. Although the influence of man on hydrology is noted in a number of chapters in the text, a final chapter contains examples selected specifically to demonstrate the effect of man on the hydrology of the surface waters of North America. There is some overlap in material covered in successive chapters, along with obvious interrelations between chapters. Some redundancy has been retained in the interest of the coherence and completeness of individual chapters and to reinforce several key elements such as the impact of man on the hydrology of North America.

Throughout, the emphasis is on the geographic distribution of hydrologic phenomena in North America. Processes that govern the hydrologic cycle, whether controlling floods, low flow, or water quality, are described in connection with their occurrence. While North America can be characterized in terms of broad climatic regions, the behavior of water influenced by climate interacting with landscape and soil creates smaller geographic units or patterns of hydrologic response. This scale is important in describing hydrologic processes and in characterizing their geography.

The editors acknowledge help and cooperation from a number of individuals not specifically listed as authors. These include A. R. Palmer, coordinator of the Decade of North American Geology Project; J. E. Slater, Environment Canada; S. L. Changnon, Jr., Illinois State Water Survey; B. B. Hanshaw, M. E. Moss, and V. R. Schneider, U.S. Geological Survey; and the many reviewers of the individual chapters.

H. C. Riggs
M. G. Wolman
June 1989

The Geology of North America
Vol. O-1, Surface Water Hydrology
The Geological Society of America, 1990

Chapter 1

Introduction

H. C. Riggs
U.S. Geological Survey, 415 National Center, Reston, Virginia 22092
M. G. Wolman
Department of Geography and Environmental Engineering, The Johns Hopkins University, Baltimore, Maryland 21218

HYDROLOGY AND GEOGRAPHY

The waters of the Earth circulate from the oceans through the atmosphere to the land, from which a part is returned to the atmosphere and the residual is returned to the ocean either by streams or underground flow. The paths of water on the land include surface flow to stream channels, infiltration to the soil, movement through the ground to a ground-water body and thence to a stream channel, and eventual return to the ocean. Water is extracted and returned t the atmosphere by evapotranspiration (evaporation and transpiration) throughout much of the land phase.

Surface water is the water in stream channels and in lakes, reservoirs, and wetlands. The amount of surface water is continually changing in response to precipitation and other elements of weather. The surface water of a region or drainage basin is characterized by the streamflow or runoff; it is commonly reported as the long-term mean discharge rate or the average volume discharged per year expressed as equivalent depth on the drainage area. The terms are used interchangeably in much of the following text. Streamflow is also an appropriate term for instantaneous flow, but runoff is not.

In its movement on and through the soil and rocks, water dissolves inorganic materials, moves sediments, and is the medium for development of biota. Consequently, the quality of surface water ranges widely according to the character of the land, and to some extent is influenced by constitutents contributed from the atmosphere. A description of a surface-water body includes its quality as well as its magnitude and variability.

The surface waters of North America drain to three oceans, as shown in Figure 1. That figure also shows the largest closed basins, regions from which no flow reaches an ocean; the runoff at some sites in such basins may be considerable but is disposed of entirely by evaporation.

The principal rivers of North America and those river channels that have mean flows exceeding 1,000 cubic meters per second (m^3/s) are shown in Figure 2. Major political subdivisions of North America are shown in Figure 3 to help the reader identify sites or regions described in this and subsequent chapters.

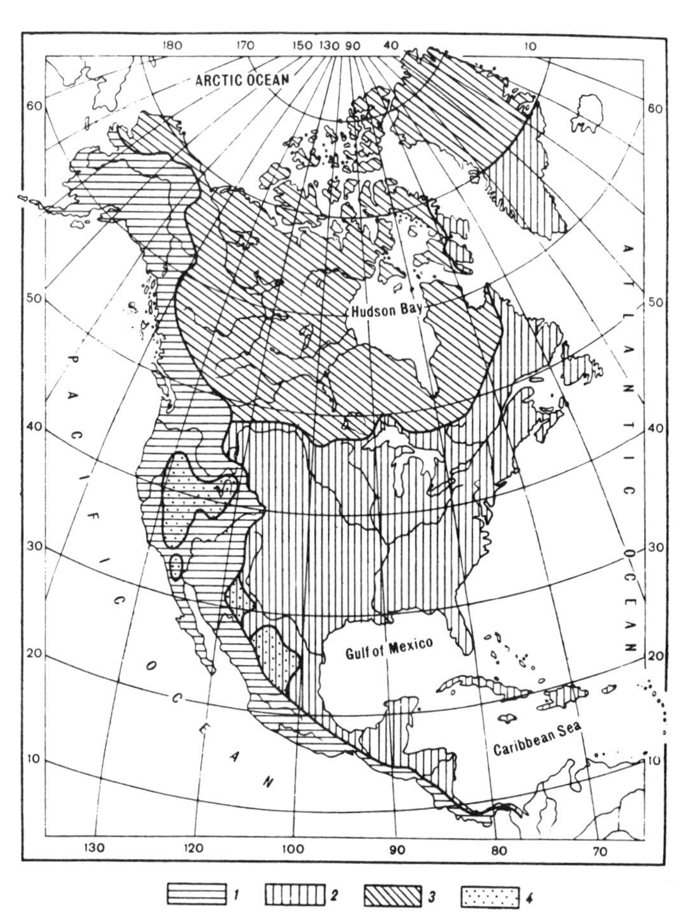

Figure 1. Areas of internal drainage and areas that drain to each of three oceans: 1, Pacific; 2, Atlantic; 3, Arctic; 4, Internal (after UNESCO, 1978).

Runoff of a major river is the integration of the many different tributary inflows, each of which is dependent on the climate and physiography of its drainage basin. The variability of runoff among these subbasins can be shown by expressing each annual runoff volume as the depth on its drainage area. This is equivalent to the yield per unit area and is commensurate with its causal precipitation.

Riggs, H. C., and Wolman, M. G., 1990, Introduction, *in* Wolman, M. G., and Riggs, H. C., eds., Surface water hydrology: Boulder, Colorado, Geological Society of America, The Geology of North America, v. O-1.

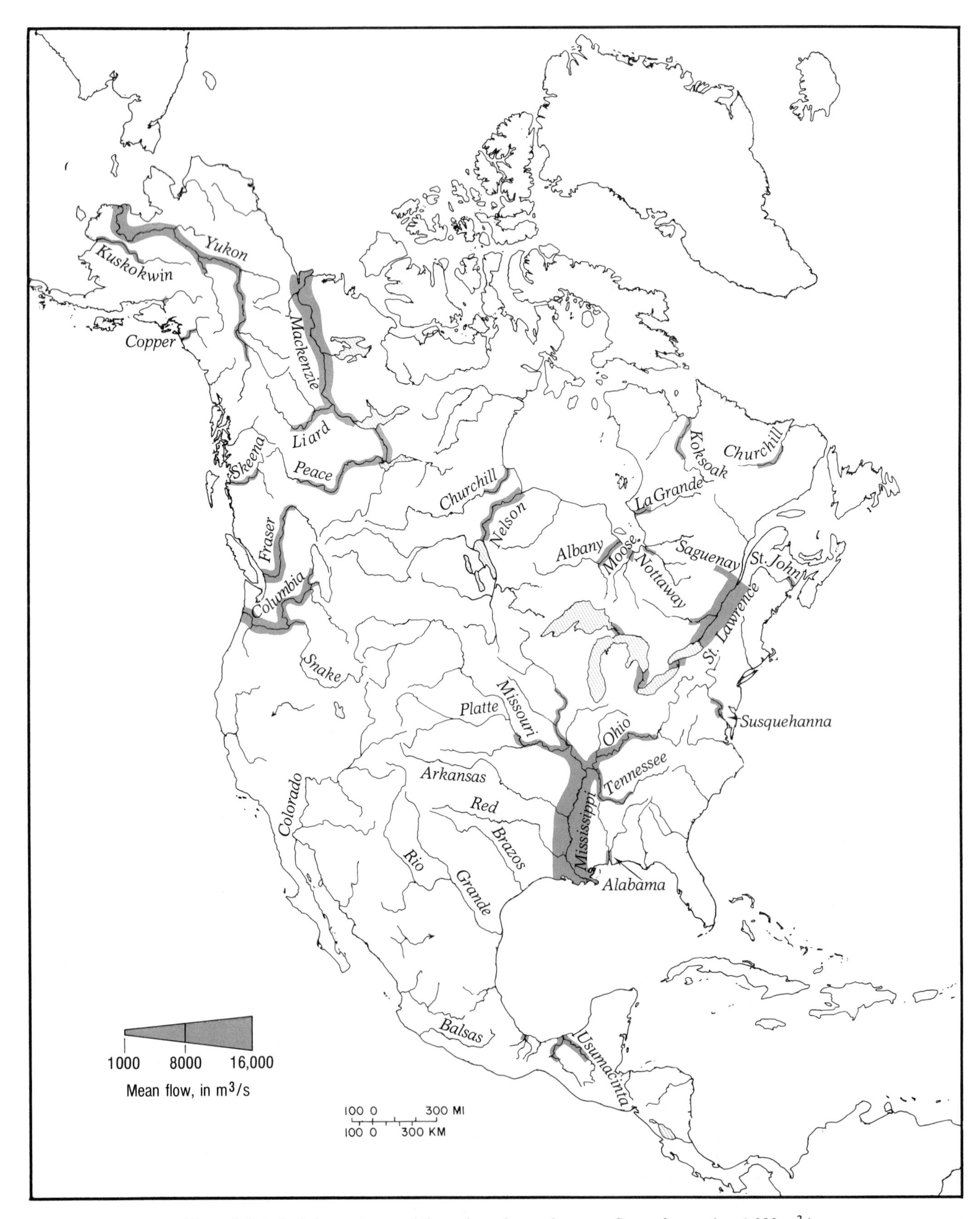

Figure 2. Principal river systems and those channels carrying mean flows of more than 1,000 m^3/s.

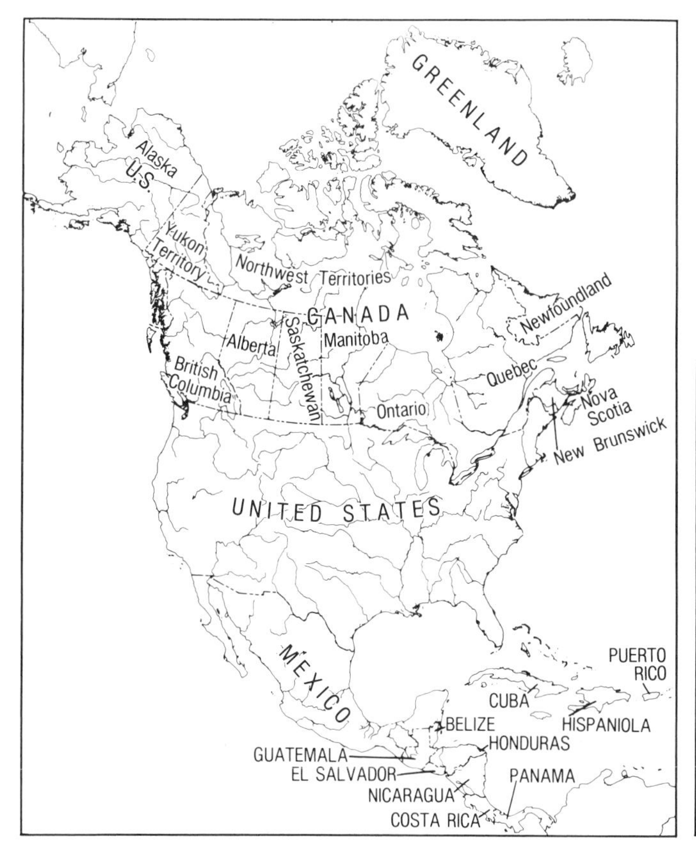

Figure 3. Major political subdivisions of North America.

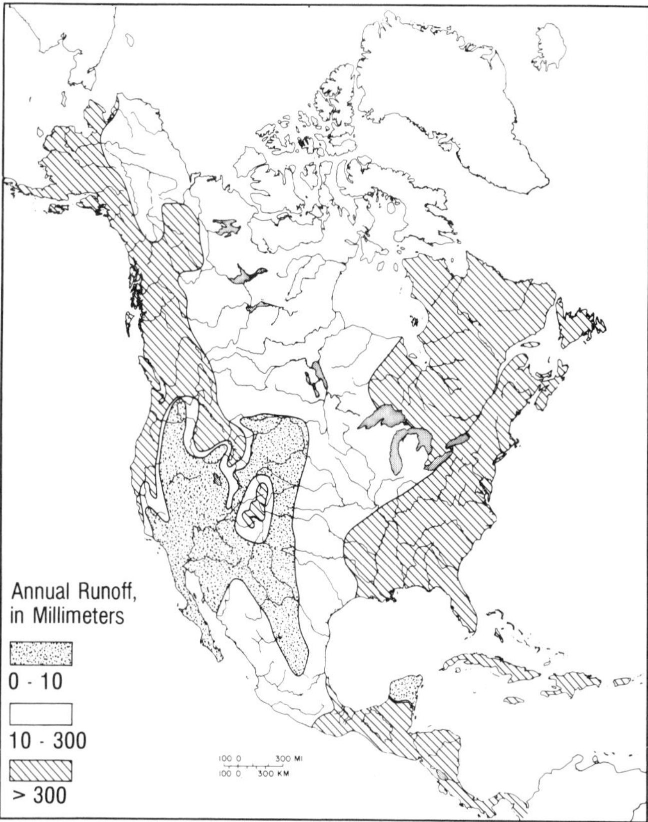

Figure 4. Ranges of annual runoff in millimeters. Runoff exceeds 1,000 mm along the northwest coast and in parts of Nova Scotia, Newfoundland, and central America. Runoff from the northern Canadian islands and from Greenland is undefined. See Plate 3A for more detail.

The great geographic variability of precipitation over North America, and to a lesser extent the variability of temperature and physiography, produces annual runoff ranging from a few millimeters in west-central United States and central Mexico to runoff on the order of 4,000 mm along the Pacific coast, from southeastern Alaska to northern Washington, and in eastern Costa Rica. The approximate locations of regions of high and low runoff are shown in Figure 4. A more detailed runoff map is included in Chapter 4.

Runoff amount and variability are modified by the temperature regime. High temperatures increase water losses from the land. Low temperatures produce glaciers, snow fields, permafrost, and seasonal snow cover; all of these modify the amount and timing of runoff. Glaciers and snowfields occur at high latitudes and at high elevations at lower latitudes (see Chapter 7). Permafrost, defined as permanently frozen ground, restricts the amount and movement of water in the ground. The approximate extent of continuous, discontinuous, and scattered permafrost is shown in Figure 5. Permafrost also exists in the western Cordillera at high elevations as far south as lat 49°. Seasonal snow cover occurs as far south as central United States and farther south in the western mountains.

Various combinations of climate, topography, and geology produce lakes and wetlands, the subject of Chapter 8. The largest

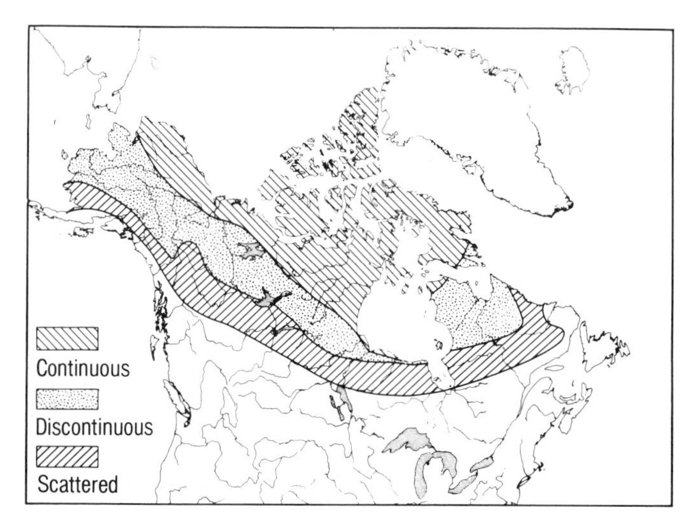

Figure 5. Approximate extent of permafrost (modified from Energy, Mines and Resources Canada, 1974, and Ferrians et al., 1969). Scattered permafrost also occurs in the western Cordillera as far south as lat 49°. Not defined in Greenland.

TABLE 1. MAJOR LAKES OF NORTH AMERICA

Name	Location	Area (km²)	Name	Location	Area (km²)
Superior	Canada-U.S.	84,500	Reindeer	Canada	6,640
Huron	Canada-U.S.	63,500	Netilling	Canada	5,530
Michigan	U.S.	58,000	Winnipegosis	Canada	5,360
Great Bear	Canada	31,400	Wipigon	Canada	4,850
Great Slave	Canada	28,600	Manitoba	Canada	4,700
Erie	Canada-U.S.	25,800	Lake of the Woods	Canada	~4,000
Winnepeg	Canada	24,400	Dubawnt	Canada	3,830
Ontario	Canada-U.S.	19,300	Amadjuak	Canada	3,120
Nicaragua	Nicaragua	8,030	Iliamna	Alaska	2,590
Athabaska	Canada	7,940	Great Salt	U.S.	2,500-6,200

North American lakes are listed in Table 1. All but four are in or border on Canada, and all but Great Salt Lake contain fresh water. Most freshwater lakes are connected to a stream system either by surface or underground drainage.

Wetlands are poorly drained areas of low relief in which the soil is seasonally or perennially saturated and which may include shallow ponds or swamps. There are major wetlands in Northwest Territories and northern Alberta, and in eastern Manitoba and northern Ontario. Wetlands exist along some of the major rivers of Alaska and at various sites along the east coast of North America, from New Jersey to Panama. Small areas of wetland occur at many inland sites.

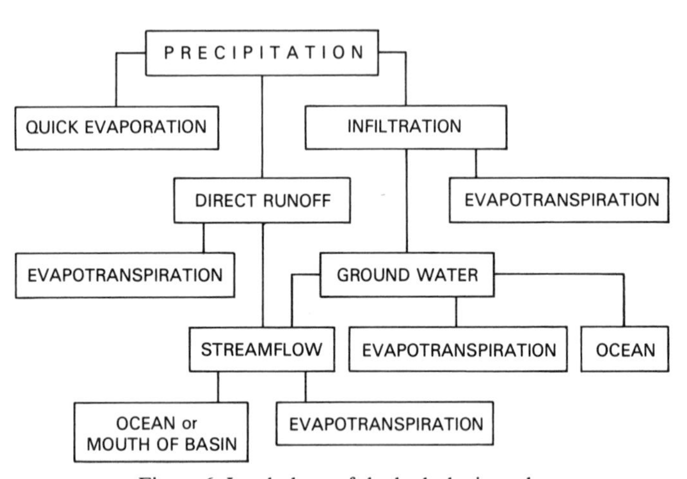

Figure 6. Land phase of the hydrologic cycle.

EFFECTS OF LAND AND CLIMATE

The generation of streamflow from precipitation can be generalized, as shown by Figure 6. That part of the precipitation that is intercepted by vegetation is reevaporated quickly, and the direct runoff that moves on or near the land surface reaches a stream channel promptly. In contrast, the water that infiltrates more deeply into the soil moves slowly to the ground-water table and thence to a stream channel. Evapotranspiration, the water returned to the atmosphere by evaporation from ground and water surfaces and by transpiration by plants, occurs throughout the land phase of the cycle, except from deep ground water.

The disposal of a small amount of precipitation on a basin may be considerably simplified over that shown by the diagram; most will be disposed of by quick evaporation and by infiltration and little or no direct runoff may occur. Most of the infiltrated water will be returned to the atmosphere by evaporation and by transpiration of plants, little or none passing on to the ground-water table.

The character of the land also influences the relative amounts of water in each of the components of Figure 6. In some highly permeable soils, direct runoff never occurs. In less permeable soils, little or none of the infiltrated water reaches a water table; most is returned to the atmosphere by evapotranspiration, particularly if the storm precipitation is small. The way precipitation is disposed of depends on the geologic, topographic, and vegetative characteristics of the drainage basin and on the pattern of precipitation and temperature.

Runoff is the residual part of precipitation after evapotranspiration has been extracted. The amount of evapotranspiration is not proportional to the precipitation because it depends on the available moisture, temperature, solar radiation, wind, and land cover. The average annual water loss (evaporation and transpiration) from a basin can be approximated by subtracting the measured runoff from the average basin precipitation if there is no significant subsurface flow from the basin. Table 2 shows that annual water loss in temperate regions is a much greater part of annual precipitation where annual precipitation is low, and conversely, that runoff is a much smaller part of precipitation.

The greater loss from Neches River than from James River, both of which have about the same precipitation, is attributable to the higher mean annual temperature in the Neches River basin. The time distribution of precipitation also could be a factor by altering the amount of time when moisture was available for evaporation.

Williams (1940) showed that annual water loss is closely

TABLE 2. ANNUAL WATER LOSS COMPARED TO PRECIPITATION

Stream	Mean Annual Water Loss (mm)			WL/P	R/P
	Precipitation (P)	Water Loss (WL)	Runoff (R)		
South Fork Skykomish River, Washington	2,940	580	2,360	0.20	0.80
Oconee River, Georgia	1,290	770	520	0.60	0.40
Neches River, Texas	1,090	860	230	0.79	0.21
James River, Missouri	1,080	740	340	0.69	0.31
Scioto River, Ohio	930	650	280	0.70	0.30
Republican River, Kansas	630	595	35	0.94	0.06
Moreau River, South Dakota	370	357	13	0.97	0.03

related to mean annual temperature (Fig. 7). He used basins with mean annual precipitations greater than 500 mm, those for which the amount of precipitation did not limit the water loss.

Runoff approaches zero from basins on which the potential evapotranspiration (the amount that would occur if moisture were always available) is much higher than the annual precipitation. However, mean annual runoff is rarely zero because heavy rainfalls always exceed the losses over short periods. Between runoff-producing storms, streamflow is zero in such regions. This condition exists in the arid parts of North America where all but the major rivers are ephemeral; that is, they flow only after substantial precipitation. Perennial streams flowing through arid regions originate in mountainous regions where the annual precipitation is higher.

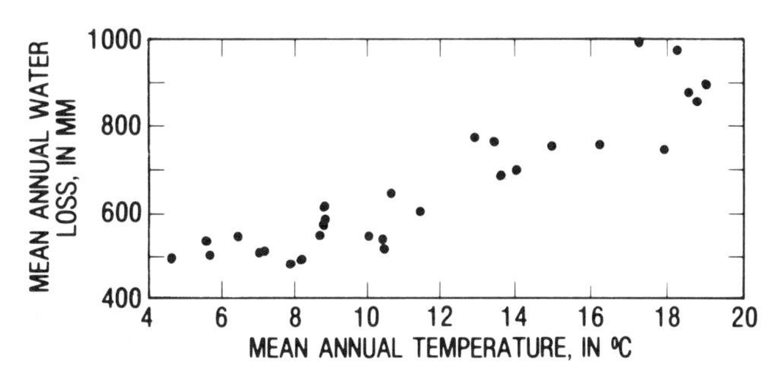

Figure 7. Basin water loss related to temperature, eastern United States (after Williams, 1940).

In arctic regions underlain by permafrost, the paths by which water moves on and through the land are more limited than those in Figure 6. Here infiltration is negligible, movement of ground water is limited, and evaporation losses are low because of the moderate temperatures during the short ice-free season.

Runoff maps such as that of Figure 4 are based on information obtained throughout a drainage basin. The existence of runoff at some interior point does not imply that it (or part of it) will eventually be discharged to an ocean. Thus, there is considerable runoff in the Great Basin of western United States (shown as internal drainage in Fig. 1), although the streams in that basin all terminate in lakes, sinks, or playas, and all the water leaves the basin by evaporation. Storage in these terminal lakes may be substantial and quite variable over a period of years. See, for example, the range in area of Great Salt Lake in Table 1. Playas are dry most of the time.

The discharge of a stream follows an annual pattern in response to the annual climatic cycle. Figure 8 shows a hydrograph of daily discharges for a year; the cyclic pattern is apparent. The hydrographs at this site for other years would show similar patterns, but the discharges for particular days or seasons might be greatly different because of variations in weather. The extent of this year-to-year variation is given in Figure 9, which shows the median and measures of the range in discharge for each day, based on 29 years of record.

Annual mean runoff also differs from year to year, as shown by the plot in Figure 10. Mean runoff (mean for all complete years of record) is shown in Figure 10 as the mean of the annual means. The variability of streamflow depends not only on the climate but also on the character of the drainage basin. A large amount of storage either in the ground or on the surface will reduce the fluctuations in runoff over those fluctuations in a basin that does not have such storage. For example, storage in the Great Lakes makes the annual outflow, the flow of St. Lawrence River, less variable than the inflows.

Annual mean flows of streams in humid regions are usually less variable than those in arid regions. Similarly, the larger the basin, the less likely that extreme weather conditions will occur over the whole basin in one period.

As mentioned above, an important element of the surface-

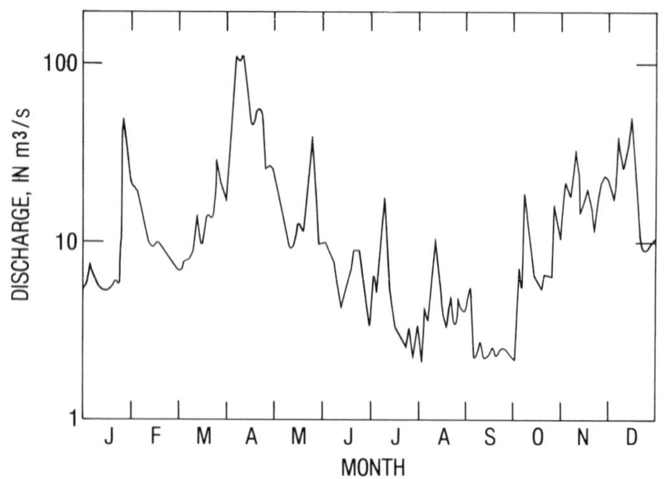

Figure 8. Hydrograph of daily discharges of west branch, Oswegatchie River, New York, 1959.

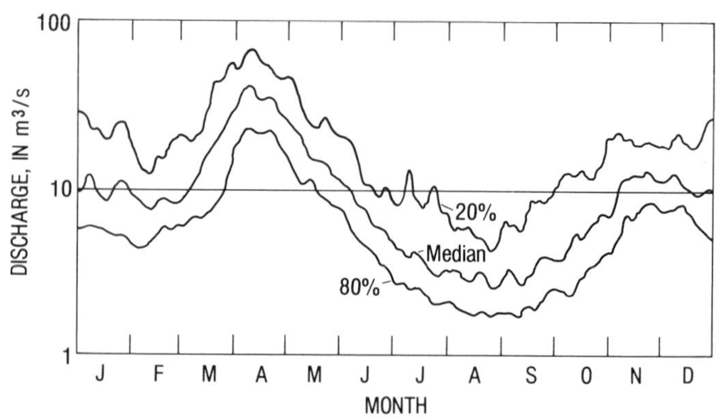

Figure 9. Median daily discharges and discharges having exceedance probabilities of 20% and 80%, based on 29 years of record, west branch, Oswegatchie River, New York (after Robert M. Beall, U.S. Geological Survey, 1963, written commun.).

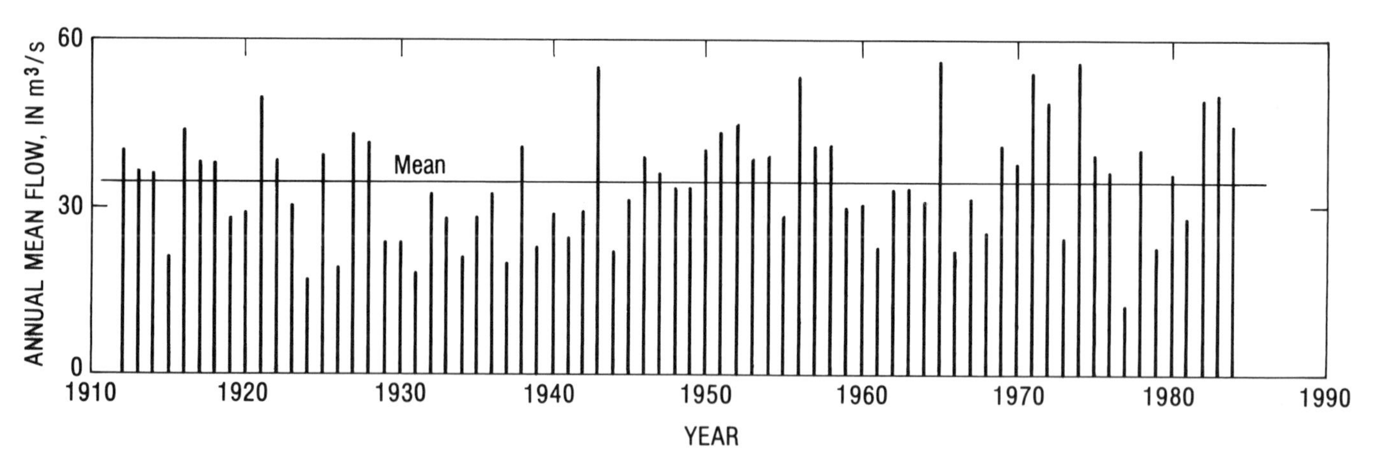

Figure 10. Annual mean flows of Boise River, Idaho, showing variability.

water supply in some regions is the water in lakes and wetlands. These modify the streamflow regime and in some cases maintain a supply of surface water when nearby streams are dry. If extensive, they also modify the climate of the region. An understanding of the movement of water into and out of lakes helps the interpretation of the variations of flows in individual streams and among streams.

As water moves over and through the ground and along stream channels, it erodes the soil, moves the eroded material (sediment), dissolves inorganic and organic materials, and accumulates biota. The result is to modify the land and to change the quality of the water. The amount of erosion and the amount of sediment moved depend on the types of soils and rocks and on the topography. Rapid surface flows, associated with steep topography, produce more land erosion and usually carry more sediment than comparable flows in regions with low land slopes. Conversely, the solution of chemical constituents of the rocks usually proceeds slowly so that a long residence time of the water in the ground is most likely to develop a high concentration of dissolved materials.

Growth of biota in a stream or lake is enhanced by low velocities and low variability of flow or lake stage. However, the oxygen exchange is speeded up by turbulent flow.

The present drainage patterns now being modified by the movement of water were determined by previous hydrologic and climatic conditions, associated for the most part with the Pleistocene period. More recently, man has had a major influence on the hydrology of some regions.

HYDROLOGIC MEASUREMENTS

Surface water of an area or drainage basin is considered to be the flow in stream channels. Water in storage is not included because over a period of a year or more the changes in the amounts of water in storage are reflected in streamflow. Continuous records of stream stage, which can be converted to discharge,

Figure 11. Staff gages.

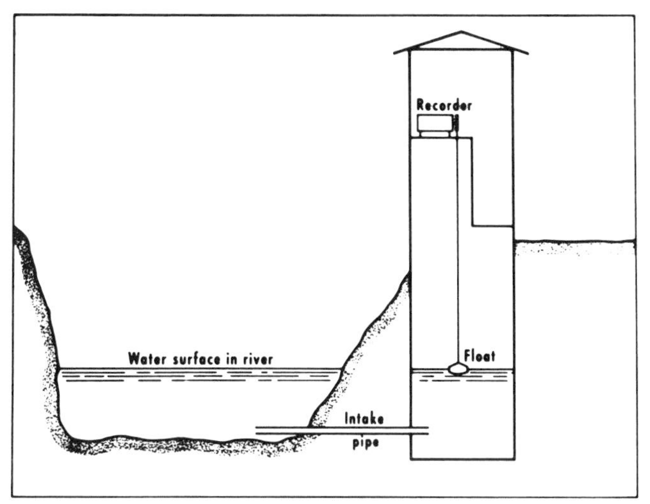

Figure 12. Diagram of a gaging station using a stilling well (after Riggs, 1967).

are obtained at gaging stations located at selected points on these channels. A gaging station consists of a structure containing instruments for sensing and recording the water-surface elevation (stage) periodically or continuously. The simplest station consists of a vertical staff gage, usually set to an arbitrary datum, on which an observer reads the water stage daily or more often (Fig. 11). More commonly the stage is sensed by a float in a stilling well connected to the stream, or by a gas-purge system, and is recorded by an instrument. A diagram of a gaging station with a stilling well is shown in Figure 12. In the gas-purge system, a gas is fed through a tube to an orifice that is permanently mounted in the stream. The gas pressure in the tube is equal to the pressure exerted by the head of water on the orifice. This pressure is measured by a manometer, which drives the stage recorder. The installation requires only an instrument shelter and a tube to the orifice in the stream.

Either analog or digital recorders have been used to record stage. The analog record is a continuous line on a chart. The digital record is a punched tape on which stage is recorded at intervals of 15 minutes to an hour depending on the "flashiness" of the stream. More recently, stage has been recorded electronically.

A discharge measurement is commonly made by the velocity-area method. Velocity in each of many vertical subsections of a stream cross section is measured by current meter at points in the vertical, as shown in Figure 13. The total discharge is the sum of the products of mean velocity and area in each of the subsections. Figure 14 shows the equipment for making a discharge measurement from a bridge. The weight is used for sounding depth and for holding the current meter in the proper position for velocity observations. Horizontal distances are measured along the bridge rail. Discharge measurements are also made from cableways, from boats, and by wading (Rantz, 1982). Methods are also available for computing the peak discharge after the passage of a flood. They utilize hydraulic equations that relate discharge to water-surface profile, channel geometry, and channel roughness (Barnes and Davidian, 1978).

The stage record at a gaging station is converted to a dis-

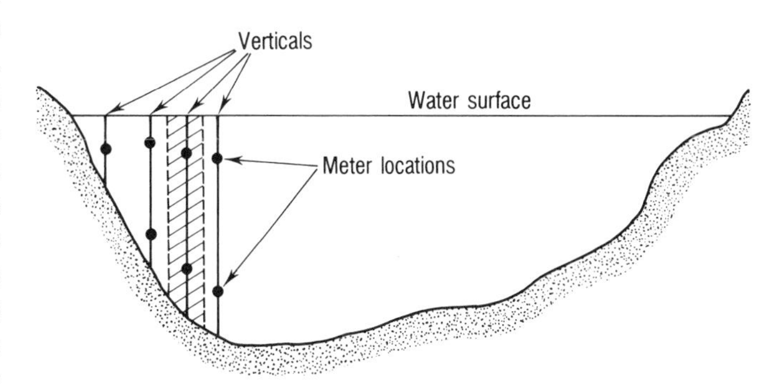

Figure 13. Stream cross section showing meter locations for a discharge measurement (after Riggs, 1985).

Figure 14. Equipment for making a discharge measurement from a bridge (after Riggs, 1985).

charge record by means of a rating curve (Fig. 15), defined by discharge measurements throughout the range in stage. The reliability of a rating curve is controlled largely by the character and permanence of the channel downstream from the gage. The rating may reflect the control by a riffle, such as that of Figure 16, or by a long channel reach, or both. In less-stable channels, particularly sand channels, the rating curve, or part of it, may be changed because of passage of a flood. And even in stable channels some temporary changes in rating occur because of aquatic growth, ice, or debris on the control.

Daily mean discharges and annual maxima and minima are computed from the stage record and the rating curve, taking into account any changes in the rating curve as indicated by discharge measurements and other information during the period of analysis (Kennedy, 1983). The principal end product is a table of daily mean discharges from which various flow characteristics can be computed, as described subsequently. Daily mean discharges are rated for reliability by the United States and Canada on the bases of the quality of the stage and discharge records and the stability of the rating curve.

The amount of sediment transported over a period of time by a stream is determined from sediment-discharge measurements and a continuous water-discharge record. A sediment-discharge measurement is made by collecting depth-integrated water samples at several verticals in the stream cross section. From analyses of these samples and the water discharge, the average sediment concentration and the rate of sediment transport at the time of measurement can be computed. Sediment-discharge measurements at various times and over a range of water discharges are the basis for a sediment rating curve (Fig. 17). Because the quantity of suspended load is not solely a function of discharge, sediment-rating curves may display considerable scatter. The rating curve can be used with the continuous water-discharge record to compute a continuous sediment-discharge record and the sediment load for a period of time obtained from that record.

The above method measures primarily the suspended sediment. The bed load, the sediment that rolls or bounds along the bottom of the channel, may be measured separately or may be estimated. Guy and Norman (1970) reviewed techniques for field measurements of fluvial sediment, and Porterfield (1972) described methods of computing fluvial-sediment discharge.

Surface waters are characterized by their chemical and biological constituents as well as by their water and sediment discharge. The identification and concentrations of each of these various constituents are obtained by laboratory analyses of water

Figure 16. Natural section control.

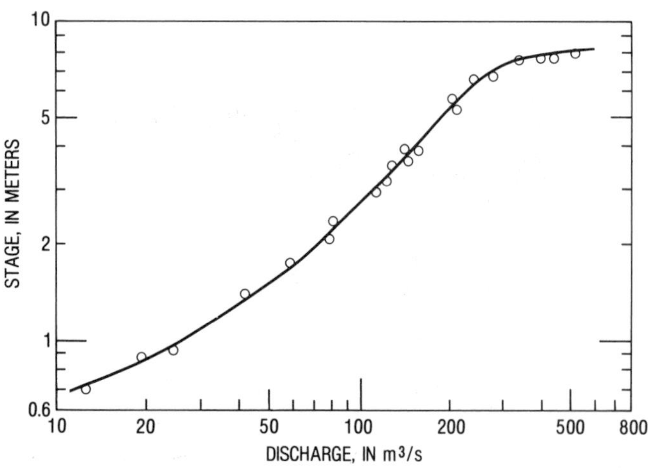

Figure 15. Rating curve for a gaging station.

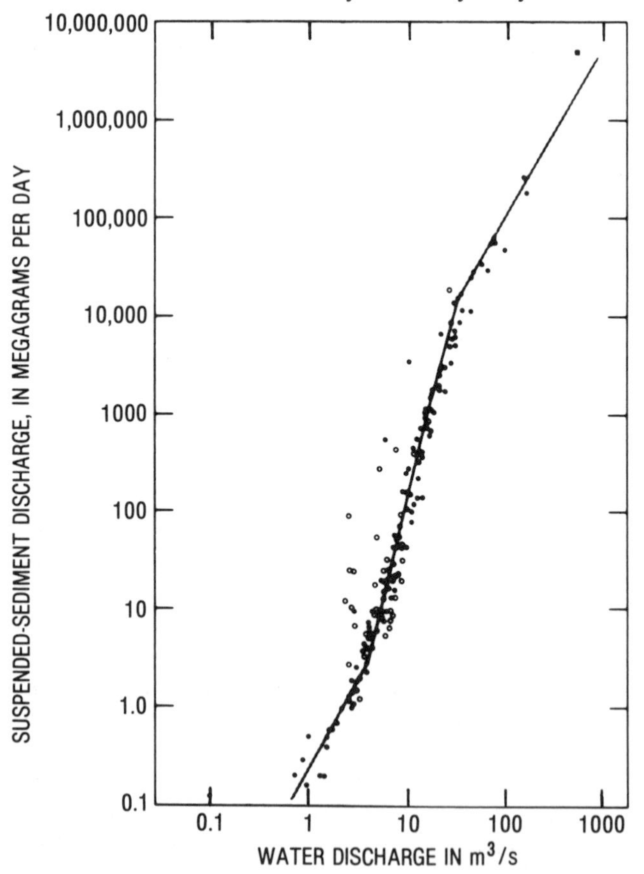

Figure 17. Sediment rating curve for Thomes Creek at Paskenta, California (after Porterfield, 1972).

samples taken through the range of stream discharge. Other indicators of water quality, which can be measured at the stream site, include water temperature, dissolved-oxygen concentration, specific conductance (a measure of total dissolved solids), and the hydrogen-ion concentration, pH. The presence, or absence, of certain biota are other indicators of water quality.

DESCRIBING SURFACE WATERS

The record of daily discharges of a stream can be generalized to a few manageable figures for describing streamflow at a site and for comparisons among streams. The hydrograph for each year can be described by the mean discharge for each month, the mean for the year, the maximum instantaneous discharge, and the lowest average discharges for various numbers of consecutive days. The overall mean and the calendar mean monthlies are widely used. Averaging of annual extremes does not produce a very useful statistic. Consequently, annual extremes are used to define a frequency curve from which particular statistics are selected. Chapters 5 and 6 on Floods and on Low Flows and Hydrologic Droughts include examples and explanations of frequency curves.

Some descriptive hydrologic statistics can also be estimated for sites that have no streamflow records. Such estimates are based on data from gaged sites and data on drainage basin and climatic characteristics, and sometimes on field information at the ungaged site. The reliability of such estimates ranges from very good to poor, depending on the amount and character of the precipitation, the hydrologic homogeneity of the region, the technique used, and on the amount of field data at the site.

Fluvial sediment in a river is commonly measured and reported as sediment concentration and as load of sediment transported per day. Both are closely related to water discharge and each can be described by its relation to water discharge. However, the cumulative load of sediment is often of more interest than the day-to-day loads, because those loads will be deposited somewhere downstream. Consequently, sediment transport at a site is usually described as average tonnes per year.

The chemical quality of a water is the composite of the concentrations of the many constitutents. The simplest descriptor is the average concentration of total dissolved solids. Average concentration of the major chemical constituents is an alternative descriptor, but it would be less useful in comparing water quality among streams that had different major constituents.

Averages are gross approximations because the concentrations are related to water discharge that is not normally distributed. Plots of concentration versus water discharge and of probability characteristics of daily mean concentrations provide much additional information. Such curves can also be provided for individual constituents.

Biological character of a water is usually described qualitatively, although useful indicators of its suitability for aquatic life are pH and the concentration of dissolved oxygen.

REFERENCES CITED

Barnes, H. H., Jr. and Davidian, Jacob, 1978, Indirect methods, *in* Hydrometry: Herschey, R. W., ed., New York, John Wiley and Sons, p. 149–204.

Energy, Mines and Resources Canada, 1974, The national atlas of Canada: Ottawa, McMillan Company of Canada, 254 p.

Ferrians, O. J., Jr., Kachadoorian, R., and Green, G. W., 1969, Permafrost and related engineering problems in Alaska: U.S. Geological Survey Professional Paper 678, 37 p.

Guy, H. P., and Norman, V. W., 1970, Field methods for measurement of fluvial sediment: U.S. Geological Survey Techniques of Water-Resources Investigations, book 3, chapter C2, 59 p.

Kennedy, E. J., 1983, Computation of continuous records of streamflow: U.S. Geological Survey Techniques of Water-Resources Investigations, book 3, chapter A13, 53 p.

Porterfield, G., 1972, Computation of fluvial discharge: U.S. Geological Survey Techniques of Water-Resources Investigations, book 3, chapter C3, 66 p.

Rantz, S. E., 1982, Measurement and computation of streamflow: U.S. Geological Survey Water-Supply Paper 2175, 631 p.

Riggs, H. C., 1967, Hydrologic bench marks in the National Parks: National Parks Magazine, v. 41, no. 232, p. 17–19.

——, 1985, Streamflow characteristics: Amsterdam, Elsevier Science Publishers B.V., 249 p.

UNESCO, 1978, World water balance and water resources of the Earth, including atlas of world water balance: Paris, UNESCO Press, 663 p., 65 maps.

Williams, G. R., 1940, Natural water loss in selected drainage basins: U.S. Geological Survey Water-Supply Paper 846, 62 p.

MANUSCRIPT ACCEPTED BY THE SOCIETY MAY 14, 1987

Printed in U.S.A.

The Geology of North America
Vol. O-1, Surface Water Hydrology
The Geological Society of America, 1990

Chapter 2

Influence of the atmosphere

Harry F. Lins
U.S. Geological Survey, 436 National Center, Reston, Virginia 22092
F. Kenneth Hare
301 Lakeshore Rd West, Oakville, Ontario, Canada L6K 1G2
Krishan P. Singh
Illinois State Geological Survey, Natural Resources Building, 615 Peabody Drive, Champaign, Illinois 61820

INTRODUCTION

Surface water hydrologic processes represent the dynamic expression of the flux of moisture to, across, and from the land surface. The primary control on this moisture flux is climate, which in turn, is controlled by the general circulation of the atmosphere. Thus, in order to understand the nature and characteristics of surface water hydrology, it is first necessary to understand the atmospheric context within which surface water processes occur. In this chapter we present this context in a conceptually broad, but topically systematic way. The chapter content includes empirical information developed over many decades of observation. We develop a more contemporary perspective, however, by characterizing these observations in a framework that draws on the current thinking associated with global change, especially with respect to climate variability and change.

The chapter begins with an overview of the concepts and elements associated with the fluxes of energy and moisture to, across, through, and from the land surface. In particular, emphasis is placed on the characteristics of the surface energy balance, especially the fluxes of latent and sensible heat. Additionally, the primary terrestrial components of the hydrologic cycle—evapotranspiration, runoff, and soil moisture—are discussed.

From this description of the atmospheric "engine" we proceed to a characterization of the climate and hydrologic effects that result from the operation of the atmospheric "engine." This includes the general atmospheric circulation (i.e., the weather delivery system), the climatology of North American extratropical and tropical storms, patterns and trends in North American droughts, regional hydroclimatic conditions, and an overview of the hydrologic cycle over North America.

The chapter concludes with a presentation of, and perspective on, the current issues of climate variability and change and what the implications of such issues are for the surface water hydrology of North America. In particular, regional patterns and temporal trends in temperature and precipitation, as compiled in the systematic record, are presented, along with longer-term reconstructions of climatic conditions in the geologic past. A clear attempt has been made to provide insights into processes that vary differentially over a variety of time scales. This has been done because the current emphasis on secular climatic change has obscured and overwhelmed the critical importance of understanding the relation between shorter-term climatic variations and hydrologic processes. Finally, a discussion of modeling is included that focuses on both hydrologic and climatic models. The latter is aimed specifically at the information content and appropriate uses and limitations of general circulation models and their output vis-a-vis the study of surface water hydrology.

THE EXCHANGE OF ENERGY AND MOISTURE

The idea of the *water balance* is simply that the losses and gains of water from a given locality must balance one another, unless there is an equivalent change in local water storage. The same conservation principle applies to *energy* in all its forms. These balances are examined briefly in principle in this section, beginning with energy.

The sun provides the energy that drives the hydrologic cycle—the fluxes of water into and out of the various reservoirs in which it may rest. The well-known radiation balance shown in Figure 1 illustrates the disposition of incoming solar radiation: absorption in the Earth's atmosphere, backscatter from the atmosphere and surface, re-radiation of longwave radiation from the Earth's surface, including from water, soil, and vegetation, and emission of longwave radiation from the atmosphere.

Much of the solar energy is absorbed at the Earth's surface where it warms the soil or water. This *sensible heat* is then transferred downward (the soil heat flux Q_G), or upward to the atmosphere. The upward flux of heat is in two forms: (1) a flux of long infrared radiation, $L\uparrow$, emitted by the soil, plants, or sur-

Lins, H. F., Hare, F. K., and Singh, K. P., 1990 *in* Wolman, M. G., and Riggs, H. C., eds., Surface water hydrology: Boulder, Colorado, Geological Society of America, The Geology of North America, v. O-1.

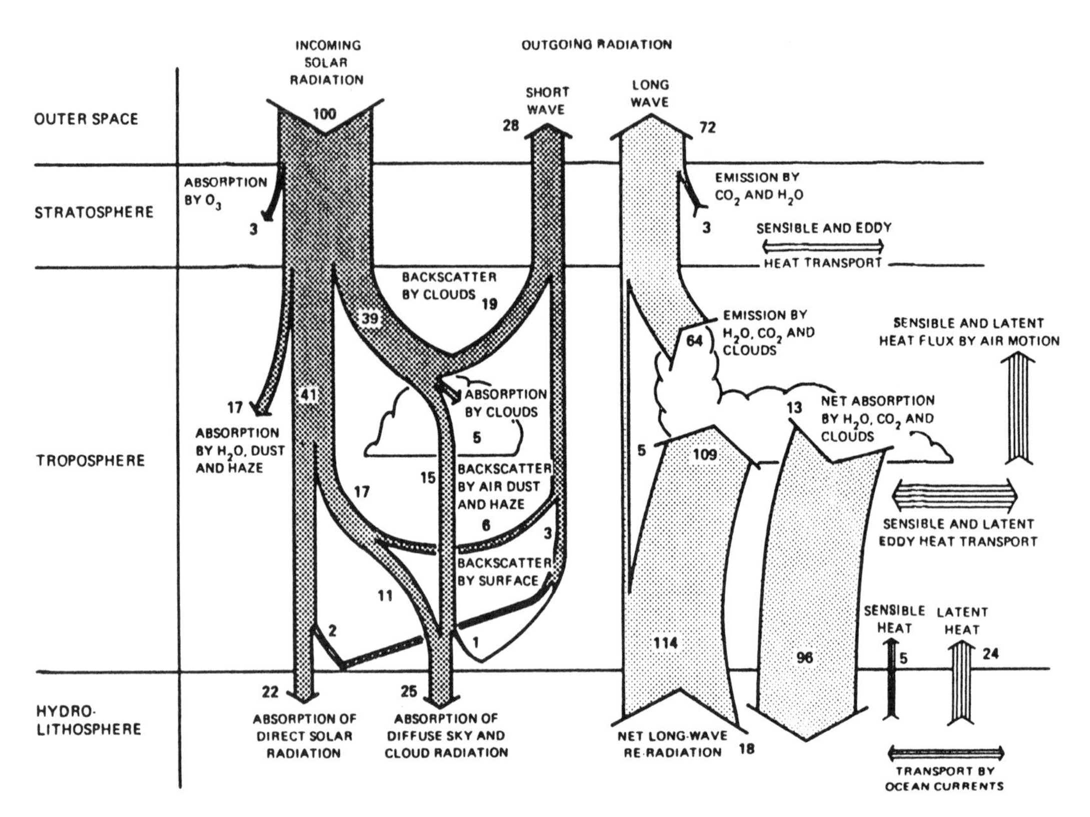

Figure 1. Energy balance of the Earth's surface and atmosphere, in percent (Schneider and Mesirow, 1976).

face water; and (2) convective fluxes of sensible heat, Q_H, or latent heat, Q_E, (the latter due to upward transport of evaporated water).

The surface also receives from the air compensating downward fluxes of long infrared radiation, $L\downarrow$, and sometimes of condensing water (dew formation). It also receives intermittent precipitation, P, of rain, hail, or snow. These fluxes must balance. Considering exchange across a surface only, there can be no storage. In reality, heat or water may be stored in the volume of soil beneath the surface of exchange. Thus a storage term ΔS may be included in the heat or water balance. By definition, at equilibrium over time the sum of additions and subtractions from storage equals zero.

The surface *radiation balance* or *net radiation* (Q_*) *is the sum of all the radiative energy receipts or losses per unit time and per unit area:*

$$Q_* = K(1 - a) - L\downarrow + L\uparrow \tag{1}$$

where K = incoming solar radiation flux (W m^{-2}), a = fraction of K reflected back (albedo, dimensionless), and $L\downarrow$ and $L\uparrow$ are the downward and upward fluxes, respectively, of long-wave (infrared) radiation at the Earth's surface (W m^{-2}).

The surface energy balance, neglecting the small amounts of heat used in photosynthesis and respiration, is then

$$Q_* - Q_G = Q_H + Q_E. \tag{2}$$

All the quantities in (2) may be intrinsically positive or negative, or may vanish. The quantity $Q_* - Q_G$ is sometimes called the *net heating* at the surface. It is clearly the source of energy that drives the convection of heat (Q_H) and latent heat (Q_E) from the surface to the atmosphere.

In hydrological and climatological work it is often more useful to use dimensionless expressions in place of the physical terms in equations (1) and (2). This is done in a standard way, using the following ratios:

$$\begin{aligned} B &= Q_H/Q_E, \textit{ the Bowen ratio;} \\ C &= N/P, \textit{ the runoff ratio; and} \\ D &= Q_*/LP, \textit{ the dryness ratio} \end{aligned} \tag{3}$$

where P = precipitation (mm s^{-1}) and N = runoff (mm s^{-1}).

For periods in which there is no change in the stored heat or water, these ratios combine, as in

$$D = (1 + B)(1 - C). \tag{4}$$

This law, introduced by Lettau (1969), means that if any two of the ratios can be determined, the third is also known. B and C can be applied to instantaneous observations, but all three can be

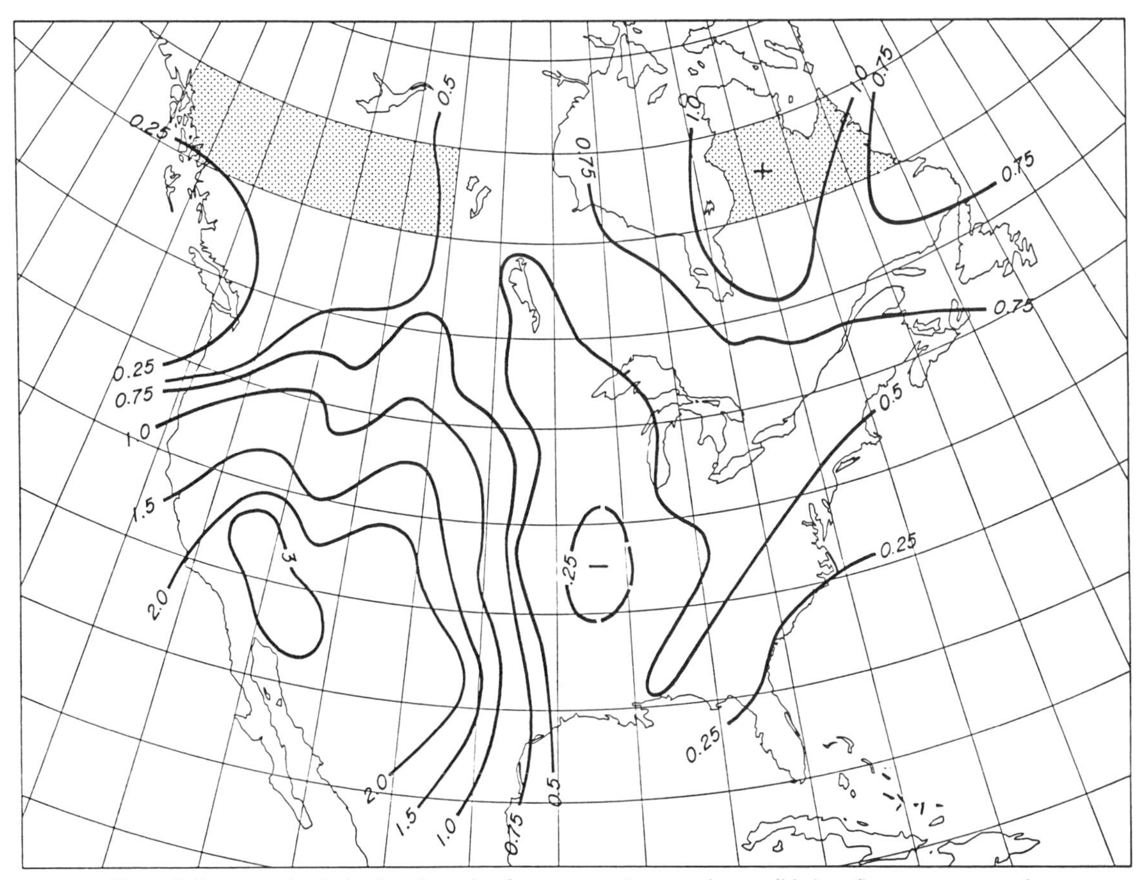

Figure 2. Bowen ratio, derived as the ratio of mean annual convective sensible heat flux to mean annual latent heat flux, $\bar{Q}_H/\bar{Q}_E$ (Hare, 1980).

defined for longer periods, or to long-term averages (such as annual means). Equation (4) actually expresses over such periods of constant heat and water storage the principles of conservation of mass and energy. Maps of the annual spatially averaged Bowen, runoff, and dryness ratios are given in Figures 2 to 4.

The surface net radiation is, on the average, strongly positive during the day and negative at night. Hence, the Earth's surface is usually warmer than the atmosphere by day and colder at night. On the average, the Earth warms the atmosphere. Temperature decreases upward through the troposphere. The resulting vertical temperature gradient is sufficient to allow convection, and hence the upward transport of water vapor and sensible heat. The convection is also sufficient, at certain times, to allow clouds to form, from which precipitation may fall. The convection may be on small scales, resulting in cumulus cloud formation, or much larger scales. The intensity of these exchanges depends on the radiative heating of the Earth's surface.

Unequal heating of the surface provides another driving element in the hydrologic cycle. It creates large-scale vertical and horizontal fluxes of water vapor from oceans to continents, from tropical to polar latitudes, and from the Earth's surface to the atmosphere, chiefly to the lowest 15 km. The general circulation of the atmosphere, operating at a global power of 10^{14} to 10^{15} watts, redistributes transported vapor worldwide, subject to the limitation that polar airstreams can carry little moisture because of their low temperatures. The upper troposphere and stratosphere are similarly limited in water-holding capacity.

Evapotranspiration

Evapotranspiration off the land surface of North America is the largest consumer of the net heating. In part it consists of evaporation off lakes, streams, and wet surfaces (notably leaves). The rest is transpiration, i.e., water absorbed from the available soil moisture, carried up to the leaf, and there evaporated via stomatal openings. This latter process is complex and subject to control by the availability of soil moisture, and also by the physiology of the plant cover (stomatal resistance).

Hydrologists and climatologists, faced with the absence of reliable measurements of evapotranspiration on a daily basis, have developed the concept of potential evapotranspiration, E_p, the evapotranspiration that may be expected off a wet surface, i.e., one in which stomatal resistance is zero and water supply unlimited. They have generally relied on empirical formulae to estimate this quantity. One such estimate can be obtained from an equation due to Slatyer and McIlroy (1961):

$$LE_p = \underset{(1)}{SQ_*} + \underset{(2)}{\rho c_p h D_z} \tag{5}$$

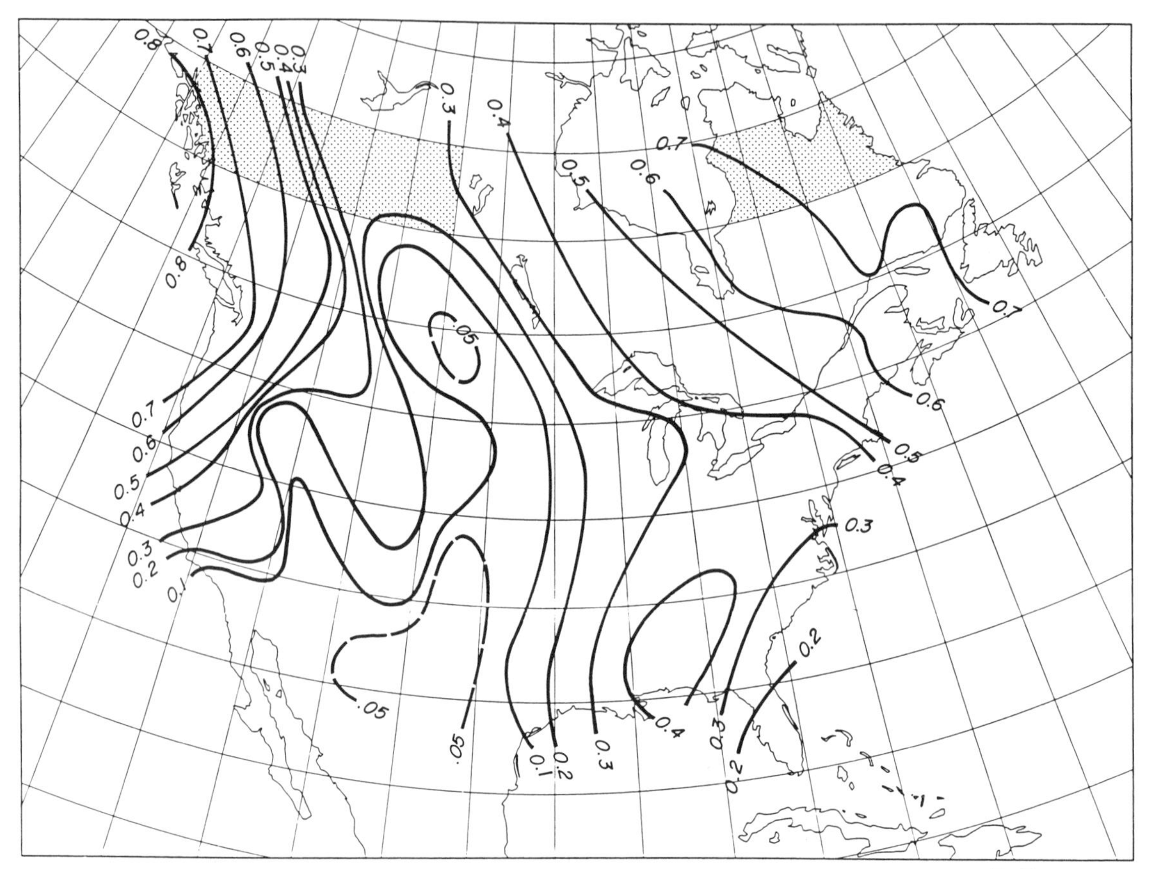

Figure 3. Runoff ratio, derived as the ratio of mean annual runoff to mean annual precipitation, $\bar{N}/\bar{P}$ (Hare, 1980).

where L is the latent heat of evaporation (about 247 MJ kg^{-1}); S = $(s + \tau)/s$, where s is the change of saturation vapor pressure per Kelvin appropriate to the temperature, and τ is 0.64 mbar ρ = air density (kg m^{-3}); c_p = specific heat of dry air at constant pressure; h = a transfer coefficient ($\sim 10^{-2}$ m s^{-1}) between the surface and the height z; and D_z = wet bulb temperature (K) depression at the standard level of observation z.

Term (1) in equation (5) is the radiative contribution to the heat needed for the evaporation. Term (2) is the net contribution from turbulent heat transfer from air to ground. Term (1) is generally much larger; therefore, absorbed radiation drives most of the evapotranspiration.

Evapotranspiration from actual surfaces is then estimated as some fraction of the potential value. Priestley and Taylor (1972) showed, and it has since been confirmed for a wide variety of moist surfaces, that term (2) tends to a stable maximum value of 26 percent of term (1) (see Hare, 1980, for a review), at least when the relation is applied to daily or monthly averages.

If $D_z = 0$, or if it equals the wet bulb depression at the surface, LE = SQ_*, defining the equilibrium evapotranspiration, E_q (Priestley and Taylor, 1972). Over much of North America, regional evaporation rates are close to the equilibrium value over a wide range of surfaces (Hare, 1980).

In detail, evapotranspiration is the loss of water to the atmosphere through evaporation from all surfaces, including free-water surfaces, soil, and manmade surfaces, and through transpiration from plants (Knapp, 1985). Satisfactory determination of daily and seasonal evapotranspiration is elusive for many reasons, including unavailability of instruments for correct measurement of evaporation from natural surfaces, dependence on prevailing energy inputs and winds, and spatial variability in recharge of soil moisture by precipitation. In practice, evapotranspiration is usually considered under two categories: lake (or free-water surface) evaporation, and evapotranspiration from plants and soil. The latter is more complex; its determination is subject to substantial uncertainties and errors. The rate of transpiration depends on the evaporative power of the air as determined by energy input, wind, saturation deficit, the amount of light (which partly controls opening of the stomata), and the availability of moisture in the leaf tissues, which in turn depends on soil-moisture availability (Lull, 1964).

Net radiation, together with the heat stored within the lake, provides the energy for lake evaporation. The evaporation rate is closely related to the vapor-pressure difference between the water surface and the atmosphere. These relations are only noted here but are described at length by Winter and others in this volume. Lake evaporation is estimated by multiplying measurements of evaporation from standard evaporation pans with the pan coeffi-

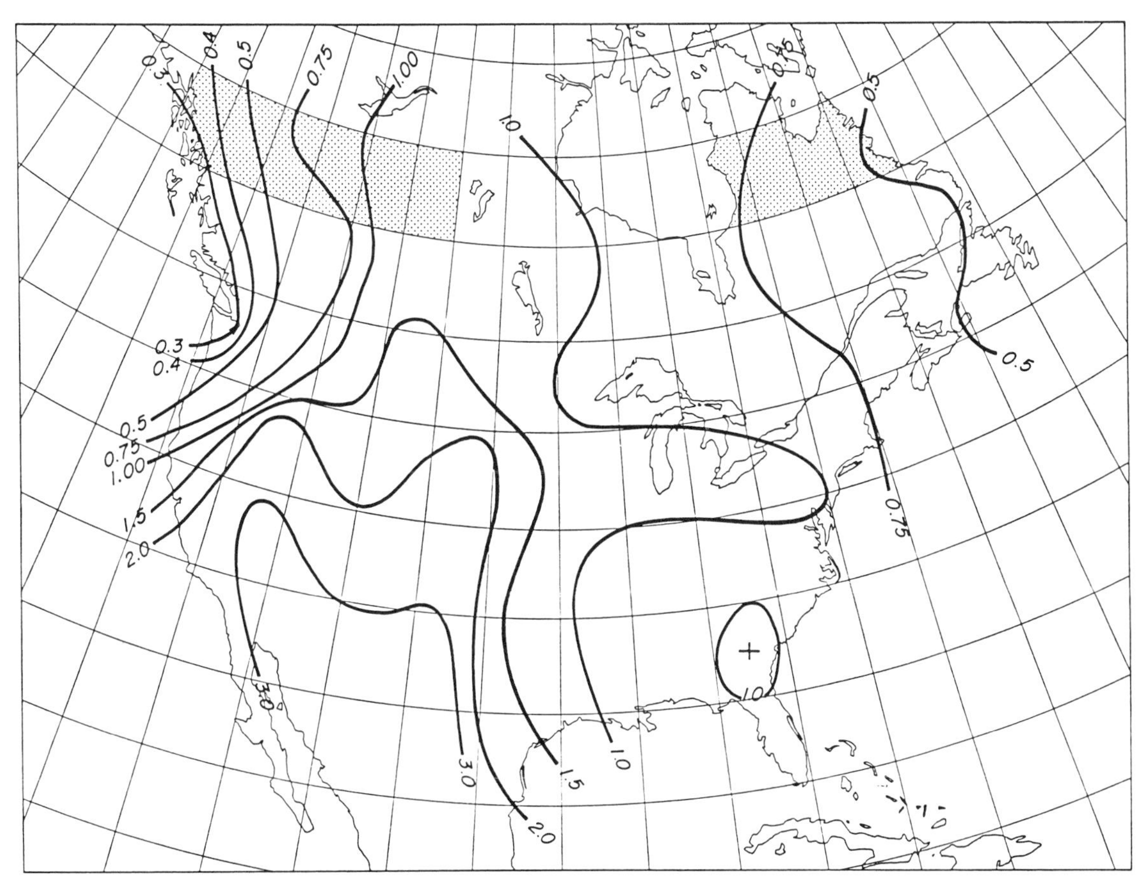

Figure 4. Dryness ratio, derived as the ratio of mean annual net radiation to the heat required to evaporate the mean annual precipitation, $\bar{Q}_*/L\bar{P}$ (Hare, 1980).

cient, which varies from season to season. In the Lake Hefner study (U.S. Geological Survey, 1954), the average monthly pan coefficients from January to December 1950–51 were estimated as 0.76, 0.13, 0.51, 0.39, 0.44, 0.62, 0.68, 0.77, 0.93, 0.90, 1.32, and 1.06. Coefficient values change because of changes in the distribution of incoming energy into latent heat (evaporation), sensible heat, and heat storage. Pan evaporation data averaged over several observation sites and 5 to 10 years provide a reliable estimate of annual lake evaporation.

For shallow lakes, evaporation is higher during hot months when the winds are also relatively strong. For deeper lakes, heat storage becomes important and evaporation is not determined solely by the input of radiant energy. For very large or very small lakes, size can also be a factor.

Figure 5 shows the mean annual lake evaporation for Canada and the United States. The values for Canada were obtained from the *Hydrologic Atlas of Canada* (CNC/IHD, 1978), and those for the United States follow the contours prepared by Kohler and others (1955). The mean annual lake evaporation decreases from a maximum of about 200 cm in the central southern border area of the United States to 20 cm or less in northern Canada. The spatial distribution of lake evaporation differs greatly from that of precipitation.

Actual evapotranspiration off the total surface of the continent cannot be derived from any single observational base. Figure 6 shows an estimate (Hare, 1980) based on spatially averaged differences between mean annual precipitation and mean annual runoff.

Runoff

Plate 3A (see Riggs and Harvey, this volume) shows ranges of runoff for North America. The mean annual runoff increases from less than 10 mm in the west-central United States and Mexico and northern Canada to more than 1,000 mm in the far northwest. Seasonal or monthly variations in runoff are marked over much of the continent, as illustrated in Plate 3B.

Soil moisture

There are no reliable statistics concerning soil moisture storage, which is measured routinely at research laboratories, but not at most climatological stations. Nevertheless, soil moisture retention, recharge, and withdrawal play crucial roles in the climatic aspects of the hydrologic cycle.

When a soil is wetted by precipitation, infiltration occurs

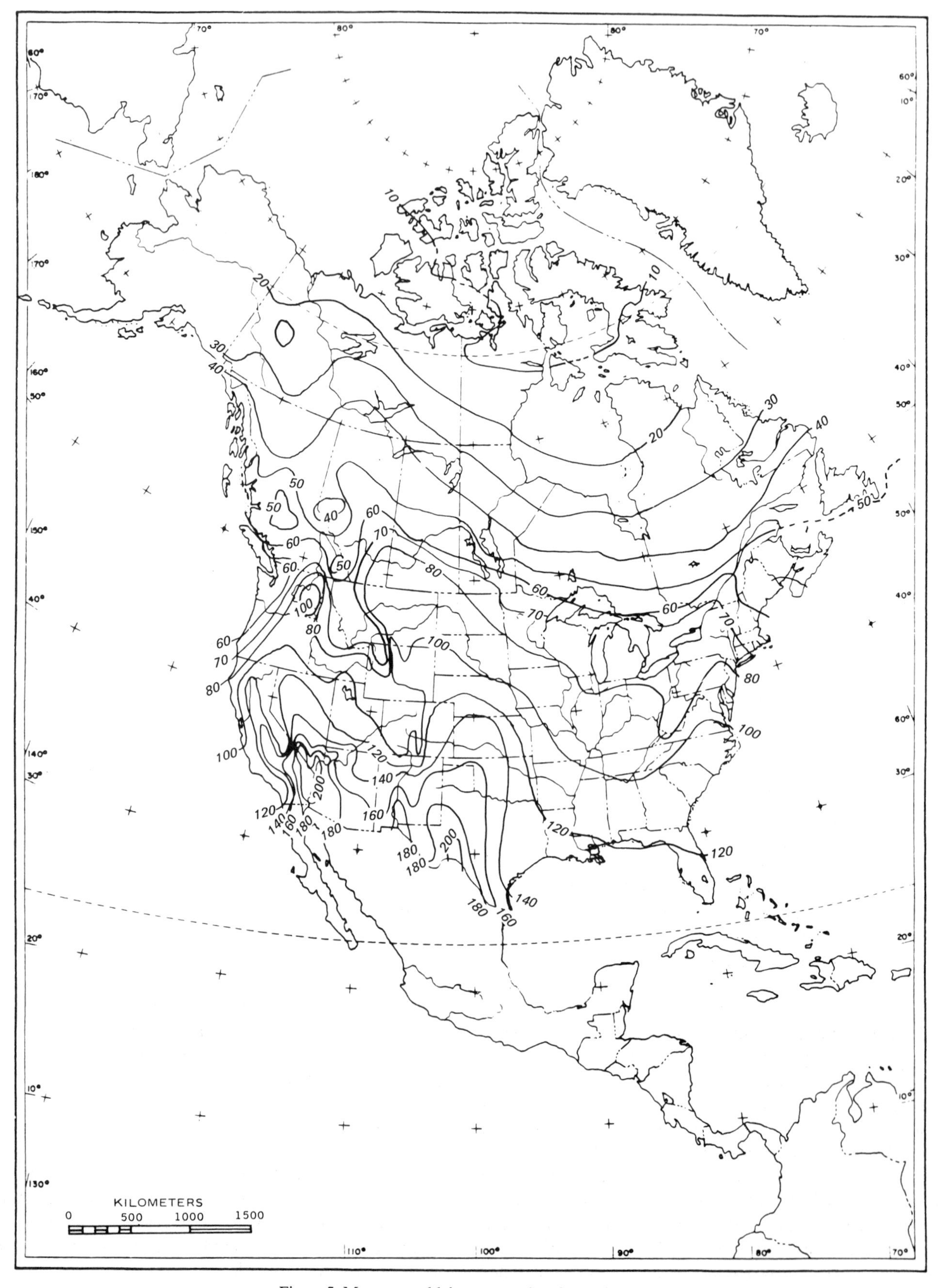

Figure 5. Mean annual lake evaporation, in centimeters.

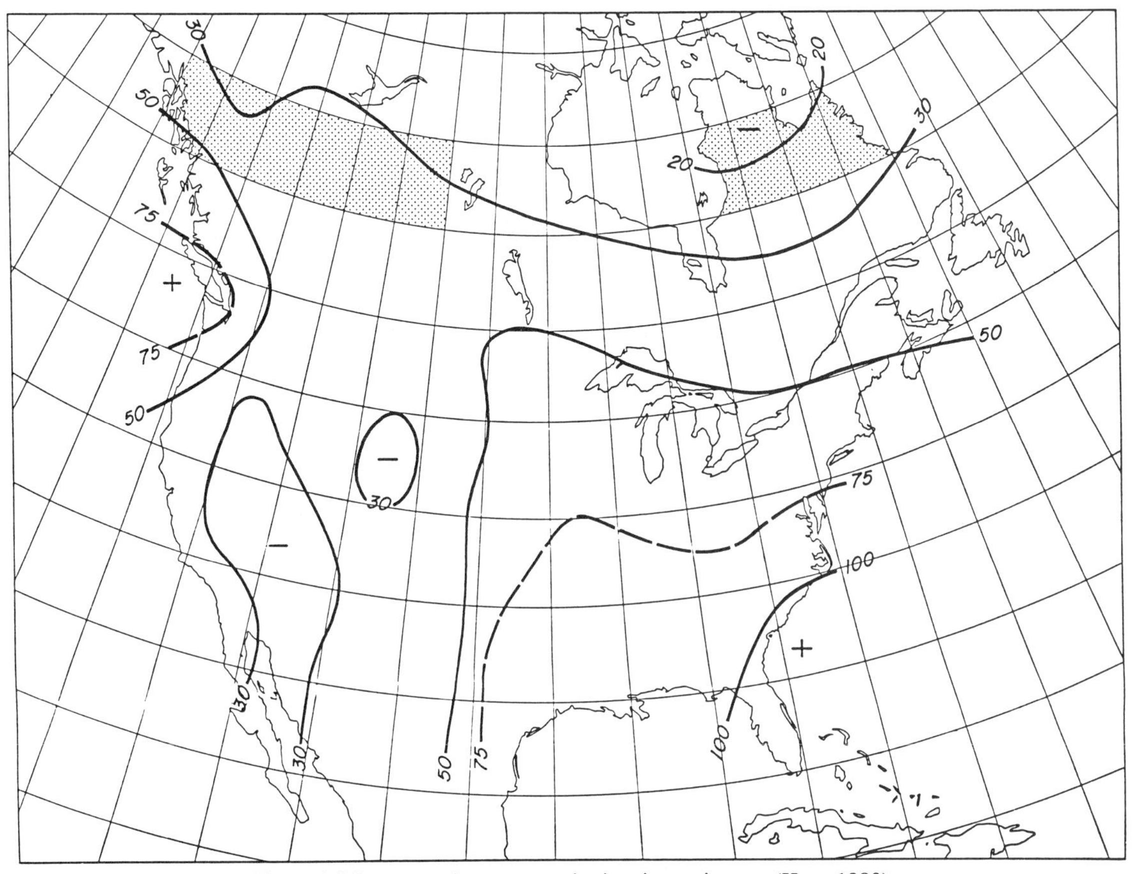

Figure 6. Mean annual evapotranspiration, in centimeters (Hare, 1980).

into the unsaturated zone and to subsoil and ground water. Water that passes below the root zone of the deepest rooted plants is removed temporarily from interchange with the atmosphere. The soil nevertheless retains within the root zone large amounts of precipitation held at various degrees of tension. When downwind percolation essentially ceases, and before evaporative withdrawals start, the soil is said to be at *field capacity.* At that point the portion of soil water available for surface evaporation, or for withdrawal by roots for transpiration, is called the *available water.* It can be withdrawn by surface evaporation or root systems, and can be replaced by percolating rain or snowmelt. The behavior of soil moisture in soils of different composition and physical characteristics is discussed by Saxton and Shiau in this volume.

Estimates vary as to the spatially averaged magnitude of available soil water. In deep, loamy soils with a high organic and clay content the storage is very high, whereas in light sands it is low. Deeply rooted forest vegetation in humid areas may have access to more than 200 mm precipitation equivalent, amounting to a third to a tenth of the annual precipitation. Tall prairie grass swards on deep black soils may also have abundant amounts of available water. Semi-desert vegetation on light soils may have less than 20 mm, even after moderate rain. Dry soils may retard or even supress the impact of heavy rainstorms on streamflow. Wet soils, by contrast, may create floods out of moderate rainfalls.

In modeling the local moisture and energy balances, one may use empirically determined values of available water, and of the soil-moisture tension increase that follows progressive withdrawals by the root system plus surface evaporation. None of these quantities is routinely observed on regional and continental scales. Hence, large-scale budgeting exercises, including numerical weather prediction or atmospheric general circulation models, have to incorporate intelligent guesses as to these vital parameters. This is a major problem for climatologists, hydrologists, and others concerned with analysis of the hydrologic cycle and it is the subject of much inquiry in connection with efforts to model the hydrologic effects of climatic change, as noted at the end of this chapter.

CLIMATIC AND HYDROCLIMATIC CHARACTERISTICS OF NORTH AMERICA

Circulation patterns affecting North America

The global wind system, more specifically referred to as the general circulation of the atmosphere, functions to transport warm air from equatorial regions toward the poles, and to maintain a return flow of cold air from polar to tropical latitudes. It

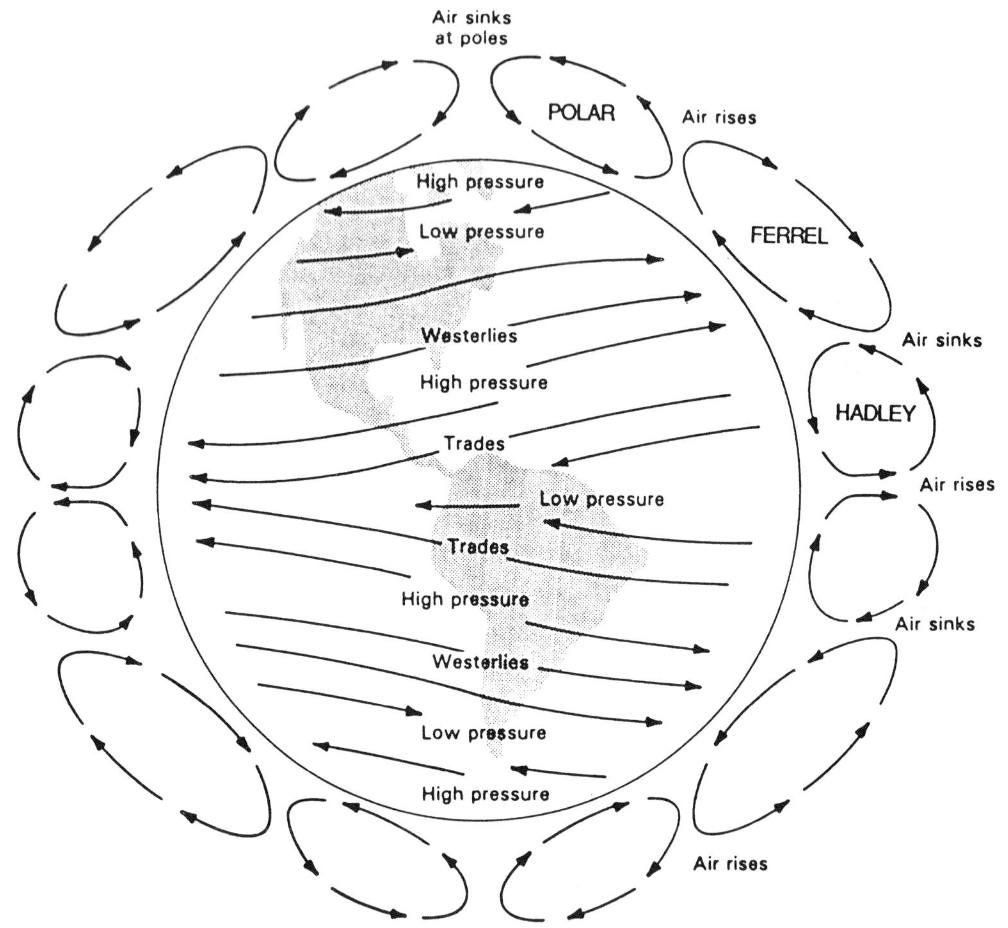

Figure 7. Schematic depiction of the general circulation of the Earth's atmosphere (modified from Washington and Parkinson, 1986).

results from four primary conditions: solar radiation, the rotation of the Earth, the distribution of land masses and seas across the Earth's surface, and tidal forces. Solar radiation and the Earth's rotation combine to form the "brute force" driver of the system. They are, by far, the dominant factors. Topographic and thermodynamic differences arising from the spatial distribution of land and sea are major modulators of the brute-force component. Tidal forces are minor modulators of brute force.

The general circulation has both vertical and horizontal structure (Fig. 7). The vertical structure, composed of the tropical Hadley, mid-latitude Ferrel, and a very weak and intermittent polar cell, mixes heat and moisture through the depth of the atmosphere. The horizontal structure, manifest as the tropical trade winds, mid-latitude westerlies, and polar easterlies, performs the same function across the Earth's surface. These large-scale features primarily determine the broad pattern of global climate. Embedded within this large-scale system are smaller circulations that tend to perturb the global flow. These smaller-scale cyclonic (low pressure) and anticyclonic (high pressure) circulations are responsible for transitory, short-term variations in atmospheric conditions, which we refer to as "the weather."

Large-scale movements of warm air poleward and cold air equatorward provide a balance between energy surpluses at low latitudes and energy deficits at high latitudes. When the temperature difference between high and low latitudes is relatively small, as is typical in summer, the circulation of the atmosphere is weaker and more zonal, or more nearly west to east (Fig. 8). When the temperature contrast is high, as in winter, the circulation may become stronger and more meridional, or north-south. Thus, seasonal variations in weather patterns can typically be characterized by the persistence of one or the other of these two circulation regimes.

Changes in the amplitude of these planetary-scale circulation "waves" reflect the positions and intensities of high- and low-pressure systems around the globe. The region beneath a wave crest is characterized by high pressure. It is typically referred to as a high-pressure "ridge." Such ridges serve to move relatively warm air poleward and are generally associated with fair weather conditions. The wave depressions are regions of low pressure, and are referred to as low-pressure "troughs." Troughs of low pressure within the general circulation move relatively cold air equatorward and are generally associated with stormy conditions. Occasionally the amplitude of a low-pressure trough becomes so great that the circulation folds back onto itself, thereby cutting the low-pressure system off from the general circulation. Without the momentum provided by the general circulation, such "cut-off" lows tend to persist over an area, sometimes bringing storminess for a week or more. In contrast, protracted periods of fair weather can accompany a high-pressure ridge cut off from the general circulation. Such ridges are frequently referred to as "blocking" highs, because they tend to block the movement of storm systems into an area.

Considerable research has focused on the characteristic and recurrent structure of atmospheric circulation because of its importance in weather and climate forecasting (Paegle and Kierulff, 1974; Wallace and Gutzler, 1981; Horel, 1981; Blackmon and others, 1984a, b; and Barnston and Livezey, 1987). Most studies have investigated mid-tropospheric circulation (i.e., between 10,000 and 20,000 ft) because circulation in this altitude range is most closely associated with surface weather patterns. The findings of these numerous investigations have demonstrated that the upper-air planetary-scale circulation is amazingly well organized at monthly to seasonal to annual time scales. The seasonal behavior of atmospheric circulation, and its patterns of persistence, have been well-documented in recent years as the systematic record of upper-air observations has lengthened to nearly 40 years. It is now known, for example, that more than a dozen distinct, coherent circulation regimes exist for the Northern Hemisphere. Some only occur during the winter season, some only in the summer, while others are restricted to the transition seasons. Most, however, tend to bridge at least a couple of seaons with varying degrees of influence. A synthesis of these findings, with emphasis on the circulation patterns affecting weather and climate in the North American sector, follows.

The most persistent mode of atmospheric circulation in the Northern Hemisphere, the only pattern to unambiguously occur in all 12 months, is the North Atlantic Oscillation (Fig. 9a). Although the location of the centers of high or low atmospheric pressure associated with this regime (i.e., its "centers of action") vary considerably throughout the year, its dominant characteristics are a strong pressure center over or just west of Greenland, with a steep north-south gradient to a subtropical pressure center of opposite sign over the Atlantic, Europe, or the United States. This opposing center can occur in any one, two, or all three of

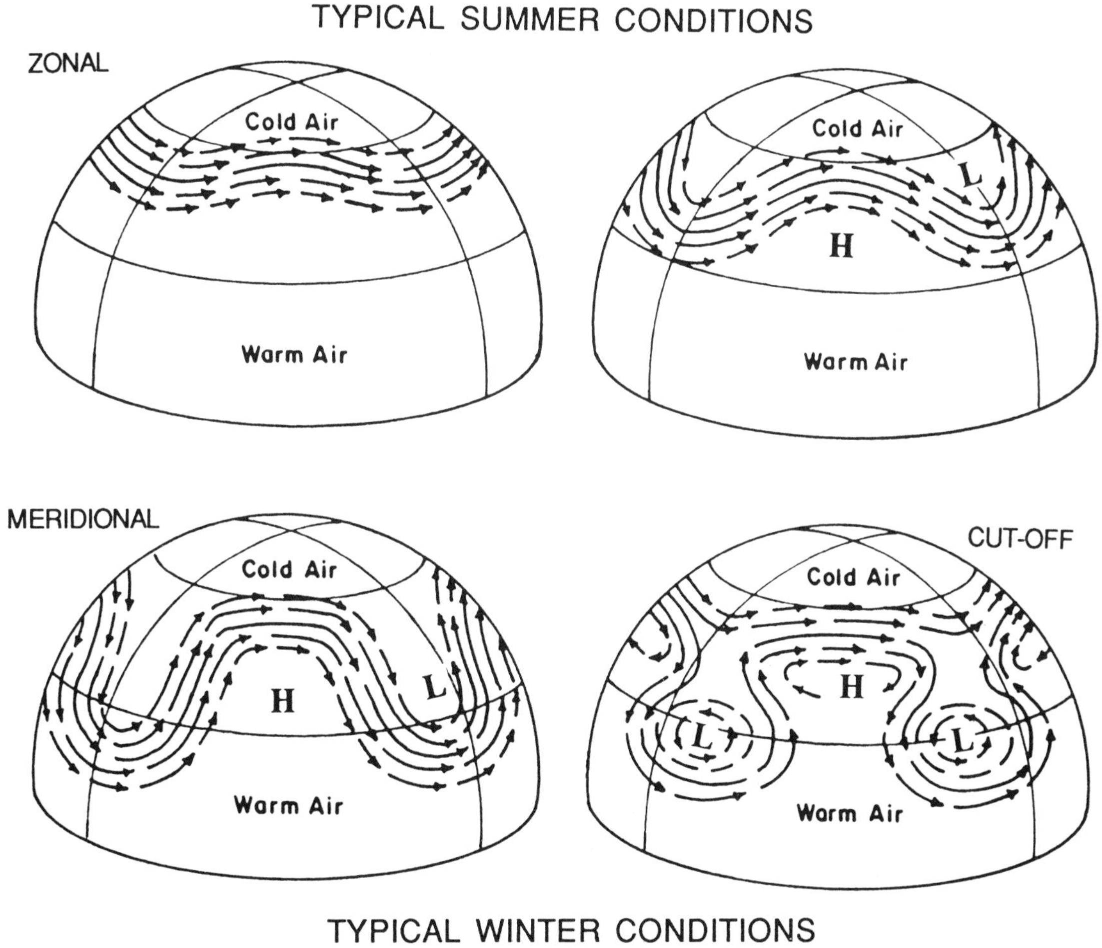

Figure 8. Zonal and meridional components of the horizontal circulation of the atmosphere. The letter L refers to low pressure and H to high pressure (Lins and others, 1988).

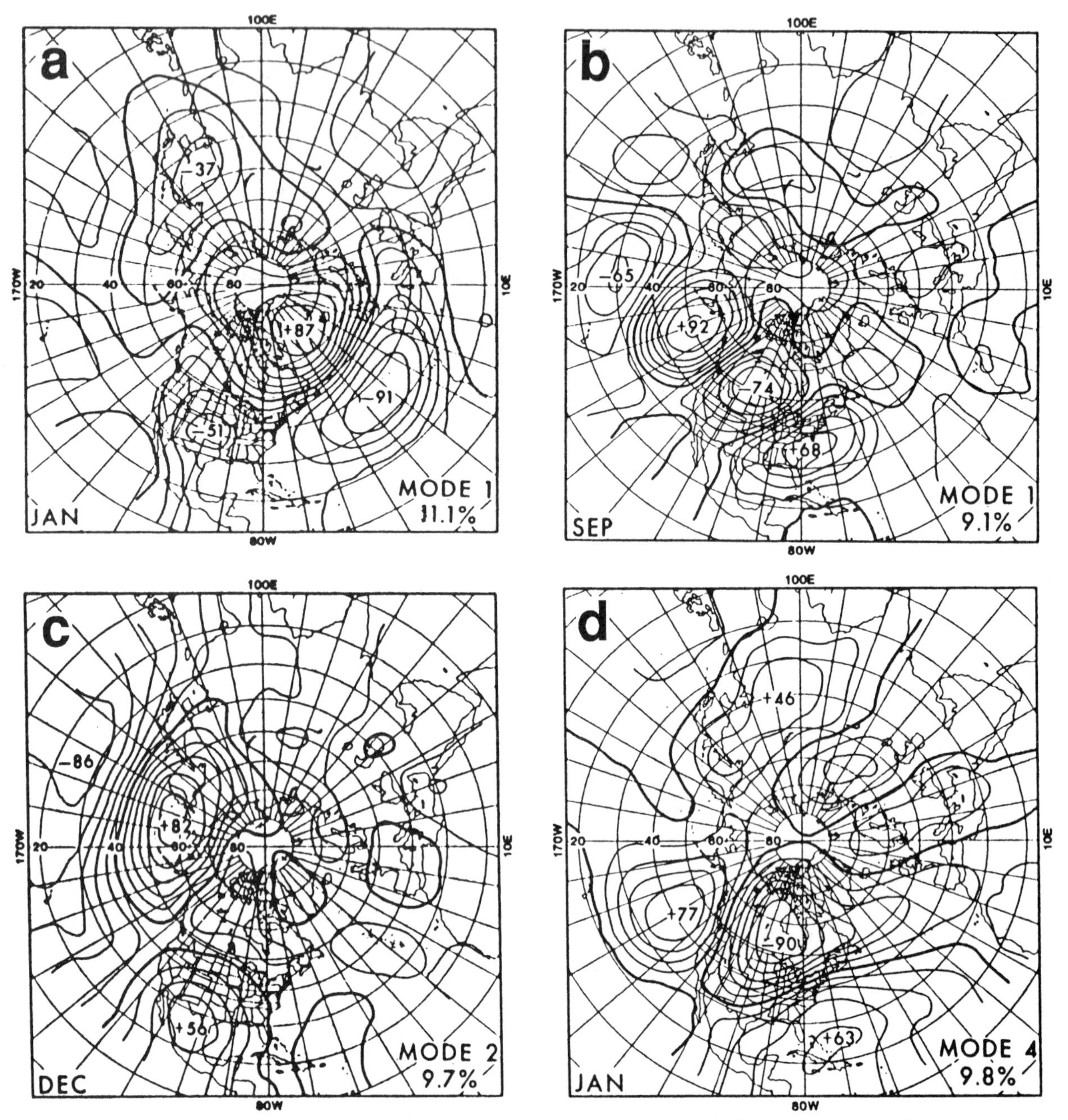

Figure 9. Characteristic shape of the 700-mbar geopotential height anomaly field for (a) the North Atlantic Oscillation (NAO) teleconnection pattern in January; (b) the Pacific/North American (PNA) pattern in September; (c) the West Pacific Oscillation (WPO) in December; (d) the Tropical/Northern Hemisphere (TNH) pattern in January; (e) the East Pacific (EP) pattern in March; (f) the North Pacific (NP) pattern in April; (g) the Pacific Transition (PT) pattern in May; and (h) the Subtropical Zonal (SZ) pattern in July (Barnston and Livezey, 1987).

these regions, or in an elongated band engulfing any two or all three zones. A less consistent element of this circulation pattern is a somewhat weaker center of action positioned over eastern Asia having a sign like that of the Atlantic. Barnston and Livezey (1987) point out that the "NAO's seasonal north-south contraction/expansion parallels that of the other basic meteorological features in the hemisphere (the profile of the westerlies, the 700 mb isotherms, etc.) in response to the annual cycle in solar insolation. The pattern is most closely associated with fluctuations in the strength of the climatological mean jet stream over the western Atlantic during the winter season; hence, the meteorological patterns affecting Western Europe. It is also the dominant influence on summer weather conditions over most of Canada and the United States. Since the NAO is an active regime throughout the year, being the most frequent or second most frequent pattern occurring in 9 out of 12 months, it is by far the

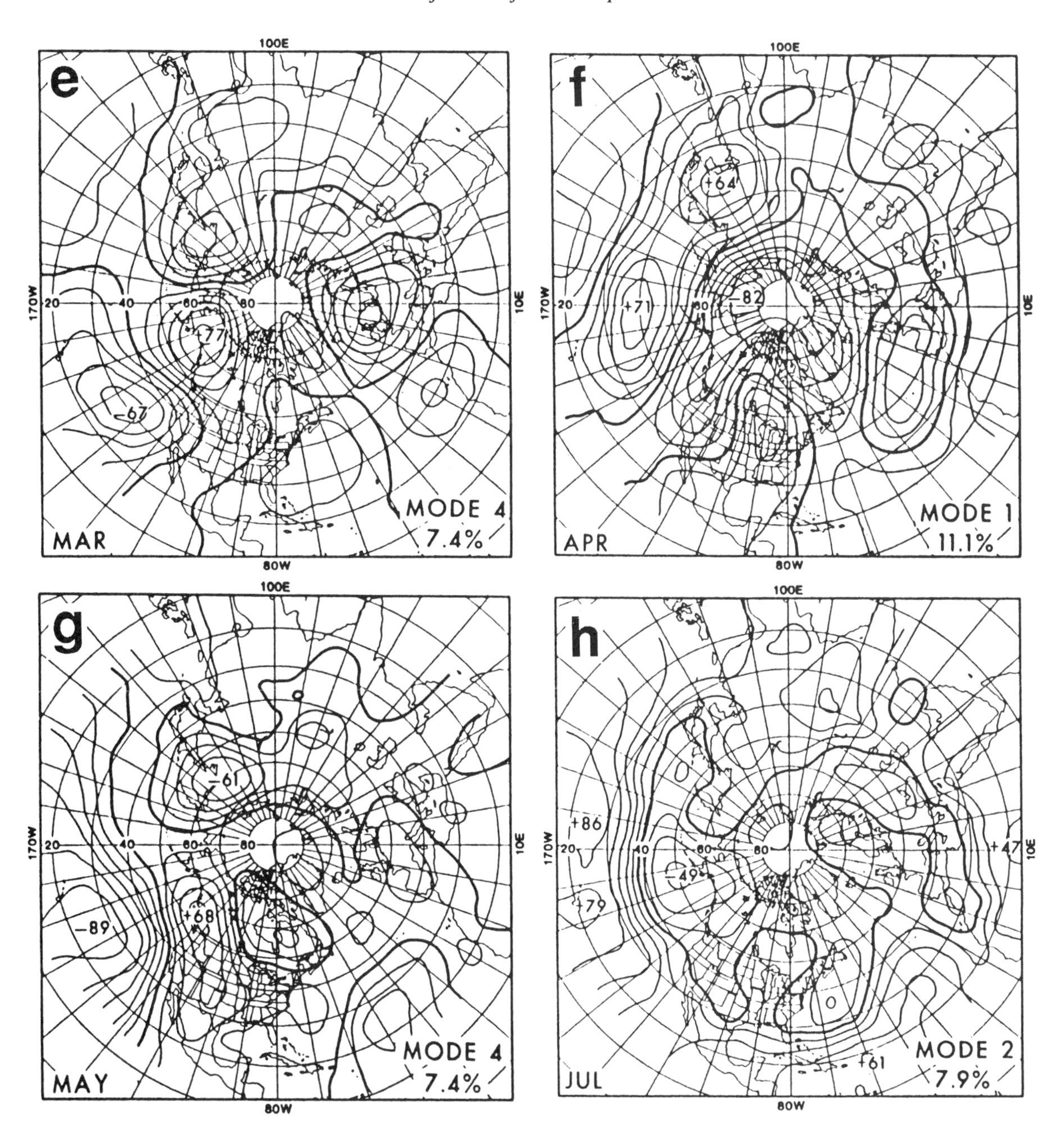

most significant contributor to low-frequency variance in mid-tropospheric pressure in the Northern Hemisphere.

The next most influential regime, at least during the winter half-year, is the Pacific/North American (PNA) pattern. In its purest form, the PNA consists of four cells with centers in, and forming an arc stretching from, the central Pacific to the Gulf of Alaska, into Alberta, and continuing southeastward across the United States to the eastern Gulf of Mexico (Fig. 9b). Each adjacent cell is of opposite sign, with the steepest gradient generally occurring from west to east between the two center cells. The PAN is almost certainly the primary determinant of winter weather for most of the North American continent. It is commonly associated with relatively dry longwave ridges over the western United States and southwestern Canada occurring in conjunction with relatively wet troughs over the eastern United States and southeastern Canada, and vice versa. This bimodal pattern is one of the most characteristic forms of winter season atmospheric variability, and is evident in the systematic record of temperature, precipitation, and streamflow, all of which are associated with mid-tropospheric circulation.

After the PNA, the next most influential form of circulation on North American winter season weather and climate is the West Pacific Oscillation (WPO). This regime, which most affects North America in November, December, February, March, and April, is characterized by a strong center over or just east of Kamchatka with an oppositely signed anomaly band spanning the Pacific Ocean between 20° and 30°N, and 160°W to 130° E (Fig. 9c). The pattern also includes a moderately strong pressure

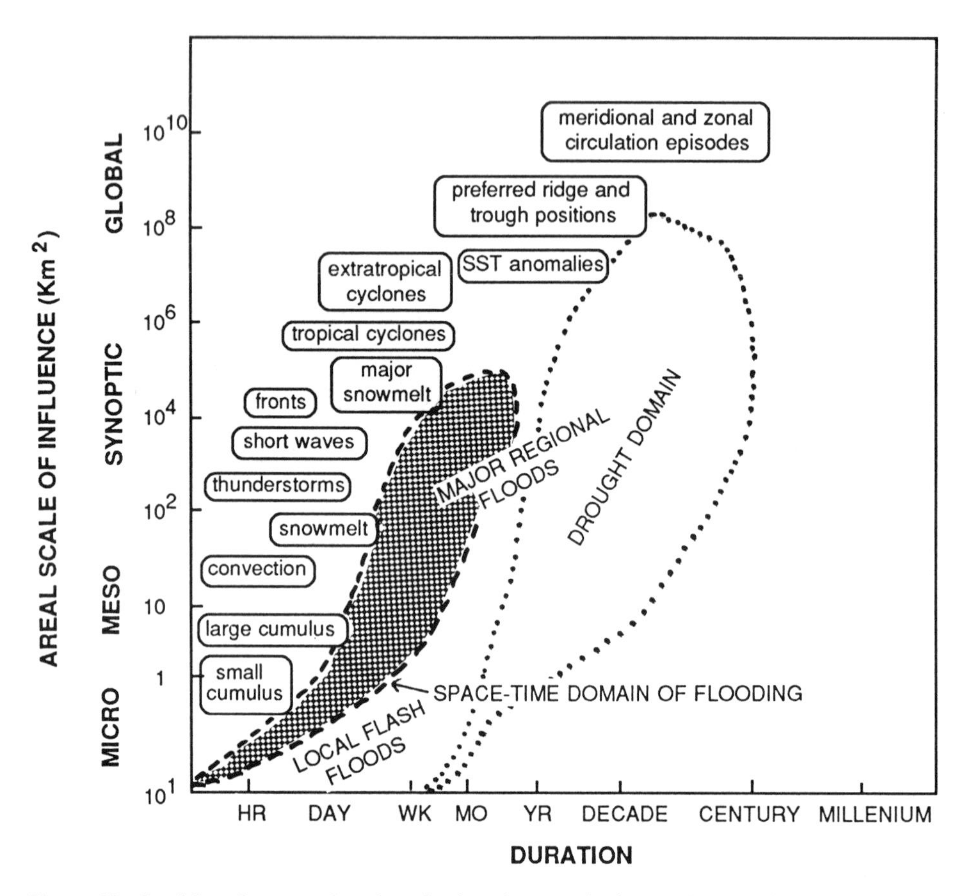

Figure 10. Spatial and temporal scales of selected atmospheric and hydrologic conditions (after Hirschboeck, 1988 and Orlanski, 1975).

center over the southern and central United States with a sign like that of the Kamchatka center. This is a regime that tends to produce a contiguous pattern of either wet or dry conditions across much of the conterminous United States and Mexico.

Another cold-season pattern, occurring primarily between November and February, is the Tropical/Northern Hemisphere (TNH) pattern (Fig. 9d). The TNH bears a resemblance to the PNA, but with only three centers of action that are displaced eastward so as to be out of phase with it. The three centers are positioned just off the Pacific Northwest coast of the U.S., with an oppositely signed center near or just north of the Great Lakes, and a broad center of sign like that of the Pacific center near or just east of Cuba.

The final cold-season pattern of any consequence for North America is the East Pacific (EP) pattern. This regime has a strong center near eastern Alaska with a steep gradient to the south and an oppositely signed center between Hawaii and the coast of California (Fig. 9e). It is generally observed only between January and March.

The transition seasons, spring and autumn, over North America are most influenced by two primary circulation patterns: the North Pacific (NP) and Pacific Transition (PT) (Figs. 9f and 9g). Indeed, the North Pacific pattern is the dominant Northern Hemisphere circulation regime during the months of April and October.

Finally, the most conspicuous summer-season regime after the North Atlantic Oscillation is the Subtropical Zonal (SZ) pattern. The SZ is characterized by an east-west band with like-signed pressure conditions in three preferred regions; the Pacific, the Caribbean and western Atlantic, and North Africa (Fig. 9h). It is an index of zonal flow conditions in the latitudinal band between 25° and 35° N, and is the only pattern having quasi-hemispheric longitudinal extent (Barnston and Livezey, 1987). The SZ is associated with the SPO pattern during the spring, and with the PT pattern in June and July.

CLIMATOLOGY OF NORTH AMERICAN STORMS

The most significant and direct factor influencing the character of surface water hydrology is the precipitating storm. Precipitation in the form of rain or snow occurs over a wide variety of scales, from localized showers covering a few square kilometers to very large mid-latitude storms that spread moisture over several million square kilometers in a period of several days. The

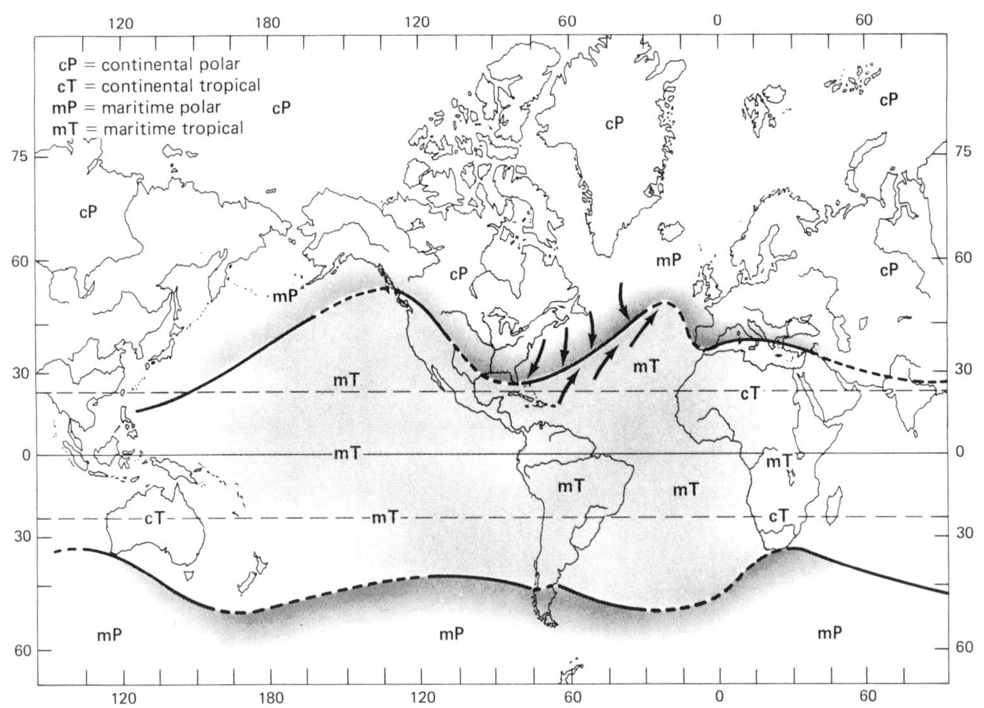

Figure 11. Average position of the polar front in January. Dashed line refers to an area where the front is not well defined. Arrows depict wind direction (Gedzelman, 1985).

ranges of spatial and temporal scales of the most important atmospheric conditions affecting surface water hydrology are depicted in Figure 10. Although it is clear from the figure that there are many discrete meteorological factors of relevance, it is nevertheless possible to reduce the number of factors to a couple of generic systems that describe much of the climatology of North American storms. Thus, for the following discussion, we focus on the primary precipitation delivery mechanisms for most of North America: extratropical and tropical cyclones.

Extratropical cyclones

Extratropical cyclones are the principal precipitation delivery mechanism for North America. They are produced by horizontal temperature contrasts in the lower troposphere and are generally associated with considerable vertical wind shear. Extratropical cyclones (along with anticyclones) are responsible for much of the meridional heat transport occurring outside the tropics. In essence, they tend to reduce the temperature gradients that give rise to them in the first place. By reducing the horizontal temperature contrasts, the intensity of the general atmospheric circulation is also reduced. Then, radiational and turbulent exchange processes begin to restore local radiational balance whereby tropical temperatures increase and polar temperatures decrease.

The air residing over large uniform areas (e.g., oceans or snow-covered surfaces) gradually acquires the character of the surface beneath it; in this way, distinct "air masses" are formed. In general, air masses are described as being either warm or cold (tropical or polar) and moist or dry (maritime or continental). As air masses develop, the meridional temperature gradient and the vertical wind shear increase. Ultimately the flow becomes baroclinically unstable, at which time large waves begin to develop. Usually these waves take the form of closed cyclonic or anticyclonic circulations at the surface, with associated closed circulations or open waves at higher elevations in the troposphere.

Frequently a large part of the temperature contrast is focused along narrow zones or fronts. Fronts are boundaries that separate different air masses. The dominant frontal zones form a nearly continuous boundary between polar and tropical air that encircles the globe. These zones constantly move, evolve and dissipate, but during each season there exists an average position that is generally referred to as the polar front (Fig. 11). Extratropical cyclones typically develop along the polar front. The steps involved in cyclone development are characterized in Figure 12. A cyclone wave thus consists of a region of cold air and a region of warm air, separated by a frontal boundary. On the west side of the cyclone, cold polar air moves equatorward, while to the east of the cyclone, warm tropical air pushes poleward.

Characteristically, the winds in the warm sector of the storm have very little curvature and, except where they converge at the storm's low-pressure center, generally flow parallel to the overall motion of the cyclone itself and to the upper-level winds of the general atmospheric circulation. As the cyclone wave grows, the area occupied by the warm sector decreases, indicating that much of the warm air is rising. Eventually, the position of the storm's low-pressure center becomes increasingly distant from the warm air, and the storm becomes occluded; this is usually the point of greatest development or intensity, essentially the same as the breaking point of an ocean wave. Once a storm occludes, it begins to dissipate or die. The life cycle of an extratropical cyclone typically lasts several days to a week.

Occasionally several cyclone waves will develop in a series along the polar front. Such a sequence of storms is referred to as a cyclone family and is most commonly observed over the North Pacific Ocean. The easternmost storm in the sequence represents the most advanced stage in the system's development, while behind the westernmost storm, cold air is being drawn equatorward to a point where it will eventually terminate the family.

The motion of the cyclones at the surface is related to the wind flow aloft. In general, extratropical cyclones tend to move essentially parallel to the adjacent mid-tropospheric winds. Gedzelman (1985) refers to this directional tendency as the steering rule, and suggests that a convenient formula governing such movements is that low-pressure systems at the surface tend to move parallel to the 500-mb winds (i.e., those at about 18,000 ft) that are directly above them, and at about half the wind speed.

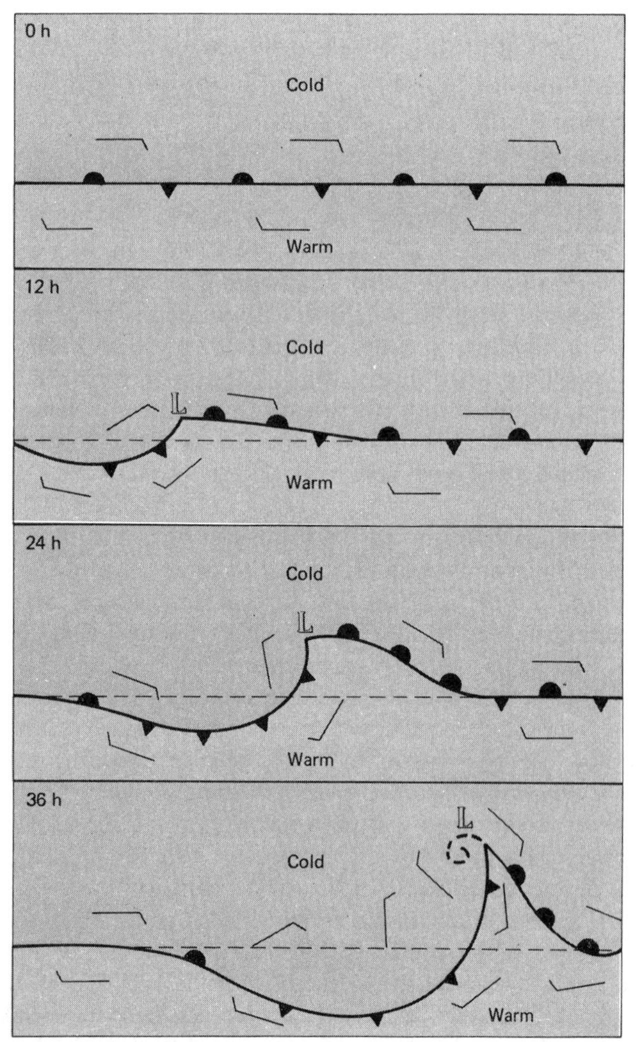

Figure 12. Graphical depiction of the formation, growth, and movement of an extratropical cyclone wave along the polar front (Gedzelman, 1985).

It is important to note that the mid-tropospheric flow pattern frequently consists of two distinct scales of waves: long waves and short waves. The long waves commonly span continental distances and tend to move quite slowly or be quasi-stationary. The short waves are of smaller amplitude and are superimposed on the long waves. Short waves are generally connected to the migratory surface cyclones and anticyclones. They progress eastward more rapidly than the long waves. In general, cyclone families tend to occur downstream from long-wave troughs.

From this general discussion of extratropical cyclone characteristics, it is possible to describe more specifically some of the patterns associated with cyclone development and movement across North America. The primary zone of North American cyclogenesis in all but the summer season is between 35° and 45°N., while during the summer this zone shifts northward to between 45° and 55°N. (Whittaker and Horn, 1981). The most active cyclogenetic areas are depicted in Figure 13. These include the East Coast, the northwestern Gulf of Mexico in winter, Colorado, the Great Basin, Alberta, and the Northwest Territories. A pronounced maximum exists in the Colorado and Great Basin areas in March, contemporaneous with a minimum in the Alberta and Northwest Territories areas. During the spring-to-summer transition, the Colorado activity declines significantly while activity in the Alberta–Northwest Territories region increases. On an annual basis, the greatest concentration of North American cyclogenesis occurs in the Colorado region. Whittaker and Horn (1981) attribute this to the sharp topographic decline going from west to east across this area and to the fact that the polar jet stream is present near this region through much of the year.

In addition to the seasonal variations in regional extratropical cyclogenesis, the tracks followed by these storms after development exhibit seasonal differences. The primary and secondary cyclone tracks from the major cyclogenetic source regions for the four mid-season months of January, April, July, and October are depicted in Figure 14. These mean tracks and source areas are based on the period 1958 to 1977 (Whittaker and Horn, 1984).

In January, storms affecting western North America follow a primary track that moves from the central Pacific into the Gulf of Alaska, where most cyclones fail to penetrate beyond the coastal mountains of Alaska and British Columbia (Fig. 14A). A secondary track then begins in the Gulf of Alaska and progresses southward along the British Columbia, Washington, and Oregon coasts. Some of these cyclones turn eastward in the vicinity of the Canada–U.S. border and track eastward, remaining intact across the Rocky Mountains. Cyclones emanating from the Alberta source region follow a primary track that initially runs southeastward from the source area and then eastward toward the Great Lakes. Colorado cyclones generally move northeastward into the Great Lakes region. Storms then tend to take one of two

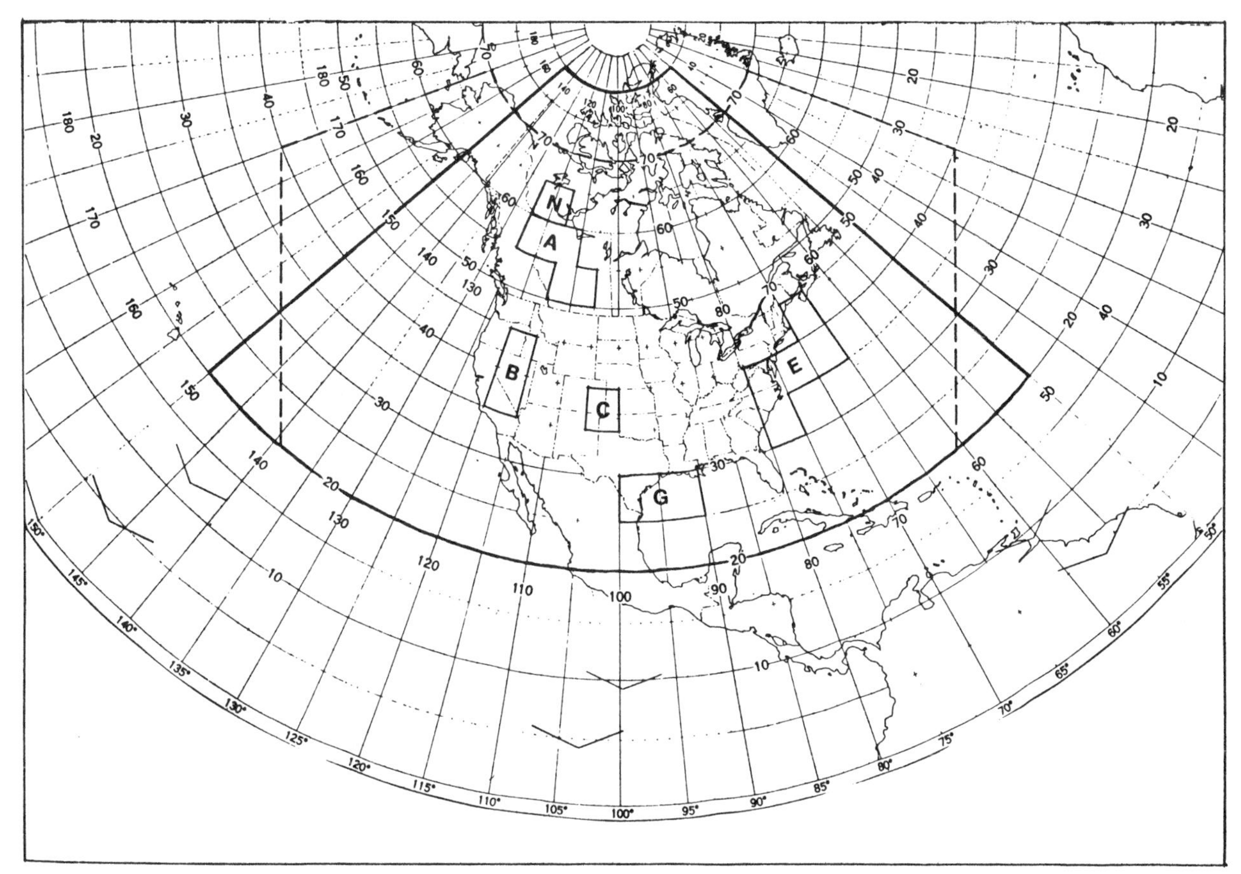

Figure 13. Principal regions of North American cyclogenesis by 5° latitude-longitude grid cells according to Whittaker and Horn (1981). Letter-designated areas refer to (A) Alberta, (B) Great Basin, (C) Colorado, (E) East Coast, (G) Gulf of Mexico, and (N) Northwest Territories. Solid line encloses the Whittaker and Horn study area.

major tracks out of the Great Lakes region: one track moves eastward down the St. Lawrence Valley and then northward into the Davis Strait, and the other extends northward along the eastern edge of Hudson Bay. The cyclogenetic region in the western Gulf of Mexico is the source of a secondary track that extends northeastward across northern Florida and into the western Atlantic. Another major track then begins off the mid-Atlantic coast in response to that area's prevailing baroclinic instability. From a synoptic viewpoint, the cyclone tracks in eastern North America and the western Atlantic appear to rotate around a point in Hudson Bay which, as Whittaker and Horn (1984) point out, is near the center of the polar vortex at 500 mb in January.

The transition from winter to spring conditions is characterized by a decrease in cyclone frequency over large portions of the Northern Hemisphere (Fig. 14B). During April the primary track of the central and eastern Pacific Ocean shifts northward from its position in January, although a nearly parallel secondary track is evident farther south. On the North American continent, two secondary storm tracks emanating from the Northwest Territories diverge as they traverse eastward across northern and central Canada. The most prominent North American storm track during this month originates in the Great Basin and Colorado cyclogenetic areas and extends east-northeastward into southern New England. A major branch of this track turns almost due north from Michigan and moves along the east shore of Hudson Bay in a pattern similar to that occurring in January. From the East Coast source area a primary track extends northeastward along the mid-Atlantic and northeast coast of the United States, continuing on across Newfoundland and eventually extending as a secondary track into the Davis Strait.

With the onset of summer, thermal gradients continue to weaken and cyclonic activity becomes more confined to continental regions. In July, the month with the weakest gradients, most of the primary cyclone tracks are poleward of 50° N (Fig. 14C). Two primary storm tracks approach North America from the North Pacific; one moving into the Bering Sea, the other into the Gulf of Alaska. Primary tracks originating in Alberta and Montana extend to James Bay. From that region a major track progresses eastward across eastern Canada and into the North Atlantic. A long secondary track stretches from Alaska across northern Canada and southeastward toward James Bay. Secondary tracks also extend northeastward from the mid-Atlantic coast, which lies in the same approximate position as the primary Atlantic track during April, and from the Lake Michigan area.

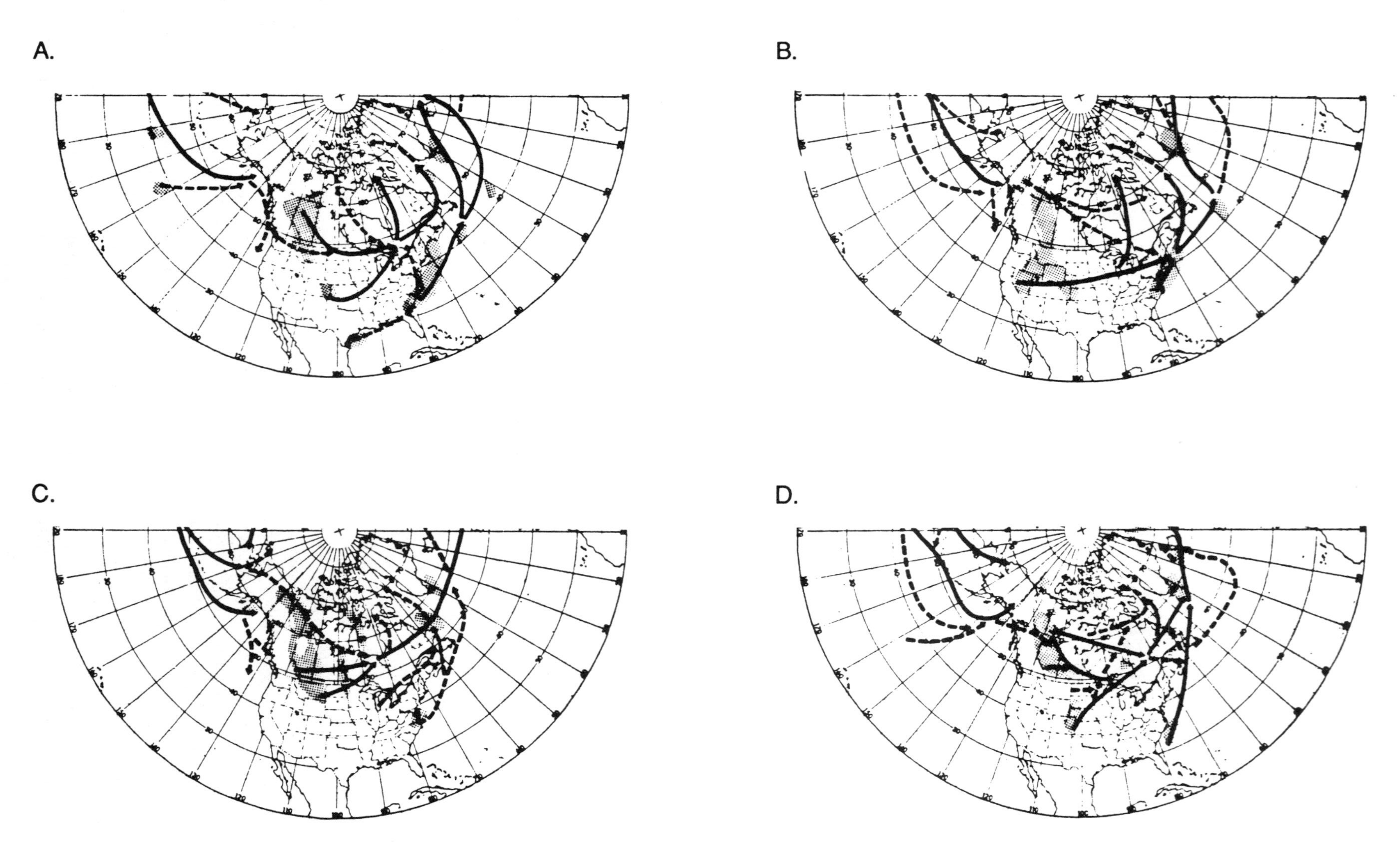

Figure 14. Primary (solid) and secondary (dashed) extratropical cyclone tracks and areas of cyclogenesis (stippled) for (A) January, (B) April, (C) July, and (D) October (from Whittaker and Horn, 1984).

Oceanic cyclone frequency increases during the transition from summer to autumn, and a southerly shift in the position of storm tracks can be noted in the pattern for October shown in Figure 14D. The primary Pacific track now moves northeastward from the central Pacific into the south coast of Alaska. Two secondary tracks are also present in the Pacific south of the primary path. On the continent, the primary track stemming from the Colorado source area reappears. This track extends northeastward across Lake Superior into Quebec where it joins a primary track originating in the Northwest Territories. Another track, originating in northern Alberta, curves southeastward to the western Great Lakes. From a confluence in Quebec, primary tracks extend northeastward to the area just south of Greenland and northward into the Davis Strait. A primary track from the source area off the East Coast of the United States moves northeastward across Newfoundland and traverses the North Atlantic.

Several studies have focused on trends in North American cyclogenesis, and each has arrived at the same general conclusion. For example, Reitan (1979), using data from the four mid-season months (January, April, July, and October), identified a mean downward trend of about 20 percent in cyclone frequency during the period 1949 to 1976. Similarly, Zishka and Smith (1980), using data from January and July, noted a significant decline in extratropical cyclones over the period 1950 to 1977. This included a 45 percent decrease in January cyclones and a 24 percent decrease in July cyclones. Whittaker and Horn (1981), who investigated the period 1958 to 1977, also identified a statistically significant downward trend in North American cyclones. They note, however, that the trend varies over different parts of the North American sector: the most notable decline occurs in the Alberta cyclogenetic source region, a less significant decline occurs in the East Coast source area, and no significant trend occurred in the Colorado source region (Fig. 15).

It is important to note, however, that this consistent assessment of trend in North American cyclone frequency is linked to the similar study periods used by the various investigators. The situation changes for different time periods. For example, Hosler and Gamage (1956) studied cyclone frequency in the United States for the period 1905 to 1954 and found no significant trend. Dolan and others (1988) evaluated trends in cyclone frequency along the East Coast source area for the period 1943 to 1984 and found no statistically significant trend. Notably, in examining Dolan and others' storm-frequency trend plot (Fig. 16) it is possible to discern a clear decline in cyclone frequency between the late 1940s and the mid- to late-1970s. This part of their record is consistent with the findings in the three other studies. This apparent trend, however, is actually associated with the fact that the years 1973 to 1977 had unusually few cyclones in comparison with the rest of the temporal record. Beginning in 1978, and continuing through 1984, cyclone frequency along the East Coast returned to more typical levels and, as a result, led to the lack of any long-term trend in the data series. Such a situation raises an important cautionary point when investigating trends in geophysical phenomena. It is imperative that the evaluation of trends,

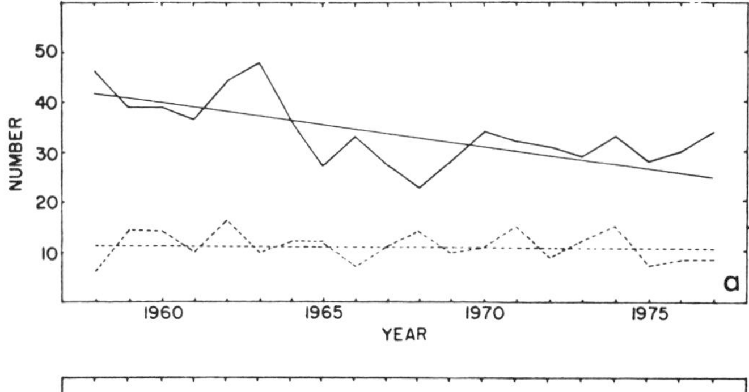

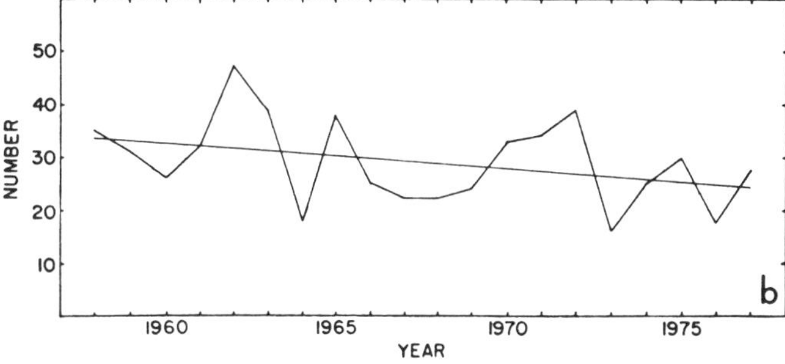

Figure 15. Frequency of cyclogenesis for (a) Alberta (solid line) and Colorado (dashed line), and (b) East Coast. In each case the straight line represents the regression line fitted to the data series (from Whittaker and Horn, 1981).

especially decadal to secular trends, be considered in the context of even longer time series of the phenomenon under study. In other words, a 15-year downward trend in cyclone frequency at some given location could be embedded within a 45-year upward trend at the same location. Care must therefore be exercised in any discussion of temporal trends not to ascribe more meaning to a temporal pattern than the data actually justify.

Tropical cyclones

Tropical storms, in general, exert considerably less influence on most North American regions than do their extratropical counterparts. There is however, one notable exception. This is the area surrounding the western and northern Gulf of Mexico and adjacent western Atlantic, including the U.S. Gulf and southeast Atlantic coasts and those of eastern Mexico. In these areas, many, if not most, extreme hydrologic events occur in conjunction with tropical cyclones (Fig. 17). To a lesser degree, the region encompassing the U.S. Southwest, Baja California, and northwestern and southwestern Mexico is also subject to occasional tropical cyclone effects.

In very basic terms, tropical cyclones are well-organized systems of thunderstorms. They rotate counterclockwise (cyclonically) in the northern hemisphere; this rotation is initially forced by the Coriolis effect. Thus, tropical cyclones never form over the equator. They are warm-core systems, and the source of this warmth is through the release of latent heat. The general rising motion in the core region is then enhanced by the warmth such

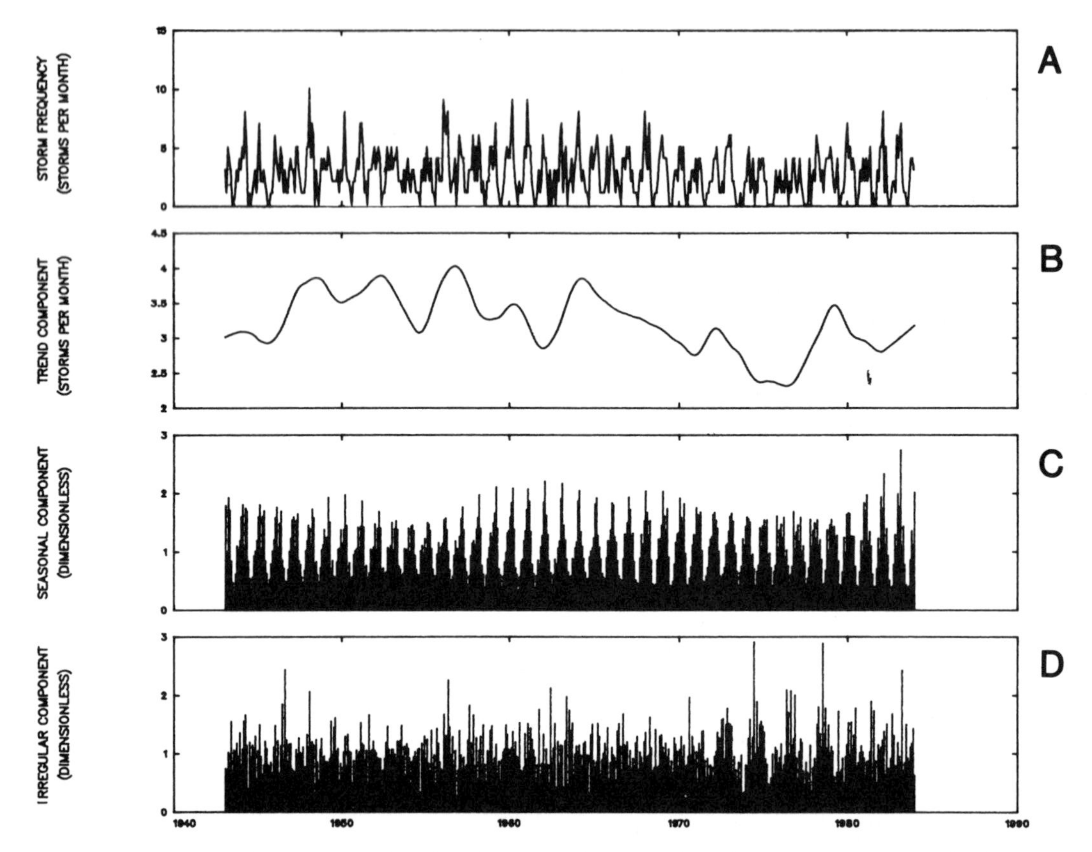

Figure 16. Plot of the raw monthly cyclone frequency data for the East Coast cyclone source region (a), and its decomposed components: trend (b), seasonal (c), and random (d) (from Dolan and others, 1988).

that a positive feedback mechanism is established that maintains and intensifies these storms. The rising air is replaced by converging air from surrounding regions, and the velocity of this air increases as it approaches the cyclone's core. The absence of large-scale vertical wind shear facilitates the vertical buildup of a warm core and is as necessary to the development and maintenance of tropical cyclones as the presence of a large thermal wind is to extratropical cyclones.

Each tropical cyclone contains bands that spiral cyclonically inward to the eye wall. These spiral bands are basically composed of lines of thunderstorms. Within the bands, precipitation and wind conditions are more severe than in the regions between. The bands appear to function as necessary mechanisms for the inward transport of kinetic energy and angular momentum.

During the period 1886 to 1977, more than 300 relatively intense Atlantic tropical cyclones affected coastal and inland areas of Mexico, Canada, and the United States (Fig. 18). Frequently, much of the major late summer and autumn rainfall over parts of the U.S. Southeast is associated with tropical disturbances. Indeed, although these storms may not be responsible for the majority of flooding events in North America, they are responsible for most of the large flood peaks occurring across the southern and southeastern U.S. and coastal regions of Mexico between June and November. Based on the tracks of recent Atlantic and eastern Pacific tropical cyclones, most of the southwestern, south-central, and eastern parts of the United States, as well as Baja California and western and eastern Mexico, are susceptible to occasional extreme flooding from the direct or indirect effects of tropical cyclones (Fig. 19).

Hirschboeck (1990) identified four distinct ways in which tropical cyclones can generate flooding. First, individual thunderstorms embedded within the spiraling cloud bands of a tropical cyclone can produce intense localized rainfall and generate flash flooding in urban areas and small drainage basins. Second, widespread heavy rainfall generated by instability throughout a tropical storm system can produce major riverine flooding over large areas that are located along the path of the storm, especially in the case of a slow-moving system. Third, storm surge, waves, and high water levels produced by low atmospheric pressure and onshore winds can produce coastal flooding, and shoreline erosion, even when the tropical cyclone does not make landfall. Finally, residual moisture fluxes from a tropical disturbance, even if in a dissipating stage, can produce unusually large amounts of precipitable water vapor, and this moisture, in the presence of strong surface heating, orographic lifting, or upper atmospheric instability, can generate heavy rainfall and subsequent flooding.

There is a flip side to the effects of tropical cyclones, however: these storms have frequently been responsible for the termi-

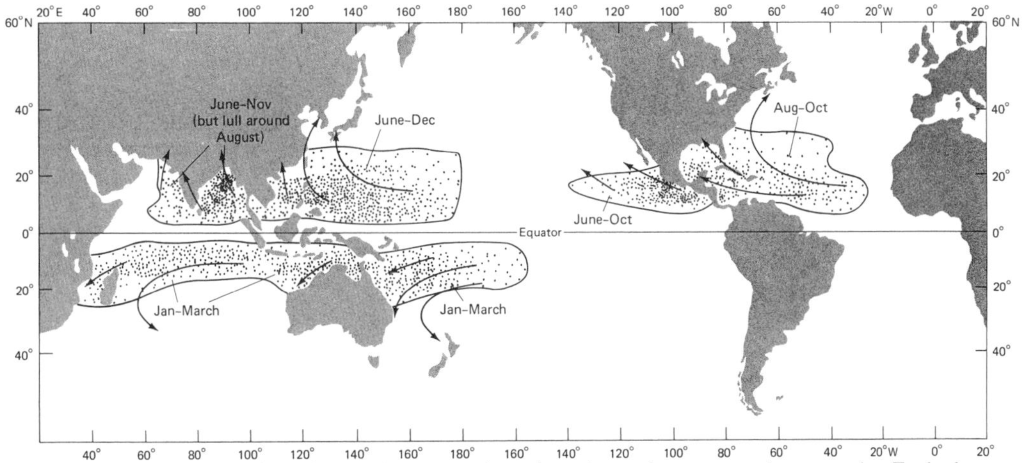

Figure 17. Regions of tropical cyclone genesis, trajectories and most prevalent months. Each dot represents a point where a tropical cyclone was initially spotted (from Gedzelman, 1985).

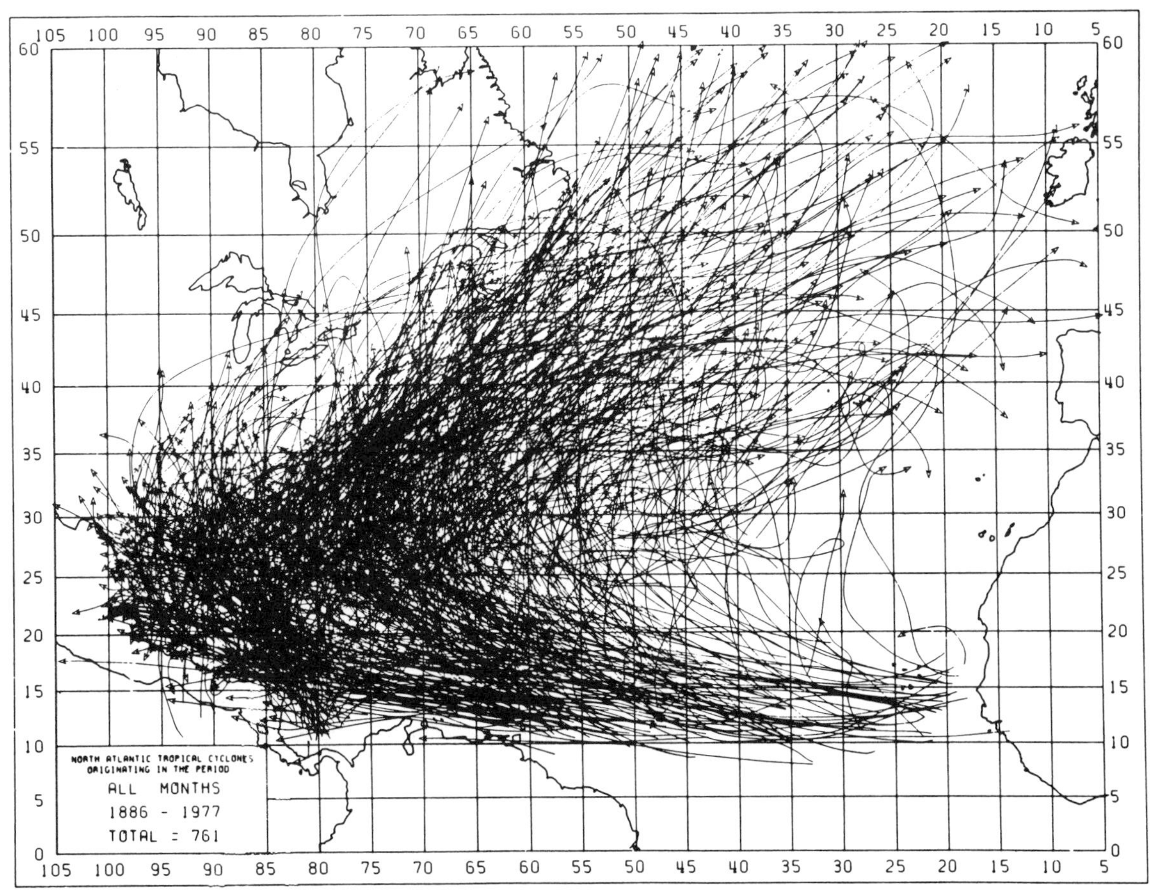

Figure 18. Plot showing the tracks of the 761 known Atlantic tropical cyclones that achieved at least tropical storm strength (i.e., maximum sustained winds between 34 and 63 knots) during the 92-year period 1886 to 1977 (from Neumann and others, 1978).

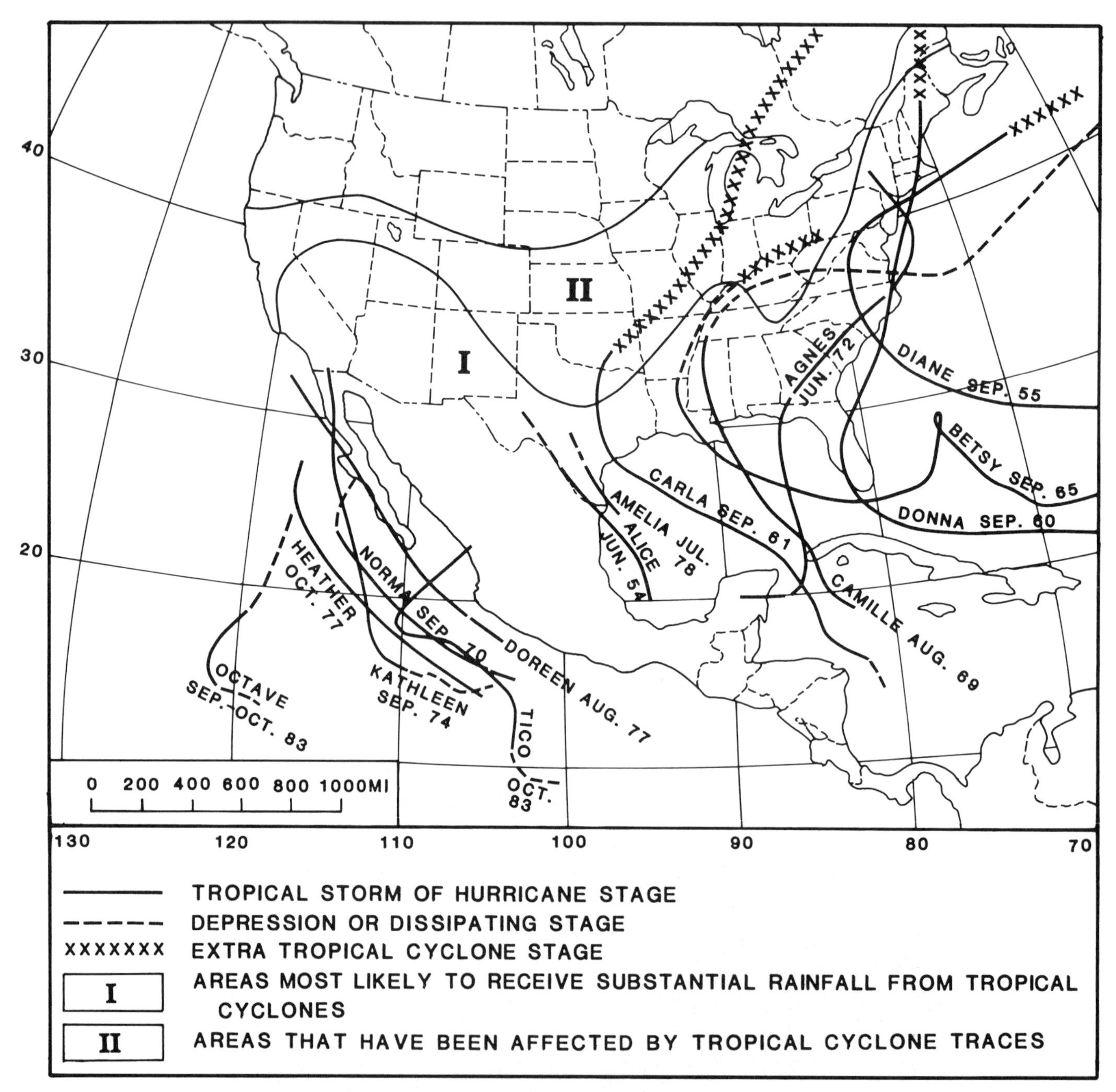

Figure 19. Tracks of flood-producing tropical cyclones and regions affected by tropical cyclone precipitation (from Hirschboeck, 1990).

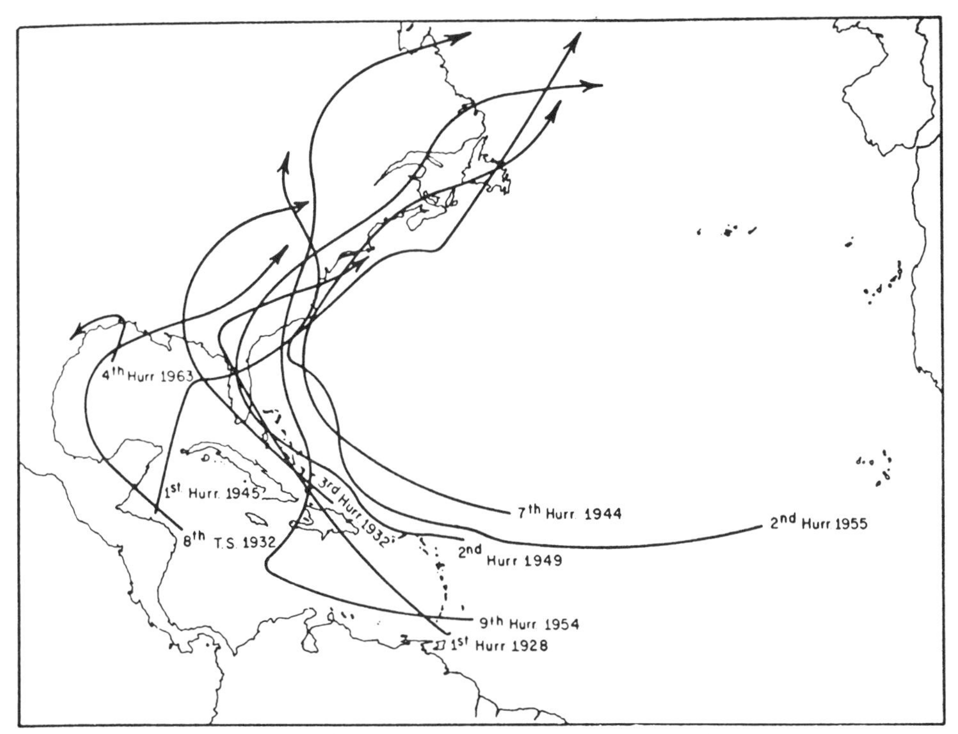

Figure 20. Tracks of nine tropical cyclones that terminated drought conditions in the southeastern United States between 1900 and 1967 (from Sugg, 1968).

nation of drought conditions in the southeastern United States (Fig. 20). Sugg (1968) has suggested that although tropical storms are generally associated with widespread destruction and even death, any evaluation of the social and economic impacts of these events should also consider the economic gains that they produce.

CLIMATOLOGY OF NORTH AMERICAN DROUGHTS

Precipitation is highly erratic in behavior. There is a striking difference between precipitation, as observed locally, and streamflow, except on the very local scale. Precipitation is fundamentally discrete in incidence; it is made up, at origin, of individual droplets or snowflakes, and these combine into discrete storm areas that have limited lives and cover detached areas. Runoff, by contrast, tends to be continuous in time, except in arid areas, and when integrated over large drainage basins, behaves in a much smoother and more predictable fashion. The disciplines of climatology and hydrology have built up different bodies of methods of dealing with these two components of the hydrologic cycle.

The climate record shows much evidence of the inherent variability of precipitation. Some of this evidence was reviewed earlier. What emerges is that in all parts of North America, a tendency exists for drought periods lasting from a few weeks to a few years. Even in well-watered areas there are periods significantly drier than others. This climatic feature is well known and sometimes exaggerated. It has a marked bearing on the design of engineered structures, notably storage dams and channelization projects, and also on navigation and irrigation requirements.

Less well known but of equal importance are the protracted wet periods present in most rainfall records. These have far-reaching consequences. The Great Lakes, for example, were at very high levels in 1987 because of several consecutive years of high rainfall and snowfall.

The incidence of drought and wet periods has been examined in many parts of North America (e.g., Diaz, 1983). Particular attention has been given to the U.S. Great Plains and Midwest and the prairies of Canada because of their importance to the world food system. Such studies normally recognize meteorological drought, defined by Palmer (1965, p. 3) as "... an interval of time, generally of the order of months or years in duration, during which the actual moisture supply at a given place rather consistently falls short of climatically appropriate moisture supply." Agricultural drought, a related concept, occurs "... only when available soil moisture is inadequate to meet evaporative demand by plants" (Oladipo, 1982). Hydrologic drought refers to periods of below-normal streamflow.

Obviously, these drought categories all spring from variations in precipitation, with differing time lags. Various objective indices have been proposed to give realistic measures of drought conditions. These indices take account, in varying measure, of the mitigating effects of soil moisture storage and consumption (e.g., Palmer, 1965; Bhalme and Mooley, 1979; Oladipo, 1982). They also conveniently identify wet periods.

Figure 21 (after Oladipo, 1982) shows for the entire Great Plains region of North America the incidence of various categories of growing-season rainfall anomalies (percent of total area affected as indicated). Table 1 shows the percentage of the area characterized by drought or wetness in the given year. The data indicate that drought or wetness rarely persists on the Great Plains as a whole for two or more consecutive years. There was a decrease in the incidence of extreme drought over the century, and an apparent slow rise in the incidence of near-normal years. The 1980s have not, however, continued these tendencies; drought was severe in several of the first 6 years of the decade over many areas, notably the spring wheat belt. In addition, drought or wetness rarely extends over very large areas (Table 1). In the extreme dry year of 1934, 87 percent of the Great Plains had drought. In the wet year of 1942, 77 percent of the area was wet. But in most dry or wet years, less than 60 percent of the area is especially wet or dry. Similar conclusions can be drawn for the Midwest and Great Lakes regions. Spatially, persistent rainfall anomalies generally occur in regions smaller than the areas of large drainage basins of North America.

Variations in precipitation over time have led to the never-ending search for periodicities in rainfall and streamflow. When demonstrated in streamflow they are usually then looked for in precipitation records, and vice-versa. In practice, it is difficult to

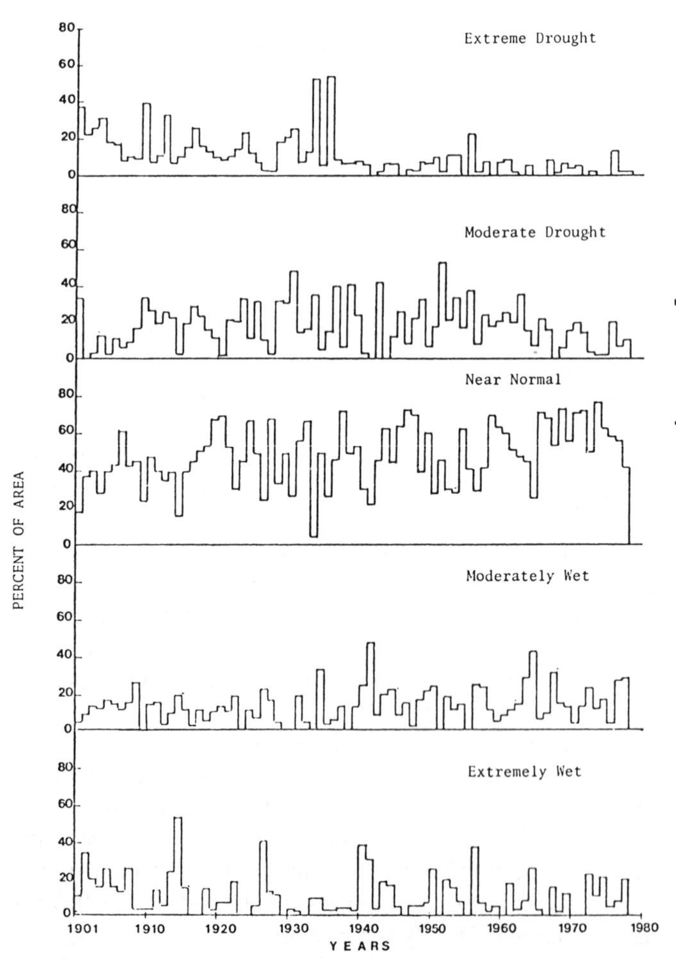

Figure 21. Percent of area within the Prairies and Great Plains in various precipitation categories for the period 1901 to 1978 (after Oladipo, 1982).

find convincing periodicities. A widely held view has been that rainfall in the Great Plains and Great Lakes basin is subject to an approximately 20- or 22-year cycle (the latter with a period similar to the Hale sunspot cycle) (Mitchell and others, 1978). This view was supported by tree-ring analysis from floodplain sites. More recent analysis (Stockton and Meko, 1983) suggests that this periodicity is weak, and is actually distributed over a range of periods from 15 to 25 years; in other words, it is at best quasi-periodic. Similar doubts have been raised by Diaz (1983). A comparable quasi-periodicity affects Great Lakes water levels (Hare, 1984). In his spectral analysis of a modified Bhalme-Mooley drought index over the entire Great Plains from southern Canada to northern Texas, Oladipo (1982) found several weak periods in the 2- to 6-year range, but concluded that drought was essentially aperiodic, adding that ". . . short-lived perioditicies will disappear when a spectrum is made of an expanded drought series."

To illustrate the characteristic look of these quasi-periodic variations, we present Figure 22 (after Changnon, 1984), which shows five-year moving averages of annual precipitation at 4 adjacent stations in the Corn Belt of Illinois. Strong variations are visible. It is natural to look for periodicities in what appears as a powerful, fluctuating response. In practice, neither rainfall nor streamflow usually contains strong periodic signals except those associated with the daily and annual rhythms.

Complicating the issue even more is the recent work of Currie (1981; 1984) indicating that droughts and floods in west-

TABLE 1. AREAL EXTENT OF ANOMALOUS CONDITIONS DURING DROUGHT AND WET YEARS IN THE GREAT PLAINS*

Year	Area of Drought (%)	Year	Area of Wetness (%)
1901	69	1902	42
1904	44	1905	40
1910	71	1908	40
1913	57	1915	73
1917	53	1923	40
1924	56	1927	62
1929	51	1935	40
1930	51	1941	61
1931	71	1942	77
1934	87	1951	47
1936	67	1957	61
1937	48	1965	67
1939	48	1968	45
1943	43	1973	44
1949	40	1978	47
1952	54		
1954	44		
1956	59		

*After Oladipo, 1982.

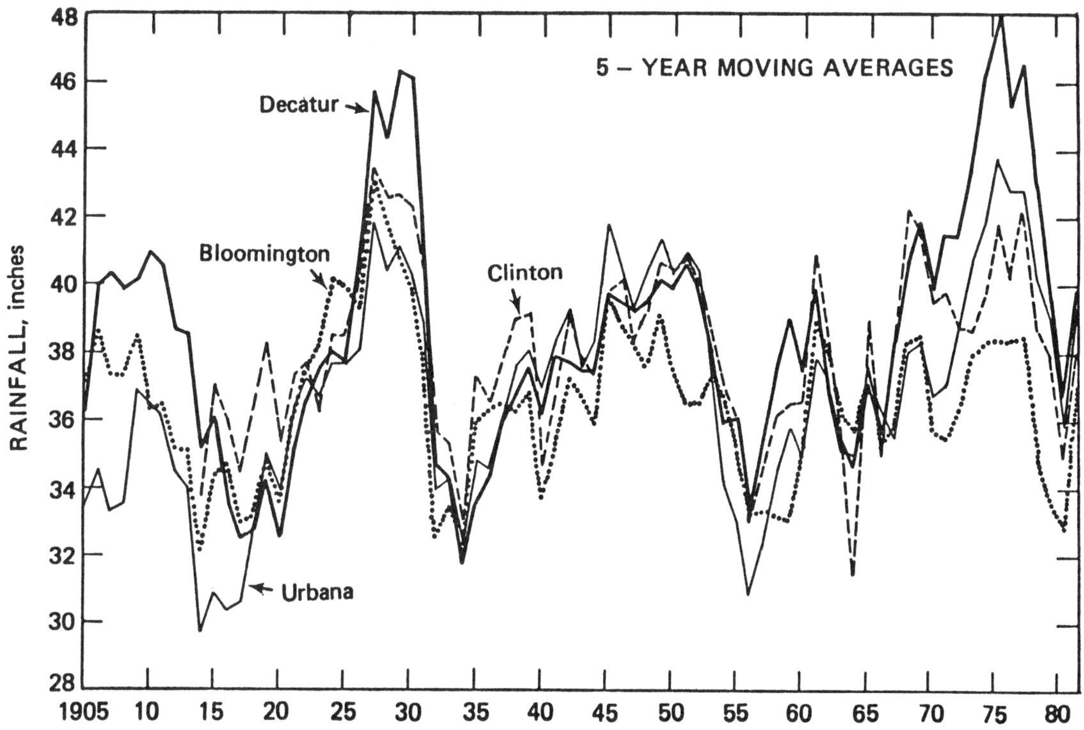

Figure 22. Variations in five-year moving average of annual precipitation at four adjacent stations in Illinois, showing characteristic quasi-periodic appearance of rainfall time series (after Changnon, 1984).

ern North America are periodic, with a period of 18.6 years. This 18.6-year term, apparent in tree-ring chronologies, conforms to the period of the lunar nodal tide. Currie's analysis indicates that during the past two centuries, droughts in the western United States and northern Mexico, and floods in western Canada, have occurred in phase with epochs of maxima in the lunar nodal tide (Table 2). This "bistable phasing," in terms of geography, implies that during the early 1990s, the time of the next nodal tide maximum, drought should be more prevalent in the western U.S. and northern Mexico, and flooding more common in western Canada.

Although the issue of drought periodicities continues to be debated by proponents and opponents, there is concensus among atmospheric scientists and hydrologists that drought is highly variable both in time and space. An impression of this variability is apparent in a study of drought in the United States by Karl and Koscielny (1982). That study focused on determining the spatial patterns and homogeneity of drought, as well as its temporal behavior. The data used in the study were monthly values of the Palmer Drought Severity Index for the period 1895 to 1981. These data were analyzed using principal components analysis; a commonly used technique for isolating modes of spatial and temporal variability in meteorological and climatological data.

From their analysis, Karl and Koscielny (1982) identified nine modes of regionally coherent drought variability, explaining 80 percent of the data variance (Fig. 23). Geographically, the nine modes, in descending order of variance explained, have their foci in the south (14%), west north-central (14%), central (11%), northwest (8%), southwest (8%), west (7%), southeast (7%), east north-central (7%), and northeast (5%) United States. Significantly, these nine regions of drought are distinguishable in terms of their intra-annual precipitation variability (Fig. 24). For example, the late spring–early summer precipitation maximum in the west north-central region is clearly distinct from the bimodal spring-autumn maximum occurring in the south region. Karl and Koscielny (1982) ascribe these inter-regional differences to the unique synoptic climatology controlling the precipitation regimes of each region. They point out that "it would be surprising if the synoptic patterns, which inhibit or enhance the normal oscillatory progression of seasonal rainfall for each of the nine regions were not also responsible for drought initiation, development, and termination."

Moreover, the temporal pattern of drought occurrence within each region appears to be distinct. Cross correlations of the time series of principal component scores for each region are generally weak, suggesting that changes in atmospheric circulation are affecting regions the size of those depicted in Figure 23. It is reasonable to assume that the timing of circulation changes would differentially affect the precipitation and, hence, the droughts in each of the nine regions. To highlight the temporal variability in drought occurrence from region to region, time series plots of the principal component scores of the PDSI are presented in Figure 25. Several points are apparent from these plots. First, the occurrence of dry years varies considerably be-

TABLE 2. SUMMARY OF NODAL 18.6-YEAR EPOCHS FOR SELECTED DATA SETS*

Nodal Tide[a]	Canada[b]	United States and Mexico[c]	China[d]	South America[e]	Africa[f]	India[g]
1508.1			D 1509.1 -1.6			
26.7			D 27.1 -0.4			
45.3			D 45.8 -0.5			
63.9			D 64.2 -0.3			
82.5	D 1580.6 1.9		D 81.5 1.0			
1601.1	D 1601.5 -0.4		D 99.7 1.4	F 1605.6 -4.5		
19.7	D 22.1 -2.4		D 1620.4 -0.7	F 21.3 -1.6		
38.3	D 40.0 -1.7		D 40.5 -2.2	F 37.5 0.8		
57.0	F 52.1 4.9		D 59.3 -2.3	F 56.4 0.5		
75.6	F 74.0 1.6		D 77.6 -2.0	F 75.1 0.4		
94.2	F 93.9 0.3			F 93.1 1.0		
1712.8	F 1712.0 0.8			F 1710.2 2.5	D 1713.4 -0.6	
31.4	F 29.8 1.6			F 27.5 3.8	D 33.4 -2.0	
50.0	F 51.5 -1.5			D 51.9 -1.9	D 50.3 -0.3	
68.6	F 67.8 0.8			D 67.9 0.7	D 66.5 2.1	
87.2	F 85.9 1.3			D 84.4 2.8	D 88.6 -1.4	
1805.8	F 1804.8 1.0	D 1805.0 0.8	F 1805.5 0.3	D 1802.3 3.5	D 1805.8 0.0	
24.4	F 23.3 1.1	D 23.6 0.8	F 22.7 1.7	D 21.6 2.8	D 21.9 2.5	
43.0	F 42.3 0.7	D 44.4 -1.4	D 46.2 -3.2	D 41.3 1.7	D 37.3 5.7	
61.6	F 58.8 2.8	D 61.3 0.3	D 62.0 -0.4	D 63.6 -2.0	?	
80.3	F 80.5 -0.2	D 78.7 1.6	D 80.6 -0.3	D 80.4 -0.1	D 84.5 4.2	
98.9	F 99.9 -1.0	D 1900.9 -2.0	D 1900.1 -1.2	D 95.1 3.8	D 1901.6 -2.7	D 1901.9 -0.3
1917.5	F 1916.0 1.5	D 18.8 -1.3	D 18.1 -0.6	F 1918.3 -0.8	?	F 13.2 4.3
36.1	F 31.8 4.3	D 34.5 1.6	F 35.3 0.8	F 36.5 -0.4	F 35.7 0.4	F 38.7 -2.6
54.7		D 54.8 -0.1	F 53.9 0.8		F 53.0 1.7	F 58.1 -3.4
73.3						F 76.0 -2.7
91.9	0.9± 1.8	0.0± 1.3	-0.5± 1.3	0.7 ± 2.2	0.8 ± 2.5	-1.5 ± 3.2
2010.5						

*From Currie, 1984

tween regions. Second, interannual variability in nearly all regions is relatively high. There does appear, however, to be some tendency for severe drought to persist over several years in the west north-central, southwest, and south regions. Third, although some regions appear to exhibit decadal-scale trends toward both wetness and dryness, no region exhibits any systematic trend over secular time scales or over the 85-year period of record. In other words, the frequency of drought across the United States does not appear to have changed during the past century

REGIONAL HYDROCLIMATIC CONDITIONS

Regional-scale hydrologic conditions, especially those that persist over months to seasons, are reflective of large-scale patterns in atmospheric circulation. To understand fully how and why certain patterns of hydrologic variability occur and recur requires an examination of the contemporaneous atmospheric conditions. This process of analyzing the spatial and temporal characteristics of hydrologic events and conditions within their climatologic context is referred to as hydroclimatology. By focusing on how the atmosphere forces conditions in the hydrosphere, hydroclimatology provides a mechanism for integrating the physical sources of variability in a hydrologic time series with the statistical properties of the varying driving force itself. This integration has the benefit of enhancing our understanding of hydrologic processes and also the quantitative assessment of their variability (Hirschboeck, 1988).

Recent research indicates that specific regions of the globe exhibit characteristic climatic and hydrologic responses to large-scale atmospheric and oceanic conditions. These characteristic responses, or teleconnections, provide a basis for predicting areas of water deficits and surpluses attendant to climate variability and change. The concept of teleconnections, introduced by Walker (1923) and refined by Namias (1981), actually refers to the statistical cross-correlation between atmospheric anomaly patterns separated by synoptic (1,000 to 2,500 km) to hemispheric distances. Some of the best documented teleconnections, the Pacific/North American, the Central North Pacific, and the Southern

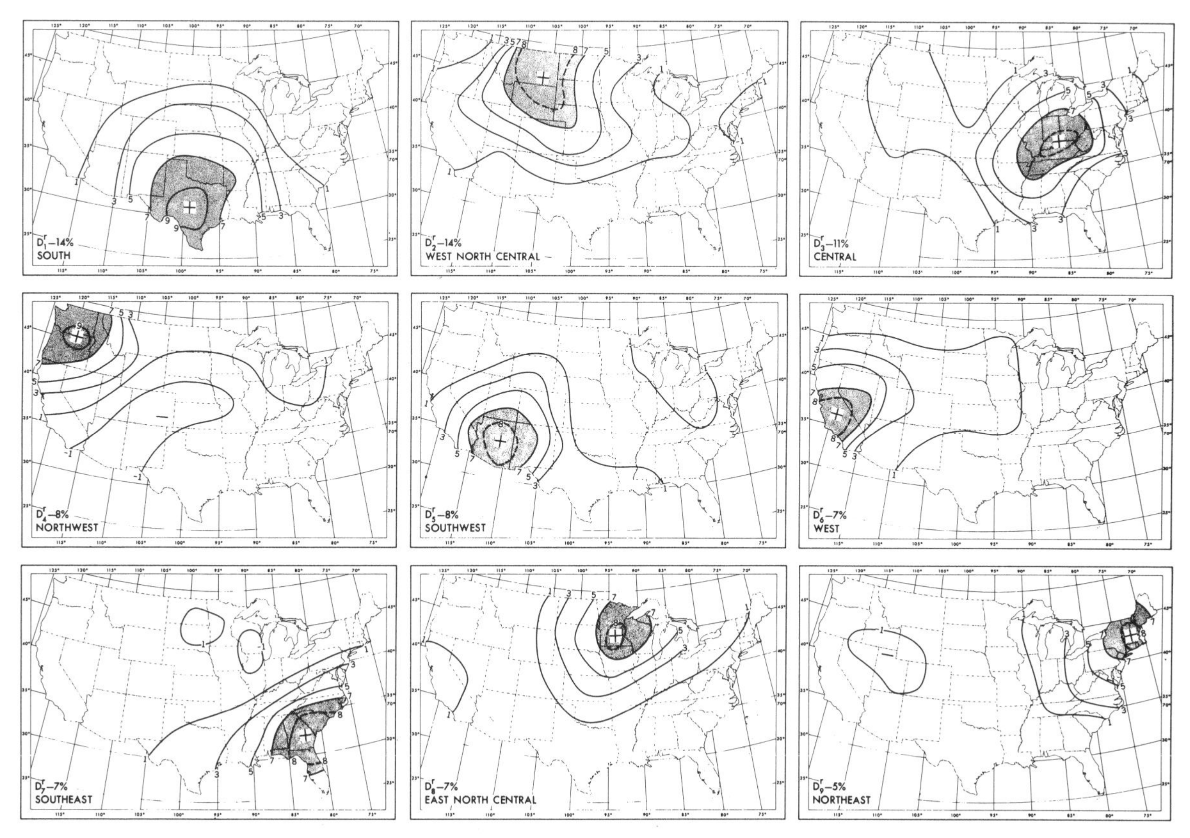

Figure 23. The first nine orthogonally rotated principal component loadings (i.e., correlation coefficients times 100) for the Palmer Drought Severity Index during the period 1895 to 1981 (Karl and Koscielny, 1982).

Oscillation patterns, for example, have been clearly associated with distinct patterns of streamflow variability in western North America, as shown in Figure 26. Refinement of these associations and expanded hydroclimatic analysis of regional patterns nationwide, and at monthly time scales, offer potential for water-resource planning and assessment.

An initial step toward expanding such analyses has been undertaken by Lins (1985a, b) through the identification of systematic patterns of interannual streamflow variability across the conterminous United States. Working with annual mean values of streamflow at 106 sites across the U.S. during the 48-year period 1931 to 1978, Lins reported the existence of five statistically significant regions of systematic discharge variability (Fig. 27). These five regions, or modes of variation, accounted for more than 56 percent of the total variance in the entire streamflow data set.

The most consistent region of flow variability is centered on the central Mississippi River valley (Fig. 27a). This area, accounting for approximately 14 percent of the total variance, tends to be representative of uniform conditions of either above- or below-normal streamflow across most of the adjacent United States. This region is also frequently a center or focal point for both flooding and hydrologic drought. The reason this region exhibits a pattern of streamflow more generally characteristic of the nation is that streamflow, as with temperature and precipitation, is closely linked with the large- (i.e., continental) scale circulation of the atmosphere. This is especially true for seasonal to annual time scales and for below-normal flow conditions that are invariability associated with geographically broad and persistent high pressure ridges centered over the central United States. Moreover, upper atmospheric pressure is commonly either above or below normal across much of the U.S., especially when averaged over seasonal to annual time scales.

A second regionally consistent pattern of streamflow variability occurs in the Pacific Northwest (Fig. 27b), and undoubtedly extends into southern British Columbia. This pattern accounts for approximately 13 percent of the total variance in annual flow nationwide. An interesting aspect of this pattern is the sign opposition existing in the correlation coefficients between the Pacific Northwest and the Southwest. What this indicates is that when streamflow is above (below) normal in the Pacific Northwest, it is frequently below (above) normal in the Southwest. This characteristic structure in streamflow variability has been noted previously in the work of Meko and Stockton (1984) as well as in precipitation variability by Walsh and Mostek (1980) and in drought variability by Karl and Koscielny (1982).

A related regional pattern is characterized by streamflow coherence across a broad section of the western United States, as depicted in Figure 27c. The core of this pattern forms a wedge spanning the California coast and stretching eastward across the Great Basin and into the Four Corners region. This mode of streamflow variability exhibited considerable interannual persistence during the early 1940s and again between 1965 and 1975, both periods of normal to above-normal flow in the western U.S.

These two western streamflow regimes primarily reflect the influence of atmospheric circulation over the eastern Pacific Ocean and western North America. Variations in the position and intensity of the Central North Pacific (CNP) atmospheric pressure center and the Pacific/North American (PNA) pressure pattern, of which the CNP is a part, affect the path of Pacific storms and the location of storm landfalls along the North American coastline. The PNA pattern is formed by the contemporaneous occurrence of four large pressure centers over the North Pacific Ocean and the North American Continent. This pattern is generally configured with a pressure anomaly of one sign situated over the subtropical North Pacific, another center of like sign over southern British Columbia and the U.S. Pacific Northwest, a strong center of opposite sign south of the Aleutian Islands, and a fourth center of opposite sign over the southeast Gulf Coast of the United States.

Several important characteristics of the relation between atmospheric circulation over the North Pacific in the winter season and streamflow in western North America have recently been reported by Cayan and Peterson (1989). First, when the low pressure center over the Central North Pacific is weak, Pacific storms tend to cross North America along the U.S.–Canada border, bringing heightened precipitation and elevated streamflows to the Pacific Northwest, from Washington through Montana and into northern Utah. Such a trajectory in Pacific storms effectively deprives the southern Alaska coast of moisture, thus reducing streamflows there to below-normal levels.

Second, when the Central North Pacific low is very strong and high pressure is situated over western Canada (i.e., PNA), storms are displaced northward such that the Pacific coast of Alaska experiences elevated streamflows while British Columbia and the U.S. Pacific Northwest typically experience below-normal streamflows. A variation of this condition often occurs during El Niño–Southern Oscillation (ENSO) events when, in addition to the distribution of high and low streamflows just described, another region of above-normal flows occurs across the southwestern United States. ENSO episodes are correlated with the PNA; hence, the same general pattern of wetness in

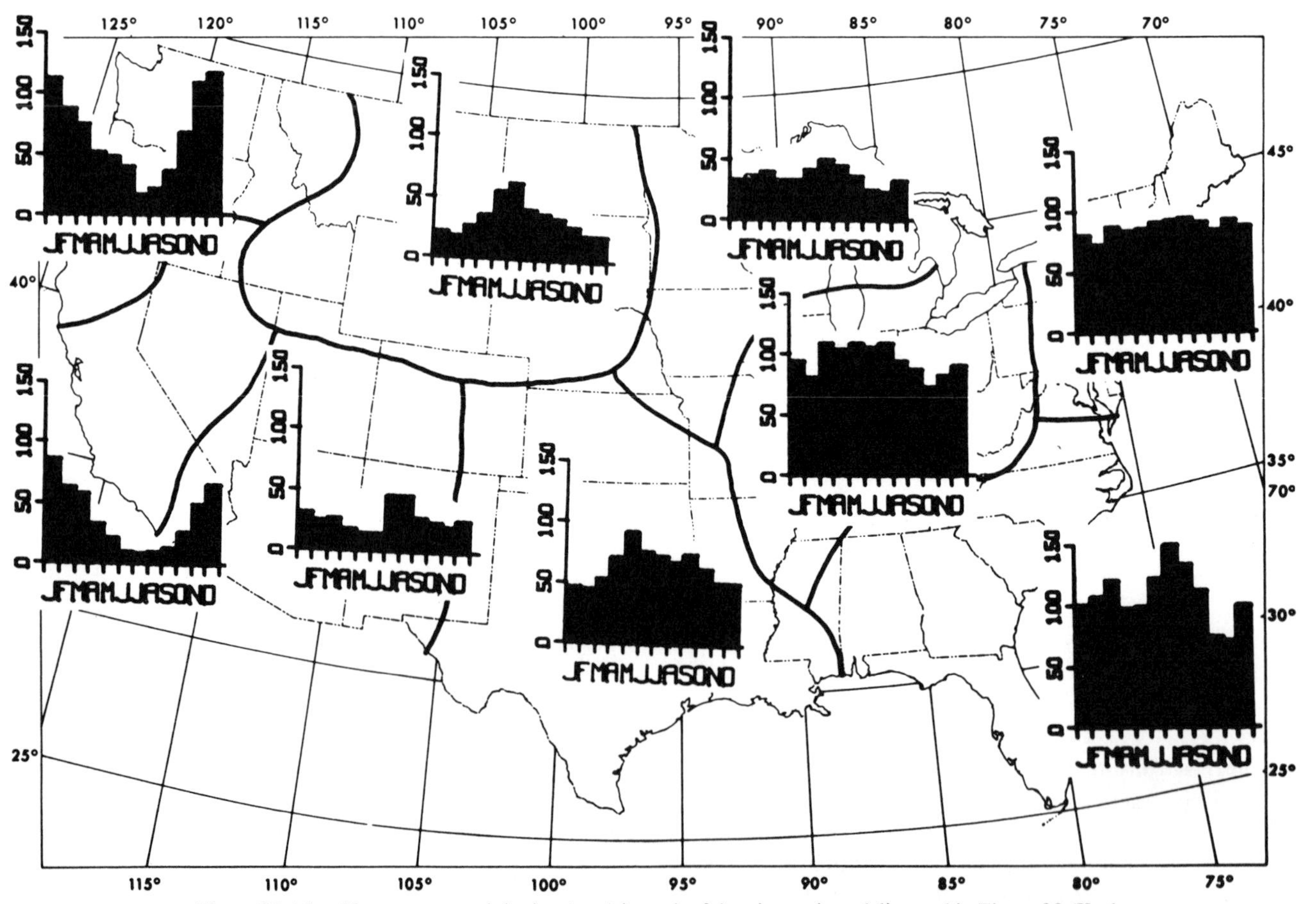

Figure 24. Monthly average precipitation (mm) in each of the nine regions delineated in Figure 23 (Karl and Koscielny, 1982).

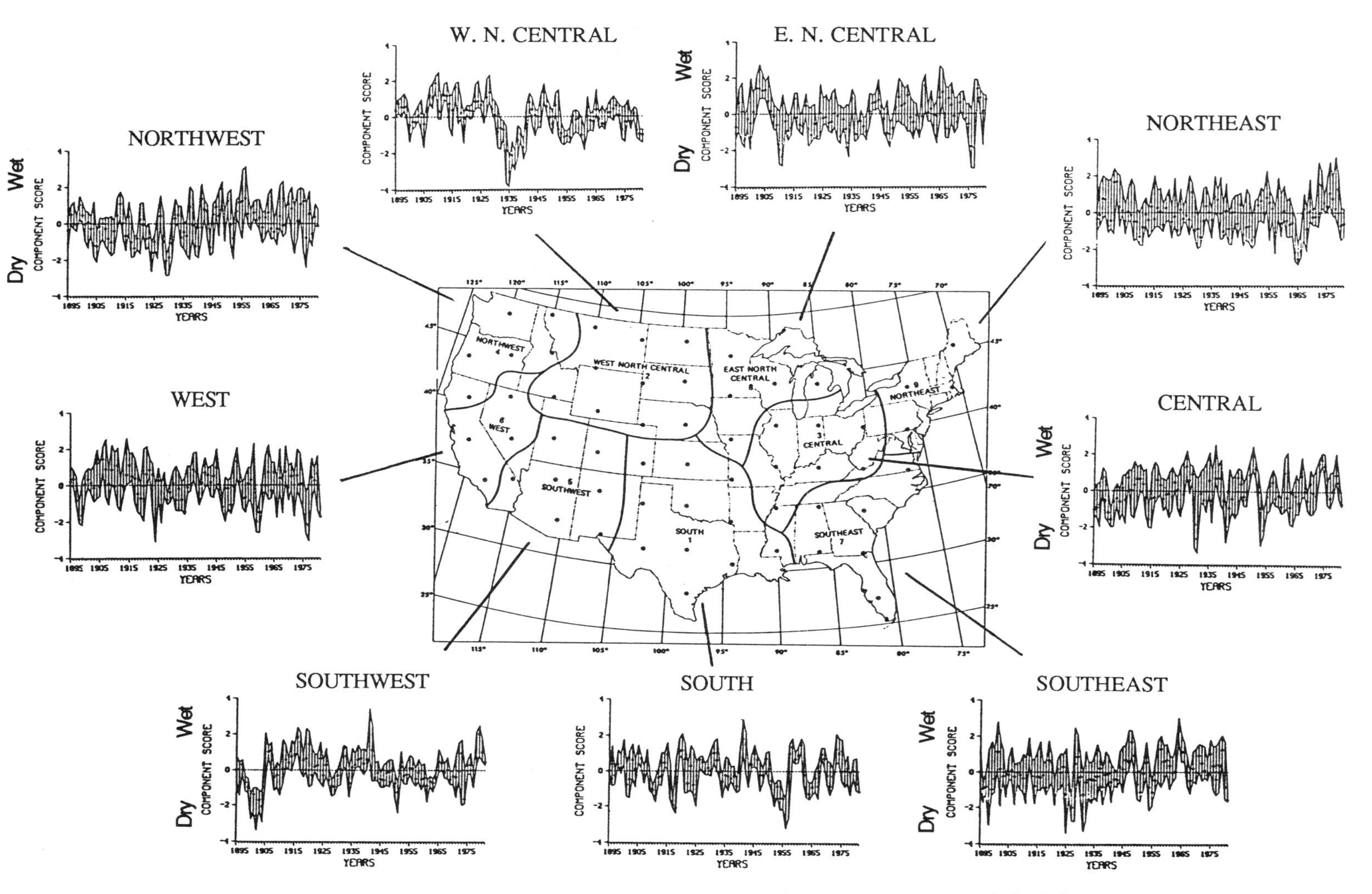

Figure 25. Annual maximum, minimum, and median principal component scores of the Palmer Drought Severity Index for the period 1895 to 1981. The plotted scores contrast years of relative wetness (above the zero line) with those of relative dryness (below the zero line) (from Lins and others, 1988, after Karl and Koscielny, 1982).

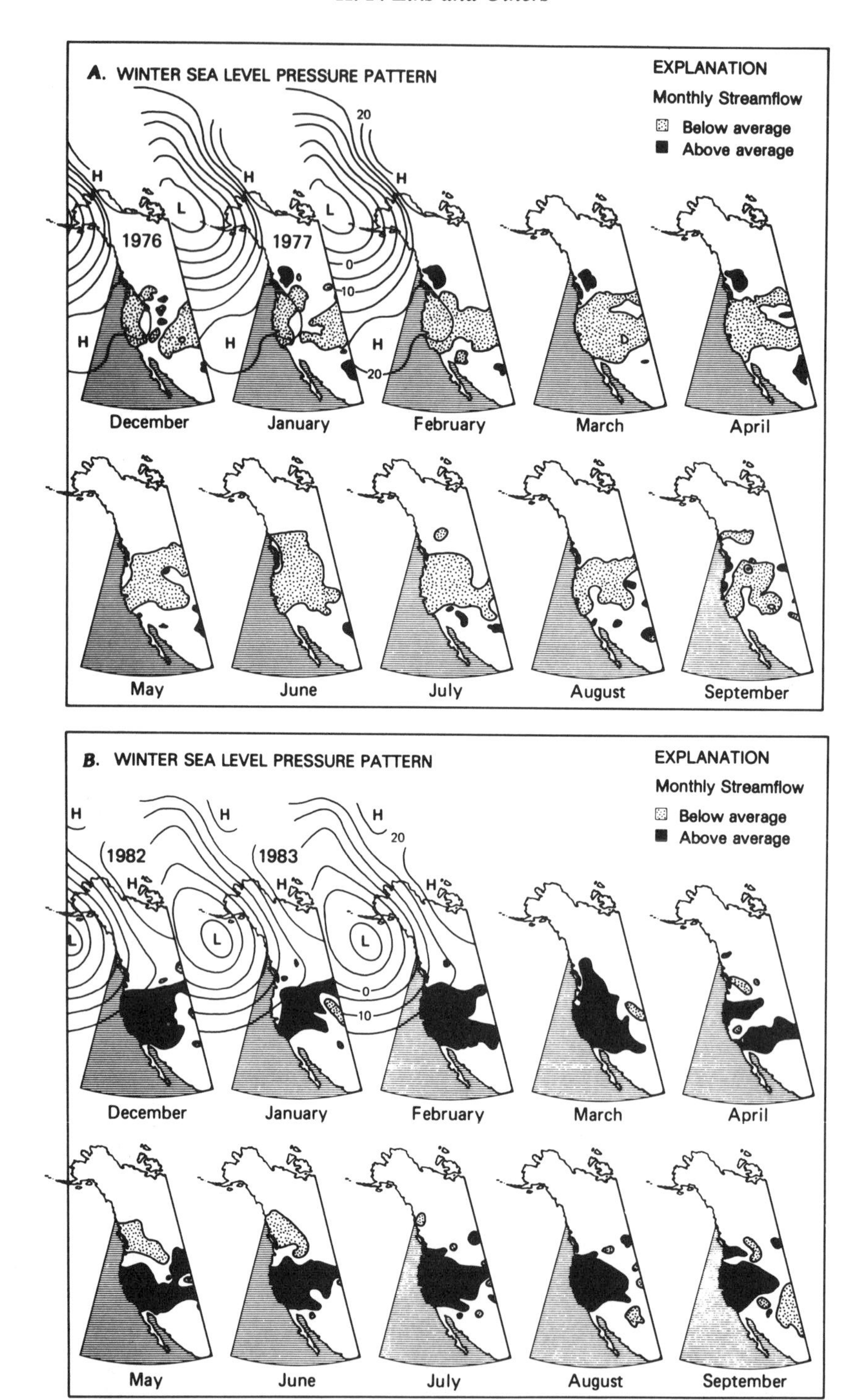

Figure 26. Mean sea-level pressure (in millibars minus 1,000) and monthly streamflow patterns for the winter of 1976–77 (A), and the winter of 1982–83 (B). "Below average" refers to the lower quartile of flow, and "above average" refers to the upper quartile of flow (Peterson and others, 1987).

Alaska and dryness in the Pacific Northwest is commonly observed. However, ENSO events also appear to give rise to a strong subtropical jet stream which travels northeastward from the Pacific across the coast of Southern California. This jet brings moist maritime air into the Southwest, producing abundant precipitation and elevated streamflows.

A fourth regional streamflow pattern, appearing in Figure 27d, encompasses the Northeastern United States and very likely continues into southern Quebec. This pattern exhibited considerable interannual persistence during the early to mid-1960s, when mildly to extremely deficient streamflows characterized much of the Northeast region. Interestingly, the early to mid-1970s were

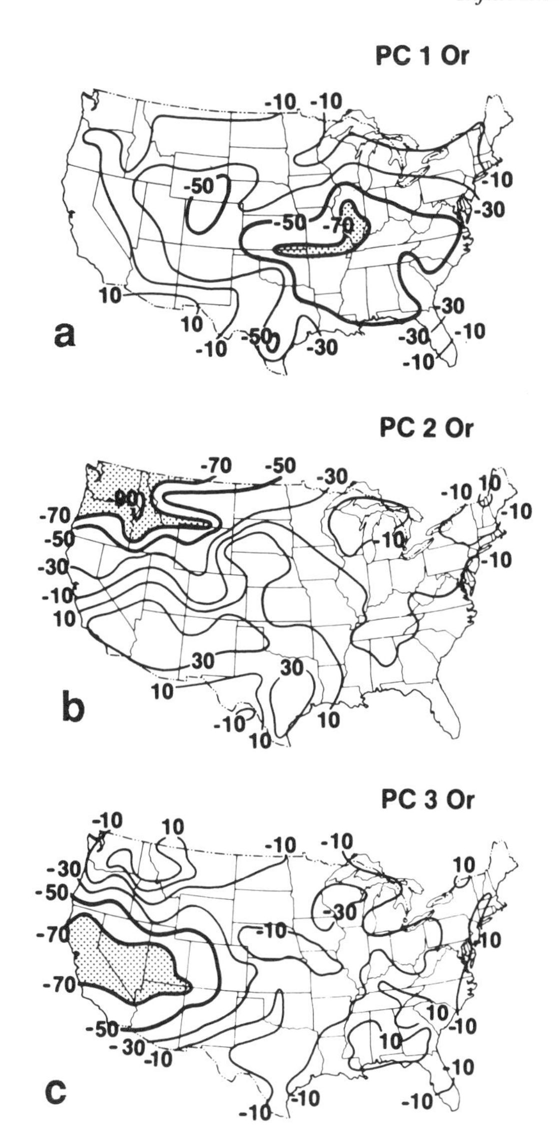

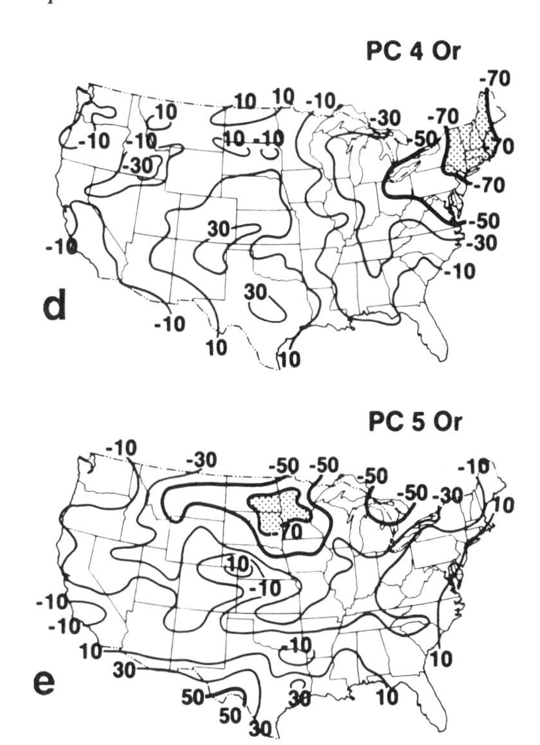

Figure 27. Loadings (×100) on the first (a), second (b), third (c), fourth (d), and fifth (e) orthogonally rotated principal components of annual streamflow.

also years of persistence in this annual pattern, although flows during this time period were generally in the above-normal range.

The final region of broad-scale streamflow variability described by Lins centers on the northern Great Plains of the United States (Fig. 27e). This is also a region of relatively high interannual persistence in streamflow anomalies. For example, between 1931 and 1941, normal to much-below-normal flows dominated this region. Then, from 1942 until 1953, normal to above-normal runoff persisted. In all but one year between 1954 and 1961, streamflow again exhibited normal to below-normal conditions. Finally, in all but three years, including the record dry year of 1977, did streams in the northern Great Plains between 1962 and 1978 flow in the above-normal range.

An additional characteristic of this northern Plains pattern is noteworthy. The area around the central Rio Grande basin in southwestern Texas contains moderately high negative correlations with those in the northern Great Plains. This pattern indicates that there is some tendency for streamflow anomalies of one extreme (drought or flood) to occur in southwestern Texas contemporaneously with streamflows of the opposite extreme in the northern Great Plains. The occurrence of such a pattern has been documented for many years (Busby, 1963; Nace and Pluhowski, 1965) and is undoubtedly associated with the shape and position of the atmospheric long-wave circulation over central North America.

THE HYDROLOGIC CYCLE OVER NORTH AMERICA

Principal controls: Net radiation and atmospheric circulation

The energy needed to drive evaporation and transpiration over the continent, as well as to melt snow and ice, is provided by incoming solar and emitted longwave terrestrial radiation. Net radiation represents the difference between radiation to the Earth from these sources and losses of infrared radiation from the surface. The source of water to the continent is controlled by the circulation of the atmosphere, which carries a large amount of water vapor from the ocean across the continent, and also contains the storm systems that precipitate part of this flux. The precipitation provides the mass input to soil and ground water and to streamflow.

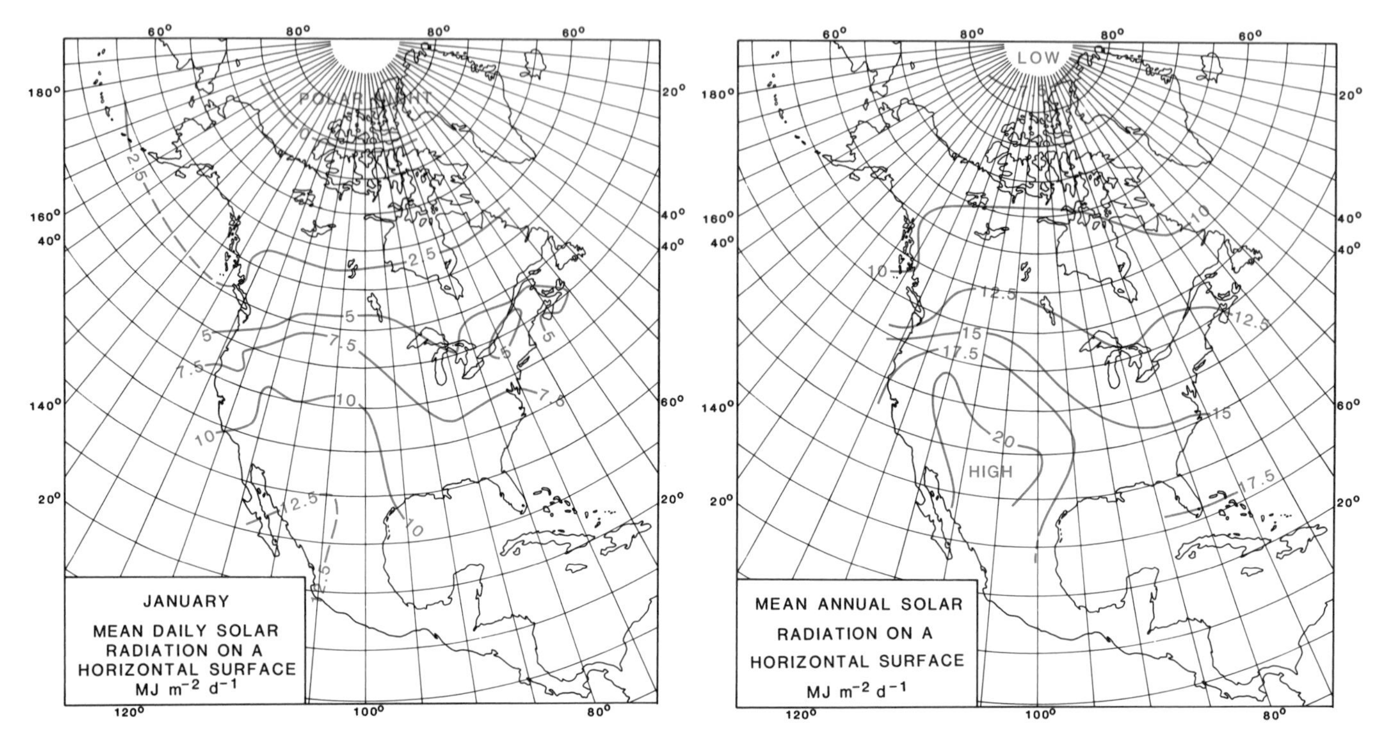

Figure 28. Mean daily global solar radiation on a horizontal surface (in megajoules m^{-2} d^{-1}) for January (Solar Energy Research Institute, 1981; McKay and Morris, 1985).

Figure 30. Mean annual global solar radiation on a horizontal surface (in megajoules m^{-2} d^{-1}) (Solar Energy Research Institute, 1981; McKay and Morris, 1985).

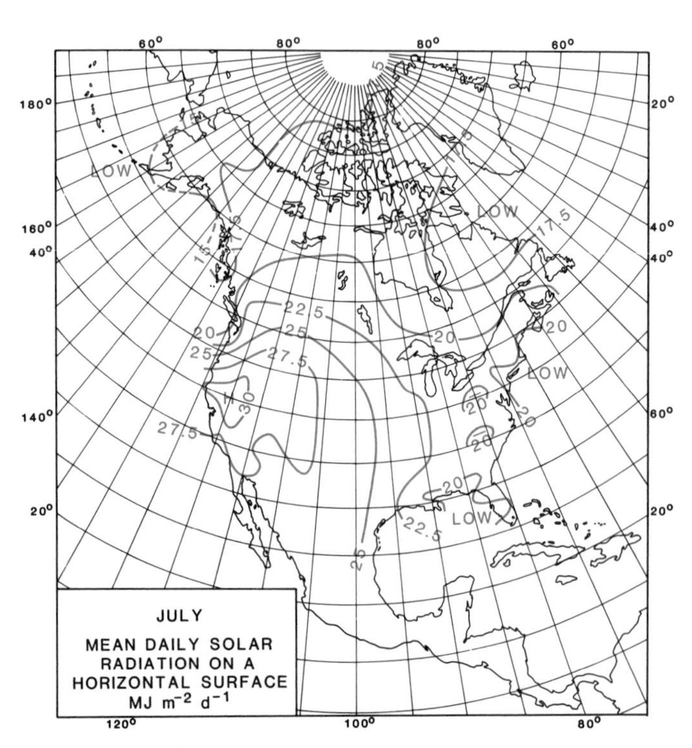

Figure 29. As for Figure 28, but for July.

Figures 28 and 29 show estimates of the solar radiation over North America in January and July, near to the minimum and maximum of the seasonal cycle. Annual averages are given in Figure 30. The radiation received depends on (1) the solar input at the top of the atmosphere; (2) absorption of solar radiation by the atmosphere; (3) the attenuation of the downward beam by backscattering and reflection by aerosols, clouds, and molecules; and (4) forward-scattered diffuse radiation from sky, aerosols, and clouds. Direct and diffuse components of the radiation are of comparable magnitude at the surface. Unfortunately the solar irradiance of the Earth's surface (the rate at which energy is received in watts per square meter, W m^{-2}, or megajoules per square meter per day, MJ m^{-2} d^{-1}) has been monitored in a haphazard fashion. Hence, Figures 28 to 30 are approximate.

As shown earlier, the local energy and water balances are dominated by the net radiation, Q_*, the sum of all radiative gains and losses. Net radiation is measured directly only at research stations, and at a small number of Canadian climatological stations. It can be calculated from the solar irradiance from known characteristics of the surface, cloudiness, aerosol distribution, and atmospheric composition. Figure 31 shows a spatially-generalized analysis of the mean annual net radiation over North America based on calculations by Henning (1978). Extensive use of this estimate by one of the present authors (FKH) suggests that is probably correct within ± 5 percent, with large uncertainties in mountain areas.

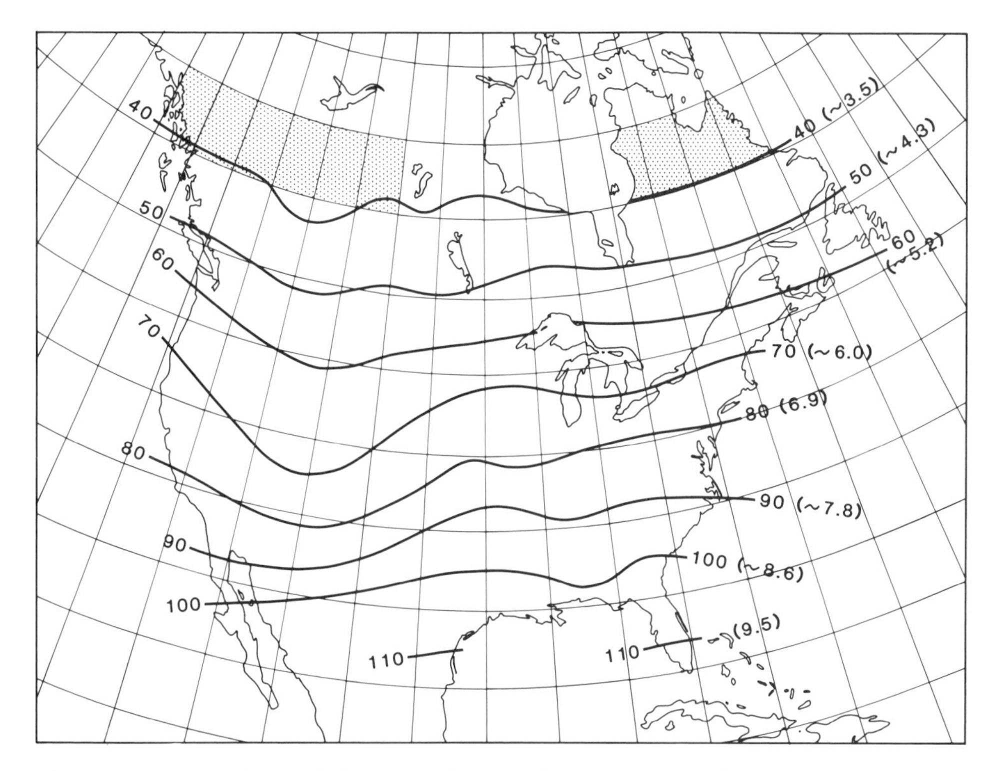

Figure 31. Mean annual net radiation on a horizontal surface (all-wave), that is, the net radiative heating averaged over 5° latitude × 5° longitude grid cells in watts m^{-2} (and megajoules m^{-2} d^{-1}). Stippling indicates uncertain values (Henning, 1978).

The input to the hydrologic cycle is precipitation as rain, snow, hail, and small amounts of dew. The sources and direction of movement of atmospheric moisture were described earlier. Figure 32 shows mean annual precipitation for North America.

Major water bodies and wetlands

The continent of North America is unique in the extent to which it is penetrated by the sea, or overlain by large lakes and areas of wet terrain. These facts have a marked bearing on the continental climate.

The northern seas include an interlocking network of channels, many quite deep, that divide arctic Canada into an archipelago of large and small islands and peninsulas. North of Barrow Strait and Lancaster Sound, some of these are mountainous and extensively glacierized. So is the eastern side of Baffin Island. The western islands are generally lower, and have only a few small glaciers. Foxe Basin and Hudson Bay occupy the floor of a huge depression in the Canadian Shield, communicating with the Atlantic via Hudson Strait (which divides Baffin Island from Labrador-Ungava). All these seas and straits are frozen in winter; in fact, arctic Canada is a vast sheet of snow and ice from January to June, with little to obstruct the free flow of the wind. A long-standing myth that Hudson Bay remained opened in winter was dispelled years ago (Hare, 1951). Hudson Bay and Strait are largely clear of ice in late summer, allowing a brief navigation season.

Foxe Basin and Hudson Bay occupy geologically ancient depressions that were again depressed repeatedly in Pleistocene times by the weight of continental ice sheets. Hudson Bay is still recovering isostatically from the disappearance of the Wisconsinan ice sheet. The resulting rise of the land is on the order of one meter per century. Raised shorelines extend inland well over a hundred kilometers in much of Ontario and westernmost Quebec.

Extensive lakes and wetlands are developed on the Canadian shield, especially in a vast ring surrounding Hudson Bay. Figures 33 and 34 show the extent of these wet surfaces, expressed as a percentage fraction of the total surface (smoothed by averaging over squares 100 × 100 km, i.e., 10^4 km^2). There is a distinct latitudinal gradient in the extent of wetland surfaces (Fig. 35; Hare, 1984). (See Winter and others, this volume, for generalized map of wetlands and Plate 2 for map of precipitation minus lake evaporation.)

The outer rim of the Canadian shield is marked by many big lakes. Along the Mackenzie system, Great Bear, Great Slave, and Athabasca lakes drain to the Arctic. The Manitoba lakes (Winnipeg, Winnipegosis, Manitoba, and others) are part of the Saskatchewan-Nelson basin, draining to Hudson Bay. The Great Lakes sit right on the continental divide between Hudson Bay and

Figure 32. Mean annual precipitation for North America (mm) (from UNESCO, 1978).

Gulf of Mexico drainages, but continue to drain to the Atlantic via the St. Lawrence and its huge Gulf. Nowhere else on the world map is there anything akin to this enormous crescent of water-bodies (see Winter and others, this volume).

Although there is no water-body of comparable size south of the Great Lakes and the Canadian border west to the Pacific, in many ways the Mississippi basin, Appalachian system, and the Atlantic–Gulf of Mexico coastal plains east of about 100°W function as moist surfaces. Through winter, spring, and early summer this vast area maintains high soil moisture; even in winter the temperatures permit some evaporation. In spring and early summer, evapotranspiration is very high. West of the Mississippi itself, freely transpiring crops may use advected heat from drier lands to the southwest, and achieve evapotranspiration rates far

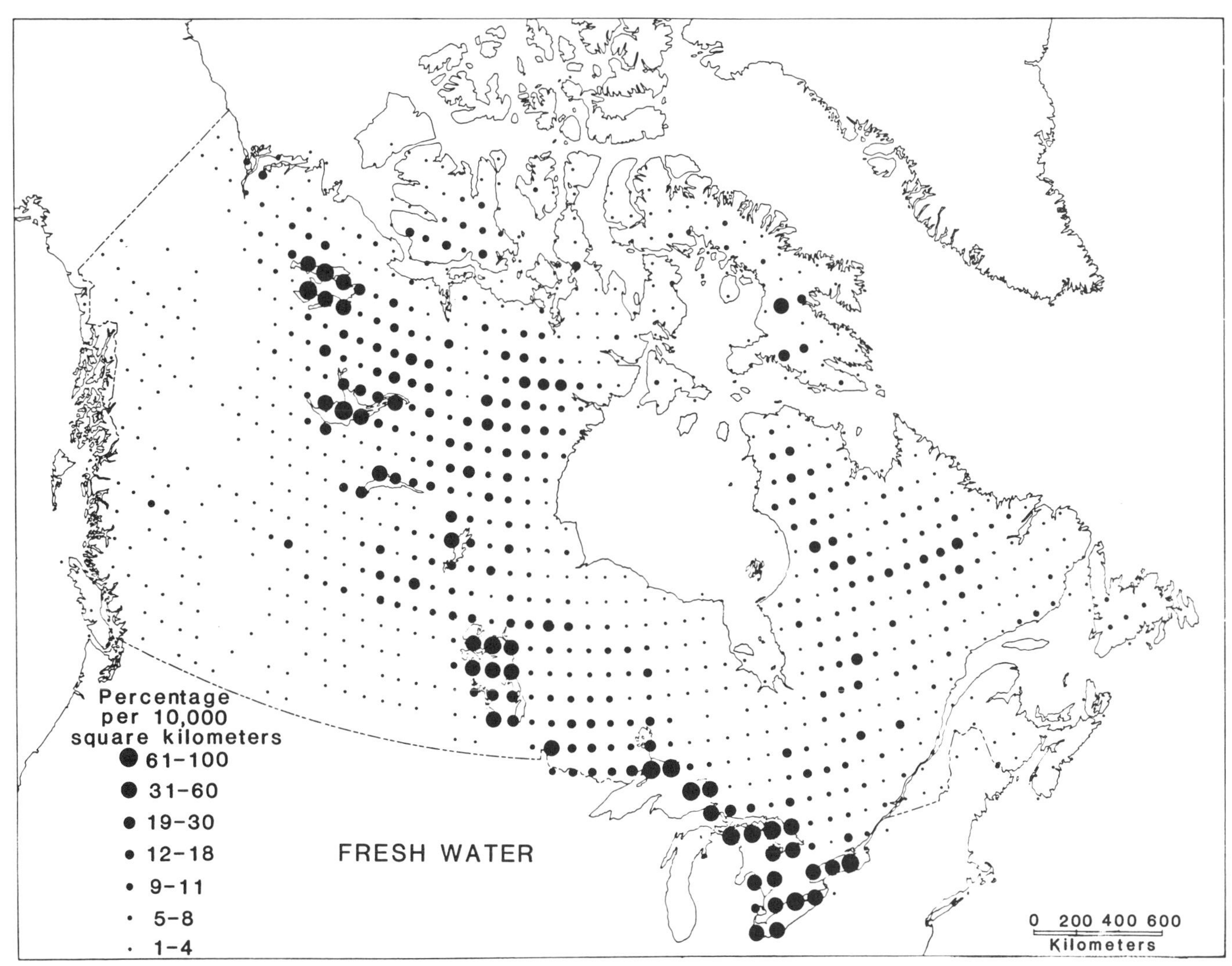

Figure 33. The area of fresh water lakes expressed as a percentage of 10,000 km^2 grid cells along latitude circles for Canada (Hare, 1984).

above the potential rates defined earlier (Rosenberg and Verma, 1978).

Airmasses traveling northward and northeastward across this area tend to remain moist even if substantial losses due to rainfall are encountered. The heat and humidity of southerly airstreams in the Midwest are usually credited to the Gulf of Mexico, with good reason, but part of the total moisture contained in such airstreams (which often yield heavy rainfalls) is derived in early summer from transpiration over the continental land surface.

A different process affects cold airstreams travelling southeastward from the Canadian sub-arctic. In fall and winter these airstreams pass over unfrozen lakes, whose stored heat sets up low-level convection. Widespread stratocumulus cloud and snow flurries occur over a belt extending some hundreds of kilometers south of the southern limit of frozen lakes. The Great Lakes themselves create much deeper convection. On the lee shores of all the lakes, and for up to 80 km inland, heavy snow flurries may fall from shallow but dense cumuliform clouds created by the very large temperature contrast between frigid air and unfrozen waters. Great snow belts occur around the southern and eastern shores of the Lakes. The areas to the east of Lake Erie and Lake Ontario are especially vulnerable to this kind of storm (see Meier, this volume).

The remaining major water body affecting North America's climate, oceans excepted, is the Gulf of Mexico. For much of the year, as noted earlier, but especially in spring and summer (extending at least through September), warm, humid, tropical airstreams penetrate the Mississippi valley, and may extend inland in the warmer months as far as the Canadian border or even beyond. These airstreams, often present aloft even in the cooler seasons, bring large amounts of water vapor into the continent. Many heavy rainstorms fall from such air—at all seasons in the deep south and in spring and summer farther north. Most of the severe weather common to the region involves these Gulf airstreams (the surface layers of which often, in point of fact, come from far beyond the Gulf—the Caribbean, and beyond that, the tropical Atlantic).

Major relief features

The gross relief of North America has a striking influence on precipitation and runoff. The Western Cordillera, extending from Alaska to the Isthmus of Panama, dominates the precipitation map, and is almost everywhere the divide between Pacific and Atlantic or Arctic drainage. The great sweep of open terrain between the Canadian Arctic and the Gulf of Mexico—the Canadian shield, the Great Plains, Midwest, and South—encourages free airstream exchange between north and south. The Appalachian system modifies the overall pattern, but is much less significant as a large-scale climate control (Bryson and Hare, 1974).

The Western Cordillera exerts its influence in several ways. Heavy rains and winter storms fall on the western sides of the northern Cordilleran ranges, whereas to the east and south, rain-shadow effects create the Mexican, U.S., and Canadian dry belt. Especially in the summer, the Cordillera tends to divert low-level onshore airstreams northward toward the Gulf of Alaska, and southward toward Mexico and the eastern Pacific. Higher-level westerlies are able to traverse the mountains, but due to the mountain barrier they tend to move poleward as they approach the coast from the Pacific, and then to move southeastward as far as the longitude of the Great Lakes—a standing ridge and trough that fluctuate in intensity and longitude, but are rarely absent. These perturbations in the westerlies greatly influence the climate of the whole continent (and to some extent that of the Northern Hemisphere).

Over the vast interior plains, conditions favor the creation of strong fronts, i.e., frontogenesis. Arctic and tropical airstreams come into contact (with many less intense conjunctions as well), and numerous synoptic-scale cyclones form in association with the strong poleward temperature gradient. At times, Pacific airstreams may also cross the Cordillera and descend to the High Plains and Prairies. This descending air increases in temperature

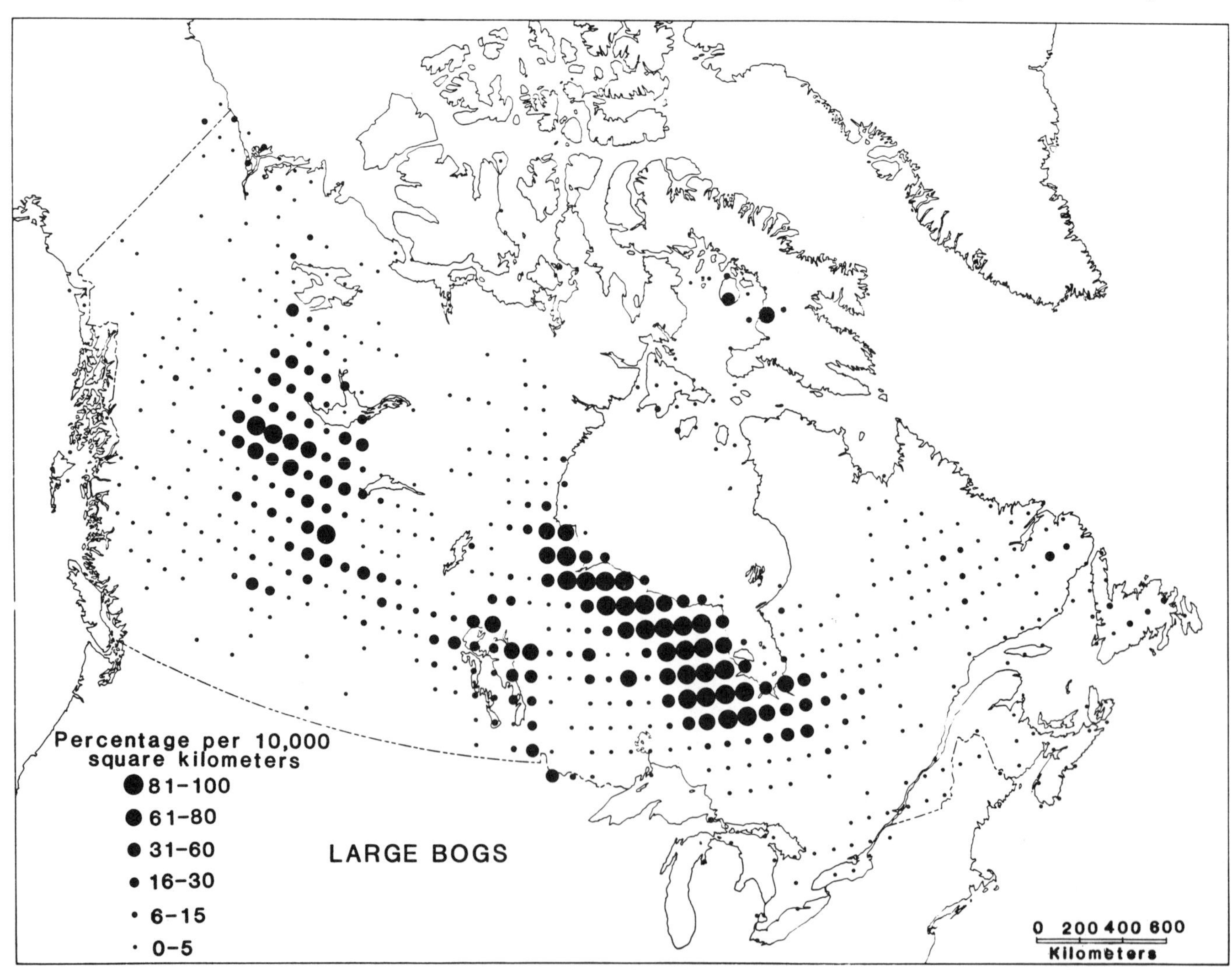

Figure 34. As Figure 33, but for large bogs.

as it moves to lower elevations (i.e., higher pressures), and the warm air often interacts with Arctic airstreams to create cyclones. Hence, the interior of North America acts as one of the northern hemisphere's main theaters of cyclogenesis (cyclone formation and intensification).

In the winter months the contrasts introduced into the North American climate by gross relief features are remarkable. For weeks at a time, frigid continental arctic air over the interior may be separated from mild Pacific air by the mountains themselves. Over the Canadian Arctic, and in some winters as far south as the mid- and eastern United States, a veritable northwesterly winter monsoon of arctic air covers the entire region; in parts of northern and eastern Canada, the persistence of these bitter winds is higher than that of the celebrated winter monsoon of China and Japan.

CLIMATIC VARIABILITY AND CHANGE

Patterns and trends in the systematic record

During the decade of the 1980s, considerable scientific attention focused on the identification of an enhanced greenhouse-induced signal in the systematic record of climate observations. This attention was stimulated by climate-model simulations that were consistently indicating that increasing atmospheric concentrations of such radiatively active trace gases as carbon dioxide would lead to higher global temperatures (Hansen and others, 1981; Manabe and Weatherald, 1975, 1980; Schlesinger, 1986; Wilson and Mitchell, 1987). Moreover, these simulations were suggesting that the build-up of CO_2 in the atmosphere during the 1980s would reach the point where a clear effect on the observed record of surface-air temperature should be imminent.

By 1989, there was a concensus among climatologists that the global mean surface-air temperature was increasing, although erratically, and that it had been doing so since the peak of the Little Ice Age and certainly through the period of systematic observations (i.e., since about 1750). Supporting this conclusion, moreover, were a number of supplementary observations from systematic data sets. These included the shrinkage of glaciers, recession of sea ice, lifting of snow and tree lines, lengthening of ice- and frost-free summers, surface decreases in slopes of temperature profiles in boreholes in dry media, and reduced fractionation in isotopes of hydrogen and oxygen in permanent ice and biospheric precipitates in the upper layers of sedimentary deposits.

However, the amount of warming has been difficult to determine with precision since the preparation of representative time series of temperature is influenced by numerous uncertainties. These include biases introduced by changes in instruments, methods of observation, exposure and height of the instruments, time of observation, relocation of stations, and urbanization (Oort and others, 1987). When these uncertainties are considered in evaluating the observed temperature record, most investigations indicate that there has been a 0.3- to 0.7-degree Celsius increase in the mean global temperature during the past century (Hansen and Lebedeff, 1987, 1988; Jones and others, 1986a, b; Oort and others, 1987; and Wigley and others, 1985). The general pattern of this increase is evident in Figure 36. Notably, however, this temperature increase is not uniformly evident around the planet. Hanson and others (1989) recently analyzed temperature records for nearly 6,000 stations across the conter-

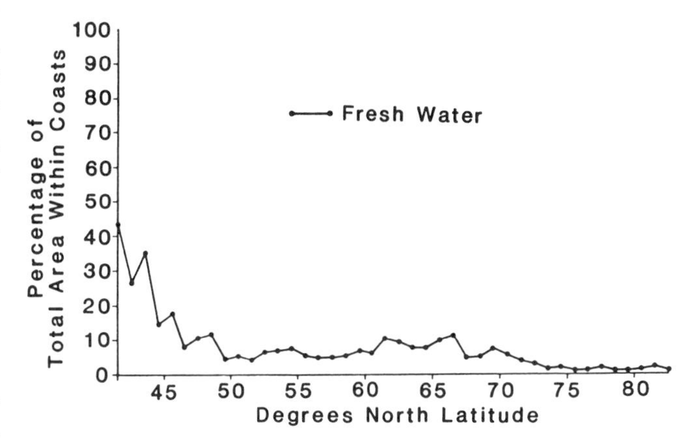

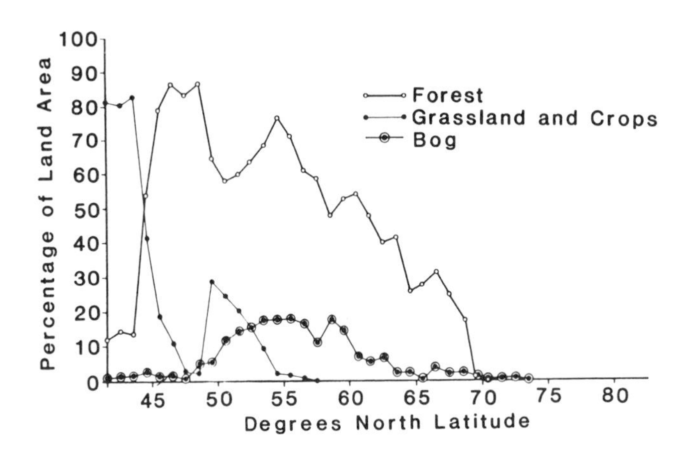

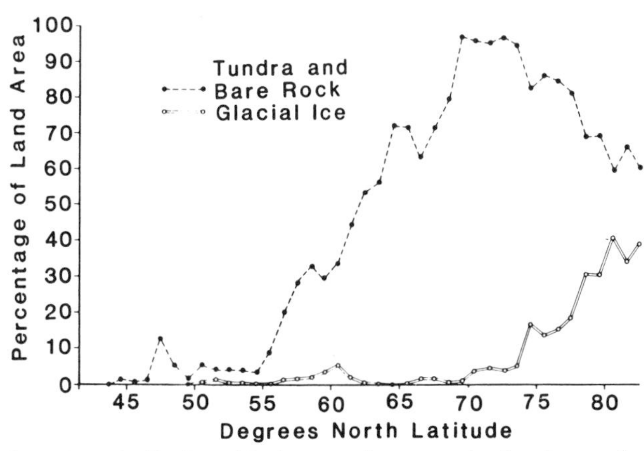

Figure 35. Distribution of fresh water, forest, grassland and crops, bog, tundra, and bare rock, and glacial ice in Canada. Note uniformity with figures 34 and 35 (Hare, 1984).

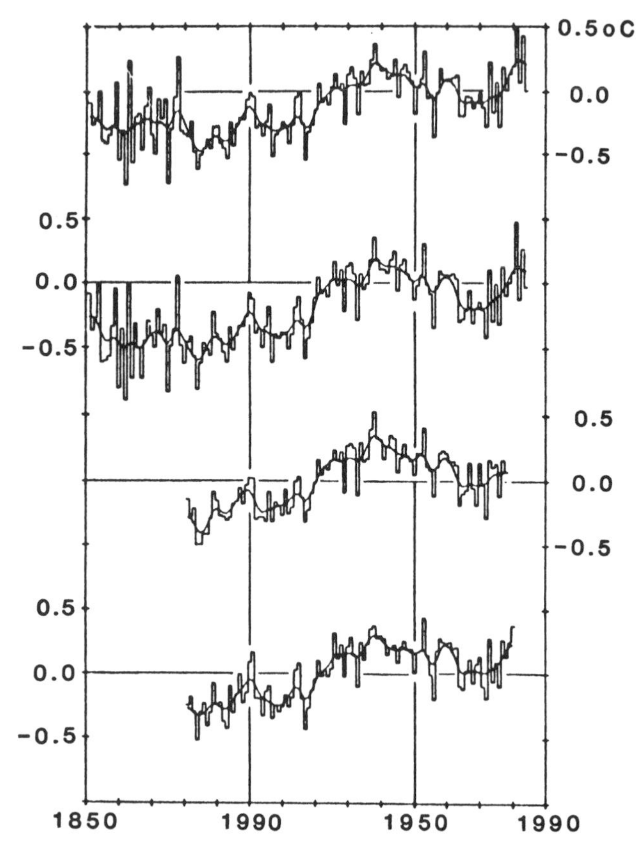

Figure 36. Comparison of Northern Hemisphere land-based air temperature anomalies (from top to bottom): Jones and others, 1986a; Kelly and others, 1985; Vinnikov and others, 1980; and Hansen and others, 1981. The filtered curve is a 13-term Gaussian filter designed to suppress variations on time scales less than 10 years. In order to estimate filtered values at the ends of each curve, six extra years are used at each end with values equal to the mean of the six years at the beginning/end of each curve (Jones and others, 1986a).

minous United States covering the period from 1895 to 1987. After adjusting for observational bias (time of observation and urbanization effects) and spatially averaging individual station data with proper spatial weighting, this group found no statistically significant overall increase in annual temperature. This finding does not negate the existence of a world-wide temperature increase since the United States occupies but a small percentage of the Earth's surface. What it does suggest, however, is that significant regional differences exist in the pattern of decadal to century-scale temperature variations.

Given that a global temperature rise has occurred, and that the amount of this increase is roughly consistent with that expected in response to the increase in greenhouse gases during the past century, most scientists agree that it is not yet possible to detect a climatic change caused by the greenhouse effect. Several reasons account for this lack of certainty regarding causality. Natural variability in temperature is relatively large on an interannual and interdecadal basis and is uncertain on secular time scales. Moreover, El Niño–Southern Oscillation (ENSO) events exert a strong influence on temperature variability. Also, a large warming has been observed in the southern hemisphere, which is difficult to explain since conventional wisdom suggests that, because of the large ocean area with its greater thermal lag, warming should occur more gradually there.

Moreover, to attribute the roughly 0.5°C rise in global mean temperature over the past century entirely to increasing atmospheric CO_2 would require ignoring the observational record, which also suggests that the upward trend began at a minimum in the temperature curve that occurred late in the Little Ice Age (Ellsaesser and others, 1986). It would also require ignoring obvious differences in the spatial and seasonal distributions of observed temperature changes from those computed by general circulation models for recent increases in CO_2. Although the observed warming does display a latitudinal amplification, a characteristic pattern in all general circulation model simulations, no other similarities with model-predicted patterns are apparent in the observed temperature record (Ellsaesser and others, 1986).

Still, at the time of this writing, there appears to be a predisposition toward causally linking the current warming to increased atmospheric CO_2 primarily because of the simultaneity of the two occurrences and also because of the inability to identify any other warming mechanism. Ellsaesser and others (1986) suggest a couple of reasons why this predisposition should be resisted. First, paleoclimatic reconstructions, even if only approximately correct, indicate that the terrestrial climate has undergone significant variations in the past. During both the altithermal and medieval little climatic optimum (ca. A.D. 900 to 1300), it appears that the atmosphere achieved and maintained mean surface-air temperatures above those of the present without increased levels of CO_2. Second, increased CO_2 is not the only process affected by man that is capable of causing the planetary temperature to increase. Modern agriculture and energy consumption practices have altered the water balance by increasing the water available to fields, parks, and landscaped living areas throughout the summer season. In combination with increased power generation for air conditioning, such practices have systematically converted energy from sensible to latent forms, thereby delaying its return to space and increasing the water vapor content of the atmosphere, and hence, the greenhouse effect. The energy balance is also affected by these practices, primarily through a decrease in surface albedo in winter when fields are left bare and snow covers are disturbed.

The conclusion to be drawn from this discussion is that although the global mean surface-air temperature is increasing, it is not yet possible to identify the reason or reasons for it. Since higher temperatures appear to have existed in earlier (pre-industrial) times through natural processes, the current warming may also be natural. It may, however, be anthropogenic, but even if it is, it need not be due entirely to increased CO_2. If CO_2 is the culprit, though, then considerable improvements in the current generation of climate models are necessary in order to reconcile their output with the instrumental record of surface-air temperature change.

Figure 37. Annual precipitation index (mean of percentiles of the gamma distribution for Northern Hemisphere continental (A) and constituent regions (B). Vertical dashed line shows year in which 50 percent of the grid points became available for analysis. Curved line shows the smooth trend line fitted through indvidual values (from Lins and others, 1988, after Bradley and others, 1987).

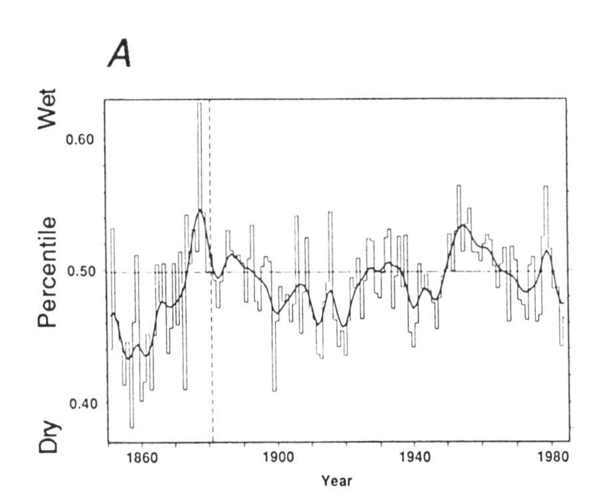

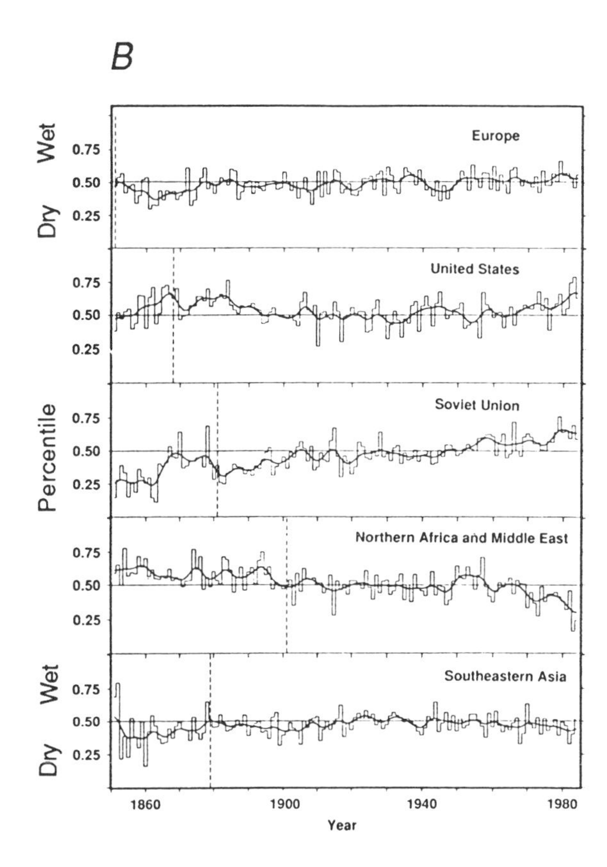

Precipitation is also a highly variable quantity in space and time. Analyses of precipitation data for the Northern Hemisphere from the mid-19th century to the present (Bradley and others, 1987), and of much longer proxy series reconstructed from tree rings, indicate that significant regional variations have occurred over decadal and longer time scales. Hemispherically, decadal and longer fluctuations in precipitation are also evident, although there does not appear to be any systematic trend over the past century (Fig. 37A).

Regional time series variations are evident in Figure 37B. Europe, for example, has undergone a small but steady increase in precipitation since the 1860s. Precipitation over the Soviet Union has increased significantly since the 1880s, with most of the change occurring before 1900 and after 1940. In northern Africa and the Middle East, precipitation exhibited little trend between 1900 and the late 1940s. Since then, however, precipitation has declined dramatically. In Southeast Asia, a relatively wet period in the 1920s and early 1930s separated two protracted periods of relative dryness. The first was centered around 1900 and the second extended from the mid-1960s to the present. In the United States, precipitation declined between the 1880s and the 1930s. Since the 1930s, however, there has been a general increase with the most marked rise occurring since the mid-1950s. Overall, no long-term trend is apparent in the U.S. precipitation record. This pattern is supported by the findings of Hanson and others (1989) who found no statistically significant trend in annual average U.S. precipitation, although a significant upward trend in autumn precipitation was indicated for the period 1895 to 1987.

During the past 30 to 40 years there have been some notable differences in the temporal distribution of precipitation on a latitudinal basis. In the middle latitudes (35° to 70°N), for example, a significant precipitation increase has occurred since 1940. In the subtropical latitudes (5° to 35°N), however, a significant decrease is evident since the mid-1950s. In the north equatorial region (0° to 5°N), little or no trend is evident in the record during the past century.

Perspectives on Climatic Change

Climatic change means a change in the mean value of a climatic parameter such as temperature, or a group of variables, over a significant period of time. In recognition of the mounting evidence of the potential impact of human activities on the global climate, climatic change is touched on briefly here because it is, in turn, closely related to prospective alterations in the hydrologic cycle of North America.

Throughout most of geologic time, the climate has been warmer than it is at present. During the last 130 m.y., temperature has been declining, as suggested by reconstructions of deep-sea temperatures based on foraminifera (Fig. 38). At the same time, major oscillations of 3 to 5°C are also reflected in the record. Climatic changes throughout geologic time are related to a complex variety of factors whose relative importance has varied

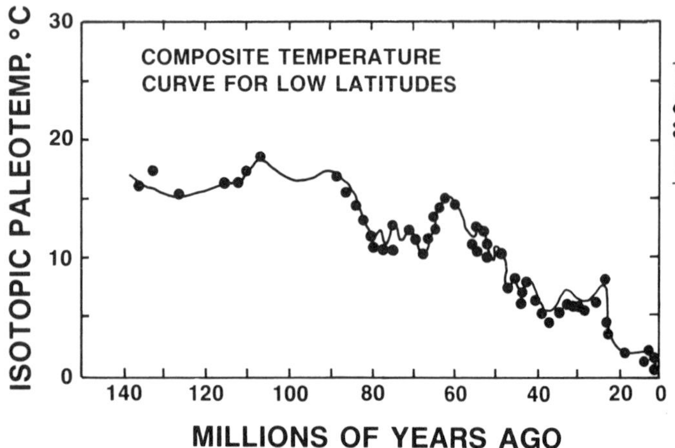

Figure 38. Deep sea water temperature history for the past 130 m.y. (Douglas and Woodruff, 1981).

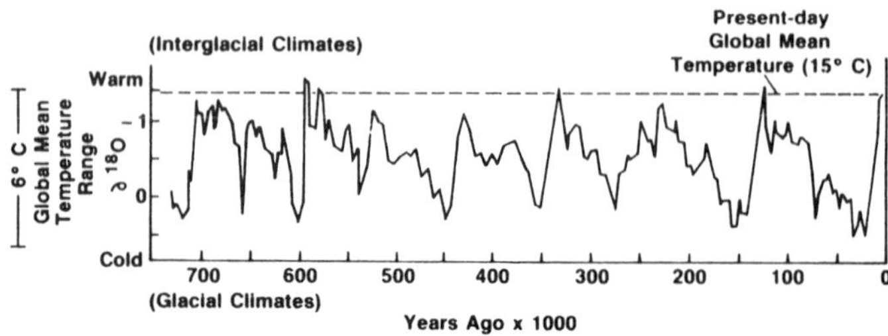

Figure 39. Composite oxygen isotope curve for the past 750,000 years of Earth history (Emiliani, 1978).

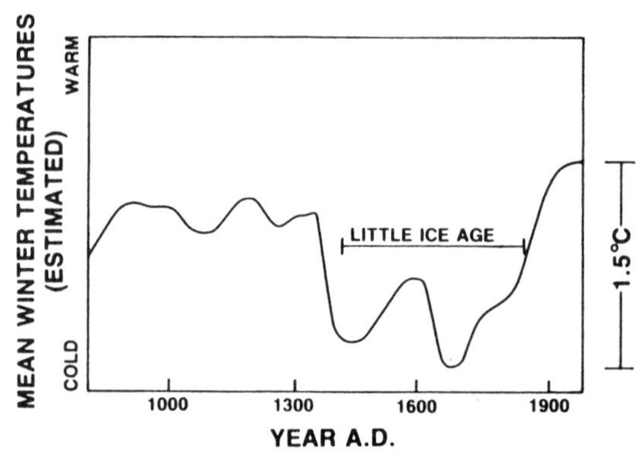

Figure 40. Winter temperature estimates for Eastern Europe during the past 1,000 years, based largely on historic non-instrument records (Lamb, 1977).

during different geologic time intervals. Thus, perturbations in the astronomical positions and dynamics of the sun and the planets, breakup and clustering of continental plates and land masses, concentrations of volcanic dust in the atmosphere, the existence of large areas of snow and glaciers, and changes in the concentration and composition of gases in the atmosphere can all be shown to have affected the climate of the globe.

The magnitude of temperature variations, including a portion of the Pleistocene glacial period, shows a range of temperature variations on the order of 6°C in the last 750,000 years, with the present global mean temperature comparable to the warmest interglacial periods (Fig. 39). Winter temperatures in Europe over the last 1,000 years (Fig. 40) again show fluctuations but of a smaller range (~1.5°C), including the much noted Little Ice Age during the Middle Ages.

Present concern over global warming is related to three sets of observations coupled with models of the energy balance of the globe incorporating the likely effect of increasing concentration of CO_2 and trace gases such as methane and chlorfluorocarbons in the atmosphere. Both CO_2 and methane concentrations have been increasing over the past 100 years, with CO_2 increasing from 280 ppm to 360 ppm and methane roughly doubling (Fig. 41). In addition, mean global temperatures have increased over the same period roughly 0.7°C. (Fig. 42). The principal contributors to these increased concentrations appear to be the burning of fossil fuels which has roughly doubled over the past century. Given the variety of factors influencing the global climate and the marked variability in temperature from year to year, projections of a continuing increase in global warming due to the "greenhouse effect" of anthropogenic sources of greenhouse gases remains uncertain. The statistical uncertainty precludes recognition of the signal of climate change within the noisy variability of the data. In addition, model projections of atmospheric warming lead to projected increases in the magnitude of cloud cover at unknown elevations. Because cloud cover at differing elevations mitigates the impact of warming (by changing the balance of short and longwave radiation reaching the earth), cloud cover may mitigate the warming effect of increasing concentration of gases emitted to the atmosphere. On balance, many investigators presently believe that global warming is likely to continue, given the probable increase in the burning of fossil fuels; they differ only in their projections of the probable magnitude of the temperature increase over the next 5 or 6 decades. A discussion of the postulated potential impact on the hydrology of selected regions of global warming due to the "greenhouse effect" follows.

MODELING

General circulation models

In recent years, the increasing concern over the social and economic impacts of both natural and anthropogenic climatic perturbations has focused considerable attention on models of the climate system. The most sophisticated climate models are those known as general circulation models, or GCMs. General circulation models depict the circulation of mass and energy in the atmosphere in three spatial dimensions, on a global scale, and at varying levels of detail that include components of oceanic and land-surface processes (Washington and Parkinson, 1986). In

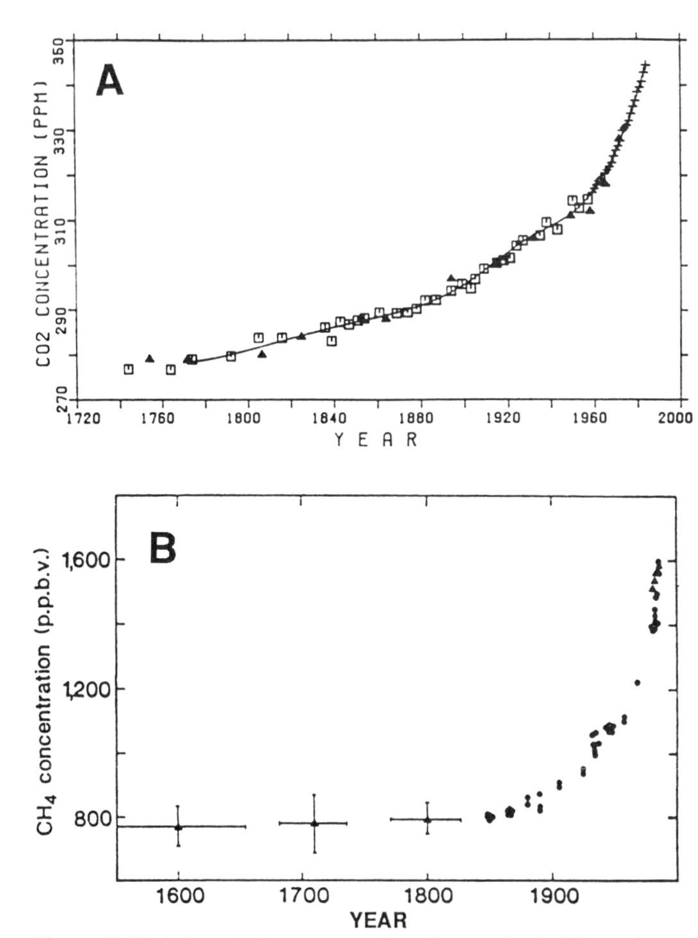

Figure 41. Variations in the concentration of atmospheric CO_2 as determined from ice cores in Greenland and Antarctica (A; Siegenthaler and Oeschger, 1987), and atmospheric methane (B; Pearman and Fraser, 1988).

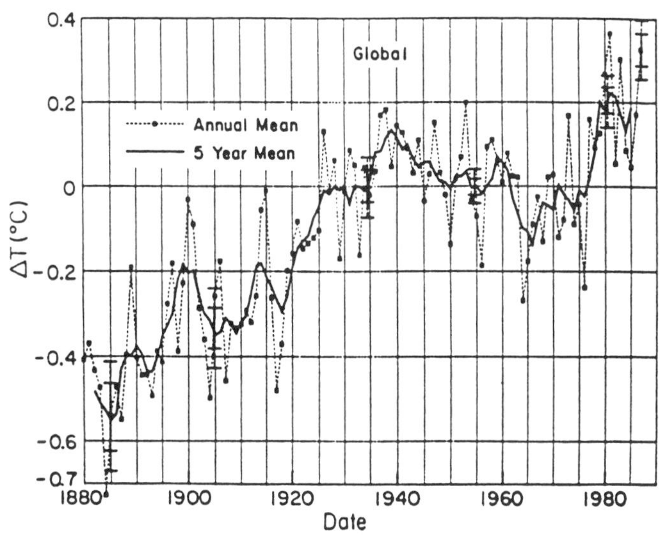

Figure 42. Normalized mean global surface air temperatures, 1880 to 1987 (Hansen and Lebedeff, 1988).

general, attempts are made in the model to represent all physical processes believed to be important. The variables used to drive GCMs include components of the wind field or vorticity, temperature, and water-vapor content. Typically, the data are aggregated into cells of several degrees latitude and longitude around the globe, and at 6 to 10 levels vertically through the lower atmosphere. The stored quantities are then stepped forward in time by applying the equations of motion and the first law of thermodynamics, so that mass, energy, and momentum are conserved. The equations also include terms describing surface friction, radiative and latent heating, and other physical processes.

One of the most crucial aspects of GCMs involves the mathematical description of the interactions between the Earth's surface and its atmosphere. Typically, the surface fluxes of heat, momentum, and moisture are described using bulk aerodynamic formulas. Land-surface temperature and soil moisture are commonly calculated by equations incorporating daily and seasonal heat and moisture storage and their vertical fluxes. Significantly, although current GCMs are structurally complex and computationally demanding even for the latest generation of supercomputers, in many respects their representations of the physical processes and conditions at and immediately below the land surface are exceedingly simple and inaccurate. For this reason, the general view among climate modelers is that GCMs are "simply not yet ready to be used for quantitative prediction at anything approaching even a multi-state region, let alone a single surrogate gridpoint representing a particular state, county or city. Over such small scales, a wide range of responses is currently predicted" (Grotch, 1988, p. 253). Thus, at regional scales, where the important water resources planning and management decisions need to be made, the data from GCMs are currently unsuitable for use in management and policy development.

A critical aspect of general circulation modeling involves parameterizing the interactions between the surface and the atmosphere. Parameterization refers to the expression of the statistical effects of various small-scale (subgrid-scale) transport processes in terms of the large-scale (grid-scale) variables explicitly resolved by the model (Global Atmospheric Research Program, 1975). Because of the wide range of scales of interacting atmospheric processes in relation to the limited spatial resolution of computational grids and observation systems, it is technically and economically impractical to observe or explicitly calculate the effects of small-scale processes in detail. However, these subgrid processes are significant, and it is important to relate their statistical effects to measurable or computable conditions at larger scales. Because the arbitrary specification of small-scale effects does not accommodate full interaction with the larger scale resolvable state of the atmosphere, parameterization is required. To illustrate the problem associated with specifying subgrid processes at the grid scale, consider how a typical GCM treats soil moisture (Fig. 43). At the Earth's surface, the moisture content of the top 15 cm of soil is a highly variable quantity, responding to differing

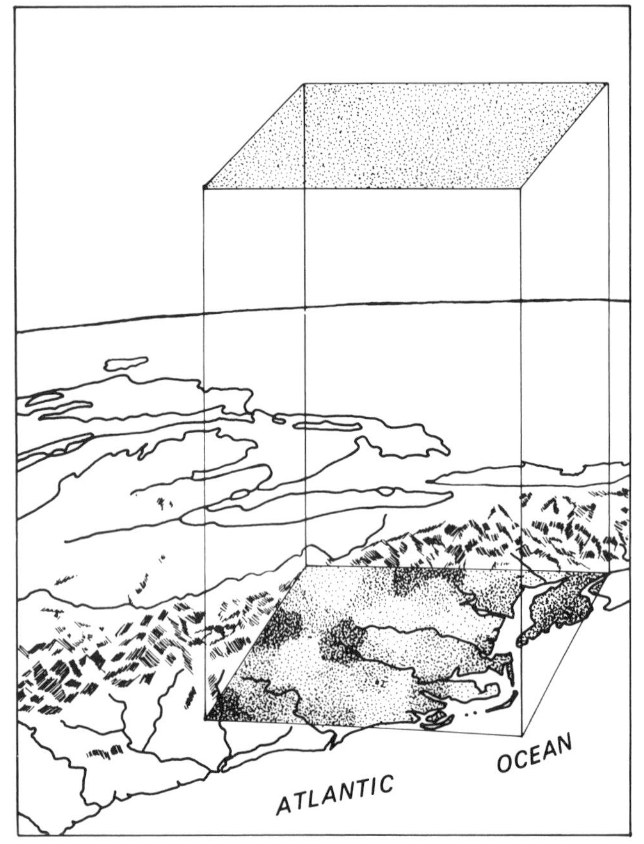

Figure 43. Characterization of how the spatial variability of a surface condition such as soil moisture, which varies over a scale of meters, is generalized to a single value through parameterization (Moss and Lins, 1989).

TABLE 3. CHANGES IN THE GLOBAL MEAN SURFACE AIR TEMPERATURE (T_S) AND PRECIPITATION RATE (P) SIMULATED BY VARIOUS ATMOSPHERIC GENERAL CIRCULATION/MIXED-LAYER OCEAN MODELS FOR A CO_2 DOUBLING*

Model Study	Change in Temperature (°C)	 Precipitation (%)
Geophysical Fluid Dynamics Laboratory Wetherald and Manabe (1988)	4.0	8.7
Goddard Institute for Space Studies Hansen and others (1984)	4.2	11.0
National Center for Atmospheric Research Washington and Meehl (1984)	3.5	7.1
Oregon State University Schlesinger and Zhao (1988)	2.8	7.8
United Kingdom Meterological Office Wilson and Mitchell (1987)	5.2	15.0

*From Schlesinger, 1988.

vegetation, soils, and geomorphological characteristics. Over an area the size of a 5° latitude by 5° longitude grid square, a map of soil moisture content would appear as a complex mosaic. Yet, the input to the GCM is a single value for the entire grid square. Clearly, for most of the globe's land surface, this is a poor representation of soil moisture conditions.

Linking hydrology and climate through GCMs is very difficult. Accounting for the myriad of interacting processes, including precipitation, interception, infiltration, runoff, and evapotranspiration, with any degree of precision or realism is very complicated, and procedures for handling these intricacies are largely underdeveloped or totally lacking.

Significantly, hydrologists, especially those involved in surficial process and ground water modeling, have been addressing many of these same problems. Whereas atmospheric modelers have focused their attention primarily on processes over characteristic scales ranging from 10,000 to 100,000 km^2, hydrologic modelers have dealt with smaller scale processes—those occurring over field (<1 km^2) or catchment (100 to 1,000 km^2) areas. Importantly, hydrologists have had to confront many of the same basic issues to parameterize field-scale processes in catchment-scale models. Although adequate solutions have not always been found, those processes significantly affecting catchment response have been identified. Thus, the hydrological community has acquired important insights necessary in guiding parameterizations from catchment scales up to GCM grid scales and is now beginning to apply them.

An intermediate step of great significance has been undertaken by atmospheric modelers to enhance the realism of subgrid-scale processes by application of a finer mesh grid. This work, generally referred to as mesoscale modeling, is a spatially and temporally limited form of general circulation modeling. The basic physical and mathematical foundations are the same, although more detail is incorporated at the mesoscale. Pielke (1984) defines mesoscale as having a characteristic horizontal spatial scale of 10 km^2 to 50,000 km^2 with a time scale of about 1 to 12 hours. Clearly, this scale of analysis is much more consistent with that used in hydrology and affords a sound basis for substantial interaction between the atmospheric and hydrologic modeling communities. The mesoscale class of models, with the inherent ability to nest within the broader GCMs, provides a likely pathway for incorporating more realistic hydrologic conditions into climate simulations at the global scale. Perhaps more important for hydrologists, however, is the potential for improving the output of catchment models by using the output of mesoscale models as input.

Hydrologic models

Given the cautions indicated above regarding both the regional resolution of general circulation models and the problems associated with modeling local and regional water balances, projecting the hydrologic consequences of global warming is highly speculative. Nevertheless, because of the potential importance of these effects on Earth surface processes, on ecology, and on water management, some of these speculative projections warrant inclusion here.

For simplicity, many projections of future precipitation regimes are based on projections of a doubling of atmospheric carbon dioxide. Results of five such projections based on different models are shown in Table 3 from which it can be seen that projected precipitation rates are roughly proportional to the magnitude of the projected temperature increase.

While potential changes in temperature and water balance have been projected for much of the United States and parts of Canada, one example is given here to illustrate the possible relation between runoff and increases in precipitation. For the southeast quadrant of the United States, Schaake and Liu (1989), using a water balance model, suggested changes in runoff ranging from 22 to 44 percent, based on a 10 percent change in annual precipitation. Their results are given in Table 4. Such percentage changes in runoff are quite large, but it is important to note that the highest of these is, as expected, associated with small magnitudes of runoff. Given the great uncertainties inherent in all of the model projections, it is clear that much work is needed to quantify the potential impacts of climatic change on the water balance of different regions of the North American continent.

TABLE 4. SENSITIVITY OF RUNOFF TO CLIMATIC CHANGE IN THE SOUTHEASTERN UNITED STATES*

Mean Annual Runoff (in)	Percent change in runoff with 10% increase in precipitation	in potential evapotranspiration
2	44	-20
5	36	-16
10	28	-12
15	22	-10

*From Schaake and Liu, 1989.

REFERENCES CITED

Barnston, A. C., and Livezey, R. E., 1987, Classification, seasonality, and persistence of low-frequency atmospheric circulation patterns: Monthly Weather Review, v. 115, p. 1083–1126.

Bhalme, H. N., and Mooley, D. A., 1979, On the performance of modified Palmer index: Archiv fur Meteorologie, Geophysik und Bioklimatologie, series B, v. 27, p. 281–295.

Blackmon, M. L., Lee, Y. H., and Wallace, J. M., 1984a, Horizontal structure of 500 mb height fluctuations with long, intermediate, and short time scales: Journal of Atmospheric Sciences, v. 41, p. 961–979.

Blackmon, M. L., Lee, Y. H., Wallace, J. M., and Hsu, H. H., 1984b, Time variation of 500 mb height fluctuations with long, intermediate, and short time scales as deduced from lag-correlation statistics: Journal of the Atmospheric Sciences, v. 41, p. 981–991.

Bradley, R. S., and 5 others, 1987, Precipitation fluctuations over northern hemisphere land areas since the mid-19th century: Science, v. 237, p. 171–175.

Bryson, R. A., and Hare, F. K., 1974, Climates of North America: Amsterdam, Elsevier, 420 p.

Busby, M. W., 1963, Yearly variations in runoff for the conterminous United States, 1931–60: U.S. Geological Survey Water Supply Paper 1669-S, 49 p.

CNC/IHD, 1978, Hydrological atlas of Canada: Ottawa, Ontario, Canadian National Committee for the International Hydrologic Decade, Supply and Services Canada, 76 p.

Cayan, D. R., and Peterson, D. H., 1989, The influence of North Pacific atmospheric circulation on streamflow in the west, *in* Peterson, D., ed., Aspects of climate variability in the Pacific and western Americas: American Geophysical Union Geophysical Monograph 55, p. 375–397.

Changnon, S. A., 1984, Climate fluctuations in Illinois; 1901–1980: Champaign, Illinois State Water Survey, 73 p.

Currie, R. G., 1981, Evidence for 18.6 year signal in temperature and drought conditions in North America since A.D. 1800: Journal of Geophysical Research, v. 86, p. 11055–11064.

——, 1984, Periodic (18.6-year) and cyclic (11-year) induced drought and flood in western North America: Journal of Geophysical Research, v. 89, p. 7215–7230.

Diaz, H. F., 1983, Some aspects of major dry and wet periods in the contiguous United States, 1895–1981: Journal of Climate and Applied Meteorology, v. 22, p. 3–16.

Dolan, R., Lins, H., and Hayden, B., 1988, Mid-Atlantic coastal storms: Journal of Coastal Research, v. 4, p. 417–433.

Douglas, R. G., and Woodruff, F., 1981, Deep sea benthic foraminifer, *in* Emiliana, C., ed., The sea, v. 7: New York, John Wiley and Sons, p. 1233–1327.

Ellsaesser, H. W., MacCracken, M. D., Walton, J. J., and Grotch, S. L., 1986, Global climatic trends as revealed by the recorded data: Reviews of Geophysics, v. 24, p. 745–792.

Emiliani, C., 1978, The cause of the ice ages: Earth and Planetary Science Letters, v. 37, p. 349–354.

Gedzelman, S. D., 1985, Atmospheric circulation systems, *in* Houghton, D. D., ed., Handbook of applied meteorology: New York, John Wiley and Sons, p. 3–61.

Global Atmospheric Research Program, 1975, The physical basis of climate and climate modeling: Global Atmospheric Research Program Publication Series 16, 265 p.

Grotch, S. L., 1988, Regional intercomparisons of general circulation model predictions and historical climate data: U.S. Department of Energy TR041, 291 p.

Hansen, J. E., and Lebedeff, S., 1987, Global trends of measured surface air temperature: Journal of Geophysical Research, v. 92, p. 13345–13372.

——, 1988, Global surface air temperatures; Update through 1987: Geophysical Research Letters, v. 15, p. 323–326.

Hansen, J. E., and 6 others, 1981, Climate impact of increasing atmospheric carbon dioxide: Science, v. 213, p. 957–966.

Hanson, K., Maul, G. A., and Karl, T. R., 1989, Are atmospheric "greenhouse" effects apparent in the climatic record of the contiguous U.S. (1895–1987)?: Geophysical Research Letters, v. 16, p. 49–52.

Hare, F. K., 1951, Some climatological problems of the Arctic and sub-Arctic, *in* Malone, T. F., ed., Compendium of meteorology: Boston, Massachusetts, American Meteorological Society, p. 952–964.

——, 1980, Long-term annual surface heat and water balances over Canada and the United States south of 60°N; Reconciliation of precipitation, runoff, and temperature fields: Atmosphere-Ocean, v. 18, p. 127–153.

——, 1984, The impact of human activities on water in Canada; Inquiry on federal water policy: Ottawa, Ontario, Environment Canada Inland Waters Directorate Research Paper 2, 102 p.

Henning, D., 1978, Charts of net radiation prepared for U.S. Environment Program, U.N. Conference on Desertification; Chart for North America, *in* Proceedings of the Reading Symposium World Water Balance: Internation Association of Scientific Hydrology, p. 361–365.

Hirschboeck, K. K., 1988, Flood hydroclimatology, *in* Baker, V. R., Kochel, R. C., and Patton, P. C., eds., Flood geomorphology: New York, John Wiley and Sons, 503 p.

——, 1990, The role of climate in the generation of floods, *in* National water summary 1988–89: U.S. Geological Survey Water Supply Paper (in press).

Horel, J. D., 1981, A rotated principal component analysis of the interannual variability of the northern hemisphere 500 mb height field: Monthly Weather Review, v. 109, p. 2080–2092.

Hosler, D. L., and Gamage, L. A., 1956, Cyclone frequences in the United States for the period 1905–1954: Monthly weather review, v. 84, p. 388–390.

Jones, P. D., and 5 others, 1986a, Northern Hemisphere surface air temperature variations; 1851–1984: Journal of Climate and Applied Meteorology, v. 25, p. 161–179.

Jones, P. D., Widley, T.M.L., and Wright, P. B., 1986b, Global temperature variations between 1861 and 1984: Nature, v. 322, p. 430–434.

Karl, T. R., and Koscielny, A. J., 1982, Drought in the United States; 1895–1981: Journal of Climatology, v. 2, p. 313–329.

Kelly, P. M., and 5 others, 1985, The extended northern hemisphere air temperature record; 1851–1984: U.S. Department of Energy Carbon Dioxide Research Division, DOE Technical Report TR022, 251 p.

Knapp, H. V., 1985, Evaporation and transpiration, *in* Houghton, D. C., ed., Handbook of applied meteorology: New York, John Wiley and Sons, p. 537–554.

Kohler, M. A., Nordenson, T. J., and Fox, W. E., 1955, Evaporation from pans and lakes: U.S. Weather Bureau Research Paper 46, 109 p.

Lamb, H. H., 1977, Climate; Present, past, and future: London, Methuen and Co., Ltd., 835 p.

Lettau, H., 1969, Evapotranspiration climatonomy; 1, A new approach to numerical prediction of monthly evapotranspiration, runoff, and soil moisture storage: Monthly Weather Review, v. 97, p. 691–699.

Lins, H. F., 1985a, Interannual streamflow variability in the United States based on principal components: Water Resources Research, v. 21, p. 691–701.

——, 1985b, Streamflow variability in the United States, 1931–1978: Journal of Climate and Applied Meteorology, v. 24, p. 463–471.

Lins, H. F., Sundquist, E. T., and Ager, T. A., 1988, Information on selected climate and climage-change issues: U.S. Geological Survey Open-File Report 88-718, 26 p.

Lull, H. W., 1964, Ecological and silvicultural aspects, *in* Chow, V. T., ed., Handbook of applied hydrology: New York, McGraw-Hill, p. 6–19.

Manabe, S., and Weatherald, R. T., 1975, The effects of doubling the CO_2 concentration on a general circulation model: Journal of the Atmospheric Sciences, v. 32, p. 3–15.

——, 1980, On the distribution of climate change resulting from an increase in the CO_2 content of the atmosphere: Journal of the Atmospheric Sciences, v. 37, p. 99–118.

McKay, D. C., and Morris, R. J., 1985, Solar radiation data analyses for Canada, 1967–1976: Environment Canada, Ottawa, v. 5, figs. 4, 5, 11.

Meko, D. M., and Stockton, C. W., 1984, Secular variations in streamflow in the western United States: Journal of Climate and Applied Meteorology, v. 23, p. 889–897.

Mitchell, J. M., Stockton, C. W., and Meko, D. M., 1978, Evidence of a 22-year rhythm of drought in the western United States related to the Hale solar cycle since the 17th century, *in* McCormac, B. M., and Seliga, T. A., eds., Solar-terrestrial influences on weather and climate: Dordrecht, Reidel, p. 125–143.

Moss, M. E., and Lins, H. F., 1989, Water resources in the twenty-first century; A study of the implications of climate uncertainty: U.S. Geological Survey Circular 1030, 25 p.

Nace, R. L., and Pluhowski, E. J., 1965, Drought of the 1950's with special reference to the midcontinent: U.S. Geological Survey Water Supply Paper 1804, 88 p.

Namias, J., 1981, Teleconnections of 700 mb height anomalies for the northern hemisphere; CALCOFI atlas 29: La Jolla, California, Scripps Institution of Oceanography, 265 p.

Neumann, C. J., Cry, G. W., Caso, E. L., and Jarvinen, B. R., 1978, Tropical cyclones of the North Atlantic Ocean, 1871–1977; Asheville, North Carolina, National Climate Center, 170 p.

Oladipo, E. O., 1982, On the spatial and temporal characteristics of drought in the interior plains of North America; A statistical analysis [Ph.D. thesis]: Toronto, Ontario, University of Toronto, 271 p.

Oort, A. H., Pan, Y. H., Reynolds, R. W., and Ropelewski, C. F., 1987, Historical trends in the surface temperature over the oceans based on the COADS: Climate Dynamics, v. 2, p. 29–38.

Orlansky, I., 1975, A rational subdivision of scales for atmospheric processes: Bulletin of the American Meteorological Society, v. 56, p. 507–510.

Paegle, J. N., and Kierulff, L. P., 1974, Synoptic climatology of 500-mb winter flow types: Journal of Applied Meteorology, v. 13, p. 205–212.

Palmer, W. C., 1965, Meteorological drought: U.S. Weather Bureau Research Paper 45, 58 p.

Pearman, G. I., and Fraser, P. J., 1988, Sources of increased methane: Nature, v. 332, p. 489–490.

Peterson, D. H., Cayan, D. R., Dileo-Stevens, J., and Ross, T. G., 1987, Some effects of climate variability on hydrology in western North America, *in* Solomon, S. I., Beran, M., and Hogg, W., eds., The influence of climate change and climate variability on the hydrologic regime and water resources: International Association of Hydrological Sciences Publication 168, p. 45–62.

Pielke, R. A., 1984, Mesocale meteorological modeling: New York, Academic Press, 612 p.

Priestley, C.H.B., and Taylor, R. A., 1972, On the assessment of surface heat flux and evaporation using large-scale parameters: Monthly Weather Review, v. 100, p. 81–92.

Reitan, C. H., 1979, Trends in frequencies of cyclone activity over North America: Monthly Weather Review, v. 107, p. 1684–1688.

Rosenberg, N. J., and Verma, S. B., 1978, Extreme evapotranspiration by irrigated alfalfa; A consequence of the 1976 midwestern drought: Journal of Applied Meteorology, v. 17, p. 934–941.

Schaake, J. C., and Liu, C-Z, 1989, Development and application of simple water balance models to understand the relation between climate and water resources, *in* Kavvas, M. L., ed., New directions for sueface water modeling, Proceedings of the Baltimore Symposium, May, 1989: Wallingford, Connecticut, IAHS Press, International Association of Hydrologic Sciences Publication 181, p. 343–352.

Schlesinger, M. E., 1986, Equilibrium and transient climatic warming induced by increased atmospheric CO_2: Climate Dynamics, v. 1, p. 35–51.

——, 1988, Model projections of the climatic changes induced by increased atmospheric CO_2, *in* Berger, A., and others, eds., Climate and Geosciences, NATO Advanced Research Workshop Proceedings: Louvain-la-Neuve, Belgium, Kluwer Academic Publications.

Schneider, S. H., and Mesirow, L. E., 1976, The genesis stratigy; Climate and global survival: New York, Plenum Press, 419 p.

Siegenthaler, U., and Oeschger, H., 1987, Biospheric CO_2 emissions during the past 200 years reconstructed by deconvolution of ice core data: Tellus,

v. 39B, no. 1-2, p. 140–154.

Slatyer, R. O., and McIlroy, I. C., 1961, Practical micrometeorology: Paris, UNESCO, p. 3-50–3-81.

Solar Energy Research Institute, 1981, Solar radiation energy resource atlas of the United States: Washington, D.C., U.S. Government Printing Office, Figs. 7, 8, 9.

Stockton, C. W., and Meko, D. M., 1983, Drought recurrence in the Great Plains as reconstructed from long-term tree-ring records: Journal of Climate and Applied Meteorology, v. 22, p. 17–29.

Sugg, A. L., 1968, Beneficial aspects of the tropical cyclone: Journal of Applied Meteorology, v. 7, p. 39–45.

UNESCO, 1978, World water balance and water resources of the Earth, including atlas of world water balance: Paris, UNESCO Press, 663 p.

U.S. Geological Survey, 1954, Water-loss investigations; Lake Hefner studies: U.S. Geological Survey Professional Paper 269, 158 p.

Vinnikov, K., Ya., and 5 others, 1980, Current climatic changes in the northern hemisphere: Meteorol: Meteorol, Gidrol, v. 6, p. 5–17 (in Russian).

Walker, G., 1923, Correlations in seasonal variations of weather VIII and IX: Memoranda of the India Meteorological Department, v. 24, p. 75–131.

Wallace, J. M., and Gutzler, D. S., 1981, Teleconnections in the geopotential height field during the northern hemisphere winter: Monthly Weather Review, v. 109, p. 784–812.

Walsh, J. E., and Mostek, A., 1980, A quantitative analysis of meteorological anomaly patterns over the United States, 1900–1977: Monthly Weather Review, v. 108, p. 615–630.

Washington, W. M., and Parkinson, C. L., 1986, An introduction to three-dimensional climate modeling: Mill Valley, California, University Science Books, 422 p.

Whittaker, L. M., and Horn, L. H., 1981, Geographical and seasonal distribution of North American cyclogenesis: Monthly Weather Review, v. 109, p. 2312–2322.

——, 1984, Northern Hemisphere extratropical cyclone activity for four mid-season months: Journal of Climatology, v. 4, p. 297–310.

Wigley, T.M.L., Angell, J., and Jones, P. D., 1985, Analysis of the temperature record; Detecting the climatic effects of increasing carbon dioxide: U.S. Department of Energy Report ER-0235, 198 p.

Wilson, C. A., and Mitchell, C.F.B., 1987, Simulated climate and CO_2-induced climate change over western Europe: Climatic Change, v. 10, p. 11–42.

Zishka, K. M., and Smith, P. J., 1980, The climatology of cyclones and anticyclones over North America and surrounding ocean environs for January and July, 1950–77: Monthly Weather Review, v. 108, p. 387–401.

MANUSCRIPT ACCEPTED BY THE SOCIETY JANUARY 5, 1990

Printed in U.S.A.

The Geology of North America
Vol. O-1, Surface Water Hydrology
The Geological Society of America, 1990

Chapter 3

Surface waters of North America; Influence of land and vegetation on streamflow

K. E. Saxton
U.S. Department of Agriculture, Agricultural Research Service, Pullman, Washington 99164-6120
S. Y. Shiau
WRB Environment Canada, Ottawa, Ontario K1A 0Y7, Canada

INTRODUCTION

Surface waters originate as precipitation determined by local climate, but they must pass over or through the land surface before accumulating as streamflow. Climatic considerations were presented in the previous chapter. In this chapter, major factors that influence the transformation of precipitation into streamflow are presented.

Although precipitation patterns and intensities largely determine the surface runoff from a particular drainage basin, two drainage basins subjected to the same climate may have dissimilar runoff characteristics due to differences in topography, geology, soil and vegetation. Striking contrasts at once come to mind:

> Rugged mountain areas with their torrential streams, contrasted with aggrading deltas and sluggish waters; maturely dissected areas with their prompt and complete drainage, contrasted with areas of recent glacial drift in which lakes and swamps abound; well-forested lands that tend to hold the rain and snow, contrasted with denuded lands that allow quick runoff; areas of cavernous limestone with their extensive subterranean drainage and big springs, contrasted with areas of relative impermeable granitic rock that have well-developed systems of surface streams and numerous small springs. (Meinzer, 1942, p. 6)

Many examples of diverse hydrologic responses are shown in subsequent chapters. In this chapter the hydrologic processes that cause such responses are described and examined for major sections of the North American continent.

To consider surface waters on a continental scale, it is only possible to present a broad scale of features for this land phase of the hydrologic cycle. But large watershed land areas are the integration and interaction of many smaller watersheds. Thus, the hydrologic processes of unit land areas and small watersheds provide understanding for larger areas. Both broad-scale descriptions of the North American continent and more detailed hydrologic processes are useful to describe surface-water interactions with the land and vegetation.

Surface soils constitute one of the most influential determinants of streamflow. A large percentage of precipitation is infiltrated into the upper soil mantle and then returned to the atmosphere by evapotranspiration. The infiltration rate is highly dependent on soil type and its moisture status. In turn, the soil characteristics are influenced by climatic and geologic processes and parent materials.

Vegetation on the landscape also plays an important role in precipitation dissipation, interception, soil protection, and thus, infiltration and evapotranspiration. This vegetation often has high economic importance to mankind as principal foodstuff and building material. However, it significantly affects surface waters through its impact on surface runoff and streamflow withdrawal for irrigation. A broad range of native and established vegetation types reflects the range of climate and soils. The North American continent contains examples of virtually every major vegetative system of the world.

Man can significantly alter the impact of land and vegetation on streamflow. Tillage, agricultural crops, forestation, grazing, mechanical impedences, and water-management systems are a few examples. Through man's development of the landscape over the past 200 to 300 years, significantly different land uses have been established over large areas of the continent with important hydrologic implications.

The topography of a watershed affects streamflow through both the hydraulics of overland and channel flow and ground-water flow. Both the total water and its time rate of discharge in rivers are dependent on the physiography of the contributing watershed. Topography is highly reflective of the local geology and geomorphic processes that have developed the various landscapes.

The several major factors that influence the downstream water flow derived from climatic precipitation all combine to form patterns of hydrologic response units. Even though there is a high degree of interaction among the soils, vegetation, and geomorphology, regions can be delineated and described that pro-

Saxton, K. E., and Shiau, S. Y., 1990, Surface waters of North America; Influence of land and vegetation on streamflow, *in* Wolman, M. G., and Riggs, H. C., eds., Surface water hydrology: Boulder, Colorado, Geological Society of America, The Geology of North America, v. O-1.

duce discernible features of hydrologic response. Such regions range from high-moisture tundra to flat farmlands to deserts.

In this chapter, several major land and vegetation associations and their physical impacts on downstream surface water are examined. Although the descriptions are limited to broad regions, the hydrologic principles described determine specific processes and watershed responses that are necessary to understand the hydrology of both small and large land units. First, a brief introduction of the broad physiographic regions of North America is presented, which suggests the range, diversity, and distribution of land features to be found. Then, the hydrologic impact of land features on surface water is discussed through the impact of soils, vegetation, and topography.

PHYSIOGRAPHY

To understand the hydrologic land influence of a large land mass such as the North American continent, it is useful to begin with a broad overview of that region's physical geography. The physiography we see today is the result of eons of crustal formation involving contractions, mountain uplifts, erosion, glaciation, and many other forces. This has left mountain ranges, plateaus, depositions, and exposed surfaces of a wide variety to interact with the current climate.

Physiography and streamflow

The streamflow emanating from any watershed is the result of many integrated influences of that watershed's physical composition. It is intuitive to know that each land mass has characteristics that relate to the amount of water received through precipitation and how this water is then distributed in space and time. Streamflow is one of the major water processes of interest in any physiographic region.

The North American continent has a diverse range of climate and physiography, which results in an even broader range of water-related characteristics. In the far north are regions locked in frozen conditions most of the year, with snowfall, snowmelt, and glaciation over surface topographies ranging from smooth to rugged. In contrast are areas of rain forest with abundant precipitation and streamflow, and the deserts with very little water of any kind.

Superimposed on the natural physical setting of the landscape are the changes brought about by man's relatively recent activity. Other than some major reservoirs and water-transport schemes, most large streams have not been significantly impacted by physical changes. Most of the man-related effects have resulted from converting the vegetation. Mountain forests have been harvested, grasslands used for annual cereal grains, and flat woodlands cleared for farming. Many areas, of course, have remained relatively unchanged by man, such as the vast, cold northlands, high mountain ranges, and dry desert lands.

The physical impact of land on streamflow can be delineated into major classifications that provide guidelines to the general nature of the regional hydrology. Regions dominated by mountains and steep slopes will have significantly different streamflow amounts and time distributions than those with flat topography, large rains, and no snow. This broad-scale perspective is useful to provide the general setting of streamflow origin and general characteristics to be expected. Inspection of the watershed geology, soils, and vegetation will further define the relationship between the Earth's surface and watershed streamflow.

Major physiographic regions of North America

North America has been divided into 12 major physiographic regions, as shown in Figure 1 (FAO-UNESCO, 1975). While recognizing that each of these regions contains much variety and includes areas of a different nature, a general understanding of each is useful for the more detailed descriptions of soils, vegetation, and topography that follow in this chapter. These regions are generally described as follows:

The Innuitian region. This region is the northernmost part of North America and comprises the northern half of the Canadian archipelago. The landscape includes mountain ranges of varying degrees of ruggedness, plateaus with subdued relief, and flat-floored valleys. At present, mountain glaciation, ice fields, and valley glaciers are still predominant features.

The Arctic Lowland and Coastal Plain region. This region includes the southern part of the Canadian Archipelago and the northern margin of the North American mainland north of the Canadian Shield region. All of the Arctic Lowland and Coastal Plain region was glaciated, leaving landforms that have controlled the terrain development over wide areas. An intricate combination of rivers, lakes, and glaciers persists today.

The Canadian Shield region. This region is a vast area centered on Hudson Bay and extending as far south as Lake Superior. The shield is a huge saucer with the depressed center at Hudson Bay. Once a mountainous region, it has been planed down by long periods of erosion and glaciation so that its present surface is a huge region of relatively low relief, a vast peneplain. This eroded surface contains only a few flat-topped knobs and ridges ranging from several meters to 135 m in relief, with most of the surface between 70 and 100 m. Erosion has disrupted all previous drainage networks and left enormous numbers of freshwater lakes in basins scraped out by glacial quarrying, moraine-dammed depressions, and erratic river systems. Today's surface waters must follow this complex river-lake system.

The Appalachian Highlands region. This region is a long narrow band of folded low mountains and valleys extending from the Canadian border to near the Gulf of Mexico and parallel to the Atlantic Coast. The northern third of the region consists mainly of metamorphosed formations with intrusions of igneous rocks in domes, while the southern two-thirds are folded sedimentary mountains grading into plateaus toward both the east and west. The gently sloping plateaus are dissected by numerous streams originating in the more complex ridge-and-valley core of the region.

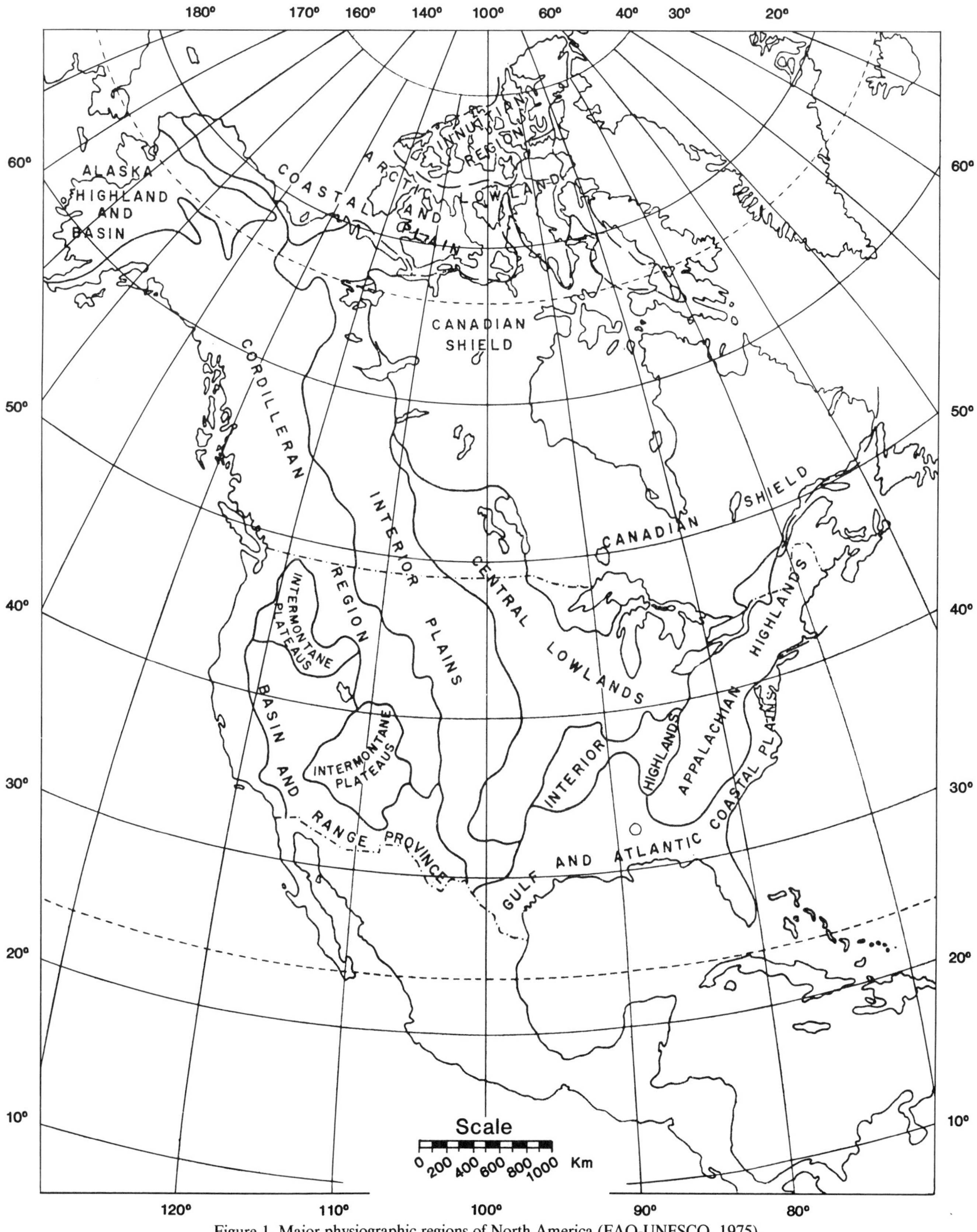

Figure 1. Major physiographic regions of North America (FAO-UNESCO, 1975).

The Interior Highlands region. This region is somewhat similar to the Appalachian Highlands. It is composed primarily of sedimentary rocks with some folds and truncations. Both sandstone and limestone are exposed in this broad-dissected plateau. The eastern part has several large basins cut by meandering streams in steep-walled valleys. West of the Mississippi River, the streams have created broad, gently sloping interfluves between deep, steep-walled valleys, and sinkholes are common in the limestone areas such as the Ozark Mountains.

The Gulf and Atlantic Coastal Plains region. This region contains an inland band of marginal marine plains along the Atlantic and Gulf of Mexico coasts, including the Florida peninsula. It is generally 150- to 300-km wide with more than half of the area at elevations less than 35 m. The extensive sedimentary deposits dip gently seaward, and extensive areas are nearly flat and featureless. Streams generally originate in the Interior Highlands region and dissect the plains as they flow toward the oceans.

The Central Lowlands region. This region contains the broad, low-lying plains in the central part of the continent. These plains are underlain by nearly horizontal beds of sedimentary rocks deposited on the floor of a shallow Paleozoic sea. Uplifts in the adjacent highlands and Canadian Shield caused erosion cycles, which produced the modern topography. The southern part contains scraped plains with streams entrenched in gorges. The northern portion has been repeatedly glaciated to varying degrees leaving a complex landscape of erosional features and debris. Slopes are gentle, there are many lakes and ponds, and the drainage is not well developed. Loess mantles a significant part of the west-central lowlands making it an important agricultural area but highly subject to runoff and erosion.

The Interior Plains region. This region is a long narrow band east of the Rocky Mountains sloping gently eastward to the Central Lowlands. The region is underlain by sedimentary rocks, and glacial drift covers the Interior Plains from the Missouri River north to the Arctic Ocean. The topography is generally rolling plains and hills with some terraced stream valleys, ponds, and bogs. South of the glaciated area, the topography is nearly level, with occasional sand dunes, and generally blanketed with a thin layer of wind-blown loess.

The Cordilleran region. This region is a belt of major mountain ranges parallel to much of the North American west coast. North of the conterminous United States, closely spaced linear mountain ranges characterize western Canada. In Alaska, the mountain belt separates into those along the southern coast and the Brooks Range in the north. In the conterminous United States, two separate mountain belts also occur; one inland containing the Rocky Mountains and extending as far south as New Mexico, and a second adjacent to the Pacific Coast containing the coastal ranges, Cascades, and Sierra Nevada mountain ranges. Uplifts, glaciation, and erosion give this region a complex topography of steep mountains and sediment-filled valleys.

The Alaskan Highland and Basin region. This region is at the northwestern terminus of the Cordilleran region and includes much of the Yukon River drainage basin in central and southwestern Alaska. The complex folded and faulted basin floor is generally covered by glacial drift and partially mantled by loess and volcanic ash. Dissected uplands are dominant in the eastern part, and broad lowland basins are typical of the western part.

The Intermontane Plateaus region. This region, located between the Rocky Mountains and the Pacific Coast ranges, is in two parts. The Colorado Plateau, in the south, has a nearly level upper surface but also has many deeply cut, steep-walled canyons created by erosion of hard and soft rocks, which provide spectacular escarpments, benches, and pediments. The Columbia Plateau, in the north, is underlain primarily by thick lava flows interbedded with some sedimentary formations. The partially glaciated terrain includes smooth upland surfaces bearing a variety of sedimentary deposits, and canyons incised by erosion from adjacent mountain streamflow.

The Basin and Range region. This region occupies the remainder of the region between the Rocky Mountains and the Pacific Coast ranges and extends well into Mexico. It consists of many intermediate to small mountain ranges and ridges separated by broad basins. The lowest elevation is some 100 m below sea level, and the highest peak about 4,000 m. The basins have varying amounts of alluvial sediments transected by stream channels. Precipitation and runoff are among the lowest on the continent, and highly variable. Topographic development over geologic time has been sporadic and incomplete due to catastrophic activity and limited hydrologic weathering. This has resulted in numerous closed basins and ephemeral lakes.

Other physiographic descriptions are given by Atwood (1940), Hunt (1967), and Gerlach (1970). Examples of other flow regimes found in various physiographic regions, such as those underlain by limestone, volcanics, alluvial deposits, and glacial till, are given by Riggs and Harvey (this volume).

SOILS

Soils are the upper mantle of the landscape and are as variable as most naturally occurring materials. Soil material, structure, chemistry, depth, and layering may all have significant hydrologic influence. Significant variation in soil character often occurs from the top to the bottom of small, rolling hills due to differences in parent material, weathering, and perhaps recent erosion and deposition.

Soils play a dominant role in the hydrologic cycle because they usually have the first opportunity to absorb, store, and release precipitation. Thus, their characteristics and evolution are important for understanding hydrologic responses of watersheds and the generation of streamflow.

Once precipitation makes an impact on the soil surface, several physical processes may occur. Infiltration through the soil surface predominates in most cases. As long as the infiltration rate exceeds the precipitation rate, no surface ponding or runoff will occur. But, as explained later, soil wetting itself reduces the infiltration rate; thus, as precipitation and infiltration continue, runoff will eventually begin.

The infiltrated water is first stored in the upper soil, and later redistributed within the profile. It is eventually either returned to the atmosphere by evapotranspiration or percolated downward to ground water. Soil characteristics, vegetation, and climate all interact in these hydrologic processes of storage and release. Soil-water flow primarily occurs one-dimensionally in the vertical. Two- and three-dimensional flow in the soil may at times become significant due to layering of impeding zones and steep slopes.

Soil science has developed many classifications and descriptions of soils. Obviously these characterizations serve many purposes other than hydrology, but the basic descriptions provide good insight about hydrologic performance. Thus, it is necessary to know the fundamentals of soil descriptions to apply this vast knowledge of soils to the understanding of surface-water hydrology.

Soil formation and classification

Soils over a specific watershed began forming in geologic time; and many have been highly influenced during millions of years by climate and vegetation. The four principal determinants of soil formation are climate, vegetation, physiography and geology. With wide variations of these soil-forming factors across a landscape as diverse as North America, large variations in soil characteristics are to be expected.

Describing and mapping soils has been a major effort in North America almost from the time of modern settlement. It is a continuing process as new knowledge and delineations become apparent and necessary. Much useful information for hydrologic inferences can be gleaned from this large knowledge bank regardless of its vintage.

The pedon is the basic soil unit that is described and mapped. Most soil scientists envision the pedon unit as an irregularly shaped solid or polyhedron that begins at the surface and extends downward to include the full set of soil horizons, or to some arbitrary depth corresponding approximately to the vertical dimensions of a set of horizons. The pedon perimeters are gradational. The actual size of the pedon is definable only relatively in that it must be large enough to be observed, sampled, and described, yet be unique among others in the scheme. Typical dimensions range from tens of meters to kilometers per side. The amount of spatial variation allowed within the descriptors of any single pedon before it is reclassified varies among systems being used.

Although modern soil classification is based upon scientific principles, it is not nearly as precise as that developed for plants and animals. In order of decreasing rank and increasing number, the hierarchy of categories is (1) order, (2) suborder, (3) great group, (4) subgroup, (5) family, and (6) series. Using this scheme over the United States requires 10 orders to be subdivided consecutively into some 10,500 series. Expanding over the North American continent would add even more (Finkl, 1982).

Modern naming of the categories has followed a root-word scheme somewhat based on the generic characteristics of the pedons being included. Examples of orders are aridisols (lack water, mesophytic plants, low humus) and mollisols (dark, friable, high calcium, formed under grass, the extensive soils of the steppes of Europe, Asia, North America, and South America). The series names are commonly associated with local terrain or previously named features. A comprehensive description of soil classification and nomenclature used throughout North America can be found in Soil Survey Staff (1960), with an overview by Muir (1962).

Most of North America has been mapped for the existing soil pedons to a greater or lesser detail depending upon local variation and economic importance. The smallest delineations are on maps with scales of 1:10,000 and show within-field soil series. Broader scales are more common in range, forests, and tundra regions. Very limited mapping or categorization has been reported in Mexico and Central America. Soil maps are generally available through local, state, and national offices of the United States and Canadian governmental divisions of agricultural and land resources.

Descriptions of the representative median profiles of the mapped pedons are published in conjunction with soil maps. While general in nature, many hydrologic characteristics can be inferred from these descriptions. Landscape locations and soil associations such as those in Figure 2 are usually shown.

To estimate the soil hydrologic influence, most soil series in the United States and those in Canada have been given a hydrologic soil classification of A, B, C, or D, and are published as part of the USDA-SCS (U.S. Department of Agriculture, Soil Conservation Service) Engineering Handbook (1955) and described in Chow (1964, p. 21.11). This broad classification of surface runoff potential varies from quite permeable sands (A) to rather impermeable clay (D).

Additional data for many soil series are available through published and computerized data banks provided by the USDA-SCS. Texture, structure, chemistry, and water-holding characteristics are enumerated by soil horizons. This detailed soils information can be most useful for small upland watersheds where specific surface-water effects and interpretations are required.

Soil-water impacts on streamflows

Hydrologically, soils are a porous mantle over the landscape; they can absorb, store, and release water. Essentially all precipitation first encounters the soil surface before it becomes a contribution to streamflow; thus, the soil has first opportunity to absorb this water. Only after satisfying the soil storage capacity or exceeding its infiltration rate will the precipitation become overland flow and direct runoff, which contributes to streamflow.

A schematic representation of the soil-plant-air system is shown in Figure 3. Soil profiles are usually not uniform with depth but have one or more somewhat distinct layers differing in texture, structure, and chemical composition due to soil-forming processes or tillage. Plant root penetration plays a very important

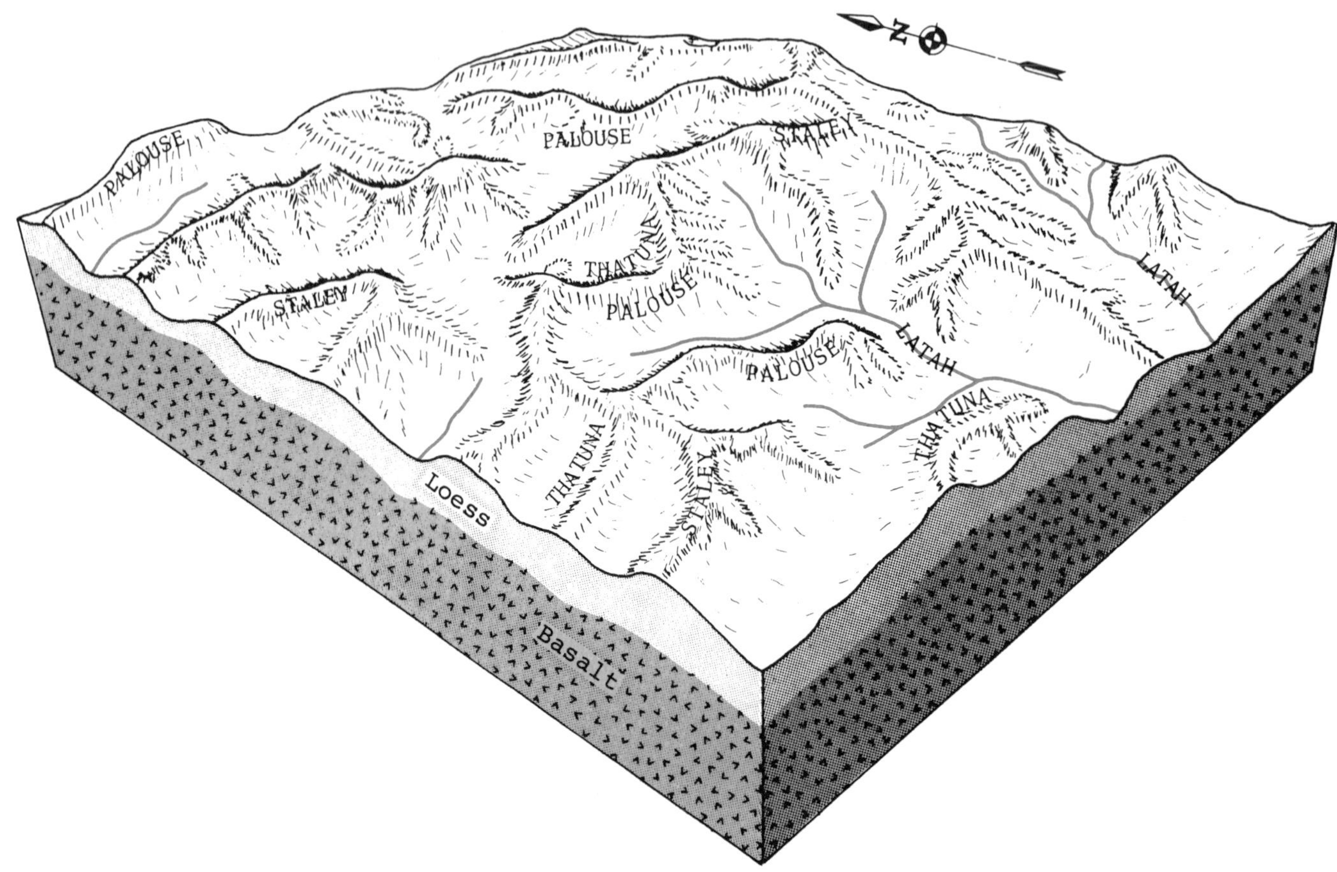

Figure 2. An example of soil associations in Washington State, United States.

role in extracting water from the soil profile by providing a flow path for evaporation back to the atmosphere. The several major hydrologic processes of water movement on and within the soil-plant-air system are shown by arrows in Figure 3.

Infiltration and the occurrence of soil water and its movement into and out of the soil are important components of hydrologic budget processes and one of the most significant determinants of crop production on agricultural lands. Thus, an understanding of the processes of soil-water storage, infiltration, and water budgets is important for understanding the impacts on watershed hydrology and streamflow.

Storage and measurement. Soils are complex compositions of granular and platy particles usually derived from decomposed surface geologic materials plus dead and living organic materials. Through complex physical and chemical processes related to climate and vegetation, they develop distinguishable soil horizons and structures in a layered fashion with depth.

Typically, some 50 percent of any soil volume is void space in the form of pores and interstices between particles. Water is held and moved through these pores largely as the result of surface tension similar to that in a capillary tube. Thus, pore size, shape, and chemicals play a major role in the rate and volume of soil-water movement and storage.

The volume of water held within any unit of soil also deter-

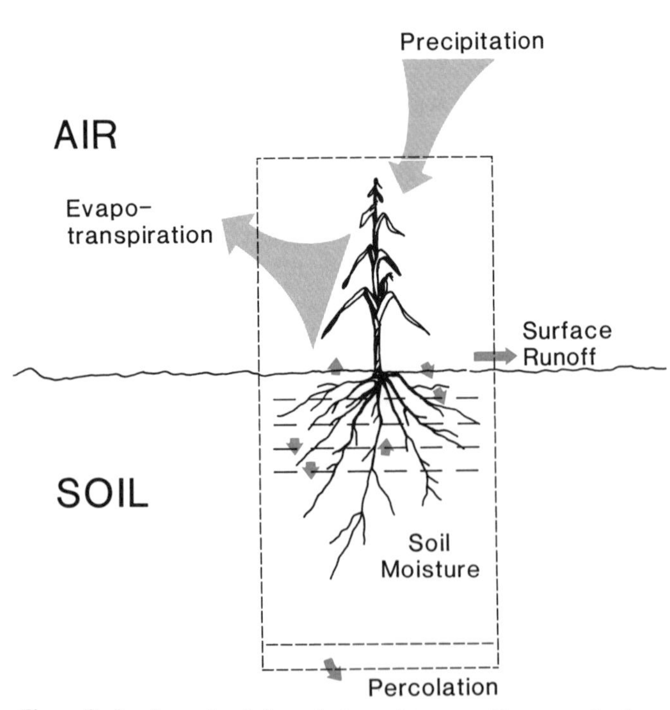

Figure 3. A schematic of the soil-plant-air system (Saxton and others, 1974).

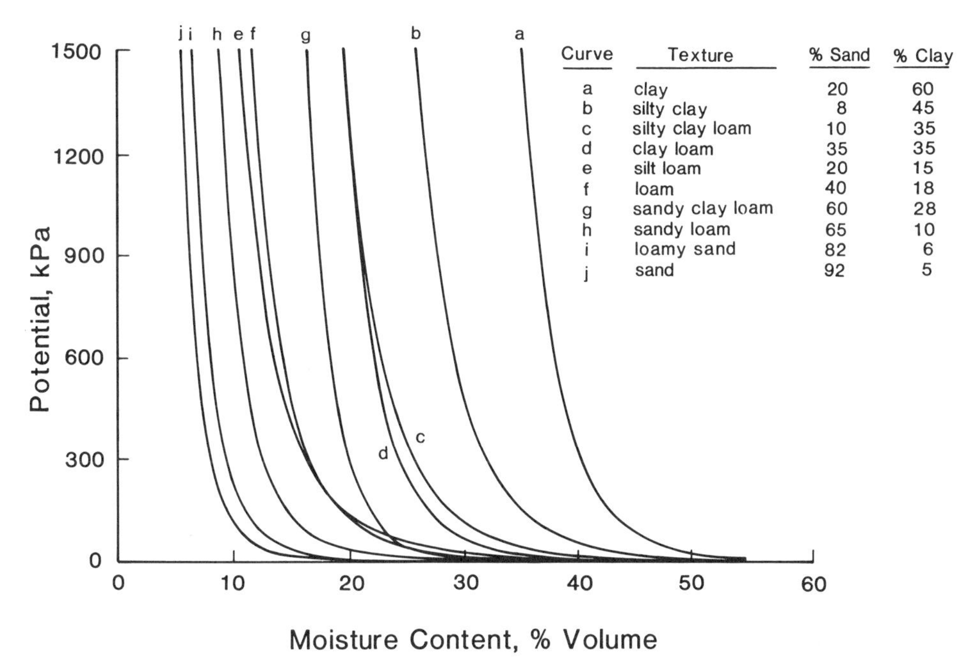

Figure 4. Soil-water potential versus moisture content characteristics of selected soil textures (Saxton and others, 1986).

mines the strength with which the capillary and surface forces hold that water. These forces increase from near zero for saturated soil to some 1,500 to 2,000 kPa (15 to 20 bars) at the dry limit where plant water extraction essentially ends. This soil characteristic largely determines the availability of soil water to plants. The pattern of these potential relationships depends on the pore-space configuration, which is, in turn, largely determined by soil texture. Figure 4 shows some typical relationships for various textured soils (Saxton and others, 1986).

Similarly, the potential rate of water flow, hydraulic conductivity, depends strongly on the soil pores and moisture content. The curves of Figure 5 are examples of hydraulic-conductivity relationships. The upper limits are saturated conductivities. Because additional soil characteristics other than texture, such as organic matter, clay type, and worm and root activity, also partially determine both soil-water potential and hydraulic conductivity, the actual relationshiops for any specific soil horizon will deviate somewhat from those of Figures 4 and 5 (Saxton and others, 1986). Further details of soil-water movement and definition can be found in Nielsen and others (1972), Hillel (1980), and Campbell (1985).

Because soil water and its movement significantly affect the hydrology of a landscape, data on soil-water volume at any particular time and location are commonly required by hydrologists. Soil water is spatially variable due to differences in precipitation, infiltration, and soil characteristics. Different locations have distinctive vertical water-content profiles due to infiltration, evaporation, transpiration, and percolation at that site during the previous several months.

The measurement of soil water for any given land area requires sampling several depth increments at several locations to arrive at an average soil-water content profile, which can then be integrated for some representative depth to determine actual quantities present. For many hydrologic purposes, the change of soil water over some time period is more important than the absolute quantities present.

The two most common measurement techniques used to determine soil water are gravimetric samples and neutron meter measurements. Gravimetric samples are soil samples obtained by auger or sampling tube. The samples are taken to the laboratory where they are weighed, oven dried, and weighed again. The water can be expressed directly as that lost, as a percent of the soil dry weight, or as a percent of total soil volume.

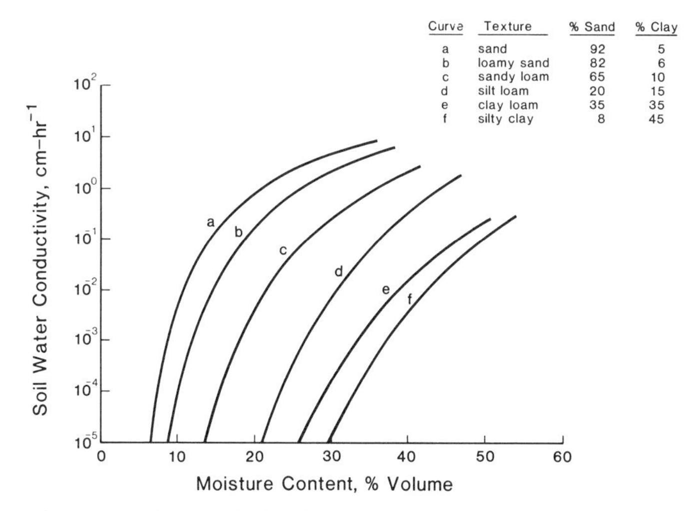

Figure 5. Soil-water hydraulic conductivity versus moisture-content characteristics of selected soil textures (Saxton and others, 1986).

The percent by volume is most useful hydrologically because, when multiplied by the represented soil depth, it provides the water depth contained in that horizon. This requires sampling a known volume so that bulk density can be calculated using the soil dry weight, or estimating an average density profile. By sampling the same locations and applying the same average density profile, changes in soil water can be determined within useful accuracies for many hydrologic purposes.

The neutron probe is the most modern and accurate soil-water meter. A radioactive neutron source emits neutrons outward into a soil mass, and a detector tube responds to those reflected back to the probe. The hydrogen atom is the principal reflecting element in the soil mass; thus, a good calibration between electronic counts from the detector and soil-water content can be established. The neutron source and detector are mounted in a cylindrical probe and separated by a lead shield. This probe is lowered into a tube placed in the soil and determines the soil water in a soil mass 15- to 25-cm in diameter about the tube at the depth positioned. The readings can be converted directly to percent by volume for water volume calculations.

The neutron method has the advantage that the measurement tubes can often be left in place; thus, the same soil volumes are measured each time, which makes the measurement of moisture changes even more accurate. A number of locations are still required to average spatial variation, and the soil surface in the immediate tube vicinity must be carefully protected if the location is to be representative. Neutron probes only operate correctly 30 to 40 cm or more below the soil surface because there is insufficient soil volume above the probe at shallower depths. Thus, these shallow layers must always be sampled gravimetrically near each tube, but no closer than about 0.5 m to avoid disturbance of soil adjacent to the tube.

Much surface hydrology is related to the quantity and time distribution of soil water. As suggested by the several simultaneous hydrologic processes in Figure 3, the soil-water content is quite variable in time, location, and depth. Replenishment is dependent on the precipitation pattern and amount, and depletion depends on the climatic evaporative demand and the vegetation growth and characteristics.

The quantity of water a soil will absorb and hold depends heavily on the texture and other characteristics. There is an upper limit, commonly referred to as field capacity, at which any further water added will likely drain downward due to gravity to wet lower layers or become a contribution to ground water. If there is an impeding layer, the water may accumulate to the point of filling nearly all void space, and thus, approach saturation. And there is a lower limit to which evaporation and transpiration usually dry the soil. This permanent wilting point is principally affected by the percent clay content and is at a high negative potential of about 1,500 to 2,000 kPa as represented in Figure 4.

Some representative water-holding capacities of soils are given in Table 1. These can be used to compute precipitation storage, plant-available water, and ground-water recharge potentials for various locations across North America. The sand plains of Georgia, with high precipitation and moderate evaporation potentials (see Plate 2), obviously have low storage capability and high ground-water recharge potential. Clay soils in hot, dry Texas are likely to store and evaporate nearly all precipitation

TABLE 1. MEAN SOIL-WATER CHARACTERISTICS OF SELECTED SOIL TEXTURES

Texture Class	Size Percentage*			Moisture Content, m^3/m^3†			Plant avail.§ mm/2m	Conductivities** x 10^{-8}m-S^{-1}	
	Sand	Silt	Clay	Saturation	Field Capacity	Wilting Point		Saturation	Field Capacity
Silt loam	20	65	15	0.47	0.29	0.11	360	550	1.00
Loam	45	40	15	0.45	0.24	0.11	260	410	0.37
Sandy loam	65	20	15	0.43	0.21	0.11	200	340	0.24
Silty clay loam	12	53	35	0.52	0.37	0.19	360	110	0.70
Clay loam	30	35	35	0.51	0.34	0.19	300	84	0.30
Sandy clay loam	55	10	35	0.49	0.29	0.19	200	53	0.08
Silty clay	10	45	45	0.54	0.42	0.26	320	85	2.50
Clay	30	20	50	0.53	0.41	0.28	260	46	1.40
Sandy Clay	50	5	45	0.51	0.34	0.24	200	36	0.19

*Defined by the USDA system.
†Defined by curves of Figure 4 (saturation = 0 kpa; field capacity = 33 kpa; and wilting point = 1,500 kpa [100 kpa = 1 atm]).
§The difference between drainable water at field capacity and plant extractable water at wilting point, in mm of water depth per 2m of soil depth.
**Defined at the selected moisture contents from Figure 5.

with very little ground-water recharge. And the thin, sandy, gravelly soils of the tundra in the north have little storage or evaporative demand, and thus, high potential for runoff and ground-water recharge.

An example of annual distribution of soil water is shown in Figure 6. This example is for a silt-loam soil with winter-wheat cover in which occasional measurements of soil water and daily estimates by computer simulation were made. The pattern of winter wetting and summer drying is obvious. The upper layers dry first and more than deeper layers, and summer rains occasionally rewet only the upper layers. Just as with climate, each year and location have unique distributions as illustrated in Figure 6; however, means and variations can be defined over longer periods. Recent developments in computer simulations of soil-water hydrology have aided these definitions (Saxton and McGuiness, 1982; Saxton and Bluhm, 1982).

Annual soil-water budgets for several locations in the United States are shown in Table 2. These values are from measured climatic data and simulated daily budgets by the Soil-Plant-Air-Water (SPAW) model of Saxton and others (1974), the same method used for the Figure 6 simulations. The profile percolation is an estimate of ground-water recharge. Both the annual precipitation and potential evaporation vary considerably with time and location. The quantity of actual evapotranspiration is dependent on the distribution of both precipitatoin and potential evaporation. When the soil water is nearly depleted during the high evaporative period, actual evaporation will be considerably less than the potential due to water availability limitations (e.g., compare central Indiana and west Texas, Table 2). Percolation and surface runoff quantities are similarly affected by soil-water availability.

Only generalized attempts to characterize soil-water budgets over the broad continental regions of North America have been made. Thornthwaite Associates (1964) developed some approximate estimates largely based on monthly mean air temperature and precipitation. Differences in mean precipitation and evaporation maps (such as Plate 2) can be used to infer water budgets such as Figure 10c (in Winter and Woo, this volume). Only by knowing the climate, soil, and vegetation patterns and their interactions with time can more accurate quantitative statements be made about the soil-water hydrology and its impact on the surface and ground-water hydrology. Lins and others (this volume) describe more precise techniques to estimate potential evapotranspiration, but these potential values need to be significantly modified in accord with soil and vegetation effects to estimate actual evapotranspiration of soil water.

Infiltration and runoff. Infiltration is the process that moves water vertically across the soil surface for subsequent transmission to the soil mass; and it is the process that determines the amount of the precipitation to become runoff, the overland

TABLE 2. ANNUAL SOIL-WATER BALANCE BY SIMULATION FOR SELECTED LOCATIONS IN THE UNITED STATES

Location*	Year	Precipitation (mm)	Potential ET† (mm)	Actual ET§ (mm)	Runoff** (mm)	Percolation‡ (mm)	Change in Soil Water‡ (mm)
Eastern	1980	494	919	474	0	-13	33
Washington	1981	410	975	390	0	-14	34
(46 - 44 x 117 - 48)	1982	470	1034	398	0	-8	80
	1983	463	1055	426	0	-9	45
	1984	391	1081	393	0	-14	12
Central	1980	424	1057	370	15	8	31
North Dakota	1981	421	911	432	14	5	-29
(46 - 51 x 100 - 38)	1982	534	856	478	25	4	26
	1983	336	1093	363	1	13	-41
Central	1982	1226	870	796	147	96	187
Indiana	1983	1036	971	855	69	60	51
(40 - 08 x 86 - 38)	1984	1037	895	813	107	50	68
West Texas	1981	444	1619	528	4	-18	-71
(35 - 09 x 102 - 07)	1982	424	1708	417	2	-2	7
	1983	317	1829	294	42	12	-31
	1984	347	1961	251	13	15	68

*State sector, latitude x longitude.
†Observed class A pan evaporation reduced by monthly lake-to-pan ratios.
§Computed evapotranspiration from established grass.
**Estimated from daily precipitation using USDA-SCS curve number method.
‡End-of-year net change.

flow component to streamflow. With this degree of importance, infiltration deserves special attention in the description of surface hydrology. Only by careful definition of the infiltration process can we estimate the quantity and time distribution of runoff generation.

Infiltration is controlled and modified by a large number of variables, all of which vary significantly in time and space. Thus, it is extremely difficult to define the potential and actual infiltration for even a small piece of landscape, and these small units integrate to form the larger watersheds. However, understanding the principal variables provides estimating methods that allow rational estimates and judgments.

The soil itself is the most important and obvious determinant of the infiltration rate. Water moves vertically into the soil by two physical forces. Gravity forces the liquid water toward a deeper soil position. Capillary action, however, is the dominant force in all but nearly saturated soil. The soil pores act as a myriad of capillary tubes where wettable surfaces powerfully attract the water molecules through chemical bonding. This capillary potential varies largely with both the size and number of soil pores; thus, it is highly influenced by soil texture, as shown in Figure 4.

The degree of soil wetness is highly influential because wetter soils have lower capillary potentials to provide the moving force. For completely saturated soils, the capillary potential is zero, and only the weak gravity forces remain to cause infiltration. When rain occurs on less than saturated soil, the capillary water potential at the soil surface becomes low due to the increased wetness. With dryer soil below, capillary forces create a strong downward gradient, and infiltration readily occurs. As the soil wets with continued infiltration, the downward gradient lessens and the infiltration rate decreases.

A potential infiltration rate can be defined for a given soil and its status based on the flow gradients and rates if water were instantaneously applied to the surface. As water infiltrates, this potential gradient begins to decline because of increased wetting of the deeper soil. With continued wetting, the infiltration rate declines until only the rate supported by the gravity potential and saturated hydraulic conductivity remains. This rate can continue indefinitely unless subsurface restrictions cause soil saturation. This effect of soil wetness is the physical reason that antecedent precipitation is a significant factor in the expected infiltration and runoff rate.

Surface runoff is generated when the infiltration rate is less than the rainfall rate. Figure 7 shows typical time distributions of a variable rainfall and infiltration rate. The area between the infiltration and rainfall rates represents surface runoff, which occurs only after the infiltration rate bcomes sufficiently low to cause surface ponding and runoff.

Infiltration is highly dependent on soil surface conditions, and these vary with the soil characteristics, land use, and biological processes. Undisturbed soil tends to form a slightly compacted surface by weathering, which is then loosened by biological activity such as worms, plant growth, and by soil freezing and thawing. Plants protect the surface from the pounding of raindrops, and thus reduce this dispersing effect. On agricultural land, tillage can either enhance or reduce infiltration depending on the roughness and compaction of the soil surface. Animal grazing often causes compaction by trampling, particularly when the soils are wet.

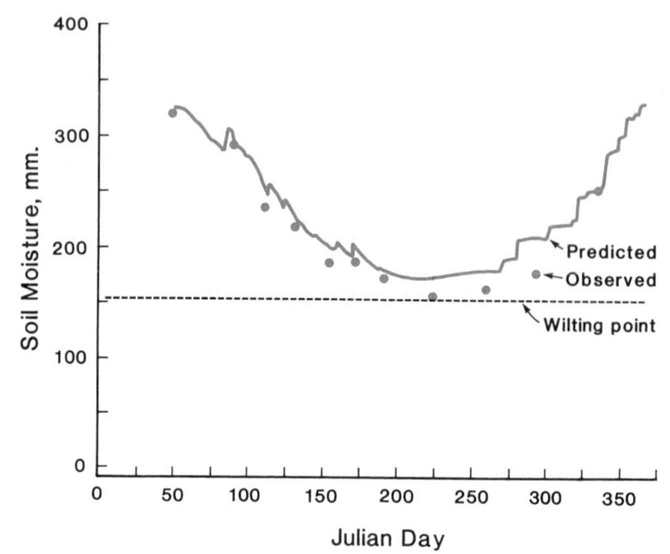

Figure 6. Soil moisture content in a 2.0-m profile beneath wheat as measured (●) and predicted (–) for an example year in eastern Washington State.

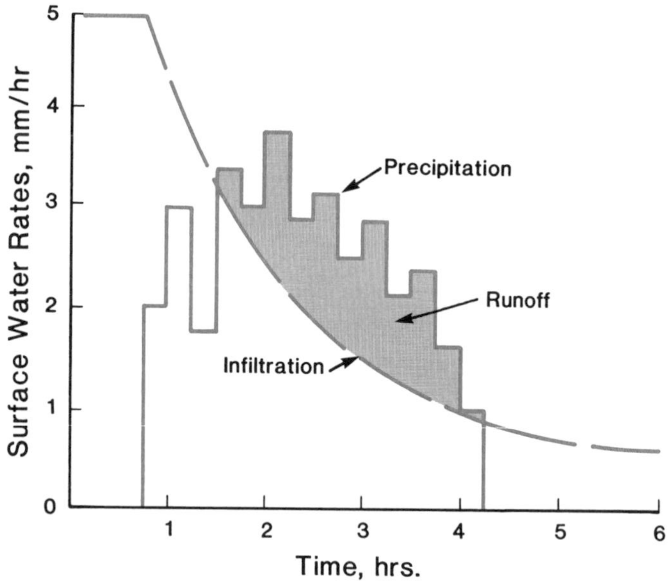

Figure 7. An illustration of infiltration, precipitation, and surface-water runoff rates.

In many cases, infiltration is not a simple, vertical movement of water into the soil profile but involves a complex two- or three-dimensional flow net within the soil and its various layers.

Where the soil layers have much different permeabilities, perched water tables can occur, resulting in lateral, downslope flow and causing nonuniform wetness patterns over the landscape, diverse infiltration rates, and seepage from the soil to the surface. Many steep forest soils are examples of this more complex case where nearly all precipitation infiltrates into the very permeable surface layers of duff and gravely soils; but this water soon emerges downslope to provide sustained streamflow and fast-response hydrographs.

All of these infiltration effects must be integrated to define the infiltration over a complete watershed or larger land area. Then these further integrate with the space-time distribution of the precipitation to form a rather complex set of physical processes, which determine the generation of surface water.

Once surface runoff is generated at any point on the landscape, it begins to flow overland in a downslope direction from that point. The direction of flow is approximately perpendicular to the local elevation contours, and the flow velocity depends primarily on the slope steepness and roughness. Overland flow from higher elevations accumulates with that from lower elevations as it progresses toward the watershed outlet.

Not all overland flow necessarily reaches the stream. Surface runoff from one land element may well infiltrate into the soil as it passes over a downslope element if that soil has less precipitation or a higher infiltration rate. This is particularly true if land use changes in the downstream direction, or precipitation suddenly lessens while the runoff is in progress. In arid regions and where rainstorms are localized, runoff from one area may well traverse to dry land and pervious streambeds and be readily infiltrated. Large sand-bed streams are particularly notorious for these transmission losses (Sharp and Saxton, 1963).

The topographic map of Figure 8 represents a typical small upland watershed in a generally rolling topographic region that has been glaciated and subsequently eroded by water. Surface runoff flow paths are from the ridge tops, which form the watershed boundaries, toward the dissecting channels in which the flow acculumates toward the downstream discharge point. Conversely, any point on the stream has a unique watershed boundary and runoff contributing area regardless of scale; thus, a similar topographic map could as easily represent a larger or smaller basin with the same principles.

The surface runoff from each element of land surface has its own unique flow path and velocities by which it finally reaches any designated downstream point. Thus, the time distribution of the accumulated flow at a stream location varies according to the integration of the upland flow generation distribution and the travel time. This effect of overland and upland channel hydraulics produces the customary somewhat bell-shaped hydrographs illustrated in Figure 9 for a simple and complex case of rainfall distribution (Burford, 1972; Burford and Clark, 1973).

Hydrograph shapes and volumes vary for each storm and watershed depending on many variables, such as precipitation intensities, space and time distributions, infiltration capacities and variabilities, and watershed topography. Although the estimation of surface runoff and hydrographs is complicated, hydrologic computations are essential for many purposes, such as roadway culvert design, flood protection, erosion control, and navigation.

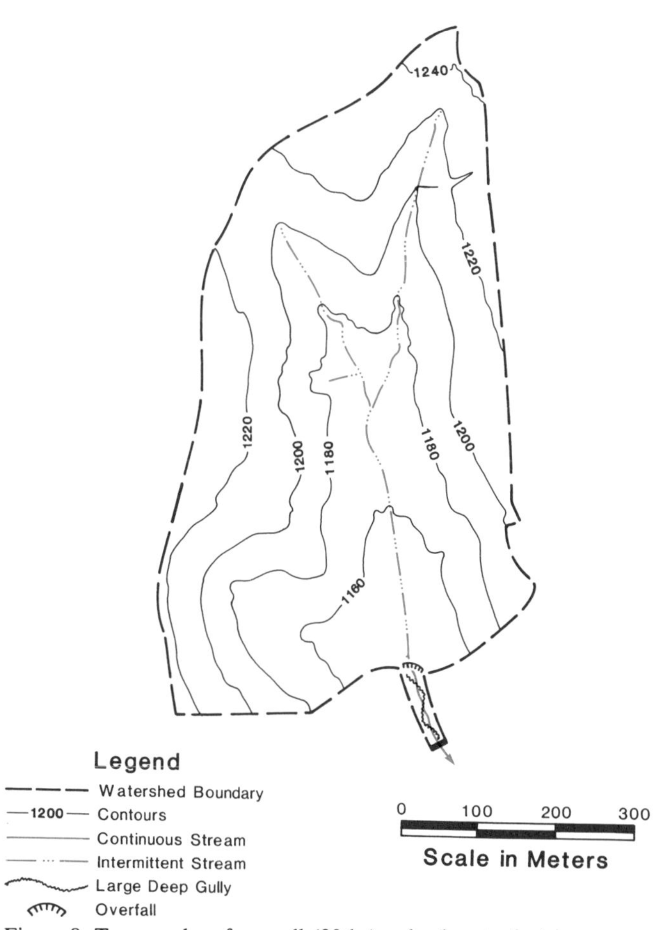

Figure 8. Topography of a small (30 ha) upland watershed in western Iowa.

As watershed size increases, the hydraulics of flow in the channel system becomes more important and the upland runoff generation less important. The time of interest also changes from hours in the small watershed to many days or months for the passage of a hydrograph in large watersheds. In very large watersheds such as the Missouri, Mississippi, and Columbia rivers, the streamflow variation becomes a seasonal phenomenon because of the integrated effect of large land areas, the large channel and reservoir storages, and seasonal climatic occurrences. Chapters in this volume by Riggs and Harvey, and by Matthai, describe these variations of surface runoff more specifically.

Percolation and ground-water recharge. Water infiltrated into the uppermost soil profile is primarily disposed of by evapotranspiration (ET) back into the atmosphere. The physics of the ET process is discussed in Lins and others (this volume) and in the section of this chapter concerning soil water. There are common situations, however, when some of the soil water percolates deeper than can be reached by plant roots for the ET

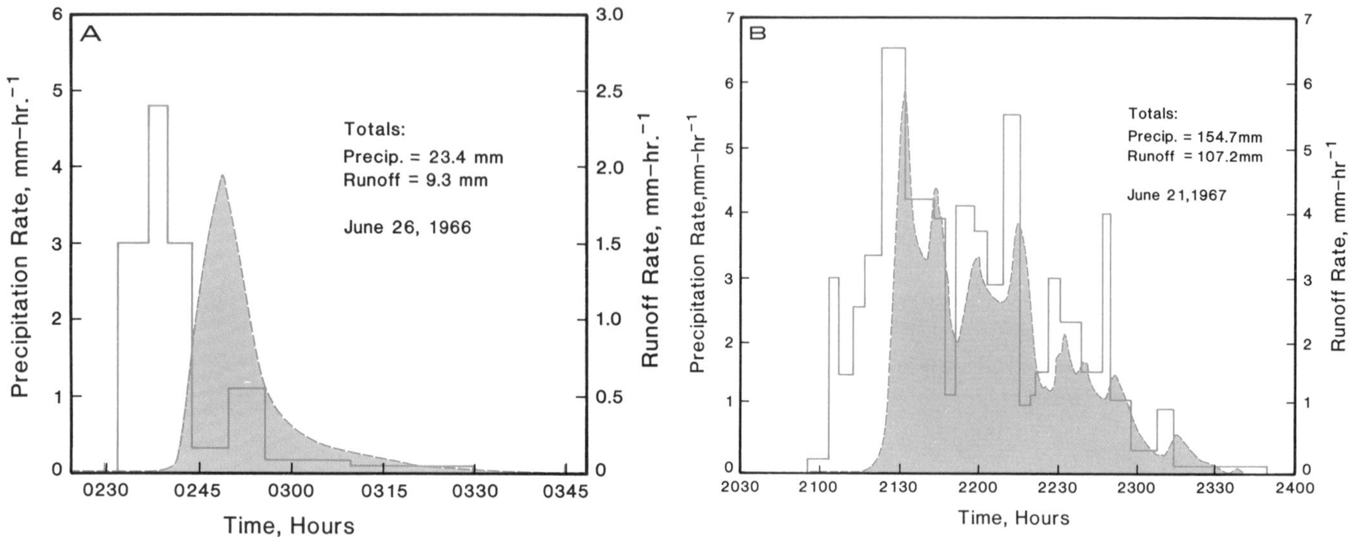

Figure 9. Examples of surface-water runoff rates from a small (30 ha) upland watershed (Fig. 8) from rainfall: (A) simple case; (B) complex case. (Burford, 1972; Burford and Clark, 1973.)

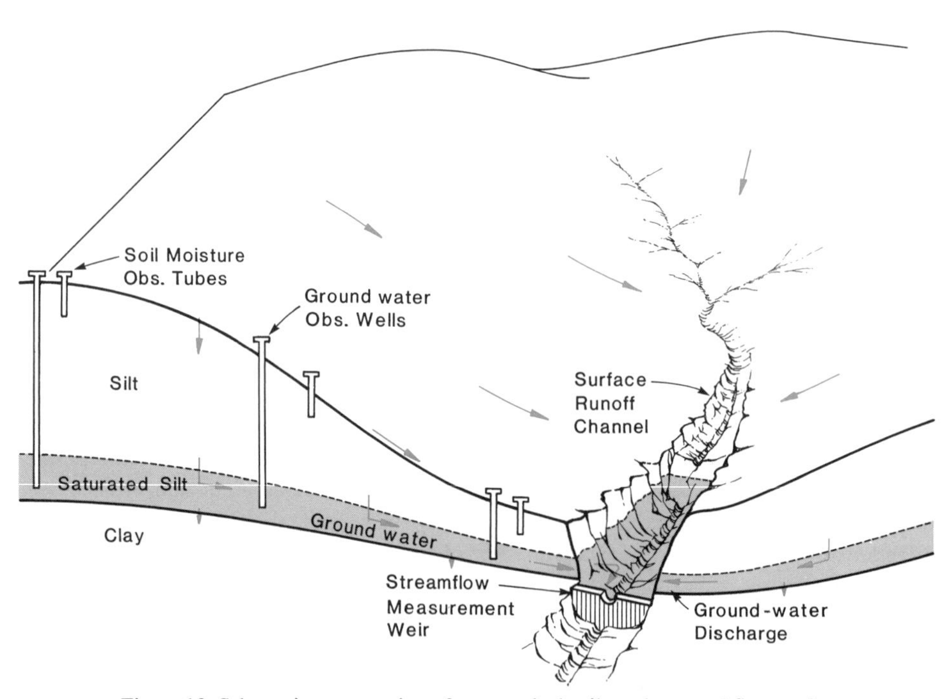

Figure 10. Schematic cross section of a watershed soil, geology, and flow paths.

process. This water is usually destined to move even deeper and become recharge to an underlying ground-water system.

The schematic watershed cross section in Figure 10 represents a simple subsurface geology. In this case, a moderately permeable upper mantle of loess overlies a very slowly permeable, clay-textured glacial till. The topography is such that the upland stream channel has become incised to the level of the glacial till. Water percolating downward through the loess forms a saturated ground-water layer above the silt-clay boundary and then moves horizontally toward the channel. The ground-water discharge occurs as a continuous seepage face along the incised channel to provide a base level of streamflow even in times of no recent precipitation.

During periods of surface runoff, the two streamflow sources—surface runoff and ground-water discharge—occur simultaneously. For larger watersheds with more complex geology, the separation of ground-water discharge from surface runoff becomes considerably more difficult than in small upland watersheds where surface runoff rapidly ceases after precipitation stops.

The interaction of these several hydrologic flow paths and storages is complex and variable depending upon the hydrologic and geologic settings. The soil acts as a large sponge over the landscape to hold infiltrated water. Percolation increases as the soil profile becomes wetter, until it approaches the saturation point when most subsequent infiltration percolates toward ground water. In drier circumstances, the soil mantle has the first opportunity to intercept the precipitation and little or no ground-water recharge occurs.

The discharge of ground water to streams is a slow and diffuse process. In most cases, the general rule that ground-water discharge rates are in some exponential proportion to the amount of water in the underground storage is a reasonable approximation (Chow, 1964, p. 14.9). With the amount in storage being the balance between discharge and percolation (recharge), this storage volume is a very nebulous quantity and is rarely known except by inference, and then only in simple geologic settings. Nevertheless, this continuum among surface runoff, soil water, and ground water is fundamental to much of our hydrologic understanding.

The data in Figure 11 demonstrate how this continuum operates. These data were obtained on the small upland loess-mantled watershed in western Iowa (United States), generally represented by the surface topography of Figure 8 and the subsurface geography of Figure 10. The graphs show daily rainfall and streamflow for a complete year with occasional soil moisture measurements to a depth of 5.5 m and frequent ground-water levels at three distances from the stream. During days of no precipitation, the streamflow was only from ground-water discharge since surface runoff ceased on this small watershed within one to two hours after any rainfall, as shown in Figure 9.

The year represented in Figure 11 had a relatively dry period until the end of May, which was followed by a very wet period in June, then quite dry in July and August. The three layers of the soil profile show the upper layers wetting with the first precipitation, then the deeper layers wetting by percolation from the upper layers as they reached their field capacity. The complete soil profile became very wet in June. This caused several days of significant surface runoff and considerable deep percolation to the ground water. The shallowest ground-water well responded first because it had less loess mantle for the percolating water to traverse, followed by the midslope well, and then the deepest well. The daily streamflow increased markedly following this ground-water recharge and continued so for several succeeding months. The ground-water levels very slowly declined over this long discharge period.

The graphs of Figure 11 show the dynamic nature and hydrologic interactions of these several processes. Driven by time-variable precipitation, the soil water and ground-water storages recharge sporadically, and ground water does so only after soil water is satisfied. Meanwhile, surface flow occurs when generated by the precipitation and infiltration regimes. While these data represent a very simple and well-documented case, these principles operate in similar fashion but in varying degrees on most of the North American landscape to create the surface runoff documented by Plate 3A and discussed in subsequent chapters.

Major soil regions of North America

The North American continent stretches from about 10°N to 80°N, is bounded by two major oceans, and has been subjected to numerous glaciations and geologic upheavals. Its soils, then, can be expected to be as diverse as the combinations of these many influencing phenomena. Virtually every major soil category of the world exists in North America to some extent, and much of the diversity of observed surface waters and their constituents can be attributed to the soils that these waters passed over and through on their way to becoming streamflow.

Although highly diverse, major soil regions of North America can be described as shown in Figure 12; each region generally is dominated by one or a few soil types as a result of common parent material and soil-forming processes. Obviously, some strong local differences can occur, but nonetheless, broad regional descriptions are very informative to begin understanding the continental soils and their hydrologic impacts.

Using the numbering scheme of Figure 12, the following are brief explanations of these regional soils, defined in FAO-

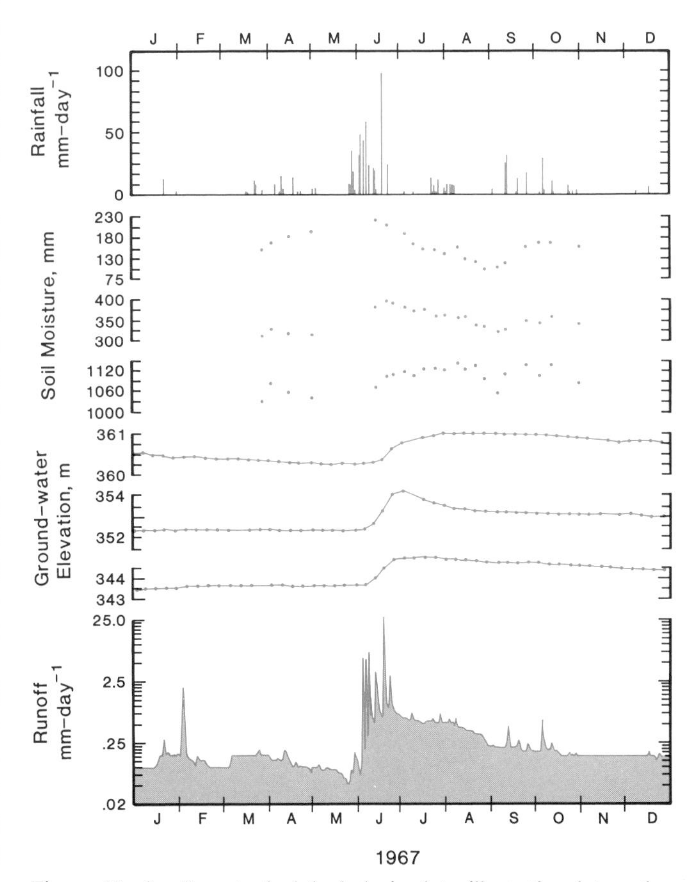

Figure 11. Small watershed hydrologic data illustrating interaction among soil water, ground water, and streamflow.

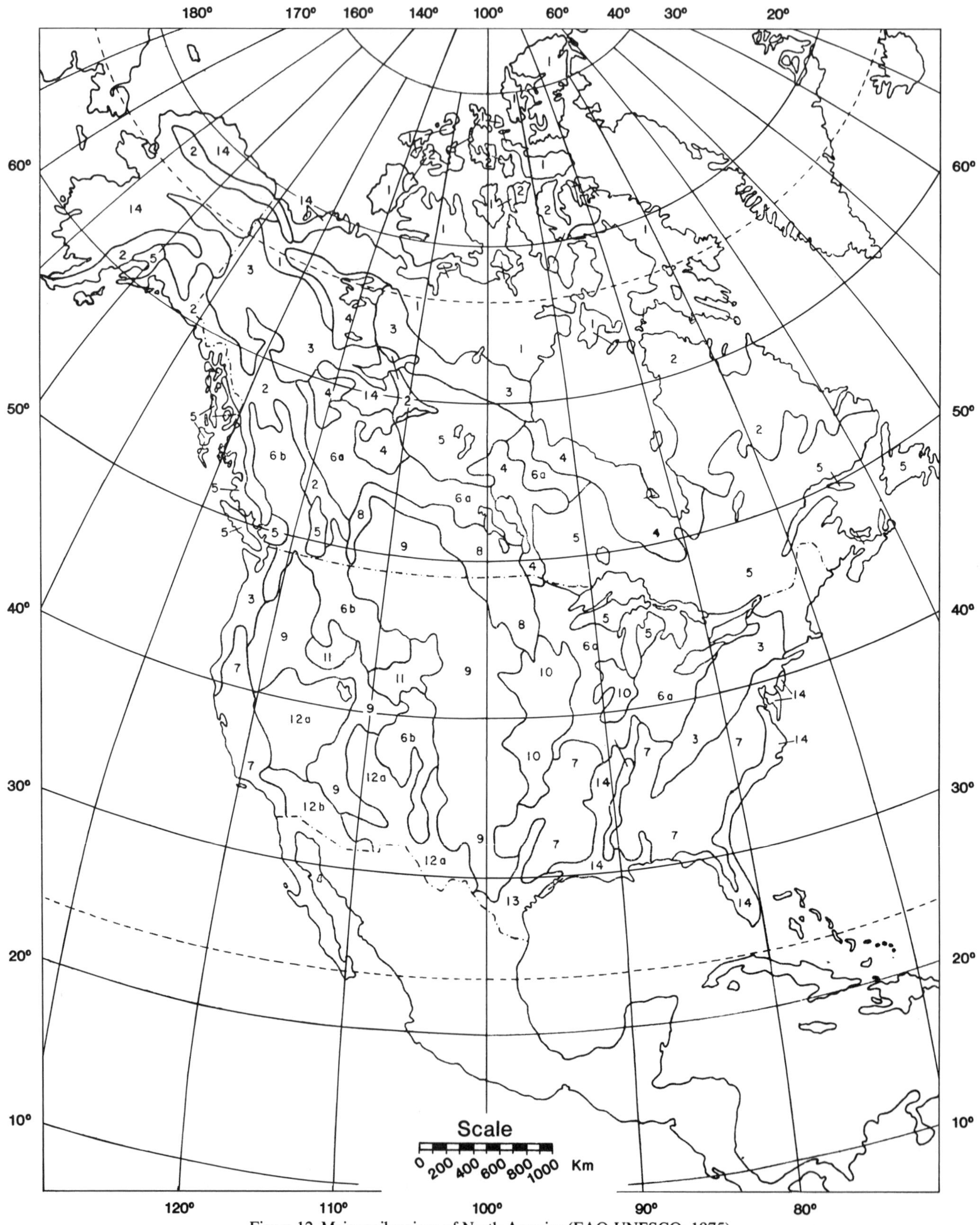

Figure 12. Major soil regions of North America (FAO-UNESCO, 1975).

UNESCO (1975). Most regions include a wide range of latitude and elevation, so any single statement represents a broad generalization. The general hydrologic performance of each soil region is given, but local deviations can be expected.

1. *Regosols* are the dominant soils across much of northern Canada, including the northern islands. They are cold-region soils where permafrost is continuous. Natural vegetation is tundra and some areas are barren. Other than limited grazing, these soils provide little economic value. They have limited infiltration potential and cause major areas to have significant surface runoff largely due to snowmelt.

2. *Lithosols* are extensive in the northern half of North America. They occur on steep mountain slopes in the western and northwestern part of the continent and on gentle and moderate slopes in northern Quebec, the coast of Labrador, and the islands in the far north. Much of the region has a sparse cover of forest and shrubs with virtually no economically productive activities. The infiltration potential is highly variable; however, with limited vegetation and evapotranspiration, much of the precipitation becomes runoff through surface or subsurface flow paths.

3. *Cambisols* are widely distributed across mountains, hills, and plains in northern and northwestern Canada and Alaska. They also are the dominant soils in the Appalachian Mountains in the eastern United States and coastal mountain ranges along the central Pacific coast. Some parts of this region in the north may have permafrost near the surface. These northern parts largely support boreal forest and tundra with some barren mountains. Agriculture is very limited. In more southern and warmer areas located in the United States, the natural vegetation is predominantly forest with some areas now converted to grazing and dairying. These soils often have high infiltration rates but relatively low storage capacities; thus, total water yield can be quite variable depending upon precipitation and vegetation patterns.

4. *Histosols* occur predominantly in the Canadian plains, which are nearly level or very gently sloping. Northern parts contain permafrost. Vegetation is generally forest or bog with little farming. These regions provide significant surface water but also large volumes of storage in lakes and ponds with slow drainage and some evaporation loss.

5. *Podzols* occur in the cool and moist eastern and northeastern parts of the continent with lesser areas on the mountain slopes of the northwest. Soil temperatures are cool to cold except in the southern areas near the United States-Canadian border; thus, natural vegetation is forest and the land is only cleared for farming in the warmer regions. Forage and grains are the principal crops. Regions with these soils are important to the streamflow in eastern Canada. They tend to have moderate to good infiltration and are important for forest production.

6. *Luvisols* are widely distributed soils and occur on both plains (6a) and mountain (6b) areas. The plains are generally moist but range from cold in Canada to warm in the southern United States. Natural vegetation is forests, but these have been extensively cleared for cultivation. The mountain areas tend to be stony, cool, and forested except for small valley areas. The quantity of surface runoff and erosion in this region is significantly influenced by farming methods and rainfall distribution and intensity. In the warmer, flatter regions, the soils have high agricultural production and are farmed intensively. In contrast, the mountain area (6b) is largely forested and at high elevation where winter snowpacks and spring melt create very important sources of surface runoff for municipal and agricultural use downstream.

7. *Acrisols* are the dominant soils in the plains and hills of the southeastern United States and on the mountain and valley slopes along the southwest coast. They generally occur in warm to hot and moist climates. Natural vegetation was both coniferous and deciduous forest, but most is now cleared for farmland. About half of the farmland is pasture and woods in the southeastern United States. Along the southwest coast, the soils are moist only during the winter due to the precipitation pattern (see Lins and others, this volume, Figure 24). The natural vegetation here varied from forests to shrubs and grasses, but most is now cleared and intensively farmed where topography permits. Valley irrigation is extensive. Infiltration is usually good to moderate in these soils. Precipitation patterns are a major factor in surface runoff since vegetation is usually quite dense and vigorous.

8. *Chernozems* occur on the northern and northeastern perimeter of the plains of North America. They are cool soils that are dry for appreciable periods during the warmer part of the year. The native vegetation was grasses, but virtually all of this area is now farmed. Crops are generally limited to cereal grains such as wheat and barley because of the limited moisture and growing season. With moderate infiltration capacities and lower precipitation, surface runoff is also limited.

9. *Kastanozems* are extensive in a broad area along the eastern side of the Rocky Mountains from southern Canada almost to the Gulf of Mexico. They also occupy extensive areas in the intermountain region to the west of the Rocky Mountains from Washington south to Arizona. East of the mountains, the soils are cool to warm, progressing from north to south, and dry to moist depending on precipitation patterns. Native vegetation was grasslands, some with brush. Where precipitation permits (generally greater than about 300 mm), much of this area is now cropped to small grains with the remainder grazed. In the intermountain area, the moisture ranges from seasonally moist to dry, slopes are steeper, and larger variations are common. Natural vegetation was shrubs and grasses and some has been converted to dry land farming. With the limited precipitation, this region does not produce large surface runoff amounts, but locally intense summer storms do cause occasional flooding.

10. *Phaeozems* occur on plains and small hills in the central United States. They are generally moist, temperate, and highly productive; thus, much of the region is farmed intensively. Dominant crops are maize, soybeans, and small grains, with some pasture and woods along steep streambanks. These soils have moderate to slow infiltration rates, and this area is subject to intense summer rains, which combine to produce local and regional floods and occasional severe erosion.

11. *Xerosols* occupy smaller plateau areas of the Rocky

Mountains in the western United States. They are generally dry and cool due to high elevation. The natural vegetation is sagebrush and sparse grass, and most of the area is now used for low-intensity grazing. In contrast to the adjacent region 6b, this dry area produces much less surface runoff even though most of the soils have low infiltration rates.

12. *Yermosols* are the dominant soils of the driest part of the continent—the basin-and-range area of the southwestern United States. They are generally hot and dry, vegetation is sparse desert shrubs and grasses with frequent barren areas, and crops are grown only in limited areas of irrigation. The topography is quite variable, ranging from about equal amounts of plains and sharp mountain ranges (12a) to regions largely dominated by plains and alluvial fans with only occasional mountain ranges (12b). Surface runoff in this region is generally limited to local floods caused by intense rains and is often absorbed by drier regions downstream.

13. *Vertisols* occur on limestones, calcareous shales, and marl in the southern part of the United States. They are warm or hot soils and dry for appreciable periods of the year. Some one-third of the region is underlain by relatively impervious sandstone. Natural vegetation was tall grass and forest. Most of the area is now used for growing cotton, rice, and soybeans or is grazed. Surface runoff is often significant because of low infiltration rates and large storm systems moving inland from the Gulf of Mexico.

14. *Gleysols* and gleyic soils occur in very wet locations due to precipitation and topography. They are found in diverse locations, being particularly extensive in Alaska and northwestern Canada, the Mississippi Valley, and along the Gulf of Mexico and Atlantic coasts. The northernmost locations have permafrost and tundra, while other northerly sites have forests and some have been cleared for farming. The Mississippi Valley and coastal areas are intensively farmed to row crops and grains. They are often on flat lowlands, which require surface and subsurface drainage. Surface runoff can be significant and of long duration in these humid, flat areas. Considerable amounts of the coatal region remain in lowland forests, swamps, and tidal marshes. These areas are high producers of runoff during wet periods.

VEGETATION

Vegetation on a landscape has a significant impact on the hydrology of a watershed and region. And, like soils, the vegetation is itself selected or adapted depending upon the climate and soils of the landscape. Despite the strong interaction of climate, soils, vegetation, and hydrology, it is useful to describe the characteristics and individual impacts of each separately.

Vegetation is described for hydrologic regions only in a community sense. Using broad generalizations, we can describe the mix of plant genera one would expect to find over significant areas of any region. While this is often not sufficient for interpreting specific hydrological observations, the broader views can be refined with additional information and surveys as appropriate to the watershed scale involved.

Impacts of vegetation on streamflow

Vegetation over the watershed contributing to streamflow will influence the water content and distribution primarily by one of two mechanisms. First, through transpiration, plants largely determine the soil-water status at any particular time, which in turn, plays a significant role in setting the infiltration rate for any subsequent rain. The mechanism is complex and closely integrated with the atmospheric and soil characteristics.

The potential for evaporation from either a water, soil, or plant surface is largely determined by the atmospheric drying capacity. Solar radiation and air humidity are the primary atmospheric variables. When water is available at the respective surface, it will evaporate at near this potential rate. Soils dry quickly at their surface and begin limiting the water loss to a reduced rate within hours of recent wetting. Plants, on the other hand, maintain a supply of evaporable water at their leaves much longer, depending on their root depth and associated soil water content. Thus, vegetated surfaces continue to dry the soil profile to a greater extent than partially canopied or bare soil. Deep-rooted plants, such as some trees and desert shrubs, are more effective in soil-water extraction than shallow-rooted plants, such as most grasses and forbs.

Vegetative transpiration can indirectly affect ground water as well. Through the transpiration of soil water, plants can increase the potential soil-water storage capacity for subsequent rainfall. With this increased storage, less of the precipitation is likely to percolate to become ground-water recharge, thus long-term ground-water storage and streamflow discharge can be significantly reduced.

Secondly, the plant canopy and residues are important for providing soil-surface protection from rain-drop impact, which can significantly influence the infiltration rate. For water to infiltrate the soil, it must be drawn below the surface. Compaction can reduce this rate to much less than that from similar but rough, friable surfaces. Thus, surface ponding and runoff will occur much sooner and to a greater volume during rainfall where the surface is compacted. The result will be greater surface streamflow and associated land erosion.

Several lesser effects of vegetation can also be important, such as root channels, and residue or duff layers. Plants encourage biological activity in the soil profile through their supply of organic matter from roots and surface residues. This activity is nearly always conducive to enhancing water infiltration and redistribution in the soil, thus reducing surface runoff and erosion. Root channels that remain in the soil after decomposition of the root sometimes form a useful avenue for water movement.

Vegetation growing adjacent to streams or in shallow, wet soils associated with ground-water tables where the roots have unrestricted access to the water are called phreatophytes. These plants can markedly reduce the volume of streamflow and ground water through transpiration. Some major river basins, particularly broad sandy desert streams, will lose a major part of their potential streamflow to phreatophytes. Usually, this is not a productive use of water for man's support, but large control schemes such as that attempted for salt cedar in the southwestern United States have generally not been highly successful. Phreatophytes often cause a diurnal pattern in streamflow rates due to changes in evaporative potential from night to day. This variation is one measure of the water quantity utilized by this vegetation.

Streamflow can be altered by large vegetative changes over the contributing watershed. These changes can be caused by any number of methods such as logging, disease, fire, insects, or revegetation by man. The effect expected on streamflow is not always obvious or predictable. Effects will be caused by a combination of changes in transpiration rates, soil-surface protection, and residues. Often the vegetation changes will be relatively temporary because the replacement vegetation has somewhat similar characteristics; thus, the hydrologic budget and streamflow will remain relatively unchanged. Documented examples where forested watersheds were harvested, then allowed to readily regrow in brush, have shown a return to near-forested hydrologic behavior within relatively few years after tree removal.

To assess the effect of vegetation on streamflow for any particular region or watershed requires knowing the water relationships of the species within the plant communities to be expected. Generally, plants have adapted to a particular climatic regime so that they behave in a particular way. Forests are often deep-rooted and have moderate ability to evaporate water compared to open water surfaces (climatic potential). Most grasses have dense canopies and readily transpire when water is available, but they tend to be shallow-rooted and reach limitations due to soil drying more quickly. Desert shrubs have sparse canopies and carefully mete out the evaporated water in this region of high evaporative potential. Each such environment and plant characteristic combination must be individually evaluated.

Major vegetation regions of North America

For a broad perspective of the natural vegetation patterns over North America, 24 regions have been delineated and are shown in Figure 13 (FAO-UNESCO, 1975). These regions can be compared with physiographic (Fig. 1), soils (Fig. 12), land use (Fig. 14), and climatic (Plate 2) regions to form an overall impression of the interrelationships that determine the vegetation distribution. Major changes by man and modern land-use regions are shown in the next section. Brief descriptions of each regional plant community before agricultural revision are as follows:

1. The *Boreal Forest* region is the most extensive vegetative region on the continent. It occupies a broad arc extending from Newfoundland west to the Rocky Mountains and then northwest through central Alaska to the shore of the Bering Sea. It is a cool to cold region with a generally moist climate where moisture saturation or deficits occur infrequently. The forest is dominantly coniferous, with spruce the most common species, but there are some deciduous trees, largely aspen and balsam poplar. A large part of these forests would be classified as productive (growth rates of about 2 to 6 cubic meters per hectare per year), but this is site dependent because the soil, topography, and climate vary. Only small parts of this area have been converted to agricultural uses.

2. The *Boreal Forest and Tall-Grass Prairie* region forms a belt between the closed cover of the boreal forest and the treeless grasslands of the tall-grass prairie. It extends in a broad arc from south-central and eastern Manitoba northwestward through Saskatchewan, north-central Alberta, and ends at the Rocky Mountains. Moisture generally limits forest growth but sustains productive grass growth. A large part of this region has been cleared and cultivated and now supports a dominantly small-grain agriculture with some grazing.

3. The *Subalpine Forest* is a dominantly coniferous forest region on the uplands of the Cordilleran region. It is similar to the boreal forest but occupies higher elevations. The region generally occurs on the eastern slopes and upper foothills of the Rocky Mountains from about the Canadian–United States border northward. These productive forests are generally spruce, fir, and pine and support an active timber industry.

4. The *Interior Montane Forest* region occurs generally between the eastern slope of the Rocky Mountains and the seaward-facing slopes of Washington, Oregon, California, and British Columbia. Its climate varies widely with rainfall, which ranges from about 175 to 850 mm/yr. Vegetation also is quite variable, ranging from sagebrush and grasslands in the lower and drier elevations to dense forests of spruce, fir, and pine at upper elevations. Some alpine meadows exist above treeline. Little farming occurs in this region; most activity is related to lumber or grazing.

5. The *Columbia Forest* occurs on the steep mountain slopes of Idaho, northeastern Oregon, and central British Columbia. The area generally has moderate rainfall of 500 to 1,250 mm/yr, occurring largely as winter snowfall, with dry summers. Dominant forest species are ponderosa pine, western white pine, western red cedar, hemlock, and fir. Essentially no farming is done, but there is extensive commercial forestry.

6. The *Western Coastal Forest* extends from southern Alaska along the west coasts of British Columbia, Washington, Oregon, and northern California. It is generally cool and humid with local areas of "rain forest" like that in northwest Washington. The region supports dense forests of red cedar, hemlock, and fir. The region remains largely in forests and has some of the richest forest productivity on the continent.

7. The *Redwood Forest* is a small but unique area in northern California and southern Oregon where the large redwood is the dominant species. The climate is temperate to warm with 800 to 2,000 mm/yr of well-distributed precipitation. The region supports a significant timber industry and little farming.

8. The *Rocky Mountain Coniferous Forest* region includes the mountain ranges of Arizona, New Mexico, and Colorado. With large elevation variations (1,300 to 3,200 m with peaks of 4,300 m), the climate is quite variable and the vegetation species noticeably vertically zoned. Rainfall ranges from 250 mm/yr in the valleys to 750 to 1,000 mm/yr on the mountain slopes. Species range from sagebrush and ponderosa pine at the lower

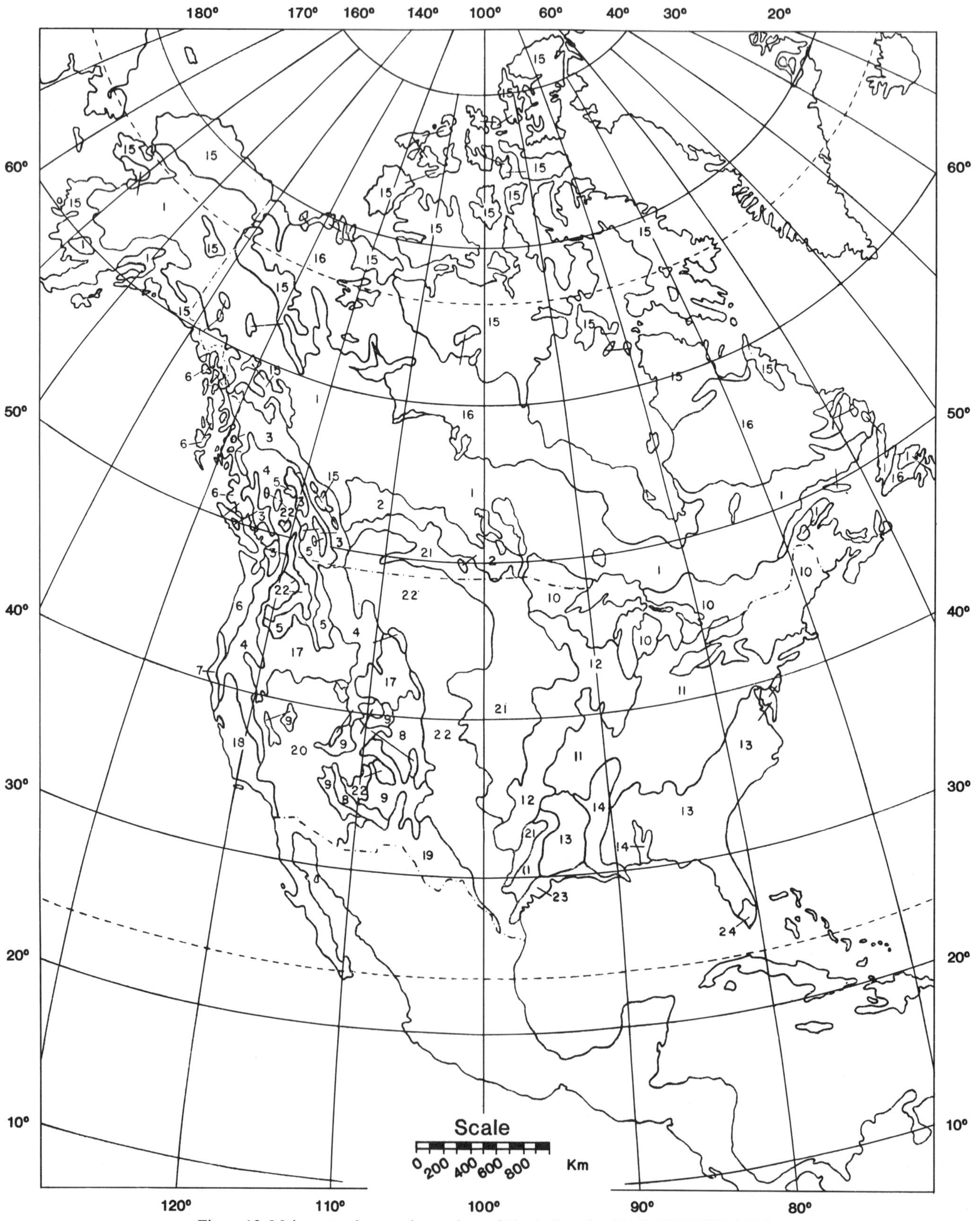

Figure 13. Major natural vegetation regions of North America (FAO-UNESCO, 1975).

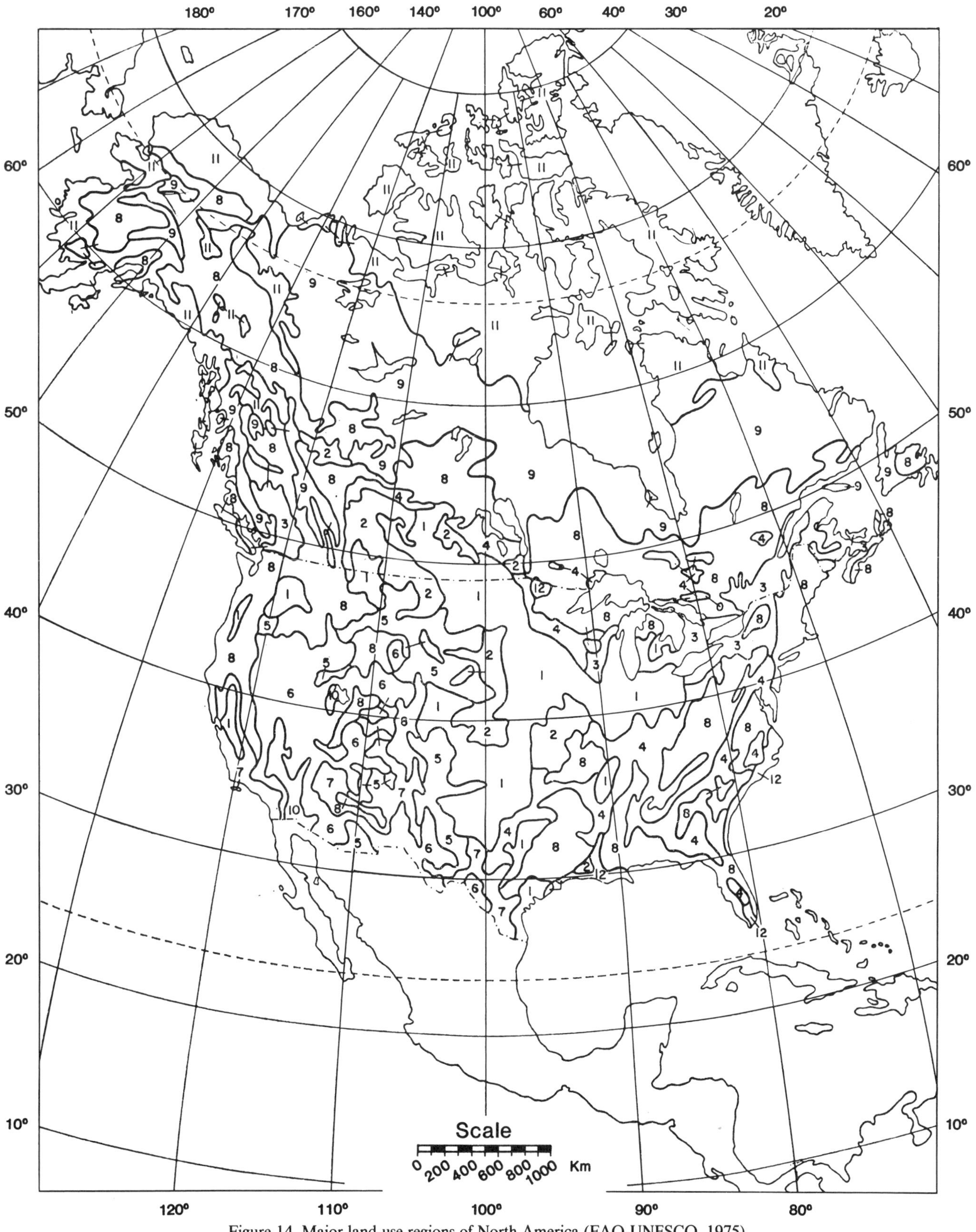

Figure 14. Major land-use regions of North America (FAO-UNESCO, 1975).

elevations to fir and spruce at higher sites, and alpine species above treeline. Small areas are cultivated; however, about half of this region is used for grazing and hay meadows.

9. The *Pinyon-Juniper Woodland* occupies the high tablelands of the southwestern United States, mostly in New Mexico, Arizona, and Utah. It is a dry region with precipitation of only 150 to 375 mm/yr. The principal vegetation is open groves of low needleleaf evergreens with a variety of shrubs and grasses. Oneseed juniper and pinyon pine are typical species. Significant areas of nearly bare ground are common. Less than half of the area is used for cattle and sheep grazing, with essentially no cultivation except small irrigated valleys.

10. The *Eastern Hardwood-Conifer Forest* in the east-central part of the continent lies in a broad arc from Ontario to the Atlantic coast. It is a transitional vegetative region between the evergreen needleleaf forest to the north and the deciduous broadleaf forest to the south; thus, the forests are mixed. Typical species are pine, hemlock, birch, beech, and maple. Because of shallow soils and poor climate, only part of this region has a productive forest capacity. Some clearing for cultivation and grazing has occurred in the south.

11. The *Southeastern Broadleaf Hardwood Forest* region is an extensive and geographically important region from southern Ontario to the northern parts of Alabama, Mississippi, and Arkansas, and eastern Texas. The region includes the temperate moist climate of the southern Great Lakes region; the warmer, moist climate of the Mississippi Valley; and the warm climate of eastern Texas with moisture-deficit summers. The common forest species range from maple and beech in the north to butternut, hickories, and oaks in the south. Most of this region is now intensively farmed to crops and pasture. Small woodlots remain on steep slopes and along valleys. There are several large metropolitan cities in this region with a generally high population density throughout.

12. The *Broadleaf Hardwood Forest and Tall-Grass Prairie* is a transition region between the eastern forests (no. 11) and more western prairies (no. 21). The original mosaic of forest and grassland has now been essentially removed by intense agricultural activity in this very productive region. Nearly all of this region is now under cultivation of annual crops, with some limited grazing on steeper slopes.

13. The *Southeastern Mixed Forest* region is another highly important vegetative area, which extends from southern New Jersey south to Florida and west to the eastern edge of Texas. This region contains much of the eastern and southern coastal plain plus the eastern part of the Piedmont of the Appalachian Mountains. With a moist climate, it contains many rivers and flood plains. The forests were largely of hickory, pines, and oak on the Piedmont slopes. From central South Carolina south to Florida and west to Louisiana, the loblolly pine and white oak were dominant. Although much of the area was farmed or deforested in the past, one-half to three-fourths of this region is now in forest or woodlots, with some cleared land for cropping and pasture, mostly in the east.

14. The *River Bottom Forest* region extends from the confluence of the Ohio and Mississippi Rivers southward on the Mississippi flood plain to the Gulf of Mexico, plus flood plains of lesser rivers. This region is nearly entirely flat flood plains with warm wet soils. The original forest was water tupelo, oaks, and cypress, but less than 10 percent now remains forested. Virtually all of the northern part has been cleared and drained for intensive agriculture, with somewhat less clearing in the wetlands of the southern part.

15. The *Tundra and Alpine Meadow* region is in high contrast to the river-bottom forest (14). This region of tundra vegetation extends across the northernmost part of the mainland of the continent from the Labrador coast of eastern Canada to the arctic and Pacific coasts of Alaska plus alpine regions above treeline in the Cordilleran region. Tundra is essentially a treeless vegetation characterized by the absence of tall woody species. Only dwarfed trees are likely to be present. The principal tundra vegetation associations are Arctic desert, lichen-moss, heath, sedge grass, and bush or scrub. Virtually no major change has been imposed in this region by man, and most of the area is devoted to wildlife or recreation.

16. The *Tundra and Boreal Forest* region is another transition zone extending in a narrow band from the Yukon-Alaskan border southeasterly to the Newfoundland highlands. The region is generally a mosaic of open scrub and tundra with patches of largely unproductive forests. This provides good wildlife shelter and a limited amount of domestic grazing.

17. The *Western Sagebrush Steppe* occurs on plains and plateaus in the northwestern part of the United States. Situated north of the very dry southwest United States and between mountain ranges, the precipitation varies from 150 mm/yr in some dry valleys to 1,500 mm/yr on windward mountain slopes. The vegetation is predominantly big sagebrush and varying amounts of native grasses interspersed with areas of bare soil. The vegetation amount and species are quite variable depending on elevation and local climate due to orographic effects. Some three-fourths of this area is used for livestock grazing, with some irrigated farming in valley bottoms.

18. The *California Steppe* region in the southwestern third of California has wide variations in topography, temperature, rainfall, and soils within short distances. Thus, the vegetation is equally diverse. The precipitation is 250 to 1,000 mm/yr, but largely occurs in the winter months with very dry summers. The valleys were originally tall grasslands, but essentially all are now in agricultural crops. The hills and low mountains have a mixture of evergreen deciduous forest consisting mostly of various pines and oaks. Many of the hills are now cleared for dry land grain farming, and the steeper slopes are used for grazing.

19. The *Desert Shrubs and Grasslands* region in the southwestern United States is a transition zone between the short-grass prairie of northern and central Texas and the desert of western Arizona. Precipitation is generally 200 to 500 mm/yr with precipitation, temperature, and vegetation quite elevation-dependent. The grasses are sparse and short and the shrubs are

mainly creosote bush and tarbrush. One-half to three-fourths of this region is used for low-intensity grazing with some irrigated and dryland farming interspersed.

20. The *Southwestern Desert Shrubs* region is the driest part of North America with precipitation of 150 to 300 mm/yr. The vegetation is an open stand of shrubs such as sage, creosote bush, sagebrush, and greasewood. Intervening spaces are occupied by short grasses or desert cacti, or are bare. Some hills in the northern part have sparse stands of juniper and pinyon pine. About 90 percent of this region is still owned by the U.S. government and is leased for light grazing. But there are also some intensely irrigated areas and some important, rapidly developing residential and commercial cities that are located where water is available.

21. The *Tall-Grass Prairie* region is a large and important region extending from southern Canada to the southern United States, parallel to the Rocky Mountains. Temperatures range from cool to warm in the north-south direction. Precipitation is generally favorable, 350 to 750 mm/yr, with a significant amount occurring during the growing season; however, it is highly variable from year to year. This region is naturally nearly treeless except along watercourses, and almost entirely covered by tall-grass species such as western wheat grass, needle grass, Indian grass, and bluestems. About three-fourths or more of this region is now cultivated, mostly to dryland grains, but some irrigated row crops are grown where water is available from streams or ground water.

22. The *Short-Grass Prairie* region is extensive. It extends from southern Alberta to the tip of Texas and contains all of the easternly foothills and outwash plains of the Rocky Mountains. A small region occurs in Washington and Oregon. The climate is quite variable and elevation-dependent. Precipitation is generally 250 to 400 mm/yr. The native vegetation was mostly grasses with some sedges, forbs, and shrubs. The dominant species were western wheat grass, blue gramma, and buffalo grass with varying amounts of sagebrush and mesquite. Essentially all of this region in Canada and Washington-Oregon is now cultivated with dryland small grains, principally wheat. The eastern Rocky Mountain slope is about half dry-land cultivated and the remainder grazed, although these proportions shift with weather and economic patterns.

23. The *Gulf Coast Prairie* region is a small, lowland area extending inland of the Gulf of Mexico in Texas and Louisiana. With a hot, humid climate, the native vegetation was tall, dense grasses such as seacoast, bluestem, and smooth cordgrass. Essentially, all of this region is now cultivated in cash crops with large areas of rice and sugarcane.

24. The *Florida Marsh and Swamp* region contains the very wet southern tip of Florida. The climate is hot and humid, and much of the land is nearly saturated most of the year, with some areas inundated all or part of the year. The vegetation is tall grasses and hydrophytic trees. Dominant grasses are sawgrass and three-awn grasses and the trees are those such as bay, cypress, and mangroves. Only about one-fifth of this region is farmed, mostly in improved pasture. The remainder is in game reserves, parks, and wetland reservations.

Major land-use regions of North America

Over much of North America, man has significantly changed the vegetation from that found before modern civilization to that of an intense agrarian society. In many cases, this land-use change brought about a significant hydrologic change due to the significant differences of water demand by the vegetation before and after settlement. In other cases, hydrologic changes have been small and temporary. In the absence of good monitoring networks, many changes are undetectable or only documentable in retrospect based on recent research and analysis of trends. Hirsch and others (this volume) discuss several examples of such changes and their hydrologic impact.

A good relationship often exists between natural vegetation and the plants man chooses to grow. This is so because the soils and climate have not significantly shifted; thus, a major shift in genera is unlikely. Some typical examples are forestlands where elevation and poor soils make plants other than trees unfeasible; or tundra, which remains unchanged. Similarly, much of the native rangeland has been left intact, although sometimes significantly altered by grazing intensity or by completely foreign activities such as mining or irrigation.

In other areas, the vegetation has been completely transformed by modern man. The "corn-belt" and the "cotton-belt" states of the central and southern United States are prime examples. These areas were originally covered by tall grass and forest, which provided good erosion protection, high evaporative demands throughout the growing season, and stable streambeds. Today, the land is nearly bare much of the year, and floods and erosion are more serious problems.

In the dry-land wheat belt of the central United States and Canada, man's impact has been at an intermediate scale. This region of short prairie grasses was replaced largely with another grass species: wheat. Other crops are intermixed and even large areas of natural terrain remain; thus, the actual impact is quite variable and not spatially coherent. While the surface waters are significantly less in this dryland region than more humid areas, they often are of high economic and social importance; even minor changes in availability may have serious consequences.

The soils, climate, and topography of each region define the suitability of that landscape to one or more uses by mankind. Other social, economic, and geographic criteria also influence the use man makes of a region, and in turn, the impact of that use on the region. Austin (1965) presents a detailed summary of land-resource regions of the United States largely based on soil and climate.

The major land-use regions of North America, as defined by current man-related (1960 to 1970) activity, are outlined in Figure 14 (FAO-UNESCO, 1975). These regions, by necessity, are broad and general delineations, but the hydrology and surface-water data and information presented in subsequent chapters can be put in perspective by relating to these regions. More detail regarding specific areas is available from sources such as the agricultural statistics of each country. A brief description of the delineated regions in Figure 14 is as follows:

Cropland. The cropland regions have been delineated as (1) mainly cropland; (2) cropland and grazing; (3) specialty cropland, pasture, and forest; and (4) cropland, pasture, and forest. These regions are generally extensively used for tillage crops such as maize, soybeans, sorghum, cotton, and wheat. Except for winter wheat, these crops leave the soil partly bare much of the year, with intensive growth in the summer period. This often creates more runoff and erosion potential than when natural vegetation was present. For example, native tall-grass prairies in the middle sections of the United States (region 1) are now extensively farmed with maize, soybeans, and other clean-tilled row crops, which leaves the soil surface unprotected by vegetation much of the year. Both runoff and erosion are some two to four times more than the native situation. However, modern farming techniques using less tillage and maintaining more surface residues are reversing this trend. Croplands are commonly interspersed with pastureland and forests, depending on the local climate, soils, and topography.

Rangeland. Grazing of natural vegetation by livestock occupies a vast part of the drier climatic regions of North America. Much of the western United States, both west and east of the Rocky Mountains, is used in this manner. The intensity and mix of grazing with forest and desert shrubland varies with precipitation and vegetation. The rangeland regions of Figure 14 have been subdivided into (5) grassland, (6) shrubland, and (7) open woodland.

Except for locally concentrated or severely over-grazed areas, these regions are most likely hydrologically very similar to their long-term history. Some increased runoff and erosion occurs due to heavy cattle trampling and vegetation removal, but the volumes of both are very low compared to those from cropland.

Forestland. Forests occupy large regions of North America, principally near the coasts and in the mountainous areas. Much forest has been harvested and several regenerations have occurred. For other areas, principally in the western mountain ranges of the United States and Canada, first harvest of the forests has only occurred in the past 50 years when extensive road systems were built to permit access. The hydrologic impact of this timber harvesting has been debated and studied; the results generally show that there are increased runoff and sediment volumes during and immediately following the road building and tree removal, but reforestation returns the areas to near-natural conditions within a matter of years. With only small percentages of a region affected at any one time, the overall impact of this human activity is generally slight. Locally, however, the changes can be significant. A vast area, principally across northern Canada, has forest cover of no commercial value in today's market; thus, this region remains virtually in its natural vegetation state. The forest regions in Figure 14 are delineated as (8) forestland, with commercial value; and (9) forestland, with no commercial value.

Other land. The remainder of North America remains largely unaffected by man's activities, except for some occasional incursions such as irrigated deserts, drained swamps, or cities. But these effects are usually local. These other lands in Figure 14 have been delineated as (10) desert shrubland, (11) tundra, and (12) swampland and marshland.

TOPOGRAPHY

A drainage basin is the topographic area providing runoff to a selected point along a stream. It is an area limited by a drainage divide and occupied by a drainage network of channels that transport upstream drainage water and sediment to the lower parts. Drainage basins are of all sizes, and form a nested hierarchy as points are selected downstream. For example, the Mississippi River has drainage basins ranging from the entire Mississippi-Ohio-Missouri river system, one of the largest river systems on Earth, to a single, tiny, unbranched tributary in the badlands of the upper Missouri River basin.

The erosional drainage basin has been recognized as a viable process-response unit since the beginning of the last century. A significant appreciation of the drainage basin unit arose with the gradual understanding of the hydrologic cycle and the basin function for conveying precipitation runoff to the river.

To relate basin topographic features with hydrologic characteristics, the topographic features must be quantitatively represented. Under the impetus supplied by Horton (1932, 1945), the description of drainage basins and channel networks was transformed from a qualitative and dedutive study to a rigorous quantitative science capable of providing hydrologists with numerical data of practical value. Horton's work was supplemented by Langbein (1947), then developed in detail by Strahler (1964). Thornbury (1965) provided a regional description of the geomorphology of the United States. Standard methods for the quantitative derivation of the various indices of basin characteristics for hydrologic analysis have been recommended by the U.S. Geological Survey (1978).

There are many ways in which the physical characteristics of drainage basins may be expressed and an equal variety of methods whereby the drainage processes may be represented. An empirical parametric approach has been employed most often to relate hydrologic characteristics—such as peak flow, mean annual runoff, and low flow—to specific quantitative drainage basin characteristics. Mathematical modeling is used by hydrologists as a basic tool to derive such relations. Models vary from the simple relation of two variables to a number of variables analyzed by multivariate techniques to very complex mathematical deterministic models.

Taylor and Schwartz (1952) used the data of 20 basins ranging in area from 50 to 4,100 km^2 located in the northern and middle Atlantic states to study the effect of drainage basin characteristics upon unit-hydrograph lag and peak flow. Drainage area, length of longest water course, mainstream length to centroid of

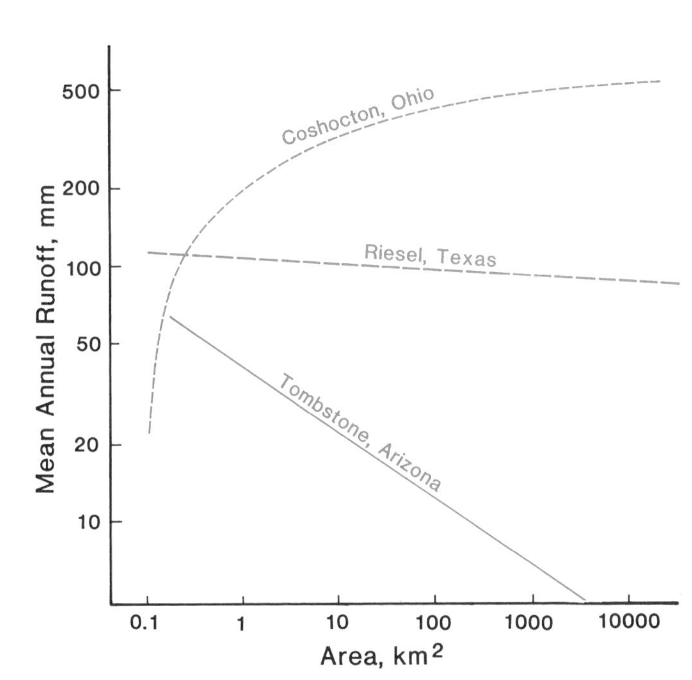

Figure 15. Relationship between mean annual runoff and basin area (Glymph and Holton, 1969).

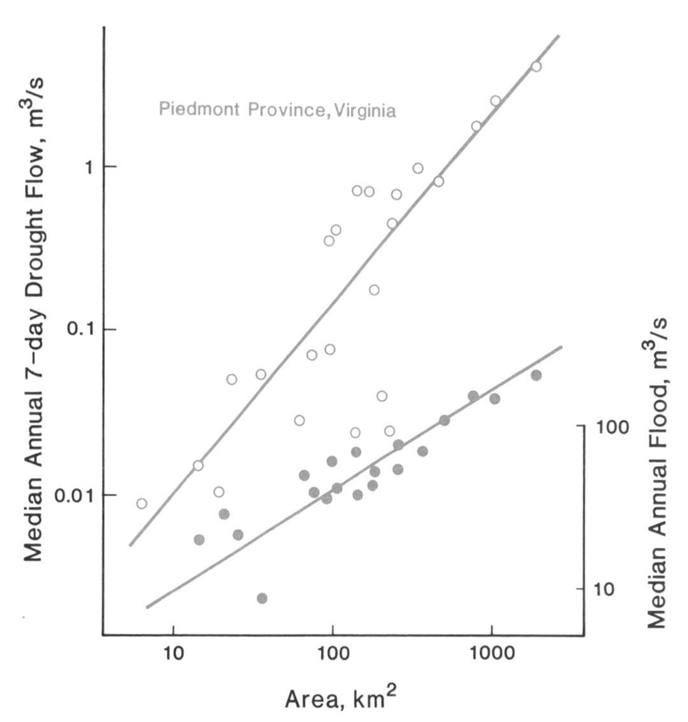

Figure 16. Relationship between flood flow and basin area (●) and low flow and basin area (○). (Giusti, 1962.)

area, and equivalent mainstream slope were judged the most significant physiographic variables with hydrologic influence.

Drainage basin area is the simplest basin characteristic related to drainage-basin processes. However, it is often difficult to interpret the significance of area because it is, in turn, correlated with many other basin characteristics. Glymph and Holton (1969) found three different relationships between mean annual runoff and basin area as shown in Figure 15. In the humid Appalachian region represented by Coshocton, Ohio, upland infiltration returns in part to channels downstream, causing a gain in runoff as basin area increases. In the arid regions such as Tombstone, Arizona, where channels absorb streamflow, runoff decreases with increasing basin size. In other areas where interflow is small and where channel gains and losses nearly balance, the runoff may not vary significantly with basin size, such as streams in the Texas blacklands near Riesel.

Giusti (1962) illustrated differences in flood flow and low-flow relationships with basin area in the Piedmont Province, Virginia, as shown in Figure 16. Both the 7-day low-flows and median annual flood flows are well correlated with the watershed area. However, where underlying geologic formations are distinctly different (e.g., limestone and shale), streamflow characteristics may not be closely correlated with drainage area.

The slope of the ground surface is a factor in the overland-flow process and, hence, is a parameter of hydrologic interest, especially for small basins where overland flow may be a dominant factor in determining hydrograph shape. The influence of basin slope is most relevant to indices of peak streamflow and to the hydrograph shape, as illustrated in Figure 17 for two basins with different longitudinal surface profiles.

The shape of the basin affects the hydrograph characteristics of lag time, the time of rise, and the peak-flow rates. Figure 18 shows the corresponding relationships between the basin shape and the hydrograph shape and time of rise. The significance of basin shape is often expressed through the pattern of the drainage network. As shown on Figure 19, the more compact efficient network (A) gives a slower rise but a higher peak, whereas a more dispersed network (B) produces a quick rise but a lower and more protracted hydrograph peak. These basin characteristic

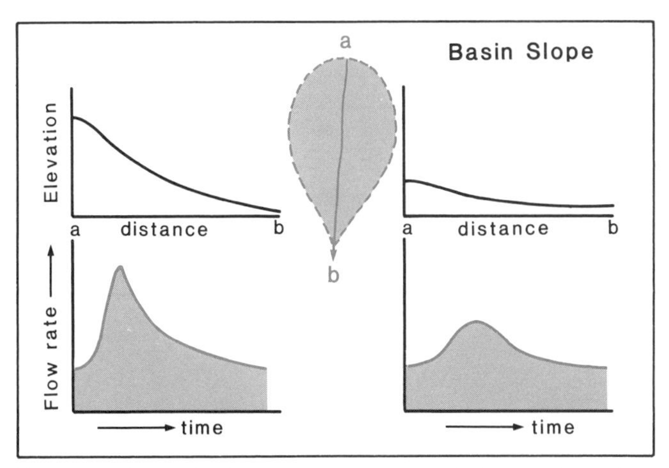

Figure 17. Effect of basin slope on the hydrograph shape and peak-flow rate.

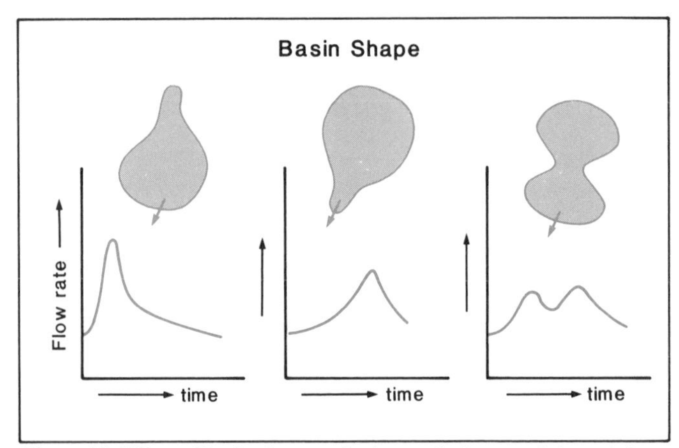

Figure 18. Effect of basin shape on hydrograph shape and time of rise.

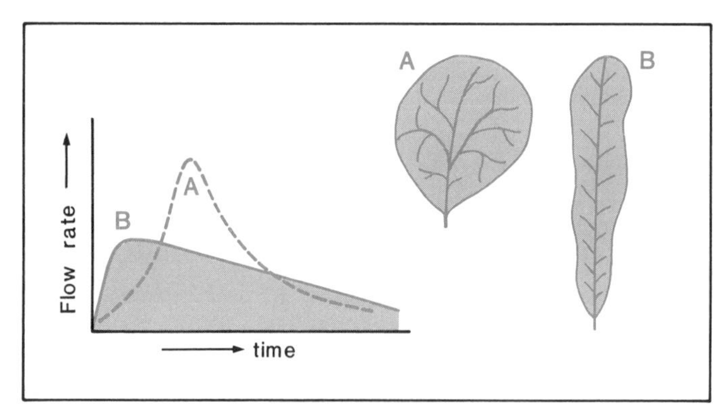

Figure 19. Effect of drainage network with contrasted bifurcation ratio on the hydrograph shape and peak-flow rate.

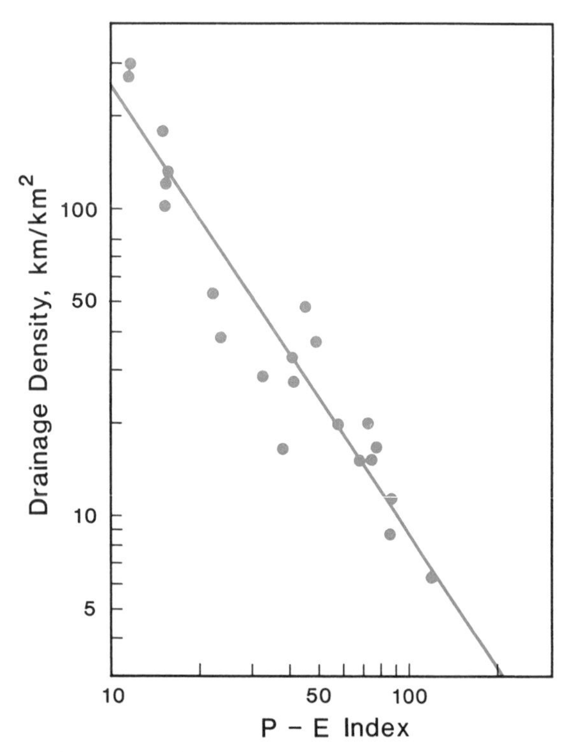

Figure 20. Relationship between drainage density and precipitation minus evaporation (P–E) index (Melton, 1957).

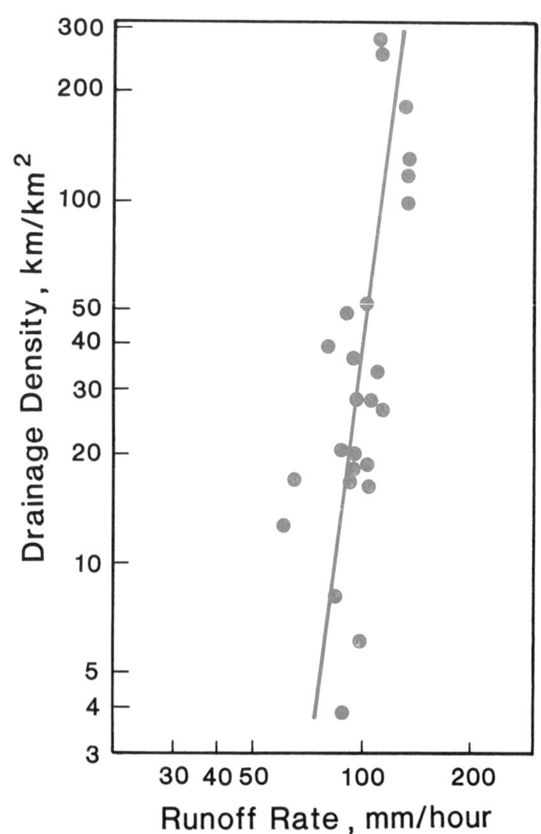

Figure 21. Relationship between runoff rate and drainage density (Melton, 1957).

effects are added to the precipitation and infiltration effects illustrated in Figure 7 to provide the complete hydrologic response of any specific watershed.

A large channel length per unit of watershed area (high drainage density) reflects a highly dissected basin that usually responds rapidly to a rainfall input. A low drainage density reflects a poorly drained basin with slow hydrologic responses. Observed values of drainage density range from as low as 3 km/km^2 in parts of the Appalachian area to 400 km/km^2 or more in Badlands National Monument, South Dakota.

Low drainage densities occur where soil materials are resistant to erosion, or conversely, are very permeable, and where the relief is small as in the Coastal Plain region of the southeastern United States. High values may be expected where soils are easily eroded or relatively impermeable, slopes are steep, and vegetal cover is scanty as in the Dakota badlands. The hydrologic significance of drainage density results from the fact that water and sediment yields are very much influenced by the length of water courses per unit area. Drainage density is developed by long-term hydrology and erosion, but it in turn affects the amount and rate of watershed discharge. Accordingly, attempts have been made to relate drainage density to climatic inputs and also to drainage-basin outputs.

A number of examples illustrate the relationships between drainage density and hydrologic response. Melton (1957) used climatic indices to show that drainage density is inversely related to the Thornthwaite precipitation effectiveness (P–E) index (precipitation minus evaporation) for 23 basins in Colorado, Utah, Arizona, and New Mexico (Fig. 20). Melton (1957) also found that runoff rate is directly related to drainage density for the same 23 basins (Fig. 21). In his study of drainage density and streamflow for the northeastern United States, Carlston (1963) found that basins with higher drainage density produce higher floodflow, as shown in Figure 22.

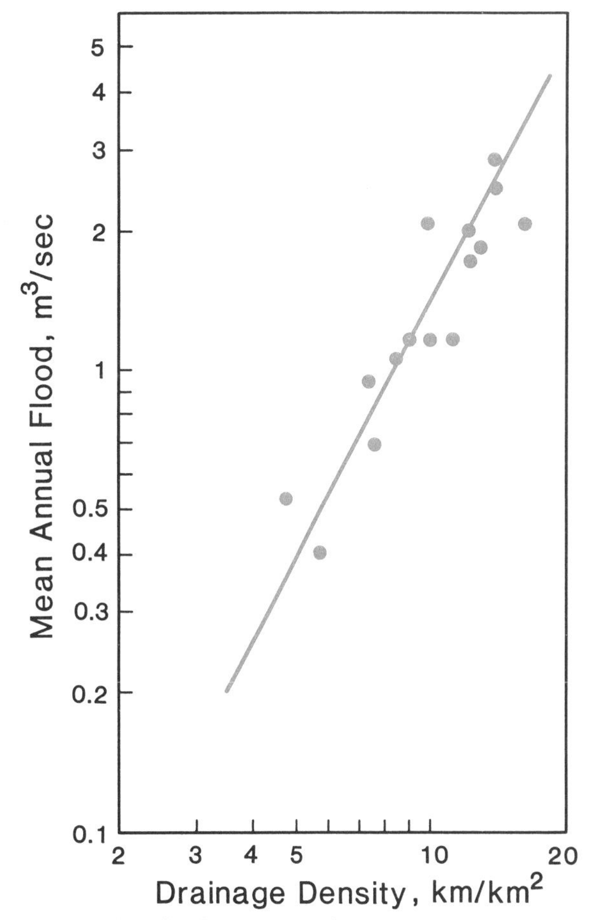

Figure 22. Relationship of mean annual flood to drainage density (Carlston, 1963).

REFERENCES CITED

Atwood, W. W., 1940, The physiographic provinces of North America: New York, Ginn Publishers, 536 p.

Austin, M. E., 1965, Land resource regions and major land resource areas of the United States: U.S. Department of Agriculture Soil Conservation Service Agricultural Handbook no. 296, 82 p.

Burford, J. B., 1972, Hydrologic data for experimental watersheds in the United States, 1966: Washington, D.C., U.S. Government Printing Office, U.S. Department of Agriculture Miscellaneous Publication 1226, p. 71.1–2.

Burford, J. B., and Clark, J. M., 1973, Hydrologic data for experimental watersheds in the United States, 1967: Washington, D.C., U.S. Government Printing Office, U.S. Department of Agriculture Miscellaneous Publication no. 1262, p. 71.1–4.

Campbell, G. S., 1985, Soil physics with BASIC; Transport models for soil-plant systems, *in* Developments in soil science: Amsterdam, Netherlands, Elsevier Scientific Publishing Company, 150 p.

Carlston, C. W., 1963, Drainage density and streamflow: U.S. Geological Survey Professional Paper 422C, p. C1–C8.

Chow, V. T., 1964, Runoff, *in* Handbook of applied hydrology: New York, McGraw-Hill Book Company, p. 14.1–14.54.

FAO-UNESCO, 1975, Soil map of the world; Volume II, North America: Paris, United Nations Educational, Scientific, and Cultural Organization, 220 p.

Finkl, C. W., ed., 1982, Soil classification, *in* Benchmark papers in soil science, v. 1: Stroudsburg, Pennsylvania, Hutchinson Ross Publishing Company, 391 p.

Gerlach, A. C., ed., 1970, The national atlas of the United States of America: Washington, D.C., U.S. Geological Survey, scale 1:7,500,000 and 1:17,000,000.

Giusti, E. V., 1962, A relation between floods and drought flows in the Piedmont province in Virginia: U.S. Geological Survey Professional Paper 450C, p. C1–C146.

Glymph, L. M., and Holton, H. N., 1969, Land treatment in agricultural watershed hydrology research, *in* Moore, W. L., and Morgan, C. W., eds., Effects of watershed change on streamflow: American Society of Civil Engineers, p. 44–68.

Hillel, D., 1980, Fundamentals of soil physics: New York, Academic Press, 413 p.

Horton, R. E., 1932, Drainage basin characteristics: EOS American Geophysical Union Transactions, v. 13, p. 350–361.

——, 1945, Erosional development of streams and their drainage basins; Hydrophysical approach to quantitative morphology: Geological Society of America Bulletin, v. 56, p. 275–370.

Hunt, C. B., 1967, Physiography of the United States: San Francisco, California, Freeman Publishers, 480 p.

Langbein, W. B., 1947, Topographic characteristics of drainage basins: U.S. Geological Survey Water-Supply Paper 968-C, p. 125–157.

Meinzer, O. E., ed., 1942, Hydrology: New York, Dover Publications, 712 p.

Melton, M. A., 1957, An analysis of the relations among elements of climate, surface properties, and geomorphology: Office of Naval Research, Geography Branks Project NR 389-042, Technical Report 11, Columbia University.

Muir, J. W., 1962, The general principles of classification with reference to soils: Journal of Soil Sciences, v. 13, p. 22–30.

Nielsen, D. R., Jackson, R. D., Cary, J. W., and Evans, D. D., eds., 1972, Soil water: Madison, Wisconsin, Soil Science Society of America, 175 p.

Saxton, K. E., and Bluhm, G. C., 1982, Regional prediction of crop water stress by soil water budgets and climatic demand: American Society of Agricultural Engineers Transactions, v. 25, no. 1, p. 105–115.

Saxton, K. E., and McGuiness, J. L., Evapotranspiration, *in* Haan, C. T., Johnson, H. P., and Brakensick, D. L., eds., Hydrologic modeling of small watersheds: American Society of Agricultural Engineers Monograph 5, p. 229–273.

Saxton, K. E., Johnson, H. P., Shaw, R. H., 1974, Modeling evapotranspiration and soil moisture: American Association of Agricultural Engineers Transactions, v. 17, no. 4, p. 673–677.

Saxton, K. E., Rawls, W. J., Romberger, J., and Papendick, R. I., 1986, Estimating generalized soil water characteristics from texture: Soil Science of America Journal, v. 50, no. 4, p. 1031–1036.

Sharp, A. L., and Saxton, K. E., 1963, Transmission losses in natural stream valleys: American Society of Civil Engineers Hydraulic Division Journal, v. 88, no. HY4, p. 121–142.

Soil Survey Staff, 1960, U.S. Department of Agriculture soil classification; A comprehensive system, 7th approximation: Washington, D.C., U.S. Government Printing Office.

Strahler, A. N., 1964, Quantitative geomorphology of drainage basins and channel networks, *in* Chow, V. T., ed., Handbook of applied hydrology, section 4-II: New York, McGraw-Hill, p. 4.39–4.76.

Taylor, A. B., and Schwartz, H. E., 1952, Unit-hydrograph lag and peak flow related to basin characteristics: EOS American Geophysical Union Transactions, v. 33, p. 235–246.

Thornbury, W. D., 1965, Regional geomorphology of the United States: New York, John Wiley and Sons, 609 p.

Thornthwaite Associates, 1964, Average climatic water balance data of the continents; Part VII, United States: Publications in Climatology, v. 17, no. 3, 615 p.

U.S. Geological Survey, 1978 (1981), National handbook of recommended methods for water-data acquisition; Chapter 7, Physical basin characteristics for hydrologic analyses: U.S. Geological Survey Office of water Data Coordination, loose-leaf format, 38 p.

USDA-SCS, 1955, Engineering handbook; Section 4, Hydrology: U.S. Department of Agriculture Soil Conservation Service, 608 p.

MANUSCRIPT ACCEPTED BY THE SOCIETY NOVEMBER 15, 1987

Printed in U.S.A.

The Geology of North America
Vol. O-1, Surface Water Hydrology
The Geological Society of America, 1990

Chapter 4

Temporal and spatial variability of streamflow

H. C. Riggs
U.S. Geological Survey, 415 National Center, Reston, Virginia 22092
K. D. Harvey
Environment Canada, Place Vincent Massey, 8th Floor, Ottawa, Ontario K1A 0E7, Canada

INTRODUCTION

Streamflow occurs as a result of the interaction of the many components of the hydrologic cycle, as shown in Figure 5 of Riggs and Wolman (this volume). The streamflow part of the cycle is driven by precipitation in the form of rain or snow. The complex processes of interception, infiltration, evaporation, and transpiration (Saxton and Shiau, this volume) serve to reduce the amount of water available for runoff at any one time and at any one location. These proceses are influenced by various climatic factors, such as precipitation, temperature, wind, and solar radiation; and by physiographic factors related to geology and topography. As a result, streamflow varies considerably both in time and space. For example, see the variation of daily streamflows in Figure 8 of Riggs and Wolman (this volume). Lins and others (this volume) and Saxton and Shiau (this volume) describe the climatic and physiographic variations throughout the continent.

This chapter describes the variation of runoff (the volume of flow for a given time) in North America and explains the major climatic and physiographic factors underlying this variation. Following a description of common terminology related to runoff, the variation of runoff in various parts of the continent is presented on maps and graphs. The next section describes the effects of climatic, topographic, and geologic factors on runoff using appropriate examples. A final section documents the integrated effects of climate and physiography within a specific watershed as runoff drains from the headwaters to the basin outlet.

Comprehensive treatments of extreme streamflows—floods at the one extreme and low flows and droughts at the other—are presented in Matthai (this volume) and Rogers and Armbruster (this volume), respectively.

FLOW DESCRIPTORS

The streamflow record computed from the data collected at a gaging station consists principally of mean flows for each day. These are called daily mean flows. The following other descriptors are computed from them.

Mean flow is the average flow for all complete years of record, expressed as a rate of discharge (m^3/s). Mean annual flow, or mean annual runoff, is equal to mean flow when expressed as a rate of discharge. Alternatively, mean annual runoff can be expressed as the depth of water over the drainage area in the average year. For example, a mean flow of 117 m^3/s would, in 1 year, cover a 3,351 km^2 drainage basin to a depth of about 1,100 mm. Therefore, the mean annual flow for this basin is 117 m^3/s or 1,100 mm. The expression of runoff as a depth of water facilitates the assessment of the area distribution of runoff, as well as the comparison of runoff with causal precipitation.

Annual mean flow is the mean flow for a particular year. A sequence of annual mean flows is shown in Figure 1. Taking the mean of these annual mean flows gives the "mean annual flow" discussed previously. Typical variation of annual mean flows is shown by the frequency curve of Figure 2. Variations of annual mean flows with time can be seen by a curve of moving averages such as that drawn on Figure 1. A moving average sometimes indicates a time trend such as might result from some physiographic change within the basin or from climatic change. More commonly the evidence provided by the moving average is inconclusive, particularly if the streamflow record is short.

Monthly mean flow is the mean flow for a particular calendar month of a specific year. Mean monthly flow is the mean of all recorded monthly mean flows for a particular month.

Annual maximum flows are also important descriptors of streamflow. Annual maximum daily mean flow and annual maximum instantaneous flow are used in the analysis of floods. Annual minimum n-day flow (where n = 1 or more) is a characteristic used in the study of low flows. These are treated extensively in Matthai (this volume) and Rogers and Armbruster (this volume), respectively.

OCCURRENCE AND DISTRIBUTION OF ANNUAL RUNOFF

North America is characterized by many different climatic and physiographic regions, with the result that runoff varies con-

Riggs, H. C., and Harvey, K. D., 1990, Temporal and spatial variability of streamflow, *in* Wolman, M. G., and Riggs, H. C., eds., Surface water hydrology: Boulder, Colorado, Geological Society of America, The Geology of North America, v. O-1.

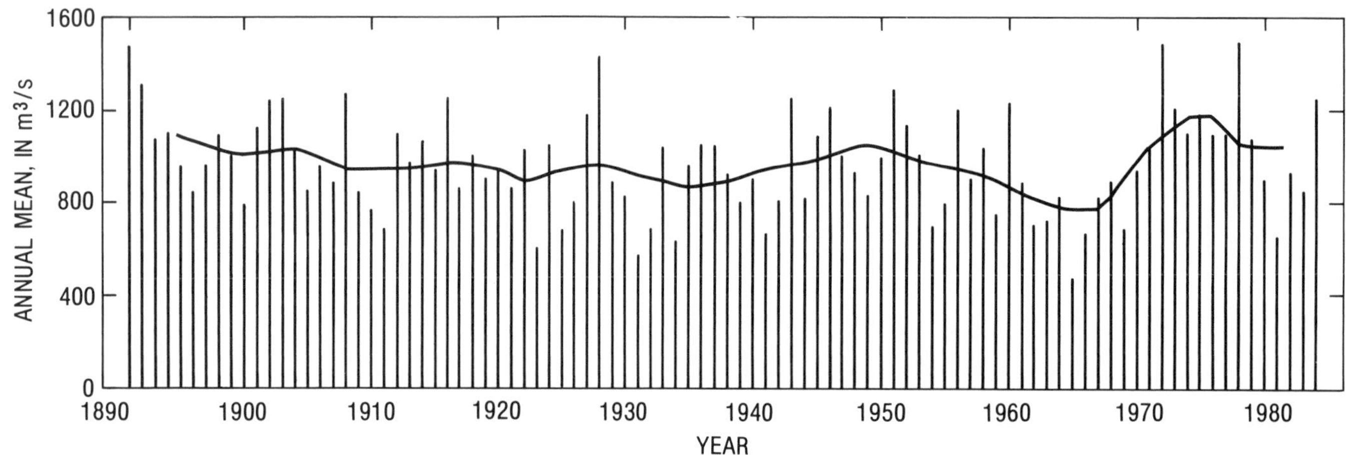

Figure 1. Annual mean flows of Susquehanna River at Harrisburg, Pennsylvania, and the 9-year moving average.

siderably. Plate 3A shows ranges of mean annual runoff, in millimeters, throughout the continent. These ranges are highly generalized because the causal climatic and physiographic factors can be highly variable over short distances, particularly in mountainous regions. Annual runoff ranges from near zero in central Mexico and Yucatan to more than 4,000 mm in eastern Costa Rica and along the western coast from Alaska to Washington State. Runoff generally is low in the western portion of the interior of the continent, and it becomes increasingly abundant toward the eastern coast. The spatial variability of runoff is greatest in mountainous regions because of the influence of topography and prevailing wind direction on precipitation and resulting runoff. Similarities of Plate 3A to the precipitation map of Lins and others (this volume) are apparent, and this underlines the importance of precipitation to surface runoff.

TEMPORAL VARIATION OF ANNUAL RUNOFF

The runoff map of Plate 3A is based on data from many sites and represents long-term average conditions. Annual mean flows at each site vary from year to year in response to year-to-year changes in the weather—particularly precipitation—and the hydrologic response of the drainage basin to these changes. The variability of precipitation from year to year, described in Lins and others (this volume), is greatest in semiarid and arid regions. Annual runoff is more variable than annual precipitation because it is the residual after the less-variable evapotranspiration losses have been subtracted as described in Riggs and Wolman (this volume).

Runoff variabilities of Pascagoula River in humid Mississippi and of Brazos River in a semiarid region of Texas are shown in Figure 3; frequency curves based on these data are shown in Figure 4. The slopes of these curves provide a good indication of

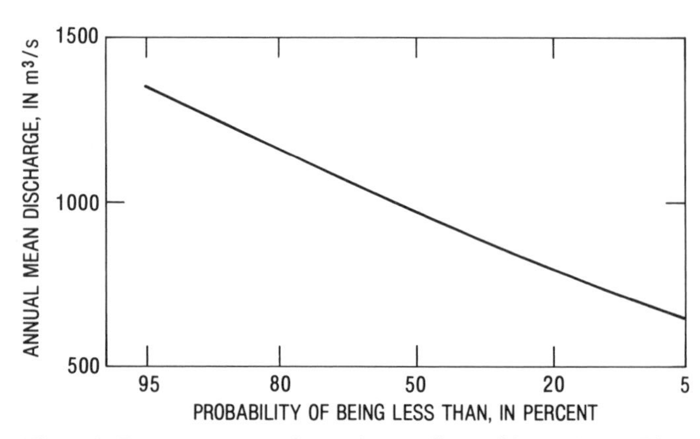

Figure 2. Frequency curve of annual mean flows of Susquehanna River at Harrisburg, Pennsylvania, 1891 through 1984.

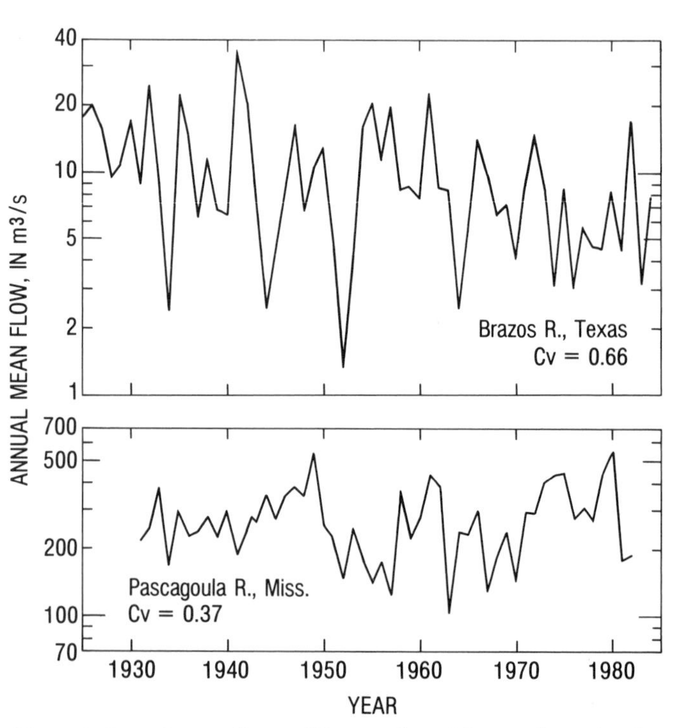

Figure 3. Annual mean flows of Pascagoula and Brazos rivers showing the higher variability, and the higher coefficient of variation (CV), of the flows of Brazos River, which drains a semiarid basin.

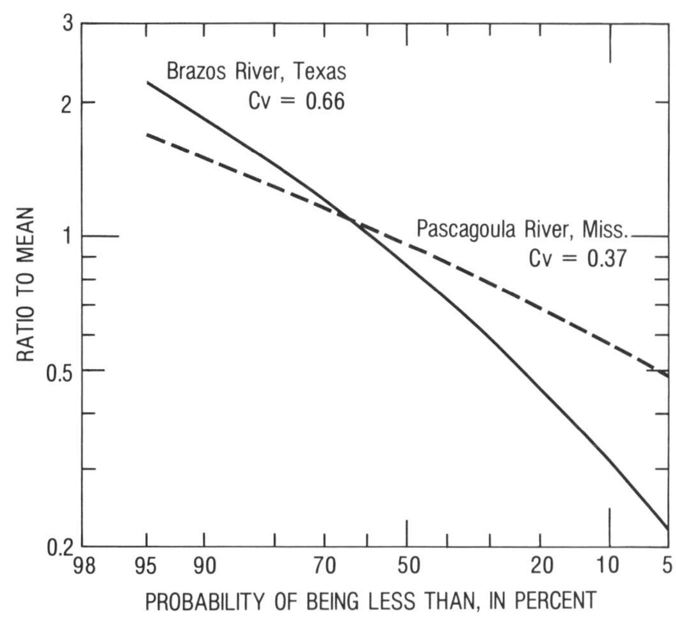

Figure 4. Frequency curves based on the annual mean flows graphed in Figure 3.

the variability of flow; the steeper the slope, the greater the variability. Thus the annual flow of Pascagoula River is less variable than that of Brazos River.

A statistical measure of variation is the coefficient of variation, defined as the ratio of the standard deviation of a variable to its mean. This parameter ranges from zero, for no variation, to a number greater than one, for extreme variation. The coefficients of variation of the annual flows of the Pascagoula and Brazos Rivers are 0.37 and 0.66, respectively, indicating, as before, that the annual flow of Pascagoula River is less variable in time than that of Brazos River.

The variabilities of annual flows of many rivers in North America are shown by the coefficients of variation in Figure 5. Although in general, low variability is associated with high annual precipitation, and high variability with low precipitation, many exceptions are shown. These exceptions are due to local conditions, usually the existence of a large ground-water body or a large surface storage, both of which tend to reduce flow variability. Consequently the coefficients shown are not necessarily representative of the immediately surrounding regions in which they appear. Also, the coefficients for rivers in Mexico and Central America are based on limited information and therefore may not indicate long-term characteristics.

FLOW REGIMES

The distribution of streamflow throughout the year at a given point—the flow regime—can be shown graphically as the ratio of mean monthly flow to mean annual flow, plotted by months. The flow regimes of selected rivers of North America are shown on Plate 3B.

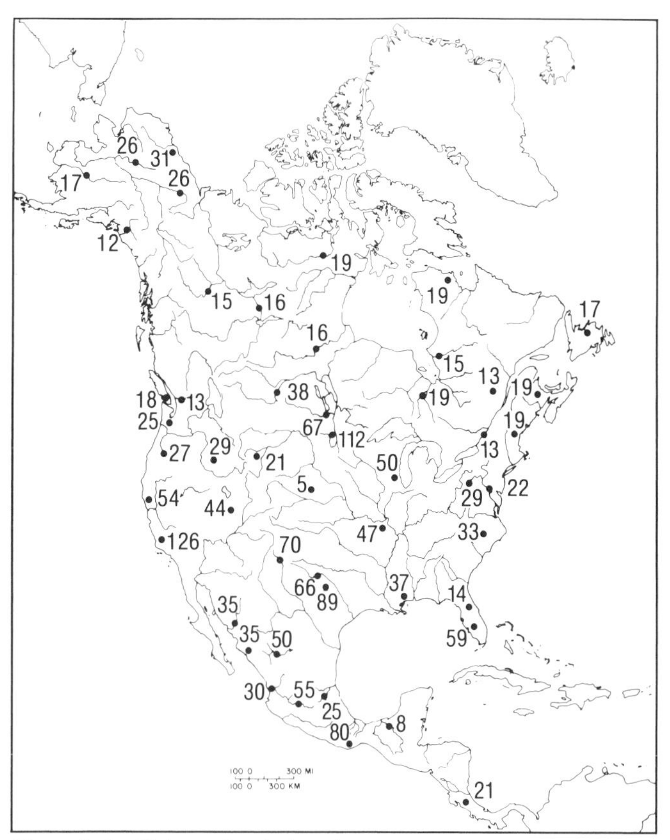

Figure 5. Coefficients of variation of annual mean flows, in percent, for selected streams.

The variation of the flow regimes across the continent is largely due to the variability of the climatic factors of precipitation and temperature. In temperate regions there is a striking similarity of the runoff regime to the precipitation regime. In colder regions this similarity is limited to the snow- and ice-free months of the year. Some of the regimes shown reflect hydrologic conditions nontypical of the climatic region. Examples of these will be described later.

Since the graphs of Plate 3B are based upon mean monthly flows, they best represent flow regimes of streams for which the coefficients of variation of the monthly flows are small. In general, the flow regime observed in a given year will be most similar to the mean flow regime if the precipitation regime is consistent from year to year, or if the major portion of annual runoff results from late spring to summer snowmelt. The flow regime observed in a given year may be significantly different from the mean flow regime if the basin is subject to extreme climatic events, such as hurricanes or droughts.

A study of Plate 3B reveals similar runoff regimes in certain regions and gradual changes in regime with distance to other regions. For instance, coastal streams in the northwestern United States and Canada experience high runoff in winter and relatively low runoff in summer. As one proceeds eastward to the midcontinent, the maximum monthly runoff trends toward spring and summer, the result of colder temperatures, different rainfall patterns, and spring snowmelt. From Mexico south, precipitation reaches its peak in late summer, and this is reflected by the runoff regimes. Regimes of far northern streams are dominated by temperature; maximum runoff is due to snowmelt in early summer.

A detailed classification of flow regimes was developed by L'vovich (1979) who used three categories of each of four types of major inputs (snow, rain, glacier, and ground water) by each of four seasons. This resulted in 40 types of identified flow regimes. His classification was used by UNESCO (1978) as the basis for outlining regions having similar flow regimes throughout the world. The UNESCO generalization for North America is shown in Figure 6. These areas of similar regime are not coincident with areas having similar rates of annual runoff, as can be seen by comparing Figure 6 with Plate 3A. No attempt was made to define a regional classification of flow regime on Plate 3B. Both flow regime and runoff magnitude are affected by climate, geology, and topography, but the effects differ in time and space because of the wide variations and complex nature of the hydrologic relationships involved.

The following sections explain some of the effects of major climatic and physiographic factors that underlie the variation in runoff regimes in North America.

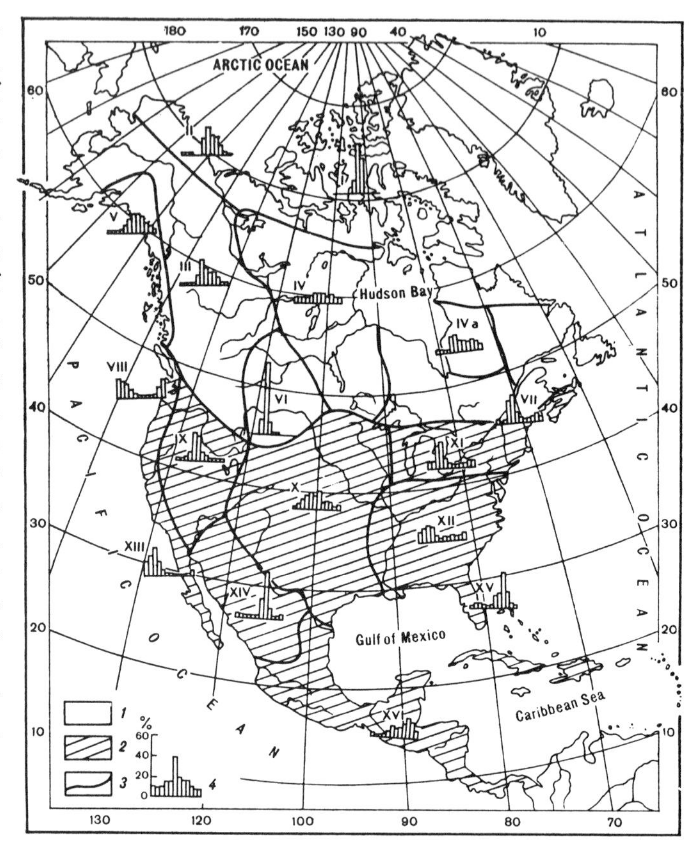

Figure 6. Regions with different flow patterns (from UNESCO, 1978). 1, rivers with floods; 2, rivers with a freshet cycle; 3, boundaries of regions with the same annual runoff pattern; 4, runoff hydrograph typical of the region.

CLIMATIC FACTORS

Precipitation and temperature are the two major climatic factors that influence runoff. Precipitation is the input to the runoff-generating process; and temperature acts as a process regulator, controlling the form of precipitation as rain or snow and the natural storage of water on and in the ground as snow and ice. Temperature also greatly influences the rate of evaporation and transpiration of water back into the atmosphere. Both spatial and temporal variations of runoff result from the interaction of these factors.

Precipitation

Annual precipitation is always the major climatic factor contributing to annual runoff. Figures in Lins and others (this volume) show the variation of mean annual precipitation across North America and the variation of precipitation regimes. In regions where temperatures are normally above freezing year round, the precipitation regime is the major determinant of the runoff regime. Such regions occur across the cotninent south of about 32° N latitude and along the west coast as far north as 60° N latitude.

Nanaimo River in British Columbia and Rio Cocle del Norte in Panama are but two examples where the annual runoff regime (Plate 3B) strongly reflects the seasonal patterns of precipitation. Both of these rivers flow in humid regions that experience abundant and consistent seasonal precipitation. Generally, the rainfall-runoff relationship becomes less pronounced as climate tends from humid to arid.

Streamflow responds promptly to precipitation in arid and semiarid regions when storm rainfall is intense. However, such regions usually experience large losses from evapotranspiration and infiltration during and following rainfall events. Less intense rainfalls produce little or no runoff. Accordingly, a runoff regime, such as that of San Pedro River, Arizona, (Plate 3B), normally reflects the precipitaton regime only during wetter periods.

In humid regions, relatively minor rates and amounts of precipitation can increase or maintain streamflow during periods of nearly continuous cloud cover when evapotranspiration losses are low and the soil is continuously saturated. Even "fog drip," resulting from the condensation of fog on vegetation, can increase or maintain flow in such climates.

Precipitation varies considerably from year to year in some regions, humid or arid. Consequently, the runoff regime for a given year can be very different from the average regime. For example, Figure 7 shows the mean monthly flows of Canoochee River, Georgia, for the period 1951 to 1960 and the monthly mean flows for a high and low runoff year in that period. Similar deviations of individual-year regimes from the average are shown in Figure 8 for Wakarusa River, which drains a semiarid region in Kansas. This stream ceases to flow in occasional years. Where

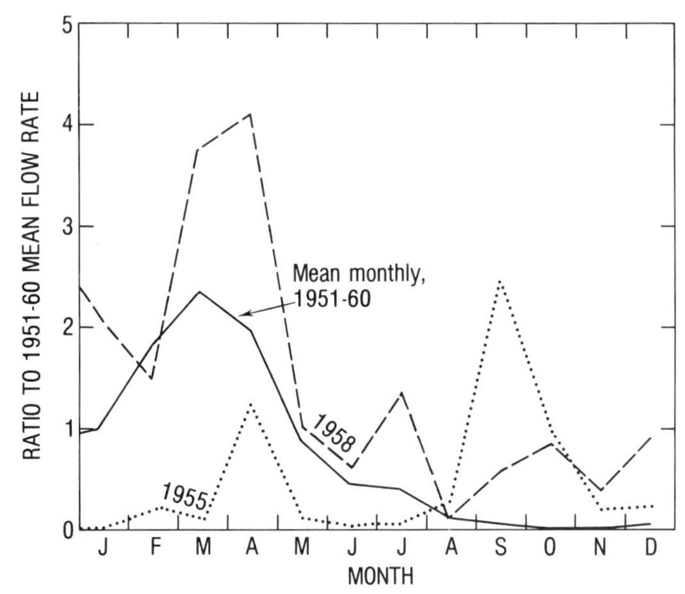

Figure 7. Mean flow regime and monthly mean flows for two individual years, Canoochee River, Georgia.

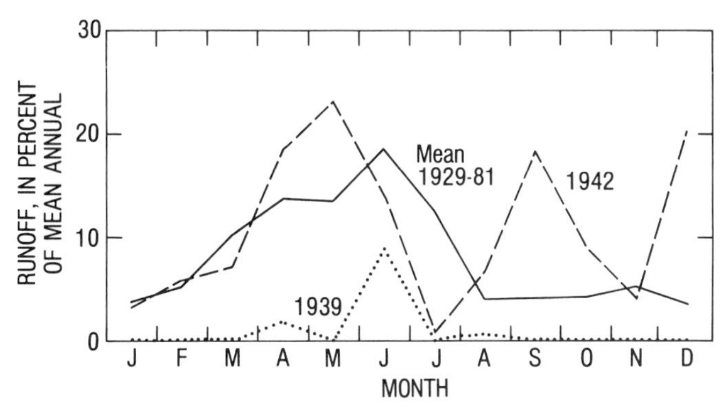

Figure 8. Mean flow regime and monthly mean flows for two individual years, Wakarusa River, Kansas.

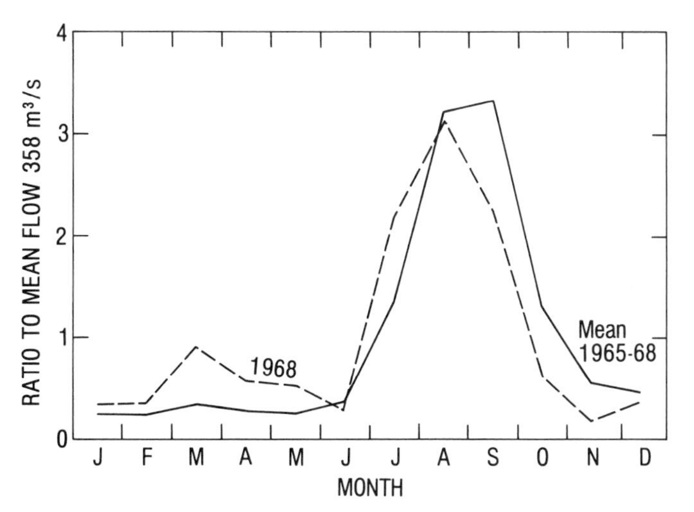

Figure 9. Mean flow regime of Rio Santiago, Mexico, and regime for 1968.

the precipitation regime is more consistent, however, the variation of the flow regime can be significantly less. Figure 9 shows such a case for Rio Santiago in Mexico.

The total flow in a year can also vary significantly from the mean annual flow, as was shown earlier by frequency curves of annual flow (Fig. 4) and by coefficients of variation (Fig. 5). The variability of both annual and monthly flows is greatest where the drainage basins are subject to occasional heavy precipitation, such as that associated with hurricanes.

In arid and semiarid regions, many streams are ephemeral, experiencing extended periods of no streamflow. In such streams, flow may not always be zero in the same months each year. Consequently, the runoff regime, as defined using mean monthly flows, could be misleading. The regime of Goose Creek, North Dakota, in Plate 3B, shows that all mean monthly flows are greater than zero; but monthly mean flows of zero did occur from October through February of three consecutive years and have occurred in July, August, and September.

In regions north of about 35° N latitude, excluding the west coast as far north as 60° N latitude, a considerable proportion of annual precipitation falls during the winter months in the form of snow. The occurrence of precipitation as snow has a profound effect on flow regimes. Runoff can be delayed for many months before the snow melts and runs off in the spring or early summer. In these regions, the spring snowmelt runoff usually accounts for more than 60 percent of the total annual runoff. In the semiarid midwestern plains, the proportion is commonly more than 80 percent. Plate 3B shows how the snowmelt runoff period dominates the flow regimes in the northern regions.

Temperature

Temperature affects every phase of the hydrologic cycle and, therefore, greatly influences surface runoff. The form of precipitation as rain or snow is governed mainly by temperature; the storage and movement of water on or in the ground is highly dependent upon temperature; and the return of water to the atmosphere through evapotranspiration is largely a function of temperature. This section focuses on the effects of temperature on runoff with respect to snow and ice, permafrost, glaciers, and evapotranspiration.

Snow and ice. Regions north of about 35° N latitude (excluding the lowlands of the west coast) experience extended periods of subzero temperatures. Of course, the higher the latitude, the longer are these extended periods. In arctic regions, freezing temperatures are common for 8 months of the year.

The major impact of temperature on runoff in such regions results from the temporary storage of water as snow and ice. Precipitation occurs as snow, and ice forms on stream channels, lakes, and ponds, and to a limited extent in the soil. The flow regimes of streams so affected show low runoffs during winter and major runoffs in spring or early summer when the snow and ice melt. The high summer runoffs shown for northern streams in Plate 3B do not result entirely from snowmelt; considerable summer precipitation occurs in that region.

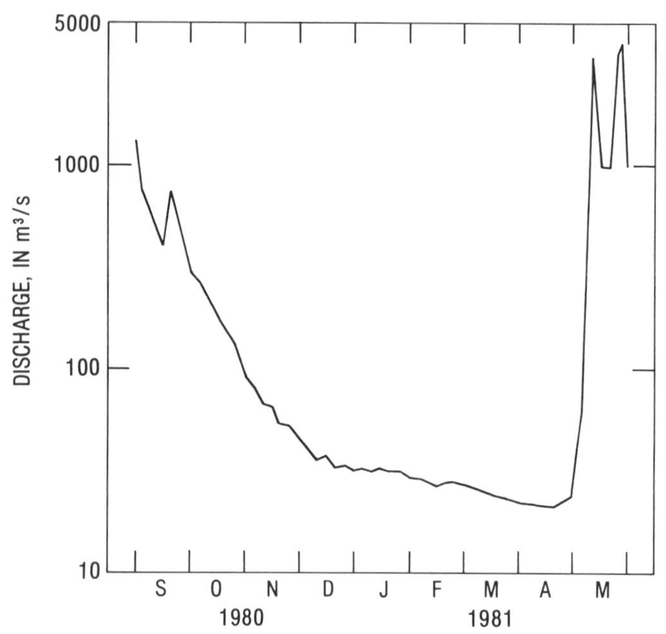

Figure 10. Winter flow of Porcupine River, Yukon Territory, showing recession during period of ice cover.

During the freezeup period, the flow in stream channels is decreased as surface ice is formed, surface runoff becomes zero, and ground-water movement is diminished because of the increased viscosity at lower temperatures. Consequently, when a channel is completely ice covered, streamflow derived entirely from ground water recedes gradually throughout the winter, as shown by the hydrograph of winter flow of Porcupine River (Fig. 10).

In some small streams such as Kuparuk River, Alaska, (Plate 3B), ground-water flow to the channel ceases during winter. In others, ground-water inflow continues throughout the winter, but the water freezes when it reaches the stream channel, resulting in an accumulation of aufeis that may fill the channel and extend to riparian areas. Figure 11 is a photo of this phenomenon.

In less extreme climates, winter flows may fluctuate considerably because periods of freezeup are interspersed with periods of melting. The effect of temperature fluctuations on flow of Salmon River, Idaho, is demonstrated by the data shown in Figure 12. The flow regimes for such streams often show a low-flow period in January or February, however.

Large daily fluctuations in discharge of small streams result from the formation and release of anchor ice, which occurs where temperature differences between night and day are great, usually at high elevations at moderate latitudes. Radiation from the earth during clear cold nights causes ice to form on the beds of streams, but never under ice cover or under bridges. Radiation from the

Figure 11. Ice formation due to ground-water inflow has encroached on car and building along an Alaskan stream (photo by Fairbanks Daily News-Miner).

sun releases the anchor ice, which floats away (Hoyt, 1913). Typical fluctuations of stream stage are shown in Figure 13.

The melting of the snowpack of a watershed can occur in a matter of days or months, depending mainly upon the temperature gradients experienced in the watershed. Temperatures are relatively uniform over a basin of low relief, such as the Goose Creek watershed in North Dakota, and melting of the snowpack occurs throughout the entire basin at about the same time. Resulting runoff is concentrated in a short period, as shown by the Goose Creek regime in Plate 3B. More typically, however, a basin has a sufficiently large range in elevation or is large enough in area such that temperatures are not uniform over the basin. As a result, snowmelt runoff is prolonged, with the areas at the lower elevations contributing runoff first and those at the higher elevations contributing last. The North Saskatchewan River in the Canadian Prairies (Plate 3B) provides an example of this phenomenon in the extreme: an early runoff event occurs in March or April due to snow melting from the Prairie portion of the watershed, and a second runoff event occurs in May or June as a result of snow melting in the headwaters of the watershed, the Rocky Mountains.

Snow accumulates for periods of 5 months or more at high elevations as far south as 35° latitude in western North America, and on the high volcanic peaks in Mexico. The spring and summer flows of streams draining such areas are augmented by snowmelt runoff only from the areas at high elevation. Any snow at lower elevations will have melted earlier. Melt of the high-elevation snowpack may continue into August. Late-summer melt usually indicates the presence of a glacier or permanent snowfield in the headwaters.

The timing and the amount of runoff from a snowpack in a particular basin depend on the water content of the pack, the sequence of temperatures during spring and summer, and the amount of rainfall, if any, during the melt period. The extent of the variation in runoff from snowmelt is shown by the hydrographs of Lamoille Creek, Nevada, for 3 years (Fig. 14). Lamoille Creek has its source above 3,000 m elevation and is gaged at an elevation of 1,940 m. Its drainage area is 65 km^2.

Runoff from a melting snowpack may fluctuate throughout the day during periods of warm days and cold nights. Daily mean fluctuations may also be considerable, as shown by data for Boise River, Idaho, in Figure 15. The basin has a mean elevation of about 1,900 m, and ranges from nearly 3,000 m to 960 m.

The average timing of annual spring snowmelt events is very much dependent upon latitude, since warm temperatures arrive earlier in the southern regions. Accordingly, the Raystown Branch, Juniata River, Pennsylvania (Plate 3B), normally experiences its major snowmelt runoff event in March, whereas, on Moose River, Ontario, this event usually occurs in late April or May. In the lowlands and plains of the arctic regions, the maximum snowmelt runoff usually occurs in June when melting temperatures are assisted by up to 24 hours of solar radiation.

Permafrost. The geographic extent of permafrost in the polar and alpine regions of North America is shown in Riggs and Wolman (this volume). Most of Alaska and half of Canada is underlain by this frozen soil condition, which is generally continuous in the north and discontinuous in the south.

Little or no free ground water exists or moves in regions of continuous permafrost conditions. An active layer of soil or rock above the permafrost thaws in the summer and freezes in the winter. Consequently, the most significant effect of permafrost on

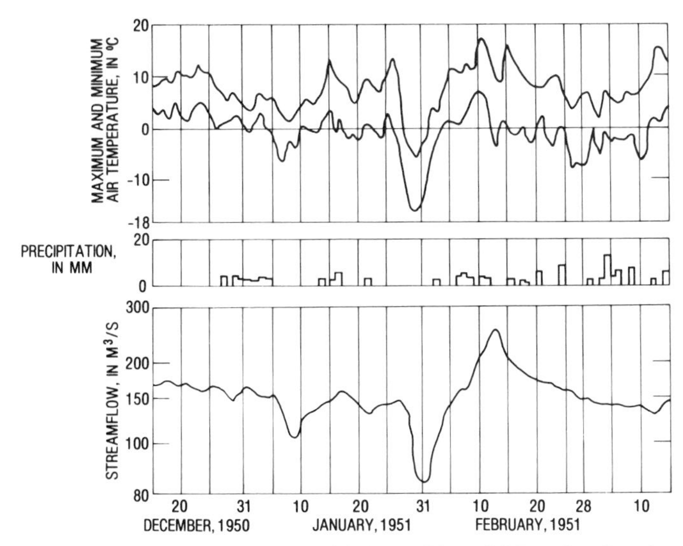

Figure 12. Winter hydrograph of Salmon River, Idaho, showing the effect of temperature on streamflow (after Simons, 1953).

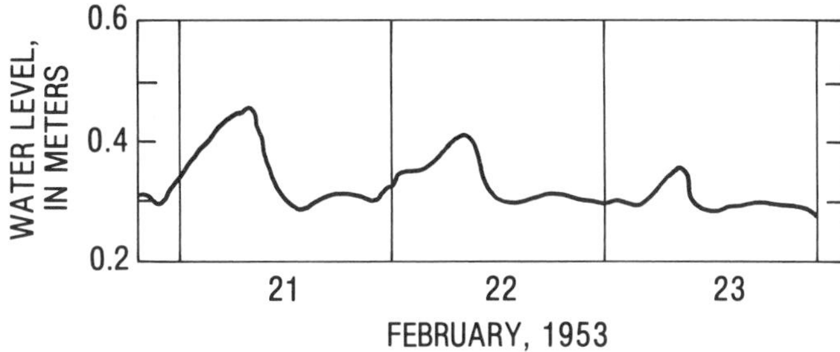

Figure 13. Diurnal fluctuation of stream stage due to formation and release of anchor ice (after Moore, 1957).

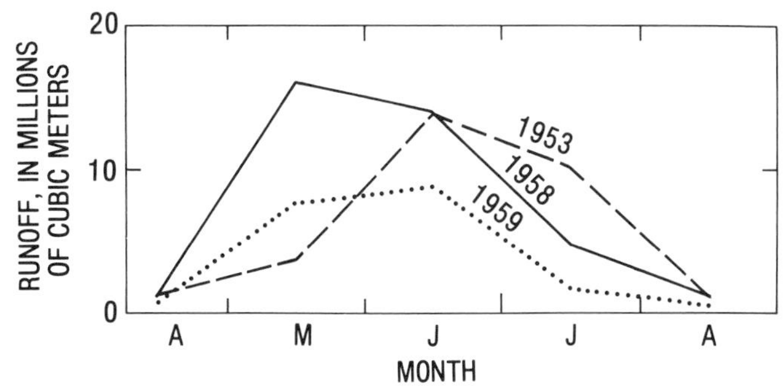

Figure 14. Snowmelt runoff of Lamoille Creek, Nevada, depends on the water content of the snowpack in April and on the subsequent weather.

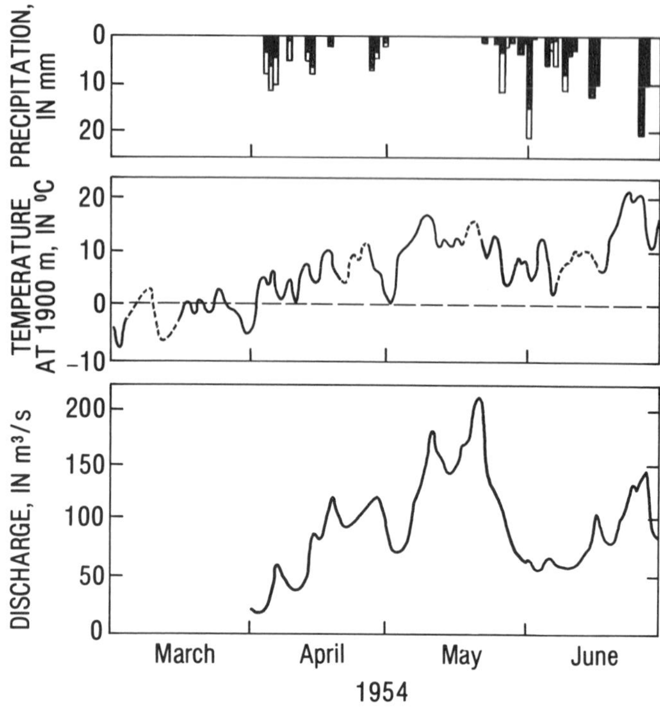

Figure 15. Snowmelt runoff of Boise River, Idaho, and concurrent temperature and precipitation (after Corps of Engineers, 1960).

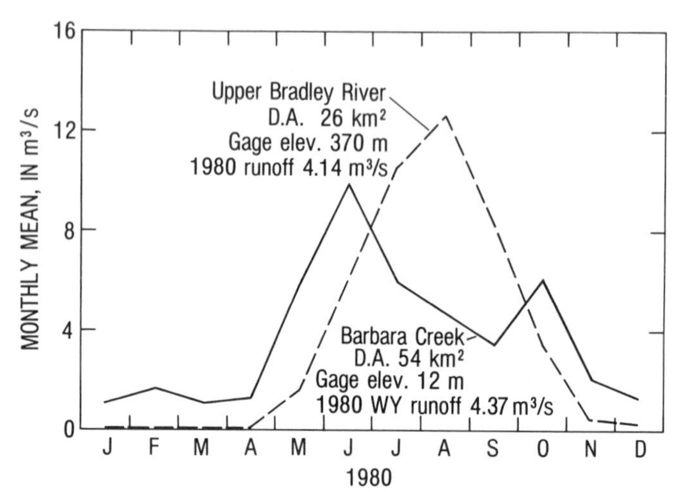

Figure 16. Monthly mean flows of glacial Upper Bradley River and of nonglacial Barbara Creek, Kenai Peninsula, Alaska.

streamflow is the lack of ground-water-derived baseflow: small streams typically cease to flow at the onset of freezing temperatures in the fall and do not flow again until warmer temperatures return in late spring or early summer. The regime of Kuparuk River in Alaska (Plate 3B) is typical.

Another important effect of permafrost on streamflow is the increased potential for surface runoff because of the reduced capacity for infiltration beneath the active layer. The result is that, during the summer, most of the rain or melting snow runs off at the surface or is evaporated.

Glaciers. Rivers having glaciers or perennial snowfields in their headwaters usually derive substantial flow from melt later in the summer than those rivers that head at lower elevations where the snowpack is exhausted each summer. The differences in the regimes of streams with and without glaciers will be particularly apparent if summer precipitation is low; clear hot weather increases runoff from a glacier. The significance of glacial melt during late summer is well illustrated by Figure 16, which compares monthly flows of two small streams on Kenai Peninsula, Alaska. The flow of Upper Bradley River is gaged 1.9 km downstream from a glacier whereas Barbara Creek drains an unglaciated basin.

Significant diurnal fluctuations in summer streamflow are characteristic of glacier-fed streams. Figure 17 illustrates the effect of Peyto Glacier on the flow of Peyto Creek in the Rocky Mountains of Alberta; the range in diurnal flow tends to increase as the summer and the melt progresses.

The influences of glaciers on streamflow are discussed in more detail in Meier (this volume).

Evapotranspiration. The evapotranspiration component of the hydrologic cycle has a major effect on streamflow throughout North America; its influence is particularly noticeable in warmer, arid regions to the south. Losses to evapotranspiration can significantly reduce streamflow throughout the year, although, for most of the continent, these losses are concentrated during the warmer "growing-season" months of late spring, summer, and early autumn. Lins and others (this volume) and Winter and Woo (this volume) describe the geographic variation in evapotranspiration

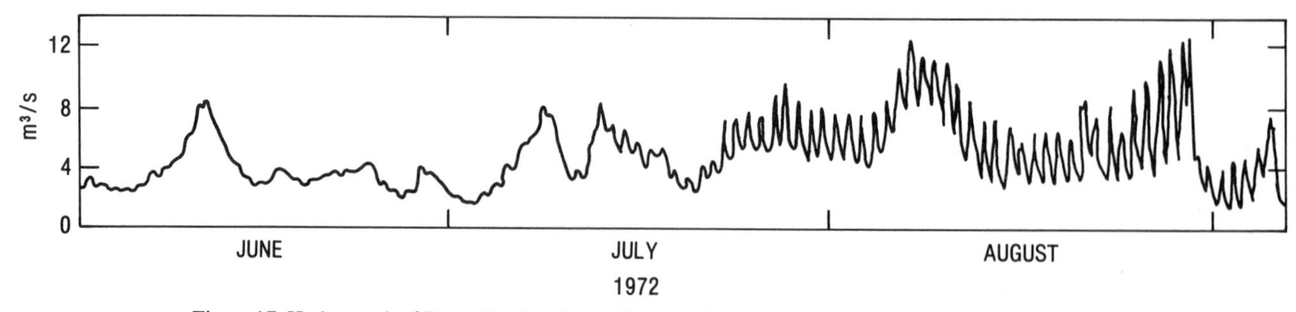

Figure 17. Hydrograph of Peyto Creek at Peyto Glacier, Alberta, showing the diurnal fluctuations due to glacial melting (after Environment Canada, 1978).

rates, and Saxton and Shiau (this volume) explain the effect of the land on these rates.

The effect of evapotranspiration on runoff is most noticeable where summer precipitation is light and ground-water contributions to flow are small. Some streams draining several thousand square kilometers run dry each summer, a phenomenon common to the Great Plains of North America. High evapotranspiration rates can reduce runoff even in more humid regions that usually experience frequent summer rainfalls. In these regions, the smaller streams may cease to flow during periods of deficient rainfall and high temperature because all the ground-water flow to the channel is removed by evapotranspiration.

The following specific effects of evapotranspiration on streamflow have been observed:

1. Flows of some small streams fluctuate diurnally during hot, clear weather but not if the sky is cloud covered. This fluctuation is attributed to the decrease in evapotranspiration from day to night. Hoyt (1936) reported diurnal fluctuations in discharge of a Maryland stream draining 15.5 km^2 (Fig. 18). Evapotranspiration apparently was from shallow ground water as well as from the water surface of this stream.
2. During a period of more than a month without rain in late summer, a small stream in Virginia ceased to flow. As temperatures decreased in the fall, the stream resumed flow before any rainfall occurred (Riggs, personal communication, 1963).
3. During periods of base-flow recession in late fall, the flows of small streams in New England have been observed to increase abruptly following a severe frost that killed the riparian vegetation.
4. The rate at which streamflow recedes during periods of dry weather in the absence of snowmelt is greater in summer than in winter. For example, the winter and summer base-flow recession curves for Little Beaver Kill, New York, both begin at 1.4 m^3/s, but 40 days later the winter curve has receded to 0.23 m^3/s while the summer curve has receded to 0.085 m^3/s (Parker and others, 1964, p. 136).

Further discussions of minimum and zero flows are found in Rogers and Armbruster (this volume).

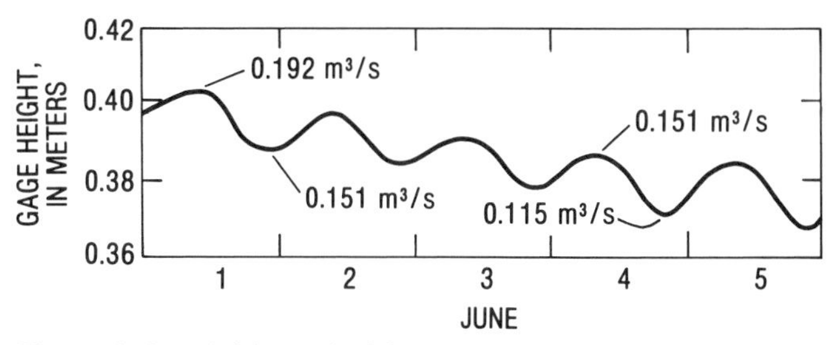

Figure 18. Gage-height graph of Owens Creek at Lantz, Maryland, 1932, showing the effect of diurnal changes in evapotranspiration on streamflow (after Hoyt, 1936).

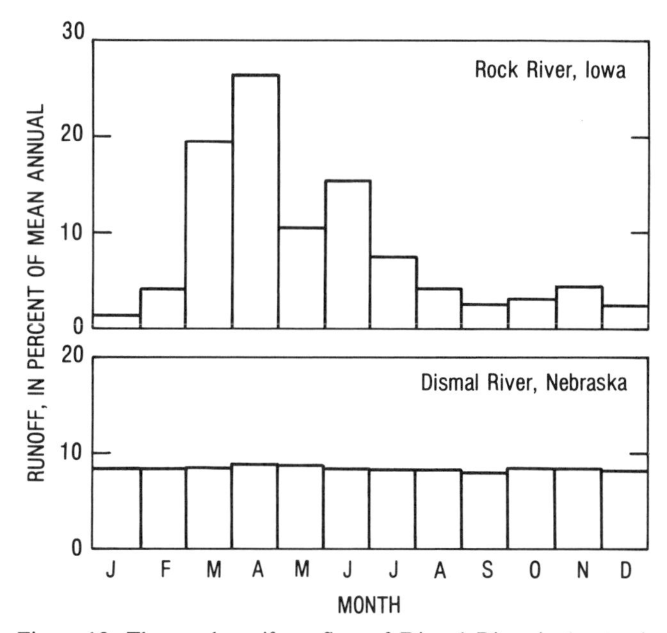

Figure 19. The nearly uniform flow of Dismal River is due to the extremely porous soil and the large ground-water storage in its basin. Rock River basin, 300 km east of Dismal River, drains a more impervious soil.

PHYSIOGRAPHIC FACTORS

The amount and distribution of surface water in North America generally reflect climatic variations, but there is also considerable influence by the physiographic characteristics of the land. Variations in geology and topography have a great impact on losses to interception and infiltration (Saxton and Shiau, this volume), on the interaction of surface water and ground water, on drainage patterns and rates of runoff, and on climate modification (Lins and others, this volume). As a result, runoff regimes within a region that is climatically homogenous may vary greatly.

Geology

The amount of precipitation that ultimately reaches the stream channels depends to some extent on the geologic materials and structure of the region. These factors control the amount and rate of infiltration of water down through the soil and its movement in the ground, often leading to its reappearance in a stream channel downstream.

High infiltration rates and large aquifer storages throughout a basin produce a runoff regime of low variability. For example, in the Sand Hills of Nebraska, all precipitation is absorbed by the extremely pervious soil. A large proportion reaches the water table which feeds regional streams. The flow regime of a typical stream in this region is shown in Figure 19. The regime of the Rock River, Iowa, which drains a more impervious soil some 300 km east of the Sand Hills, is also shown for comparison. The runoff per unit area is about the same for each basin.

Generally, regions deeply overlain by unconsolidated surficial materials, as in the Great Lakes–St. Lawrence lowlands, are subject to significant infiltration and ground-water contribution to streamflow. In areas where sand and gravel are widespread, the annual ground-water contribution to surface flow can exceed 50 percent of the mean annual flow; nearly all of the summer flow may be from ground water. Figure 20 illustrates the seasonal variation of the ground-water contribution to Big Otter Creek, which predominantly drains a sand plain area in southern Ontario. The percentage contribution is lowest during the spring snowmelt runoff period and highest during the summer and fall.

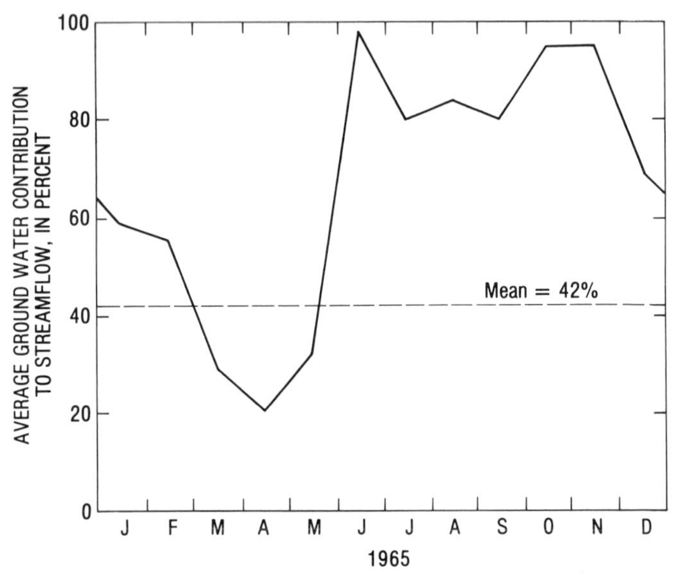

Figure 20. Ground-water contribution to Big Otter Creek, which drains a sand plain area in southern Ontario (after Ontario Ministry of Natural Resources, 1984).

Streams fed by extensive aquifers of glacial valley-train deposits in Ohio maintain high flows between rainstorms, and their annual flows are less variable than those of nearby streams lacking such productive aquifers. The Mad River regime, shown in Figure 21, is typical.

High infiltration rates may occur in consolidated material such as limestone and volcanics. In the limestone Yucatan Peninsula of Mexico, there are no stream channels; precipitation not lost to evapotranspiration recharges an aquifer, which discharges directly to the ocean. More commonly, a limestone aquifer discharges through springs rather than by effluent seepage at many points along the stream channel.

Typical of an extensive limestone terrane is the presence of sinkholes, caves, and springs and the absence of small perennial streams. The thick Mississippian limestone terrane from central Indiana through Kentucky to central Tennessee is an example. The regimes of the larger streams in limestone terranes usually differ little from those draining other geologic units. Most of the

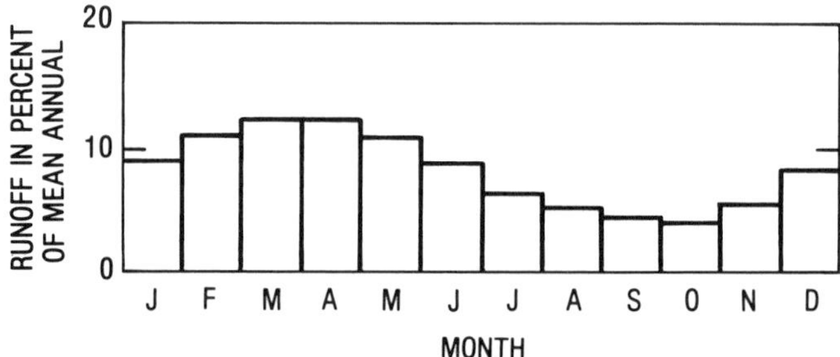

Figure 21. Regime of Mad River, Ohio, whose basin contains extensive aquifers of valley-train deposits.

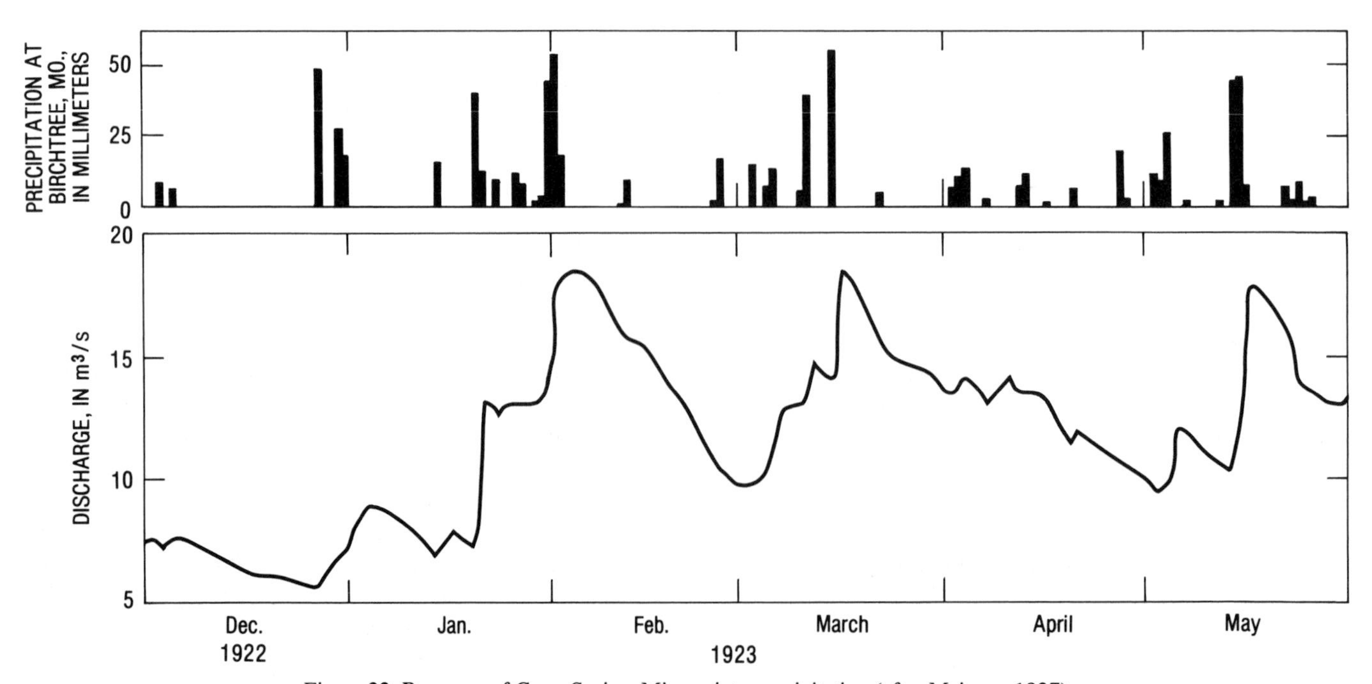

Figure 22. Response of Greer Spring, Missouri, to precipitation (after Meinzer, 1927).

major springs issuing from limestone occur in Florida, Missouri, Arkansas, and Texas. The responses of these springs to precipitation depend on the degree of development of the underground drainage system. A well-developed system with little storage will transmit precipitation fairly promptly, as shown in Figure 22 for Greer Spring in southern Missouri.

Silver Springs, Florida, is probably the largest limestone spring in North America. Its flow is derived from the large Floridan aquifer, and thus it responds only slightly to individual rainfalls. The mean monthly runoffs range only from 7.91 to 8.95 percent of the mean annual runoff. The sequence of annual flows of Silver Springs (Fig. 23) shows that the moderating effect of the aquifer extends beyond a year. Compare this annual sequence with that of Pascagoula River in Figure 3.

Drainage systems in limestone terranes may be partly on the surface and partly underground in regions of rough topography. This accounts for the "lost rivers" in such regions. An example is Lost River, a tributary to Cacapon River in West Virginia, which flows underground for several miles.

In less well developed surface drainages in karst topography, the ground-water level often intersects the ground surface at intervals, producing short surface channels. If the intersection of the water level and the ground surface occurs at a sink, a pond is formed. The character of the pond depends on the movement of water. If water moves slowly through the pond, the water will be contaminated by material falling on the surface—this seems to be the case of the "wells" in Yucatan, although these are no longer being contaminated by human sacrifices. Some ponds in karst contain exceptionally clear water, such as Blue Hole in central Belize.

Aquifers in some terranes may extend beyond the drainage area defined by the surface topography. Discharge from such an aquifer in one basin may include recharge from an adjacent basin so that the unit flow of one stream is much grater than that of the other stream even though the climatic characteristics are similar. For example, Figure 24 shows the regimes of streams draining adjacent (southeastern Idaho) basins underlain by limestone. Bloomington Creek has a large spring in its headwaters; Cub River does not.

The movement of water into and through volcanic terranes ranges from slow to moderate for silicic rocks to rapid for basalts having large joint openings and cavities. Discharges of ground water from volcanic terranes usually are localized. Some of the major springs of North America issue from basalt. The Snake River Plain, occupying about 30,000 km^2 in southern Idaho, is underlain by basalts, and has no surface runoff. Its aquifer is recharged by rainfall, by flows of the Big and Little Lost rivers, Birch Creek, and other streams that sink at the margin of the plain, and recently by waste water from irrigation. The aquifer discharges through springs along the north side of Snake River. This discharge averaged about 114 m^3/s before irrigation of the plain began. The high infiltration in this region does not produce highly variable outflows because of the enormous storage capacity of the aquifer.

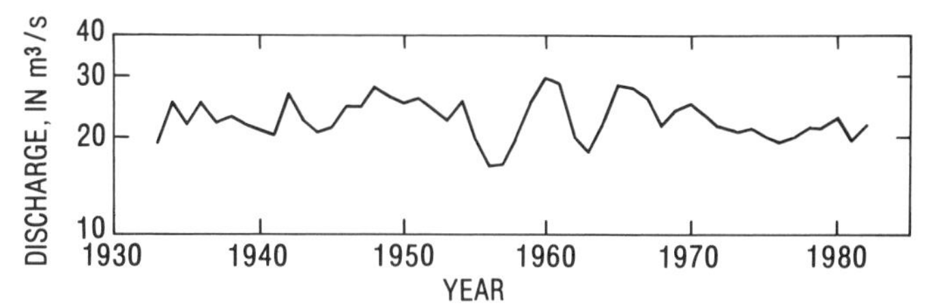

Figure 23. Annual mean flows of Silver Springs, Florida, showing the moderating effect of the Floridan aquifer.

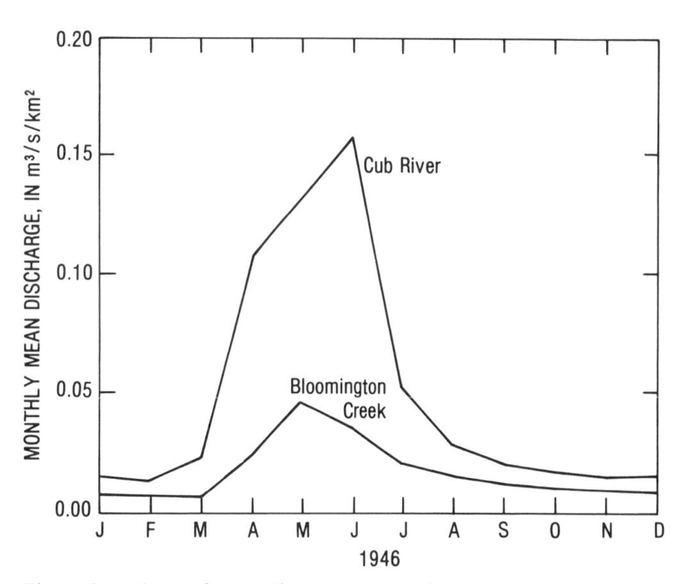

Figure 24. Yields of two adjacent Idaho drainage basins. Difference is attributed to movement of ground water across topographic boundaries.

The region of the volcanic rocks in southern Idaho extends into northern Nevada, northeastern California, and much of Oregon and eastern Washington. Large springs also occur there, particularly in the Sacramento River basin, California, and in the Deschutes River basin, Oregon. Metolius River, tributary to Deschutes River, is fed by perennial springs and has an extremely uniform flow (Fig. 25). McDonald and Langbein (1948) estimated that the ground water in storage in that basin was about three times the mean annual flow.

In addition to limestone and volcanic terranes, basins underlain by alluvium or porous sandstone, severely fractured rocks, or ones having anomalous structural features produce nontypical runoff characteristics.

Water flowing through alluvial valleys may not always be contained in the channel. Some water may move below the channel bed. Under such conditions, the surface-water flow may differ appreciably from reach to reach; in small streams, the surface flow may be zero in some reaches.

Major faults can disrupt streamflow and ground-water movement. Streams crossing the Balcones fault zone in Texas lose large amounts of water to the ground. The amount lost at a particular time depends on that available in the streams and on

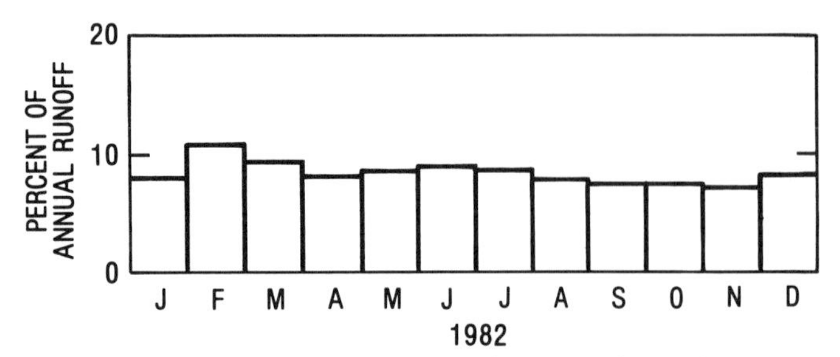

Figure 25. Regime of Metolius River, Oregon, which drains porous volcanic rocks.

the water level of the aquifer. Flow records of the Sabinal River at sites above and below the fault zone show that nearly all the low flows are lost as the stream crosses the fault (Fig. 26). Similar losses occur in other streams in the vicinity. The extensive aquifer underlying the region discharges through springs in the vicinity of San Antonio.

Topography

Topography is a major determinant of the drainage patterns of a watershed and, hence, has considerable influence on rates of runoff. Steep basin slopes and channel gradients tend to produce rapid runoff with little opportunity for loss of water by evapotranspiration or infiltration. Conversely, the hydrologic response of basins of low topographic relief is slower, tending to allow for more infiltration and for interception of surface runoff in lakes, ponds, and swamps. Such storage can signficantly delay direct runoff and increase evapotranspiration from the basin. These effects are usually more noticeable on the extremes of low—floods and low flows—than on the monthly flow regime.

Major surface storage can dramatically reduce the variability of a flow regime. This phenomenon is well demonstrated by the regime of the St. Lawrence River (Plate 3B), whose flows are highly regulated by the enormous storage of the five Great Lakes. Surface storage significantly moderates the flows of most rivers that originate in or flow through the Canadian Shield physiographic region (Energy, Mines, and Resources, Canada, 1974). For example, see Nelson River on Plate 3B.

The runoff regimes of Bear Creek and Cherry Creek, tributaries to the South Platte River near Denver, Colorado, are quite different because of major differences in both climate and topography. The headwaters of Bear Creek are at an elevation greater than 3,000 m in the Rocky Mountains, and the stream drops to about 1,500 m. In contrast, the Cherry Creek basin lies in the lower plains east of the mountains where temperatures are warmer, annual precipitation is lower, and summer thunderstorms are common. The regimes of the two streams, shown in Figure 27, reflect these significant topographic and climatic differences.

The topography of some regions of North America is such that some drainage basins are "closed"; that is, they are basins from which no surface runoff reaches an ocean. The Great Basin, most of which lies in Utah and Nevada, contains streams hydro-

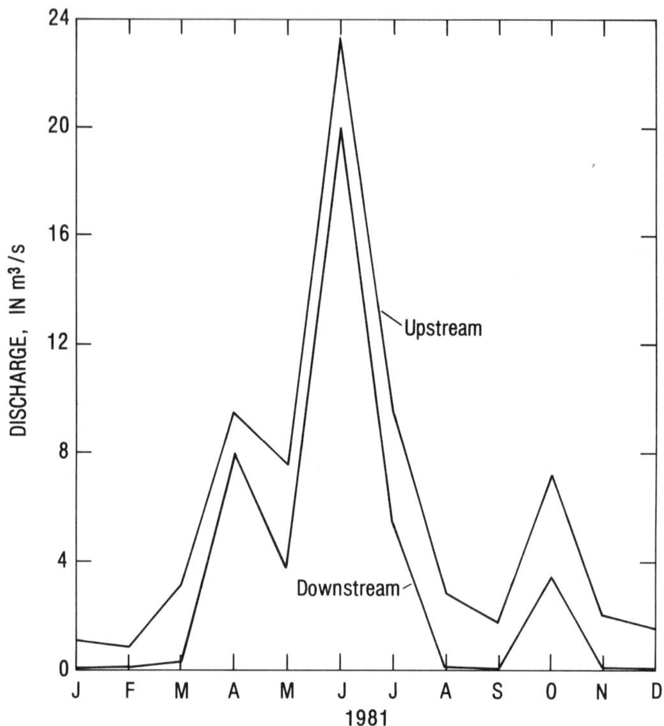

Figure 26. Monthly mean flows of Sabinal River, Texas, upstream and downstream from the Balcones fault zone showing losses to the ground.

logically similar to those in adjacent regions except that they terminate in lakes, "sinks," and playas. Part of the flows of these streams may recharge aquifers beneath the basins. The Great Basin is not well-defined topographically—it contains about 100 separate basins, each of which is a closed basin (Fenneman, 1931, p. 348). Only larger streams terminate in permanent lakes, such as Great Salt Lake, Walker Lake, and Pyramid Lake. Other large, closed basins are found in Saskatchewan, Wyoming, North Dakota, Chihuahua, and Coahuila.

An indirect effect of topography on runoff is related to climate modification: temperatures decrease as elevations increase; large water bodies supply moisture to the atmosphere and modify temperatures; and mountains produce an orographic or barrier effect on precipitation in which mild, humid air is forced aloft and compelled to give up its moisture, with the result that higher precipitation is received on the approach side of the mountain than on the lee side.

The barrier effect is very apparent along the western coast of North America and is reflected by generally decreasing runoff from west to east. Figure 28 presents the mean annual runoff of typical streams along a transect from western Washington to western Montana, which crosses the Olympic Mountains, the Cascades, the Bitterroots, and the Rockies.

In many basins, the specific effects of topography on streamflow are difficult to identify because of the dominant effect of geology. The characteristics of runoff from similar topography in the Cascade and Coast Ranges in Oregon vary considerably from

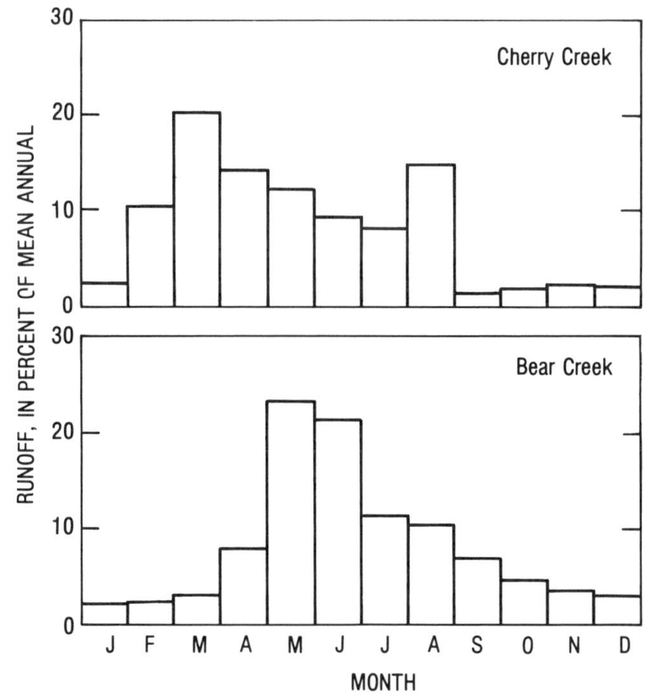

Figure 27. Regimes of two tributaries of South Platte River, Colorado. Cherry Creek basin lies in the Plains; Bear Creek heads in the Rocky Mountains.

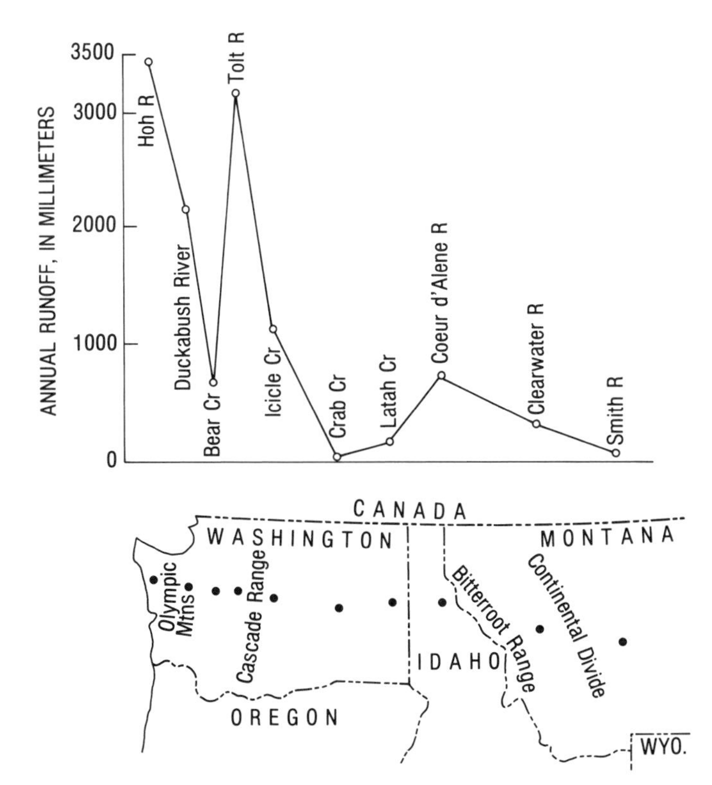

Figure 28. Mean flows of typical streams on a transect across several mountain ranges. These flows reflect the effects of the mountains on precipitation.

basin to basin; differences are due to the varying amounts of porous lava among the individual basins. For stream basins having a large range in topographic relief, there are usually significant climatic and geologic differences from the headwaters to the lower reaches of the watershed. Climatic differences may be higher precipitation in the headwaters, or the occurrence of snow in the headwaters and only rain in the lowlands. The flow regime depends on the amounts and timing of runoff from the various parts of the basin. The following section further discusses this concept.

FLOW VARIABILITY ALONG A STREAM

As described in previous sections, streamflow is a highly variable parameter in time and space, a complex function of the interaction of climate, physiography, and vegetation. The following description of the Saskatchewan River basin demonstrates the integrated effects of the various factors influencing runoff and the resulting variation of streamflow throughout a watershed.

The Saskatchewan River system (Fig. 29) originates in the Rocky Mountains, in the Columbia Ice Field, the largest glacier in Canada. The system drains eastward, eventually emptying, via Nelson River, into Hudson Bay. The Saskatchewan River is formed by the merging of its two principal tributaries, the North and South Saskatchewan rivers. At The Pas, Manitoba, the river drains 347,000 km^2 of area covering three broad physiographic regions: the Western Cordillera, consisting of the Rocky Mountains and foothills; the Canadian Shield, characterized by low, rounded hills and valleys with thin soils and numerous lakes and rivers; and the Interior Plains, a region of low relief; pervious soils, and poorly developed drainage patterns. Most of the Saskatchewan River basin lies in the Interior Plains region. Within this part of the watershed there are several areas of internal drainage—the closed basins described in the previous section.

The climate of the Saskatchewan River basin may be described as an extreme continental type, featuring warm summers and long, cold winters. Annual precipitation varies considerably over the watershed. The greatest amounts occur in the mountainous headwaters, where it averages in excess of 600 mm. Precipitation decreases markedly east of the foothills. Most of this portion of the basin averages 300 to 400 mm annually, while some areas experience less. In general, evapotranspiration exceeds precipitation east of the foothills, and so most of the Saskatchewan basin can be considered semiarid. In fact, semidesert conditions and "badlands" topography are found in the southwestern part of the basin. Grassland, parkland, and bush country are more common to the north and east. These basin characteristics are reflected in the mean flows and mean annual runoff depths shown in Table 1.

The hydrographs of mean monthly flows of Saskatchewan River and its tributaries (Fig. 30) are dominated by runoff generated during the snowmelt season, which occurs as early as March and as late as July, depending on the prevailing source of streamflow. Generally, the source moves westward as the snowmelt season progresses, from the Interior Plains in early spring, through the foothills in early summer, to the high snow and ice fields by midsummer. The occurrence of rainfall during this period can signficantly augment flow.

Figure 30a shows the mean monthly flows of Saskatchewan River at The Pas. These flows reflect the integrated runoff from all portions of the 347,000 km^2 drainage area. Contributions of the north and south branches of the river, at points not far upstream from the confluence, are shown also for comparison.

An interesting pattern emerges when tributaries on the north and south branches are studied. In Figure 30b, it is evident that most of the flow of North Saskatchewan River observed at Prince Albert originates in the Cordilleran region above Edmonton, with the Interior Plains area contributing relatively little flow. Note, in particular, the vast difference in flow observed on Battle River, which lies in the Interior Plains, from that on North Saskatchewan River at Edmonton, even though the drainage areas above each observation point are about the same.

The case is similar for the south branch (Fig. 30c). Red Deer River primarily drains the Interior Plains region and, accordingly, its flow is relatively small. Furthermore, a nearly three-fold increase in drainage area between Medicine Hat and Saskatoon results in only a modest increase in flow.

It is also interesting to look at the timing of maximum runoff in the basin. In general, streams originating on the Interior Plains, such as Battle River, peak earlier in response to snowmelt. Maximum monthly flows of the mountain-fed streams, like Clearwater River, are delayed until June and July, when precipitation and melting snow and ice combine to produce the highest rates of runoff in the Saskatchewan basin.

TABLE 1. MEAN FLOW AND MEAN ANNUAL RUNOFF AT SITES MARKED ON FIGURE 29

River	Mean Flow (m^3/s)	Mean Annual Runoff (mm)
Clearwater River	26	252
Red Deer River	48	137
Bow River	92	369
Oldman river	87	174
North Saskatchewan River at Edmonton	215	248
North Saskatchewan River at Prince Albert	242	106
Battle River	16	37
South Saskatchewan River at Medicine Hat	200	150
South Saskatchewan River at Saskatoon	262	94
Saskatchewan River at The Pas	658	93

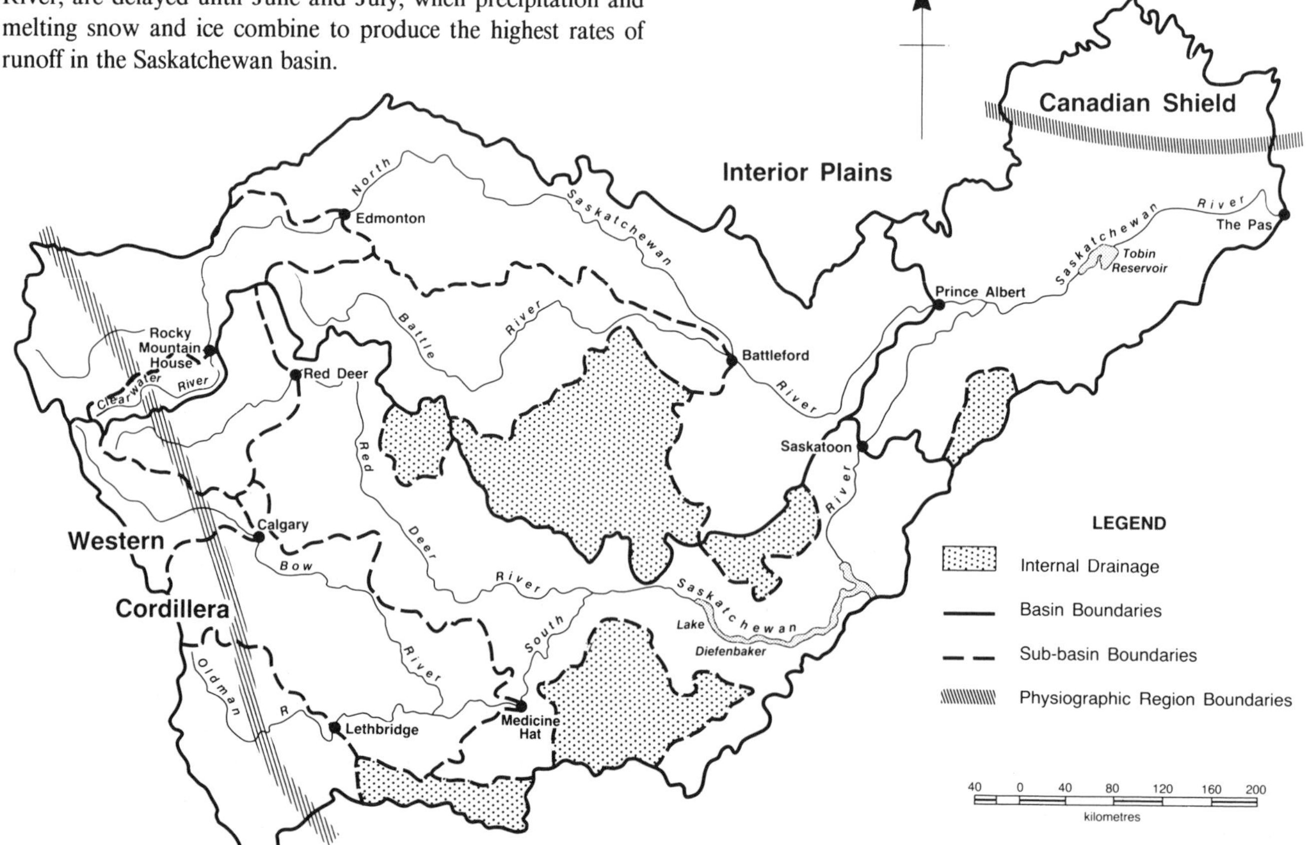

Figure 29. Saskatchewan River basin.

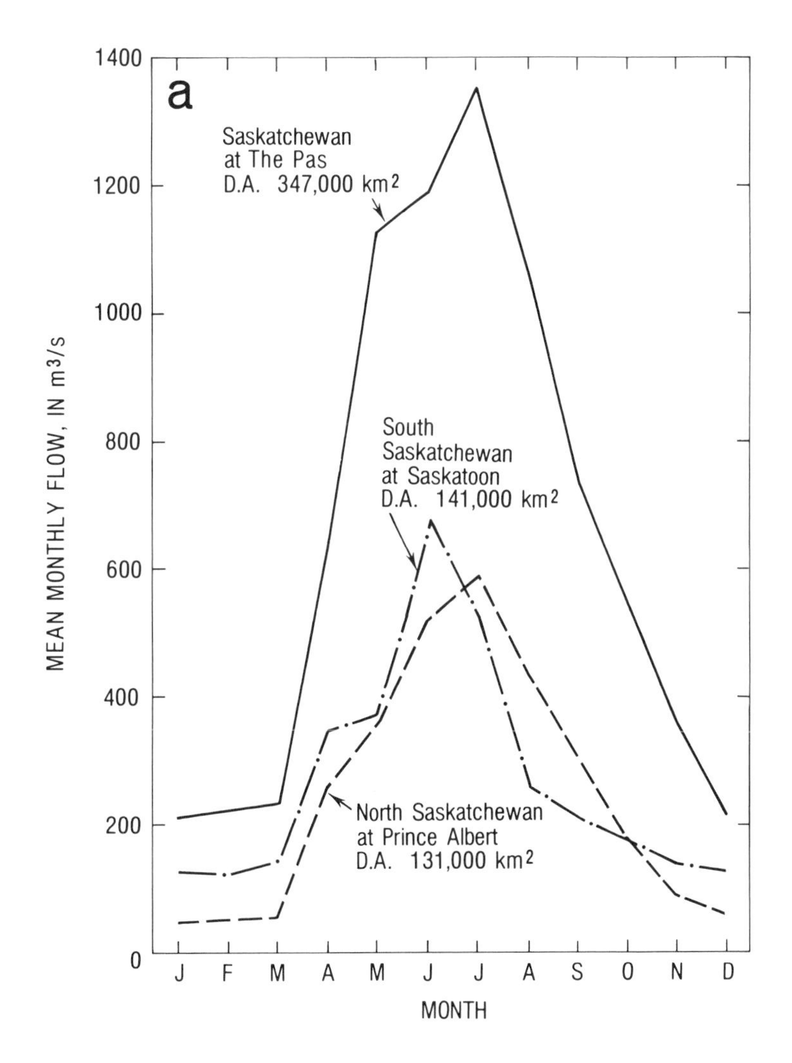

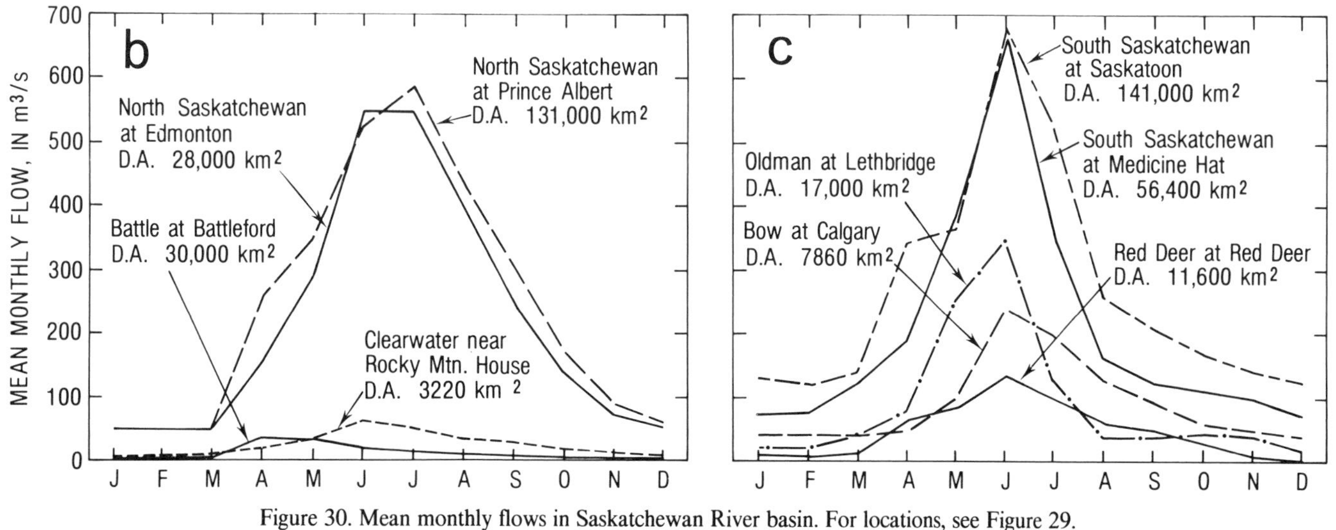

Figure 30. Mean monthly flows in Saskatchewan River basin. For locations, see Figure 29.

REFERENCES

Corps of Engineers, 1960, Runoff from snowmelt: Washington, D.C., U.S. Army Corps of Engineers Engineering and Design Manual EM1110-2-1406, U.S. Government Printing Office, p. 30.

Energy, Mines, and Resources, Canada, 1974, The national atlas of Canada: Ottawa, Macmillan Company of Canada, 254 p.

Environment Canda, 1978, Hydrological atlas of Canada, Runoff Map: Ottawa, Supply and Services Canada, scale 1:20,000,000.

Fenneman, N. M., 1931, Physiography of western United States: New York, McGraw-Hill Book Company, 534 p.

Hoyt, J. C., 1936, Droughts of 1930–34: U.S. Geological Survey Water-Supply Paper 680, 106 p.

Hoyt, W. G., 1913, The effects of ice on streamflow: U.S. Geological Survey Water-Supply Paper 337, 77 p.

L'vovich, M. I., 1979, World water resources and their future (English translation): Washington, D.C., American Geophysical Union, 207 p.

McDonald, C. C., and Langbein, W. B., 1948, Trends in runoff in the Northwest: EOS American Geophysical Union Transactions, v. 29, no. 3, p. 387–397.

Meinzer, O. E., 1927, Large springs in the United States: U.S. Geological Survey Water-Supply Paper 557, 94 p.

Moore, A. M., 1957, Measuring streamflow under ice conditions: American Society of Civil Engineers Proceedings, Journal of Hydraulics Division, v. 83, no. HY1, Paper 1162, 12 p.

Ontario Ministry of Natural Resources, 1984, Water quantity resources of Ontario: MNR Publication 5932, Toronto, 72 p.

Parker, G. G., and others, 1964, Water resources of the Delaware River basin: U.S. Geological Survey Professional Paper 381, 200 p.

Simons, W. D., 1953, Concept and characteristics of base flow in the Columbia River basin: Western Snow Conference Proceedings, 21st Meeting, Boise, Idaho, p. 57–61.

UNESCO, 1978, World water balance and water resources of the earth, including atlas of world water balance: Paris, UNESCO Press, 663 p. and 65 maps.

U.S. Department of the Interior, 1970, The national atlas of the United States of America: Washington, D.C., U.S. Department of the Interior, Geological Survey, p. 118–119.

MANUSCRIPT ACCEPTED BY THE SOCIETY MAY 1, 1987

Printed in U.S.A.

The Geology of North America
Vol. O-1, Surface Water Hydrology
The Geological Society of America, 1990

Chapter 5

Floods

Howard F. Matthai
836 Creek Drive, Menlo Park, California 94025

Figure 1. Main shopping street of Winsted, Connecticut, after flood of August 1955 (photo by Hank Murphy, Hartford Times).

GENERAL DEFINITIONS

The term "floods" can have quite different connotations to hydrologists and to laymen. Hydrologists are primarily concerned with discharge, the volume per unit of time at a point, along a stream, or throughout a drainage basin or basins, usually on a yearly basis. However, in arid regions the "annual flood" in some years may be a very small discharge or even zero—certainly not a "flood" in the lay sense. Even on a perennial stream, the annual maximum discharge or "flood" during a drought year may be a flow that is exceeded on many days in a wet year.

Laymen are seldom concerned with discharge. They want to know how high the water reached or will reach and how much damage has occurred (Figs. 1 and 2), or they may have only a passing interest in some of the spectacular aspects of the flood (Fig. 3). Although the height or stage of a flood flow is related to discharge, the residents along a stream are not interested in the relation. Therefore, some events called "floods" by hydrologists are nonevents to the citizenry.

Conversely, sustained runoff may not reach very high rates of flow, but the total volume may be sufficient to inundate low-lying areas. This type of a flood is of prime importance to local interests. Closed basins, flood plains, and land behind levees that have failed are some of the locations subject to this condition.

Matthai, H. F., 1990, Floods, *in* Wolman, M. G., and Riggs, H. C., eds., Surface water hydrology: Boulder, Colorado, Geological Society of America, The Geology of North America, v. O-1.

Figure 2. Exploits River at Bishop's Falls, Newfoundland, January 1983 (photo by Newfoundland Department of the Environment).

Figure 3. Street bridge in Putnam, Connecticut, at height of August 19, 1955, flood (photo by John Callahan, Providence Journal Bulletin).

FLOODS AT A POINT

Definitions

Flood magnitudes are determined on a regular basis at locations where a continuous record is obtained by instruments at a gaging station. At most sites, the record is of the stage or gage height of the stream, which is then converted to discharge by means of a stage-discharge relation. This relation is defined by current-meter measurements and possibly indirect measurements of the larger floods. Flood magnitudes may be determined on an intermittent basis at dams, powerhouses, and ungaged sites.

The flood of primary interest is the annual peak discharge, the greatest discharge that occurs in a year. The water year, which is October 1 to September 30, is used in the United States, but the calender year is used elsewhere in North America and herein. The greatest discharge may not coincide with the highest stage because of backwater, scour and fill in a channel, ice jams, or the works of man, such as levee construction.

An example of a record of annual peaks is that for the Pecos River near Langtry, Texas. The annual peak discharges from 1901 to 1984 are plotted in Figure 4. Large differences from year to year are evident as well as periods of several high annual floods or of several low annual floods. A period of high annual floods also may include a low annual flood and vice versa.

A flood series, other than the annual peak series, is the partial-duration series, which includes all discharges each year above a base discharge. The base is chosen so that, over several years, the number of peaks will average three per year. There are some limiting criteria regarding snowmelt periods and whether or not peaks are independent. The use of the additional data in the partial-duration flood series does not improve the reliability of flood-frequency relations if one is interested in floods with exceedance probabilities less than 0.1.

Flood frequency

A plot of the relation between annual peaks and time (Fig. 4) shows the historical record but little else, so another interpretation of the data is needed to characterize the annual floods as a guide to the future. This is done by a flood-frequency analysis, which will provide an estimate of the probability that a flood of a given magnitude will be exceeded, or conversely, the discharge of a flood with a selected exceedance probability. Both "exceedance probability" and "recurrence interval" are used in the literature; however, "exceedance probability" is the term now preferred. For annual events only, one term is the reciprocal of the other.

The instructions for flood-frequency analyses set forth by the U.S. Water Resources Council (1981) are the result of an attempt to standardize analyses. However, many recent investigators, including Wallis and Wood (1985) have proposed variations on the instructions or different techniques. Uncertainties in all methods make any analysis somewhat subjective.

Figure 5 is a flood-frequency plot of the annual peaks shown in Figure 4, except that all peaks less than 250 m^3/s (cubic meters per second) have been eliminated because they could be affected significantly by storage and diversions upstream. This figure shows how the assumption made about the major floods affects the frequency curve in the range of probabilities of 0.02 to 0.01.

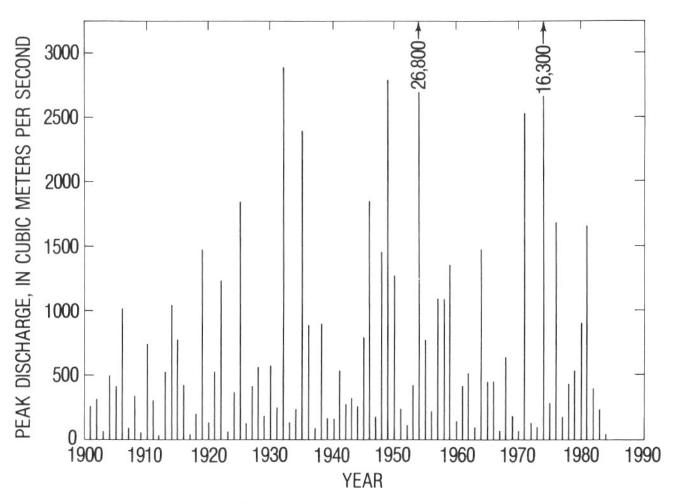

Figure 4. Annual peak discharges on Pecos River near Langtry, Texas.

The 0.01 exceedance probability or 100-year flood commonly is used as the design flood for bridges and other structures along a stream, as the basis for delineating flood plains for flood-insurance purposes, and as the basis for zoning ordinances. An inaccurate determination of this discharge, if too high, could result in the overdesign of a structure and a resulting waste of construction funds; if too low, a structure could be repeatedly damaged or destroyed, which also wastes funds replacing it, not to mention the hidden costs of delays to the public and losses incurred if a business cannot be operated.

This problem is illustrated by the flood-frequency curves in Figure 5. The 84-year record is considered to be a long one as far as streamflow records are concerned, but two rare floods occurred during this period (Fig. 4). Curve no. 1 in Figure 5 is the frequency curve obtained by using only the 84-year record directly. From this curve, the 0.01 exceedance probability flood is 10,000 m^3/s. Based on geologic evidence, Kochel and Baker (1982) estimated the recurrence intervals for the two rare floods to be at least 2,000 years and between 500 and 700 years respectively. Curve no. 2 represents this conclusion, and the 100-year flood from this curve is 6,700 m^3/s. Kochel and Baker also stated that the floods of 1954 and 1974 were higher than any other flood in at least 2,000 years. Curve 3 is based on this assumption and it shows the design flood to be 4,600 m^3/s. This large spread in discharges for the design flood is caused by the extreme difference between the two highest floods and the third highest. The author believes that the best estimate of the 100-year flood may be between the values from Curves 2 and 3 and possibly closer to the value from Curve 3.

The use of an inappropriate level of protection is not economical either. For example, golf courses, parks, parking lots, and other low-density developments could be flooded more frequently without causing undue property losses. Also, it may be more economical to design bridges and culverts on some rural roads for a more frequent flood, such as one with an exceedance probability of 0.02 or 0.04.

Factors affecting the magnitude and variability of annual floods

The magnitude of an annual flood depends on many factors; principal factors are size of drainage area, amount and intensity of the causative precipitation, and the rate of infiltration of water into the soil. The latter two may vary widely from year to year, and the effective drainage area may be less than the total if only a part of the drainage area contributes. In some regions, snowmelt may produce the annual flood or, more commonly, add to the flood produced by precipitation. Major floods also may result from transient effects such as ice jams, volcanic eruptions, glacier movements, and failures of man-made dams. Differences in geology among basins also produce different flood responses.

Effects of the various factors are described in the following sections. Each flood is the result of a particular combination of factors, but the combination differs with time and location, which makes flood phenomena very complex and difficult to analyze.

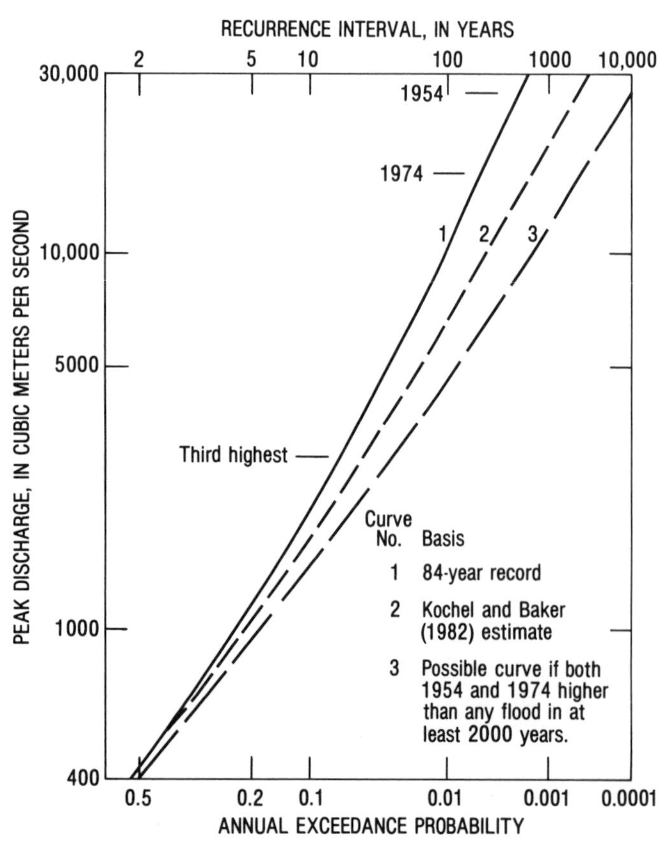

Figure 5. Flood-frequency curves for Pecos River near Langtry, Texas.

Drainage area. Drainage area has been found to be the most significant factor affecting the magnitude of floods. The magnitude of the mean annual flood, in cubic meters per second (m^3/s), varies as roughly the 0.7 power of the drainage area. This exponent tends to be less in areas of low relief, such as prairies, and greater in mountain areas. This relation, however, does not necessarily hold for very small areas, such as test plots, or for large areas for which all parts will rarely contribute to a flood. In the latter case, storms may not cover the entire drainage area, or in arid areas, the discharge downstream may decrease because of infiltration losses, overbank storage, and the low probability that the timing of the peaks on the tributaries will coincide with the peak on the main stream.

The contributing area will not be the total drainage area if part of the area receives little precipitation. This condition can occur even on relatively small areas when the length of the basin is oriented at nearly right angles to the long axis of the storm cell. Determination of the contributing area for a given storm is difficult and costly; few attempts are made to define it.

The contributing area can change with time when some of the storm precipitation falls as snow at the higher elevations in mountainous regions or on just part of a basin in the midcontinent prairies. When this occurs, the storm will contribute virtually no runoff to the flood peak from the snow area.

The drainage area may actually change when the runoff is sufficient to cause overflow from a normally closed basin or overflow from another stream usually not tributary to the point of interest.

Precipitation. Precipitation is another highly significant factor affecting the magnitude and variability of floods. Generally, the greater the amount of precipitation or the higher its intensity, the greater the flood magnitude. These relations are not always true, as discussed below and under "Antecedent conditions."

A storm of a given total amount and intensity will produce floods of different magnitudes if the storm travels up the basin, down the basin, or across the basin. The peak discharge from a given storm will be different if the storm cell lies across the basin rather than along the basin.

Though the total precipitation may be the same in two storms, the temporal distribution during the storm period can cause differences in flood peaks. When greater amounts fall near the end of a storm, the peak is apt to be greater than when they fall near the onset of the storm.

Very intense rainfall can exceed infiltration rates and surface depression storage capacity, thereby making a normally absorbent soil act like an impervious surface. The ground does not have to be saturated for this condition to exist.

Similar conditions exist in parts of southern Ontario and the northeastern United States where clay and silt swell on wetting, thus reducing infiltration, increasing the magnitude of flood peaks, and shortening the interval between the occurrence of rainfall and the time of the peak.

Precipitation from hurricanes is usually heavy and intense.

Where only the fringe of a hurricane strikes land, severe flooding can occur locally over a relatively small area. In 1969, Hurricane Camille devastated a large area along the Gulf Coast of the United States east of the Pearl River in Mississippi (Wilson and Hudson, 1969), and Hurricane Agnes caused catastrophic flooding in New York, Pennsylvania, and nearby states (Bailey and others, 1975). Hurricane floods have occurred many times from Texas to Massachusetts in the United States. Hurricane Hazel caused floods as far north as southern Ontario in 1954.

The cyclonic or hurricane season in Mexico is usually during August and September (Garcia-Quintero, 1950). The violent storms generally originate in the Caribbean Sea, and their path is westward over the Yucatan Peninsula; then, after crossing the southern part of the Gulf of Mexico, they may go inland or go northward toward Texas. The flood damage from the hurricanes negates most of the beneficial effects from the rain. Catastrophic storms occurred on August 29, 1909, and August 29, 1938, on the east coast of Mexico.

Along the west coast of Mexico, hurricanes frequently degenerate into tropical storms; but even these can cause severe flooding in northern Mexico and near San Diego and in the Imperial Valley in California.

Antecedent conditions. Flood magnitudes can be affected by the condition of the basin prior to the storms producing the floods. For each of the following pairs of conditions, the first condition will probably cause a greater peak discharge: whether the ground is near saturation from previous rainfall or is dry, whether it is frozen or not, whether vegetation has been removed or is natural, whether there is a snowpack or not, and whether or not an alluvial channel was wetted by recent flood runoff.

The relation of the first three pairs or combinations to flood magnitude should be evident. The last two need some explanation. A heavy, warm rain on a snowpack may melt much of the snow and release the water stored therein, thus increasing the peak discharge. This condition occurs at times in the Sierra Nevada in California. Light, cold rainfall can be absorbed in the snowpack, thus storing what might otherwise run off and reducing the peak discharge. An alluvial channel that has carried even modest flood runoff during the previous few days or weeks will be partly saturated, and some accumulated growth and debris in the channel will have been removed. Therefore, infiltration losses and the resistance to flow will be reduced, and the resulting flood magnitude might be greater in the downstream reaches.

If a flood of a given magnitude occurs in the winter when leaves are off the trees and cultivated areas on the flood plain have little or no vegetation, the stage for a given discharge will be lower than if the flood occurred in the summer. Crops on the flood plain and foliage on the trees that are inundated increase the resistance to the flow and raise the flood level significantly.

Snowmelt. Copious, warm rain on an existing snowpack can melt a lot of it. The conversion of water stored as snow to runoff, augmenting that from rainfall, has caused some major floods. This condition occurs, at times, in the Sierra Nevadas of California, usually in November or December; in the plains of central North America, in the Atlantic provinces of New Brunswick, Nova Scotia, and Newfoundland, as well as in southern Ontario and southern Quebec in the early spring; and in the northern Rocky Mountains in early summer.

On the west coast of North America, the warm rain usually comes from a weather system moving in from the vicinity of Hawaii. In the midcontinent, precipitation from the Gulf of Mexico or a significant increase in temperature can cause rapid snowmelt.

Much above normal snowpacks with a high water content provide natural storage of water that might cause flooding. Melting of such snowpacks may cause high-magnitude floods if the temperature increases at the right time in late spring or early summer and remains high. Alternating periods of very warm weather with periods of lower temperatures may not cause high rates of runoff, but the runoff volume may be sufficient to cause inundation of low-lying areas.

One of the outstanding snowmelt floods occurred in May 1948 in the Columbia River basin in the northwestern United States and southwestern Canada. The 28,600 m^3/s peak at The Dalles, Oregon, was only exceeded since 1858 by the 29,500 m^3/s peak in 1894. The 1948 flood had its immediate cause in a sequence of temperatures conducive to production of a flood from snowmelt (U.S. Geological Survey, 1949). During April and early May, temperatures had been subnormal, so that melting of the snow in the high mountains was delayed. Snow surveys showed that the water content of the snowpack, already above normal, contraseasonally increased during April and early May. At some locations, the water content exceeded 200 percent of normal for May 1. Temperatures rose on May 16, remained abnormally high, and were accompanied by heavy rain.

The floods of March through April, 1960, in eastern Nebraska and adjacent areas (Brice and West, 1965) were triggered by a rapid rise in temperature that melted the snow cover, which had an above normal water content. Precipitation was less than 80 mm in the 10-day flood period, though it was a contributing factor. Temperatures rose about 11°C in the headwaters of eastward-oriented streams seven days before temperatures rose in the downstream reaches. Flood flows in the headwaters had exceedance probabilities greater than 0.1, but some in the lower reaches had exceedance probabilities of roughly 0.01.

Unusually warm temperatures in February 1981, plus rainfall, caused widespread flooding in southern Ontario.

In Arctic areas, alternating periods of thawing and freezing can create a "multidecker sandwich" of ice, water, and slush lenses on the smaller streams. The final thaw may not produce much water; but when it flows over the top of all the lenses, maximum stages result.

Basin geology. Geology affects flood runoff in many complex ways. Infiltration rates and amounts, stream patterns, basin shape, land slopes, natural basin storage, and the resultant vegetation are some of the factors.

Infiltration. Infiltration is controlled by the surficial geology. Relatively impervious soils such as in granitic basins of the

Sierras and Rockies keep infiltration amounts and rates low. But, moderate to high infiltration can occur in seemingly impervious areas that are highly fractured.

A stream draining a basin composed mainly of volcanics may not flood unless there is heavy precipitation for an extended period over most of the basin. Such streams drain areas near Mount Lassen in California and other volcanic areas in the Cascades. The precipitation does not appear as runoff until some weeks or months later, and the runoff occurs at moderate rates. The Thousand Springs along the Snake River between Milner and King Hill, Idaho, drain part of the basalts of the Snake Plain where surface flooding is an infrequent event.

Alluvium can have high infiltration rates and store large quantities of water. The floods of June 1965 in the South Platte River basin of Colorado (Matthai, 1969) originated in the headwaters of several streams. The flood on Kiowa Creek peaked at 1,180 m^3/s near Elbert, Colorado, but infiltration in the channel was sufficient to reduce the flow at its mouth, about 135 km downstream, to a flow so small that it did not erode the grass growing in the stream bed. This attenuation is probably the norm as the channel at the mouth is much smaller than it is in the midreaches.

In southern California, some streams leave canyons and spill flood waters over the apexes of alluvial fans. Unless the flood is a large one, the water infiltrates into the alluvium and may disappear entirely. A few weeks or months later the water appears near the toe of the fan.

The main channels in the Tucson basin, Arizona, are efficient natural infiltration galleries (Burkham, 1970). The average annual inflow to Burkham's seven study reaches was about 81 million m^3 of which 58 million m^3, or 70 percent, was depleted by infiltration. The annual variation in infiltration, from about 30 to 90 percent of the average annual inflow, is mainly the result of variation in streamflow. Infiltration losses of this magnitude are one of the main reasons flood peaks are attenuated as the flood moves downstream in an alluvial channel.

In karst areas, water is lost into sinkholes and underground drainage channels, thus reducing both the peak discharge and flood volume. Vineyard and Feder (1982), in comparing spring-fed to non-spring-fed streams, state that though most peak discharges of spring-fed streams are lower during minor flood events, the annual runoff is generally greater. The areas of dolomitic rock of Cambrian and Ordovician age in the Ozarks province of southeastern Missouri have many springs that modify flood runoff patterns.

Big Spring, the largest in Missouri, has an average daily discharge of 12 m^3/s. The maximum daily discharge is just three times the average, and the minimum daily discharge is 55 percent of the average. This is in marked contrast to streams where the instantaneous peak discharge or the maximum daily discharge is hundreds of times the average discharge.

Stream patterns. The pattern of the streams in a basin is controlled by the basin geomorphology, which affects the number and spacing of channels, their sinuosity, slopes, and widths. Because tributary slopes are usually steeper than those of the main stream and tributary lengths are shorter, most floods on tributaries are of shorter duration with more rapid rises and recessions. Therefore, tributary peak discharges or stages seldom are concurrent with those on the main stream. Floods from downstream tributaries pass before the flood peak from the rest of the basin arrives. This sequence can cause multiple peaks or main-stream flood hydrographs with plateaus or small bumps on the rising or falling stages.

Figure 6. Erosion scars and deposition south of Castle Rock, Colorado, from flood of June 1965 (Denver Post photograph by Lowell Georgia).

Where tributary gradients are low enough, flood peaks on a large stream may cause backwater on a tributary. This might cause a higher water-surface elevation on the tributary than the one associated with its peak flood discharge.

Erosion can occur most any place along a stream, but the channel alignment of the smaller, steeper streams is controlled mainly by the geology and topography. Of course there are exceptions. Heavy, intense rainfall on a small butte south of Denver, Colorado, in June 1965 caused sufficient runoff for it to take the line of least resistance downslope. Trees were washed out and a new channel eroded, the debris being deposited near the toe of the butte (Fig. 6).

In contrast, flood flows and even nonflood flows in larger, lower gradient streams may change the topography by meandering or change the channel alignment within the flood plains. The formation of oxbows along some channels is one example. When a flood cuts off an oxbow or otherwise shortens the channel, the steeper gradient may increase erosion, especially by headcutting. The erosion may continue until the stream returns to approximately its former alignment and slope.

Prior to 1905, the Gila River in Safford Valley, Arizona, was less than 100 m wide (Burkham, 1972). Floods during the next 12 years widened the channel to about 600 m and destroyed the meanders and vegetation on the flood plain. During the period 1918 to 1970, the flood plain was reconstructed almost entirely by the accretion of sediment, and the main channel width was reduced to less than 60 m in 1964. Floods in 1965 and 1967 widened the channel to 120 m by 1968. These changes affected the timing and velocity of flood waves within the reach (Burkham, 1976).

A flood in a large basin is the integrated result of all the floods at points upstream. Such a flood may increase in magnitude as more and more tributaries contribute to the runoff. Or it may decrease as channel storage and infiltration deplete the discharge. Or it may exhibit various combinations of the factors presented above, both in time and space.

Basin shape. The basin shape is determined by geology, and the shape determines the time of concentration of the runoff, which is the time required for water to travel from the most remote point in the basin to the outlet or point of measurement. The tributaries in a long, narrow, "cigar shaped" basin have different times of concentration in relation to the outlet. When a general storm occurs, those in the lower part of the basin will peak and start receding before flow from the upstream tributaries reaches the lower basin. The peak discharge at the outlet will not be the sum of the peak discharges on the tributaries.

Tributaries in a "fan-shaped" basin will have nearly the same times of concentration; thus, tributary flows will combine to cause a faster rise to and recession from an increased peak discharge at the outlet.

Land slopes. Land slopes affect the time of concentration and the amount of surface depression storage. The effect may either augment or counteract the effect of basin shape. Streams may develop meanders where land slopes are small, thus increasing the distance floods flows must travel as well as reducing the water-surface slope.

Different land slopes produce markedly different runoff hydrographs. This is illustrated in Figure 7 by the hydrographs of two streams with comparable drainage areas in North Carolina during the same storm in August 1955. The Tar River drainage is in the Piedmont, and Swift Creek drains the Coastal Plain.

The steeper slopes on the Tar River reduce the time of concentration and are the main factor causing a sharper and higher peak earlier than on Swift Creek, which has mild slopes.

Natural basin storage. The retreat of the glaciation at the end of the last ice age left numerous depressions, some of which are now the many lakes in north-central North America. Small depressions, sometimes called "prairie potholes," may go dry in drought periods. All of these provide some natural storage of flood runoff, thus reducing the peak discharges that may have occurred without the storage. Some multiple regression analyses of flood-frequency relations show that the area of lakes and ponds in a basin is statistically significant.

Resultant vegetation. Geology determines the kinds and depths of soils in a basin and, to some degree, the types and density of vegetation. This interrelation is important because vegetation does affect flood discharges.

Vegetation and the ground litter it generates can intercept rainfall, thereby delaying the time that runoff may start and reducing the amount of water available for runoff. Vegetation in a stream channel or along the banks will increase the resistance to flood flows, which in turn increases the flood stage and the time of travel of the flood peak along the channel.

Transient effects. Transient effects can be caused by natural phenomena such as ice jams, volcanic eruptions, landslides, and glacier-dammed lakes; or as a result of man's activities such as the failure of a dam or deforestation.

Ice jams may occur during the spring breakup when the ice cover disintegrates into chunks of ice, which jam at a natural constriction in the stream or at a bridge crossing. Extreme amounts of water are not involved in each case, but the critical factor is the height to which the blockages cause the water to rise. Overbank flooding, impact of the ice on structures, and some-

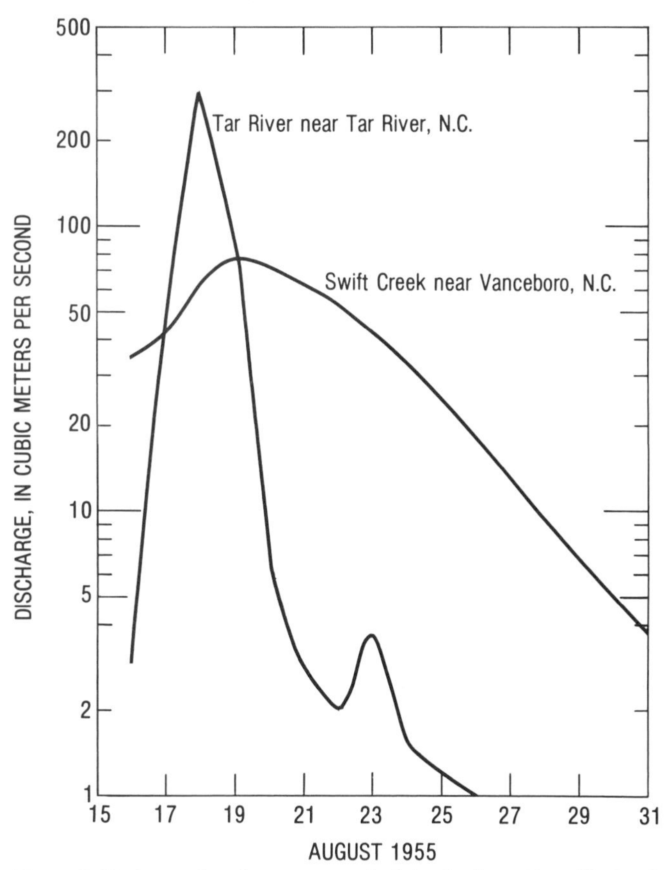

Figure 7. Hydrographs of two streams draining basins with unlike land slopes.

Figure 8. Ice jam. Rivière Saint Francois at Richmond, Quebec, May 1973 (photo by Raymond Barrett).

times channel-bed scour under the jam are the detrimental results of this transient condition.

In many northern rivers, ice-related flooding produces the highest water levels. No particular region in Canada is immune from an occasional ice-jam flood. The formation of ice jams is not limited to large rivers or particularly cold regions. Serious ice jamming has been observed on the Snake River in Idaho and on smaller streams in southern Ontario such as the Thames and Moira rivers.

Aufeis formation in streams and culverts also causes flooding. Aufeis occurs on small streams where inflow from springs during very cold weather freezes and completely fills or overfills the stream channel with ice. Many such formations form in Alaska and northern Canada.

A condition that occurs occasionally in northern lakes is the formation of a pressure ridge in the lake ice. The ridge may block the inflow from a tributary to the lake and cause severe flooding with very little flow in the stream. The town of Hay River on Great Slave Lake in the Northwest Territories has been flooded by this phenomenon in 1963, 1974, and 1985.

Ice jams are generally considered in two categories: freeze-up jams and break-up jams. In regions like the Atlantic provinces of Canada, which are subject to frequent alternations of cold and mild spells throughout the winter, a number of freeze-up and break-up events may occur during the winter. The terms freeze-up and break-up do not, therefore, necessarily denote fall and spring events.

A troublesome feature of freeze-up jams is that unless mitigative measures are taken or the discharge falls substantially, the ice accumulations may remain and the associated flooding may continue until the next thaw. Another characteristic of freeze-up jams is that, if they remain for the winter, the thick, solid ice that forms may adhere to the stream bed and be a severe impediment to the break-up ice run and hence facilitate the formation of a break-up jam. Freeze-up jamming can be aggravated downstream of a peaking hydroelectric plant if sudden increases in releases from the plant cause an advancing freeze-up accumulation to consolidate and thicken, thus producing high flood stages. Such an event has occurred on the Peace River in Alberta.

Break-up jams may form during the destruction of the ice cover, either during midwinter or more usually during spring melt (Figs. 8 and 9). The thickened ice condition produced by a midwinter break-up jam may last until spring melt, while that resulting from a spring jam normally lasts for only a short period of time. Break-up jams generally occur in rivers where the climate is cold enough to form a competent ice cover, and where substantial runoff can be generated. Jams may be more common in north-flowing streams because the thaw occurs first in the headwaters. This runoff can result from snowmelt in the spring, or from midwinter rainfall events.

Where flood levels are of primary interest in project studies, it is insufficient to carry out a flood analysis that does not include the consideration of ice jams. However, the analysis of ice-jam flood levels is more complex than that of open water levels because ice jams are essentially local phenomena. Although ice-jam floods may be frequent in a certain reach of river, other reaches a short distance upstream or downstream may be entirely free of them. Therefore, information on ice-jam floods in one reach cannot necessarily be transferred to other reaches, even on the same stream. It is also impossible to generalize ice-jam data regionally,

Figure 9. Ice shove. Rivière Saint Francois at Richmond, Quebec, May 1973 (photo by Raymond Barrett).

as is often done for open-water flood discharges; therefore, there is a stronger need for local historical data.

The eruption of Mount St. Helens in May 1980 caused outstanding floods on several streams but principally in the Toutle River basin in Washington. The intense heat melted most of the snow and ice cover, and the water transported mud, ash, sediment, blocks of ice, and other debris. Some of the material was carried as far as the Columbia River, 120 km downstream. Deposition ranged up to 5 m in the lower Toutle and Cowlitz river channels and about 3 m on adjacent flood plains (Lombard and others, 1981). Because channel changes were so extensive and because of the extremely high sediment and debris content of the flow, peak discharges were not determined. These changes increased the potential for very high stages in subsequent floods.

Landslides have been large enough at times to dam a stream temporarily. The threat of a flood caused by breaching of the slide material at some unknown future time is very real.

A landslide occurred in June 23, 1925, on the Gros Ventre River in northwestern Wyoming, impounding about 200 million m^3 (Alden, 1928). About 60 million m^3 was released 23 months later on May 18, 1927, when seepage and erosion of the slide material became sufficient to cause the failure of the "dam." The flood wave was about 5 m high near Kelly, 6 km downstream.

The phenomenon of outbursts of water from glacial-dammed lakes is sometimes called "jökulhlaup," an Icelandic term. Such events occur in Canada, Alaska, and some of the northern tier of states in the conterminous United States.

Lake George, located about 80 km east of Anchorage, Alaska, is a temporary impoundment formed by the advance of Knik Glacier. When the glacier pushes against the flank of Mount Palmer, water is impounded to form Lake George, the largest of 53 glacier-dammed lakes in southeastern Alaska and British Columbia. Water pressure and rising temperatures cause the toe of the glacier to suddenly disintegrate, and the water floods the Knik River valley in June or July. During the period 1958 to 1966, such floods occurred each year except 1963. The largest amount of water released was 2.2 km^3 in 1958, and the peak discharge reached 10,200 m^3/s at the gaging station. Since 1967, the glacier has not advanced far enough to form a lake; hence, no flooding.

Though Lake George has not formed since 1967, Lake Berg, only 200 km southeast, produced outburst floods on the lower Bering River in 1981–1982 and 1983 (L. R. Mayo, written communication, 1983). The area is about 100 km east of Cordova, Alaska, and very few people frequent the area, but the midwinter outburst of 1981–1982 may have been the first since at least 1901.

The outburst in July 1983 released approximately 2.4 km^3 of water in about three days. The peak flow was estimated as nearly 30,000 m^3/s, and about 1,000 km^2 along the Bering River were flooded to an average depth of 2.2 m. The flood also established a secondary, subglacial outburst channel into the Gandil River where the controls to a system of lakes were breached; thus compounding the flooding. The lakes are now permanently drained.

Wilson (1981) investigated historic floods from glacier-dammed Lake Warren in New York and estimated peak discharges up to 300,000 m^3/s with velocities of 5 m/s.

Nisqually Glacier and several nearby glaciers on the south side of Mount Rainier in Washington are sources of glacier outburst floods (Richardson, 1968). The probabilities of occurrence are estimated to be between 0.3 and 0.1. Floods in this area occurred at least seven times between 1926 and 1967. The flood flows included rocks, boulders, trees, and other debris. A few smaller floods have been observed at Mount Baker and Mount Hood, both volcanic mountains in the Cascades.

Tulsequah Lake, British Columbia, is on the eastern margin of the Juneau ice field and is impounded by a distributary arm of Tulsequah Glacier (Marcus, 1960). It drained annually for the 17 years, 1942 to 1958, and periodically for at least 50 years. Most of 230 m^3 was discharged in about 48 hours in 1958. Greater discharges probably occurred in the 1910 to 1920 period when lake levels were higher. Apparently, outbreaks at this location have not been documented since 1958.

Post and Mayo (1971) tabulated information on the 32 largest glacier-dammed lakes in Alaska. They also locate on maps 750 lakes, mainly in southeastern Alaska and adjacent Canada, with surface areas greater than 0.1 km^2.

Other lakes where glacier outburst floods have occurred include Hazard Lakes and Lake Donjek in Yukon Territory and Flood and Summit lakes in British Columbia.

Failures of man-made dams usually occur in a relatively short time. The resulting flood wave can destroy most anything in its path for many kilometers downstream. The high velocity of the floodwater is the main destructive force, but submersion can cause serious damage.

Some of the more outstanding dam-failure floods are the Johnstown flood in Pennsylvania in 1889, and floods resulting from failures of the St. Francis Dam in southern California in 1928 (Association Engineering Geologists, 1978), the Swift Dam in Montana in 1964 (Boner and Stermitz, 1967), the Teton Dam in Idaho in 1976 (Ray and Kjelstrom, 1978), and the Kelly Barnes Dam in Georgia (Sanders and Sauer, 1979).

The Johnstown flood is one of the better known flood events. It was caused by the failure of an unnamed earth-fill dam on the South Fork Little Conemaugh River in southwestern Pennsylvania in May 1889. The dam was completed in 1853 and impounded about 13.5 million m^3 (Hoyt and Langbein, 1955). The dam deteriorated for about 20 years until several changes were made in 1879, none of which improved the hydraulic properties of the dam and its appurtenances. The storm on May 30–31, 1889, raised streams in the area east of Johnstown to overbank conditions, and water overflowed the dam for three hours until it failed. The reservoir was emptied in 45 minutes, causing an average discharge of 5,700 m^3/s with a peak discharge possibly twice that much.

The wall of water 9 to 12 m high cleaned out the valley all the way to Johnstown, 24 km downstream. The flood reached Johnstown in one hour at an average speed of 7 m/s. More than 2,100 people lost their lives.

Johnstown was inundated by floods in 1936 and 1977 (Hoxit and others, 1982). The peak discharge in 1977 at a site near the source of the 1889 flood was about one-tenth that in 1889. The more severe floods occurred on other streams north of Johnstown.

The St. Francis Dam, completed in May 1926, near Saugus in southern California, failed on March 12, 1928, releasing the 47 million m^3 stored behind the 205-ft high, concrete, gravity-type dam (Association of Engineering Geologists, 1978). Failure seems to have been caused by seepage along the fault plane until a soft gouge was washed out. Pieces of the western abutment were carried about 2 km downstream. More than 500 people lost their lives, and damage was more than $10 million.

Swift Dam in northwestern Montana was a rockfill structure completed in 1915 that stored 42 million m^3. The entire volume was released in roughly one hour during the flood of June 1964. Dense stands of trees and brush along the channel and on the flood plains of Birch Creek were cleaned out, and some of the debris was deposited many kilometers downstream (Fig. 10). Even large blocks of the compacted earthfill were carried as far as 0.8 km below the dam site (Hadley, 1967). The first site where the channel approximated preflood conditions was 20 km downstream. A peak discharge of 25,000 m^3/s was computed by the slope-area method at this site.

The failure in June 1976 of the earthfill Teton Dam in Idaho caused damage throughout a 180 km reach downstream (Ray and Kjelstrom, 1978). Water reached depths of 23 m in the canyon just downstream and ripped out large cottonwood and willow trees. Under these changing conditions, the resistance coefficients used to compute the peak discharge of 65,000 m^3/s at a site 4 km below the dam cannot be verified. However, this peak discharge is partially substantiated by another peak discharge computation of 30,000 m^3/s at a site 10 km further downstream. Velocities in the canyon were on the order of 8 m/s.

Kelly Barnes Dam near Toccoa, Georgia failed on November 6, 1977. The rock-crib and earthfill dam impounded 777,000 m^3, which flowed out in less than 1 hour. The peak discharge at a site 1,300 m downstream from the dam was computed as 680 m^3/s. Damages were estimated at $2.8 million, and 39 people were killed. These and other dam failures are discussed in more detail by Costa (1985).

Some man-made changes in land use can affect flood peaks temporarily. Some changes will usually increase the magnitude and frequency; some will decrease them. Two of the changes that may have significant effects are: clearcutting a forested area and a change from a forest to field crops.

Clearcut logging was done on two of three small watersheds on the Oregon coast. Compared to the control basin, peak discharges and the 3-day flood runoff volume were greater after logging (Harris, 1977); but the increase in the peak discharge in one basin was not statistically significant. The mean of 16 peak discharges in the other basin was 20 percent higher than prelogging conditions would indicate.

This effect will be a transient one if the logged area is left to return to a natural condition. The growth of grasses, shrubs, and trees will in time stabilize the area so that the magnitude and

Figure 10. Looking downstream after failure of Swift Dam on Birch Creek, Montana, June 1964. Part of the right end of the upstream face of the dam is near center of photo—see arrow (photo by U.S. Bureau of Reclamation).

frequency of flood peaks will be similar to those before clearcutting.

A condition similar to clearcutting exists after a forest fire, which may be caused by lightning or by man's negligence. Heavy rains can increase the erosion and some of the peak discharges until the burned-over area is reseeded and new growth begins.

FLOODS ALONG A CHANNEL

Many of the factors affecting the magnitude and variability of floods at a point that are discussed in the previous section also apply to floods along a channel. The total drainage area, the contributing area, precipitation patterns and types, geology, infiltration, temperature effects, and tributary inflow are among those factors.

One factor not discussed above is channel storage. As soon as runoff reaches a channel, some of it goes into channel storage to be released at a later time. This temporary natural storage tends to reduce flood peaks downstream and to extend the time that the stream is above normal stages.

The passage of a flood wave down the South Platte River in Colorado and Nebraska is illustrated by the hydrographs in Figure 11. This flood originated in the Bijou Creek basin, which is tributary to the South Platte River about 40 km upstream from the Balzac gaging station. The peak discharge near the mouth of Bijou Creek was 13,200 m^3/s. The reductions in the peak discharges of 9,500 m^3/s between Bijou Creek and Balzac on the South Platte River, 2,500 m^3/s between Balzac and Paxton, and 330 m^3/s between Paxton and North Platte were due primarily to channel storage, though infiltration was also a factor.

The time of travel of the peak was about 6 hours from Bijou Creek to Balzac, 67.5 hours more to Paxton, and another 24 hours to North Platte. These represent travel rates of 6.7, 3.1, and 2.1 km/h, respectively, between the four points. Another factor causing the decreasing rate of travel is the decrease in the channel slope in the downstream direction.

A special type of channel storage is produced by wetlands. A detailed discussion of wetlands and their locations is contained in Winter and Woo (this volume). Wetlands may either decrease or increase flood peaks. The complexity of the interrelations of antecedent conditions, the amount of rainfall or snowmelt, the effect of the timing of the runoff, infiltration, evapotranspiration, and other factors makes a forecast of possible effects uncertain (Miller and Frink, 1984).

Some wetlands, such as the prairie potholes in North Dakota, are off-stream storages. If they are relatively full and inflow is sufficient for them to overflow, then the contributing area is increased, and the peak discharge may be increased. Wetlands through which a stream flows will act as reservoirs to impound runoff and reduce the flood peak, but the added resistance to the

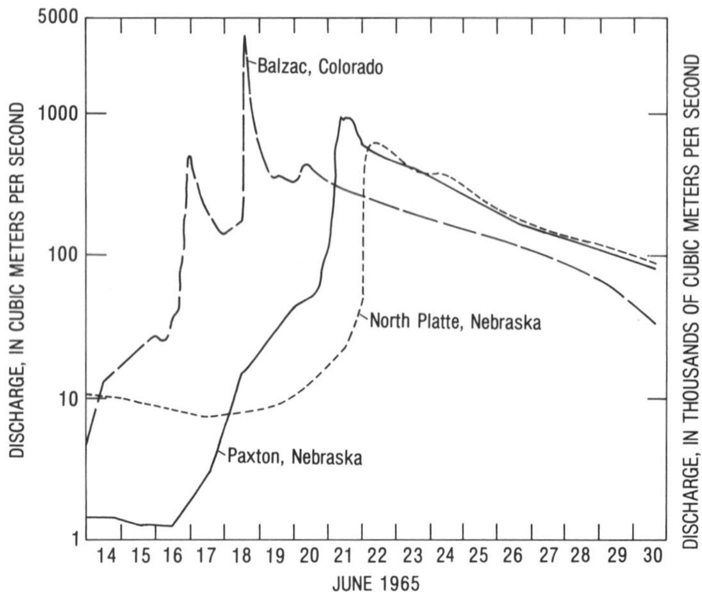

Figure 11. Hydrographs for South Platte River during flood of June 1965.

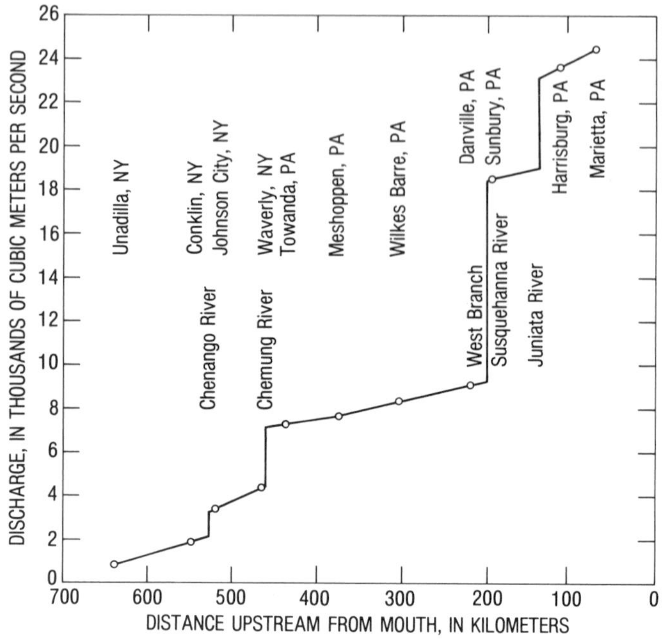

Figure 12. Discharges of floods with exceedance probabilities of 0.01 along Susquehanna River, New York and Pennsylvania.

flow from the vegetation will delay the peak and extend the time that downstream reaches are at or above flood stage.

Some individuals speculate that the reclamation of wetlands by government agencies or developers has increased the magnitude and frequency of floods. These changes are difficult to prove. The natural variation of floods with time is probably far more significant than any man-made change, and the large variation in flood discharges may mask or dwarf small changes in response in a basin. Water- and soil-conservation measures are often useful in reducing flood flows in small streams, but they have little or no influence in the protection of areas along major streams or in the control of unusually large floods. A number of studies cited by Miller and Frink (1984) are inconclusive about the effect of wetland reclamation, so the question remains unresolved.

How the 0.01 exceedance probability flood varies along a stream is illustrated in Figures 12 and 13. Flood discharges with exceedance probabilities of 0.01 on the Susquehanna River in New York and Pennsylvania are plotted versus distances upstream from the mouth in Figure 12. The discharges are for natural conditions, but there is presently (1985) enough flood-control storage and other regulation to reduce these discharges from 5 percent at Conklin to 18 percent at Sunbury. The average reduction is 13 percent at the 11 locations. Note that the contributions from four major tributaries have been included in the plot. One may approximate the 100-year recurrence-interval, or 0.01 exceedance probability, flood at other locations if the distance upstream from the mouth or from one of the points identified in the graph is known.

Figure 13 is a similar plot for the Santa Cruz River in Arizona. This plot shows a decrease in discharge in the downstream reaches—a typical condition in arid areas.

Large floods can change the topography, sometimes drastically, in a stream valley during the flood period, or they may initiate changes that are minor at first but can become progressively larger and frequently more detrimental. Matthai (1969) and Snipes and others (1974) describe some of the kinds of changes occurring during a flood (see Fig. 6). Another example of a change from a very small channel to a sizeable gully in a short time is illustrated in Figure 14.

The opposite of erosion, deposition, also occurs during floods. Stratified deposition of sand and gravel about 2 m deep is shown in Figure 15. The material was transported by a flood caused by very heavy and intense rainfall over El Dorado Canyon, Nevada (Glancy and Harmsen, 1975).

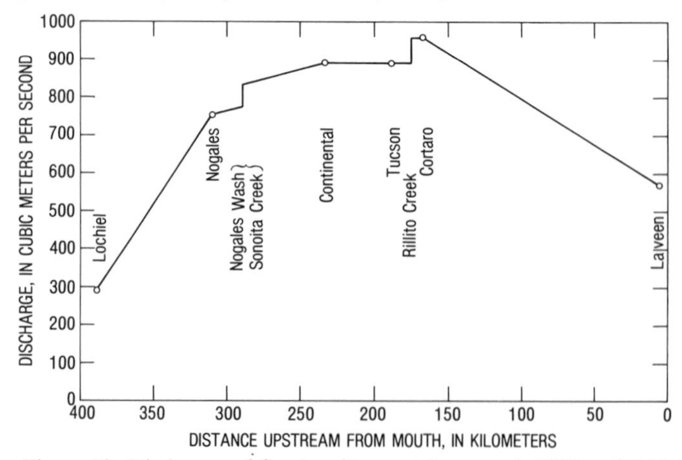

Figure 13. Discharges of floods with exceedance probabilities of 0.01 along Santa Cruz River, Arizona.

Figure 14. Channel carved by flood during hurricane Camille, Nelson County, Virginia, August 19 to 20, 1969. Before the storm the channel was only a few centimeters deep and a person could step across it easily (photo by G. P. Williams, U.S. Geological Survey).

Figure 15. Stratified sediment deposited at mouth of El Dorado Canyon, Nevada, during flood of September 14, 1974. Note alternating layers of mixed particle sizes with only a general impression of coarsest sediment in lowest strata (photo by P. A. Glancy, U.S. Geological Survey).

A different kind of depositional material is depicted in Figure 16. The largest of these granite boulders was about 0.6 × 0.7 × 1.2 m (Shroba and others, 1979). This bouldery bar obstructed part of the channel and deflected the floodwaters against the west wall of the canyon where a second channel was cut along the highway.

A progressive change is exemplified by head cutting, usually in small channels. The erosion may be severe enough to damage valuable farm or range land, marketable timber, or structures. Another progressive change occurs when streams erode the ridge separating them, and one stream captures the other. Floods may accelerate this process.

Man shifts channels at many sites to suit his convenience; however, problems similar to those mentioned above may occur. Such changes may be only temporary because streams frequently tend to return to their previous configuration because that is more stable than the new alignment. Elaborate and costly preventive measures may be necessary to maintain the change.

MAN'S INFLUENCE ON FLOOD PEAKS

Many dams have been built to provide flood control. Even dams built primarily to provide storage for irrigation or hydroelectric power usually create some capacity that can be used for flood control.

The ideal situation during a flood is the complete containment of the flood runoff into a reservoir. Then the stored water can be released at a later date as a controlled flow when the downstream channel can carry it. This situation occurred at John Martin Dam on the Arkansas River in Colorado in June 1965 (Snipes and others, 1974). The peak inflow into the reservoir was about 2,800 m^3/s, and the release was only 0.5 m^3/s. As a result, the peak discharge at Lamar, 30 km downstream, was only 2,100 m^3/s from inflow below the dam.

Another means of reducing a flood peak is by diverting some of the flow out of the channel. The flood control system in the Sacramento Valley in California utilizes several large gated weirs along the Sacramento River to divert water into bypass channels constructed specifically to carry floodwaters away from Sacramento and several other cities (Fig. 17).

The diversions from the lower Mississippi River to protect New Orleans are classic examples of man's ability to reduce a flood discharge. The Atchafalaya River receives water from the Mississippi River through Old River 75 km northwest of Baton Rouge, Louisiana (Wells, 1980). About 30 percent of the flow in the Mississippi River is diverted here, and a maximum diversion of 17,500 m^3/s can be made. In addition, the Morganza Control Structure, 50 km downstream from Old River, has a capacity of

Figure 16. Boulders deposited near south end of Drake, Colorado, during flood on Big Thompson River, July 31 to August 1, 1976. Discharge 790 m^3/s (photo by U.S. Geological Survey).

17,000 m^3/s if needed for flood protection. It has been used only once, and that was in 1973–1974 to divert flood water to the Atchafalaya River.

The Bonnet Carre spillway, 32 km upstream from New Orleans, can carry large flood flows to Lake Pontchartrain. Control is by a gated structure with a capacity of 7,100 m^3/s.

This flood control system could reduce a Mississippi River flood of 77,000 m^3/s upstream from Old River to only 35,000 m^3/s at New Orleans.

The area subject to flooding in the Red River basin in Manitoba is located in the nearly level lacustrine plain through which the Red River flows. Stages above bankfull have exceedance probabilities of 0.1 (Environment Canada, 1976); because the terrain is so flat, the flooded area can become quite extensive. In 1950, about 1,400 km^2 were flooded, and an estimate of about twice that has been made for the flood of 1852. Inadequate drainage and backwater on the tributaries extend the area inundated.

The Red River Floodway, capacity 1,700 m^3/s has diverted floodflows around Winnipeg since 1968. The Assiniboine River diversion can bypass up to 700 m^3/s around Winnipeg, and Shellmouth Reservoir can store 530.4 million m^3 to help regulate flood flows. These projects were completed in 1970. The total cost was $90 million. The first real test of the system came in 1974, and the estimated savings in flood damage was $200 million. Another flood occurred in April and May, 1979. The flood control system reduced the peak stage at Winnipeg by 3.399 m from that in 1950 for an otherwise comparable flood. Flood damages within the city were minimal, but outside Winnipeg, damages in Manitoba exceeded $30 million (McLaurin and Wedel, 1981).

Levees have been built to confine floodwaters to the main channels of numerous streams. The constricted flows will have higher water-surface elevations and possibly higher velocities than the same discharges if not contained within levees. Levees have prevented many millions of dollars of damage to crops and habitations on the flood plains. However, when a levee fails, large areas may be inundated.

Figure 17. Sacramento weir on Sacramento River, California, November 21, 1950, diverts flow into Yolo bypass. Head on weir 2.4 m; discharge 3,060 m^3/s (photo by California Water Resources Board).

The continued effectiveness of levees depends upon maintenance of the structural soundness of the levee and adequate control of vegetation growth or deposition in the channel. Dredging is sometimes warranted to maintain the channel capacity.

The paving of large areas, such as airports and shopping malls, creates impervious surfaces that can increase the magnitude of flood peaks from small drainage basins. On the other hand, a parking area with curbs, or buildings with flat roofs, can provide some temporary storage of rainwater, thus reducing peak discharges.

The effects of urban development on flood runoff have been documented by Crippen (1966). Half of a small basin with a cover of native grasses with some brush and scattered trees was developed, with streets, homes, lawns, and a golf course. About 20 percent of the developed area is impervious. Flood peaks are now more frequent than in predevelopment times. The magnitude of those occurring once or twice a year has been increased, but the higher, less frequent peaks have been increased to a much smaller degree.

How urban development affects flood peaks depends upon complex interactions of many variables. Some pertinent variables can be identified by four questions: (1) How are the impervious areas distributed with respect to one's point of interest? (2) Does the area have storm sewers? (3) Do downspouts discharge into gutters or onto lawns and flower beds? (4) Do culverts become obstructed with debris?

Anderson (1970) showed that sewer installation increased flood magnitudes by a factor of two to three independent of impervious development and that a completely impervious surface increases the smaller floods by a factor of 2.5 with a decreasing effect upon larger floods. The effect on a flood with an exceedance probability of 0.01 is insignificant. Sauer and others (1983) show similar results throughout the United States.

FLOOD FLOW UNDER SUPERCRITICAL CONDITIONS

Definitions

Critical depth in a channel is that depth at which the energy head, the actual depth plus the velocity head, is a minimum for a given discharge. When the depth is greater than critical depth, the flow is called subcritical or tranquil flow. When it is less than the critical depth, it is supercritical or rapid flow.

A change in channel geometry, an increase in channel roughness, or a flatter slope may be sufficient to cause supercritical flow to change to subcritical flow. This conversion of kinetic energy to potential energy is accomplished through a hydraulic jump. Conversely, the change from subcritical to supercritical flow usually occurs near the brink of a fall, at the head of a rapids in a channel, or at the crest of a spillway on a dam. A flood with high kinetic energy can produce either transient or permanent channel changes (sometimes in minutes or hours), erosion around structures, and considerable impact damage. The abrupt increase in depth at a hydraulic jump and the resulting turbulence can also cause damage.

Flow conditions

Supercritical flows are characterized by shallow depths, high velocities, and a wavy water surface. These may occur in steep streams or on alluvial fans, and in alluvial channels where the resistance to flow is low. Supercritical flow in part of a cross section of an alluvial channel is indicated by the occurrence of a series of sinusoidal standing waves or a series of larger waves that break up and then reform.

SHEET FLOW

Sheet flow occurs when water flows overland with no defined channel and where depths are only a few centimeters. Usually, velocities are high, and the flow may be supercritical. Water tends to follow the line of least resistance, which may change after surface irregularities are eroded or moved.

Though not a frequent event, most such events occur in arid or semiarid regions such as the southwestern United States.

MUDFLOWS

A unique type of flood is a mudflow or debris flow. These occur infrequently at any given site because many interrelated conditions must be met before such an event takes place. In most cases, a relatively small amount of water is involved because the water acts only as a lubricant. The water-solids mix can buoy and transport large rocks. Mudflows occur mainly in arid or semiarid areas after stormwater or snowmelt infiltrates the surface layers, where slopes are steep and the flow is fairly well confined in a channel.

Mudflows have been observed in southern California (Sharp and Nobles, 1953), the east side of the Sierra in Nevada (L. J. Snell, written communication, 1961), the Rocky Mountains in Colorado (Curry, 1966), in British Columbia (Miles and Kellerhals, 1981), and other locations (Wieczorek and others, 1983).

The high velocities of some mudflows can make them very destructive. Curry (1966) estimated maximum surface velocities of 16 m/s. Also, if large quantities of material are transported, man-made structures can be partially buried if they are located where the mudflow debouches out of a canyon.

Alter (1931) reports briefly on 33 mudflows in 11 counties in Utah. All occurred in a 2-week period in August 1930. Some deposits of mud and debris were 2 m deep, and a few boulders, estimated to weigh as much as 136 metric tons, were deposited on valuable farmland.

Mudflows have occurred in the city of Sierra Madre, California, in 1954, 1959, 1961, 1962, 1969, 1976, and 1978 in spite of an extensive flood control system (Los Angeles County Flood Control District, 1979). Extensive structural damage occurred in 1959, and roughly 1 m of debris was deposited in some areas in 1976 and about 2 m in 1962.

Wrightwood, California, has been damaged by debris and mudflows numerous times (Morton and Campbell, 1974). The flows are caused by short-duration, high-intensity rainfall or by a high seasonal rainfall, at least 760 mm, followed by one or two days of moderately heavy rainfall (State of California, 1976). The investigators estimated that an event with an exceedance probability of 0.01 would produce about 810,000 m^3 of debris in one of the four canyons that has had mudflows.

The term "debris torrents" instead of mudflows, is used in Canada where the flows contain a larger proportion of bigger debris, including living and dead organic material. Debris torrents typically occur in steep channels when slides from adjacent hillslopes enter the channel and move downstream during extreme discharge events or when movement of accumulated debris in the channel is initiated. Several events in British Columbia have involved more than 50,000 m^3 of debris. See Figures 18 and 19 for examples.

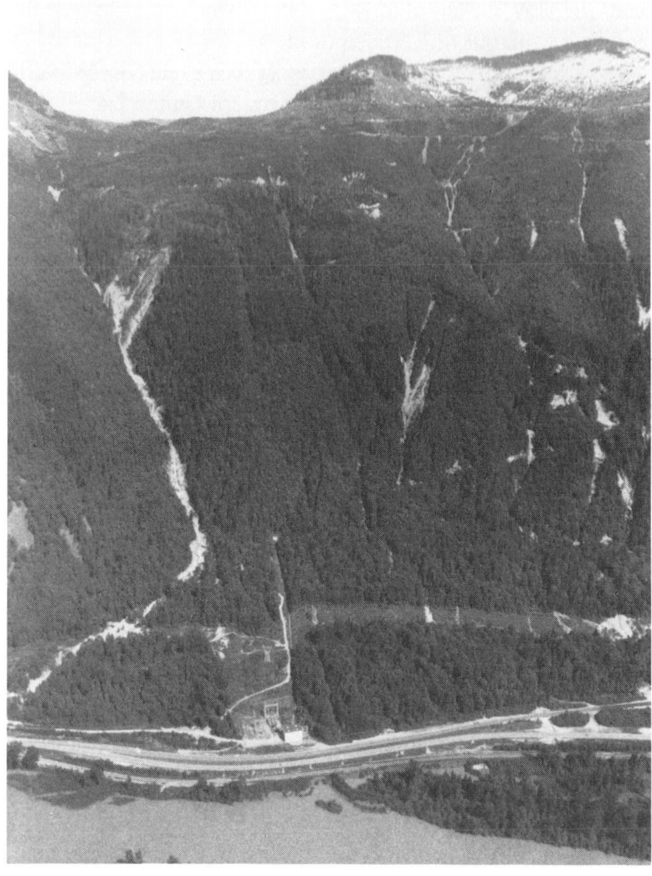

Figure 18. Inactive and recently active debris-torrent tracks near Thurber, British Columbia (photo by Thurber Consultants, Limited).

Initiation of a debris torrent requires a steep channel slope, generally at least 15° but more commonly about 25°. Once initiated, debris torrents can travel over a wide range of slope angles, depending on such factors as the degree of confinement, amount of water or cohesive materials, and channel roughness. The very coarse and noncohesive debris torrents typically seen in the mountains of western Canada will deposit on slopes of around 12° or less. If confinement ends at the exit of a gully, the torrent can come to a relatively abrupt stop on slopes as steep as 10°, leaving a lobate mass of tangled debris. On the other hand, debris torrents out of small, steep gullies sometimes discharge into larger, flatter channels and can travel long distances on slopes as low as 5° because of the added water provided by the larger stream.

Debris torrent deposits are characterized by boulder levees, irregular and often indistinct ridges or lines of boulders along either side of the runout zone. A debris torrent that stops abruptly may leave a prominent debris lobe several meters high, particularly if the debris is coarse. During transport the coarsest material in a debris torrent tends to ride to the top and to the front of the moving mass; this arrangement is likely to be preserved in the deposits (Miles and Kellerhals, 1981). Cuts into debris lobes tend to show boulders on top, supported by a matrix of finer materials with no bedding. Also, debris-torrent transport produces less abrasion and, therefore, more angular material than stream transport.

Estimates of debris-torrent frequency are only possible in the rare situations where several events have been recorded for one channel over a long period. Such records do not exist in Canada. An estimate of the time elapsed since the last one or two events is normally the best information that can be expected. It can provide a very rough estimate of the return period, provided that changes in land use are not significantly affecting the drainage basin.

Numerous studies have shown that clearcut logging and road construction on steep slopes increase the frequency of slope failures. This, in turn, increases the debris supply to stream channels; slides into streams during periods of high runoff also act to trigger debris torrents. The overall result is generally a dramatic increase in the frequency of debris torrents. For northwestern United States, one study shows increases in the spatial frequency (i.e., number of events per km^2 per year) by factors of ten due to clearcut logging, and by factors of over 100 where logging roads across steep slopes are also present.

The statistical aspects of debris-torrent frequency in a particular channel differ from normal flood frequency, in that time series of debris-torrent occurrences appear to contain various forms of long-term memory. The physical explanation for one form is that bedrock channels tend to be scoured almost clean of debris by a debris torrent, which prevents the start-up of future torrents until sufficient new debris has accumulated, a process that may take many years. Torrents originating in tributaries upstream, or massive debris avalanches, can occur and still move through such scoured reaches. An opposite effect has been reported for channels in deep unconsolidated materials. Here a major debris torrent can destabilize the channel to such a degree that relatively small floods turn into debris torrents during subsequent years, until a long period of low runoff permits restabilization.

MAXIMUM DISCHARGES VERSUS DRAINAGE AREA

An easy method for comparing a flood discharge with prior experience in other regions of North America is to plot peak discharge versus drainage area. Figure 20 is such a plot. The curve is an average of the 20 highest of many maximum known discharges; 18 sites are in the United States, and two are in Mexico. Data for these sites are in Table 1. An average curve was used rather than an envelope curve because the reliability of these high discharges varies considerably. Also, recent studies indicate the possibility that some of these discharges may be too high rather than too low.

This curve should only be used to roughly estimate a maximum flood at a site. It should not be used for engineering studies because of the complex interaction of many variables at a given

Figure 19. Debris-torrent deposits, British Columbia (source of photo undetermined).

site and also because there is no basis for assigning probabilities to the discharges.

Another problem is that the entire drainage area, which is used in this plot, may not contribute to the flood. For example, the 1954 peak discharge of 26,800 m^3/s on the Pecos River (Figs. 4 and 5) plots well below this curve when the entire drainage area, 91,100 km^2, is used. The storm that produced this flood covered only an area estimated as 9,100 km^2 in the lower part of the basin. If only the contributing area is used, the data plots just above the curve.

Another constraint is that the curve may not be representative of the geologic and climatic conditions at one's point of interest.

An interesting sidelight to the curve in Figure 20 is that the International Association of Hydrological Sciences (IAHS, 1984) lists floods greater than 70,000 m^3/s in Bangladesh, India, and Russia that exceed the curve values by as much as a factor of 3.

IAHS (1984) computes a coefficient, K, suggested by Francou (Francou and Rodier, 1967), which is given by the equation

$$K = 10\left[1 - \frac{(\log Q) - 6}{(\log A) - 8}\right]$$

where Q is the flood discharge in m^3/s and A is the drainage area in km^2. The largest floods in the world, from catchments larger than 100 km^2, plot as a straight line corresponding approximately to a K value of 6. This line is shown on Figure 20; but above 30,000 m^3/s, it is based only on floods outside North America.

Another way to compare peak discharges is to express them as unit values—in cubic meters per second per square kilometer ($m^3/s/km^2$). The main reason large areas yield low unit discharges is that the entire area may not be contributing. See the discussion above regarding Figure 20. The high unit discharges from small drainage areas pose a problem if one extrapolates unit values too far. Values derived from areas smaller than about 0.5 km^2 are questionable bases for practically all extrapolations.

A plot (not shown) on log-log coordinates of unit discharge versus drainage area for sites 6 to 16 in Table 1 and for a few other sites with high unit discharges shows that cubic meters per second per square kilometer varies approximately as $360/A^{0.54}$, A being in square kilometers. If unit discharge versus drainage area for 47 sites (IAHS, 1984) in Canada, Costa Rica, Cuba, Guatemala, Hawaii, Mexico, Panama, and the United States are plotted, a very rough relation shows unit discharge varying as $95/A^{0.4}$. A logical conclusion is that most of these floods, though they may have been outstanding locally, were not extreme hydrologic events.

Unit discharges for smaller regions are given by Crippen and Bue (1977) who divided the conterminous United States into 17 flood regions. Maximum floods experienced show considerable variation among the 17 regions.

MAXIMUM FLOOD STAGES

The maximum flood stage is very important locally. Forecasts of this stage determine whether or not existing levees will be adequate or whether sandbags or other measures will be needed to contain the flood. At some sites, the maximum stage can be

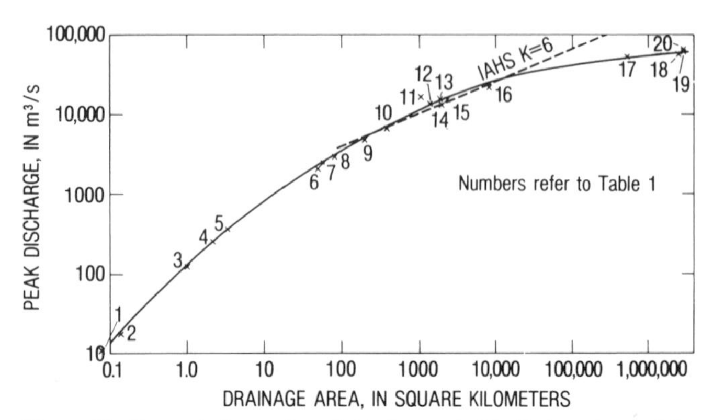

Figure 20. Greatest observed peak discharges in North America and the IAHS curve defined by world maximums.

forecast within acceptable limits; but at others the forecast can be considerably in error because of many complicating factors.

The maximum stage may not be concurrent with the maximum discharge because of ice or debris jams, backwater from another stream, or return of overbank flow. Leafy vegetation that is submerged in summer will cause a higher stage than bare branches or crops that have been plowed under in winter. If no significant floods have occurred in recent years, vegetation may have encroached in the channel; and unless the current flood washes it out, higher stages than expected will occur. Extensive gravel mining in a channel can lower the maximum stage in the vicinity until the stream has time to reestablish premining conditions.

Maximum stages can be caused by floods of long duration. The volume of runoff creates the problem. A sustained period of snowmelt runoff is one such condition.

HISTORIC FLOODS AND PALEOFLOODS OR PREHISTORIC FLOODS

Records of floods are relatively short. Qualitative information is available for some early-day floods that antedate systematic collection of flood data. Thomson and others (1964) present brief statements about floods in New England occurring in 78 of the 321 years from 1635 to 1955, as well as an allusion to an Indian legend about a Noah-like flood a long time ago. They concluded that one or more of the great New England floods of 1927, 1936, 1938, and 1955 exceeded the greatest known historical floods on most of the major rivers. Outstanding floods are known to have occurred in most of northern California in 1861–1862, 1867–1868, 1881, 1889–1890, and 1907 (McGlashan and Briggs, 1939).

Data on known floods at a few sites extend back only 150 years or less, and there are even fewer sites where fragmented information is available for earlier, isolated events. Therefore, estimates of floods based on geologic evidence and interpretation will provide knowledge about what has happened and what might happen in the future as well as helping to improve flood-frequency relations and evaluation of risks from low-probability floods.

Geologic methods often involve the use of pockets of sand and gravels on channel banks, also known as slack-water sediments, as evidence of the high flood stage. The size and distribu-

TABLE 1. DATA FOR THE 20 SITES USED IN FIGURE 20

Number on Figure 20		Date of Flood	Drainage Area (km²)	Discharge (m³/s)	K
1	N. F. Waihee Stream near Heeia, HI	02/04/65	0.08	10.6	4.53
2	Twomile Creek tributary near Point Allegheny, PA	07/18/42	0.135	18.1	4.65
3	Big Branch near Sunburst, NC	08/30/40	1.0	127	5.13
4	Humboldt River tributary near Rye Patch, NV	05/31/73	2.20	253	5.30
5	Big Creek near Waynesville, NC	08/30/40	3.42	368	5.40
6	Bronco Creek near Wikieup, AZ	08/18/71	49.2	2,080	5.75
7	S. F. Wailua River near Lihue, HI	04/15/63	58	2,470	5.82
8	Arroyo San Bartolo, Cerca Poblado, Mexico	1976	81	3,000	5.86
9	Mailtrail Creek near Loma Alta, TX	06/24/48	195	4,810	5.94
10	Seco Creek near D'Hanis, TX	05/31/35	368	6,510	5.98
11	West Nueces River near Kickapoo Springs, TX	06/14/35	1,041	16,400	6.42
12	Cihuatlan, Paso del Mojo, Mexico	1959	1,370	13,500	6.19
13	West Nueces River near Bracketville, TX	06/14/35	1,813	15,600	6.16
14	Dry Devils River near Carta Valley, TX	06/24/48	1,970	13,000	5.99
15	West Nueces River near Cline, TX	06/14/35	2,279	15,200	6.08
16	Eel River at Scotia, CA	12/23/64	8,060	21,300	5.92
17	Ohio River at Metropolis, IL	02/01/37	525,800	52,300	4.38
18	Mississippi River at New Madrid, MO	02/04/37	2,391,000	56,900	2.32
19	Mississippi River near Arkansas City, AR	02/16/37	2,928,000	61,100	2.08
20	Mississippi River at Vicksburg, MS	05/01/27	2,953,000	64,600	2.22

tion of cobbles and boulders deposited along the channel and on the flood plain, and carbon-dating procedures are also used. Observations are made where erosion from the high velocities and debris is minimal so that the present cross-sectional area is probably representative of the conditions at the time of the flood.

One type of paleoflood is caused by some momentous event. The breakout from glacial Lake Missoula that sent water down the Columbia River and across the scablands of Washington, and the breakout of Lake Bonneville at Red Rock Pass in northern Utah that caused a great flood in the Snake River basin are two examples. Volcanic eruptions are another cause of flooding.

A second type is caused by intensive rain, often on small areas but not limited to them. Some floods on the Pecos River in Texas are from this source.

The Bonneville Flood occurred in the late Pleistocene, about 30 ka (Malde, 1968). A peak discharge of 425,000 m^3/s was computed by the critical depth method at a site south of Boise, Idaho, about 550 km downstream from Red Rock Pass; therefore, this discharge was significantly attenuated by channel and overbank storage. The peak discharge at Red Rock Pass may have been about four times that computed south of Boise. Based on material deposited along the reach, maximum velocities ranged from 5 m/s to 10 m/s.

The Lake Missoula flood occurred about 13 to 18 ka and persisted about one week (Baker and Nummedal, 1978). They estimated the maximum discharge as 21.3×10^6 m^3/s, with velocities reaching 30 m/s. Discharges and velocities of these magnitudes can cause considerable scour and fill or channel realignment in a short time. Clarke and others (1984) made computer simulations of the physical equations governing the enlargement by water flow of the tunnel penetrating the ice dam impounding glacial Lake Missoula. They obtained a range of 2.74×10^6 to 13.7×10^6 m^3/s for the maximum discharge and estimated that 2,184 km^3 were released. These results are considered to be in line with paleohydrologic estimates.

Lowell Glacier in the St. Elias Mountains of southwestern Yukon Territory advanced across Alsek Valley and blocked south-flowing Alsek River many times during the Neoglacial interval (Clague and Rampton, 1982). The resulting lake, termed Neoglacial Lake Alsek, was about 200 m deep at the glacier dam and 100 km long at its maximum, the largest glacier-dammed lake formed in North America since the end of the Wisconsin glaciation. Geologic evidence indicates that the lake emptied catastrophically when the Lowell Glacier dam failed.

Historical records and radiocarbon and tree-ring dates on buried soils and driftwood indicate that Lake Alsek extended into Dezadeash Valley sometime between A.D. 1848 and 1891, between A.D. 1736 and 1832, twice between 250 and 500 years ago, and at least once between 800 and 2,900 years ago. During each ponding phase, Lake Alsek may have emptied and filled repeatedly, perhaps on a regular cycle, as is common for some existing glacier-dammed lakes.

During a future surge of Lowell Glacier, the Alsek River may be blocked again. However, Lowell Glacier has thinned and receded somewhat since Lake Alsek last extended beyond the St. Elias Mountains in the nineteenth century, thus reducing the possible flood hazard.

Studies of slack-water deposits along the lower Pecos River in Texas (Kochel and Baker, 1982) have produced a 10,000-year flood record. In this case, the floods of 1954 and 1974 were higher than any other flood in at least 2,000 years. The respective discharges of 26,800 m^3/s and 16,300 m^3/s were considered by the authors to have recurrence intervals of at least 2,000 years and 500 to 700 years, respectively. Another interpretation of the data could be made to show that these modern floods may have recurrence intervals of 10,000 and 2,500 years. Both floods were caused by heavy rain in a few hours when hurricanes occurred.

Other unusual floods are briefly noted below.

During glacial retreat about 12 to 11 ka, Lake Regina in Saskatchewan flooded southeastward to Lake Souris in North Dakota, thence to Lake Hind and Lake Agassiz in Manitoba (Kehew and Clayton, 1983). The floods, probably on the order of 10^5 m^3/s, eroded new spillway segments and enlarged existing segments.

Researchers with ongoing studies on the Salt and Verde rivers in Arizona (L. Ely, written communication, 1984) have tentative results based primarily on slack-water sediments. One interesting result is that on neither river is the largest Holocene flood more than twice the size of the largest flood during the gaged record of 40 years on the Verde River and 60 years on the Salt River. Both recent floods were storm runoff.

There is an unresolved discussion about the validity of comparing paleofloods with recent ones. Climatic conditions at the end of the last ice age were certainly different than those of the present. Should floods from Lake Missoula and Lake Bonneville, or for that matter those caused by failures of dams, be compared with floods caused by precipitation? Should hurricane-caused floods be separated from nonhurricane floods? Statisticians claim that all floods are not from the same population, but where should an analyst make the division?

Dendrochronologists have studied tree rings in a number of areas and have defined several abnormally wet or dry periods. A few tree-ring samples in southern California go back to 1385 (Schulman, 1947), and some others to 1512 (Stockton and Jacoby, 1976). Quantitative data on floods cannot be derived from tree rings, but Stockton and Jacoby conclude that the period 1906 to 1930 in the Colorado River basin was one of anomalously persistent high runoff and that it apparently was the greatest and longest high-flow period within the last 450 years. This raises the possibility that some of the floods that occurred during this period in the Colorado River basin could be rarer ones than the systematic record would indicate. Other studies have been made in western Canada, the western United States, and northern Mexico.

There have been outburst floods from glacier-dammed lakes in recent years (see above section on "Transient effects"), but the flood magnitudes have not been very large compared to those of the paleofloods.

FLOODING FROM LAKES

Harney and Malheur lakes, in a closed basin in southeastern Oregon, occupied about 182 km^2 in 1980. Snowmelt runoff since 1981 has been well above normal, and water-surface elevations have risen about 3 m. The two lakes are now one and cover about 730 km^2. Ranch homes, hay fields, and roads have been flooded. The only railroad has been inundated, so a lumber mill has had to lay off workers (Lynch, 1985). Flooding in the Malheur National Wildlife Refuge has destroyed the habitat of some species of birds and created favorable conditions for others. Cooler than normal summers the last few years have reduced evaporation losses, thus adding to the problem. Other closed basins that have had flood problems in recent years include the Great Salt Lake in Utah, Tulare Lake in California, and the Humboldt-Carson sink and Walker Lake in Nevada. In addition, flooding has occurred at Lake Tahoe, Clear Lake, and Lake Elsinor in California and around the Great Lakes.

The most recent flood problems around Great Salt Lake in Utah have occurred since 1982. The greatest recorded seasonal rise in the water surface was 1.6 m between September 18, 1982, and June 30, 1983, to an elevation of 1,281.7 m. The rise resulted from much above average rainfall in 1982, greater than normal snowfall during the autumn of 1982 through the spring of 1983, and unseasonably cool weather during that spring (Arnow, 1984). The lake surface would have been about 1.5 m higher in 1983 had there been no consumptive use by man, which began in 1847. The highest level recorded was 1,283.7 m in 1873. Prehistoric information indicates that Great Salt Lake has remained near its historic average elevation of about 1,280 m for the last 8,000 years except for two rises above about 1,290 m.

After falling 0.15 m during the summer of 1983, the lake level rose 1.5 m between September 25, 1983 and July 1, 1984 to an elevation of 1,283.0 m; and the lake covered 5,960 km^2 (Arnow, 1985). Though the rise in 1983–1984 was slightly less than that in 1982–1983, the volume of water was about 15 percent greater because of the greater surface area at the higher elevations.

Roads, railroads, recreation and industrial facilities, and wildfowl-management areas had been established on the exposed lake bed. Damage to these installations has been estimated as $212 million.

The lake level was only 0.3 m lower than the 1984 maximum elevation on October 16, 1985, when the seasonal rise started and continued to an elevation of 1,283.2 m in 1985 and to 1,283.8 m on June 6, 1986. The total volume was about 12 percent more in 1986 than in 1984.

Lake Elsinore is about 110 km southeast of Los Angeles, California; it receives most of its water from the San Jacinto River. Prior to 1965, when importation of Colorado River water began, the lake was dry for many years in succession. Since 1965, a shallow lake covering about 16 km^2 has been maintained. Prior to 1980, outflow is known to have occurred only in 1872, 1883–1884, and 1916–1917, and there probably was outflow in 1862. Outflow is northeastward to Temescal Wash (White, 1982).

As is often the case, urban development occurred around the lake shores without much consideration for flood risk. Between February 13 and 23, 1980, the lake level rose 3.65 m. This called for dredging the outlet channel that had been blocked by alluvial deposits from tributary canyons, protecting homes with sandbags and temporary levees, and evacuating people. Though the inflow decreased from a maximum of 220 m^3/s on February 22 to less than 28 m^3/s in mid-April, the maximum outflow was only 7 m^3/s; therefore, the lake rose another 2 m by March 20. The Lake Elsinore flood damaged 300 structures, flooded about 100 septic tanks, and displaced 2,000 people.

Flooding from the Great Lakes occurs in Ontario and the states adjacent to them. Maximum stages in the lakes generally occur in late spring or early summer after the annual snowmelt. A large volume of runoff into Lake Superior can affect stages in the downstream lakes during the next two or three years. Stages higher than normal may occur if there is excessive precipitation and relatively little evaporation from prolonged cloudy, cool periods. At high lake levels, property damage from flooding may be increased by erosion due to wind-driven wave action. Shallow lakes, such as Erie and St. Clair, have the additional possibility of wind setup when winds are orientated along the long axis of the lake. Records show Lake Erie to have been as much as 2 m above static level at the downwind end of the lake under high-wind conditions. This problem is usually aggravated around March when spring wind conditions are especially strong and the lake level is high. Monroe, Michigan, at the west end of Lake Erie, is susceptible to this kind of flooding. Lakes Superior and Ontario have control structures; however, these are only partially effective when conditions are severe. Lakes Huron, St. Clair, and Erie are uncontrolled and exhibit large fluctuations. Monthly mean water levels have a range of 1.2 m on Lake Superior, increasing downstream to 2.0 m on Lake Ontario. In November 1985, all of the Great Lakes except Lake Ontario reached record high monthly mean levels.

FLOODING FROM THE OCEANS

Coastal areas are also subject to oceanic flooding when hurricanes or tropical storms occur, particularly if they coincide with high tides. Hurricanes occur along the Atlantic and Gulf coasts of the United States and along both coasts of Mexico. Those along the west coast of Mexico frequently move westward and dissipate, or if they move northward they weaken to tropical storms, which have caused floods in southern California. Besides the flooding, erosion of the coast line and the battering by the waves are major hazards to structures.

Data collected and theories devised along one coast cannot be transferred to another coast, primarily because the configurations of the continental shelves are different. Also, the infrequent occurrence of oceanic flooding at a given point makes any frequency study questionable. Therefore, estimates of what might happen when a hurricane strikes, or how often, are not reliable.

Tidal waves or tsunamis are caused mainly by major earthquakes. The strong quake in Alaska in March 1964 caused identifiable seismic sea waves along the west coast of Canada and the United States and as far away as Antarctica and Wake Island (Eckel, 1970). These waves resembled high, fast-moving tides rather than breaking waves in some places, and they were generally much lower in amplitude than the locally generated waves. The local waves struck during or immediately after the earthquake and had generally subsided long before the arrival of the train of seismic sea waves.

OVERVIEW

Determination of peak discharge

Peak discharges are determined by several methods (Rantz and others, 1982). Direct measurement of a flood peak by current meter may be the most reliable method, provided the water depths and velocities can be measured without undue error. When a current-meter measurement is made soon after the peak occurred, a short extension of the stage-discharge relation is used to determine the peak discharge, and results are usually reliable.

When or where current-meter measurements cannot be or have not been made for numerous reasons, indirect measurements are made if the stage-discharge relation must be extended by at least 100 percent. They are indirect in the sense that field data are obtained from the evidence left by the flood at a bridge, a culvert, a dam, a road embankment, or a reach of channel. Surveys of high-water marks, cross sections of the channel, and the geometry of any structures are obtained. Survey data empirical coefficients based on laboratory experiments or on field verifications, and usually the Manning flow equation are used to compute the peak discharge. Results may or may not be as reliable as those from a current-meter measurement depending on the stability of the channel, the adequacy of high-water marks, and whether or not the conditions fall within the limits upon which the coefficients are based.

The stage-discharge relation

A reliable stage-discharge relation must be derived from both correct stage and discharge data. There are problems associated with the correct determination of both stage and discharge. Some of the problems are described in the following sections.

Uncertainties in flood stages

Most flood stages are determined within acceptable limits by recorders over a stilling well directly connected by intake pipes to the stream, or by pressure gages. However, the timers on the recorders may be stopped during the flood, the recorder may malfunction, and the orifice for the pressure gage or the intakes to the stilling well may be plugged, buried, or dislodged from their proper positions, thus causing erroneous readings. Even the stage from a nonrecording gage will depend upon how the observer interprets the surge on a staff gage or whether the observer is overly cautious about lowering a wire-weight gage into very rough water with lots of floating debris. Comparison of the recorded or observed stage with nearby high-water marks is essential. Where a gage is mounted on a bridge pier or a wingwall for a culvert, there may not be any high-water marks nearby, so a relation should be established between the recorded stage and a point on the stream bank where high-water marks are apt to be found.

The correct stage may be obtained at the gage, but it may not represent an average water-surface elevation because of superelevation, especially on a small stream. A gage may be located so that reliable stages are obtained for moderate-sized floods, but a very large flood might change the intake conditions so that some of the kinetic energy of the flood is converted into potential energy in the stilling well, thus causing a stage that is too high. Or, the high velocity past the open end of a dammed intake may cause drawdown, which will lower the stage below the true stage.

Uncertainties in flood discharges

Some of the problems encountered that reduce the reliability of peak discharges were mentioned in general terms above. A high-water current-meter measurement by itself may not be very reliable, but a stage-discharge relation averaging several such measurements can be reliable.

In streams less than 2 m deep with velocities more than 3 m/s, the water surface is usually very rough, with standing and breaking waves and other forms of turbulence. Under these conditions, soundings may be as much as 0.4 m in error; and the meter will swing up and downstream and transversely even with a 220 kg streamlined sounding weight. Therefore, the "measured" discharge may be in error.

In alluvial channels, the streambed can be changing constantly at every point, thus causing changes in velocity. An experienced hydrographer needs 30 to 60 minutes to make a current-meter measurement. The velocity observed at the initial point will probably not be the same at the time the velocity at the final point is obtained, but it is assumed to be when the discharge is computed.

This problem is partly overcome when using indirect methods. A cross section of an alluvial channel may not be the same shape after a flood as it was during a flood, but the area will be nearly the same because local scour and fill in the section or channel reach tend to offset each other (Colby, 1964).

Ice in the flow and layers of slush ice in northern latitude streams make measurements by current meter difficult and less reliable than those made in open water. The accuracy of periods of ice effect, which may include maximum discharges, is downgraded because of the many uncertainties.

Unstable stage-discharge relations

Even if stages and discharges are reliably determined, the stage-discharge relation may be changing almost continuously or may change every time the stream rises. Alluvial channels have unstable stage-discharge relations. An example of the large scour and fill that occurred in a short period of time on an alluvial stream is shown in Figure 21. During the flood of March 1938 (Troxell and others, 1942), the East Fork San Gabriel River near Camp Bonita, California, peaked about 5:00 p.m. on March 2. Note that the water surface that morning was lower than the stream bed on March 3, after the flood. The same relation holds for the water surface on March 3 with respect to part of the streambed on March 7. It is evident that the four measurements shown will not define a stage-discharge relation.

Ice is another factor causing unstable stage-discharge relations. The backwater from ice varies with time and with the discharge. In some streams, the ice effect is intermittent during the day, and the discharge may be related to the temperatures at the measurement site and those at higher elevations in the basin.

The magnitudes of the errors caused by the problems mentioned previously are never known. Field personnel estimate the limits between which they believe the result of a current-meter or indirect measurement should be. But when a discharge value is qualified as plus or minus 20 percent, that doesn't mean that it is 20 percent in error. It can be the correct value, and it is often the most probable value because some errors may be compensating. One must also consider the magnitude of the discharge upon which the percentage is based. A large percent error in a small figure is usually meaningless. For example, a 100-percent error in a discharge of 0.1 m^3/s is only 0.1 m^3/s. However, a 5-percent error in 1,000 m^3/s is 50 m^3/s.

Observations of field data and computations based upon them are usually recorded to three significant figures, though four significant figures are used for certain quantities. The three-figure guideline seems both practical and realistic, considering the variations in field conditions in space and time.

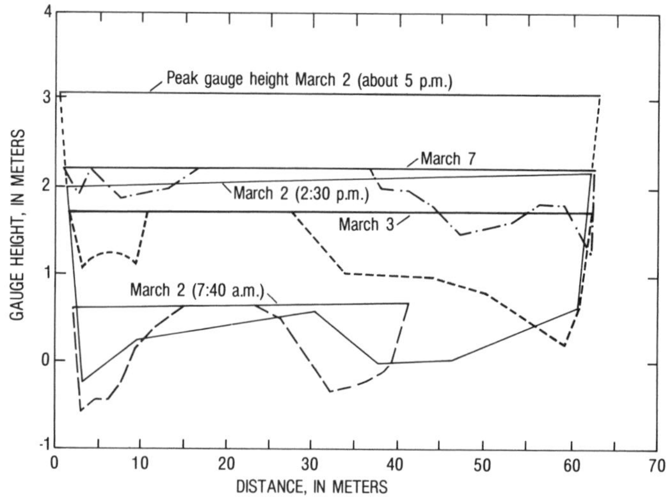

Figure 21. Changes in a cross section on East Fork San Gabriel River near Camp Bonita, California, during the flood of March 1938.

Lack of flood data or limited data, even if the latter are correct, reduce the reliability of some hydrologic analyses. No data may be available for your site of interest, so data from nearby sites must be transferred through regional relations (Thomas and Benson, 1970). Limited or short-term data may not be representative of long-term values if they happen to be collected during an abnormally wet or dry period or if they include one extreme event.

REFERENCES CITED

Alden, W. C., 1928, Landslide and flood at Gros Ventre, Wyoming: Transactions of the American Institute of Mining and Metallurgical Engineers, v. 76, p. 347–361.

Alter, J. C., 1931, Mud flows in Utah, *in* Monthly weather review, 1930: U.S. Weather Bureau, v. 58, p. 319–321.

Anderson, D. G., 1970, Effects of urban development on floods in northern Virginia: U.S. Geological Survey Water-Supply Paper 2001-C, 22 p.

Arnow, T., 1984, Water-level and water-quality changes in Great Salt Lake, Utah, 1847–1983: U.S. Geological Survey Circular 913, 22 p.

—— , 1985, Rise of Great Salt Lake, Utah, *in* National water summary, 1984: U.S. Geological Survey Water-Supply Paper 2275, p. 31–33.

Association of Engineering Geologists, 1978, Failure of the St. Francis Dam, San Francisquito Canyon, near Saugus, California: Association of Engineering Geologists, 94 p.

Bailey, J. F., Patterson, J. L., and Paulhus, J.L.H., 1975, Hurricane Agnes rainfall and floods, June–July 1972: U.S. Geological Survey Professional Paper 924, 403 p.

Baker, V. R., and Nummedal, D., eds., 1978, The Channeled Scabland: National Aeronautics and Space Administration, 186 p.

Boner, F. C., and Stermitz, F., 1967, Floods of June 1964 in northwestern Montana: U.S. Geological Survey Water-Supply Paper 1840-B, 242 p.

Brice, H. D., and West, R. E., 1965, Floods of March–April 1960 in eastern Nebraska and adjacent states: U.S. Geological Water-Supply Paper 1790-A, 144 p.

Burkham, D. E., 1970, Depletion of streamflow by infiltration in the main channels of the Tucson basin, southeastern Arizona: U.S. Geological Survey Water-Supply Paper 1939-B, 36 p.

—— , 1972, Channel changes of the Gila River in Safford Valley, Arizona 1846–1970: U.S. Geological Survey Professional Paper 655-G, 24 p.

—— , 1976, Effects of changes in an alluvial channel on the timing, magnitude, and transformation of flood waves, southeastern Arizona: U.S. Geological Survey Professional Paper 655-K, 25 p.

Clague, J. J., and Rampton, V. N., 1982, Neoglacial Lake Alsek: Canadian Journal of Earth Sciences, v. 19, p. 94–117.

Clarke, G.K.C., Mathews, W. H., and Pack, R. T., 1984, Outburst floods from glacial Lake Missoula: Quaternary Research, v. 22, no. 3, p. 289–299.

Colby, B. R., 1964, Scour and fill in sand-bed streams: U.S. Geological Survey Professional Paper 462D, 32 p.

Costa, J. E., 1985, Floods from dam failures: U.S. Geological Survey Open-File Report 85-560, 54 p.

Crippen, J. R., 1966, Selected effects of suburban development on runoff in a small basin near Palo Alto, California: U.S. Geological Survey Open-File Report, 19 p.

Crippen, J. R., and Bue, C. D., 1977, Maximum floodflows in the conterminous United States: U.S. Geological Survey Water-Supply Paper 1887, 52 p.

Curry, R. R., 1966, Observation of alpine mudflows in the Tenmile Range, central Colorado: Geological Society of America Bulletin, v. 77, p. 771–776.

Eckel, E. B., 1970, The Alaska earthquake March 27, 1964; Lessons and conclusions: U.S. Geological Survey Professional Paper 546, 57 p.

Environment Canada, 1976, Canada water yearbook, 1975: Ottawa, Inland Waters Dirctorate, p. 209–210.

Francou, J., and Rodier, J., 1967, Essai de classification des crues maximales dans le monde: Bondy, France, Cahiers ORSTOM, Serie Hydrologie, v. IV, no. 3, p. 19–46.

Garcia-Quintero, A., 1950, Hydrology of Mexico: Proceedings, American Society of Civil Engineers, v. 76, Separate no. 38, 17 p.

Glancy, P. A., and Harmsen, L., 1975, A hydrologic assessment of the September 14, 1974, flood in El Dorado Canyon, Nevada: U.S. Geological Survey Professional Paper 930, 28 p.

Hadley, R. F., 1967, Erosion and deposition caused by floods of June 1964 in northwestern Montana: U.S. Geological Survey Water-Supply Paper 1840-B, p. 115–129.

Harris, D. D., 1977, Hydrologic changes after logging in two small Oregon coastal watersheds: U.S. Geological Survey Water-Supply Paper 2037, 31 p.

Hoxit, L. R., Maddox, R. A., Chappell, C. F., and Brua, S. A., 1982, Johnstown–western Pennsylvania storm and floods of July 19–20, 1977: U.S. Geological Survey Professional Paper 1211, 68 p.

Hoyt, W. G., and Langbein, W. B., 1955, Floods: Princeton University Press, 469 p.

IAHS, 1984, World catalogue of maximum observed floods: International Association of Hydrological Sciences–Association International des Sciences Hydrogeologique, Publication no. 143, 355 p.

Kehew, A. E., and Clayton, L., 1983, Late Wisconsin floods and development of the Souris-Pembina spillway system in Saskatchewan, North Dakota, and Manitoba, in Glacial Lake Agassiz: Geological Association of Canada Special Paper 26, p. 187–209.

Kochel, R. C., and Baker, V. R., 1982, Paleoflood hydrology: Science, v. 215, no. 4531, p. 353–361.

Lombard, R. E., Miles, M. B., Nelson, L. M., Kresch, D. L., and Carpenter, P. J., 1981, Channel conditions in the lower Toutle and Cowlitz rivers resulting from the mudflows of May 18, 1980: U.S. Geological Survey Circular 850-C, 16 p.

Los Angeles County Flood Control District, 1979, Mudflow study: City of Sierra Madre, Los Angeles County, California, 8 p.

Lynch, W., 1985, Floods produce changes in waterfowl habitat: Eugene, Oregon, *The Register-Guard,* April 8, 1985, p. 12C.

Malde, H. E., 1968, The catastrophic late Pleistocene Bonneville flood in the Snake Plain, Idaho: U.S. Geological Survey Professional Paper 596, 52 p.

Marcus, M. G., 160, Periodic drainage of glacial-dammed Lake Tulsequah, British Columbia: Geographical Review, v. 50, no. 1, p. 89–106.

Matthai, H. F., 1969, Floods of June 1965 in South Platte River basin, Colorado: U.S. Geological Survey Water-Supply Paper 1850-B, 64 pp.

McGlashan, H. D., and Briggs, R. C., 1939, The floods of December 1937 in northern California: U.S. Geological Survey Water-Supply Paper 843, p. 427–479.

McLaurin, I., and Wedel, J. H., 1981, The Red River flood of 1979: Winnipeg, Manitoba, Environment Canada, Inland Waters Directorate, Water Resources Branch, 14 p.

Miles, M. J., and Kellerhals, R., 1981, Some engineering aspects of debris torrents: Proceedings of the 5th Canadian Hydrotechnical Conference, The Canadian Society for Civil Engineering, v. 1, p. 395–420.

Miller, J. E., and Frink, D. L., 1984, Changes in flood response of the Red River of the North Basin, North Dakota–Minnesota: U.S. Geological Survey Water-Supply Paper 2243, 103 p.

Morton, D. M., and Campbell, R. H., 1974, Spring mudflows at Wrightwood, southern California: Quarterly Journal of Engineering Geology, v. 7, p. 377–384.

Post, A., and Mayo, L. R., 1971, Glacier-dammed lakes and outburst floods in Alaska: U.S. Geological Survey, Hydrologic Investigations Atlas, HA 455, scale 1:1,000,000.

Rantz, S. E., and others, 1982, Measurement and computation of streamflow; Measurement of stage and discharge: U.S. Geological Survey Water-Supply Paper 2175, v. 1, p. 79–284.

Ray, H. A., and Kjelstrom, L. C., 1978, The flood in southeastern Idaho from the Teton Dam failure of June 5, 1976: U.S. Geological Survey Open-File Report 77-765, 48 p.

Richardson, D., 1968, Glacier outburst floods in the Pacific Northwest: U.S. Geological Survey Professional Paper 600-D, p. 79–86.

Sanders, C. L., and Sauer, V. B., 1979, Kelly Barnes Dam flood of November 6, 1977, near Toccoa, Georgia: U.S. Geological Survey Hydrologic Atlas, HA-613, scale 1:12,000.

Sauer, V. B., Thomas, W. O., Jr., Stricker, V. A., and Wilson, K. V., 1983, Flood characteristics of urban watersheds in the United States: U.S. Geological Survey Water-Supply Paper 2207, 63 p.

Schulman, E., 1947, Tree-ring hydrology in southern California: University of Arizona Bulletin, Laboratory of Tree-Ring Research Bulletin no. 4, 36 p.

Sharp, R. P., and Nobles, L. H., 1953, Mudflow of 1941 at Wrightwood, southern California: Geological Society of America Bulletin, v. 64, p. 547–560.

Shroba, R. R., Schmidt, P. W., Crosby, E. J., and Hansen, W. R., 1979, Geologic and geomorphic effects in the Big Thompson Canyon area, Larimer County, Colorado: U.S. Geological Survey Professional Paper 1115, Part B, p. 107–115.

Snipes, R. J., and others, 1974, Floods of June 1965 in Arkansas River basin, Colorado, Kansas, and New Mexico: U.S. Geological Survey Water-Supply Paper 1850-D, 97 p.

State of California, 1976, Wrightwood; Debris and mud flow investigation: Department of Water Resources, Southern Division, 23 p.

Stockton, C. W., and Jacoby, G. C., Jr., 1976, Long-term surface-water supply and streamflow trends in the upper Colorado River basin based on tree-ring analyses: National Science Foundation, Lake Powell Research Project Bulletin no. 18, 70 p.

Thomas, D. M., and Benson, M. A., 1970, Generalization of streamflow characteristics from drainage-basin characteristics: U.S. Geological Survey Water-Supply Paper 1975, 55 p.

Thomson, M. T., Gannon, W. B., Thomas, M. P., Hayes, G. S., and others, 1964, Historical floods in New England: U.S. Geological Survey Water-Supply Paper 1779-M, 105 p.

Troxell, H. C., and others, 1942, Floods of March 1938 in southern California: U.S. Geological Survey Water-Supply Paper 844, 399 p.

U.S. Geological Survey, 1949, Floods of May–June 1948 in Columbia River basin: U.S. Geological Survey Water-Supply Paper 1080, 476 p.

United States Water Resources Council, 1981, Guidelines for determining flood flow frequency: Bulletin 17B, 183 p.

Vineyard, J. D., and Feder, G. L., 1982, Springs of Missouri: Missouri Department of Natural Resources, Water Resources Report no. 29, 212 p.

Wallis, J. R., and Wood, E. F., 1985, Relative accuracy of log-Pearson III procedures: Journal of Hydraulic Engineering, American Society of Civil Engineers, v. 111, no. 7, p. 1043–1056.

Wells, F. C., 1980, Hydrology and water quality of the lower Mississippi River: State of Louisiana Water Resources Technical Report no. 21, p. 5–9.

White, C. R., 1982, Lake Elsinor flood disaster of March 1980, *in* Storms, floods, and debris flows in southern California and Arizona, 1978 and 1980: National Research Council, p. 387–397.

Wieczorek, G. F., Ellen, S., Lips, E. W., and Cannon, S. H., 1983, Potential for debris flows and debris flood along the Wasatch Front between Salt Lake City and Willard, Utah, and measures for their mitigation: U.S. Geological Survey Open-File Report 83-635, 45 p.

Wilson, K. V., and Hudson, J. W., 1969, Hurricane Camille tidal floods of August 1969 along the Gulf Coast: U.S. Geological Hydrologic Investigations Atlases HA-395 through HA-408, scale 1:24,000.

Wilson, M. P., 1981, Catastrophic discharge of Lake Warren in the Batavia-Genesee region [Ph.D. thesis]: Syracuse, Syracuse University, 119 p.

Manuscript Accepted by the Society May 1, 1987

Printed in U.S.A.

The Geology of North America
Vol. O-1, Surface Water Hydrology
The Geological Society of America, 1990

Chapter 6

Low flows and hydrologic droughts

J. D. Rogers
Environment Canada, Motherwell Building, 1901 Victoria, Regina, Saskatchewan S4P 3R4, Canada
J. T. Armbruster
U.S. Geological Survey, 6481-B Peachtree Industrial Boulevard, Doraville, Georgia 30360

DEFINING LOW FLOWS

Streamflow fluctuates in time in response to the annual cycles of precipitation and temperature and to random deviations from these cycles. The resulting lowest flow for each year, usually an average for some number of consecutive days, is designated as an annual low flow. That flow usually occurs during a period of no precipitation and is derived from ground-water discharge or surface storage being discharged from lakes, marshes, or melting glaciers.

The hydrograph in Figure 1 shows two low-flow periods for which the daily discharges are averaged to obtain the low flows for those durations for that year. The set of annual low flows includes those for the 1, 3, 7, 14, 30, 60, 90, 120, and 183 consecutive days. Where annual minimum flows usually occur in late summer or autumn, the climatic year, beginning April 1, is used rather than the calendar year. The magnitudes of these annual low flows at any site may range widely from year to year according to current and antecedent weather and to physical conditions of the drainage basin.

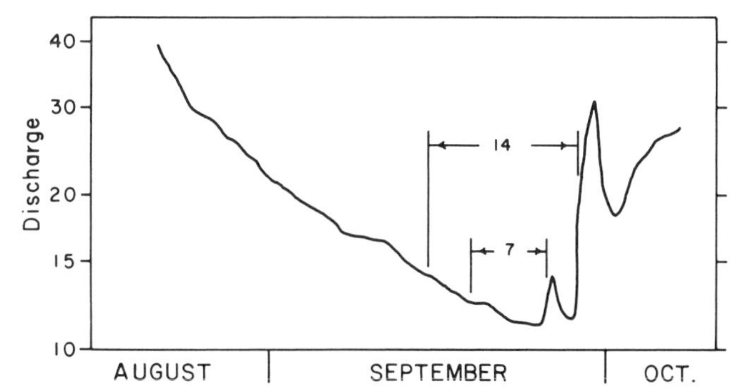

Figure 1. The lowest part of an annual hydrograph on which are shown the periods for which the daily discharges are averaged to give the 7-day and 14-day low flows for that year.

CHARACTERIZING LOW FLOWS

The lowest flows of a stream usually occur in the same season each year. Examination of the flow regimes on Plate 3B will show that in many parts of the United States and southern Canada, the lowest monthly flows generally occur in late summer when temperatures and evapotranspiration are high, and snowmelt runoff is no longer significant. In the northern part of the U.S. and southern Canada, two low-flow periods per year may occur: one during the late summer months, and another during winter when no surface runoff occurs and water in the channel freezes. Farther north the minimum flow is always in late winter or early spring; on small streams, that minimum may be zero.

Again referring to Plate 3B, spring is the season of minimum flow in Mexico, Central America, and the Caribbean Islands, that being the season of low precipitation. Southern Florida streams tend to be low in both spring and late autumn.

The magnitudes of annual low flows and their variability are expressed in various ways. A simple characteristic is the minimum flow of record. It is not commonly used because it may have been affected by some unusual condition in the channel; therefore, it may not be representative of the dependable flow of the stream, nor does it give any indication of variability.

Better ways of characterizing low flows utilize flow data for each of many years in a probability analysis from which selected points are used to represent the low-flow characteristics.

One way of examining low flows is to construct a cumulative frequency curve based on all daily discharges for all complete years of record. This type of analysis combines into one curve the flow characteristics of a stream throughout the range of discharge, without regard to the sequences of occurrence. These curves, usually called duration curves, have been in general use since 1915. Figure 2 shows examples. Low flows are represented by the lower end of the duration curve. Common indexes of low flow taken from the duration curve are the discharges exceeded 95 and 98 percent of the time. These percentages apply to the whole period of record used to define the curve. They are not probabilities of exceedance in a single year.

The slope of the lower end of the duration curve is an indicator of the variability of low flows; a flat slope indicates that flows in that range are reduced only slowly during extended periods without significant rain. However, the lower end of the

Rogers, J. D., and Armbruster, J. T., 1990, Low flows and hydrologic droughts, *in* Wolman, M. G., and Riggs, H. C., eds., Surface water hydrology: Boulder, Colorado, Geological Society of America, The Geology of North America, v. O-1.

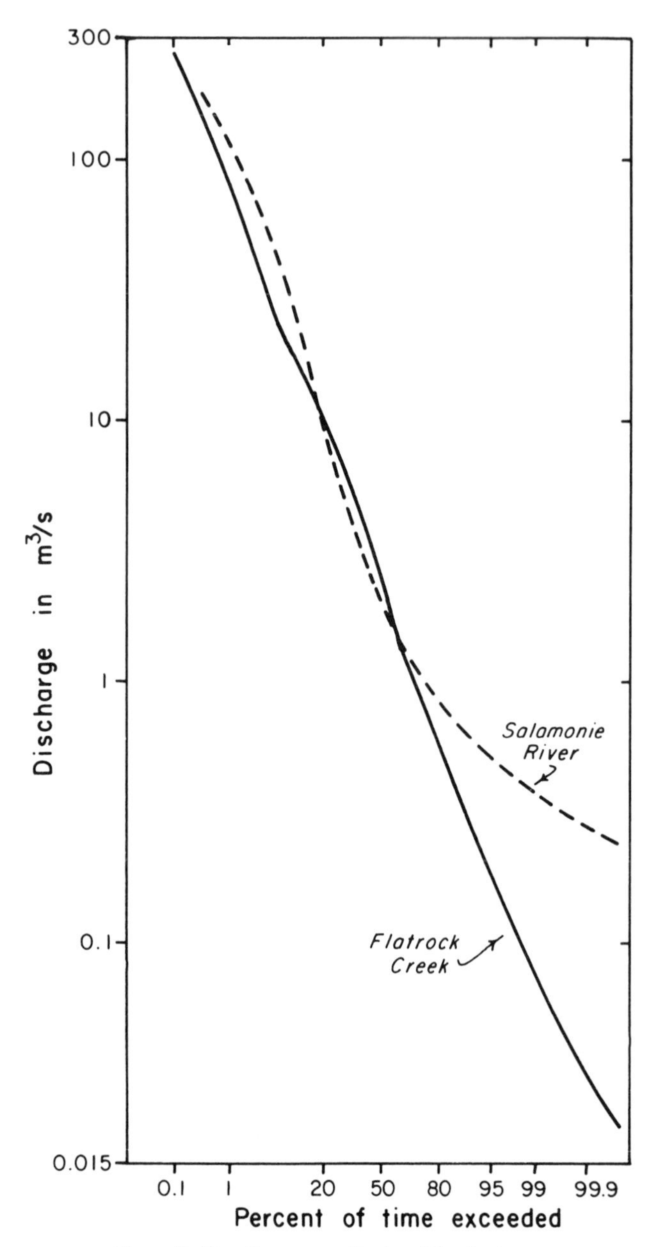

Figure 2. Duration curves for two Illinois streams.

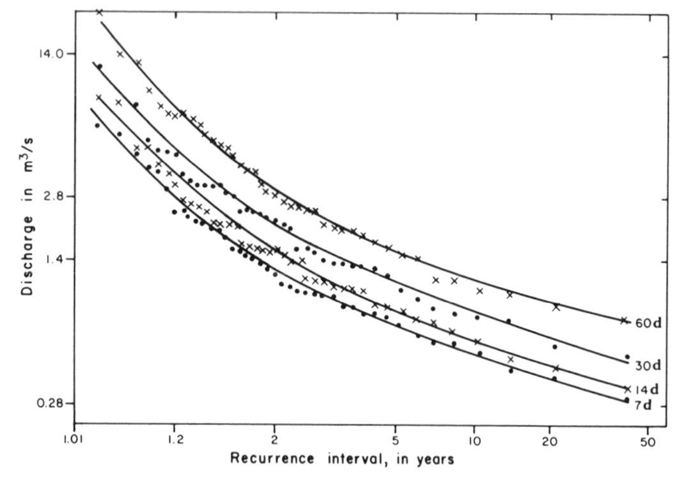

Figure 3. Typical low-flow frequency curves for 7, 14, 30, and 60-day periods, Spoon River, Illinois (after Riggs, 1972).

duration curve may have been defined by many daily means in one year of extremely low flow. Thus the slope is only an approximate indicator of low-flow variability. A steep slope of the lower end of a duration curve such as that for Flatrock Creek in Figure 2, indicates that low flows are not dependable and that they may be zero at times.

Duration curves for periods before and after some streamflow regulation, such as by a storage project, are often used to show the change in low flow due to the operation of the project.

Low-flow characteristics are more commonly described by frequency curves based on annual low flows. These curves are defined by a graphic method, or the data are fitted to a particular statistical distribution. The graphic method assumes no particular shape of the fitted curve and thus is more likely to represent the characteristics of low flows at various levels of discharge as will be shown later.

Figure 3 shows typical low-flow frequency curves. Various low-flow indexes may be taken from these. A common one is the 7-day 10-year low flow, which is the discharge at 10-year recurrence interval on the 7-day curve.

The frequency interpretation is as follows: the 7-day low flow will be less than the 7-day 10-year low flow at intervals averaging 10 years in length, or the probability is 0.1 that the 7-day low flow in any one year will be less than the 7-day 10-year low flow.

Fairly good relations between the 7-day 10-year low flow discharge from a frequency curve and the discharge exceeded 95 percent of the time from the duration curve have been found, but these relations change regionally. Indexes based on low-flow frequency curves are preferable to ones based on the duration curve, and they may be estimated more reliably at sites with little streamflow data.

FACTORS AFFECTING LOW FLOWS

Typically, streamflow is sustained by ground-water discharge during periods of low flow. The ground-water discharge and its recession rate at any time depend (1) on the basin geology, which controls the amount of storage available and the rate at which it drains to the stream; (2) on the precipitation, which determines the amount of water in storage; and (3) on temperature principally as it relates to evapotranspiration losses.

The ground-water body is recharged when water from rain or melting snow infiltrates to the zone that is saturated with water. Recharge raises the water table, the top of the zone of saturation. Where the water table is higher than the water level in the stream and the aquifer material is porous, water will drain to the stream. Without further recharge, the flow to the stream recedes gradually as the water table is lowered by reduction in the volume of water in the aquifer. The rate of recession can be described by a base-flow recession curve (Hall, 1968) derived from segments of streamflow hydrographs for periods when all the streamflow is derived from ground-water flow. A slow rate of recession indicates a large aquifer storage relative to the rate of drainage, and consequently a small variability in annual low flows.

Geologic effects

Two Illinois streams having much different low-flow characteristics are Salamonie River and Flatrock Creek. Their duration curves are shown in Figure 2, and their low-flow frequency curves are in Figure 4. Looking at the low-flow frequency curves, Flatrock Creek has considerable annual minimum flow at times, but the stored water drains rapidly until base flow is very small.

Spring-fed streams usually have dependable low flows. Silver Springs, Florida, is an example. Its flow is derived entirely from the Floridan aquifer and its annual minimum flow is rarely less than 80 percent of the mean for the year. Rivers whose flows are derived largely from ground water occur in the Sand Hills of Nebraska, in the volcanics of Oregon, and in many other locations in the U.S. and Mexico.

Not all springs flow uniformly. Some limestone aquifers drain rapidly because of the well-developed system of fractures and solution channels in the rock. See Figure 22 in Chapter 4 for an example.

Topographic effects

The principal effect of topography on low flows occurs in basins with little relief in which lakes, swamps, and wetlands are common. Low flows of streams draining such basins are derived principally from drainage of water stored on the surface. Little ground-water drainage occurs because of the flat gradients. The annual low flows of such streams are not highly variable so long as the surface storage is not depleted. But a basin containing a large percentage of shallow wetlands could have no flow in some channel reaches following an extended dry period in which the surface storage evaporated.

Climatic effects

The principal climatic effect on low flows in temperate climates is the amount and temporal distribution of precipitation. Major recharge to ground water results from continuously wetted soil such as would result from rain for several days or from the gradual melting of a snowpack. In many parts of North America, the principal recharge occurs in winter and spring, and its magnitude varies from year to year. The spring recharge may deplete gradually throughout the summer in regions having little summer precipitation. For example, the relation of spring streamflow to autumn streamflow (in Fig. 5) indicates that no appreciable recharge occurred during most summers.

More commonly, precipitation occurs throughout the summer and produces some recharge, as indicated by the hydrograph of Figure 6 showing estimates of base flow and the estimated increase in base flow (resulting from recharge derived from summer storms).

Frequent light rains during the low-flow season may produce no recharge, but they may produce a little runoff, which results in a minimum flow in excess of that derived from ground water. This condition often occurs in small streams in the eastern U.S. and accounts for high values of low flows in some years, the steep slope of the upper end of the low-flow frequency curve, and the sharp decrease in slope at about the 2-year recurrence interval. A typical curve is in Figure 7. This Florida stream has large annual minimum flows from drainage from Okefenokee Swamp in some years and none in others. Springs maintain some flow at all times.

A dogleg frequency curve might also represent a basin with

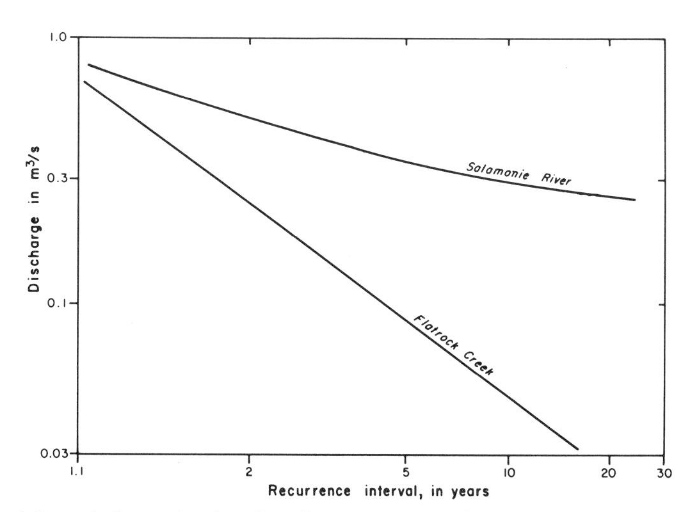

Figure 4. Seven-day low-flow frequency curves for two Illinois streams having very different low-flow characteristics. See Figure 2.

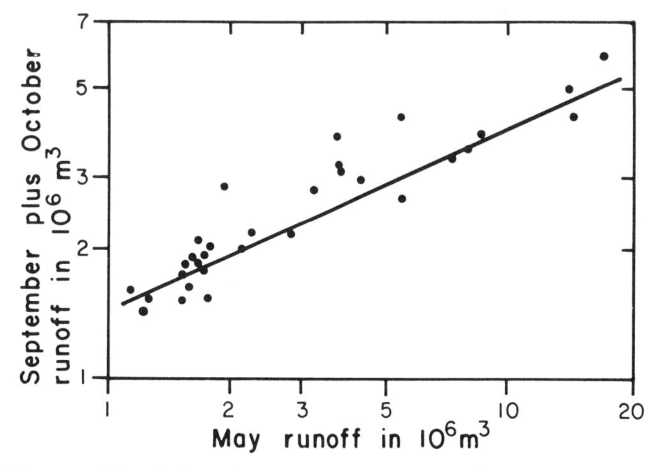

Figure 5. The relation of autumn runoff to May runoff indicates little summer recharge, Mill Creek, California (after Riggs and Hanson, 1969).

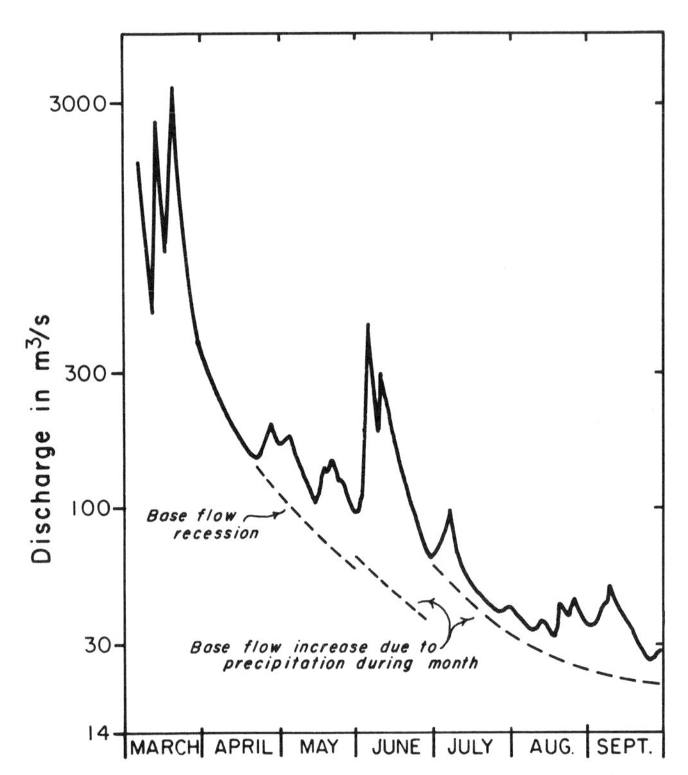

Figure 6. Hydrograph showing estimates of base flow and estimated increases due to storms in May and June, Potomac River, Maryland (after Riggs and Hanson, 1969).

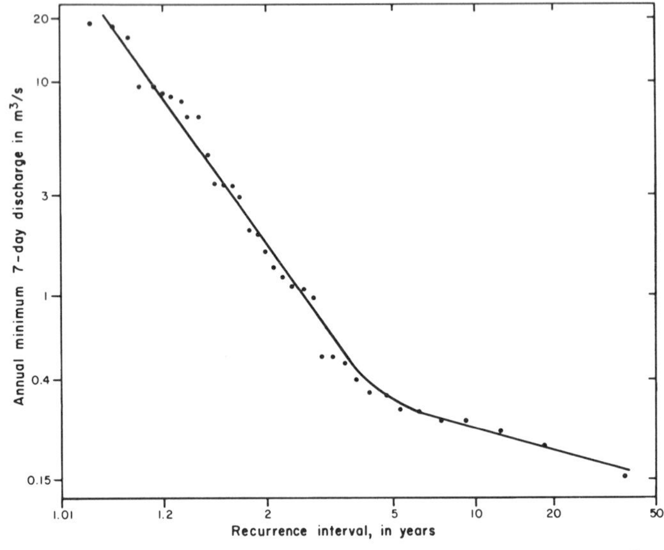

Figure 7. Low-flow frequency curve for Suwannee River, Florida (after Riggs, 1972).

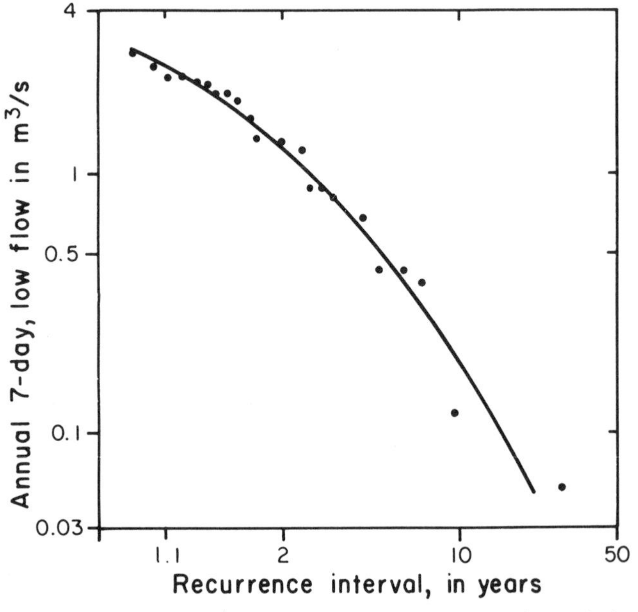

Figure 8. Low-flow frequency curve for North Anna River, Virgina, showing the effect of evapotranspiration, and possibly depletion of ground water in some years.

two aquifers, one of which is readily recharged and quickly drained, and a second one which is recharged infrequently and drains slowly. It might also represent a stream that receives a nearly constant inflow from a sewer outfall or from mine drainage.

Temperature affects the annual low flow in various ways. In temperate climates where the low-flow season coincides with hot weather, evapotranspiration from shallow ground water and from the stream channel may substantially reduce the streamflow over that drained from the aquifer. Where the ground-water contribution is small and the stream channel is wide and shallow, all of the water may be removed by evapotranspiration and the stream would have no flow—although pools of standing water may remain. This condition would result in a frequency curve similar to that of Figure 8. It is not known whether the occasional very low discharges of North Anna River are due to riparian evapotranspiration, to a small base flow from the aquifer, or both (see Riggs and Harvey, this volume).

Low temperatures reduce the rate of ground-water movement toward the stream. Under extremely low temperatures the ground-water flow will cease and the winter streamflow will be zero.

Ground water does not drain to some stream channels. In arid and semiarid regions, the water table is lower than the channel so that streamflow infiltrates through the channel bed. Flow in such a channel would occur following a rain or it might be derived from an aquifer in the headwaters. If the latter is the case, then the low flow of the stream would range from a dependable flow in the headwaters to a flow decreasing in rate and increasing in variability downstream. Streams originating in arid or semiarid regions are nearly all ephemeral—they cease to flow annually or more often.

Effects of man

Reservoirs are built and operated to modify the flow regime. The operating schedule carried out on most storage projects re-

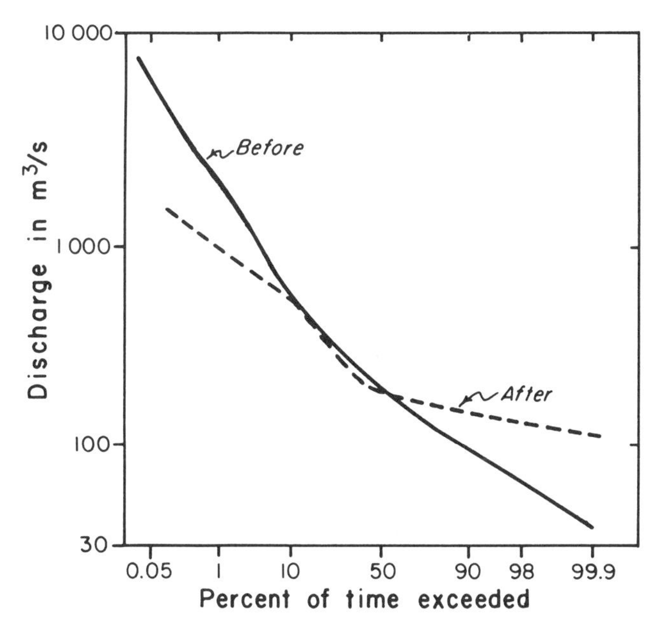

Figure 9. Duration curves showing the effect of Clark Hill Reservoir on the flows of Savannah River at Augusta, Georgia (after Stallings, 1967).

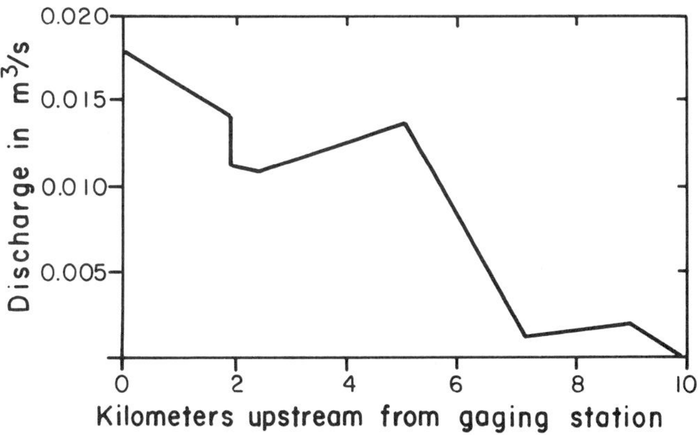

Figure 10. Discharge of Swartz Creek, Michigan, on October 12, 1966 (after Twenter and Knutilla, 1972).

sults in higher low flows and lower peak flows than the natural flows. The duration curves for flows before and after construction of a reservoir on Savannah River show the change in regimen (Fig. 9).

In addition to the obvious impact of reservoir operations on the low-flow regime of a stream, man has a fairly wide variety of other activities that can affect the low flows. One of those activities is the pumping of ground water. When a well is drilled into a ground-water body and water is pumped out of it, a cone of depression is set up in the surrounding water table (or potentiometric surface) and water begins to flow toward the well. If such a well is located near a stream that is normally fed by water discharging from the ground-water system, pumping can reduce or completely halt the discharge of ground water to the stream in that reach. In extreme cases, heavy pumping near a stream may draw water from the stream. Streams where this is most likely to happen are those in valleys filled with alluvial material or other highly permeable material.

Urbanization may affect the annual low flows of small streams. The increase in impervious surface area and the improved surface drainage both tend to inhibit the movement of water to the ground-water body, and thus reduce the base flow of the stream. On the other hand, surface discharges of waste water to streams may result in larger annual minimum flows than occurred previously, especially if the waste water originates from storage or from another stream.

The ground-water flow to a stream may also be reduced by the uptake of water by water-loving plants called phreatophytes. These occur along stream banks in many dry parts of the continent and significantly reduce the flows of streams. These plants, which are both native and imported, grow more rapidly and in greater numbers if nutrients have been added to the water from municipal waste or agricultural runoff (Culler and others, 1982).

HOW BASIN CHARACTERISTICS AFFECT LOW FLOWS

Possibly one of the best ways to show how the low-flow regime of streams are related to the ground-water system is to present an example of a case study where detailed measurements of ground-water discharge to streams in an area were made. The description that follows is from Twenter and Knutilla (1972).

Seepage measurements (discharge measurements at intervals along the channels) were made in the Swartz Creek basin (Oakland County, Michigan) on October 12, 1966; the discharge is shown in Figure 10. This was during a period of dry weather and the measurements represent base flow, wherein runoff was principally ground water. Although the measurements were made in the fall, some transpiration probably took place, and with temperatures of more than 16° C, evaporation losses were significant.

Significant differences in ground-water seepage in the upper reaches of Swartz Creek are noted, although the glacial materials throughout the basin are fairly uniform in hydrologic properties. Ground-water seepage is substantially greater in the reach of stream from kilometers 7 to 5 (Fig. 10) where the stream gradient is also greater. The steeper gradient of the water table in this area, as reflected by stream gradient, water-table contours, and the slope of the surrounding land surface, results in increased ground-water flow; the amount of flow being directly proportional to slope of the water table. In the reach of stream from about kilometers 9 to 7, a decrease in discharge is noted. The decrease is attributed, in part, to evapotranspiration; and although no data are available to substantiate this, some water probably infiltrated into the ground in this reach and returned to the stream between kilometers 7 and 5.

The decrease in streamflow between kilometers 5 and 2 can be attributed to evaporation from the several lakes in this area and to transpiration from swamps. In reaches where streamflow decreases, stream gradients are small, and the percentages of wet swampy lands are substantially greater; these conditions are conducive to greater evaporation losses.

As can be seen, gains and losses to streamflow within the upper reaches of the creek are directly attributable to the influence of land-surface configuration. Downstream, near the gage, the composition of materials at land surface comes into play and the increase in flow here can be attributed, in part, to permeable glacial materials that transmit water to the stream more readily. However, stream gradient still appears to have an important influence.

AREAL AND TEMPORAL VARIABILITY OF LOW FLOWS

One cannot generalize low-flow characteristics by regions because those characteristics are highly related to basin geology, which may be very different across topographic boundaries, or even within a single basin.

Some estimate of the relative magnitudes of low flows in various parts of the continent can be made from the monthly flow regimes of Plate 3B.

Streams whose minimums are zero include ones in the more arid regions, small ones in the far north, and small ones in humid regions draining basins underlain by limestone or by some very impervious materials.

Temporal variability is a function of climate as well as geology.

HYDROLOGIC DROUGHTS

Drought results from less than normal precipitation for an extended period of time. The effects of a drought on man's activities and interests depend on the severity, duration, and geographical extent of precipitation deficiency, and on whether precipitation is used directly (for example, to maintain soil moisture), or whether water supplies are drawn from streams or aquifers. A lack of rainfall for several consecutive summer months may not produce a drought if the normal rainfall for those months is near zero, as it is in parts of western and southern North America, because the word drought implies a deficiency from normal.

Six types of droughts have been defined by the World Meteorological Organization (Subrahmanyan, 1967):

*1. **Meteorologic drought:*** Defined only in terms of precipitation deficiencies in absolute amounts, for specific durations.

*2. **Climatologic drought:*** Defined in terms of precipitation deficiences, not in specific quantities but as ratio to mean or normal values.

*3. **Atmospheric drought:*** Definitions involve not only precipitation, but possibly temperature, humidity, or wind speed.

*4. **Agricultural drought:*** Definitions involve principally the soil moisture and plant behavior, perhaps for a specific crop.

*5. **Hydrologic drought:*** Defined in terms of reduction of streamflows, reduction in lake or reservoir storage, and lowering of ground-water levels.

*6. **Water-management drought:*** This classification is included to characterize water deficiencies that may exist because of failure of water-management practices or facilities such as integrated water-supply systems and surface or subsurface storage to bridge over normal or abnormal dry periods and equalize the water supply through the year.

An extended period of deficient precipitation might produce a drought that could be classified by any or all of the first five types. Deficiency in soil mositure develops quite promptly in response to lack of precipitation, but the effect on water supplies is delayed because streamflow is sustained by ground water. The ground-water contribution decreases with time so that as the period of deficient precipitation is extended, an agricultural drought may become a hydrologic drought.

The effects of rainfall on streamflow depend on the magnitude and intensity of the causative storms. A series of low-volume, low-intensity rainstorms may provide sufficient water to maintain soil moisture, yet may be insufficient to cause appreciable direct runoff, or any ground-water recharge. Hence, storms of this type do little to allieviate the streamflow recession; under these conditions a hydrologic drought would not be preceded by an agricultural drought.

The summer runoff of many streams in the northern half of the continent and from mountainous regions farther south is derived from snowmelt. This melt both recharges the ground-water body and produces surface runoff. Where summer precipitation is normally low, summer runoff depends on the water content of the spring snowpack and on the temperature regime during the period of melting. Thus an unusually low spring snowpack produces unusually low summer runoff, and results in an annual drought. Relief from that drought may not occur until the next year. The effect of the spring snowpack on April through July runoff is shown in Figure 11.

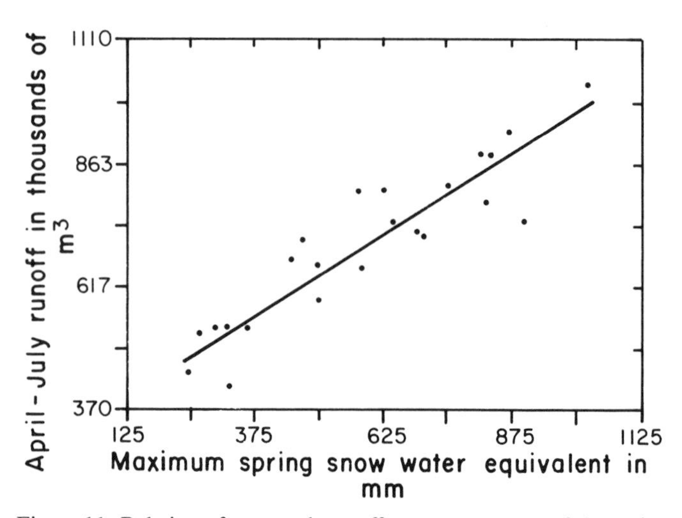

Figure 11. Relation of snowmelt runoff to water content of the spring snowpack, Rogue River, Oregon (after U.S. Soil Conservation Service, 1972).

Flows in glacier-fed streams depend on the amount of solar radiation and on other energy inputs to the glacier or the perennial snow field. High summer temperatures coupled with clear skies will increase the melt rate at the time when such conditions reduce runoff from other parts of the basin. The contribution of a headwater glacier can partially offset a deficient runoff from downstream reaches of a stream. A substantial amount of summer streamflow in the headwaters of Columbia River and Saskatchewan River is from melting of the Columbia Ice Fields.

DROUGHT CHARACTERISTICS AND HISTORY

Hydrologic droughts may occur within one year or over several consecutive years. Streamflow in a 1-year drought may be unusually deficient in the low-flow season but may not be deficient in other parts of that year. The consequences of such a drought are inadequacies of water supplies during the period when streamflow falls below demand. Such inadequacies may persist for a few months. Examples of unusually low annual runoff are shown in Figure 12 for two Canadian rivers.

A 1-year drought can be characterized as to severity on a particular river by showing its recurrence intervals on low-flow frequency curves for several periods of consecutive days. The societal effects are, of course, subjective. The areal extent of such a drought can be indicated by the locations of rivers whose flows were in the drought range.

Multiyear droughts result from below-normal runoff for several consecutive years. The hydrologic consequences of lower than normal precipitation and streamflow for several consecutive years are the lowering of the ground-water table because of lack of substantial recharge, and the depletion of soil moisture such that subsequent precipitation is less effective in producing runoff. Effects of a multiyear drought are greatest where surface storage has been provided in order to sustain a higher dependable flow. Such storage usually protects against a 1-year drought, but a succession of low annual runoffs may result in depletion of the storage to the extent that the usual water demand cannot be met. Surface storage in hot, dry climates is also reduced significantly by evaporation from the reservoir water surface.

Droughts often affect large areas. Hoyt (1936) identified from precipitation records, 38 annual droughts in the humid states of the U.S. during the period 1881 to 1934; in 12 of these droughts, over 23 percent of the area of these states was affected. In the semiarid states, Hoyt identified 28 droughts during the same period, and in 11 of these over 44 percent of the area was affected. From 1942 to 1956, precipitation was less than 85 percent of normal for 8 years or more over a region extending from southern California to western Texas and encompassing most of Arizona and New Mexico (Thomas, 1962).

Periods of drought prior to the availability of precipitation records have been estimated by various means. But droughts that may have occurred hundreds or thousands of years ago are not likely comparable to recent ones because climatic characteristics then and now may be different. Information from the nineteenth century on levels of enclosed lakes, fluctuations in tree growth (from ring widths), and precipitation seems to point to a severe drought from about 1830 to 1850 in a large part of the western U.S. More recent drought years, based on precipitation records, are 1889, 1890, 1894, 1901, 1910, and 1917 in the Interior Plains. The "Dust Bowl" drought of the 1930s was notable for its severity, length, and area extent (Hoyt, 1936, 1938; Matthai, 1979). Other major droughts occurred in the southwestern U.S. from 1942 to 1956 (Thomas, 1962), in the southern half of the U.S. in 1952 to 1956 (Nace and Pluhowski, 1965), in the eastern U.S. and Canada in the 1960s, and in central and western North America in 1976 to 1977. The latter two droughts are described in following sections.

The drought of the 1960s in the Northeast

Although climate is the underlying natural physical factor in determining the severity of a drought, other factors, such as the concentration of population and the intensity of human activity, can add to that severity. The drought of the 1960s in the northeastern United States and adjacent Canada is the longest and most

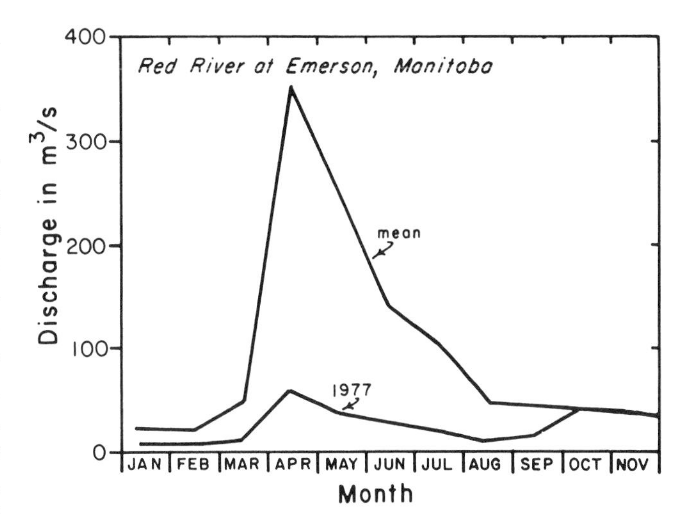

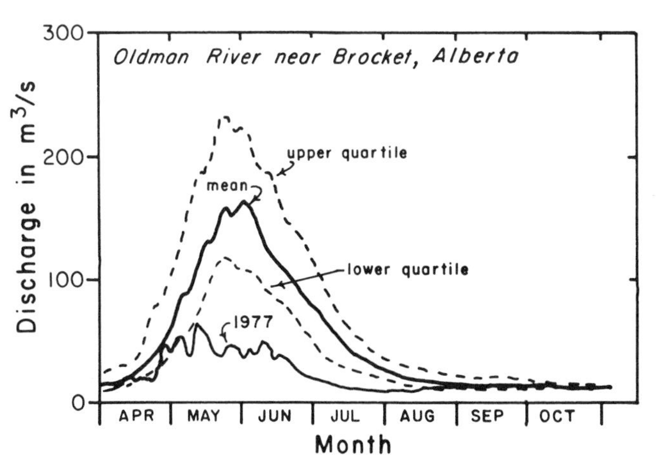

Figure 12. Unusually low runoff of two Canadian rivers in 1977.

severe in that region's recorded history. Its severity was amplified beyond that by natural causes because the northeastern U.S. is heavily populated and has high demands on its water resources. The 1960s drought began in late 1961 and extended into 1966. The progresson of the drought that follows is taken from Barksdale and others (1966).

Runoff during 1961 was near normal for the Northeast. Storage in major reservoirs generally was near or above normal through September. By the end of the following March, ground-water levels in 18 of 72 wells were more than 25 percent below the long-term average lows.

Runoff during 1962 was below normal in a wide band extending from southern Maine and western Connecticut and Massachusetts into northeastern Ohio. Streamflow in much of the Northeast was especially low from May to August. Ground-water levels were below average in most wells. Natural runoff in the Delaware River basin was inadequate to maintain the flows in the lower reaches of the Delaware River as required by the Delaware River Compact. Consequently, the Delaware River Master ordered large releases from reservoirs in the headwaters of Delaware River in New York. Storage in these reservoirs furnishes part of the water supply of New York City. Shortage of water for crops and lawns was prevalent throughout New England and New York. Serious agricultural drought in some counties in New York resulted in the declaration of an emergency to make special assistance available to farmers there.

Below-normal runoff continued through 1963. By May, storage in the New York City reservoir system (in the Delaware River basin) dropped to 42 percent of capacity, the lowest of record for that time of year. Only conservation releases were made from the Delaware River basin reservoirs. But even with these releases, the flows at the head of the Delaware River estuary were too low to prevent a sharp rise in salinity in the lower part of the estuary. Many shallow wells went dry, especially in Vermont, Connecticut, and Ohio.

Runoff was again below average in 1964 in a large part of the Northeast, from southern Vermont and New Hampshire southward through western New Jersey and eastern Pennsylvania. Drought conditions intensified in the Delaware River basin, and the river's flow in September was the lowest of record at Montague, New York. Flow at Montague had been below normal in 32 of the 37 months since August 1961. After moderate winter runoff, streamflow declined to record lows during summer. Ground-water levels indicated an extension of the drought area from the nucleus of the previous year in every direction except westward. Two-thirds of the observation wells recorded lows more than 25 percent below average. By August many small communities were experiencing water shortages. Water-use restrictions were imposed on several communities in New York, Massachusetts, and Rhode Island. New York reported a drought-caused reduction in agricultural production, and closed forests because of fire hazard resulting from extreme dryness.

Drought conditions persisted and intensified throughout much of the Northeast during 1965. Streamflow was much below normal, and during the year many streams reached record lows, partly as a result of a carryover ground-water deficiency from the previous growing season. At the year's end, the salt front of 250 parts per million of chloride was advancing upstream toward the Philadelphia water intake as a result of the exceptionally low flows in Delaware River.

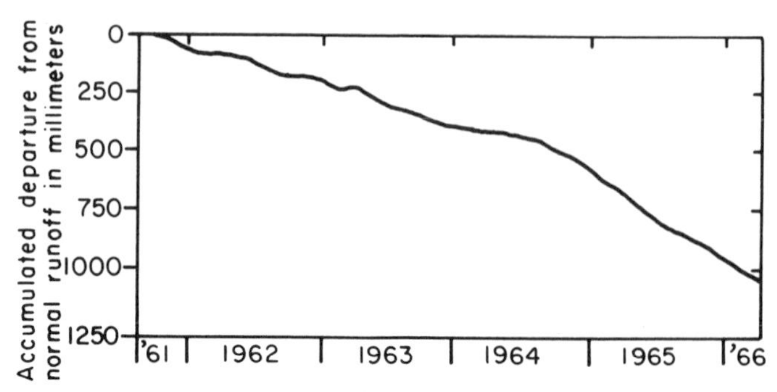

Figure 13. Runoff deficiency, South Branch Raritan River, New Jersey, during drought of 1961 to 1966 (after Barksdale and others, 1966).

In August, 1965, President Johnson declared that those parts of the states of Delaware, New Jersey, New York, and Pennsylvania included within the Delaware River basin and its service area were affected by the drought to the extent that federal assistance for certain emergency actions was warranted. Critical water shortages also were identified in almost a score of public water supplies outside the Delaware River basin, ranging from New Hampshire to Pennsylvania. Federal assistance was also provided to them. Several hundred communities in the region experienced water shortages sufficient to require restrictions on water use during the year; many of these were in Massachusetts.

Low runoff continued through 1966 so that by the end of that year, shortages were region wide and were severe over much of that region. Figure 13 shows the cumulative departure from normal of the flow of South Branch Raritan River, New Jersey, from 1961 to mid-1966. The runoff deficiency for that 4.7-year period was equivalent to 1.8 years of runoff. From September 1961, runoff was below normal for 51 of the 56 months.

The drought was ended by above-normal precipitation in 1967 and in several succeeding years, during which period, storage in the ground and in surface reservoirs was replenished.

The drought of 1976 to 1977

The drought of 1976 to 1977 was the most severe in at least 50 years in many parts of the North American continent. The following description is taken largely from Matthai (1979). The combined flow of the "Big Five" rivers of central North America—Mississippi, Columbia, St. Lawrence, Ohio, and Missouri—averaged less than normal for six consecutive months during this drought. Index streamflow stations indicated that monthly streamflow was deficient in some states and provinces

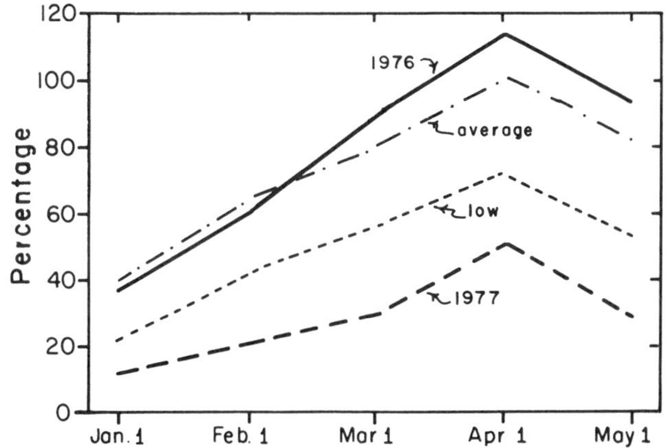

Figure 14. Water content of the spring snowpack in Columbia River basin as a percentage of the April 1 average, showing the 1977 deficiency (after Matthai, 1979).

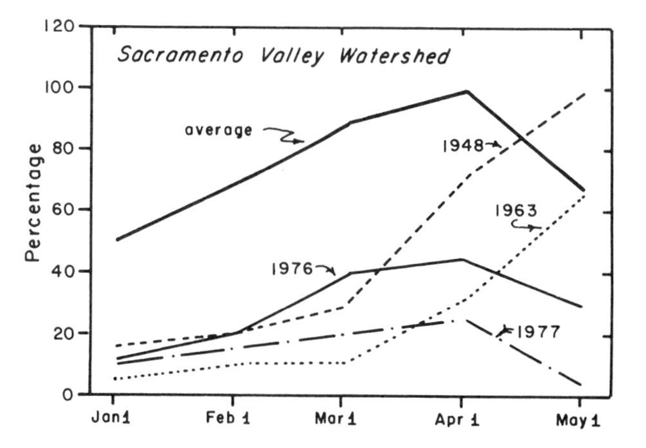

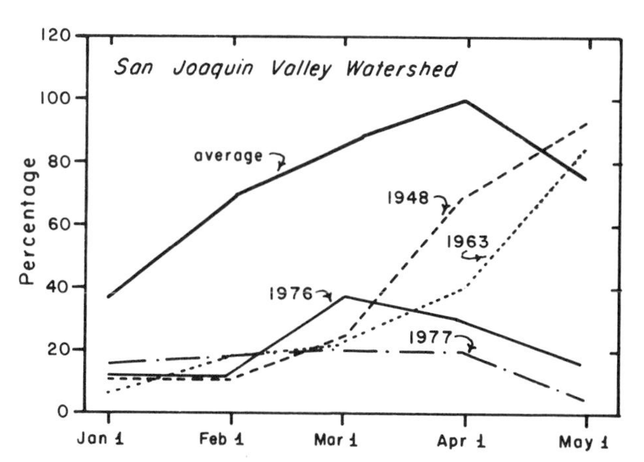

Figure 15. Water content of the spring snowpack in two California river basins as a percentage of the April average, showing the deficiencies in 1976 and 1977 (after Matthai, 1979).

for 12 months or more. Record low amounts of rainfall, snowfall, and runoff, and increased withdrawals of ground water were prevalent. Use of carryover storage in reservoirs throughout the United States during 1976 maintained streamflow at near normal levels for some streams, but some of the reservoirs went dry or dropped well below their outlet works in 1977. Carryover storage in the fall of 1977 was very low. Ground-water levels were at or near record lows in many parts of the United States, hundreds of wells went dry, thousands of wells were drilled, and water was hauled in many rural areas and to a few towns.

Although some areas were spared from the severe effects, for example much of New England and much of Louisiana, Mississippi, Arkansas, and the northern Canadian interior, probably hardest hit were the western and central parts of the continent. The water supply in the Northwest is highly dependent on snow in the Olympics Mountains of Washington, the Cascades of Oregon and Washington, and the Rockies of British Columbia, Alberta, Idaho, Montana, Wyoming, and Colorado. The drought in this region did not develop until after the winter of 1976 to 1977 when precipitation was well below normal. Water content of the 1977 spring snowpack in the Columbia River basin was a record low as shown in Figure 14. The 1977 runoff of Columbia River at The Dalles, Oregon, adjusted for change in storage, was the lowest in a record that began in 1879. Annual flows were the lowest of record at other long-term stations as well.

In California, the drought in 1977 was more severe and more widespread than in either 1924 or 1976; and in much of the state, the drought of 1976 was worse than that in 1924. Although the snowpack in 1976 in the Sierras was the lowest on record at one-third of the snow courses, the snowpack was even lower in 1977. Figure 15 compares the water content of the 1977 snowpack in the Sacramento and San Joaquin watersheds to the water contents in several other very dry years. However, heavy releases of stored water in 1976 masked the effects of the deficient snowpack in those watersheds. The stored water was depleted in 1976 so that the 1977 runoff was only that generated by melting of the deficient snowpack; the resulting runoffs of a tributary to Sacramento River are shown in Figure 16.

As a consequence of the low runoff, the generation of hydroelectric power in 1977 dropped to 30 to 40 percent of average. The annual unimpaired runoff to the Sacramento River Delta from the Sacramento and San Joaquin rivers was only 28 percent of normal. Most coastal streams in the central part of California had about 10 percent of normal runoff.

In the Upper Colorado Basin, below normal runoff occurred in all but one or two months for periods as long as 18 months, and flows were generally less than those in 1934. New record lows were set for daily, monthly, and annual flows. Storage in reservoirs was severely reduced.

In the Missouri Basin, 1976 was driest in Kansas and Nebraska, but 1977 was the driest in Colorado and Missouri. Runoff at a number of gaging stations with long-term records reached new lows, and storage at most reservoirs was well below normal. Some streams remained at these sustained low flows for many months.

High runoff in the Nelson River basin, Manitoba, in the spring of 1976 filled most lakes and reservoirs. But the summer and autumn of 1976 was warm and dry. This was followed by a warm, dry winter in which much of the snowpack was lost by evaporation. The dry antecedent conditions combined with little spring runoff from snowmelt produced record low flows in the spring of 1977.

The 1976 to 1977 drought was terminated in most of the western United States by above-normal precipitation between September 1977 and January 1978. These fall and winter rains increased streamflow to above normal and replenished reservoir storage. Normal or above-normal precipitation continued after January 1978. Ground-water levels were rising but were still below average in January.

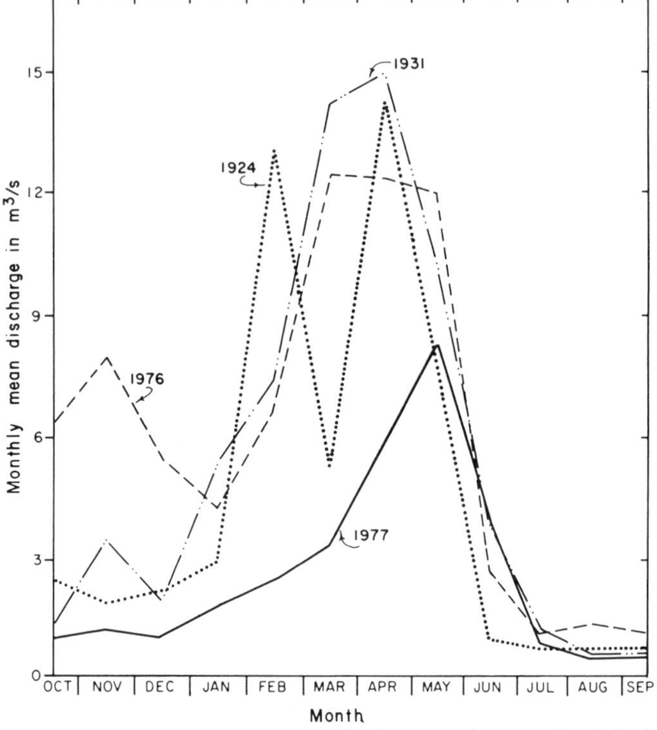

Figure 16. Monthly mean discharges for four drought years, North Fork American River, California. Discharge for 1976 was sustained by releasing water from storage. No stored water was available in 1977 (after Matthai, 1979).

REFERENCES

Barksdale, H. C., O'Bryan, D., and Schneider, W. J., 1966, Effects of drought on water resources in the Northeast: U.S. Geological Survey Hydrologic Investigations Atlas HA-243, 1 plate.

Culler, R. C., Hanson, R. L., Myrick, R. M., Turner, R. M., and Kipple, F. P., 1982, Evapotranspiration before and after clearing phreatophytes, Gila River flood plain Graham County, Arizona: U.S. Geological Survey Professional Paper 655-P, 67 p.

Hall, F. R., 1968, Base-flow recessions; A review: Water Resources Research, v. 4, no. 5, p. 973–983.

Hoyt, J. C., 1936, Droughts of 1930–34: U.S. Geological Survey Water-Supply Paper 680, 106 p.

——, 1938, Drought of 1936, with discussion on the significance of drought in relation to climate: U.S. Geological Survey Water-Supply Paper 820, 62 p.

Matthai, H. F., 1979, Hydrologic and human aspects of the 1976–77 drought: U.S. Geological Survey Professional Paper 1130, 84 p.

Nace, R. L., and Pluhowski, E. J., 1965, Drought of the 1950s; With special reference to the midcontinent: U.S. Geological Survey Water-Supply 1804, 88 p.

Riggs, H. C., 1972, Low-flow investigations: U.S. Geological Survey Techniques of Water-Resources Investigations, book 4, chapter B1, 18 p.

Riggs, H. C., and Hanson, R. L., 1969, Seasonal low-flow forecasting *in* Hydrological forecasting: Geneva, Switzerland, World Meteorological Organization Technical Note no. 92, WMO-no. 228, TP 122, p. 286–299.

Stallings, J. S., 1967, South Carolina streamflow characteristics; Low-flow frequency and flow duration: U.S. Geological Survey Open-File Report, 83 p., 1 map.

Subrahmanyam, V. P., 1967, Incidence and spread of continental drought: Geneva, Switzerland, World Meteorological Organization, International Hydrological Decade, Reports on WHO/IHD Projects, no. 2. 52 p.

Thomas, H. E., 1962, The meteorologic phenomenon of drought in the Southwest: U.S. Geological Survey Professional Paper 372-A, 42 p.

Twenter, F. R., and Knutilla, R. L., 1972, Water for a rapidly growing community; Oakland County, Michigan: U.S. Geological Survey Water-Supply Paper 2000, 150 p.

U.S. Soil Conservation Service, 1972, Snow survey and water-supply forecasting: SCS National Engineering Handbook, Section 22, USDA, Soil Conservation Service, 228 p.

Manuscript Accepted by the Society May 1, 1987

Printed in U.S.A.

The Geology of North America
Vol. O-1, Surface Water Hydrology
The Geological Society of America, 1990

Chapter 7

Snow and ice

Mark F. Meier
Institute of Arctic and Alpine Research and Department of Geological Sciences, University of Colorado, Boulder, Colorado 80309

INTRODUCTION

Snow, ice and the hydrologic cycle

The surface-water components of the hydrologic cycle usually are considered to be rainfall and subsequent runoff. Yet, in many parts of North America, for a major part of the year, the more appropriate concept is precipitation and storage as snow and ice, followed by melting, followed by runoff. The timing involved in these two concepts is very different: Runoff follows rainfall almost immediately, whereas the time between snowfall and melt can range from days to months for seasonal snowcovers, and from months to millennia for glaciers. The predictive analysis of runoff also is very different: Prediction of runoff from rainfall requires knowledge of the precipitation pattern in time and space, whereas the prediction of runoff from snow and ice melt requires knowledge of the amount and distribution of snow in storage (measurable), and the meteorologic conditions that cause melt. The fact that snow and ice accumulate and melt to produce runoff in a very different way than does rainfall commonly is ignored in a simple hydrologic analysis.

An enormous amount of water in the form of ice is seasonally exchanged with the atmosphere and oceans in North America. The amount stored for periods of years or longer as glacier ice is also huge (Fig. 1, Table 1). The seasonal snowcover fluctuates from 3 to 18 million km^2 (Ropelewski, 1986), or from 14 to 84 percent of the land surface. These seasonal variations are not as large as those in Eurasia, but do have an important effect on the Earth's reflectivity (albedo). In much of western and northern North America, most of the runoff is produced by the melting snowcover. Glacier ice covers 10 percent of the total North American land surface, almost exactly that of the global land average. However, this amounts to only 2 percent of North America exclusive of Greenland. The amount of water stored as glacier ice on the land exclusive of Greenland (about 50,000 km^3) is probably greater than that stored in all other surface water bodies in North America.

Snow, ice, and climate

Although the sun is the energy source that drives the global hydrologic cycle and the circulation of the atmosphere and oceans, our planet requires snow and ice to help maintain energy equilibrium in order to maintain a hospitable climate. In the tropical lower latitudes, more radiant energy is received than is lost; this requires other areas where more energy is lost than gained in order to maintain equilibrium and to help drive the circulation. This compensatory loss is provided in the high latitudes, where ice and snow dominate the environment. Snow has the highest reflectivity to solar radiation, in the visible-light spectrum, of any widely distributed natural material on the Earth's surface, yet it is an almost perfect emitter of energy at infrared and longer wave lengths. Thus, snow absorbs little solar energy and radiates heat to outer space efficiently.

Ice, climate, and sea level are closely related. Glaciers and ice sheets, which contain most of the world's ice (and fresh water), grow and shrink as the climate changes. It is not clear, however, whether or not the present rise in sea level is caused by ice wastage. The largest ice mass—the Antarctic Ice Sheet—is thought by most glaciologists to be growing, thereby taking water out of the ocean (National Academy of Science, Committee on Glaciology, 1985). The volume of the second-largest ice mass—the Greenland Ice Sheet—seems to be stable under present climatic conditions. However, the remaining 3 percent of Earth's glacier ice, consisting of mountain glaciers and small ice caps, generally has been wasting away since the beginning of this cen-

TABLE 1. AREA AND VOLUME OF MAJOR ICE AND SNOW MASSES IN NORTH AMERICA

	Area (thousands of km^2)	Volume (thousands of km^3)
Seasonal snow		
(Average annual coverage)*	10,700	
Glaciers and ice caps†	356	50
(Excluding the Greenland Ice Sheet)		
Greenland Ice Sheet§	1,670	2,740

*From Ropelewski (1986), does not include snow on sea ice.
†From Meier (1984)
§From Radok and others (1982)

Meier, M. F., 1990, Snow and ice, *in* Wolman, M. G., and Riggs, H. C., ed., Surface water hydrology: Boulder, Colorado, Geological Society of America, The Geology of North America, v. O-1.

Figure 1. Snow and crevasses on the upper reaches of Klawatti Glacier, North Cascades, Washington, in the heart of an area of extensive glacier cover. Photograph by Austin Post, Sept. 1, 1956.

tury; an average thickness equivalent to about 0.31 m of water per year has been lost (Meier, 1984). This water, lost from the frozen reserves, augmented streamflow during the first part of this century, and has caused between one-fourth and one-half of the observed rise in sea level. It is interesting that 40 percent of this global total of glacier wastage occurred in the mountains bordering the Gulf of Alaska (Meier, 1984).

Ice, the "greenhouse effect," and sea level

What will happen to climate and sea level in the future? Climate is being changed by fossil-fuel combustion and other human activities that add CO_2 (carbon dioxide), methane, chloroflurocarbons, and other so-called "greenhouse" gases to the atmosphere. Climate models indicate that the doubling in CO_2 concentration may cause an increase in global air temperature of 1.5° to 4.5°C; other "greenhouse" gases may cause an additional increment of warming, and the expected warming will be accentuated at high latitudes (National Academy of Sciences, Carbon Dioxide Assessment Committee, 1983). Thus, it is a question of vital concern whether this "greenhouse" warming will cause large amounts of ice to melt, thus raising global sea level and causing extensive damage to coastal and low-lying regions.

A recent study by the National Academy of Science, Committee on Glaciology (1985) analyzed the present state of knowledge on the cause of present-day changes in sea level and possible future changes that could be expected because of glacier wastage from a CO_2-induced change of climate. This study suggests that the small glaciers of the world will contribute about 0.1 to 0.3 m of sea-level rise by the year 2100. Wastage of the Greenland Ice Sheet may contribute a like amount. The effect of an altered climate on the Antarctic Ice Sheet is less certain. Two recent reports (National Academy of Science, Carbon Dioxide Assessment Committee, 1983; National Academy of Science, Committee on Glaciology, 1985) suggest that a modest rise of 0 to 0.3 m is likely, although a 1-m rise from this cause alone cannot be ruled out. Thus the total rise in sea level by the year 2100 due to changes in the global ice cover is likely to be about 0.6 ± 0.2 m, and more than half of this change will be caused by North American ice wastage.

SEASONAL SNOW COVER

Snow is a pervasive element of the environment throughout Canada, Alaska, and the northern-tier states of the conterminous United States, as well as at the higher elevations as far south as central Mexico. Globally, the seasonal snow cover of the northern hemisphere is an important factor in the heat budget and, therefore, in Earth's climate. Snow is such an efficient reflector of radiation that, if the Earth were to become completely snow covered, the mean temperature of the Earth would drop to −89°C and remain that way (Budyko and Kondratiev, 1964). Both globally and regionally, the presence of snow cover causes a positive feedback to climate. With snow on the ground, more solar radiation is reflected, causing cooling and preservation or augmentation of the snow. Extensive snow and ice cover contributes to stronger cyclogenesis, and to southward displacement of storm tracks along the East Coast of North America (Walsh and Ross, 1985). The effect of snow on humans and on the hydrologic cycle, however, is not limited to climate. For example, snowmelt runoff in populated areas supplies water for virtually all types of offstream and instream uses; snow-pack management can optimize soil moisture and minimize frost penetration for agriculture; and snowmelt floods can cause major economic losses. Snow also has major effects on transportation, construction, recreation, and other activities (Colbeck and others, 1979; McKay, 1981).

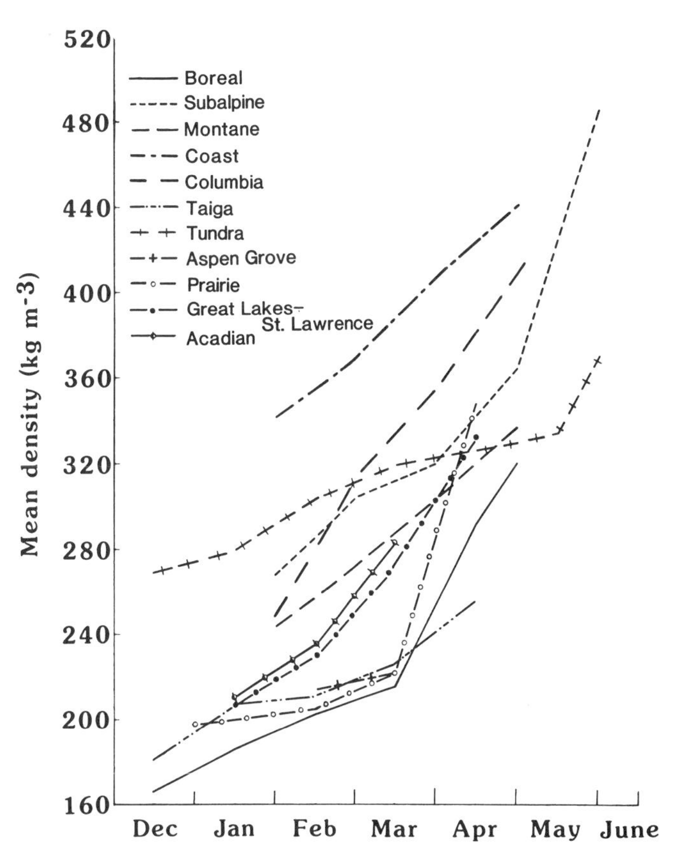

Figure 2. Variation of snow density with time for snowpacks in various areas characterized by vegetation type. After McKay and Gray (1981).

Kinds of snow and snowpacks

Snow types and methods for managing snow and snowmelt water are diverse. The Inuit recognize about two dozen different kinds of snow; certainly "Api" (snow not yet picked up and reworked by the wind) is different from "Upsik" (snow that has been reworked by wind and deposited as a firm mass), and neither should be confused with "Siqoq" (snow blown along the ground by the wind; Kirk, 1980). Throughout North America, the distribution of snow types is related to the distribution of vegetation types. The initial density of a snow cover, and its change with time, varies, as does the vegetation type from region to region (Fig. 2), reflecting the regional climate. The problems and benefits associated with snow, and methods for its management, differ greatly from region to region. These differences can be illustrated by considering snow in five contrasting regions: the Great Plains/Canadian prairies; the Cascade Range of Oregon, Washington, and British Columbia; the southern Rocky Mountains; eastern Canada/northeastern United States; and the Arctic.

Great Plains/Canadian prairies. Snowpacks in these areas typically are thin (50 to 150 mm of water equivalent), and because snow is blown into swales or gullies or into the lee of vegetation, snowpacks are extremely variable areally (Willis, 1979). The snow usually is dry, and initially has low density but becomes dense where it is windpacked ("Api" to "Upsik" in the words of the Inuit); the snow crystals may fragment and evaporate as they are blown along the ground.

Horizontal-transport studies are important for understanding the distribution, management, and hydrology of snow. Snowpacks in this region, although small in amount, are important economically because they supply critical soil moisture at the beginning of the crop-growing season, and because they insulate plants from frost damage (Steppuhn, 1981). Snow deposition is managed by the use of agricultural cropping practices, such as controlling the height of stubble and by the construction of windbreaks that concentrate the snow and delay melting until the ground thaws. Snowmelt floods, exacerbated by frozen ground, occasionally occur in some areas. Snow data are used for soil-moisture conservation studies and other purposes; a major problem in hydrologic modeling for this region is related to an incomplete understanding of evaporation/sublimation from the snowpack and infiltration into frozen ground.

Cascade Range. Snowpacks in the Cascade Range of Oregon, Washington, and British Columbia (Fig. 3) are very thick (more than 1,000 cm in places), with more than 5,000 mm water equivalent, and are almost never far below the freezing point in temperature. The amount and distribution of the snow are strongly dependent on elevation, because of the importance of the freezing level during storms and because of a marked increase of precipitation in the mountains with increasing elevation. The distribution of snow also depends, to a lesser extent, on a west-to-

Figure 3. A 10-m deep pit was needed to study a single winter's snow accumulation in the North Cascades of Washington State. Such thick heavy snowpacks are not uncommon in this region. Devices imbedded in the pit walls are being used to study the movement of meltwater through the snow. Photograph by M. F. Meier.

east decrease in precipitation across the Cascade Range. The snowpack is coarsely crystalline, very dense, and usually wet throughout. Evaporation from snow is negligible compared with snow loss caused by melting. The snowpack is not managed, except in a minor way through forestry practices. Because of extensive hydroelectric development in the region and the relatively low ratio of reservoir storage to annual streamflow, accurate predictions of snowmelt runoff are essential for meeting energy needs economically and for managing multiple uses of the river systems. A number of complex runoff-forecasting schemes, supported by an extensive network of real-time data-acquisition sensors and telemetry and snow-course observations, are used to help predict snowmelt runoff. However, because of extensive forest cover, current remote-sensing techniques are of limited value. Snow avalanches commonly damage property and occasionally cause fatalities. Rain-on-snow events can occur at any time of the year, but are most likely to cause flooding in the fall when the snowpacks are widespread but relatively thin.

Rocky Mountains. In these mountains, snowpacks are moderate in thickness, relatively cold, and extremely variable areally. Redistribution of the snow by wind causes the snow to be concentrated in cornices along ridge crests, in valleys, and in openings in timber stands. Evaporation, usually unknown in amount, often occurs when plumes of snow are blown from high ridges. The snow generally is dry and powdery when it falls but increases in density because of metamorphism and compaction over time. A form of snow termed "depth hoar" (fragile crystals that develop within a snowpack because of vapor transfer) is common.

Snow avalanches are serious hazards in many areas. Snowmelt runoff is used to irrigate crops throughout the intermountain West and Southwest; other major uses include power generation, municipal water supplies, and mitigation of river salinity. In spite of a relatively high ratio of reservoir storage to annual streamflow, the water is so valuable and so intensively utilized that complex data networks and runoff-forecasting schemes are employed, especially in the Southwestern United States. Several remote-sensing techniques are in use to inventory the snow resource, and much effort has gone into studies of snow management. Management options include cloud seeding to increase snow accumulation, construction of large snow fences on ridges to concentrate the snow into drifts so as to minimize evaporation and extend the runoff period, and forest-cutting practices to maximize the trapping of snow and reduce snowmelt-runoff peaks.

Southeastern Canada/northeastern United States. Snow in this highly populated and heavily industrialized area is, in many respects, intermediate in its characteristics between the typical snows of the western mountains and the plains. However, there are distinctive patterns and problems associated with it. The amount of snow accumulation increases with altitude, as in the western mountains, but depends in a sensitive way on moisture sources. Thus, very large local accumulations may be deposited downwind from each of the Great Lakes (Fig. 4). Acid deposition is a problem in this area, and it is exacerbated by the snowpack. Soluble pollutants are captured in the snow and accumulated for months; as the snowpack recrystallizes, these pollutants are moved to the grain boundaries, and when the snowpack warms up to freezing (ripens), the pollutants are flushed out in a sudden pulse (Fig. 5). Thus the deleterious effect is concentrated.

Arctic. The snowcover of the Arctic is most notable for its long duration of 9 to 10 months, or even more. As such, its presence is a dominant factor in the ecology. Also, the local snow distribution is primarily determined by wind drifting, with thick accumulations along riverbanks, gullies, or as cornices; elsewhere the snowpack is rarely much thicker than the tundra vegetation. The snow is often wind-packed at the surface, but is of low density below the surface due to temperature-gradient metamorphism and the formation of depth hoar, and thus is similar to the snowpacks of the Canadian prairies. Because of the pervasive effect of wind and cold temperature, and thus the difficulty of trapping precipitation in a gage, as well as a very sparse network of gages, precipitation amounts in the Arctic are poorly known.

Snow is an engineering or construction material in the Arctic for houses, for roads, and for protecting construction sites. It is also concentrated to provide local water supply.

These descriptions of various snowpacks in North America show that it is risky to generalize about snow as a water resource, or to apply snow-research results broadly without considering the local characteristics of snow accumulation, snowpack properties, snowmelt processes, or local needs for snow data. We can all learn from the Inuit.

Snow distribution in North America

Snow covers the ground in late summer or fall and remains until late spring in the north; it covers the ground only briefly during mid-winter in the south. But the actual pattern in time and space is variable within any given season and from year-to-year. Figure 6 shows the (a) buildup and (b) decay of the Northern Hemisphere snowcover at monthly intervals during 1984–1985, as mapped from meteorologic satellite images. In the Arctic, Subarctic, and Boreal latitudes the timing and pattern are relatively consistent from month to month, but in the middle and southern United States, the pattern is complex spatially and temporally and depends on the vagary of individual storm systems.

The long-term average annual snowfall (the accumulated depths of individual snowfalls, a frequently measured but somewhat meaningless statistic as the value depends on the frequency of measurement) shows a relatively simple pattern except in mountainous area (Fig. 7a), as does the long-term probability of snow cover in midwinter (Fig. 7b).

The mountain ranges bordering the Pacific Ocean from California north to south-central Alaska receive heavy snowfall, but the amount is dependent on elevation: areas near sea level from British Columbia south receive most of their winter precipitation as rain. At high elevations, the annual precipitation is greater, and the fraction that occurs as snow is greater. Annual snowfalls exceeding 29,000 mm have been recorded in western Washington, and higher values undoubtedly have occurred farther north. These areas receive heavy precipitation because they are directly exposed to moisture-bearing storms originating from the Aleutian low-pressure area. On the other hand, the lee (northeastern) slopes of these mountains receive much less snowfall, making the snow distribution a complex function of elevation, orientation, and distance from the Pacific Ocean.

Relatively heavy snowfall occurs in eastern Canada, in Newfoundland, eastern Quebec, and along the east coast of Baffin Island (Fig. 7a). These areas are often affected by storms deriving their moisture from multiple sources, mainly from the southern Atlantic and the Gulf of Mexico, and occasionally from the Pacific Ocean. Local areas of heavy snowfall also occur downwind (usually east-northeast) of the Great Lakes as indicated by the distribution of the snowpack in Figure 4.

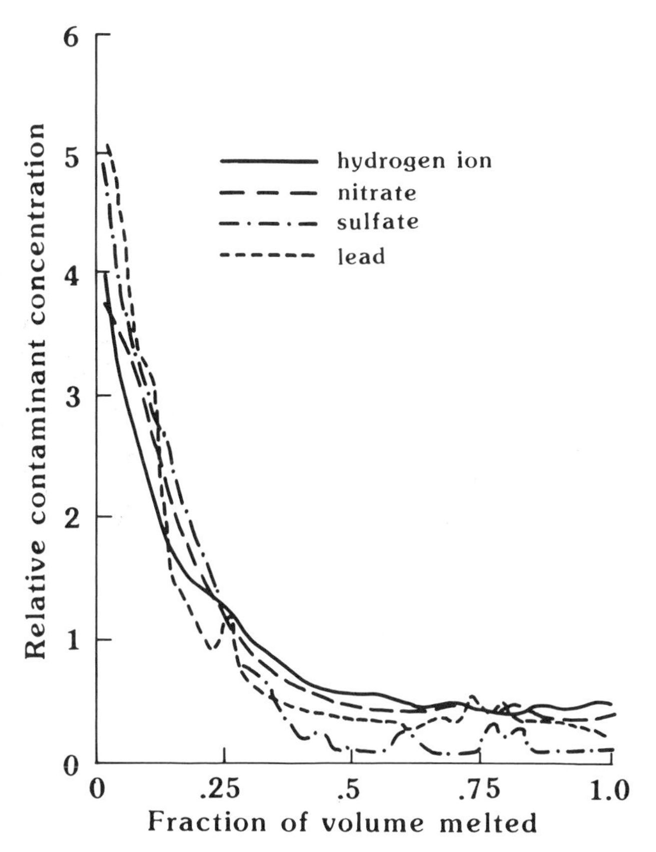

Figure 5. The ratio of concentration, C, of selected contaminants in a meltwater fraction to the average concentration, C_0, as a function of the fraction melted. Soluble impurities such as acid droplets may be accumulated and stored in a cold snowpack for up to six months. Then when the snowpack warms up and liquid water moves through it, the solutes at the grain boundaries are suddenly flushed out. This causes a rapid acidification or contamination of the meltwater, with possible resultant damage to downstream ecosystems. This typical melt curve shows that the greatest concentration of contaminants is in the initial meltwater. After Johannessen and others (1977).

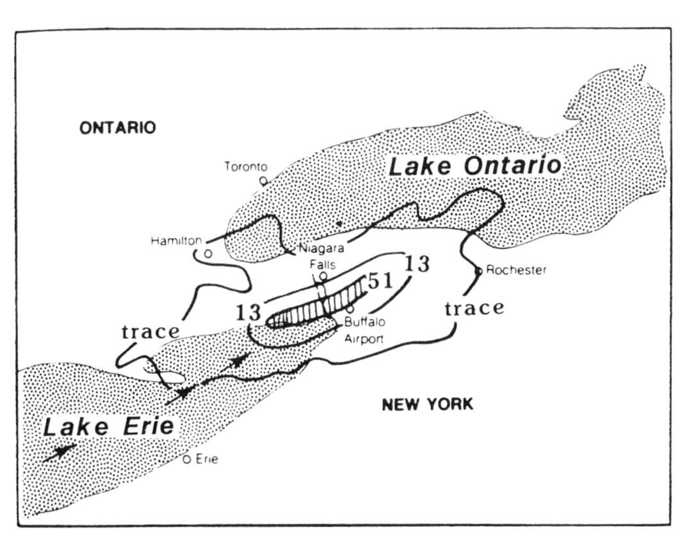

Figure 4. Typical snowfall deposition pattern, in cm, produced by the "lake effect" for a storm on November 15, 1974. After McKay and Gray (1981).

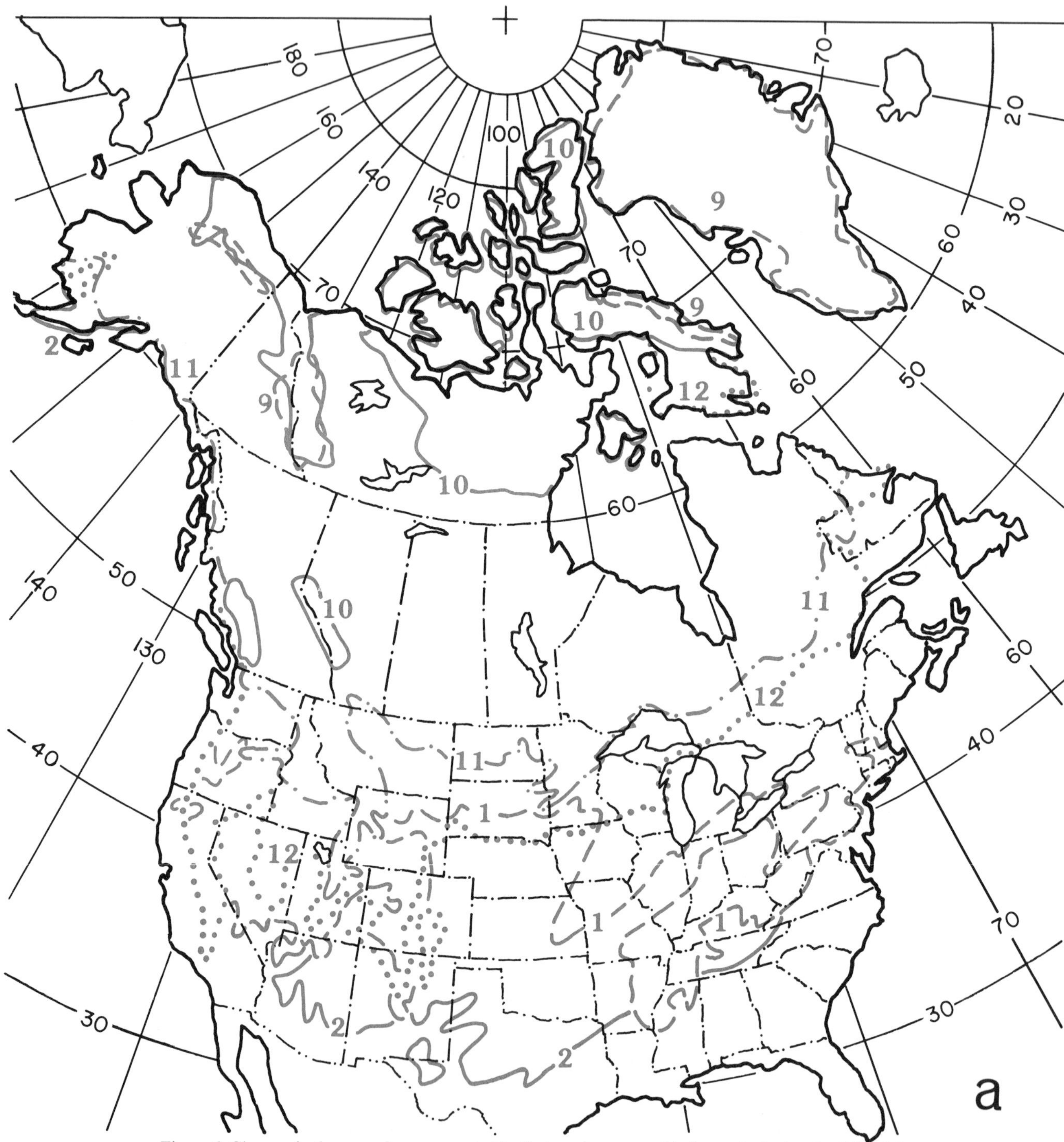

Figure 6. Changes in the area of snowcover in North America at monthly intervals during the 1984–85 winter, as compiled from meteorological satellite images. Adapted from: Northern hemisphere snow and ice boundaries (weekly charts), NOAA-NESDIS/Synoptic Analysis Section (1985–86). Maps show (a) buildup and (b) decline of the snowcover. Curves are designated by the month, but represent averages during about one week as follows: 9, Sept. 1–7, 1984; 10, Oct. 1–7, 1984; 11, Oct. 29–Nov. 4, 1984; 12, Nov. 26-Dec. 2, 1984; 1, Dec 31–Jan. 6, 1985; 2, Jan. 28–Feb. 3, 1985; 3, Feb. 25–Mar. 3, 1985; 4, April 1–7, 1985; 5, April 29–May 5, 1985; 6, May 27–June 2, 1985; 7, July 1–7, 1985.

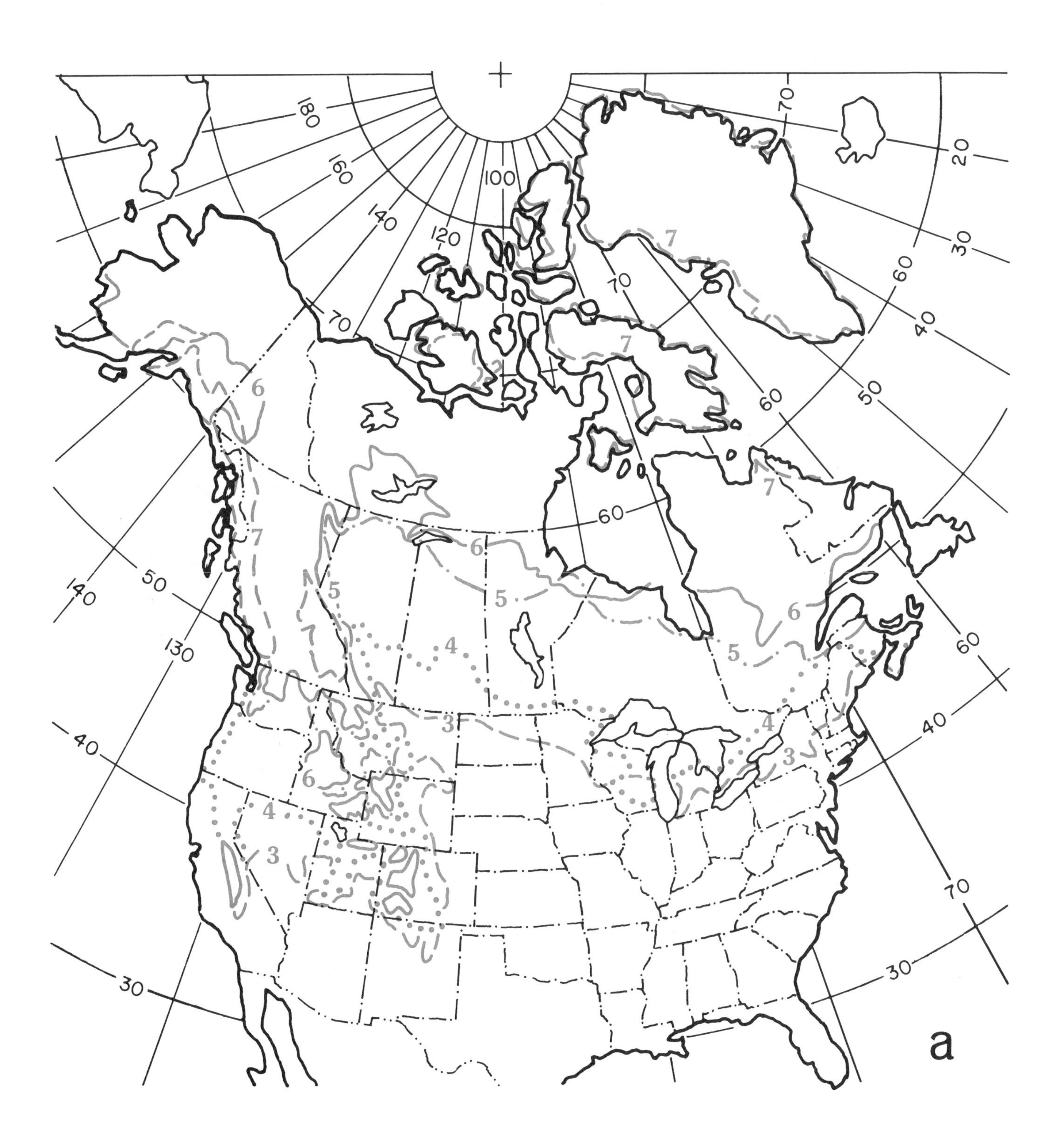
180
160
140
120
100
70
70
70
60
60
50
40
30
20
60
50
40
60
70
30
140
130
50
40
30
60
7
7
7
7
6
6
6
6
5
5
5
5
4
4
4
3
3
3
a

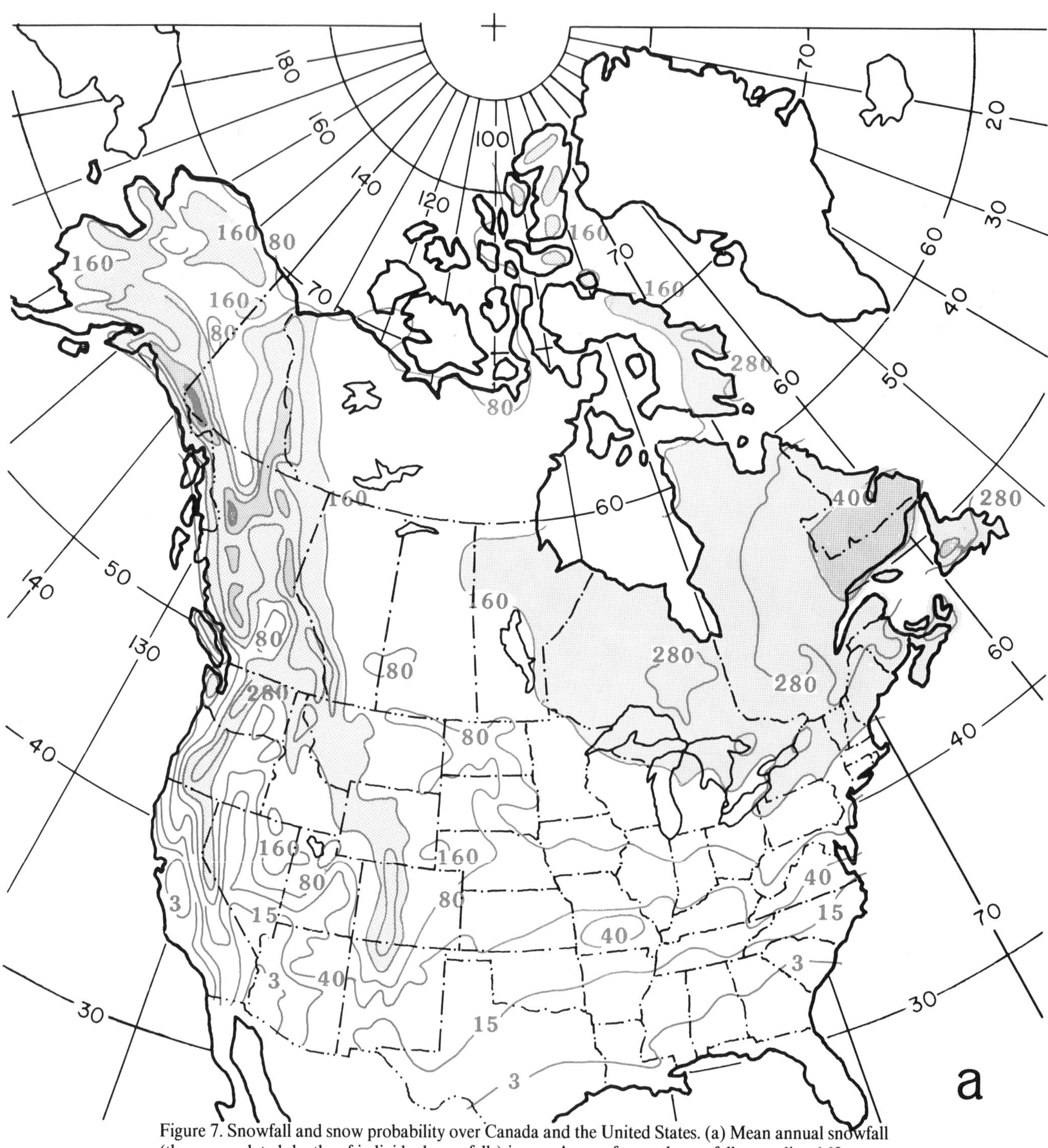

Figure 7. Snowfall and snow probability over Canada and the United States. (a) Mean annual snowfall (the accumulated depths of individual snowfalls) in cm. Areas of annual snowfall exceeding 160 m are lightly stippled, areas exceeding 400 mm are heavily stippled, and areas exceeding 800 cm are black. Note: curves are smoothed and do not show strong local variations especially in the western mountains. After: Schemenauer and others (1981). (b) Long-term probability of snow cover on January 31. After Dickson and Posey (1967).

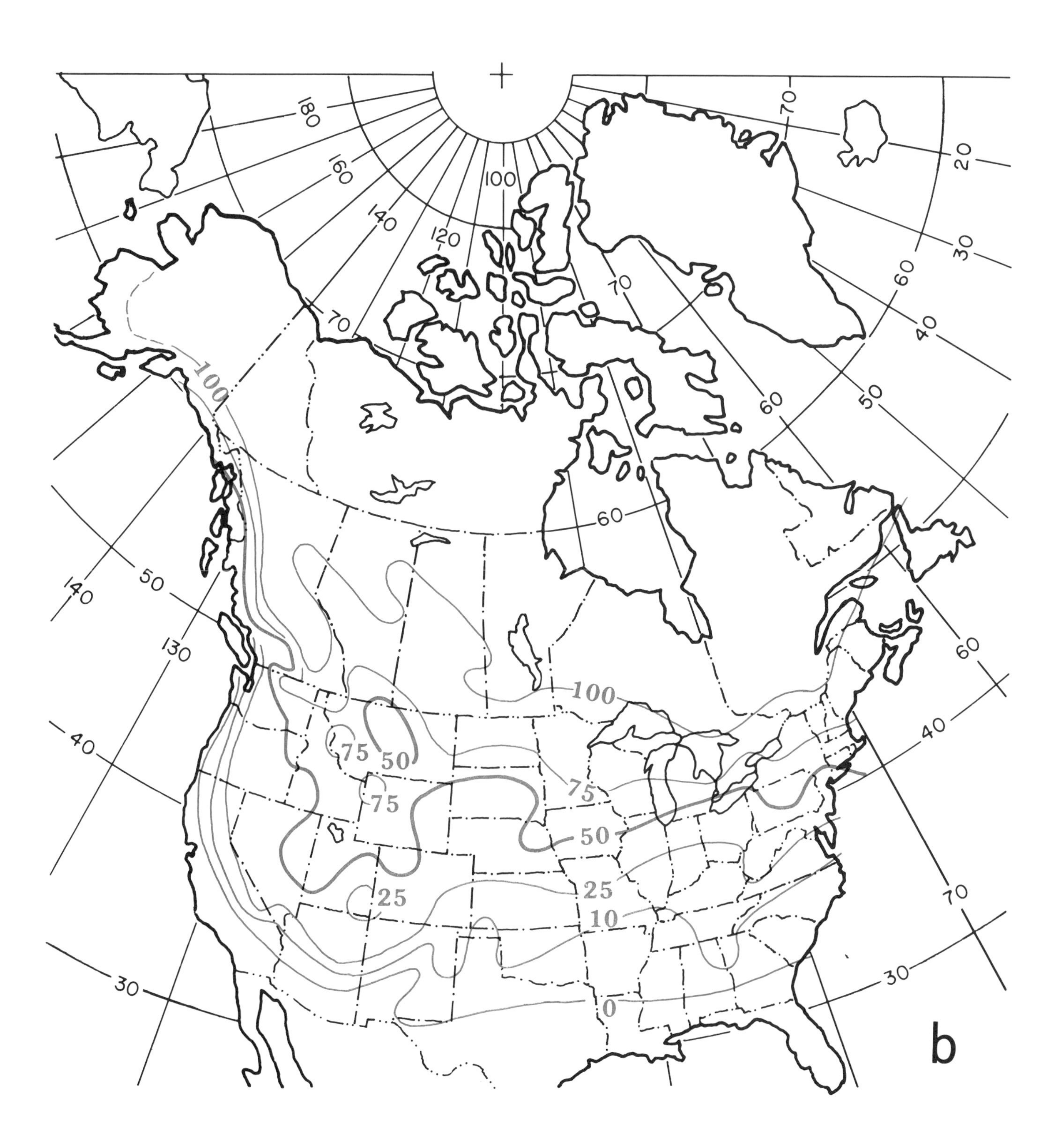
180
160
140
120
100
70
70
70
60
60
60
50
40
30
20
60
50
40
30
70
60
50
40
30
140
130
100
100
75
50
75
75
50
25
25
10
0
b

In Greenland, relatively heavy snowfall occurs along the western margin but somewhat inland from the coast, and along the southeastern coast, due to the orographic rise of moisture-laden air masses moving northeast from the North Atlantic and Baffin Bay onto the Ice Sheet or along the east coast.

Measuring snowpacks and forecasting runoff

One of the most propitious, and economically valuable, aspects of snow as a water resource is that it can be measured as it lies on the ground. From this information, the volume of subsequent runoff can be predicted. In turn, knowledge of snowmelt runoff facilitates the management of reservoirs and permits development of rivers for many uses. Forecasts of snowmelt runoff are especially important in the mountainous West, where most of the runoff is provided by snowmelt. Many agencies in the United States and Canada are involved in cooperative programs to forecast river flows from snow surveys.

A major difficulty in providing accurate forecasts of snowmelt runoff is the areal and temporal variability of snow in the mountains. This variability stems, in part, from the variability of precipitation in the mountains and, in part, from the redistribution of snow by wind and avalanches after it falls. Measurement of the snow is done with precipitation gages, by measurements made at snow courses, by snow pillows, and by techniques that measure electrical properties or the absorption of radioactivity. Precipitation gages, even those with wind-deflecting baffles, measure less than the true amount falling, depending largely on wind speed. Measurements of snow on the ground are made at snow courses, which involves the extraction and weighing of cores from the snowpack several times each winter; this technique can accurately measure the snowpack at a certain place (the snow course) but is labor intensive, expensive in mountain regions, sometimes dangerous, weather dependent, and is not adaptable to telemetry and real-time data collection. Also, it is hard to determine whether a given snow course is representative of the local terrain and land cover so that the information is an accurate index of regional snow cover. Snow pillows, which measure the weight of snow on the ground, have been integrated into automatic data-collection networks. On occasion, pillows can give suspect readings because of "bridging," whereby snow on the pillow is partly supported by ice lenses or buried crusts. These sensors generally are reliable, but snow pillows can provide only a limited sample of the variable snow resource because of the cost of installation and maintenance.

Remote sensing eventually may overcome the spatial sampling problem, but several of these techniques are still in the experimental phase. Four methods have been actively investigated: (1) airborne gamma-ray surveys—repeated aircraft flights along a designated path over relatively flat terrain to detect the attenuation of natural Earth radioactivity, whereby the degree of attenuation is proportional to the mass of snow on the ground; (2) aircraft flights in the mountains to visually or photographically record the elevation of the snow line or the fraction of the area covered by snow (commonly used to check runoff-model calculations); (3) satellite images made with visible to infrared radiation sensors that are used to monitor snow-covered areas; and (4) satellite data from passive microwave sensors, which may provide some information on snowpack mass, assuming that no liquid water is present (Hall and Martinec, 1985). The first two methods are expensive, the second is labor intensive, and the last two have numerous drawbacks, including low frequency of coverage (if high-resolution sensors, such as those on Landsat, are used), low resolution and obfuscation by clouds (if meteorological satellites with visible and infrared sensors are used), and ambiguity of interpretation (if satellite active and passive microwave sensors are used). Thus, improvement in understanding of the remotely sensed properties of natural snowpacks and in developing improved techniques for the large-scale, all-weather determination of snow mass by satellite remote sensing have been given a high priority for research (National Academy of Science, Committee on Glaciology, 1983). The trend toward privatization of high-resolution satellites, such as Landsat, may make use of their images for snow measurements too expensive for routine monitoring.

Volume forecasts use measurements of the amount of snow on the ground at a fixed date (such as April 1) as an index to the total amount of water likely to be produced by snowmelt after that date, usually in terms of what the river flow will be in relation to a specified "normal" flow. These forecasts are useful in managing reservoirs to optimize allocations to competing water users and to minimize the amount of water or hydroelectric power wasted because of uncertainty in the forecasts. However, forecasts at frequent intervals, including daily forecasts of inflows to many reservoirs, often are required for river systems that are extensively developed. Volume forecasts can be updated when runoff begins, and new forecasts can be made using new measurements of snow mass at monthly or bimonthly intervals. These volume forecasts generally are developed by use of a statistical regression analysis of snow mass against subsequent runoff as a percentage of "normal" runoff. The accuracy of the result depends, in part, on the statistical population of past snowmelt-runoff scenarios used in the regression analysis.

The prediction of extreme or unusual events, which are not represented in the historical record, is sometimes inaccurate. This can be illustrated by noting the unusual conditions in the upper Colorado River basin in 1983—conditions that occurred in many drainages in the western United States that year. The flow of the Colorado River at Lees Ferry, Arizona, is dominated by snowmelt runoff that generally peaks in June. During the 1983 water year, precipitation was only slightly (about 6 percent) higher than normal, yet the runoff volume for the year was more than twice normal (corresponding to a recurrence interval of 100 years), and the June–July flow was even more exceptional, corresponding to a 200-year event (Shafer and others, 1984). Predictions made on May 1 at many forecast sites were substantially low.

Air temperature during the snow-accumulation season and through the normal time of low-elevation melting was unusually

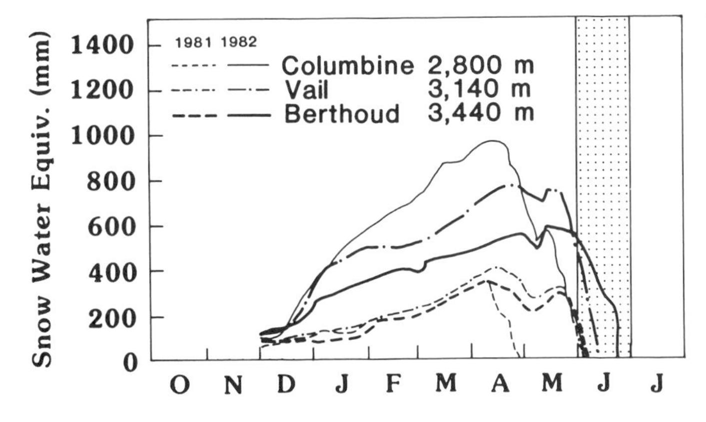

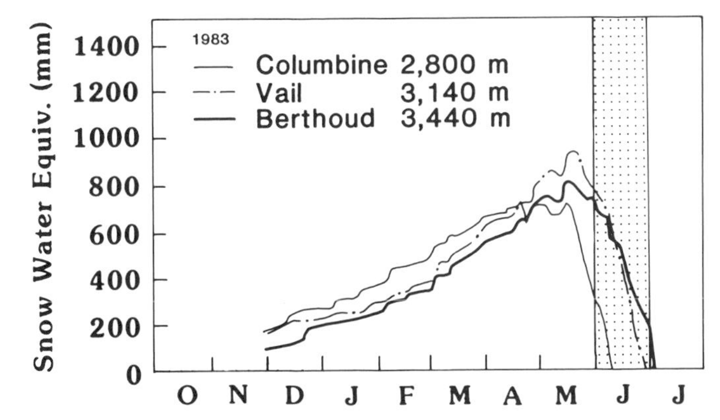

Figure 8. Variation in snow water equivalent during 1981, 1982, and 1983, upper Colorado River basin. Snow depth was measured by snow pillows at three representative snow courses for typical thick- and thin-snowpack years (1981 and 1982, respectively, top graph), and for 1983 (bottom graph). The month of June is shaded. Note that, in 1983, the maximum snowpack was reached in late May, in contrast to the earlier years, and the decline in June was rapid and nearly simultaneous at all elevations. Modified from Shafer and others (1984).

low; snow continued to accumulate at all elevations in the mountains until about May 21. Then, with very warm weather and the sun high in the sky, melt occurred at all elevations very rapidly in late May and June. This melt pattern was in marked contrast to the usual period of melting of April through June (Fig. 8). Because of the previous cool temperatures, the snow-covered area in late May also was more extensive than usual—a factor not yet included in most operational forecast models. Soil moisture had been high all year, and the shortening of the melt season into about half of its normal length reduced the opportunity for meltwater infiltration and evapotranspiration. The rate of runoff, therefore, was large, and major flooding resulted. No precedent for that scenario was found in the historical data sets used in forecasts made by statistical regression analysis. Furthermore, differences between forecast and actual runoff were uncharacteristic, based on previous experience and evaluations of the accuracy of past forecast performance (Shafer and others, 1984). Physically based conceptual models offer a better framework for understanding and predicting such events (Shafer and others, 1984).

The snowmelt process

Understanding the snowmelt process requires consideration of the energy fluxes that produce melt. Snow receives energy from the sun (short-wavelength radiation), from radiation emitted by water vapor and clouds in the air (long-wavelength radiation), from warm air, from the heat of condensation, from rainfall, and from heat conducted from the underlying ground. As previously mentioned, much of the incoming solar radiation is reflected back from the snow surface because of its high reflectivity (albedo); the snow also emits longwave radiation. The radiation balance of a snowpack depends not only on the algebraic sum of these components but also on atmospheric conditions, snow temperature, and other factors that vary with time. Heat energy that is gained or lost from the snow by the exchange of heat with the air by eddy conduction, and by evaporation or condensation, depends primarily on the wind, humidity, and temperature gradients in the air just above the snow surface.

The effects of these various energy components vary with time of day (Fig. 9) and time of year, and with elevation, latitude, and vegetation cover. In general, net radiation increases in relative importance as the season advances from winter to summer and with increasing elevation. Male and Granger (1981) summarized the different contributions to the total melt in 24 separate studies at diverse locations. They found that net radiation accounted for 17 to 100 percent of the melt, heat transfer (from air) for –42 to 79 percent, and latent-heat (evaporation-condensation) transfer to –74 to 36 percent (the minus sign indicates that the snow cooled instead of melted).

Even simple generalizations, such as that a clear day implies the highest net radiation balance, are not always true. Hubley (1957) found that the net shortwave radiation on snow near Juneau, Alaska, on a clear day in June, was 10.35 MJ m^{-2} (247 calories per square centimeter), and that the net longwave radiation was –5.11 MJ m^{-2}, leading to a net radiation balance of 5.24 MJ m^{-2}. On a totally overcast day with cloud temperatures of about 5°C, the net shortwave radiation was reduced to 5.20 MJ m^{-2}, but the net longwave-radiation balance was +2.10 MJ m^{-2}, leading to a net radiation balance of 7.29 MJ m^{-2}, which was 40 percent greater than on a clear day.

Clearly, the measurement or prediction of all the energy fluxes that determine the rate of snowmelt is impossible for operational use at most locations. Therefore, attempts are made to relate snowmelt to one or a very few simple variables that are easily measured.

Air temperature is used widely to estimate melt rate, usually by an empirical formula such as

$$M = k (T - T_0),$$

where M = melt rate, in mm of water per day; T = air temperature; T_0 = reference temperature (usually 0°C); and k = melt

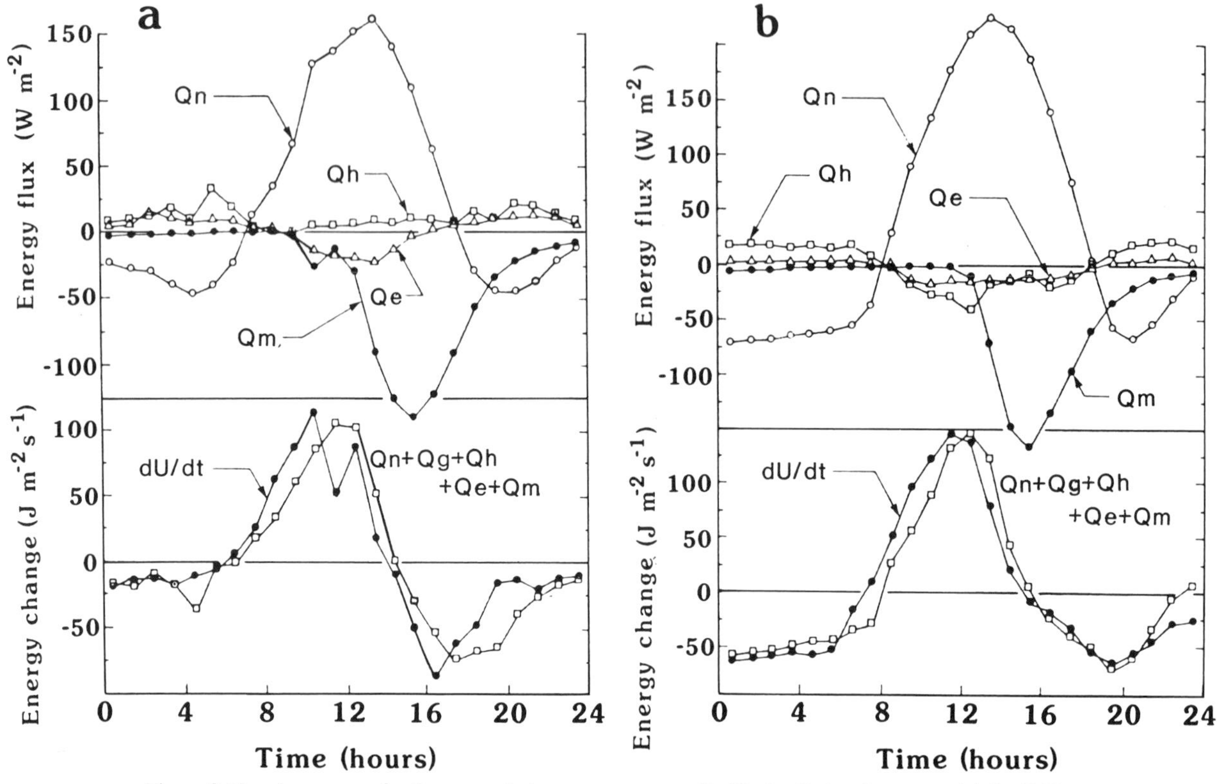

Figure 9. Hourly averages for the energy balance over snow at Bad Lake, Saskatchewan on (a) April 10, 1974, and (b) April 14, 1974. Energy balance terms are Q_n, net radiation; Q_g, ground heat flux; Q_h, sensible heat flux; Q_e, latent heat flux; Q_m, melt. The difference between the curves for the rate of change of internal energy du/dt and the sum of the heat fluxes indicates the accuracy of the independent measurements. Note that the sensible heat flux Q_h is much smaller than the radiation heat flux Q_n during the daytime, and can be either positive or negative. Thus it appears to either assist or counteract the radiation flux. After Granger and Male (1978).

factor. T can be the average or maximum daily temperature, depending on the model.

This approach, although commonly applied, is not sufficiently precise for modern water-resources management purposes for several reasons. Observations show considerable scatter in the M(T) relation (Fig. 10). Air temperature is not involved directly in the energy-balance components but is correlated with several components, such as incoming shortwave radiation. These correlations make temperature, in effect, a nonlinear function of the combination of these fluxes, and therefore, the average daily temperature may not reflect the daily average of the combination of these fluxes. Also, other important factors, such as the wind gradient near the snow surface, may not correlate with air temperature. Another problem with the simple air-temperature formula is the need to use a wide range of values for the empirical melt factor, k, when applying the formula to contrasting areas (Gray and others, 1979).

Air temperature also is commonly used to distinguish the fraction of the precipitation that falls as rain (or snow). Assuming that the precipitation is entirely as rain if the air temperature is above 0°C, and as snow if it is below 0°C, is reasonably accurate when applied instantaneously, but it is very poor when used with averages over appreciable intervals of time such as weeks or months (Fig. 11).

Many snowmelt models now reflect a compromise between the use of simple temperature indices and complete energy-balance approaches, e.g., Anderson (1973). Empirical or physical relations are used to estimate each of the important heat fluxes. Of these, the ones most difficult to estimate include the heat transfer from the air by eddy conduction and the incoming long-wave radiation from water vapor and clouds. Recent studies have shown that, under some circumstances, these fluxes may be estimated just as well from gross air-mass properties as from local measurements near the surface. For instance, Male and Granger (1981) show that the eddy-convection transfer of heat in a snow-covered area of large extent (unbroken by forests or mountains) is related to the air temperature at the 850-millibar air-pressure level—a variable that is characteristic of the air mass and one that is measured routinely several times daily at many locations. As Male and Granger point out, air-mass characteristics are more predictable than the temperature at a point near the snow surface. Consequently, research at an air-mass scale should give improved insight into the exchange of heat and mass with the snow surface, because such studies will be considering the causes of the ex-

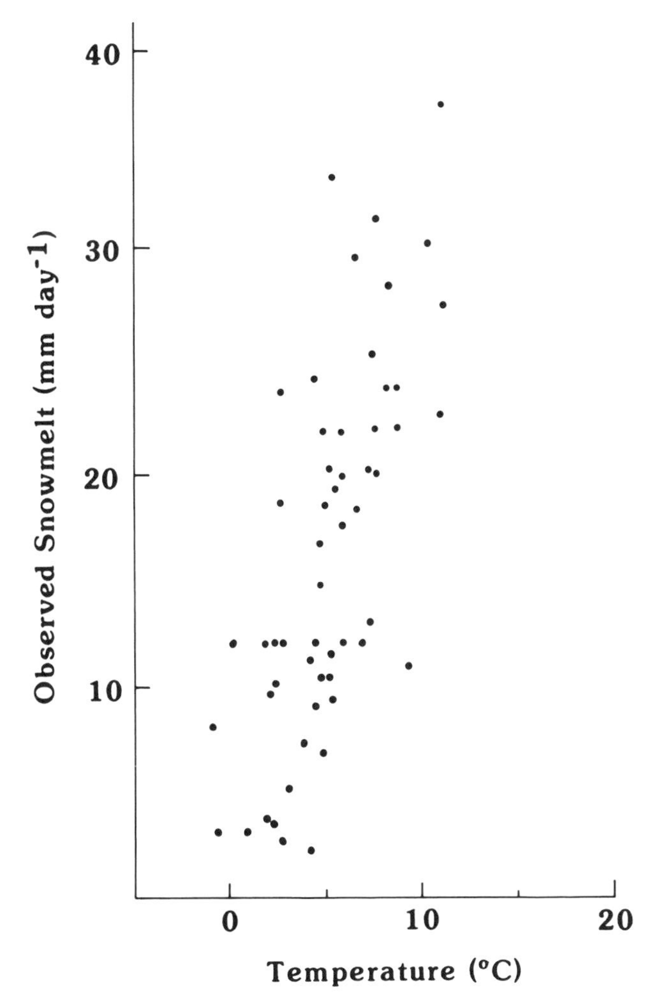

Figure 10. Observed snowmelt, in cm of water per day, as a function of average daily air temperature during the period 1973 to 1977, for the Marmot Creek drainage basin, Alberta. Several subbasins and different time intervals are included for each year. Modified from Barnaby (1980).

change and not just the effect. Clearly, improvement in snowmelt modeling is needed. Especially important are new models that will predict reliable melt rates from elementary meteorological variables.

Water flow through snow

Water flow through a mass of snow is, in the simple case, the motion of a fluid through a porous medium. In reality, it is more complex, partly because the medium may be either saturated or unsaturated, but more especially, because the "medium may be the message"—the flowing fluid may interact with its frozen phase, the medium (Colbeck, 1978; Wankiewicz, 1979).

Colbeck (1972, 1978) has analyzed the problem of the flow through an unsaturated snow mass, without interaction with the medium; in this case the flux of water, u, per unit area is

$$u = \frac{k'}{m}\left(\frac{\partial p}{\partial z} + \rho g\right),$$

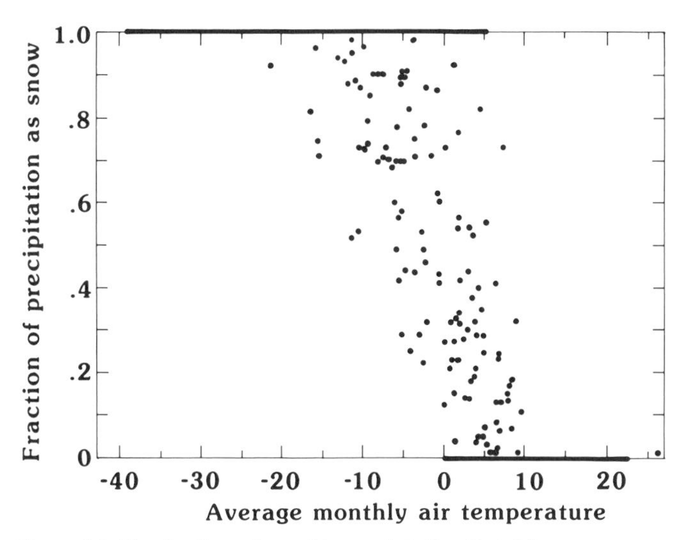

Figure 11. The fraction of monthly precipitation that fell as snow as a function of the average monthly air temperature in Canada in 1975. Modified from Ledley (1985).

where k′ is the conductivity of the system for liquid water, m is the viscosity of liquid water, p is the capillary pressure, ρ the density of water, g the acceleration of gravity, and z a vertical coordinate. The pressure gradient term, $\partial p/\partial z$, is generally small compared to the gravity term when appreciable flow is occurring, and if there are no abrupt changes in density, it is neglected subsequently. The term k′ can be estimated from

where k$k' = k\,S^n$

where k is the intrinsic conductivity of the snow mass, S is the effective liquid water saturation (the fraction of pore volume occupied by moving water), and n is an empirical coefficient (~3). Thus

$$u = \frac{\rho g}{m} k\,S^n$$

and the flow increases with the liquid water saturation, other conditions being equal. This leads to the formation of waves, as variations in liquid water flow move down through a snowpack with the peaks of concentration (= saturation) moving faster than the minima. This has been confirmed by observation (Fig. 12). In reality, the motion of water through snow is more complicated. Low-conductivity ice layers within a snow pack cause local saturated zones (perched water tables); the water leaks through gaps in these to the more uniform unsaturated snow below. A saturated zone often forms at the snow/soil interface. The effect of these saturated zones is to speed the routing of meltwater from the surface to the adjacent hydrologic system, but this increase in speed depends on the amount of water in transit. The important point is that a snow cover increases the lag time by more than an order of magnitude compared to a snow-free slope, and that both

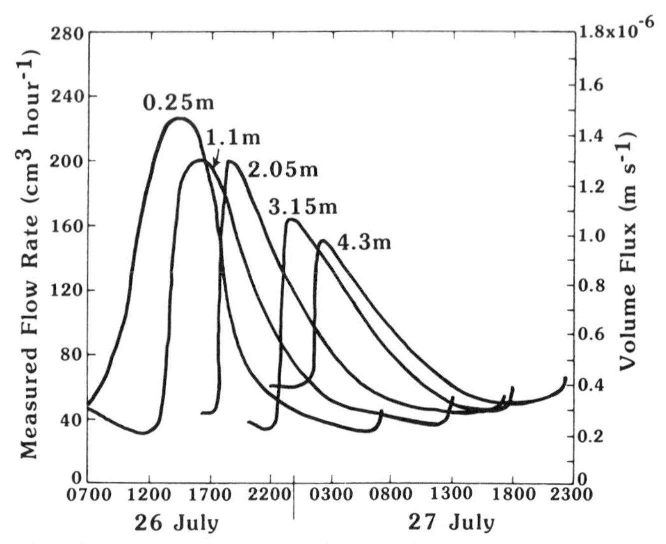

Figure 12. Changes in the diurnal wave of meltwater as measured at various depths below the snow surface. Note the development of a "shock wave." After Colbeck and Davidson (1973).

TABLE 2. THE EFFECT OF SNOW COVER ON RUNOFF LAG (HOURS)*

	Slope	Rainfall Rate ($m\ s^{-1}$) 1 x 10^{-6} Light	3 x 10^{-6} Moderate	10 x 10^{-6} Heavy
(a) Brush Slope, L = 100 m	3°	0.33 h	0.22 h	0.13 h
	10°	0.23 h	0.15 h	0.093 h
(b) Snow Slope, L = 100 m	3°	13 h	13 h	13 h
	10°	3.8 h	3.8 h	3.8 h
(c) Percolation, Snow 1 m deep	Dry Snow	12 h	5.0 h	1.9 h
	Wet Snow	3.8 h	2.2 h	1.1 h

*After Wankiewicz (1979).

vertical percolation and downslope flow contribute to this lag (Table 2).

More difficult to analyze is the motion of water into and through a cold (subfreezing) snowpack. In this case the moving water refreezes, giving up its latent heat to warm the cold snowpack. But if the water flow is slightly concentrated in some local areas relative to the surrounding areas, more warming will occur in the local areas, and flow can progress farther there, leading to the water "drilling" its way through a cold snowpack. The actual dynamics of this process are not yet adequately described.

Thus, analysis of the water flow through snow is complex. In the case of thin snowpacks, this is not an important problem to the streamflow forecaster, but it does limit the accuracy of short-term forecasts, especially in areas such as the West Coast where thick snowpacks are common.

GLACIERS

Although an enormous reserve of water, equivalent to precipitation over the entire globe for about 60 years, is stored in the form of glacier ice, most of this ice is in relatively unpopulated regions. In continental North America, however, the volume of water stored as snow and ice in glaciers probably exceeds that stored in all of the lakes, ponds, rivers, and reservoirs.

Distribution of glaciers in continental North America

The glaciers in western North America are associated with a great chain of highlands—the North American Cordillera (Fig. 13). This chain comprises two vast mountain complexes, the Pacific Mountain system and the Rocky Mountain system, that roughly parallel the Pacific coast from California and Colorado through Canada and Alaska to the Arctic; these systems are separated by a complex of basins, plateaus, and isolated mountain ranges.

Most of the precipitation that forms glaciers in this region is derived from the North Pacific Ocean. In winter, moisture-laden storms spawned from the semipermanent Aleutian low-pressure area move inland along the Pacific coast. Heavy precipitation occurs when these storms impinge on and are lifted by the Pacific Mountain system; annual values of more than 5,000 mm have been measured near the coast in Washington, British Columbia, and Alaska. Annual runoff as high as 10,000 mm has been recorded, suggesting that some coastal mountains receive much more precipitation than that recorded in the low-elevation gages. Strong precipitation-shadow effects occur where these storms move over high ridges. Movement of warm moist air into the interior is frequently blocked in winter by a persisting high-pressure area north and east of the Rocky Mountains as well as by the mountains themselves. Consequently, winter precipitation decreases markedly from the coast inland. In summer the Aleutian low shifts northward, and a North Pacific high-pressure system develops. Storms moving north of the high-pressure area bring to the Alaskan coast abundant summer precipitation, and precipitation rates in late summer and fall may be the highest of the year. Farther inland, moisture is precipitated from cyclonic or convective storms moving eastward toward low-pressure areas east of Hudson Bay. In the inland mountains of Alaska and western Canada and in the Rocky Mountains, precipitation may be greater in summer than in winter. Even so, the amount of summer snowfall is low, and the annual precipitation is much less than along the coast. North and east of the coastal mountains, annual-precipitation values less than 500 mm are common.

Most of the glacial ice and the largest glaciers in western

Figure 13. Map showing the distribution of glaciers (blue dots or shapes) and the elevation of the equilibrium line on glaciers (black lines) for western North America. Dots represent one or more glaciers and are not to scale. From Meier and others (1971).

North America occur in the Pacific Mountain system (Meier and others, 1971). The coastal Kenai and Chugach Mountains are fairly low, but snowfall is so heavy that large icefields occur; many large valley glaciers debouch into the sea; for example, see Figure 14. To the north is the long arc of the Alaska Range, with Mount McKinley (elevation 6,194 m) and major complexes of icefield and valley glaciers. A continuous structural valley (the Denali fault zone) traversing the northern part of the range is the locus of a large number of surging glaciers. Most glaciers in the Alaska Range are retreating markedly.

The eastern Chugach Mountains and the Wrangell Mountains merge into the St. Elias Mountains north of Icy Bay. Here an area of more than 50,000 km^2 is covered with ice; these are the most extensive icefields on the North America continent (Fig. 15). Noteworthy glaciers include the Bering, the largest glacier in continental North America (200 km long and 5,800 km^2 in

Figure 14. Columbia Glacier, a 65-km-long valley glacier that terminates in the sea near Valdez, Alaska. This glacier is currently disintegrating due to an iceberg-calving instability. In front of the terminus ice cliff (about 50 to 60 m high) is an area of floating icebergs and ice fragments jammed between the glacier and a submerged moraine shoal. Photograph by M. F. Meier, August 14, 1984.

area); the Seward-Malaspina (5,200 km^2 in area); and the Hubbard, a valley glacier 120 km long. On the extreme northeastern fringe of the range, subpolar ice (having temperatures considerably below the freezing point) is known to occur in the lower reaches of the glaciers. Many surging glaciers occur in the St. Elias Mountains; most other glaciers there are probably stable or are retreating slightly. Some of the tidal glaciers are advancing, but a few other tidewater glaciers are receding exceptionally fast (Fig. 14).

A vast system of icefields, valley glaciers, and isolated smaller glaciers extends along the crest of the coast Mountains of southeastern Alaska and British Columbia from the Yukon border south into the North Cascades and Olympic Mountains of Washington. Many of the smaller glaciers of the Coast Mountains were advancing in the early 1960s, but only a few had continued to do so by 1969; most large glaciers have been retreating. Glaciers in the North Cascades and Olympic Mountains advanced in the period 1950 to 1958. From southern Washington to northern California, glaciers occur only on high volcanoes. In the Sierra Nevada of California the small glaciers show both maritime and continental characteristics. Small glaciers also occur on the high volcanos of Mexico.

The Rocky Mountain system contains thousands of mountain glaciers but only a few small icefields. Many tiny cirque glaciers occur in the Brooks Range of Alaska, but moderate-size valley glaciers occur only in the eastern end of that range. In this high-latitude environment the glaciers are subpolar, and stream icings form below many of the glacial termini. Small mountain

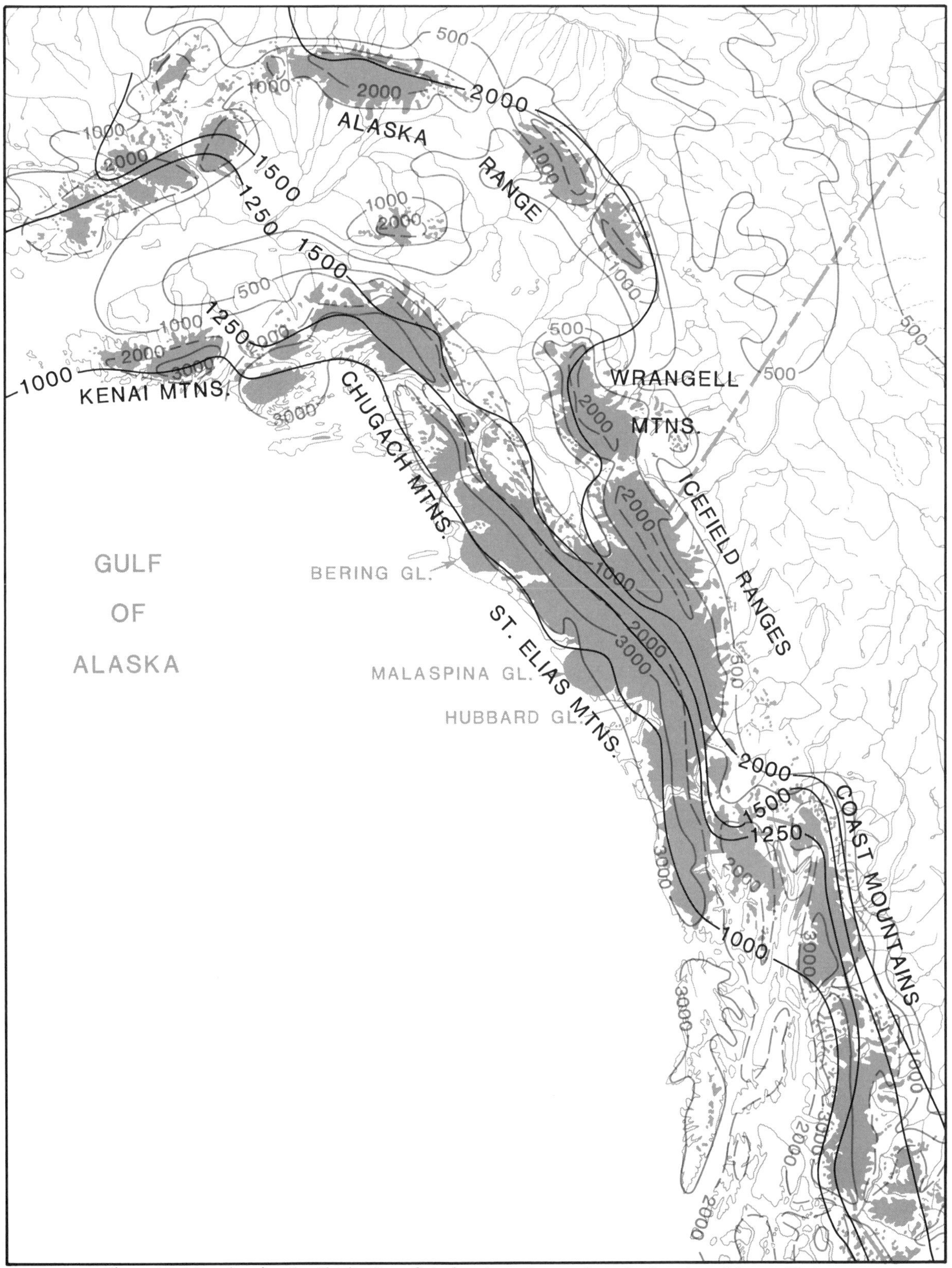

Figure 15. Map showing the distribution of glaciers (gray dots or shapes), elevation of the equilibrium line (black lines), and annual precipitation (blue lines) for the mountains bordering the Gulf of Alaska. From Meier and others (1971).

glaciers in the Mackenzie and Selwyn Mountains of Canada are undoubtedly subpolar but have not been studied in detail.

Many glaciers exist in the Canadian Rockies along the Alberta–British Columbia border. Small icefields covering up to 280 km^2 occur near the headwaters of the Columbia, Athabasca, and Saskatchewan Rivers. In the Rocky Mountains of the United States, only small cirque glaciers occur except in the Wind River Mountains of Wyoming, where there are compound cirque glaciers, valley glaciers, and ice-cap–like glaciers on high plateaus. The glaciers in Montana, Wyoming, and Colorado are found at very high altitudes (approaching 4,000 m in Colorado). However, in some areas such as the Wind River Range, the local mesoclimate results in large amounts of orographic and convective precipitation and abundant wind drifting so that considerable snow accumulation is possible.

In the mountain ranges between the Rocky Mountain system and the Pacific Mountain system, many scattered small mountain glaciers occur. Tiny vestigial glaciers occur in the intermountain ranges of the United States, such as the Wallowa, Salmon River, and Wasatch Mountains in northeastern Oregon, Idaho, central Utah, respectively, and on Wheeler Peak in eastern Nevada. Whether some of these are true glaciers or not is a matter of definition. In northwestern Alaska, small glaciers are present only in the Wood River and Kigluaik Mountains.

Glaciers of the Canadian Arctic

The Canadian Arctic Islands contain about 108,600 km^2 of perennial ice cover, the largest concentration of ice outside of the Greenland and Antarctic Ice Sheets. For a more complete description of the glaciers and ice caps in this area, see Koerner (1989). Most of the ice in the high Canadian Arctic is on Ellesmere, Devon, and Axel Heiberg Islands. Large ice caps and their outlet glaciers occur in these areas, in addition to many mountain valley and cirque glaciers, and small local ice shelves occur along the northern coast of Ellesmere Island. The glaciers are cold, flow very sluggishly, and at least the larger ice caps have existed for more than 100,000 years, according to ice cores (Paterson, and others, 1977). Accumulation rates are moderate (0.4 m/yr) on slopes facing the open waters of Baffin Bay, to less than 0.15 m/yr on slopes facing to the northwest. Mass balances during the last 20 or so years have been slightly negative (Koerner, 1985), but the annual turnover is very low.

Farther west, only a few, tiny, stagnant ice caps occur on Melville Island. These are thin ($<$100 m) and date from the late Holocene to recent. To the south, Bylot and Baffin Islands contain large ice masses that are both warmer and more active than those of the Queen Elizabeth Islands. Wisconsinan ice has been identified in ice caps on Baffin, Devon, and Ellesmere Islands, so they are relics of the Pleistocene Laurentide Ice Sheet. Small glaciers also occur south of Hudson Strait, in Labrador.

Ice in Greenland

The Greenland Ice Sheet, second only to the Antarctic Ice Sheet in size, contains about 7 percent of the fresh water on Earth. Around the margins of the ice sheet are many ice caps and other kinds of glaciers, totalling about 80,000 km^2 in area (Fig. 16). Only about 17 percent of Greenland is free of perennial ice.

The physical and climatic characteristics of the Greenland Ice Sheet have been described in detail by Fristrup (1966), Weidick (1975), Budd and others (1982), Radok and others (1982), and Reeh (1989). The surface is characterized by southern and northern domes, with elevations of 2,830 and 3,205 m respectively, connected by a broad, almost horizontal saddle with altitudes of about 2,500 m (Fig. 17). The ice divide is displaced to the east because of the higher bedrock topography along the eastern margin. The average ice thickness is about 1.79 km, and the maximum ice thickness is about 3.42 km (Radok and others, 1982). The base of the ice sheet is close to sea level in the central and northwestern regions (Fig. 17).

Relatively heavy precipitation is delivered to Greenland by cyclonic disturbances moving onto or along the coast from Baffin Bay and the north Atlantic. Precipitation is especially heavy when these storms move up the slope of the ice sheet. Thus, snow accumulation rates are high (50 to 250 cm of water equivalent per year) along a belt inland from the west coast, and along the southeast coast (Fig. 17). Elsewhere, accumulation rates are lower, and are less than 15 cm/yr in interior northeast Greenland. Ice is lost by melting along the margins and by the discharge of icebergs. Neither of these mass losses are adequately known; it is thought that the ice sheet is approximately in equilibrium at the present (Weidick, 1985), but the marginal areas appear to be thinning and the central area (at least west of the divide) appears to be thickening.

The flow of the Greenland Ice Sheet is generally sluggish, except where the flow is concentrated into ice streams. The best known of these is the Jakobshavn Ice Stream, which moves at a rate of more than 20 m/day at the terminus.

One of the most scientifically exciting aspects of the Greenland Ice Sheet has been the record of environmental history revealed in ice cores. This includes records of past air temperature, precipitation, fallout (atmospheric, cosmic, terrestrial, and volcanic), and atmospheric composition (including "greenhouse" gas concentrations; Langway and others, 1985; Reeh, 1989).

Hydrologic characteristics of glacier runoff

Runoff or no runoff? Thermal regime of glaciers. At very high elevations and at very high latitudes, the summer climate is so cold that no meltwater forms. At lower elevations and latitudes, most of the glacier ice remains at the melting temperature throughout the year, and meltwater formed in the summer (and sometimes in the winter) can flow through the ice mass and emerge as runoff. In between these two extremes, the situation is more complex. This is illustrated by the concept of snow facies (Benson, 1962; Müller, 1962), as shown on Figure 18. Runoff is possible only in portions of the wet snow, superimposed ice, and ablation-zone facies. In general, polar glaciers are characterized by dry snow facies; temperate glaciers, by the wet snow through

Figure 16. A subpolar glacier, Polaris Glacier, Hall Land, northwest Greenland. Photograph by Austin Post, July 19, 1964.

the ablation facies; and subpolar glaciers, by percolation through the ablation facies.

Mass balance as a function of elevation. In mountain regions, and along many, but not all, ice-cap and ice-sheet margins, snow accumulation rises with elevation. On the other hand, snow and ice melting decrease with elevation. Snow that does not melt off becomes new firn. At the equilibrium line, surface ablation (melting and other processes that lose mass) equals surface accumulation (mass gain) as averaged over several years. The net surplus of mass accumulated above the equilibrium line is moved down by the slow process of glacial flow. Ablation processes may produce runoff from both above and below the equilibrium line, but not in the dry snow or percolation facies. These concepts are illustrated in Figure 19.

The net mass balance (accumulation minus ablation) gradient with elevation depends on, among other things, latitude and the degree of continentality of the climate (Fig. 20). Generally, the net mass balance/elevation function is thought to be relatively consistent in a given mountain or ice cap slope area; this is a powerful tool for extrapolating local measurements to regions. The spatial variation in the equilibrium line altitude (ELA) is shown in Figures 13 and 15.

Glaciers as water resources. The importance of glacier ice to water-resource development stems largely from seasonal and long-term natural regulation of meltwater. Water is stored as ice and snow in winter when the need for water is low and becomes available during the warm summer when it is most needed for irrigation and other purposes. The estimated amount of runoff from glaciers in North America exclusive of Greenland is given in Table 3.

The rate of meltwater production during the summer is determined primarily by two factors: (1) incoming solar radiation and (2) albedo of the snow or ice surface; although eddy conduction and other processes also contribute. In May and June, the

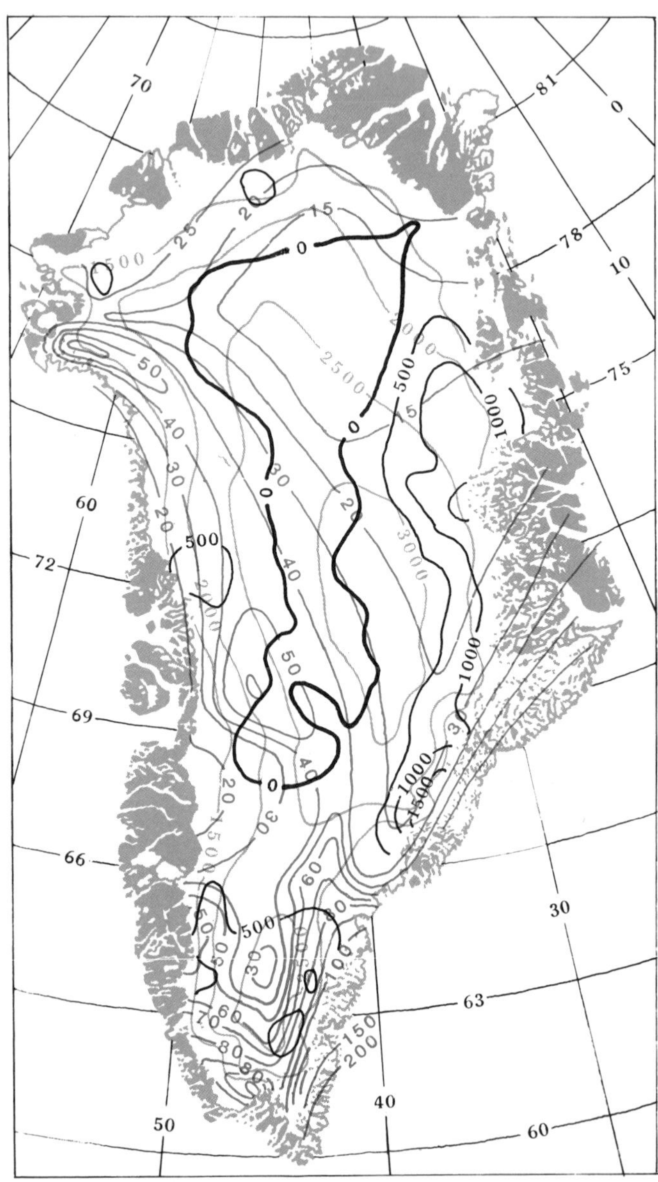

Figure 17. The Greenland Ice Sheet, showing surface topography (gray), and subglacial topography (black), both in meters, and accumulation rates (blue) in cm of water equivalent per year. Modified from Radok and others, 1982, and Reeh (1985).

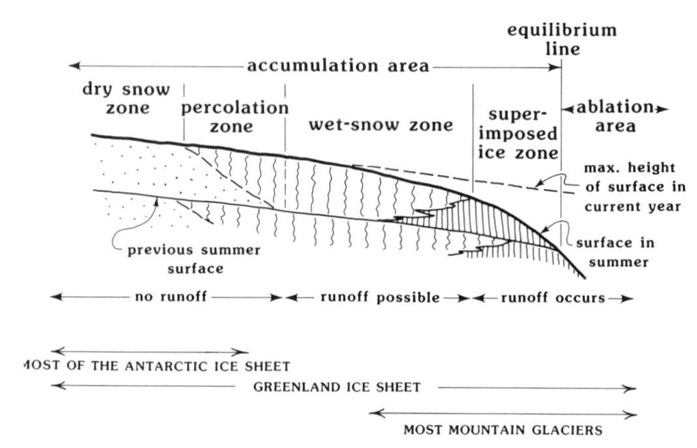

Figure 18. Diagrammatic section through snow and firn (old snow). At high altitudes no surface melting occurs. At slightly lower altitudes surface melting occurs but refreezes in that winter's cold snow layer. At lower altitudes refrozen (superimposed) ice masses develop, and some subsequent melt of snow and superimposed ice may run off depending on the temperature of the underlying snow and ice mass. At the lowest altitudes (the ablation zone, below the equilibrium line), melting exceeds snow and ice accumulation and some or all of this meltwater runs off to the sea. Modified from Benson (1962).

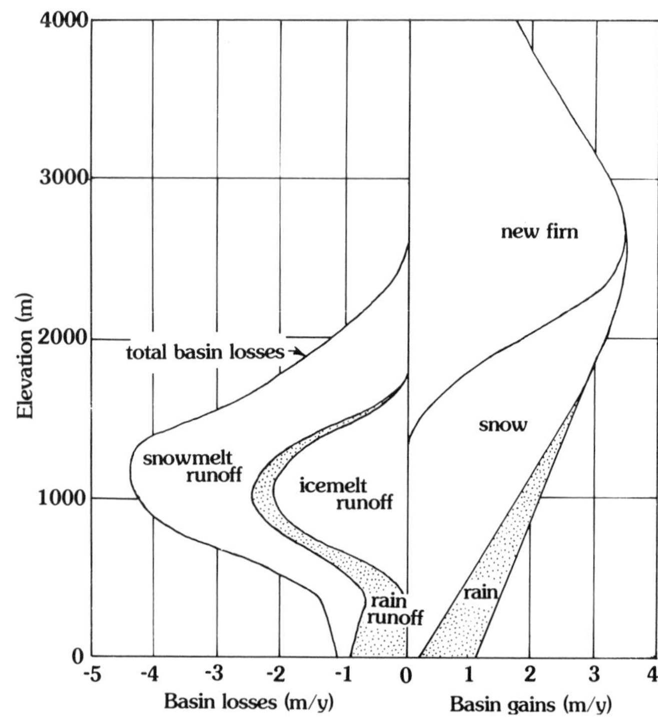

Figure 19. Diagrammatic altitude distribution of mass gains and losses from a glacierized mountain drainage basin. Note that above about 2,500 m, no runoff is lost from the basin; the excess mass gain is moved down in elevation by glacier flow. After Young (1985).

incoming radiation is intense, but the snow albedo also is high and snow covers most of the glacier, so that only moderate melt rates are possible. By July and early August, the albedo of the snow has dropped, the snow cover has become smaller, and considerable glacier ice with comparatively low albedo is exposed so that high melt rates occur even though the incoming radiation has decreased slightly. By September and October, the incident radiation has decreased markedly so that, in spite of low-albedo surfaces, the melt rates are only moderate. The amount of melt and runoff is sensitive to the duration of the snowcover, and this can be quite variable from year to year.

The amount and distribution of glacier runoff is quite different from that resulting from a lowland seasonal snow cover at the same latitude. Three effects can be noted:

1. At high elevations, precipitation is greater than at low elevations, more precipitation occurs in the form of snow, water losses to evaporation are negligible, and lower temperatures cause snowmelt to occur later in the year than at lower elevations.

2. In rugged mountains, snow blows from ridge crests or steep slopes and forms thick accumulations in hollows or basins. Some snow lies on slopes that face the sun, whereas other snow lies on shadowed slopes. Thus, snowmelt on some slopes begins earlier than on flat lands; as a result, the peak of meltwater runoff may be subdued, and floods may be rare. The runoff peak occurs much later during the year, and the runoff is sustained for a much longer time in rugged terrain than in flat terrain.

3. When a seasonal snow cover melts, the runoff is approximately equal to the water-equivalent volume of the snow. Glacier runoff, on the other hand, can be much less or much more than the water equivalent of the winter snow and can continue late into the summer, even after all traces of winter snow have disappeared.

This delayed effect, coupled with the dependence of snow and ice melt on heat balance rather than on the amount of ice available, produces a remarkable natural regulation of streamflow from year to year (Meier, 1969). The effect of an unusually hot summer is obvious. Very warm air normally is associated with highly positive radiation balances, which cause rapid melting and unusually large runoff volumes. Conversely, cool summers normally are accompanied by lower radiation balances and unusually low volumes of summer runoff from glaciers.

The effect of precipitation is less obvious. Higher-than-normal snowfall on glaciers produces less-than-normal runoff because of the reflective (high-albedo) properties of fresh snow. Snow in summer immediately raises the albedo of a glacier-ice surface from about 0.35 to about 0.75, causing a two- to threefold drop in the amount of solar radiation absorbed. Usually, thick winter snow means that the high albedo of snow will persist over a larger area of the glacier's basin for a longer period into the summer, which greatly decreases the possible melt (Figs. 21 and 22).

When glacier runoff is combined with nonglacier runoff, the effects of abnormal temperature or precipitation on the total runoff tend to balance out and produce a more stable and even flow.

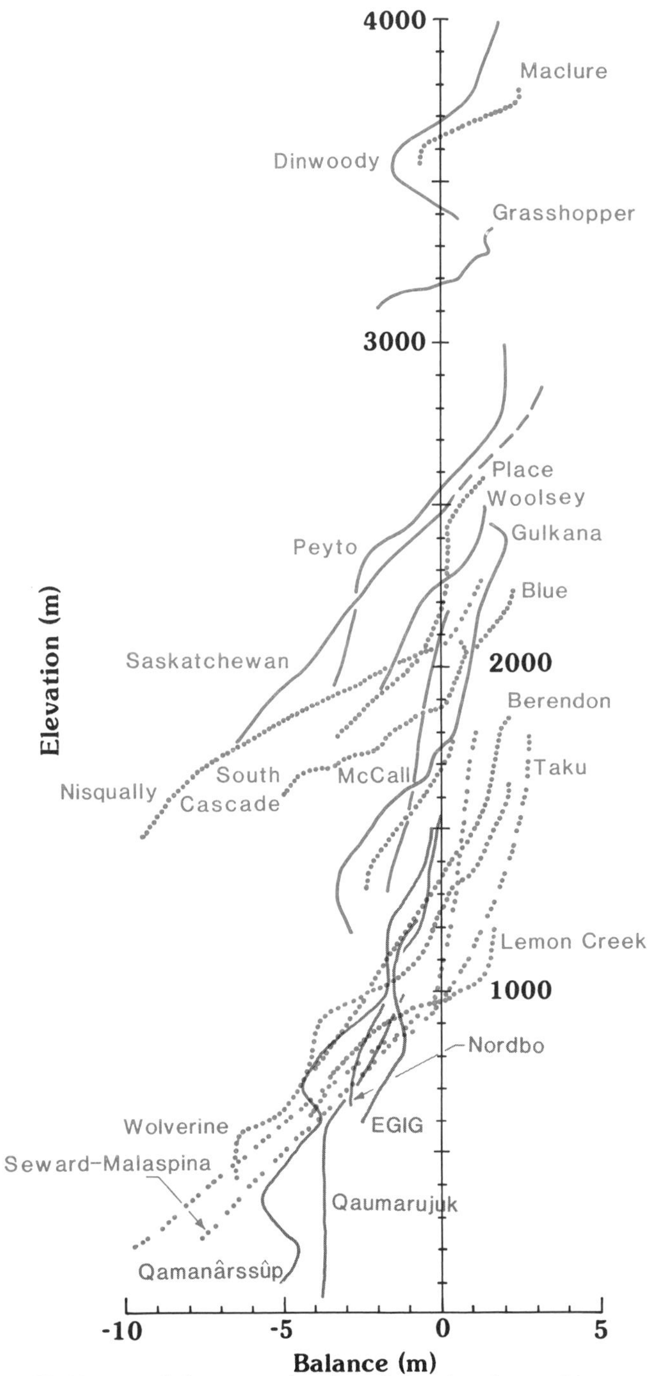

Figure 20. Net mass balance as a function of elevation, for maritime (dotted blue lines) and continental (solid blue lines) glaciers in western North America, and for Greenland (gray lines). After Meier and others (1971) and Reeh (1989).

Thus, runoff tends to be fairly uniform in regions with some glaciated basins (Fig. 23). This natural regulation has an important economic benefit; river systems can be operated more efficiently with a given amount of reservoir storage. Glacier runoff, however, cannot be forecast by ordinary procedures. Techniques based on streamflow records unaffected by glacier runoff will

TABLE 3. AREA OF GLACIERS AND ESTIMATED ANNUAL RUNOFF VOLUMES

Region	Area* (thousands of km^2)	Estimated Annual Runoff* Total per unit area (ma^{-1})	Total volume (km^3 a^{-1})	Portion due to long term ice wastage (km^3 a^{-1})
Greenland†	80	0.5	40	7
Queen Elizabeth Islands	108.6	0.3	37	7
Baffin and Bylot Islands, Labrador	42.0	0.5	20	4
Brooks Range, Kigluaik Mountains	0.7	1.1	0.8	0.1
Alaska Range; Talkeetna, Kilbuk Mountains	14.9	1.8	27	5
Aleutian Islands, Alaska Peninsula	2.2	2.0	4	0.7
Coastal Mountains, Kenai Peninsula to 52°N	88.4	3.8	340	63
Mackenzie, Selwin, Rocky Mountains north of 55°	0.9	1.2	1	0.2
Coast Mountains south of 55°N	14.0	4.0	56	10
Cascades, Olympics, Rocky Mountains south of 55°N, Selkirks (Canada), Northern Rocky Mountains	4.0	1.6	6	1
Middle and Southern Rocky Mountains, Sierra Nevada	0.1	3.0	0.3	0.1
Mexico	0.01	1.0	0.001	0.0002
Total	356	21	530	98

*Runoff is from icemelt alone, runoff from rainfall is not included. Data on glacier area and those used in the estimations of runoff are from Meier (1984).
Runoff per unit area is the same as the annual amplitude $\underline{a}$ (average of accumulation and ablation, both considered positive) plus a portion due to ice wastage. The runoff portion due to ice wastage is calculated on assumption that ratio of the mean net balance to $\underline{a}$ is constant and equal to 0.23, as in Meier (1984). This ratio undoubtedly varies from region to region, but few data are available.
†Not including the Greenland Ice Sheet.

lead to inaccurate answers because the causes of the variability are so different from those operating in glacier runoff streams. Volume-type forecasts lead to entirely wrong answers, because the runoff volume from glaciers is inversely, not directly, related to the magnitude of the spring snowpack. New techniques need to be developed to cope with this problem.

The effect of glacier recession or advance on streamflow has important consequences for long-range water-resources planning. The present period of glacier recession happens to coincide with the "base" or "normal" period of record for many important hydrologic data. About 15 percent of the present summer flow of the Columbia River at the international border between the United States and Canada is derived from about 2,600 km^2 of glaciers in Canada, and perhaps one-third of this flow is from glacier wastage (Tangborn, 1980). Glacier wastage has been very large in the Susitna River basin of Alaska, where a major hydroelectric development is planned (Harrison and others, 1983). In this and similar areas, planning for the future should take the loss of water and ice storage into consideration. This is especially important now that the reality of the "greenhouse" effect is becoming accepted.

Glacier "catastrophes"

The interaction of the melt phase (water) with the solid phase (ice) of glaciers is interesting, as it can produce several kinds of instabilities that may result in large and rapid changes in the ice mass and great hazards to people. These hazards include sudden glacier advances (surges), rapid glacial retreats due to iceberg calving, floods caused by the rapid release of water from within or alongside a glacier, and rapid melting of ice on volcanos that produce mud and debris flows.

Surges. Most glaciers flow slowly and steadily, varying their flow rate gradually in response to climatic changes. Surge-type glaciers, on the other hand, have alternating phases of very rapid flow (surges) and normal slow flow (quiescence). The duration of the surge phase typically is from less than 1 to about 3 years, and the duration of the quiescent phase typically is from 10 to 100 years. During the surge, a 10- to 1,000-fold increase in flow rate may occur, and large ice displacements and increases in glacier length may result (Meier and Post, 1969).

The changes that occur during a glacier surge can be spectacular (Fig. 24). Surges may destroy transportation lines and structures; they also may block rivers and impound lakes, and may cause major floods when these blockages open.

The cause of the alternating cycle of surge and quiescence, and the mechanism of rapid flow, have long puzzled glaciologists. The 1983 surge of Variegated Glacier, Alaska, which ended abruptly when meltwater drained out and caused a flood, was carefully observed by a team of scientists. It now appears that the rapid acceleration of motion leading to a surge is caused by the blockage of drainage conduits under the glacier, resulting in a linked-cavity system of water at the bed of the glacier and consequent rapid sliding of the glacier over its bed (Kamb and others, 1985). Many aspects of glacier surging still need explanation, however. For example, how can the buildup of water and its effect on sliding be modeled? What causes the cyclic alternation of rapid and normal behavior? Why do surge-type glaciers occur in some areas (for example, in the St. Elias Mountains) and not in other areas (for example, adjacent Chugach Mountains)? And, could parts of the Antarctic Ice Sheet surge and cause sudden rises in sea level?

Calving glacier instability. Glaciers that end in the sea by discharging ice as icebergs may also exhibit alternating phases, but of very slow advance and very rapid retreat. The advance or retreat of these tidewater glaciers depends on the difference between the rates of ice flow toward the terminus and of iceberg discharge from the terminus. Post (1975) surmised that the rate of iceberg calving depended on water depth, and that this could cause anomalous or asynchronous advances and retreats. A glacier in a deep-water fjord might be protected from a high rate of calving if it terminated on a moraine shoal. Through erosion on the upstream face and deposition on the downstream face, the shoal would be moved down the fjord and the glacier would slowly advance. But if the glacier were to retreat off the shoal so that it terminated in deep water, iceberg calving would occur at such a high rate that the glacier would retreat rapidly and irreversibly until the terminus was again in shallow water, such as at the head of a fjord. Thus a history of one of these tidewater glaciers would show an alteration between long periods of slow advance and short periods of rapid retreat.

This scenario was predicted (Meier and others, 1980) and has been confirmed (Meier and others, 1985) at Columbia Glacier, Alaska (Fig. 14), and at other glaciers (Meier and Post, 1987). The changes in the glacier that occur when rapid retreat begins are impressive. The rate of ice flow at Columbia Glacier accelerated from 3.7 m/day to 15 m/day during the period 1977 to 1984, yet iceberg calving increased far more rapidly, causing the terminus to begin retreating at a rate of 1 km/yr (Meier and Post, 1985).

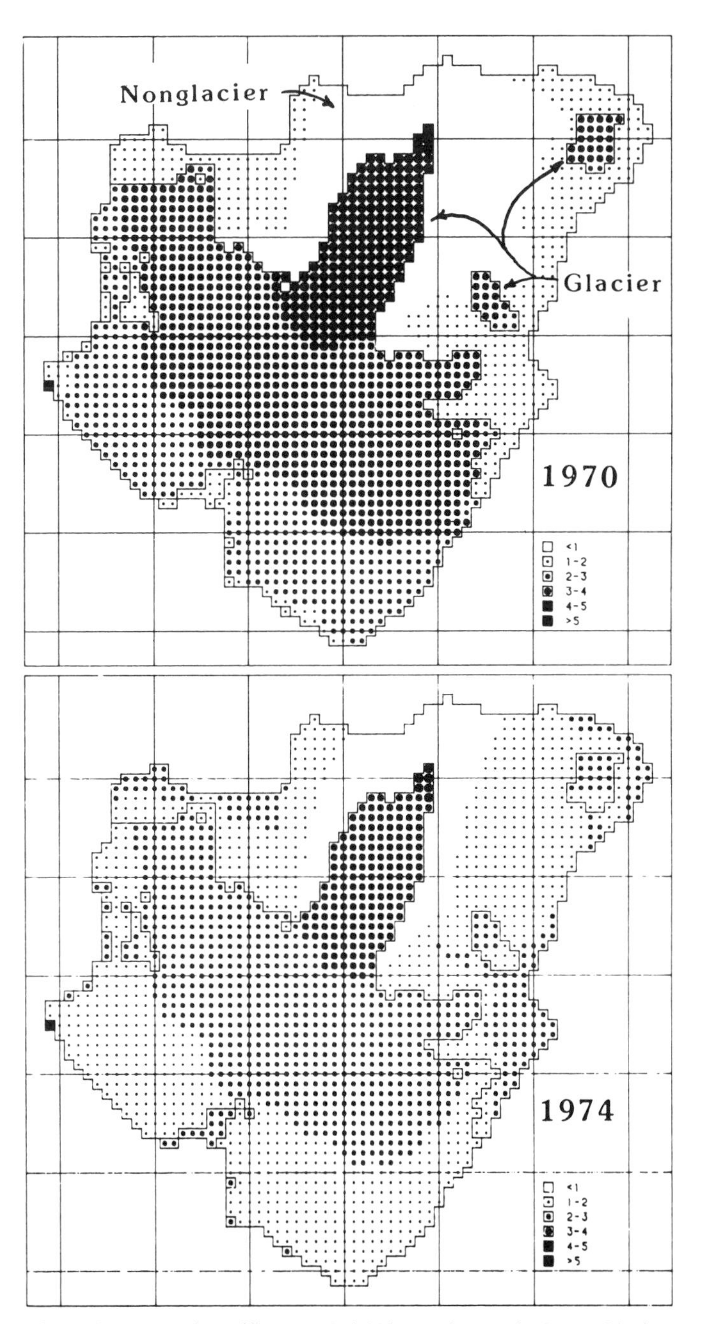

Figure 21. Maps of specific annual yield in m of water for Peyto Glacier, Alberta, and its drainage basin, in 1970 and in 1974. 1970 was a year of low winter snowfall and a warm summer, and 1974 was the opposite. In 1970, the nonglacier area produced a low specific yield (annual runoff in m^3 per m^2 of area), but the glacier produced a high specific yield; in 1974 the opposite was true. Also, specific yield increases with elevation off the glacier and decreases with elevation on the glacier. Area of glacier is separated from nonglacial area by a solid line. Values are shown on a 100 × 100 m grid. After Young (1982).

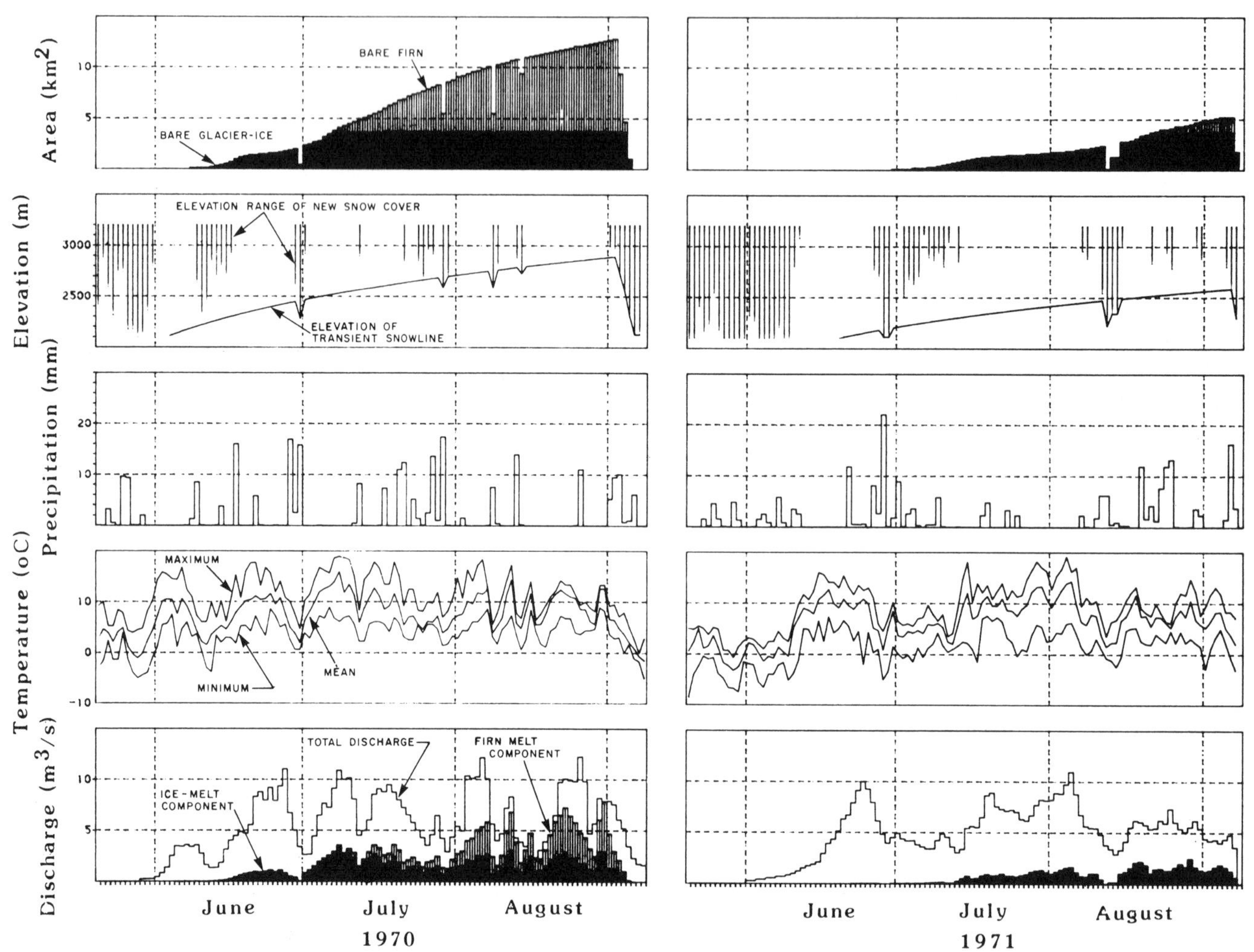

Figure 22. Change in snow-covered area, meteorology, and discharge for Peyto Glacier, Alberta, in 1970 and in 1974. The low amount of runoff from ice and firn melt in 1974 was a consequence of a delayed and slow rise in the snowline caused by heavy winter precipitation and relatively cool summer weather. After Young (1982).

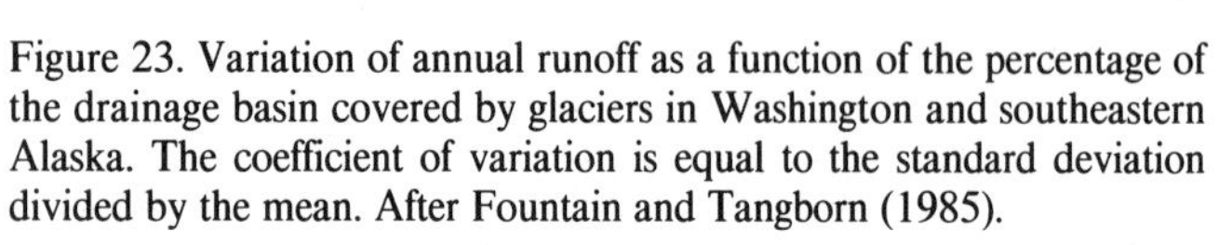

Figure 23. Variation of annual runoff as a function of the percentage of the drainage basin covered by glaciers in Washington and southeastern Alaska. The coefficient of variation is equal to the standard deviation divided by the mean. After Fountain and Tangborn (1985).

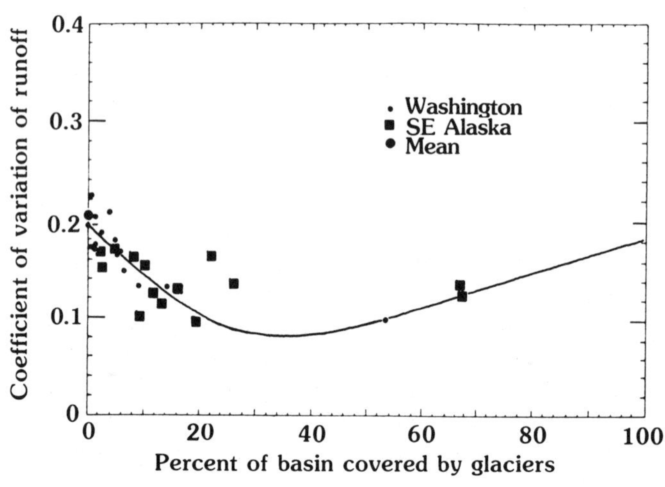

Figure 24. Two views of the 1983 surge of Variegated Glacier, Alaska. (A) June 4; (B) July 3. Most of the surge activity ended on July 4. Ice-flow velocity during the surge exceeded 60 m per day. Time-lapse photography by R. M. Krimmel, U.S. Geological Survey.

At least three implications of this glacial behavior deserve mention: (1) the advance and retreat of these glaciers depends on local conditions at the terminus and not on climate change, so one cannot infer climatic variations from study of their history; (2) when rapid retreat begins, there may be a huge increase in iceberg discharge, which could affect transportation as well as the local ecology; and (3) the steady advance of tidewater glaciers can block tributary fjords, turning them into freshwater lakes (such as the present situation at Russell Fjord/Hubbard Glacier) or suddenly release these lakes during the retreat phase.

Outburst floods and debris flows. Even in normal glaciers, the internal "plumbing" (englacial drainage channels) may become plugged, as a result of the movement of the ice, which causes water to be stored in or behind the glacier. Or, a glacier may block a valley, impounding a lake. Later, this water may be released as a sudden "outburst" flood, which is commonly known by the Icelandic term "jökulhlaup."

Outburst floods can be large (about 2,000 m^3/s on Nisqually River at Mt. Rainier, Washington in 1955), or even catastrophic in areas where large glaciers are present. Outburst floods are well known at Lake George (Knik Glacier) in Alaska; Tulsequah Lake (Tulsequah Glacier) in British Columbia; Summit Lake (Salmon Glacier) in British Columbia; and Hazard Lake (Steele Glacier) in the Yukon Territory (Post and Mayo, 1971; Clague and Mathews, 1973; Clarke, 1982). One of the largest glacier floods on record is the 1922 jökulhlaup from Grimsvotn, Iceland, which discharged about 7 km^3 of water in a 4-day period, producing a flood discharge estimated to be almost 60,000 m^3/s at its peak (Thorarinson, 1953).

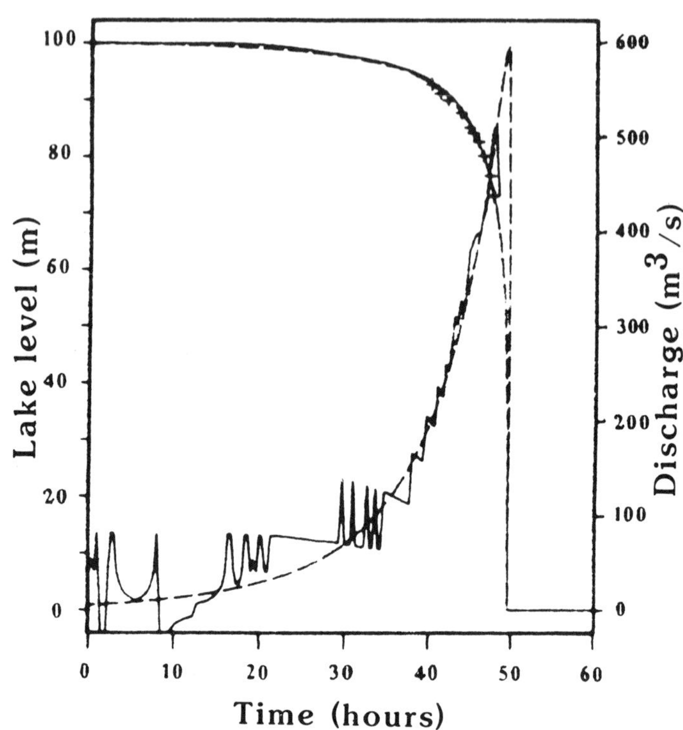

Figure 25. Discharge hydrograph and change in lake level during the outburst of a small glacier-dammed lake (Hazard Lake, Yukon Territory) in 1978. The saw-tooth component on the discharge curve is an artifact of numerical analysis. Dashed line is a model prediction. After Clarke (1982).

These floods may be caused by the release of water that had been contained within or under the glacier (the flood at Nisqually Glacier), or the release of a lake dammed by ice (the floods at Knik, Tulsequah, Salmon, and Steele Glaciers), or the outburst of accumulated meltwater because of volcanic heating (the floods at Grimsvotn). The impounded water, once it finds a minuscule channel, enlarges the channel at an ever-increasing rate as the heat produced by the loss of potential energy melts the channel walls. The flow increases at an exponential or near-exponential rate until nearly all the water is drained (Fig. 25; Bjornsson, 1975; Nye, 1976). One of the most ominous aspects of glacier outburst floods is that they cannot be predicted with the present state of knowledge.

Snow and ice on volcanos can be useful to other scientific studies by acting as a natural calorimeter, whereby changes in the ice can be used to monitor or measure the rate of heat released (Frank and others, 1977). Volcanic heat can rapidly melt ice if there is an actual eruption. The resulting flood may mobilize silt, sand, gravel, and larger objects and rush downvalley as a mudflow or debris flow. The source of water that caused the disastrous mud flows of the May 18, 1980, eruption of Mount St. Helens, Washington, has not been clearly identified, but enough water was contained in the glaciers that were destroyed to mobilize the mud flows (Brugman and Meier, 1981). To assess these debris-flow hazards, attention is now given to measuring the ice volume on active volcanos in Alaska, Washington, Oregon, and California (Dreidger and Kennard, 1984).

CONCLUSIONS

In a broad context, snow and ice are important to the global circulation of the atmosphere and ocean as well as to the global hydrologic cycle. If the amount and distribution of snow and ice were steady and unchanging, it would be a simple matter to ignore the role of snow and ice in the less-accessible high elevations and latitudes on the hydrologic cycle. But the ratio of the amount of liquid to solid water on Earth is ever changing, partly because of human activity, and the changes in these rates may have important societal consequences. We cannot understand what is happening or what is going to happen to the Earth's climate and hydrology without understanding snow and ice.

REFERENCES CITED

Anderson, E. A., 1973, National weather service river forecast system—Snow accumulation and ablation model: National Oceanic and Atmospheric Administration Technical Memorandum NWS HYDRO-17, 150 p.

Barnaby, I. E., 1980, Snowmelt observations in Alberta: Proceedings, Western Snow Conference 48th Annual Meeting, April 15–17, 1980, Laramie, Wyoming: Colorado State University, Ft. Collins, p. 128–137.

Benson, C. S., 1962, Stratigraphic studies in the snow and firn of the Greenland Ice Sheet: SIPRE (Snow, Ice, and Permafrost Research Establishment) Research Report 70.

Bjornsson, H., 1975, Explanation of jökulhlaups from Grimsvötn, Vatnajökull, Iceland: Jökul, v. 24, p. 1–26.

Brugman, M. M., and Meier, M. F., 1981, Response of glaciers to the eruptions of Mount St. Helens, *in* Lipman, P. W., and Mullineaux, D. R., eds., The eruptions of Mount St. Helens, Washington: U.S. Geological Survey Professional Paper 1250, p. 743–756.

Budd, W. F., Jacka, T. H., Jenssen, D., Radok, U., and Young, N. W., 1983, Derived physical characteristics of the Greenland Ice Sheet Mk I: University of Melbourne Meteorology Department, Publication No. 23, 103 p.

Budyko, M. E., and Kondratiev, K. Y., 1964, The heat balance of the Earth, *in* Odishaw, H., ed., Research in geophysics 2—Solid earth and interface phenomena: Cambridge, Massachusetts Institute of Technology, p. 529–554.

Clague, J. J., and Mathews, W. H., 1973, The magnitude of jökulhlaups: Journal of Glaciology, v. 12, no. 66, p. 501–504.

Clarke, G.K.C., 1982, Glacier outburst floods from "Hazard Lake," Yukon Territory, and the problem of flood magnitude prediction: Journal of Glaciology, v. 28, no. 98, p. 3–21.

Colbeck, S. C., 1972, A theory of water percolation in snow: Journal of Glaciology, v. 11, no. 63, p. 369–385.

—— , 1978, The physical aspects of water flow through snow: Advances in Hydroscience, v. 11, p. 165–206.

Colbeck, S. C., and Davidson, G., 1973, Water percolation through homogeneous snow, *in* The role of snow and ice in hydrology: Proceedings of the Banff Symposium, UNESCO and World Meteorological Organization, p. 242–256.

Colbeck, S. C., Meier, M. F., Benson, C., and others, 1979, Focus on U.S. snow research: University of Colorado, Boulder, World Data Center for Glaciology Glaciological Data Report GD–6, p. 41–52.

Dickson, R. R., and Posey, J., 1967, Maps of snow-cover probability for the Northern Hemisphere: Monthly Weather Review, v. 95, p. 347–353.

Driedger, C. L., and Kennard, P. M., 1984, Ice volumes on Cascade Volcanoes; Mount Rainier, Mount Hood, Three Sisters, and Mount Shasta: U.S. Geological Survey Open-File Report 84–851, 55 p.

Fountain, A. G., and Tangborn, W. V., 1985, The effect of glaciers on streamflow variations: Water Resources Research, v. 21, no. 4, p. 579–586.

Frank, D., Meier, M. F., and Swanson, D. A., 1977, Assessment of increased thermal activity at Mount Baker, Washington, March 1975–March 1976, with contributions by Babcock, J. W., Fretwell, M. O., Malone, S. D., Rosenfeld, C. L., Shreve, R. L., and Wilcox, R. E.: U.S. Geological Survey Professional Paper 1022–A, 49 p.

Fristrup, B., 1966, The Greenland Ice Sheet: Seattle, University of Washington Press, 312 p.

Granger, R. J., and Male, D. H., 1978, Melting of a prairie snowpack: Journal of Applied Meteorology, v. 17, no. 12, p. 1833–1842.

Gray, D. M., and others, 1979, Snow accumulation and distribution, *in* Proceedings, Modeling of Snow Cover Runoff, Hanover, New Hampshire, September 26–28, 1978: U.S. Army Cold Regions Research and Engineering Laboratory, p. 3–33.

Hall, D. D., and Martinec, J., 1985, Remote sensing of ice and snow: London and New York, Chapman and Hall, 189 p.

Harrison, W. D., Drage, B. T., Bredthauer, S., and others, 1983, Reconnaissance of the glaciers of the Susitna River basin in connection with proposed hydroelectric development, *in* Proceedings of the 2nd Symposium on Applied Glaciology, New Hampshire, August 23–27, 1982: Annals of Glaciology, v. 4, p. 99–104.

Hubley, R. C., 1957, An analysis of surface energy during the ablation season on Lemon Creek Glacier, Alaska: EOS Transactions of the American Geophysical Union, v. 38, no. 1, p. 68–85.

Johannessen, M., Dale, T., Gjessing, E. T., and others, 1977, Acid precipitation in Norway; The regional distribution of contaminants in snow and the chemical concentration processes during snowmelt; Symposium, Isotopes et impuretes dans les neiges et glaces, Actes du colloque de Grenoble, aout/septembre 1975: International Association of Hydrological Sciences Publication 118, p. 116–120.

Kamb, B., Raymond, C. F., Harrison, W. D., and others, 1985, Glacier surge mechanism: 1981–1983 surge of Variegated Glacier, Alaska: Science, v. 227, no. 4686, p. 469–479.

Kirk, R., 1980, Snow: New York, Morrow Quill Paperbacks, 320 p.

Koerner, R. M., 1985, Canadian arctic islands: glacier mass balance and global sea level, *in* Glaciers, ice sheets, and sea level: effects of a CO2-induced climatic change: Washington, D.C., National Academy Press, Report of a workshop, September 13–15, 1984, Seattle, Washington, p. 145–154.

—— , 1989, Queen Elizabeth Islands Glaciers, *in* Fulton, R. J., ed., Quaternary Canada and Greenland: Ottawa, Ontario, The Geological Survey of Canada, The Geology of North America, v. K-1 (in press).

Langway, C. C., Jr., Oeschger, H., and Dansgaard, W., eds., 1985, Greenland Ice Core—Geophysics, geochemistry, and the environment: American Geophysical Union Geophysical Monograph 33, 119 p.

Ledley, T.A.S., 1985, Sensitivity of a thermodynamic sea ice model with leads to time step size: Journal of Geophysical Research, v. 90, no. D1, p. 2251–2260.

Male, D. H., and Granger, R. J., 1981, Snow surface energy exchange: Water Resources Research, v. 17, no. 3, p. 609–627.

McKay, G. A., 1981, Snow and Living Things; Section 1, Snow and Man, *in* Gray, D. M., and Male, D. H., eds., Handbook of snow; Principles, Processes, Management, and Use: New York, Pergamon Press, p. 3–31.

McKay, G. A., and Gray, D. M., 1981, The Distribution of Snowcover, *in* Gray, D. M., and Male, D. H., eds., Handbook of Snow; Principles, Processes, Management, and Use: New York, Pergamon Press, p. 153–190.

Meier, M. F., 1969, Glaciers and water supply: American Water Works Association Journal, v. 61, no. 1, p. 8–12.

—— , 1984, Contribution of small glaciers to global sea level: Science, v. 226, no. 4681, p. 1418–1421.

Meier, M. F., and Post, A., 1969, What are glacier surges?: Canadian Journal of Earth Sciences, v. 6, no. 4, p. 801–817.

—— , 1985, Fast tidewater glaciers: Journal of Geophysical Research, v. 92, no. B9, p. 9051–9058.

Meier, M. F., Tangborn, W. V., Mayo, L. R., and Post, A., 1971, Combined ice and water balances of Gulkana and Wolverine Glaciers, Alaska, and South Cascade Glacier, Washington, 1965 and 1966 Hydrologic Years: U.S. Geological Survey Professional Paper 715–A, 23 p.

Meier, M. F., Rasmussen, L. A., Post, A., Brown, C. S., Sikonia, W. G., Bindschadler, R. A., Mayo, L. R., and Trabant, D. C., 1980, Predicted timing of the disintegration of the lower reach of Columbia Glacier, Alaska: U.S. Geological Survey Open-File Report 80–582, 58 p.

Meier, M. F., Rasmussen, L. A., and Miller, D. S., 1985, Columbia Glacier in 1984; Disintegration underway: U.S. Geological Survey Open-File Report 85–81, 17 p.

Müller, F., 1962, Zonation in the accumulation area of the glaciers of Axel Heiberg Island, N.W.T., Canada: Journal of Glaciology, v. 4, no. 33, p. 302–311.

National Academy of Sciences, Carbon Dioxide Assessment Committee, 1983, Changing Climate: Washington, D.C., National Academy Press, 496 p.

National Academy of Sciences, Committee on Glaciology, 1983, Snow and ice research—An Assessment: Washington, D.C., National Academy Press, 126 p.

——, 1985, Glaciers, ice sheets, and sea level; Effect of a CO_2-induced climatic change, Report of Workshop, September 13–15, 1984, Seattle, Washington: Washington, D.C., National Academy Press, 330 p.

Nye, J. F., 1976, Water flow in glaciers—jökulhlaups, tunnels, and veins: Journal of Glaciology, v. 17, no. 26, p. 181–207.

Paterson, W.S.B., Koerner, R. M., Fisher, D., Johnsen, S. J., Clausen, H. B., Dansgaard, W., Bucher, P., and Oeschger, 1977, An oxygen-isotope climatic record from the Devon Island ice cap, Arctic Canada: Nature, v. 266, no. 5602, p. 508–511.

Post, A., 1975, Preliminary hydrography and historical terminus changes of Columbia Glacier, Alaska: U.S. Geological Survey Hydrologic Investigations Atlas HA 559, 3 sheets.

Post, A., and Mayo, L. R., 1971, Glacier dammed lakes and outburst floods in Alaska: U.S. Geological Survey Hydrologic Investigations Atlas HA 455, 3 sheets, 10 p.

Radok, U., Barry, R. G., Jenssen, D., Keen, R. A., Kiladis, G. N., and McInnes, B., 1982, Climatic and Physical Characteristics of the Greenland Ice Sheet, Parts 1 and 2: Boulder, University of Colorado Cooperative Institute for Research in Environmental Sciences, 193 p.

Reeh, N., 1985, Greenland ice-sheet mass balance and sea-level change, *in* Glaciers, Ice sheets, and sea level: effect of a CO_2-induced climatic change: Washington, D.C., National Academy Press, Report of a Workshop held in Seattle, Washington, 1984, p. 155–171.

——, 1989, Greenland Ice Sheet, *in* Fulton, R. J., Heginbottom, J. A., and Funder, S., eds., Quaternary geology of Canada and Greenland: Ottawa, Geological Survey of Canada, Geology of Canada series, v. 1 (Geological Society of America, The Geology of North America, v. K-1) (in press).

Ropelewski, C. F., 1986, Snow Cover in Real Time Climate Monitoring, *in* Kukla, G., Barry, R. G., Hecht, A, and Wiesnet, D., eds., Proceedings, Snow Watch '85, October 1985: World Data Center A for Glaciology (Snow and Ice) Glaciological Data Report DG-18, p. 105–108.

Schemenauer, R. S., Berry, M. O., and Maxwell, J. B., 1981, *in* Gray, D. M., and Male, D. H., eds., Handbook of Snow: Principles, Processes, Management, and Use: New York, Pergamon Press, p. 129–152.

Shafer, B. A., Jensen, D. T., and Jones, K. C., 1984, Analysis of 1983 snowmelt runoff production in the upper Colorado River Basin, *in* Proceedings, Western Snow Conference, 52nd Annual Meeting, April 17–19, 1984, Sun Valley, Idaho: Ft. Collins, Colorado State University, p. 1–11.

Steppuhn, H., 1981, Snow and agriculture, *in* Gray, D. M., and Male, D. H., eds., Handbook of Snow: Principles, Processes, Management, and Use: New York, Pergamon Press, p. 60–125.

Tangborn, W. V., 1980, Contribution of glacier runoff to hydroelectric power generation on the Columbia River: Data of Glaciological Studies, *in* Proceedings of the International Symposium on Computation and Forecasts of the Runoff from Glaciers and Glacierized Areas (Tbilisi, 1978): Moscow, Academy of Sciences of the U.S.S.R., Section of Glaciology of the Soviet Geophysical Committee and Institute of Geography Publication 39, p. 62–67, 140–143.

Thorarinsson, S., 1953, Some new aspects of the Grimsvotn problem: Journal of Glaciology, v. 2, no. 14, p. 267–274.

Walsh, J. E., and Ross, B., 1985, Snow Cover, Cyclogenesis, and Cyclone Trajectories, *in* Kukla, G., Barry, R. G., Hecht, A., and Wiesnet, D., eds., Proceedings, Snow Watch '85, October 1985: World Data Center A for Glaciology (Snow and Ice) Glaciological Data Report DG-18, p. 23–35.

Wankiewicz, A., 1979, A review of water movement in snow, *in* Colbeck, S. C., and Ray, M., eds., Proceedings, Modeling of Snow Cover Runoff, September 1978: Hanover, New Hampshire, U.S. Cold Regions Research and Engineering Laboratory, p. 222–252.

Weidick, A., 1975, A review of Quaternary investigations in Greenland; Published for the Geological Survey of Greenland, Copenhagen, Denmark: Columbus, Ohio State University Institute of Polar Studies Report 55, 161 p.

——, 1985, Review of glacier changes in west Greenland: Climate and Paleoclimate of Lakes, Rivers, and Glaciers: Zeitschrift für Gletscherkunde und Glazial Geologie, v. 21, p. 301–309.

Willis, W. O., 1979, Snow on the Great Plains, *in* Colbeck, S. C., and Ray, M., eds., Proceedings, Modeling of Snow Cover Runoff, September 1978: Hanover, New Hampshire, U.S. Cold Regions Research and Engineering Laboratory, p. 56–62.

Young, G. J., 1982, Hydrological relationships in a glacierized mountain basin, *in* Hydrological Aspects of Alpine and High Mountain Areas, Proceedings of the Exeter Symposium, July, 1982: International Association of Hydrological Sciences Publication 138, p. 51–59.

——, 1985, Introduction; The need for predictive techniques, *in* Young, G. S., ed., Techniques for prediction of runoff from glacierized areas: International Association of Hydrological Sciences Publication 149, p. 3–23.

Manuscript Accepted by the Society November 17, 1988

ACKNOWLEDGMENTS

B. E. Goodison, W. T. Pfeffer, H. C. Riggs, B. A. Schafer, M. G. Wolman, and G. J. Young provided helpful comments on this manuscript, but are not to be blamed for any inadequacies or misinterpretations.

The Geology of North America
Vol. O-1, Surface Water Hydrology
The Geological Society of America, 1990

Chapter 8

Hydrology of lakes and wetlands

Thomas C. Winter
U.S. Geological Survey, MS-413, Box 25046, Denver Federal Center, Denver, Colorado 80225
Ming-Ko Woo
Department of Geography, McMaster University, Hamilton, Ontario L8S 4M1, Canada

HYDROGEOLOGIC FACTORS THAT AFFECT THE DISTRIBUTION OF LAKES AND WETLANDS

The existence of lakes and wetlands depends on the specific geologic setting that favors the ponding of water, and on the hydrologic processes that allow the body of water to persist at a given site. Lakes can occur only in topographic depressions, but wetlands occur in depressions, on flat areas, on slopes, and even on drainage divides. Lakes and wetlands have some common characteristics, but they differ in many aspects of water storage, water circulation, water loss to the atmosphere, and the thermal and chemical characteristics of their waters.

Not only do lakes and wetlands differ in many respects from each other, but there are also many differences within each group. Furthermore, because of the complex physical, chemical, and biological processes within lakes and wetlands, many of which are poorly understood, attempts to classify lakes or wetlands or both have a number of shortcomings. As an example, wetland classifications have been developed separately in Canada (Zoltai and Pollett, 1983) and the United States (Cowardin and others, 1979); the classifications cannot easily be unified, nor are they based entirely on hydrologic considerations. Because this report concentrates on geologic and hydrologic aspects of lakes and wetlands, the geologic settings in which lakes and wetlands are located are discussed first, followed by a discussion of the hydrologic processes controlling the flux of water to and from lakes and wetlands.

Geologic settings

In an extensive discussion of the origin of lake basins, Hutchinson (1957) presents a classification including 76 types. Most types are related to geologic processes, but a few types are related to biological processes. The 76 types are grouped according to 11 general processes: tectonic, volcanic, landslide, glacial, solution, fluvial, eolian, shoreline, organic accumulation, animal activity, and meteorite impact. Because the classification is of lakes only, it does not include some geologic settings favorable to formation of wetlands. Therefore, this chapter discusses geologic settings according to the following major categories: topographic depressions, slope discontinuities, subsurface stratigraphy, and permafrost.

Topographic depressions. All lakes and many wetlands occur in topographic depressions. The greatest number of lakes and types of lakes occur in glacial terrain. Of Hutchinson's (1957) 76 lake types, 20 are related to topographic depressions in the glacial-process group. The other topographic depressions most favorable for formation of lakes and some wetlands are those related to damming and to subsidence.

Glacial scour and deposition. Glacial erosion and deposition are responsible for many topographic depressions. Joints, faults, and bedding planes are enlarged by glacial action which removes blocks of rock, thereby creating lake basins. Lakes in such basins have irregular shorelines that reflect the structural pattern of the region. For example, the glaciated Canadian Shield has an enormous number of lakes linked by intricate networks of surface drainage. Glacial scour has produced this great number of lakes, which includes the greatest number of the world's largest lakes (Herdendorf, 1982). Some of the shallower lakes have subsequently been filled with sediments, but they remain saturated and sustain wetlands known as muskegs (Robison and others, 1962).

Glacial ice can sometimes erode preexisting valleys to great depths. Large lakes such as Great Bear Lake or Great Slave Lake, Northwest Territories, or Lake Winnipeg, Manitoba, were created when old valleys were deepened by glacial erosion. More spectacular examples are the Great Lakes, which were enlarged by the glacial scouring of a preexisting drainage system (Flint, 1971). The Great Lakes were even larger in the past, but altered drainage patterns and isostatic rebound of the Earth's crust decreased the lakes to their present size.

In mountainous areas, cirque lakes occur at valley heads. Where a chain of depressions has been eroded along a valley, paternoster lakes result. In some fiord areas, such as the coast of British Columbia, glaciers deepen elongated valleys at low eleva-

Winter, T. C., and Woo, M.-K., 1990, Hydrology of lakes and wetlands, *in* Wolman, M. G., and Riggs, H. C., eds., Surface water hydrology: Boulder, Colorado, Geological Society of America, The Geology of North America, v. O-1.

tions, and deep lakes fringed by steep mountain walls may result if the valleys are enclosed at the seaward end. These are known as fiord lakes (Hutchinson, 1957), but the name also applies to inland lakes that were formed by glacial erosion of elongated valleys; examples include lakes in the Pacific Northwest such as Lakes Okanagan, Shuswap, Kootenay, and the Arrow Lakes of British Columbia.

The great number of lakes in glacial terrain that occur in areas of glacial deposits were initially formed in kettle holes, depressions caused by the melting of buried ice masses. Kettle-hole lakes are common on glaciated terrain skirting the Canadian Shield in the north-central United States and the prairie provinces of Canada. They are also common in glacial terrain within mountainous areas such as the Rocky Mountains and the northern Appalachian Mountains. Other glacial-lake basins are formed by damming of valleys, as discussed in the following section.

Damming of drainage. Glacial processes: Glaciers can directly impound lakes when the ice flows up or across a valley or lowland. Glacier-dammed lakes may be partly or completely drained by conduits through the ice dam, causing large floods known by the Icelandic term jökulhlaups (Gilbert, 1971).

Fluvial processes related to glaciers can form lakes in river valleys. For example, Lake Pepin, in Minnesota and Wisconsin, was formed in the Mississippi River Valley as a result of a fan deposited by discharge from a large tributary. Conversely, aggradation by the main river can cause lakes to form in tributary valleys. This occurred along the Mississippi River during the late Pleistocene. In mountainous areas, lateral moraines in large glacial valleys may function as dams and form lakes in smaller side valleys.

Landslides: Landslide dams form in a variety of physiographic settings. The most common types are rock slumps, mud flows, and rock and debris avalanches. The most common mechanisms to initiate such landslides are earthquakes and excessive rainfall and snowmelt. Most landslide dams fail shortly after they form. Costa and Schuster (1988) indicated that, of 63 documented cases of such dams, about half failed within 10 days. However, some landslide-dammed lakes have persisted for many years, such as Hebgen Lake, Montana, west of Yellowstone National Park.

Tectonic activity: Tectonic activity has formed many lake basins. Fault blocks can be down-thrown or tilted to form the geologic dam. Examples of these types of lakes are common in the Basin and Range province in the western United States. Three of the largest lakes in the United States are in grabens: Lake Tahoe, California and Nevada; Pyramid Lake, Nevada; and Great Salt Lake, Utah. Reelfoot Lake, Tennessee, was also formed by tectonism.

Riverine bars: In many river valleys, shifting of the main channel and formation of riverine bars form dams that cause formation of oxbow lakes. Lake Chicot, Arkansas, is a well-studied example of this type of lake (Shiebe and others, 1984).

Eolian processes: Lakes and wetland basins are common in some dune fields. Hundreds of lakes and wetlands occur in the interdunal lowlands in the Nebraska sandhills, in coastal areas along the oceans, and along the Great Lakes.

Coastal processes: Marine processes can create lagoons as embayments, inlets, or estuaries are sealed off by long-shore deposits. Where the lagoons are still linked to the sea, saltwater intrusion commonly causes brackish-water conditions, but when the lagoon mouth is completely sealed, river and precipitation input commonly causes these lakes to be fresh. Wetlands can develop from shallow coastal lakes, after vegetation growth and organic deposition. The Dismal Swamp, Virginia and North Carolina, is an example of a forested wetland of this type; the peat is generally more than 1.5 m thick, and it is interspersed with open water (Kirk, 1979).

Subsidence. Depressions can be created by land subsidence. Dolines, commonly called sinkholes, are formed by the collapse of overlying rock and soil into caverns produced by dissolution of carbonate rocks. Many lakes in Florida, which commonly have circular outlines, originated as solution lakes. Occasionally, further solution and subsidence along lines of structural weakness cause the formation of large, steep-sided depressions called poljes. In some cases, dry poljes and smaller depressions may be temporarily flooded as runoff from intense rainstorms causes the discharge of large quantities of surface water and ground water to the depressions (Brook and Ford, 1980).

In volcanic areas, subsidence of volcanic craters produces calderas which, when filled with water, form approximately circular lakes that have steep sides. One well-known example is Crater Lake, Oregon, which has a surface area of 64.4 km^2 and a depth of 608 m.

Slope discontinuities. Studies of ground-water flow systems have led to considerable insight into the mechanisms of ground-water recharge and discharge (Toth, 1963, 1966; Meyboom, 1966; Freeze, 1969, Lissey, 1971; Winter, 1976). Modeling studies have been useful particularly in developing understanding of the variety of geologic boundaries that can cause ground-water discharge.

In areas of steep land slopes, such as embankments or river-valley walls, the water table sometimes intersects the land surface, and ground water discharges directly to the land surface in the lower parts of the steep slope. The constant supply of ground-water seepage keeps the soil saturated, and permits aquatic plants to grow in the lower parts of the slopes.

Modeling studies have further shown that, where the water table has an upward break in slope, ground water moves upward toward the land surface in the part that has the lower slope. For example, where steeply sloping end moraines meet flatter glacial-lake plains, till plains, or outwash plains, ground water moves upward toward the water table in the plain and the flow is concentrated near the valley side (Fig. 1A). This same pattern of ground-water movement also occurs in flood plains that have steeply sloping, adjacent valley sides.

Subsurface stratigraphy. Two types of subsurface stratigraphy affect the formation of wetlands; the stratigraphy of the peat itself and the stratigraphy of the underlying mineral sub-

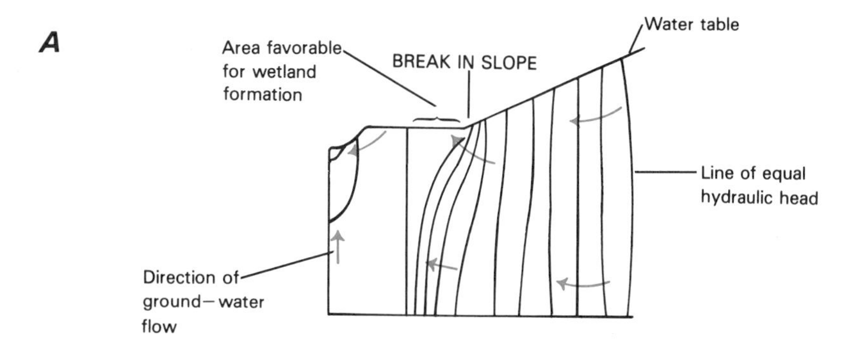

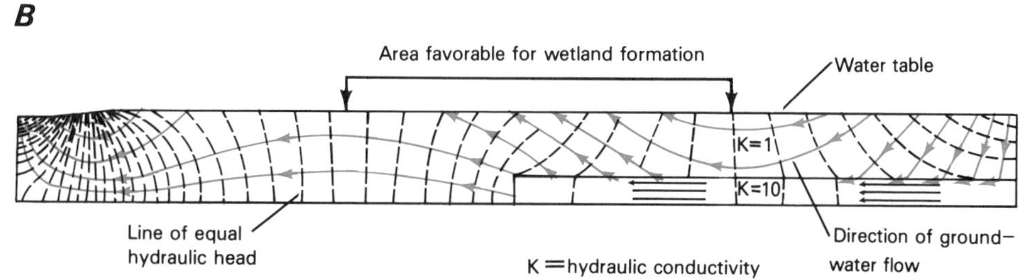

Figure 1. Formation of wetlands caused by ground-water discharge associated with: (A) break in slope of water table (modified from Winter, 1976), and (B) variation in subsurface stratigraphy (modified from Freeze, 1969). Land surface is not shown in these diagrams.

strate. Wetlands commonly occur in areas with impeded drainage. Impeded drainage produces minimal oxygen concentrations in which decomposition is greatly decreased, allowing the accumulation of dead organic matter as peat (Hofstetter, 1983). The rate of peat accumulation is controlled by the rate of vegetation production minus the rate of its decomposition. In the Arctic, both vegetation growth and decomposition are slow; in the tropics, both vegetation growth and decomposition are relatively rapid. Therefore, it is probable that the maximum thickness of peat occurs in the intermediate temperate zone.

As peat thickness increases, the lower layers become progressively compacted and humified, thus altering their hydraulic characteristics. In most peat deposits, porosity and hydraulic conductivity decrease with depth (Boelter, 1965, 1969; Radforth, 1977). The surface zone, known as the acrotelm (Ingram, 1983), is more permeable than deeper zones. The lower zone, known as the catotelm, generally permits little water to be stored and transmitted.

Unlike most other mineral substrates, peat volume decreases after drying, and the original physical properties are not fully restored after resaturation. Therefore, the hydrologic characteristics of peatlands may be irreversibly altered when the area is drained.

Subsurface stratigraphy within underlying mineral substrates affects ground-water movement. In complex, heterogeneous geologic terrain, upward ground-water movement may occur where permeable rocks pinch out, or where permeable rocks have a different dip relative to adjacent less-permeable rocks. Freeze (1969) and Winter (1976) provide several examples of ground-water movement toward the land surface within these geologic configurations (Fig. 1B).

Permafrost. Permafrost, defined as earth material that has a temperature continuously below 0°C for at least 2 consecutive years, is commonly considered impervious for practical purposes. Where permafrost is continuous, this impervious substrate causes water to be impounded in land-surface depressions.

Where permafrost is subject to local melting, the areas of ice loss result in depressions. This process creates thermokarst lakes, many of which begin as shallow ponds formed inside ice-wedge polygons. These ponds may enlarge and coalesce, and eventually expand into lakes.

Oriented lakes occur in northern Alaska (Carson and Hussey, 1962) and in the Tuktoyatuk Peninsula, and Cape Bathurst on the northern shore of Arctic mainland Canada (Mackay, 1956). The lakes are elliptical, rectangular, or irregular in outline, and their long axes are oriented perpendicular to the prevailing winds; this orientation is probably a result of wave erosion.

In areas of continuous permafrost, lake beds are commonly separated from the regional ground-water system beneath the permafrost zone. In areas of discontinuous permafrost, ground thaw can establish subsurface links between the lake water and the regional ground water. In some cases, a lake may contain water derived from outside the surficial drainage basin. Hartman and Carlson (1973) indicated that the presence of lakes can increase ground thaw in central Alaska; if they are connected to the regional ground-water flow systems, seepage to the lakes is greatly increased.

Wetland formation also can be favored by the presence of permafrost. In the northern coastal plains of the Yukon Territory and Alaska, precipitation exceeds evaporation, and extensive wetlands occur. However, for most parts of the Arctic islands, precipitation is minimal and wetlands form in pockets that have abundant local water supply; for example, below late-lying snowbanks, in front of glaciers, or at the base of concave slopes.

Hydrologic processes related to the water balance of lakes and wetlands

A favorable geologic setting is only one of the requirements for the existence of a lake or wetland. It also is necessary that an adequate and persistent supply of water be available. Lakes and wetlands are dynamic, continuously receiving and losing water through interchange with the atmosphere, streams, and ground water. Furthermore, the relative importance of each of these components in maintaining the water supply is variable in time and space. The water balance of a prairie lake might be controlled largely by interchange with atmospheric water (precipitation and evaporation), a riverine lake by streamflow in and out, and a lake in permeable rocks by ground-water seepage in and out.

Most Alaskan and Canadian lakes and wetlands receive precipitation as both snowfall and rainfall. The contribution from snow decreases southward, and becomes insignificant in the southern United States and Mexico, except in the high mountains. Evaporation accounts for a sizable water loss in some parts of North America, but in northern latitudes the presence of winter ice cover decreases the evaporation season to only several months each year.

Depending on the geologic setting, lakes have a considerable range of storage capacity. Water storage in wetlands is underground or in surface depressions. When the water table is low, considerable storage capacity is available in the nonsaturated peat. When the peat is saturated, only surface depressions provide storage; if these are also filled, any additional water will run off the wetland quickly. Thus, during certain periods in the dry season, wetlands are effective in retarding or preventing runoff, but water is generally not retained for a sufficiently long period to regulate seasonal streamflow.

Interaction with atmospheric water. For most lakes and wetlands, precipitation and evaporation or transpiration, or both, are the major components of water gain and loss. The only common exceptions are those lakes that have large streams entering and leaving. Therefore, an understanding of the distribution of various aspects of climate (Lins and others, this volume) is useful in understanding the distribution of lakes in North America.

Large quantities of precipitation correspond with the windward slopes of large mountain ranges where orographic effects are substantial, and with regions where air-mass frontal activities are pronounced. Tropical areas, which have frequent convectional storms, also receive large quantities of rainfall. Conversely, precipitation is minimal in continental interiors where the atmosphere is dry. The precipitation map of North America shows that maximum quantities occur on the western slopes of high mountain ranges in the West. A secondary belt of substantial precipitation parallels the east coast of North America. In contrast, the driest part of North America is the desert in the southwestern United States and northern Mexico.

Snow falling on drainage basins containing wetlands and lakes does not provide water recharge until the melt season. Large quantities of rainfall and snowmelt invariably raise lake levels and commonly flood wetlands.

Lakes and many wetlands have the maximum potential supply of moisture available for evaporation and transpiration. Because of this, the potential water loss to the atmosphere is affected largely by the availability of energy for evaporation. Therefore, actual evaporation and transpiration from lakes and wetlands are controlled largely by wind and vapor fluxes above the lake or wetland. Areas that have persistent winds, warm water, and dry overlying air masses have maximum evaporation and transpiration. Therefore, water loss to the atmosphere is greatest in desert regions and least in cold regions of the north and in high mountains.

In colder regions, transpiration—and especially evaporation—are further retarded by the duration of the snow and ice cover that intercepts the energy fluxes. In addition, substantial variations in evaporation can occur daily. Evidence of active evaporation is reflected in water-level changes that commonly have diurnal steplike declines in dry periods, indicating faster evaporation rates during the day and slower rates at night (Dolan and others, 1984). To summarize the general gains and losses of atmospheric water with respect to surface water in North America, a map showing the distribution of the balance of precipitation minus lake evaporation is given in Plate 2.

Interaction with surface water. Streams and lakes have somewhat similar relations with respect to the atmosphere. Therefore, like lakes, streams are more common where precipitation exceeds evaporation, and lakes in these wetter regions have more interaction with streams. Geologic terrain also has a significant effect on characteristics of streamflow (Riggs and Harvey, this volume). Streams flowing on relatively impervious rocks, or permafrost, have variable discharge, whereas streams flowing on permeable rocks have lower peak flows and more sustained base flow. These streamflow characteristics are reflected in patterns of lake-level fluctuations.

For some soil conditions, water flows in sheets on the land surface, but surface water is more commonly channelized in streams and gullies. Surface water that reaches lakes mixes with lake water and is stored and subsequently evaporated or discharged through the outlet or as seepage to ground water, or both. In wetlands, surface flow may seep underground, be retained in hollows, or continue to travel across the wetlands in sheets or along channels.

Lakes and wetlands commonly retain part of surface inflow

and release the water during an extended period. The resulting outflow hydrographs are generally smooth, and the time of peak flow lags behind the initial peak runoff into the lake or wetland. The degree of flow modification related to storage depends on the size of the lake, the extent of the wetland, and the characteristics of flow through them. Where streams disappear as they enter the wetland, but reappear at lower elevation, there is thorough mixing of surface water with the water in the acrotelm of the wetland, and the flow pattern is greatly modified. In contrast, streams flowing through a wetland along well-defined channels have less exchange with ground water and the streamflow regime is little changed by the wetland (Woo and Valverde, 1981).

Interaction with ground water. Studies of ground-water flow near surface water have indicated that seepage is controlled by such nondynamic factors as overall geometry and complexity of the geologic framework, anisotropy of the porous media, and depth of the surface water (Winter, 1976) and by dynamic factors such as the continually changing configuration of the water table (Winter, 1983). These studies indicate that movement of ground water to and from surface water is affected by the magnitude of the underlying local flow systems. A local ground-water flow system includes the water that recharges at a water-table high and discharges in an adjacent lowland, which commonly is occupied by a lake or wetland. Local flow systems are bounded by flow lines that separate the local flow system from underlying intermediate or regional flow systems. If the local flow systems are small, they may not completely enclose the lake or wetland, and water will seep out of the lake or wetland in the part of the bed that does not receive seepage inflow from the local flow systems.

If the local ground-water flow system is large enough to completely enclose the lake or wetland, water cannot seep out of the lake or wetland because hydraulic heads within the local flow system are greater than the hydraulic head represented by the lake level (Winter, 1978).

Therefore, understanding the interaction of surface water with ground water requires understanding the dynamic character of the distribution of hydraulic head in the contiguous ground-water system. This distribution of hydraulic head is directly related to infiltration and water movement through the unsaturated zone. Therefore, to thoroughly understand the interaction of ground water and surface water, the entire subsurface-water system (including infiltration and water movement through the unsaturated zone, movement into and through the ground-water zone, and movement through the lake or wetland bed) needs to be considered as a single continuous system.

Recent field and numerical-modeling studies of variably saturated subsurface flow have indicated that ground water moves into surface water more quickly than previously believed. These studies indicate that ground-water mounds, or ridges, form rapidly directly adjacent to surface water immediately after infiltration (Winter, 1983). This focused recharge could account for the rapid chemical changes detected in surface water after precipitation or snowmelt (Sklash and Farvolden, 1979).

From the above studies it is apparent that climate is a major factor affecting the ground-water component of lake and wetland water balances. For example, because of the dynamic response of ground-water flow systems to variations in time and space of recharge, which is related to climate, lakes and wetlands in exactly the same geologic settings could have different interaction with ground water in different climatic settings.

DISTRIBUTION AND HYDROLOGIC CHARACTERISTICS OF LAKES AND WETLANDS IN NORTH AMERICA

Lakes occupy about 8 percent of the surface area of Canada (0.7×10^6 km^2). The surface area of lakes in the United States and Mexico has not been determined, but it is much less than Canada. Areas that have numerous lakes include the low Arctic, Canadian Shield, Atlantic provinces of Canada, northeastern United States, north-central United States, and Florida (Fig. 2). Many very large lakes occur in North America. In a survey of the world's largest natural lakes, defined as being greater than 500 km^2 in surface area, Herdendorf (1982) indicated that 122—48 percent of the world's total—are located in North America, and most of these are on the Canadian Shield.

Wetlands cover 18 percent of Canada (1.7×10^6 km^2) and less than 4 percent of the United States (0.4×10^6 km^2). Most wetlands are found in the subarctic, the north-central plains, and the Atlantic coastal zone from Labrador to Florida (Hofstetter, 1983; U.S. Geological survey, 1983; Zoltai and Pollett, 1983). These wetlands include bogs, fens, swamps, marshes, muskegs, and sloughs. A basic hydrological distinction can be made between ombrotrophic wetlands (bogs), which are sustained primarily by precipitation, and minerotrophic wetlands (fens), which are sustained by mineral-rich water from lateral drainage and ground-water inflow.

In the following sections, the hydrologic characteristics of lakes and wetlands in different parts of North America, as shown in Figure 2, are described. Comparison is based on the water balance, because it offers a useful summary of the processes that control the gains, losses, and storage changes that depend on the environmental controls of the regions. Examples of water-balance studies, however, are available mainly for small drainage basins. For large lakes and extensive wetlands, an insight into the regional variability of their hydrologic elements can be gained by examining the water level or runoff hydrographs, which manifest the combined effects of the water-balance components.

Arctic

The Arctic is considered here as the area north of the tree line, where the land is barren or covered only with tundra vegetation. Prolonged cold results in the development of permafrost beneath all lands and most water bodies. Permafrost occurs at depths of no more than several meters below all lakes and less than 1 m in the wetlands. For 8 to 10 months each year the ground surface is frozen and the lakes and wetlands are covered

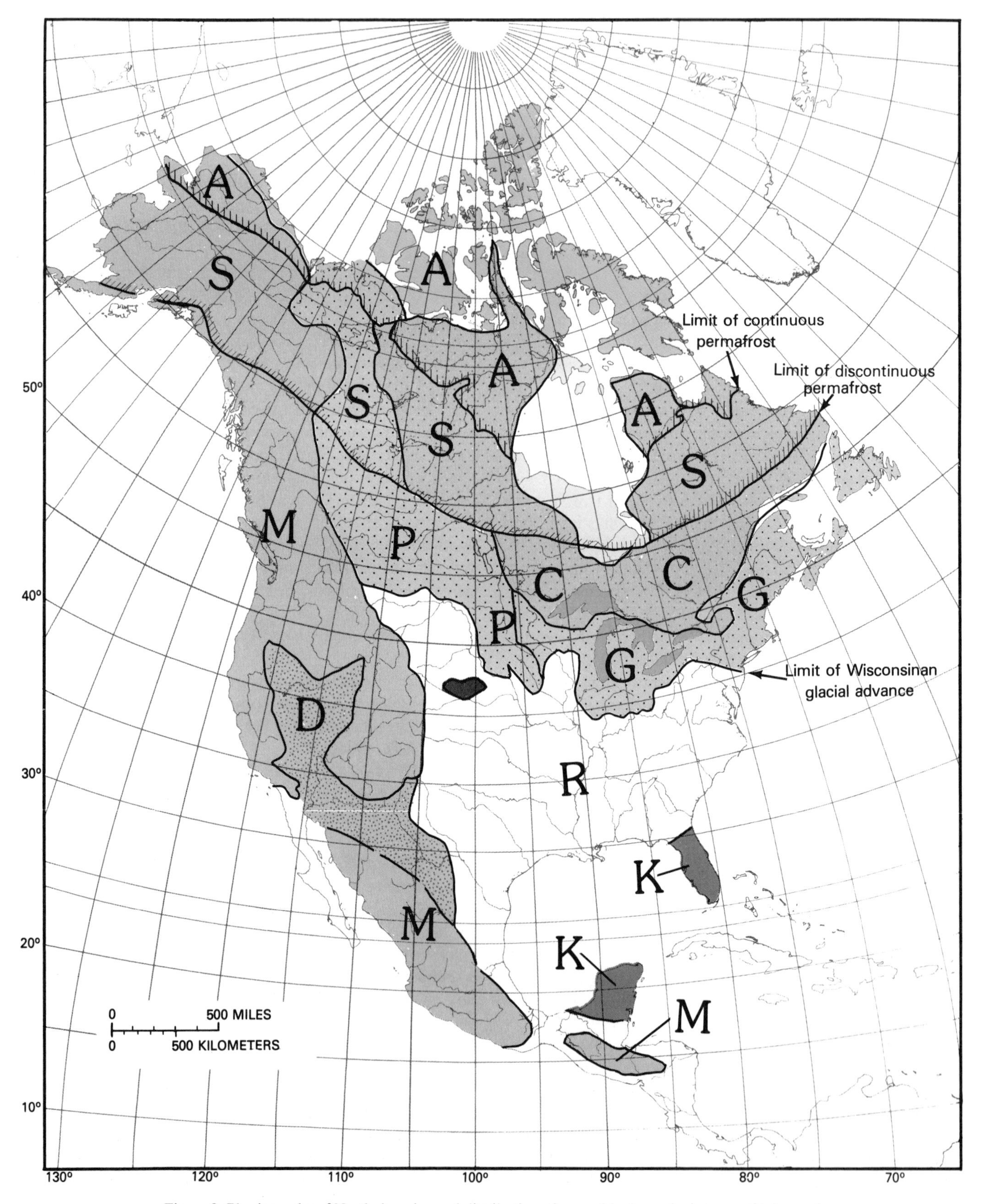

Figure 2. Physiography of North America and distribution of general hydrogeologic types of lakes and wetlands. (Explanation on facing page.)

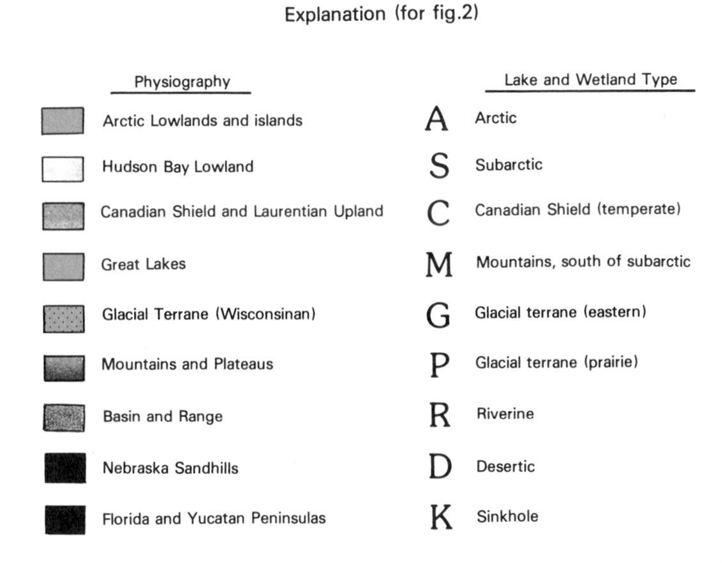

with snow and ice. During the long periods of darkness in winter, solar radiation is minimal. Even during the prolonged periods of sunlight in summer, solar radiation is minimal because of the extreme angle of solar radiation at northern latitudes. Therefore, the winter snowpack usually does not melt until May to July.

The water balance of small Arctic lakes is characterized by: (1) a large influx of snowmelt, which produces the largest quantity of the annual water input, (2) the lack of water exchange with deep ground water due to the presence of permafrost, and (3) a relatively short evaporation season due to the lengthy duration of ice cover. Several seasonal hydrologic periods can be distinguished, each having a distinct water balance (Woo, 1980).

Winter in the Arctic Islands is characterized by the formation and maintenance of an ice cover that commonly is 2 m thick (Schindler and others, 1974). Depending on the latitude and elevation, this ice cover may persist for 9 months in the coastal zone, but some lakes at higher elevations may be continuously ice covered in certain years. There is minimal water loss from the lakes because the ice cover prevents evaporation and outflow is minimal or ceases for small lakes throughout winter. The snow cover is subject to continual redistribution because of active drifting in the open terrain, which commonly blocks lake outlets.

When melting occurs in spring, water from the basin floods the lake ice, as indicated by the hydrograph of Small Lake near Resolute, Cornwallis Island (Fig. 3). The water level rises continuously until the snow dam is breached; this invariably produces the annual peak outflow and the rapid decline in water level. The presence of permafrost at shallow depths permits exchange of lake water with only the shallow thawed layer in the basin. In addition, snowmelt and rain directly on the lake surface represent only a small fraction of the total water input. Therefore, Small Lake, as well as other small Arctic lakes, receives most inflow from the associated drainage basin (Woo and others, 1981). The 1978 summer water balance of Small Lake indicated that rainfall was about the same as evaporation, and outflow was about 2.5 times greater than evaporation. Inflow from the drainage basin approximately balanced the outflow (Fig. 3).

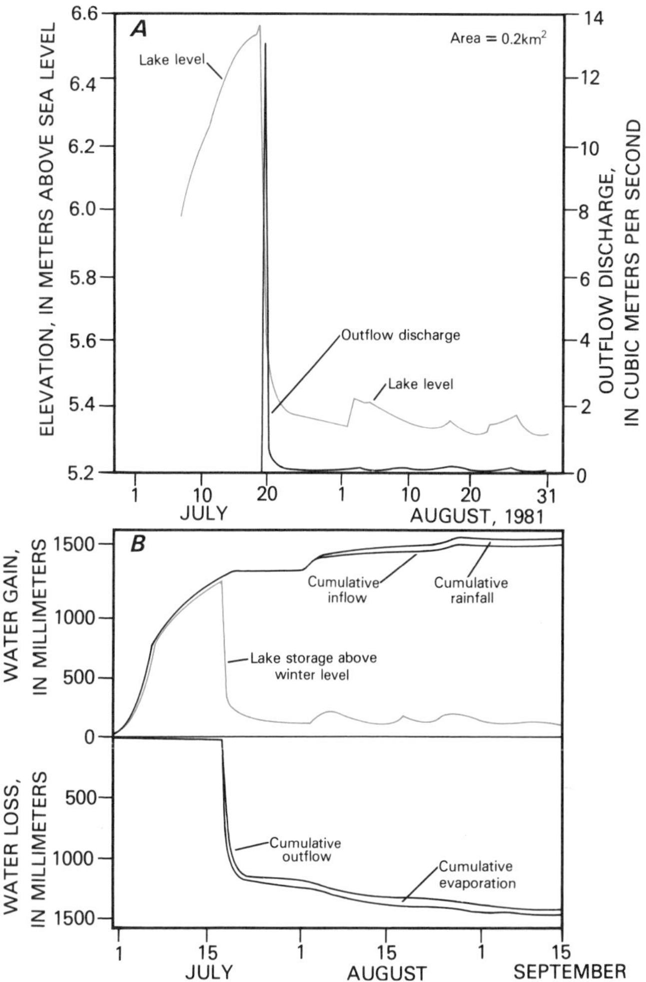

Figure 3. Fluctuation of (A) lake level and magnitude of outflow and (B) water balance of Small Lake near Resolute, Cornwallis Island, for the summer of 1978 (data from Woo, 1980).

Large Arctic lakes have different hydrologic characteristics than small lakes, but inadequate information prevents detailed water-balance analyses for the many lakes that exceed 100 km^2 in area. However, the hydrologic behavior of these lakes can be inferred from their hydrographs. Several lakes of the Dubawnt River drainage system in central Keewatin, Northwest Territories, provide examples. Dubawnt Lake, Marjorie Lake, and other minor lakes have declining water levels in winter as water is drained through the lake outlets (Fig. 4). Snowmelt in June causes rapid rises in lake level, accompanied by a sudden increase in outflow. Because it is much larger than Marjorie Lake, Dubawnt Lake has greater storage capacity relative to its inflow, and its level continues to rise as water inflow is temporarily stored. Thus, the peak water level for Dubawnt Lake is in late summer and early autumn, but the peak for Marjorie Lake corresponds to the end of snowmelt, in June or July. Autumn rainfall and de-

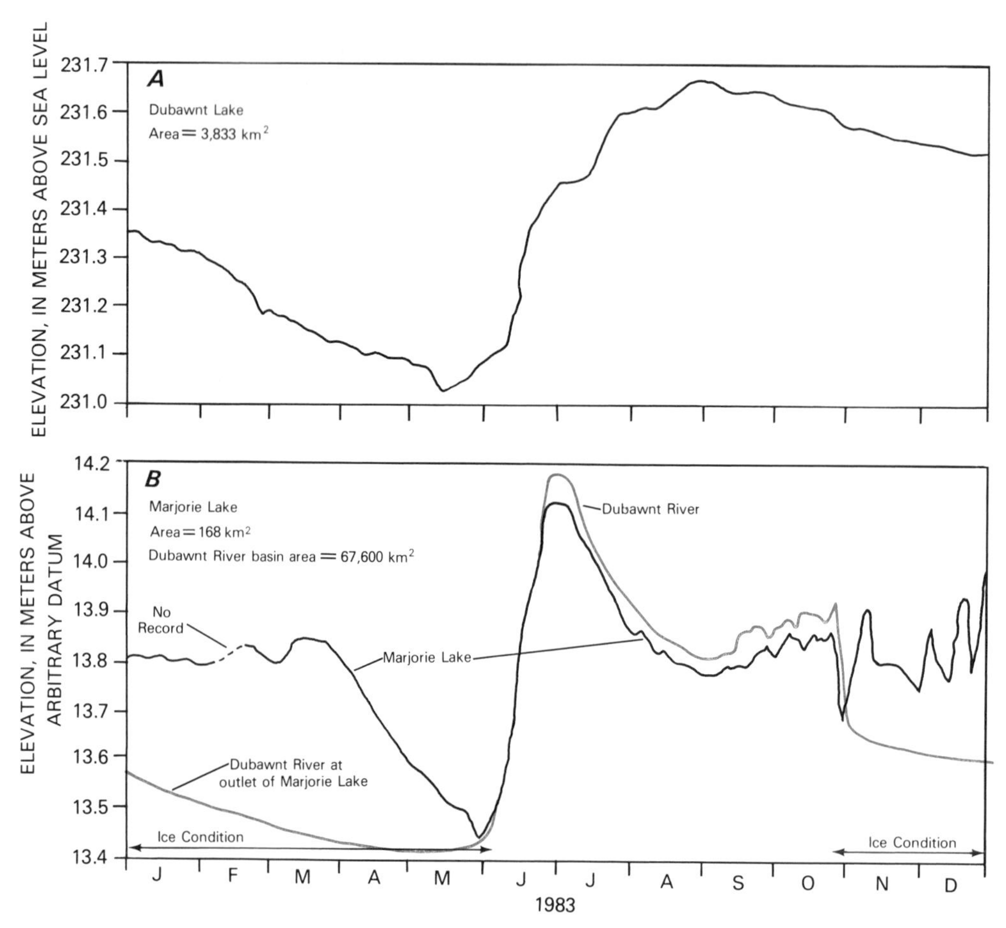

Figure 4. Fluctuations of lake level and outflow from two large Arctic lakes. A: Dubawnt Lake; B: Marjorie Lake.

creased evaporation results in a greater net water gain for Marjorie Lake, which rises, whereas the high water level in Dubawnt Lake is generally sustained. Outflow discharge parallels the level of Marjorie Lake until freezeup, when ice is formed at the lake outlet. Outflow then decreases suddenly, followed by a gradual recession, but the lake level may continue to fluctuate or decline at a gradual rate.

Most Arctic wetlands are sustained by water from lateral sources rather than by precipitation. Thus these wetlands are associated with lakes, streams, and late-lying snowbanks that provide abundant water for most of the thaw season. The cold climate permits only the growth of sedges, mosses, lichens, and a limited variety of stunted vascular plants. The peat layer is generally no more than 0.5 m thick, except under isolated peat mounds. During the snowmelt period, these wetlands remain frozen just below the surface. Because little water infiltrates the frozen soil, the bulk of the melt runoff is discharged within a short period. Examples of discharge from wetland basins at Barrow, Alaska (basin area, 1.6 km^2), and at Truelove Lowland, Devon Island (basin area, 0.4 km^2) (Brown and others, 1968; Ryden, 1977) are shown by the hydrographs in Figure 5. Both hydrographs indicate pronounced peak flow due to snowmelt. In summer, water may be stored in the thawed zone and in surface depressions. Evaporation during the summer approximates rainfall in total quantity, but for Truelove Lowland, snow sublimation is another significant water loss.

The magnitude of various water-balance components for a small wetland in central Keewatin provides a quantitative assessment of its water gains and losses (Roulet and Woo, 1986). This wetland is adjacent to a small lake that discharges most of its peak snowmelt runoff to the wetland. During the study (mid-May to early August, 1983) the lake provided 430 mm of inflow to the wetland; the snow on the wetland itself yielded 146 mm of water, and rainfall contributed 34 mm. Outflow from the wetland was 331 mm and it occurred as surface runoff mainly during the spring melt. Ground-water outflow was minimal because of the low gradient and the low hydraulic conductivity of the compacted peat. Evapotranspiration removed 217 mm of water, but melting of ground ice released 250 mm of water. The result of these gains and losses was a water storage increase in the wetland of about 310 mm. Most of this surplus in storage reverted back to ice or was drained from the wetland prior to ice formation.

Subarctic

The subarctic is directly south of the tree line; it is underlain by discontinuous permafrost and has abundant lakes and wetlands. Although small ponds may have no hydraulic connection with the ground-water system below the permafrost (Hartman and Carlson, 1973), most lakes are probably linked to the ground-water system below the permafrost through a network of thawed zones (Kane and Slaughter, 1973). It is inferred that permafrost does not exist beneath large lakes, such as Great Bear or Great Slave Lakes, because the large water bodies prevent the lake bed from freezing. However, during severe winters all the lakes have a continuous ice cover, and wetland surfaces are frozen.

Small lakes in the subarctic respond quickly to water inflow because of their limited storage capacity. FitzGibbon and Dunne (1981) indicated that peak outflow for several lakes in Schefferville, Quebec, commonly occurred during the spring melt period. These peaks occurred about 2 weeks after the beginning of melt, but most of the lag time was due to water storage in the drainage basin rather than in the lakes. The effect of lake storage in smoothing and lagging the hydrographs becomes less apparent when the lakes constitute less than 5 percent of the basin area. In this situation, outflow peaks have the same magnitude as lake inflow.

As with Arctic lakes, some subarctic lakes have their outlets blocked by snow and ice so that water is impounded behind the blockages during the melt season. A steeply rising limb in the outflow hydrograph commonly is produced when the snow and

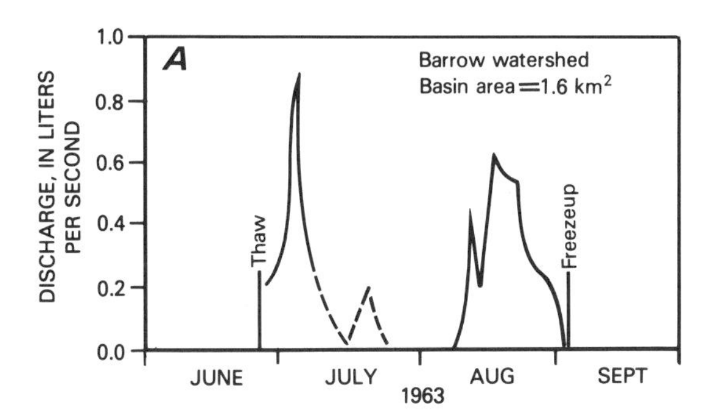

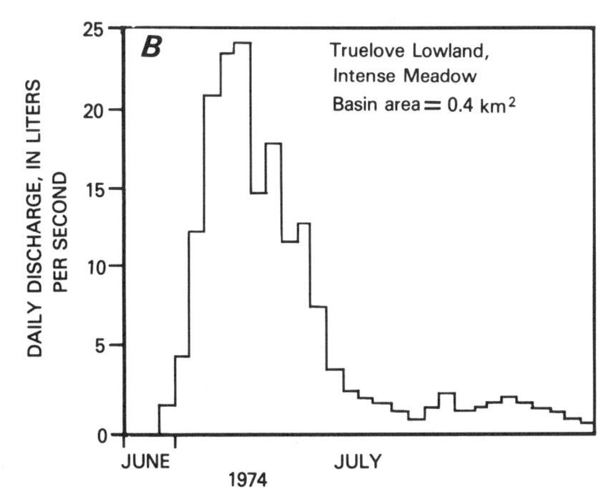

Figure 5. Discharge from two small wetland drainage basins at (A) Barrow, Alaska (data from Brown and others, 1968); and (B) Truelove Lowland, Devon Island (data from Ryden, 1977).

ice blockages are suddenly cleared. The annual peak outflow for drainage basins dominated by lakes also occurs in spring, as indicated by the Knob Lake basin of Schefferville, Quebec (Fig. 6). There, Findlay (1966) indicated that meltwater contributed most of the runoff, and that autumn rainfall caused secondary high flows.

The storage effect is more apparent in the larger lakes, which can sustain winter outflow through a gradual release of water held in storage. For example, Great Bear Lake maintains a steady discharge of 400 to 500 m^3/s throughout the year (Fig. 7). Smaller lakes, such as Kakiska, are more affected by seasonal and yearly variation of inflow and outflow. Daily fluctuations in water level and in discharge are apparent for both Great Bear and Great Slave Lakes, but the magnitude of the fluctuations is small. Water-level rises in Great Bear Lake occur later in the spring than those in Great Slave Lake because the former is farther north and the melting of snow and ice is delayed. Although high water level and lake outflow are usually caused by the melting of snow and ice in the spring, early winter rainstorms can produce peaks late in the year. The discharges from Kakiska and Great Bear Lakes in late 1981 illustrate this phenomenon.

The presence of permafrost in the low-lying areas of the subarctic results in impeded drainage and the formation of many wetlands. In central Alaska, wetlands in valley bottoms have near-continuous moss cover, and vascular flora dominated by black spruce, low-bush cranberry, blueberry, and Labrador tea. Studies at Glenn Creek near Fairbanks indicate that rainstorms generate rapid runoff from the wet bottomlands, where the high water table provides little storage capacity (Dingman, 1973). In northern Saskatchewan and at the southern edge of the subarctic, FitzGibbon (1982) noted that the runoff response of the wetlands to rainfall depends on the antecedent moisture storage in the organic layer. Discharge from the wetlands responds most quickly when the water table is close to the land surface. Freezing in winter can exert pressure on the underlying peat because it is squeezed between the impermeable frost and the underlying till. Thus, the hydraulic head below frost level in fens may be higher than the ground surface. If the peat is not saturated, as in the bogs that continue to drain in winter, the frost penetration does not affect the water table (Price and FitzGibbon, 1982).

The Hudson Bay Lowland is the most extensive wetland area in North America and spans northeastern Manitoba and northwestern Ontario. The area is underlain mostly by Paleozoic bedrock. During deglaciation, the Tyrrell Sea deposited marine sediments on the till, while fluvial deposition by the northward-draining rivers became active (Adshead, 1983). The flat lowland has since undergone uplift, and raised beaches and former mud-flats are now covered with peat for hundreds of kilometers inland. Peat thickness ranges from several centimeters along the coast to several meters inland. Vast areas of impeded drainage result in coastal marshes, bogs, and fens interspersed with numerous ponds and small lakes. Most raised beaches and streambanks are vegetated by black spruce; the wetter sites are covered with sedges and cattail.

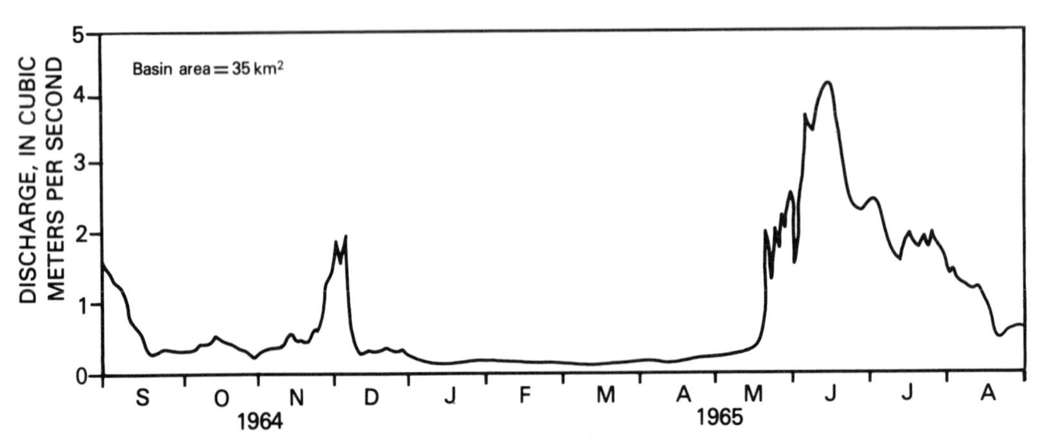

Figure 6. Discharge from Knob Lake, which is in a small subarctic basin containing several lakes (data from Findlay, 1966).

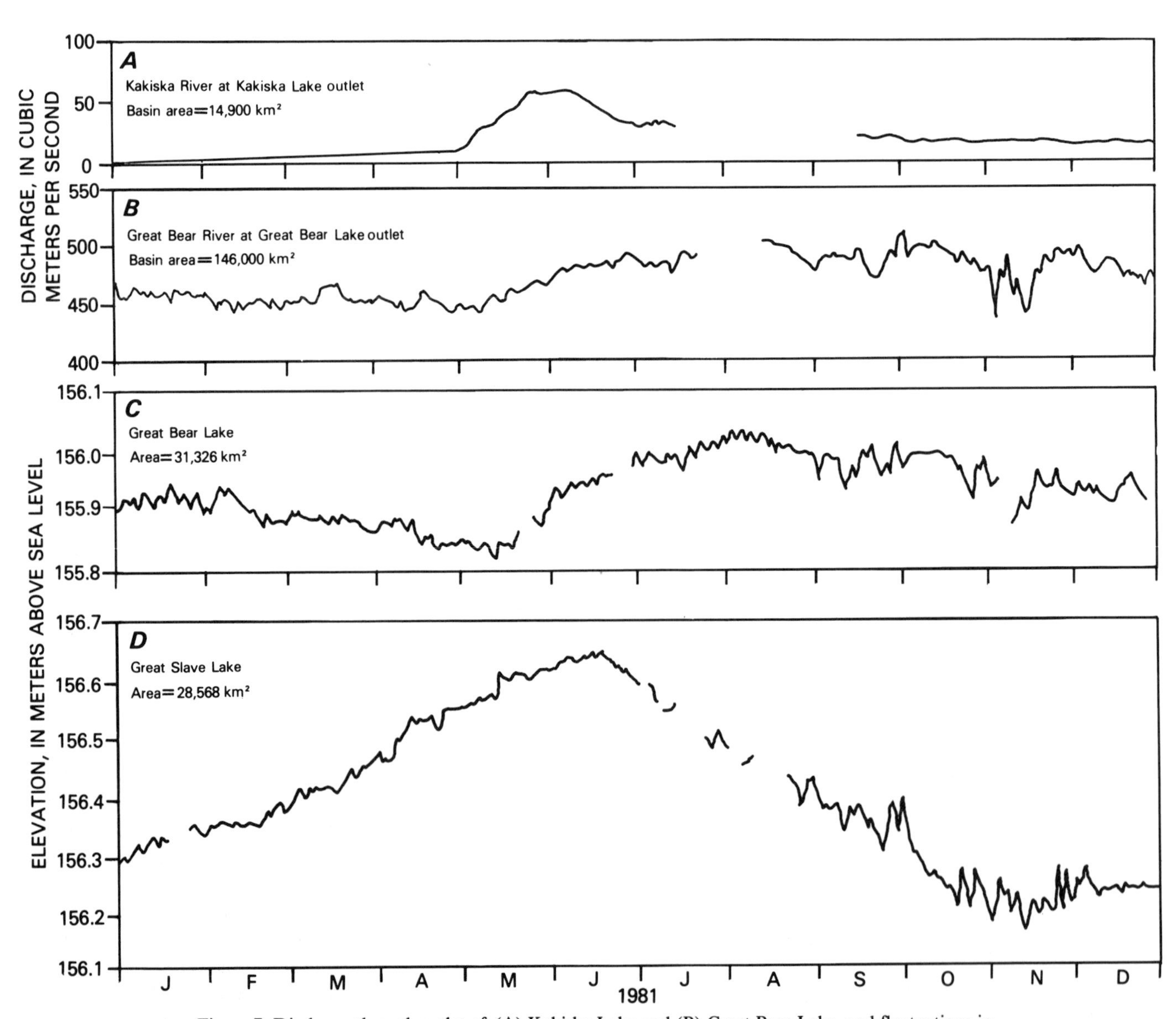

Figure 7. Discharge through outlet of: (A) Kakiska Lake and (B) Great Bear Lake, and fluctuations in level of (C) Great Bear Lake and (D) Great Slave Lake for 1981.

Permafrost is continuous along the coast of Hudson Bay, and becomes thicker in the north. The depth of frost decreases where snow accumulates to greater thickness, because of the insulating effect of snow. During spring snowmelt, most of the water cannot penetrate the frozen zone, and surface runoff as much as a meter deep sweeps across the wetlands toward the rivers and the coast. The drainage network is poorly integrated except adjacent to the major rivers. Many rivulets and rills that are dry or contain stagnant water in summer become active channels during the spring flow. Although the ground-water level remains high throughout summer, subsurface flow is not as important as surface runoff, mainly because of the low hydraulic gradient (0.4–1.5 m/km). Evaporation occurs in the frost-free period, which lasts from 5 to 7 months. The water table declines in response to evaporation, but summer rainstorms can quickly raise the water level, occasionally reactivating surface flow. Late-autumn rainstorms are common, and these storms, together with decreased evaporation, increase the discharge. Flow recedes in winter, but water can seep above the land surface and freeze. Icing is especially noticeable along streambanks, where seepage faces are common.

Drainage basins entirely covered with wetlands and small lakes generally have little stream discharge under ice conditions in winter. An example is the Ekwan River in northern Ontario (Fig. 8), where annual peak discharge is the result of snowmelt during the time when the channel remains partly ice covered. The abundance of wetlands delays streamflow response to rainfall but not to snowmelt, partly because of the presence of frost that hinders infiltration. Streamflow generally increases after a rainstorm, but there is a long recession as water is released from wetland storage. Streamflow decreases gradually in summer as evaporation increases, but frequent autumn rains cause a secondary peak flow before the river attains its winter pattern of low discharge.

Canadian Shield in temperate latitudes

The Canadian Shield consists largely of crystalline bedrock extensively scoured by glaciers Most of the uplands consist of barren rock or rock covered with a thin soil, but the depressions commonly contain glacial deposits. Many lakes and wetlands occupy the depressions, and many are connected by an intricate drainage network. Much of the Canadian Shield is forested; conifers are dominant in the north and mixed coniferous-deciduous forests are dominant in the south.

Snowfall constitutes a substantial part of annual precipitation. Lakes are ice covered in winter, but the depth of frost penetration in wetlands varies inversely with the depth of snow. Snowmelt occurs between March and May, yielding substantial surface runoff to lakes and wetlands. The rate of evaporation is largely controlled by the incident radiative energy, which is greatest between June and September.

Some of the most comprehensive studies of lake hydrology have been of small lakes located on the Canadian Shield. Rawson Lake, located within the Canadian Experimental Lakes Area in western Ontario, has an area of 0.54 km^2 within a 3.96-km^2 drainage basin (Schindler and others, 1976). Results from several years of study indicate that stream inflow from the basin and lake outflow are the major components of the water budget (Fig. 9). Runoff is great in spring as well as in the early fall when evaporation is decreased. Evaporation is limited to about 7 months each year; the maximum rates occur in mid-summer. Although the general pattern holds for all of the years studied, there is a large year-to-year variation in the magnitude of the various water-balance components. Similar results were obtained for Harp Lake east of Georgian Bay in southern Ontario (Scheider and others, 1979), particularly with respect to the dominance of streamflow in the water balance of the lake.

Although the two above studies would indicate ground water is insignificant in the water balance of lakes on the Canadian Shield, the study of Perch Lake (Barry, 1979), also in southern Ontario, indicates that ground water can be an important part of the water balance of some lakes on the Canadian Shield. Perch Lake, 0.45 km^2 in area, has a drainage basin 8 km^2 in area. The lake is underlain by 20 to 30 m of sand and clay overlying the crystalline rocks of the Canadian Shield. Barry (1979) indicated that the water balance of Perch Lake is dominated by surface-water interchange with the lake, but that ground-water

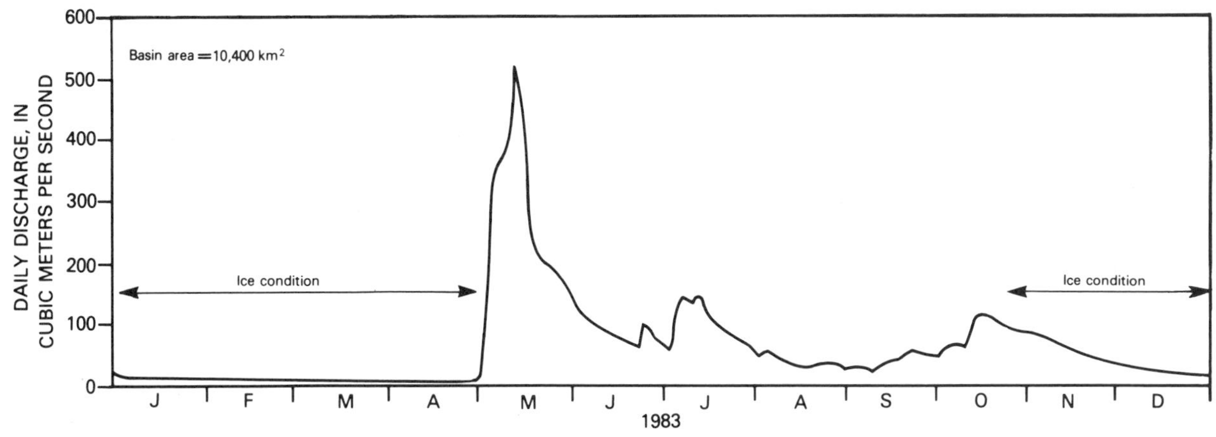

Figure 8. Discharge of the Ekwan River, which drains a subarctic wetland basin.

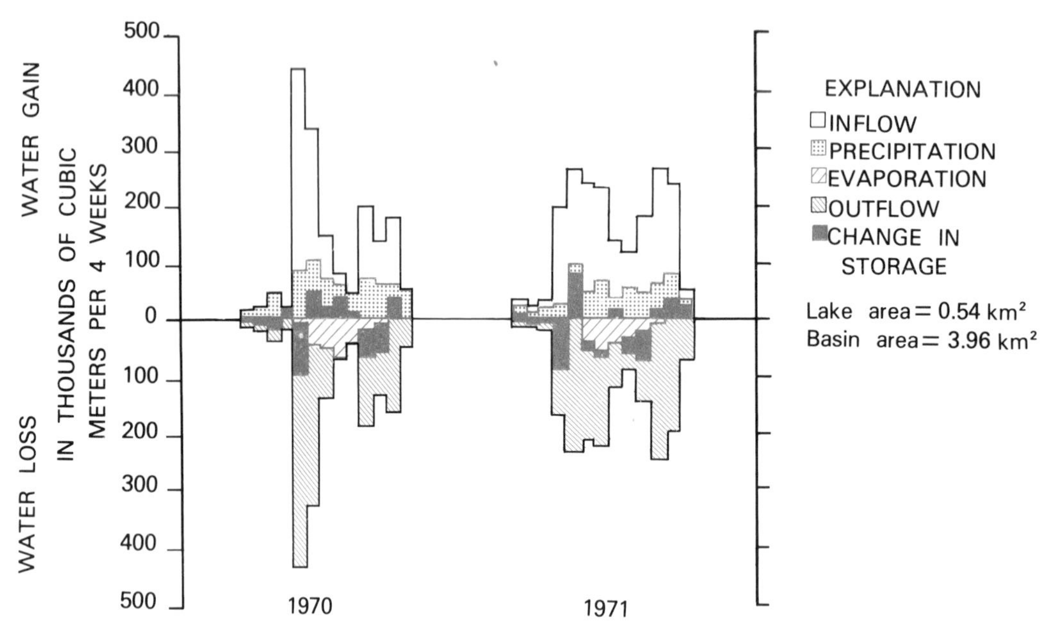

Figure 9. Water balance of Rawson Lake, a small lake in the Canadian Shield (data from Schindler and others, 1976).

inflow is approximately equal in quantity to precipitation and evaporation.

In physiographic settings similar to the Canadian Shield, that is, glacially scoured crystalline rocks having glacial deposits in some depressions, results similar to the above studies are obtained. For example, in a hydrologic study of three lakes in the Adirondack Mountains of New York, Murdock and others (1986) indicated that the effect of ground water on lake-water chemistry is directly related to the quantity of glacial drift in the drainage basins of the lakes. At Mirror Lake in the White Mountains of New Hampshire, Winter (1984) found drift as thick as 55 m in an area thought to be underlain largely by bedrock. Although the water budget of Mirror Lake also is dominated by surface-water inflow and outflow, seepage losses from the lake are similar in magnitude to inflow from one of the principal streams, except during periods of peak flow.

In general, surface water dominates the water budgets of small lakes on the Canadian Shield. However, ground-water inflow and outflow can be appreciable, depending on the quantity of glacial deposits within the drainage basin of specific lakes.

Although comprehensive water-balance studies of large lakes on the Canadian Shield have not been done, some general hydrologic characteristics have been determined. Unregulated lakes tend to have declining water levels after their surfaces freeze, and reach their lowest level just before snowmelt. Melting of snow and ice in the spring causes a rapid water-level rise, and in autumn another water-level rise is caused by increased rainfall. Therefore, these lakes generally have high water levels immediately before freezeup. In the Canadian part of this lake region, many of the larger lakes are regulated to generate hydroelectric power.

The Great Lakes

The five Great Lakes are among the most notable geographic features of the North American continent. They cover a total area of 241,000 km^2. Unlike most large lakes, the Great Lakes have a relatively small drainage basin compared to the area of the lakes. The northern part of the drainage basin is part of the Canadian Shield and has rugged topography and densely wooded areas. The southwestern part of the drainage basin is located on the central lowland, where lakes and wetlands abound. The eastern part of the basin is mainly farmland, and the Appalachian Mountains at the southeastern corner provide high relief. The Great Lakes are the largest bodies of water in the continental interior, and tend to modify local climate. One notable effect is the deeper snowfall on the leeward (eastern) side of the lakes, caused by cooling of moisture-laden air as it reaches the land.

The hydrologic characterization of the Great Lakes can best be described by a water balance (Fig. 10). Annual precipitation (Fig. 10A) tends to increase southeastward, from about 650 to 1,200 mm. Superimposed on this general trend are the many local anomalies of greater precipitation induced by the local snowfall. Particularly evident is the area of maximum precipitation in the southeastern corner, where the lake effect is reinforced by orographic uplift as the prevailing westerly winds hit the Applachian Plateau and the Adirondack Mountains. Precipitation is uniformly distributed throughout the year in the east but shows a slight summer maximum in the west. Summer precipitation occurs as rainfall and winter precipitation is a mixture of snow and rain. On the average, snow cover lasts for about 5 months in the area around Lake Superior, on the north shore of Lake Huron, and in the mountains east of Lake Ontario. Snow

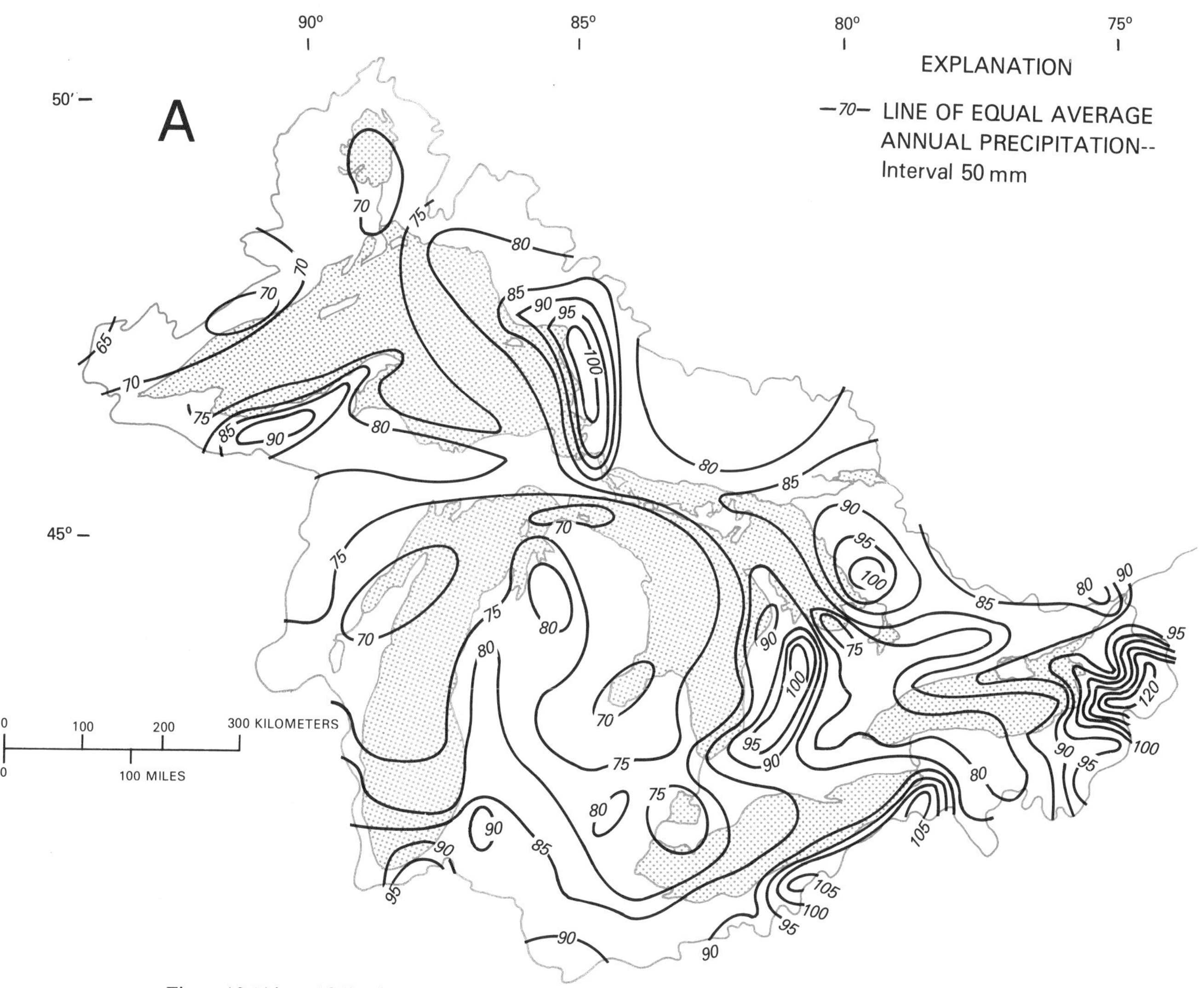

Figure 10 (this and following 2 pages). Atmospheric water components of the Great Lakes. A: precipitation; B: evaporation; and C: difference between precipitation minus evaporation (data from Philips and McCulloch, 1972).

cover lasts for about 2 months around southern Lake Michigan and western Lake Erie, and 3 to 4 months elsewhere.

Every winter, ice covers part of the Great Lakes (Assel and others, 1983), depending on the heat storage of the lake water and the severity of the winter. Ice begins to form in early November in the protected inlets and embayments where the water is shallow and has less heat stored, and the wave action is less intense. In a normal winter, the relatively shallow Lake Erie is completely covered by ice, whereas the deeper Lake Superior freezes much later and is never completely ice covered, even in the most severe winters. Lake Michigan is about 40 percent ice covered in a normal winter, and ice coverage on Lake Ontario seldom exceeds 20 percent because of its large heat storage. The shallow bays of Lake Huron usually are ice covered in winter, but under high wind and wave conditions the extent of the ice cover is greatly decreased. Ice begins to melt in March, although remnants may persist for 2 more months.

Annual evaporation decreases northward across the Great Lakes, ranging from more than 600 mm in southern Lake Michigan and Lake Erie to about 450 mm in northern Lake Superior (Fig. 10B). Lake Erie has the greatest evaporation because, being the shallowest, it has the highest water temperature. The least evaporation is from Lake Superior, which is the coldest and deepest lake. Evaporation is least in spring, when the lake water is cold relative to the air (Phillips and McCulloch, 1972). Because of their large heat storage capacity, the Great Lakes are commonly warmest in the fall; therefore, it is not unusual for their greatest evaporation rate to be in the fall. Evaporation from small lakes in the land around the Great Lakes has a different seasonal regime; maximum loss is in midsummer.

The difference between annual precipitation and evaporation (Fig. 10C) indicates that most of the available runoff is derived from the northern and eastern sections of the basin. Depending on location, runoff represents about 40 to 50 percent of

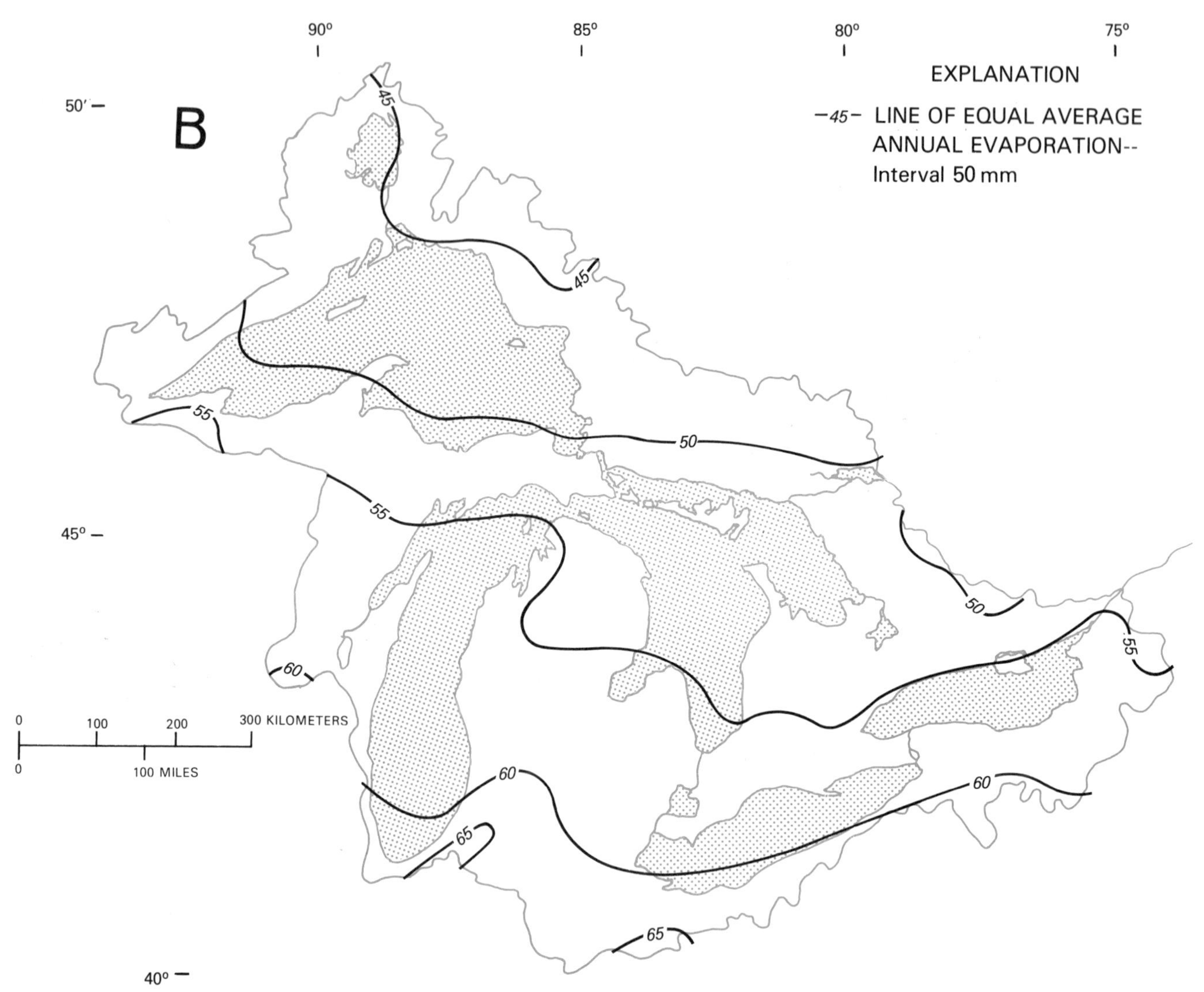

annual precipitation. Water flows from Lake Superior and Lake Michigan to the outlet of Lake Ontario, from which the St. Lawrence River conveys the flow to the Atlantic Ocean. Lake Superior outflow peaks in late summer (August–September), but the lower Great Lakes have peak outflow in June. Mean discharge from the Great Lakes basin as recorded at Cornwall, Ontario, is 72,000 m^3/s, equivalent to 290 mm per year.

The level of the Great Lakes fluctuates in response to long-term storage, annual water balance, and wind effects (Fig. 11). Several types of water-level variations have been recognized according to their durations (Richards, 1967). Short-term (hours-to-days) variations caused by wind-induced seiche or by rapid changes in atmospheric pressure may result in lake-level changes of several meters along the shore line. Annual cycles of low water in late winter and high levels in summer reflect the seasonal changes in water gains and losses. Long-term variations include a downward trend for the levels of Lakes Huron and Michigan that is partly caused by human activities, such as the deepening of the shipping channel in the St. Clair River. However, at present (1986), lake levels are rising throughout the Great Lakes.

Wetlands occur sporadically along the shore of the Great Lakes. These areas are occasionally inundated, depending on the lake-level fluctuation and the height of waves.

Mountainous regions outside the Arctic

Lakes and wetlands in mountainous regions outside the Arctic occupy river valleys and depressions on the plateaus. Although some lakes and wetlands occur at low elevations, exemplified by the Pacific coastal region, these water bodies commonly have significant water input from the higher elevations in the drainage basins. The sources of water supply include rainfall, snowmelt, and glacier melt. The basin response to such inputs is rapid because of the steep terrain and thin soil mantle on most slopes.

Although comprehensive studies of the water balance of lakes in mountainous regions have not been done, several lakes in the Western Cordillera provide examples of the annual patterns of lake-level fluctuation in this type of terrain. Bowron Lake (elevation 910 m), a small lake in central British Comumbia, lies in the rugged headwater area of Bowron River basin; local relief

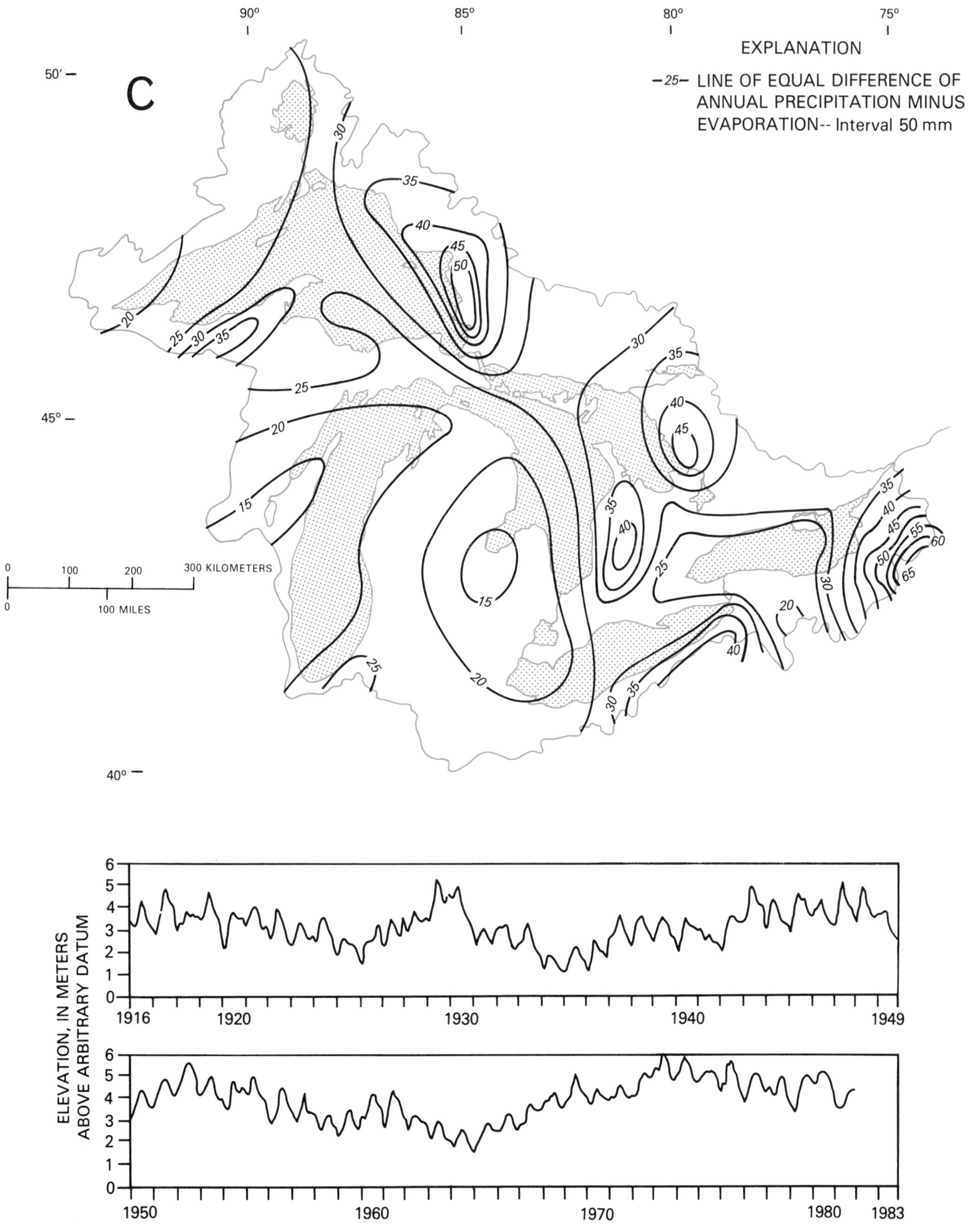

Figure 11. Annual and long-term water-level fluctuations of Lake Erie.

of the drainage basin exceeds 1,200 m. Streams have low winter flow that increases abruptly from snowmelt, resulting in a marked lake-level rise in April or May (Fig. 12A). Throughout summer, several lake-level rises occur, but a general reccession begins in August. Minor lake-level rises in autumn are caused by rainstorms; when the precipitation becomes snowfall, the lake returns to its winter low level.

Horse Lake (elevatoin 990 m), located on the interior plateau of British Columbia 185 km south of Bowron Lake, also has low water levels in winter. However, here snowmelt causes a gradual rise to high water levels in late July (Fig. 12B) because of the rolling topography and low local relief (less than 400 m), as well as the presence of a series of lakes upstream.

Harrison Lake, about 100 km east of Vancouver, British Columbia, has a lake area of 237 km^2 and a drainage basin of 7,870 km^2. Although this lake is about 10 m above sea level, its upper basin extends above 2,000 m and includes several small ice caps and many valley glaciers. Winter is usually a period of low lake level, but winter rain at lower elevations can cause minor lake-level rises. Rapid lake-level increase in spring corresponds with snowmelt in the mountains (Fig. 12C). The snowmelt period can be quite long in those years when the snowline recedes slowly up the mountain slopes. Glacier melt also contributes flow in summer and is proportional to the percentage of basin area containing glaciers. Lake level is low in early fall, which is at the

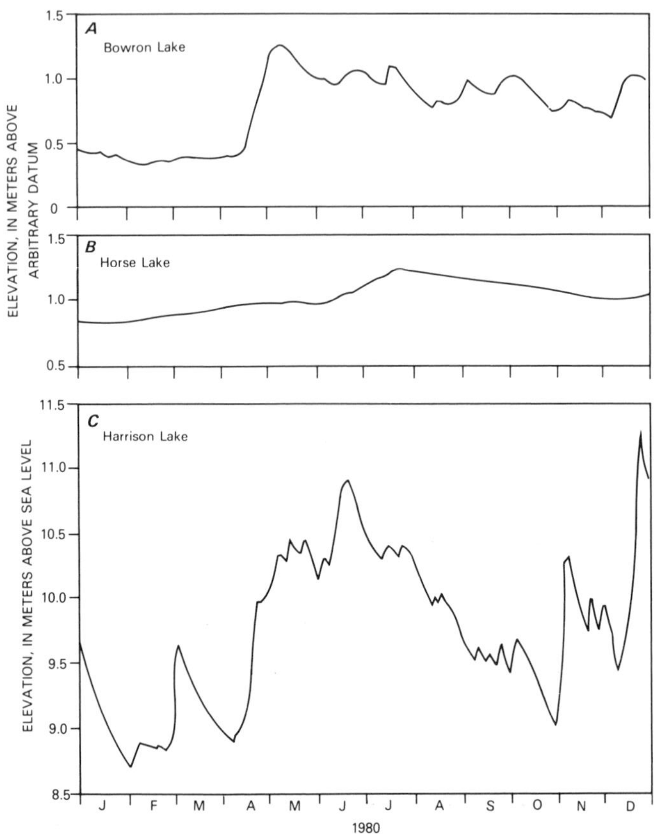

Figure 12. Annual water-level fluctuations of three mountain lakes in Canada. A: Bowron Lake; B: Horse Lake; and C: Harrison Lake.

end of the glacier-melt season but before the arrival of frequent winter rain. The annual hydrograph of this lake in mountainous terrain is not as smooth as that of lakes on the interior plateau. Throughout the year the lake-level peak corresponds to specific snowmelt and precipitation periods.

In general, studies of streamflow to and from natural lakes in mountainous terrain indicate that water balances of these lakes are dominated by streamflow, although precipitation to these lakes also is significant. Water loss to evaporation can be minimal, as evidenced by an evaporation study of a high-altitude lake in Colorado (Spahr and Turk, 1985). In a study of a small basin (Copper Lake) in the North Cascade Range, Washington, Dethier (1979) determined that precipitation and runoff were about equal, 380 to 560 cm annually. He also determined that evapotranspiration was minimal, less than 30 cm annually.

The extent of wetlands in mountainous areas is limited by the lack of flat ground except in the intermountain plateaus. However, small wet meadows are common at the base of steep slopes. In the plateaus of the Canadian Cordillera, severe winters prevent snowmelt until late spring, and the rivers draining from wetlands have low discharge maintained by base flow. Like the Arctic and subarctic wetlands, spring floods are prominent, as is evidenced by the Muskeg River in north-central British Columbia (Fig. 13). The magnitude of floods in this area depends on the total winter snowfall and the rapidity of snowmelt. After the melt period, a steady level of runoff is sustained, and summer rainstorms produce mostly minor water-level rises.

Because the supply of moisture is abundant, evaporation from mountain wetlands proceeds at approximately potential rates (Sturges, 1968). A faster rate of evapotranspiration compared to evaporation from open-water bodies is attributed to the peat surface being a good sink for solar radiation, and the irregular microrelief of the wetland increasing the surface area for evapotranspiration.

Glacial terrain (continental)

The region of lakes and wetlands in North America within glacial terrain is best defined by the extent of Wisconsin glacial deposits. Pre-Wisconsin glacial deposits have integrated drainage networks, and nearly all lakes and wetlands have been drained.

Most lakes and wetlands are kettle-hole type, and are located within moraines and outwash plains. Wetlands are common in kettle holes as well as in glacial lake plains, such as the glacial Lake Agassiz plain, particularly north and east of the Red Lakes in northern Minnesota.

As stated in the first part of this paper, lakes in depositional glacial terrain occur south of the Canadian Shield in a broad band from the St. Lawrence Valley to the prairie provinces of Canada (Fig. 2). This entire region has a great diversity of bedrock and climate. Because both of these factors have pronounced effects on lake hydrology, the discussion of lakes in glacial terrain is in two parts. Lakes in the eastern part are generally in glacial drift overlying a variety of bedrock types, and are in a climate where precipi-

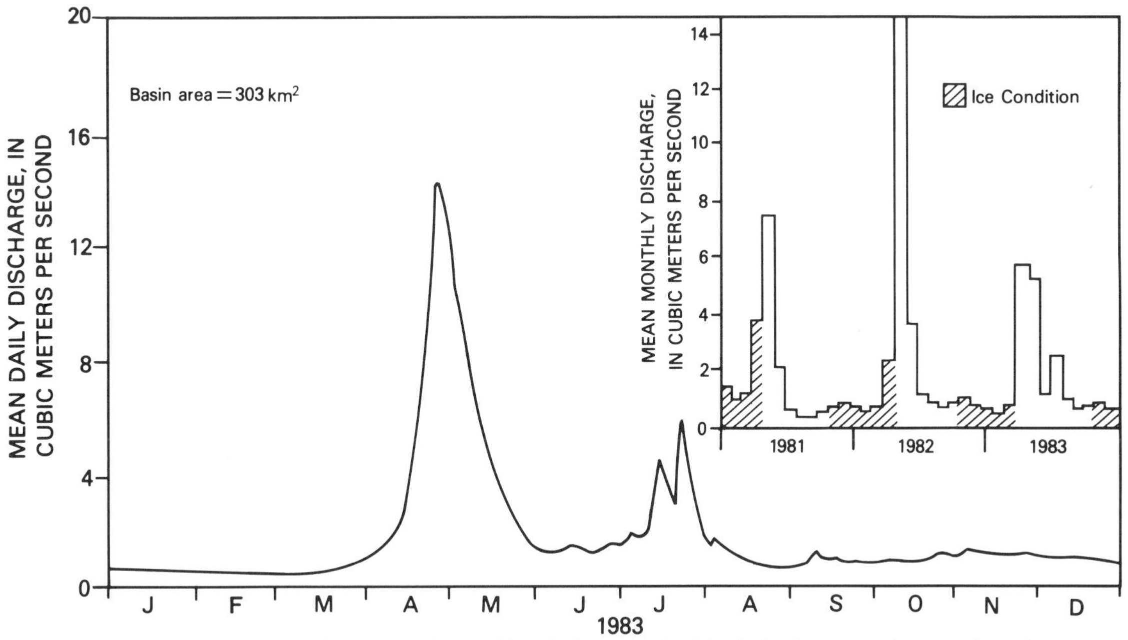

Figure 13. Discharge of Muskeg River, which drains a wetland basin in the mountainous region of British Columbia.

tation exceeds evaporation, whereas prairie lakes and wetlands are in glacial drift overlying relatively uniform, fine-grained bedrock, in a climate where evaporation exceeds precipitation.

Eastern United States and Canada. In general, glacial drift in the eastern part of glaciated North America is more permeable than in the prairies because the bedrock from which the drift is derived is more permeable. For example, in the White Mountains of central New Hampshire the glacial till consists mostly of silt and sand and it contains numerous cobbles and boulders (Winter, 1984). Glacial till in the vicinity of the Great Lakes, however, contains Paleozoic shale as well as lake clay and silt, so that glacial till to the south is commonly clayey (Scott, 1976; Frye and others, 1965).

Although general differences between glacial drift in the east and glacial drift in the west exist, the principal differences in the hydrology of these lakes are related to climate. Not only does precipitation increase to the east, reaching greatest quantities in the Applachian Mountains, but evaporation rates in eastern North America are among the least on the continent. Because of the excess of precipitation over evaporation, lakes and wetlands in the eastern glaciated region not only have a positive water balance with respect to atmospheric water, but they also tend to have more interchange with streamflow.

The water balance of glacial-terrain lakes in eastern North America that have streams entering and leaving them is greatly affected by that streamflow, as evidenced by Lake Sallie, Minnesota (Mann and McBride, 1972). Water gains and losses to the atmosphere are also substantial for many lakes in this region. Ground-water inflow and outflow can also be substantial in lakes situated in relatively permeable glacial drift. In the case of Lake Sallie, ground water contributed about 13 to 16 percent of the inflow to the lake; this percentage is equal in volume to precipitation and evaporation for the lake. Precipitation and evaporation were about equal for East Twin Lake, Ohio (Cooke and others, 1973), but there the ground water constituted about 25 percent of inflow to the lake.

Many lakes in glacial terrain are not integrated into a drainage network, and interact only with atmospheric water and ground water. The water balance of such lakes is generally dominated by atmospheric-water interchange, and the relative volume of water exchange with ground water varies widely. For Williams Lake, Minnesota, ground water contributed about 15 percent of water input to the lake, and an equal percentage of lake water seeped to ground water (Siegel and Winter, 1980). For Mirror Lake, Wisconsin, ground water accounted for about 20 percent of the water gain and 16 percent of the water loss (Possin, 1973). At Shadow Lake, Wisconsin, ground water contributed about 40 percent of the water gain and only 2 percent of the water loss (Possin, 1973). Both Mirror and Shadow lakes have some interchange with streams, but interchange does not dominate the water balance. Perhaps the greatest reported interaction of ground water with a lake is for Pickerel Lake, Wisconsin, where ground-water inflow and outflow are in the range of 70 to 75 percent (Hennings, 1978).

The glaciated regions of north-central and eastern United States and eastern Canada have extensive wetlands that commonly occur in former glacial lake basins, along rivers, lake margins, and coastal zones. Vegetation ranges from cattail and sedge marshes to shrub and forested swamps. In addition to having poor drainage, wetlands in this region have an abundant water supply. Snowfall is particularly important in the central area, and generates large quanitites of meltwater that floods many wetlands in spring. For example, in northern Minnesota, spring runoff accounts for more than 60 percent of annual water yield (Bay, 1969; Winter and others, 1967). Snowfall and rainfall are sufficient in the Maritime Provinces and Newfoundland to support complexes of bogs and fens that blanket the slopes, small depressions, and even some high ground. These wetlands have a high water table throughout most years.

Both ombrotrophic and minerotrophic wetlands are abundant in this region. Bay (1967) studied two groups of wetlands in Minnesota. The raised bog had a water table higher than the contiguous ground water, whereas the ground-water "bog" (which by definition is a fen) received some ground-water inflow. Bay (1967) determined that the raised bog had greater annual variation in the fluctuation of its water table than did the fen (Fig. 14). However, during summer the patterns of water-table fluctuations for both types of wetland were similar.

Although wetlands of the interior normally have low water levels in winter, the Altantic coastal zone receives winter rainfall and sporadic snowmelt. Minimal evapotranspiration and a high water table are commonly maintained in the cold period (Vecchioli and others, 1962).

In the interior lowland, maximum evapotranspiration occurs in summer, but as the water table declines, potential evapotranspiration is not attained (Munro, 1979). Bay (1967) determined that, throughout the entire growing season between May 1 and November 1, transpiration from wetlands in northern Minnesota was less than the potential rate in certain years but more than the potential rate in other years, depending on how wet or dry the summer is.

During the entire summer season, evapotranspiration is the major mechanism by which water is lost from wetlands. Woo and Valverde (1981) determined that evapotranspiration from a wetland in southern Ontario totaled 554 mm, approximately balancing the total rainfall of 541 mm. This evapotranspiration was more than double the inflow (255 mm) and outflow (242 mm).

Runoff from wetlands is logarithmically proportional to water level in the peat (Bay, 1968); increasingly large runoff occurs when the water level is within the hydrologically active layer, the acrotelm. Runoff responds readily to summer rainstorms because the water table is commonly within the permeable acrotelm. However, peak runoff rates are generally slow, and the recessions are long, indicating the effectiveness of wetlands for short-term storage (Bay, 1968). On a long-term basis, wetlands do not provide adequate perennial storage to sustain low flow in the dry season, and they are therefore ineffective as a seasonal regulator of streamflows (Verry and Boelter, 1972).

Recent studies (Siegel, 1981) within the Red Lake peatland, Minnesota, indicate that ground water is a more important factor than previously believed. Although the wetland is hundreds of square kilometers in area, and most of it is distant from contiguous uplands, some local ground-water flow systems may extend into the underlying minerogenic soil, making the upward-moving ground water a source of chemical elements that nourish certain minerotrophic plants.

Prairie lakes and wetlands. Most prairie lakes and wetlands are within the region of North America that extends northward from central Iowa through western Minnesota, eastern South Dakota and North Dakota, southwestern Manitoba, and southern Saskatchewan to southeastern Alberta (Fig. 2).

Because the glaciers overrode relatively fine-grained sedimentary rock in the prairie regions, the glacial till there is generally characterized as silty and clayey. In general, the glacial till is silty in Canada and western North Dakota, and overlies Tertiary siltstone and sandstone (Scott, 1976); the till is more clayey in eastern North Dakota and South Dakota, Minnesota, and Iowa, where it overlies Cretaceous shale. The widespread fine-grained texture of glacial till in the prairie region of North America has a substantial effect on the surface-water and ground-water hydrology of this region.

Overall, the topography of the north-central prairie is flat. Where it is not extremely flat, such as in morainal areas, the natural drainage network is not well developed, and the many depressions are not connected by an integrated drainage system.

Major topographic features of the glaciated prairie are the relatively rugged end moraines that extend for hundreds of ki-

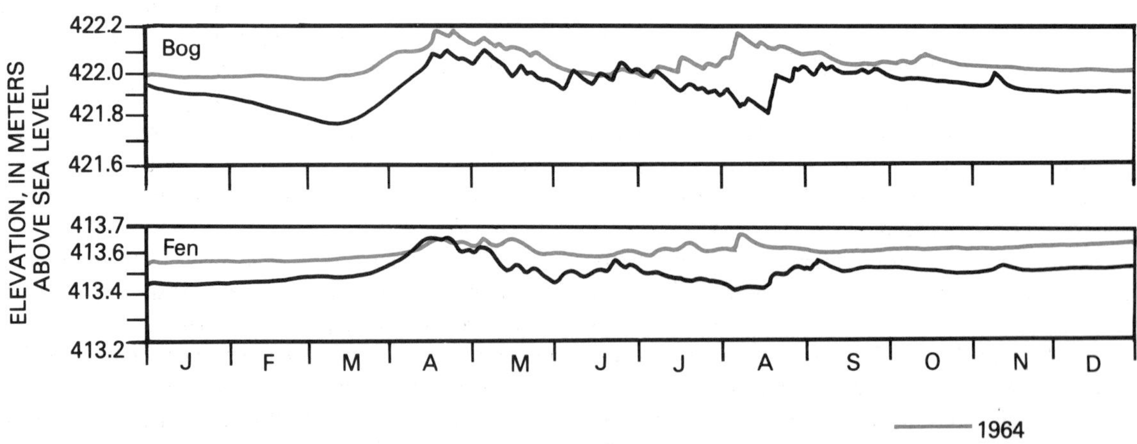

Figure 14. Water-table fluctuation in a raised bog and a fen in northern Minnesota (data from Bay, 1968).

lometers northwest through the area. These end moraines are characterized by steep slopes at their edges, and they rise for more than 100 m above the surrounding flatter plains. Within the end moraines, local relief is commonly 15 to 45 m. Outside the areas of end moraines, gorund moraines have local relief of only a few meters. The majority of prairie wetlands occur within depressions in end moraines and ground moraines.

Other features of glacial terrain are lake plains and outwash plains. The lake plains are flat and consist largely of clay and silt; there are sandy beach ridges along the margins. Outwash plains also generally are flat, but they commonly contain lakes and wetlands in ice-block depressions. Outwash deposits generally consist of stratified sand and gravel.

The climate of the northern prairie of North America is continental; there are hot summers (maximum temperatures as much as 40°C) and long cold winters (minimum temperatures as low as –50°C). Evaporation exceeds precipitation throughout the region. The deficit of precipitation compared to evaporation ranges from –50 mm in Iowa to –750 mm in western Nebraska. Because of the northern position of the prairies, it is not unusual for lakes and wetlands to be frozen for more than 6 months. Many small lakes and wetlands are frozen to the bottom part of this time.

Studies of the hydrology of prairie lakes and wetlands indicate that the water balances of these water bodies are dominated by precipitation and evaporation. Stream inflow and outflow are not generally major components of the water balance, except for the overland flow of snowmelt into the depressions prior to thawing of soil frost. For example, Shjeflo (1968) indicates that runoff from snowmelt contributes about 30 percent of inflow to wetlands in North Dakota. Ground-water interaction, although only 5 to 25 percent of the water exchange, has a significant effect on the chemistry of these systems. Studies by LaBaugh and others (1987) and Swanson and others (1988) indicate that the chemical character of prairie lakes and wetlands is determined by their position relative to ground-water flow systems. Therefore, the following discussion concentrates on the ground-water component of prairie lakes and wetlands.

Using concepts of integrated flow systems, Canadian hydrologists studied a number of areas in the prairie, many of which contain lakes and wetlands. For example, Meyboom and others (1966) mapped patterns of ground-water flow and discharge characteristics for a wide range of scales, from individual wetlands to major regional systems. Intensive study of ground-water movement near individual wetlands indicated seasonal reversals of ground-water flow directions, depending on the magnitude of recharge to ground water relative to water levels in wetlands (Meyboom, 1966, 1967). Studies by Lissey (1971) in the Canadian prairie resulted in additional concepts of recharge. Lissey (1971) indicated that both ground-water recharge and discharge occur largely in depressions in the prairie, many of which contain lakes and wetlands. Therefore, wetlands could serve both functions of recharge and discharge, but he did not elaborate on the seasonality of these functions.

In a recent study, the hydrogeologic settings of about 180 lakes and wetlands were examined with respect to regional ground-water flow systems in Kidder and Stutsman counties, North Dakota (Swanson and others, 1988). The study included modeling of ground-water flow systems relative to lakes and wetlands along two transects across the Missouri Coteau in the two-county area. Results indicate that, on a regional scale, ground water is recharged at topographic highs and is discharged to regional lowlands, such as lakes in the Kidder County outwash plain. Lakes in the Crystal Springs area receive discharge from large ground-water flow systems in the glacial drift, and the hydrologic section indicates that some of the lakes here might even be receiving discharge from bedrock water that moves through the glacial drift before entering the lakes. The Crystal Springs area is particularly interesting because it also lies in a depression between major topographic highs to the north and south. Because of these topographic highs, it is likely that ground-water flow systems, even larger than those shown to the east of Crystal Springs in Figure 15, discharge ground water to the area from the north and south.

Although hydrologic sections are useful in developing concepts of the interaction of ground water and lakes on a regional scale, local flow systems are more complex and the assumption that water-table highs underlie land-surface highs is not always true. Furthermore, ground-water recharge occurs in numerous locations throughout the landscape, regardless of regional topographic position.

For example, LaBaugh and others (1987) indicated that the configuration of the water table between prairie wetlands in North Dakota is spatially complex and temporally dynamic; seasonal and annual variations of seepage direction and chemistry can be large. The wetlands selected for intensive study lie within an integrated ground-water flow system (Fig. 16). Wetland T8 is an area of recharge to ground water. Snowmelt accumulates in wetland T8 in the spring, and the water table slopes away from the wetland nearly all the time. Between wetlands T8 and T3, configuration of the water table does not follow the topography of the land surface, indicating that recharge beneath hills in this highest part of the study area is less than in the adjacent depressions. Wetlands P1 and P8 are in areas of ground-water discharge. In the low areas between wetlands T3, P1, and P8, configuration of the water table is dynamic. Reversals of ground-water flow are common between wetlands T3 and P1 in the vicinity of well 13. Timing and duration of these reversals are affected by seasonal and annual fluctuations in recharge by snowmelt and by rainfall. Less pronounced and less frequent reversals of flow occur between wetlands P1 and P8.

In general, most ground water recharged locally near any surface-water body will discharge into the lake or stream nearest the shore, and seepage will decrease in a nonlinear rate toward the middle of the water body (Fig. 17) (Pfannkuch and Winter, 1984). Furthermore, as discussed above for the North Dakota site, constantly changing recharge conditions adjacent to surface water can cause complex ground-water flow and seepage patterns

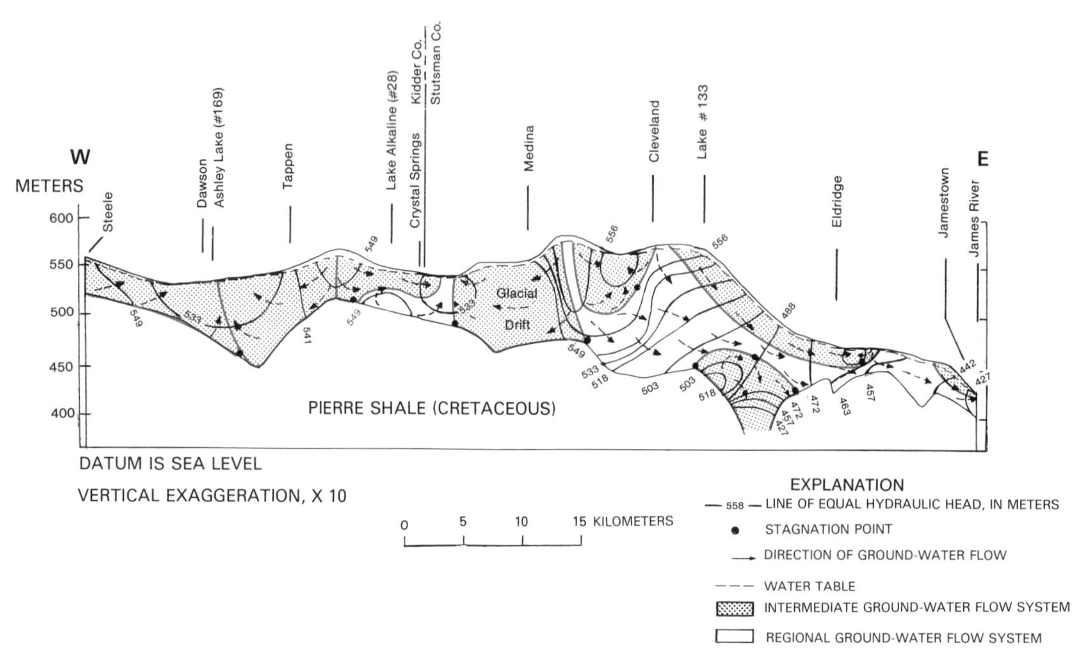

Figure 15. Hydrologic section showing intermediate and regional ground-water movement in Kidder and Stutsman counties, North Dakota (modified from Swanson and others, 1988).

in lakes and wetlands (Fig. 18) (Winter, 1983). Therefore, it is not unusual to observe freshwater springs discharging into saline lakes. For example, ground water discharging into Chase Lake, North Dakota, was consistently more dilute than the lake water. In addition, the relative contribution of ground-water discharge from local ground-water flow systems and that from regional systems can can vary considerably between lakes, which will cause the overall chemistry of the lakes to vary considerably.

Differences in topographic position, though small, present different opportunities for seepage from lakes within any regional topographic position. For example, within a regional topographic low, a lake that is slightly higher than an adjacent lake could have seepage from the higher lake part of the year, whereas the lower lake might not have seepage from it. This opportunity to lose water from one of the lakes and not the other, together with different relative mixes of ground water from local and regional flow systems, could explain greatly different water chemistry between adjacent lakes or wetlands.

In the Sandhills of Nebraska, the dunes are commonly 30 to 90 m in height, and they overlie sandstone largely of Tertiary age. Therefore, from the standpoint of ground-water hydrology, the sand dunes and bedrock together constitute a single hydrogeologic unit. Because of this, thickness of the ground-water system interacting with lakes and wetlands in the Sandhills is as much as 300 m in western Nebraska.

Analysis of water-level fluctuations in observation wells near five lakes in the Crescent Lake National Wildlife Refuge in the Sandhills indicates that water-table configuration beneath sand dunes in this area varies considerably in time and space, depending on the topographic configuration of the dunes. If the topography between lakes is hummocky, ground-water recharge is focused at topographic lows, causing formation of water-table mounds. These mounds prevent ground-water movement from topographically high lakes to adjacent lower lakes. If a dune ridge is sharp, the opportunity for focused recharge does not exist, resulting in water-table troughs between lakes. Lakes aligned in descending altitude parallel to the principal direction of regional ground-water movement generally have seepage from higher lakes toward lower lakes (Winter, 1986).

Riverine lakes and wetlands

Lakes and wetlands are common features in the flood plains of large rivers in many environments; therefore they are not outlined areally in Figure 2. In some areas, wetlands cover the entire flood plain. Extensive wetlands are to be expected partly because of the close proximity of the river and, therefore, the wetland's vulnerability to flooding. Surface inflow from the river is commonly the the dominant component of water input to lakes and wetlands located on flood plains. For wetlands that are virtually shallow parts of rivers, the water balance, both input and output of water, is totally dominated by river flow.

The water balance of lakes and wetlands located on flood plains is somewhat different from other lake and wetland types because of the dominating effect of the river. For example, lakes and wetlands in glacial terrain are greatly affected by local climate and ground-water conditions. In contrast, lakes and wetlands in flood plains, because they are so closely associated with the river, are commonly affected by the climatic conditions affecting the entire river basin, which might be a considerable distance

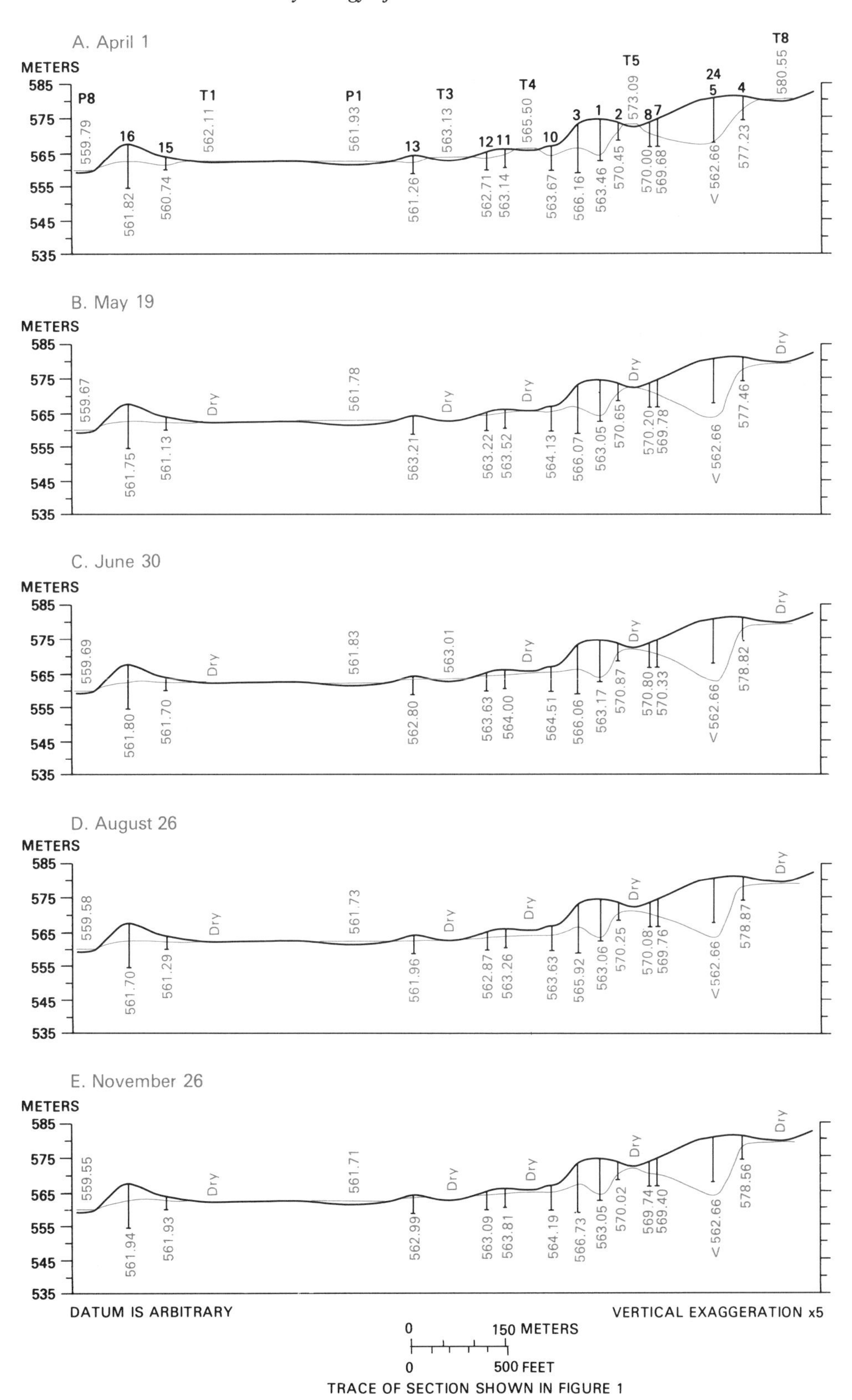

Figure 16. Altitudes of water levels of wetlands and of the water table for five dates in 1980, in a section through the Cottonwood Lake area, North Dakota (modified from LaBaugh and others, 1987).

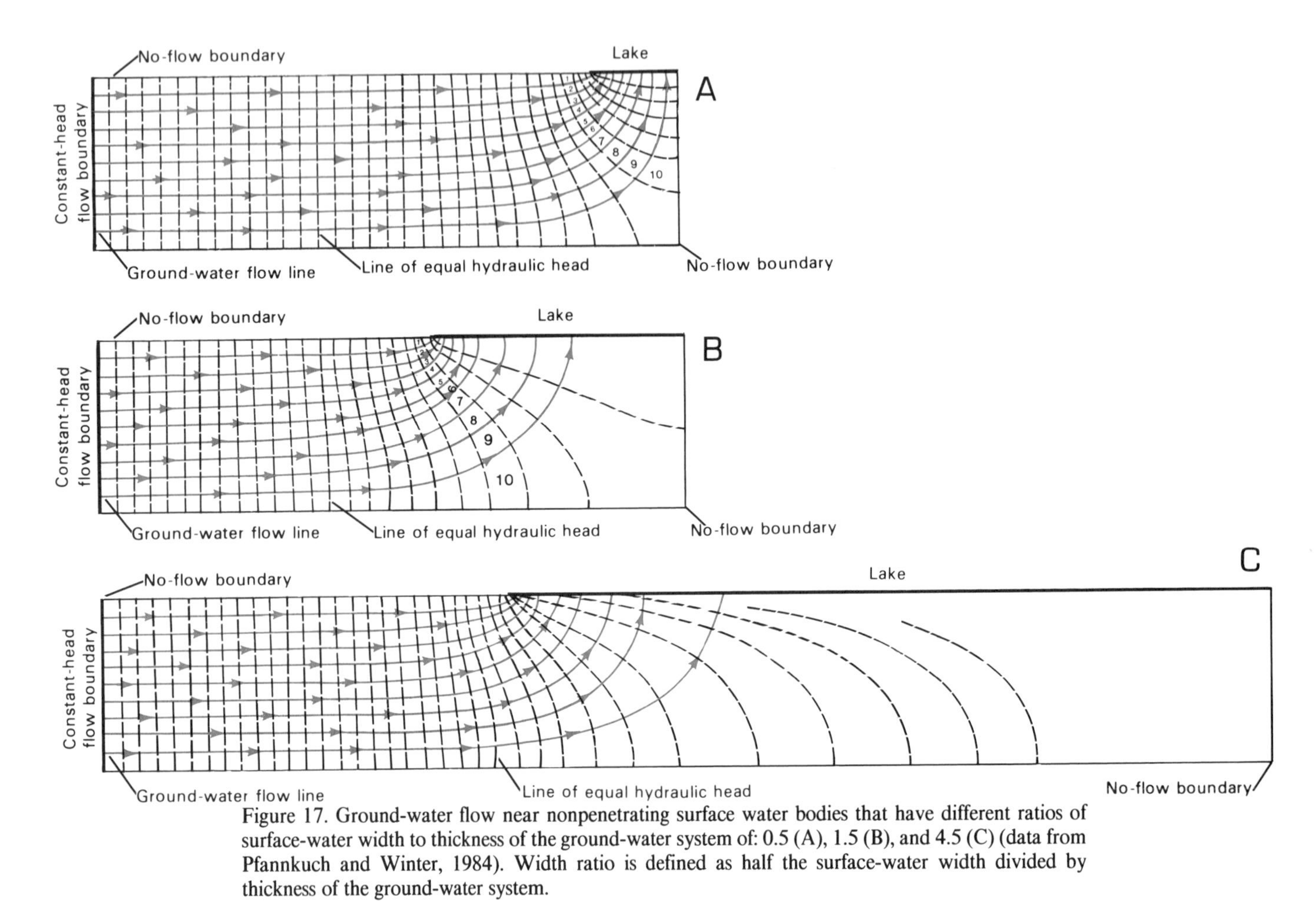

Figure 17. Ground-water flow near nonpenetrating surface water bodies that have different ratios of surface-water width to thickness of the ground-water system of: 0.5 (A), 1.5 (B), and 4.5 (C) (data from Pfannkuch and Winter, 1984). Width ratio is defined as half the surface-water width divided by thickness of the ground-water system.

away from a given lake. Two contrasting examples are found in the largest river systems on the continent.

Spring runoff, largely from snowmelt in the headwaters, commonly floods lakes and wetlands on the flood plains of the Mississippi River system in the southern United States, which is thousands of kilometers away. Spring snowmelt in north-flowing rivers, such as the McKenzie River and Red River of the North, begins in the southern headwaters areas while the northern reaches of the rivers are still icebound. The damming effect of the river ice in the north on the rising discharges from the south results in extensive flooding in the northern flood plains.

Although surface flooding constitutes a major water gain to lakes and wetlands located on flood plains, the nearby river also has significant effects on ground-water movement in flood plains. Because alluvial deposits in flood plains commonly have highly permeable zones, a significant hydraulic connection between the river and lakes and wetlands on the flood plain usually exists. Rises in river stage cause river water to go into bank storage, which results in rises in levels in the lakes and wetlands, as has been shown for the flood-plain wetlands in the Platte River Valley near Grand Island, Nebraska (Hurr, 1983), and Lake Chicot, an oxbow lake on the Mississippi River flood plain in Arkansas (Schiebe and others, 1984).

Ground water from adjacent uplands can also be an important source of water to lakes and wetlands on flood plains. Because flood plains are topographically low, they are natural places for discharge from ground-water flow systems of all magnitudes, from local to regional. The pattern of ground-water discharge in flood plains is discussed in the "Slope discontinuities" part of the first section of this chapter.

Water loss from lakes and wetlands on flood plains, where wetlands are not contiguous with the river, is usually dominated by evapotranspiration and ground water. Transpiration losses are particularly substantial in flood-plain settings because the wetness of the surface and shallow ground water result in plants being able to transpire water at maximum potential. Water loss to ground water also can be substantial for the same reason that bank storage can be a major source, that is, the high permeability of alluvial deposits.

Desert lakes

A number of large lakes occur in the Basin and Range physiographic province of western United States (Fig. 2); most are in fault-block depressions and have no surface outflow. Because of their large drainage basins and the desert climate, the

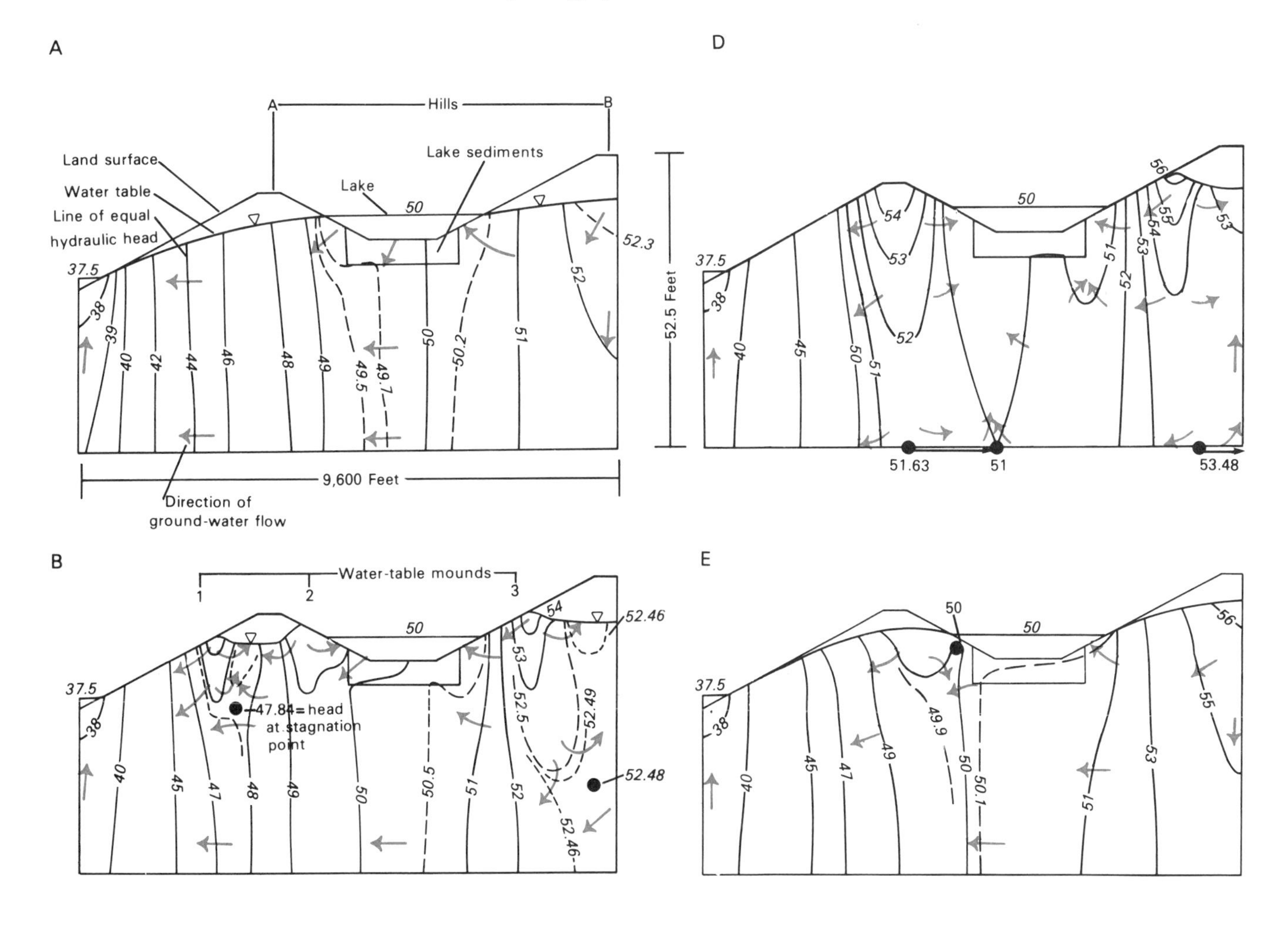

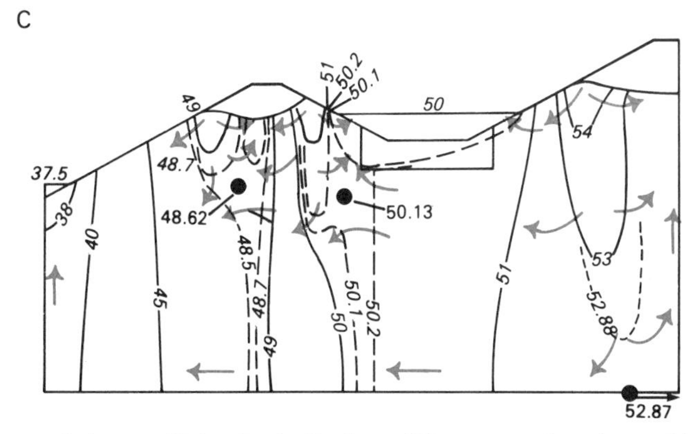

Figure 18. Distribution of hydraulic head and direction of ground-water flow for variably saturated porous media near surface water. Anistropy of the porous media is 500. Beginning with a steady-state water table (A), results are of conditions following 5 (B), 10 (C), and 15 (D) days of infiltration from snowmelt and its effect on growth of water-table mounds and local ground-water flow systems. The final part, (E), shows conditions after 7 months of redistribution. Note that seepage from surface water is about to resume at this time (data from Winter, 1983).

water balance of lakes in the Basin and Range province is considerably different from other lakes.

Direct input of water from rainfall is minimal in most desert lakes. Nearly all water input is derived from streamflow. Because these lakes commonly occupy the lowest topographic position in a drainage basin, it generally is assumed that they receive ground-water inflow, but that it is minimal (Langbein, 1961). Because seepage losses to ground water are also assumed to be minimal, the only significant loss of water from these lakes is to the atmosphere.

The relatively simple water budget of streamflow gain and evaporation loss has been reported in several studies of closed-basin lakes in Oregon (Phillips and Van Denburgh, 1971), although some water may come from springs (Van Denburgh, 1975; Jones, 1965). Even though the ground-water contribution of springs to desert lakes has been determined in a few studies, it should be noted that a thorough study of ground-water flow near desert lakes has never been made. Such studies are needed not only to determine the seepage into lakes but also to test the assumption that the lakes have minimal to no seepage outflow.

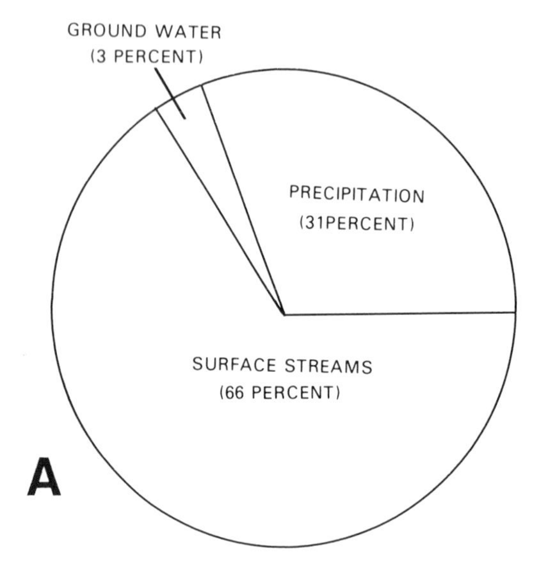

Figure 19. Magnitude of (A) inflow components to Great Salt Lake, and (B) fluctuation of the level of Great Salt Lake, 1846 to 1985 (modified from Arnow, 1985).

The largest desert lake, and the one that receives the most public attention, is Great Salt Lake, Utah. The water balance of Great Salt Lake is dominated by stream inflow (Fig. 19A) and evaporation losses. Great Salt Lake has been the subject of geological studies of its history, and records of level changes have been maintained since the 1840s (Fig. 19B).

As stated by Arrow (1984), Great Salt Lake is the modern remnant of a much larger water body, Lake Bonneville, which covered about 51,800 km^2 in Utah, Nevada, and Idaho during the most recent ice age of the Pleistocene Epoch. Lake Bonneville reached its maximum level, about 305 m above the present surface of Great Salt Lake, about 16 to 17 ka. About 11 ka, the lake declined to its current level of about 1,280 m above sea level.

On July 1, 1984, the lake level was at an altitude of 1,282.98 m above sea level, and the lake covered an area of about 5,957 km^2. This level was still below the historic high level in 1873 at 1,283.67 m above sea level. At that time, the lake surface covered about 6,475 km^2. At the other extreme, the lowest lake level was recorded in 1963 at 1,277.52 m above sea level, when the lake covered less than 2,590 km^2.

The lake level has a yearly cycle; it begins to decline in the spring or summer, when loss of water by evaporation from the lake surface is greater than the combined inflow from surface streams, ground water, and precipitation directly on the lake. The level begins to rise in autumn, when the loss of water by evaporation is exceeded by inflow. According to past records, the rise can begin at any time between September and December, and the decline any time between March and July. Thus, the level and volume of the lake reflect a dynamic equilibrium between inflow and evaporation. The surface area and brine concentration are the major aspects of the lake that affect the volume of evaporation. During dry years the water level declines, causing a decrease in surface area; consequently, the volume of evaporation decreases. Moreover, as the lake level declines, the brine generally becomes more concentrated, which also decreases the rate of evaporation. During wet years, the water level rises, causing an increase of surface area; consequently, the volume of evaporation increases. As the lake rises, the brine generally becomes less concentrated, which also increases the rate of evaporation (Arnow, 1984).

Sinkhole lakes

Lakes commonly occur in karst terrain where the water table is at shallow depth. Although lakes of this type occur in many places in North America, the areas are small and scattered. Florida is an exception; there, nearly all the lakes are related directly or indirectly to sinkholes. Some lakes in Florida lie directly within the limestone, but most actually occur within surficial clastic sediment overlying the limestone. The depression containing the lake is a consequence of the collapse of limestone at depth.

The water balance of lakes in Florida generally depends little on gains and losses from streamflow. Precipitation is a major source of water, and evaporation accounts for the major loss (Hughes, 1974, 1979). Because of the sandy geologic setting of most lakes in Florida, ground water is also a significant source of water (Lichtler and others, 1976). Although recent work in Florida on ground-water flow near lakes indicates ground-water inflow is common, seasonal reversals of seepage can occur, and lake water seeps to ground water in some areas of the lakebeds.

It is commonly assumed that lakes in Florida are hydraulically connected to the Floridan aquifer and that the hydraulic head of the Floridan aquifer is the primary control on lake levels. However, numerical modeling indicates that the lakes could be affected most by flow systems in the surficial sand aquifer, which is commonly separated from the Floridan aquifer by the confining Tertiary Hawthorne Formation. Hydraulic heads in the Floridan aquifer have a significant effect on seepage only if the confining bed at the base of the surficial sand is discontinuous.

A different type of sinkhole lake occurs in the Yucatan Peninsula of Mexico. These lakes, called cenotes, lie directly within the limestone, and the lakes are simply a surface expression of ground water within the limestone (De la O Carreño, 1950).

EFFECT OF HYDROGEOLOGIC AND CLIMATIC SETTING ON CHEMICAL CHARACTERISTICS OF LAKES AND WETLANDS

Chemical processes within lakes and wetlands are complex, and many occur without much heed to subtle changes in the physical world around them. That is, once an element or compound exists within an aquatic ecosystem, much of the concentration of different forms of the element at any given time depends on internal cycling, largely through biochemical processes within that ecosystem.

However, many processes depend on the flux of chemicals to and from the ecosystem; these fluxes are the cause of elements and compounds existing in the ecosystem. Therefore, to understand an ecosystem thoroughly it is necessary to understand the transport of water and chemicals to and from that ecosystem. As discussed thus far in this chapter, the water balance of lakes and wetlands is variable in space and time with respect to which components are dominant and which are not.

Lakes and wetlands dominated by surface-water inflow and outflow largely reflect the chemistry of the river. Lakes and wetlands that gain most water from surface-water inflow and lose most water by evapotranspiration are an ultimate sink for chemicals. Materials brought into the water body have no way of leaving, and these water bodies become increasingly saline.

Desert lakes are the best example of these types of water bodies, and they are typically brackish or saline. Furthermore, the salinity of desert lakes directly reflects the variations in the hydrologic balance at the lake. For example, Arnow (1984, p. 31–33) indicated that:

> The salinity of the brine in Great Salt Lake before 1959 varied inversely with the lake level. . . . During 1869, for example, when the lake was within a few feet of its historic high level, the concentration of dissolved minerals was 15 percent of the brine weight. During 1930, however, when the lake was about 10 ft lower, the mineral concentration was 21 percent.
>
> Between 1957 and 1959, the Southern Pacific Transportation Co. built a railroad causeway, which divided the lake and restricted the movement of the brine. The southern part of the lake receives more than 90 percent of the freshwater inflow, whereas the inflow to the northern part is nearly all brine that moves through the causeway from the southern part. Thus, the water in the southern part always was higher and fresher than the water in the northern part of the lake. From 1959 to 1982, the brine concentration north of the causeway remained relatively constant at or close to saturation regardless of changes in lake levels. The concentration decreased somewhat, however, during the large lake-level rises of 1983 and 1984. The brine south of the causeway was close to saturation during the historic low lake level in 1963. As the lake rose, the salinity of the brine south of the causeway continued to change inversely with the lake level, but the salinity was less than it would have been before the construction of the causeway. In 1977, for example, at a lake level of about 4,200 ft, the mineral concentration was approximately 12 percent, whereas before 1957 at the same level, it would have been more than 20 percent.

The only significant way desert lakes can lose material is by deflation from dry areas of the lakebeds. If a desert lake does not have high salinity, the lower salinity could be an indication that some water might be seeping to ground water, or that the lake could have intermittent overflows to adjacent topographically lower areas during high stages.

Lakes and wetlands dominated by water gains from the atmosphere are generally chemically dilute. Concentrations of chemicals in precipitation are usually minimal; therefore, there is minimal chemical input from the atmosphere. It is likely that dry deposition of chemicals into lakes and wetlands is greater than wet deposition. However, reliable data on inputs from dry deposition are rarely available. Even though evaporation should concentrate the chemicals existing in water bodies that have precipitation as the major input, the fact that these water bodies have minimal surface water or ground-water inflow usually indicates that they are topographically high and are therefore likely to have water and chemical loss to seepage (see wetland T8, Fig. 16).

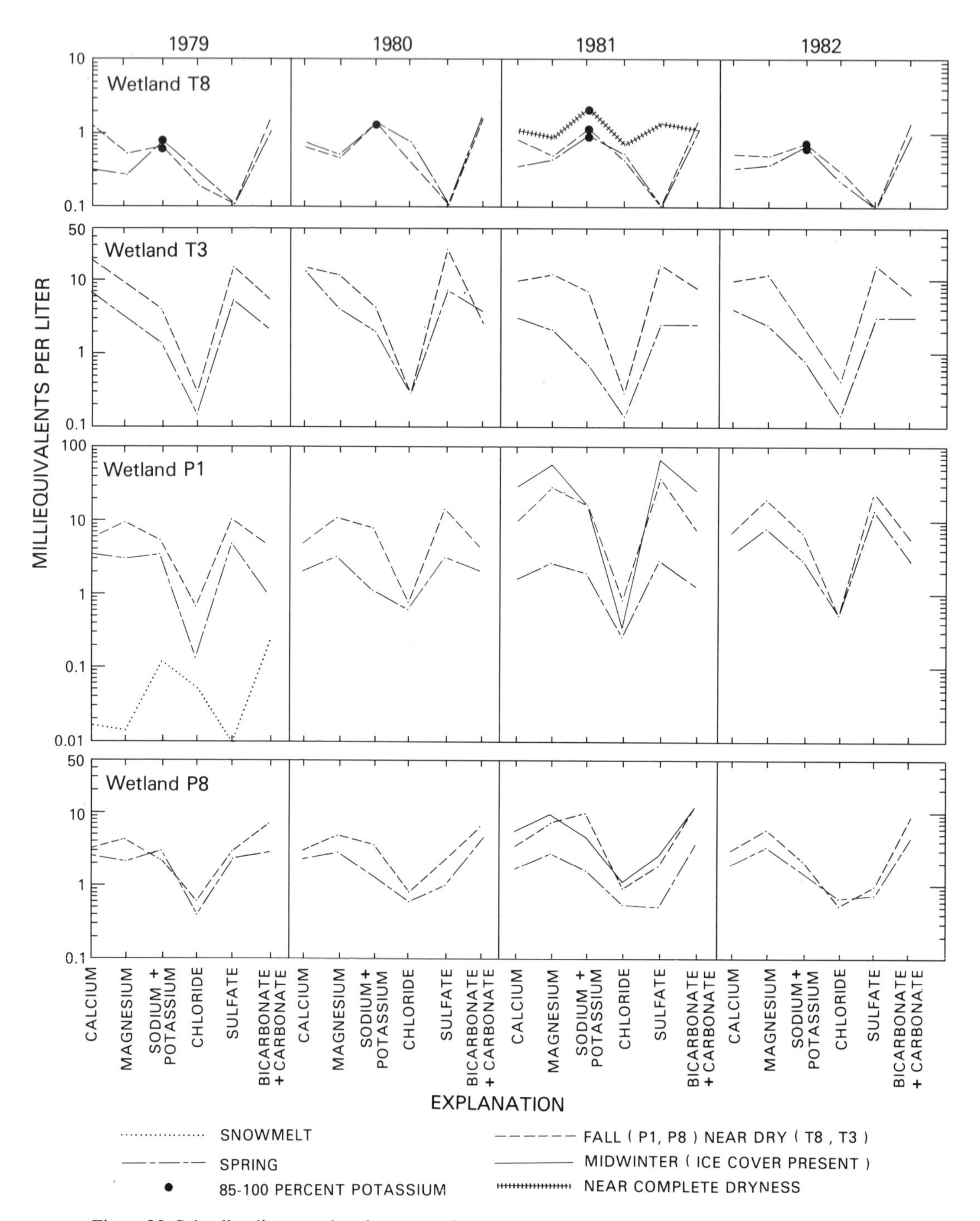

Figure 20. Schoeller diagrams showing seasonal and annual changes in the major-ion composition of the water in wetlands in the Cottonwood Lake area, North Dakota (modified from LaBaugh and others, 1987).

Lakes and wetlands dominated by ground-water inflow generally have chemical characteristics similar to ground water. However, even if lakes and wetlands are dominated by atmospheric-water components, if ground water is only as much as 10 to 20 percent of the water input, the water body commonly reflects the chemistry of ground water because ground water is the dominant source of chemicals.

The relation of ground-water flow systems, ground-water chemistry, and lake and wetland chemistry is most evident in prairie lakes and wetlands. In the regional hydrologic section through central North Dakota shown in Figure 15, flow systems of different magnitudes interacting with lakes and wetlands result in a variety of water chemical types and salinities, even in adjacent lakes. The lakes that are topographically low are extremely saline (dissolved solids as much as 70,000 mg/L) because they receive input from local and regional ground-water flow systems and they lose water only to the atmosphere.

However, even in small areas that are topographically high, lakes and wetlands have a range in hydrologic function, and their chemical characteristics reflect their position relative to ground-water flow systems. As pointed out previously (Fig. 16), the Cottonwood Lake area, North Dakota, is an excellent example of adjacent small water bodies having different hydrologic functions. At this site, the relation of hydrologic setting and temporal variability in hydrology to nutrient content and geochemical characteristics of a group of prairie wetlands and adjacent ground water was studied during 1979–1982 (LaBaugh and others, 1987). Although data were collected from many wetlands at the study site, emphasis in that report is primarily on four wetlands, two seasonally flooded and two permanently flooded. Two of the wetlands, T8 and T3, contained water only for a few weeks to months after filling in spring and early summer; both were completely dry by August. The permanent wetlands, P1 and P8, are in areas of ground-water discharge. None of the wetlands received water by channelized surface-water inlets. Only wetland P8 has a channelized, intermittent surface-water outlet. Ground-water level data indicated that high points of the water table did not always occur beneath land-surface highs. Reversals of ground-water flow occasionally occurred between two of the wetlands.

Significant differences existed in the chemical composition of the wetlands based on their hydrologic setting. In general, the dominant cations and anions in the wetlands were potassium and bicarbonate in wetland T8, calcium and sulfate in wetland T3, magnesium and sulfate in wetland P1, and magnesium and bicarbonate in wetland P8 (Fig. 20). Significant seasonal differences existed in the water chemistry of the wetlands in ground-water discharge areas. Water in three of the wetlands, T3, P1, and P8, was most dilute while they filled in spring after ice melt. Concentration increased during the open-water period, and two of the wetlands, P1 and P8, became most concentrated under ice cover.

Concentrations of total phosphorus and total nitrogen in wetlands were greatest in areas of ground-water recharge and least in areas of ground-water discharge. Differences in the chemistry of water from wells in the adjacent ground water resulted primarily from position of the wells in the ground-water flow system (Fig. 16). The chemical type of water from well 12, which was located in a ground-water recharge area, was calcium sodium bicarbonate. Water from well 4, located downgradient from wetland T8, and water from well 16, located downgradient from wetland P1, were typically calcium-sulfate type. Water from well 13, located between wetlands T3 and P1 in an area of changing ground-water flow direction, was a magnesium-sulfate type.

Although the above relations are particularly evident in prairie environments, the mutual effects of lake and ground-water interchange also are evident in larger lakes and in more humid environments. For example, Groschen (1981) indicated that Williams Lake, Minnesota, has chemical characteristics related to ground-water input. Conversely, lake water seeping into the ground-water system creates a plume of ground water that has a chemistry reflecting the chemistry of the lake water.

The dynamic reversals in seepage direction related to near-shore focused recharge can have an effect on lake and wetland water chemstry and, in some cases, on biology. For example, Hurley and others (1985) have determined that the flux of silica related to the increased ground-water inflow in the spring results in a diatom bloom that would not be expected, based on the silica concentration in the lake the remainder of the year.

Because a lake or wetland and its related ground-water system meet at the lakebed, it is conceivable that benthic organisms could be key indicators of ground-water interaction with these surface-water bodies. In some cases the indication is clear. For example, cattails (a freshwater plant) growing in scattered areas of saline lakes indicates fresh ground-water inflow. Conversely, as seen in wetland P1 in the Cottonwood Lake area, the absence of cattails along part of the shoreline is an indication of highly mineralized ground-water inflow in that particular segment of the wetland.

The above brief discussion is primarily intended to indicate a few important hydrogeologic and climatic processes that affect lake and wetland chemistry. A more complete discussion of the chemistry of lakes in North America is given by Hem and others (this volume), and the most complete synthesis on the limnology of North America is given in Frey (1966).

REFERENCES CITED

Adshead, J. D., 1983, Hudson Bay river sediments and regional glaciation; III, Implications of mineralogical studies for Wisconsin and earlier ice-flow patterns: Canadian Journal of Earth Sciences, v. 20, p. 313–321.

Arnow, T. 1984, Rise of Great Salt Lake, Utah: U.S. Geological Survey Water-Supply Paper 2275, p. 31–33.

—— , 1985, Water-level and water-quality changes in Great Salt Lake, Utah, 1843–1985: Utah Geological Association Publication 14, p. 227–235.

Assel, R. A., Quinn, F. H., Leshkevich, G. A., and Bolsenga, S. J., 1983, Great Lakes ice atlas: Ann Arbor, Michigan, National Oceanic and Atmospheric Administration, Great Lakes Environmental Research Laboratory, 115 p.

Barry, P. J., 1979, Hydrological and geochemical studies in the Perch Lake basin; A second report of progress: Chalk River, Ontario, Atomic Energy of Canada Ltd., 295 p.

Bay, R. R., 1967, Evaporation from two peatland watersheds: 14th General Assembly, International Union of Geodesy and Geophysics, Bern, Switzerland, p. 300–307.

—— , 1968, The hydrology of several peat deposits in northern Minnesota, U.S.A.: Third International Peat Congress, Ottawa, Runge Press, p. 212–218.

—— , 1969, Runoff from small peatland watersheds: Journal of Hydrology, v. 9, p. 90–102.

Boelter, D. H., 1965, Hydraulic conductivity of peat: Soil Science, v. 100, p. 227–231.

—— , 1969, Physical properties of peats as related to degree of decomposition: Soil Science Society of America Proceedings, v. 33, p. 606–609.

Brook, G. A., and Ford, D. C., 1980, Hydrogeology of the Nahanni karst, northern Canada, and the importance of extreme summer storms: Journal of Hydrology, v. 46, p. 103–121.

Brown, J., Dingman, S. L., and Lewellen, R. I., 1968, Hydrology of a drainage basin on the Alaskan coastal plain: U.S. Army Cold Regions Research and Engineering Laboratory Research Report 240, 18 p.

Carson, C. E., and Hussey, K. M., 1962, The oriented lakes of Arctic Alaska: Journal of Geology, v. 70, p. 417–439.

Cooke, G. G., McComas, M., Bhargava, T. N., and Heath, R., 1973, Monitoring and nutrient inactivation on two glacial lakes before and after nutrient diversion: Kent State University, Center for Urban Regionalism Interim Research Report, 92 p.

Costa, J. E., and Schuster, R. L., 1988, The formation and failure of natural dams: Geological Society of America Bulletin, v. 100, p. 1054–1068.

Cowardin, L. M., Carter, V., Golet, F. C., and LaRoe, E. T., 1979, Classification of wetlands and deepwater habitats of the United States: Washington, D. C., U.S. Fish and Wildlife Service, 103 p.

De la O Carreño, A., ed., 1950, Los recursos naturales de Yucatan: Boletín de la Sociedad Mexicana de Geografiá Estadista, v. 59, 377 p.

Dethier, D. P., 1979, Atmospheric contributions to stream water chemistry in the North Cascade Range, Washington: Water Resources Research, v. 15, no. 4, p. 787–794.

Dingman, S. L., 1973, Effects of permafrost on stream flow characteristics in the discontinous permafrost zone of central Alaska, *in* Proceedings, North American Contribution to Second International Conference on Permafrost: Washington, D. C., National Academy of Sciences, p. 447–453.

Dolan, T. J., Hermann, A. J., Bayley, S. E., and Zoltek, J., 1984, Evapotranspiration of a Florida, U.S.A., freshwater wetland: Journal of Hydrology, v. 74, p. 355–371.

Findlay, B. F., 1966, The water budget of the Knob Lake area, *in* Hydrological studies in Labrador-Ungava: McGill Subarctic Research Paper 22, p. 1–95.

FitzGibbon, J. E., 1982, The hydrologic response of a bog-fen complex to rainfall, *in* Canadian Hydrology Symposium 82; Hydrological processes of forested areas: Ottawa, National Research Council of Canada, p. 333–346.

FitzGibbon, J. E., and Dunne, T., 1981, Land surface and lake storage during snowmelt runoff in a subarctic drainage system: Arctic and Alpine Research, v. 13, p. 277–285.

Flint, R. F., 1971, Glacial and Quaternary geology: New York, John Wiley, 892 p.

Freeze, R. A., 1969, Theoretical analysis of regional ground-water flow: Canadian Department of Energy, Mines, and Resources, Inland Waters Branch, Scientific Series no. 3, 147 p.

Frey, D. G., editor, 1966, Limnology in North America: Madison, University of Wisconsin Press, 734 p.

Frye, J. C., Willman, H. B., and Black, R. F., 1965, Outline of glacial geology of Illinois and Wisconsin, *in* Wright, H. E., Jr., and Frey, D. G., eds., The quaternary of the United States: Princeton, New Jersey, Princeton University Press, p. 43–61.

Gilbert, R., 1971, Observation on ice-dammed Summit Lake, British Columbia, Canada: Journal of Glaciology, v. 10, p. 351–356.

Groschen, G. E., 1981, Geochemistry of Williams Lake, Hubbard County, Minnesota [M.S. thesis]: Minneapolis, University of Minnesota, 136 p.

Hartman, C. W., and Carlson, R. F., 1973, Water balance of a small lake in a permafrost region: University of Alaska Institute of Water Resources Report IWR-42, 23 p.

Hennings, R. G., 1978, The hydrogeology of a sand plain seepage lake, Portage County, Wisconsin [M.S. thesis]: Madison, University of Wisconsin, 69 p.

Herdendorf, C. E., 1982, Large lakes of the world: Journal of Great Lakes Research, v. 8, no. 3, p. 379–412.

Hofstetter, R. H., 1983, Wetlands in the United States, *in* Gore, A.J.P., ed., Ecosystems of the world; 4B, Mires: Swamp, bog, fen and moor: Regional studies: Amsterdam, Elsevier, p. 201–244.

Hughes, G. H., 1974, Water balance of Lake Kerr; A deductive study of a landlocked lake in north-central Florida: Florida Bureau of Geology Report of Investigations no. 73, 49 p.

—— , 1979, Analysis of water-level fluctuations of Lakes Winona and Winnemissett: U.S. Geological Survey Water-Resources Investigations 79-55; 30 p.

Hurley, J. P., Armstrong, D. E., Kenoyer, G. J., and Bowser, C. J., 1985, Groundwater as a silica source for diatom production in a precipitation-dominated lake: Science, v. 277, p. 1576–1578.

Hurr, R. T., 1983, Ground-water hydrology of the Morman Island Crane Meadows Wildlife Area near Grand Island, Hall County, Nebraska: U.S. Geological Survey Professional Paper 1277-H, 12 p.

Hutchinson, G. E., 1957, A treatise on limnology; v. 1, Geography, physics, and chemistry: New York, John Wiley, 1015 p.

Ingram, H.A.P., 1983, Hydrology, *in* Gore, A.J.P., ed., Ecosystems of the World. 4A, Mires: Swamp, bog, fen and moor; General studies: Amsterdam, Elsevier, p. 67–158.

Jones, B. F., 1965, The hydrology and mineralogy of Deep Springs Lake, Inyo County, California: U.S. Geological Survey Professional Paper 502-A, p. 1–56.

Kane, D. L., and Slaughter, C. W., 1973, Recharge of a central Alaska lake by subpermafrost groundwater, *in* Proceedings, North American Contribution to Second International Conference on Permafrost: Washington, D. C., National Academy of Science, p. 458–462.

Kirk, P. W., editor, 1979, The Great Dismal Swamp: Charlottesville, University of Virginia Press, 427 p.

LaBaugh, J. W., Winter, T. C., Adomaitis, V., and Swanson, G. A., 1987, Geohydrology and chemistry of prairie wetlands, Stutsman County, North Dakota: U.S. Geological Survey Professional Paper 1431 (in press).

Langbein, W. B., 1961, Salinity and hydrology of closed lakes: U.S. Geological Survey Professional Paper 412, 20 p.

Lichtler, W. F., Hughes, G. H., and Pfischner, F. L., 1976, Hydrologic relations between lakes and aquifers in a recharge area near Orlando, Florida: U.S. Geological Survey Water-Resources Investigations 76-65, 61 p.

Lissey, A., 1971, Depression-focused transient groundwater flow patterns in Manitoba: Geological Association of Canada Special Paper 9, p. 333–341.

Mackay, J. R., 1956, Notes on oriented lakes of the Liverpool Bay area, Northwest Territories: Review of Canadian Geography, v. 10, p. 169–173.

Mann, W. B. IV, and McBride, M. S., 1972, The hydrologic balance of Lake Sallie, Becker County, Minnesota: U.S. Geological Survey Professional Paper 800-D, p. D189–D191.

Meyboom, P., 1966, Unsteady groundwater flow near a willow ring in hummocky moraine: Journal of Hydrology, v. 4, p. 38–62.

——, 1967, Ground water studies in the Assiniboine River drainage basin; Part II, Hydrologic characteristics of phreatophytic vegetation in south-central Saskatchewan: Geological Survey of Canada Bulletin 139, 64 p.

Meyboom, P., van Everdingen, R. O., and Freeze, R. A., 1966, Patterns of ground water flow in seven discharge areas in Saskatchewan and Manitoba: Geological Survey of Canada Bulletin 147, 57 p.

Munro, D. S., 1979, Daytime energy exchange and evaporation from a wooded swamp: Water Resources Research, v. 15, p. 1259–1265.

Murdock, P. S., Peters, N. E., and Newton, R., 1986, Hydrologic analysis of two headwater lake basins of differing lake pH in the west-central Adirondack Mountains of New York: U.S. Geological Survey Water-Resources Investigations Report 84-4313, 49 p.

Pfannkuch, H. O., and Winter, T. C., 1984, Effect of anisotropy and groundwater system geometry on seepage through lakebeds; 1. Analog and dimensional analysis: Journal of Hydrology, v. 75, p. 213–237.

Philips, D. W., and McCulloch, J.A.W., 1972, The climate of the Great Lakes basin: Toronto, Atmospheric Environment, 40 p.

Phillips, K. N., and Van Denburgh, A. S., 1971, Hydrology and geochemistry of Abert, Summer, and Goose lakes, and other closed-basin lakes in south-central Oregon: U.S. Geological Survey Professional Paper 502-B, 86 p.

Possin, B. N., 1973, Hydrogeology of Mirror and Shadow lakes in Waupaca, Wisconsin [M.S. thesis]: Madison, University of Wisconsin, 85 p.

Price, J. S., and FitzGibbon, J. E., 1982, Winter hydrology of a forested drainage basin, *in* Canadian Hydrology Symposium 82; Hydrological processs of forested areas: Ottawa, National Research Council of Canada: p. 347–360.

Radforth, N. W., 1977, Muskeg hydrology, *in* Radforth, N. W., and Brawner, C. O., eds., Muskeg and the northern environment in Canada: Toronto, University of Toronto Press, p. 130–147.

Richards, T. L., 1967, Meteorological problems on the Great Lakes, *in* Dolman, C. E., ed., Water resources of Canada: Toronto, University of Toronto Press, p. 96–107.

Robison, W. C., Dodd, A. V., and Thompson, W. F., 1962, Muskeg, review of research: U.S. Army Quartermaster Research and Engineering Command, Earth Sciences Division Technical Report ES-5, 32 p.

Roulet, N. T., and Woo, M. K., 1986, Hydrology of a wetland in the continuous permafrost region: Journal of Hydrology, v. 89, p. 73–91.

Ryden, B. E., 1977, Hydrology of Truelove Lowland, *in* Bliss, L. C., ed., Truelove Lowland, Devon Island, Canada: A high arctic ecosystem: Edmonton, University of Alberta Press, p. 107–136.

Scheider, W. A., Moss, J. J., and Dillon, P. J., 1979, Measurement and uses of hydraulic and nutrient budgets, *in* Proceedings of the National Conference on Lake Restoration: Minneapolis, U.S. Environmental Protection Agency 440/5-79-001, p. 77–83.

Schiebe, F. R., Swain, A., Cooper, C. M., and Richie, J. L., 1984, Material budgets of Lake Chicot, *in* Nix, J. F., and Schiebe, F. R., eds., Limnological studies of Lake Chicot, Arkansas, Arkansas Lake Symposium: Arkadelphia, Arkansas, Ouachita Baptist University, p. 1–17.

Schindler, D. W., Welch, H. E., Kalff, J., Brunskill, G. J., and Kritsch, N., 1974, Physical and chemical limnology of Char Lake, Cornwallis Island (74°N Lat): Journal of Fisheries Research Board of Canada, v. 31, p. 585–607.

Schindler, D. W., Newbury, R. W., Beaty, K. G., and Campbell, P., 1976, Natural water and chemical budgets for a small Precambrian lake basin in central Canada: Journal of Fisheries Research Board of Canada, v. 33, p. 2525–2543.

Scott, J. S., 1976, Geology of Canadian tills, *in* Glacial till; an interdisciplinary study: Royal Society of Canada Special Publication no 12, p. 50–66.

Shjeflo, J. B., 1968, Evapotranspiration and the water budget of prairie potholes in North Dakota: U.S. Geological Survey Professional Paper 585-B, 49 p.

Siegel, D. I., 1981, Hydrogeologic setting of the glacial Lake Agassiz peatlands, northern Minnesota: U.S. Geological Survey Water-Resources Investigations 81-24, 32 p.

Siegel, D. I., and Winter, T. C., 1980, Hydrologic setting of Williams Lake, Hubbard County, Minnesota: U.S. Geological Survey Open-File Report 80-403, 56 p.

Sklash, M. G., and Farvolden, R. N., 1979, The role of groundwater in storm runoff, *in* Back, W., and Stephenson, D. A., eds., Contemporary hydrogeology: New York, Elsevier, p. 45–65.

Spahr, N. E., and Turk, J. T., 1985, Estimation of evaporation from Ned Williams Lake, Flat Tops Wilderness Area, Colorado: U.S. Geological Survey Water-Resources Investigations Report 85-4244, 13 p.

Sturges, D. L., 1968, Evapotranspiration at a Wyoming mountain bog: Journal of Soil and Water Conservation, v. 23, p. 23–25.

Swanson, G. A., Winter, T. C., Adomaitis, V. A., and LaBaugh, J. W., 1988, Chemical characteristics of prairie lakes in south-central North Dakota; Their potential for impacting fish and wildlife: U.S. Fish and Wildlife Service Technical Report 18, 44 p.

Toth, J., 1963, A theoretical analysis of groundwater flow in small drainage basins, *in* Proceedings of Hydrology Symposium no. 3, Groundwater: Ottawa, Queen's Printer, p. 75–96.

——, 1966, Mapping and interpretation of field phenomena for ground water reconnaissance in a prairie environment, Alberta, Canada: Bulletin of International Association of Scientific Hydrology, v. 11, no. 2, p. 1–49.

U.S. Geological Survey, 1983, Hydrological events and issues: U.S. Geological Survey Water-Supply Paper 2250, 243 p.

Van Denburgh, A. S., 1975, Solute balance at Abert and Summer lakes, south-central Oregon: U.S. Geological Survey Professional Paper 502-C, 29 p.

Vecchioli, J., Gill, H. E., and Lang, S. M., 1962, Hydrologic role of the Great Swamp and other marshland in upper Passaic River basin: Journal of American Water Works Association, v. 54, p. 695–701.

Verry, E. S., and Boelter, D. H., 1972, The influence of bogs on the distribution of streamflow from small bog-upland catchments; Hydrology of marsh-ridden areas, *in* Proceedings of Minsk Symposium: Paris, Unesco Press, p. 469–478.

Winter, T. C., 1976, Numerical simulation analysis of the interaction of lakes and ground water: U.S. Geological Survey Professional Paper 1001, 45 p.

——, 1978, Numerical simulation of steady state three-dimensional groundwater flow near lakes: Water Resources Research, v. 14, no. 2, p. 245–254.

——, 1983, The interaction of lakes with variably saturated porous media: Water Resources Research, v. 19, no. 5, p. 1203–1218.

——, 1984, Geohydrologic setting of Mirror Lake, West Thornton, New Hampshire, U.S. Geological Survey Water-Resources Investigations Report 84-4266, 61 p.

——, 1986, Effect of ground-water recharge on configuration of the water table beneath sand dunes and on seepage in lakes in the sandhills of Nebraska: Journal of Hydrology, v. 86, p. 221–237.

Winter, T. C., Maclay, R. W., and Pike, G. M., 1967, Water resources of the Roseau River watershed, northwestern Minnesota: U.S. Geological Survey Hydrologic Investigation Atlas HA-241, scale 1:250,000.

Woo, M. K., 1980, Hydrology of a small lake in the Canadian High Arctic: Arctic and Alpine Research, v. 12, p. 227–235.

Woo, M. K., and Valverde, J., 1981, Summer streamflow and water level in a mid-latitude forested swamp: Forest Science, v. 27, p. 177–189.

Woo, M. K., Heron, R., and Steer, P., 1981, Catchment hydrology of a high Arctic lake: Cold Regions Science and Technology, v. 5, p. 29–41.

Zoltai, S. C., and Pollett, F. C., 1983, Wetlands in Canada; Their classification, distribution, and use, *in* Gore, A.J.P., ed., Ecosystems of the world; 4B, Mires: Swamp, bog, fen and moor: Regional studies: Amsterdam, Elsevier, p. 245–269.

MANUSCRIPT ACCEPTED BY THE SOCIETY MAY 14, 1987

Printed in U.S.A.

The Geology of North America
Vol. O-1, Surface Water Hydrology
The Geological Society of America, 1990

Chapter 9

Hydrogeochemistry of rivers and lakes

John D. Hem
U.S. Geological Survey, MS 427, 345 Middlefield Road, Menlo Park, California 94025
Adrian Demayo
Environment Canada, Place Vincent Massey, 8th Floor, Ottawa, Ontario K1A 0E7, Canada
Richard A. Smith
U.S. Geological Survey, 412 National Center, Reston, Virginia 22092

INTRODUCTION

This chapter has three principal objectives: (1) to summarize the present chemical composition of North American surface waters and point out any discernible trends with time; (2) to review chemical and biochemical principles and processes that control natural water composition, and the ways in which these may be involved in attaining the particular chemical compositions and trends that we can observe; and (3) to point out some of the more important factors that must be considered in collecting surface-water-quality data.

This discussion is concerned principally with inorganic chemistry and geochemistry. However, biochemical processes in river and lake water influence their chemical composition extensively, and specific effects are pointed out. Aquatic biology is discussed in another chapter. The physical processes occurring in lakes also have very important effects on water chemistry; these aspects of limnology are covered more fully in a separate chapter.

Data on which this chapter is based relate mostly to waters of Canada, the United States, and Mexico. For most of Central America, very little water-quality information is available. Tables 1 through 3 contain analytical data for river waters that illustrate some of the principles discussed, and summarize major features of the chemical composition of North American surface waters.

To a certain extent, at least, one would expect the average concentrations of dissolved elements in surface fresh water to reflect the relative abundance of the elements in the crustal rocks exposed at and near the land surface. Such a broad generalization has some validity for the four major cations of most natural water, calcium (Ca^{2+}), magnesium (Mg^{2+}), sodium (Na^+), and potassium (K^+), and for silicon combined with oxygen as dissolved silica. These five elements are among the seven most abundant in igneous and sedimentary rocks, after oxygen (Hem, 1985). The other two of the seven elements, Al and Fe, form oxides or hydroxides of very low solubility at near-neutral pH and therefore do not normally go into solution in major amounts in river or lake water.

Major anions in surface water, on the other hand, display a more complex relation to rock composition. In the average river water the five most abundant anions are bicarbonate (HCO_3^-), sulfate (SO_4^{2-}), chloride (Cl^-), fluoride (F^-), and nitrate (NO_3^-), primarily representing the nonmetals carbon, sulfur, chlorine, fluorine, and nitrogen, but also including oxygen in three of the anions.

Oxygen is by far the most abundant element in crustal rocks, composing 46.6 percent of the lithosphere (Goldschmidt, 1954). In rock minerals the predominant anionic species is O^{2-}, and water itself is almost 90 percent oxygen by weight. The other nonmetalic elements listed above are less widely distributed in rocks. Phosphorus is an abundant anionic element in igneous rocks; fluorine, sulfur, carbon, and chlorine are present in lesser amounts. These elements all play essential roles in life processes and, except for phosphorus and fluorine, they commonly occur in earth-surface environments in gaseous form, or forms readily soluble in water.

In a very broad general sense then, the major cationic components of river and lake waters tend to reflect composition of associated rocks and the relative resistance of minerals to weathering attack. The anions, which must be present in these water solutions in electrochemical balance with the cations, tend to reflect the importance of various chemical and biochemical processes that have broken down the rock minerals as well as the chemical, biochemical, and physical processes taking place in the aquatic and the surrounding environments. Thus, the dominance of bicarbonate anions in most river water is related to carbon dioxide from air and biological processes in soil. Dissolution of CO_2 in water produces the weak acid H_2CO_3, which attacks rock minerals.

Both sulfur and nitrogen participate in biologically mediated oxidation reactions producing H^+ that becomes available for weathering attack on rock minerals. For example, pyrite, FeS_2, may be converted to Fe^{2+} and SO_4^{2-} by oxidation. Sedimentary

Hem, J. D., Demayo, A., Smith, R. A., 1990, Hydrogeochemistry of rivers and lakes, *in* Wolman, M. G., and Riggs, H. C., eds., Surface water hydrology: Boulder, Colorado, Geological Society of America, The Geology of North America, v. O-1.

TABLE 1. MINIMUM AND MAXIMUM VALUES OF MAJOR ION CONCENTRATIONS AND RELATED PROPERTIES OBSERVED IN SOME CANADIAN STREAMS, 1976 to 1982
(GEMS/WATER Data Summary, 1983, Environment Canada data records, and U.S. Geological Survey annual data reports. Analyses in mg L^{-1} except specific conductance and pH.)

Description	1		2		3		4		5		6		7	
	Min	Max	Min	Max	Min	Max	Min	Max	Min	Max	Min	Max	Min	Max
Silica (SiO_2)	1.9	2.8	3.0	5.2	2.0	3.3	3.9	6.9	1.5	2.9	4.5	7.3	1.4	4.8
Calcium (Ca)	1.1	2.3	15.7	27	1.5	2.9	12	21	12	19	27.5	62.3	29	43
Magnesium (Mg)	.41	.79	3.2	6.1	.4	.7	---	---	6	9	6.6	16.3	4	12
Sodium (Na)	.5	.7	1.0	2.4	1.4	3.5	1	5	2	5	1.4	4.2	2	14
Potassium (K)	<.1	.2	.1	.8	<.1	.4	.5	.9	.6	3.3	.5	.8	.7	1.5
Bicarbonate (HCO_3)	3.2	9.9	56	77	2.8	11	40	77	52	110	77	202	89	137
Sulfate (SO_4)	1.7	2.0	9.0	25	2.5	5.1	6	12	9	16	15	40	22	42
Chloride (Cl)	.2	.8	.5	1.7	1.6	3.0	1	5	2	8	.5	3.5	2	19
Fluoride (F)	.02	.04	.1	2.6	<.02	---	.0	.1	.1	.1	<.05	.13	.1	.1
Nitrate + Nitrite (as NO_3)	<.04	.44	.13	.75	<.04	.53	.00	.66	.13	.93	.09	.75	.09	.8
Dissolved solids	---	---	---	---	---	---	---	---	---	---	104	225	---	---
Discharge ($m^3 sec^{-1}$)	---	---	811	3,350	362.4	484.2	---	---	---	---	---	---	---	---
Specific Conductance ($\mu S cm^{-1}$, 25°C)	12	23	116	180	22	35	---	160	110	206	165	435	210	350
pH	5.1	7.3	7.2	8.3	6.7	6.8	7.5	8.0	7.5	7.9	7.6	8.4	7.4	8.5

Description	8		9		10 (date of sample)		11		12 (date of sample)		13 (date of sample)	
	Min	Max	Min	Max	3/6/80	2/11/76	Min	Max	20/7/82	24/3/81	27/4/78	8/9/78
Silica (SiO_2)	3.2	6.9	.6	2.6	.0	.4	---	---	2.0	4.0	14	12
Calcium (Ca)	6.0	27.6	24	32	35	39	---	---	7.3	13	46	70
Magnesium (Mg)	2.3	10.7	8	12	8.0	8.7	---	---	2.1	3.4	17	28
Sodium (Na)	3.4	6.4	9	18	11	10	---	---	1.8	3.8	14	59
Potassium (K)	.3	.8	1.4	2.9	1.1	1.4	---	---	.5	1.2	5.5	6.3
Bicarbonate (HCO_3)	24	119	93	132	110	117	---	---	28	49	60	232
Sulfate (SO_4)	4.5	9.0	11	27	25	28	---	---	4.0	6.1	63	100
Chloride (Cl)	4.7	12	3	19	18	23	---	---	1.9	4.7	12	84
Fluoride (F)	.03	.05	.1	.2	.2	.1	---	---	.1	<.1	.2	.2
Nitrate + Nitrite (as NO_3)	1.0	7.0	.04	.84	1.9	.75	.25	.97	<.4	.40	5.8	.09
Dissolved solids	---	---	---	---	154	168	---	---	34	60	251	474
Discharge ($m^3 sec^{-1}$)	.2	1.92	---	---	6,881	6,371	---	---	668	220	411	30.9
Specific Conductance ($\mu S cm^{-1}$, 25°C)	67	256	205	401	285	320	30	605	65	124	410	805
pH	6.5	8.2	7.7	8.6	7.8	7.5	6.6	8.1	7.4	7.4	8.1	8.4

See key to column heads at end of Table 1 (continued on next page).

rocks containing carbonate or sulfate (limestone and gypsum, for example) add substantially to the HCO_3^- and SO_4^{2-} contents of some waters.

Chloride is much less involved in weathering processes than are sulfur and carbon. Most geochemists believe that much, or most, of the chloride in river water is derived from sea salt carried landward and deposited by rainfall and dry fallout. A considerable part of the river chloride loads, however, can be the result of human activities.

Some of the more important processes that influence the chemical composition of river water will be briefly reviewed here, and some of the techniques that have been used to study them will be described. The summary of chemical composition of North America's surface waters, which concludes this chapter, will then be more readily understandable as the end result of the interaction of many processes and the influence upon them of human activities on this continent that intensified greatly after the first European colonists arrived.

TABLE 1. (CONTINUED)

Description	14		15 (date of sample)		16		17		18		19		20	
	Min	Max	29/9/80	27/6/77	Min	Max	Min	Max	Min	Max	Min	Max	Min	Max
Silica (SiO_2)	2.3	4.5	.5	.2	---	---	0.3	4.8	---	---	4.4	5.1	<1	6.0
Calcium (Ca)	7.1	13	36	38	---	---	34	57	4.0	4.2	24	42	23.5	65.8
Magnesium (Mg)	1	2	7.8	8.0	---	---	12	20	.7	.8	6	9	9.7	23.7
Sodium (Na)	1	2	13	13	---	---	12	23	6.8	6.8	4	10	6.5	31
Potassium (K)	.1	.3	1.4	1.6	---	---	1.8	3.2	.4	.5	.4	1.3	1.0	5.2
Bicarbonate (HCO_3)	18	36	99	110	---	---	126	215	.6	2.6	76	119	110	219
Sulfate (SO_4)	5	8	26	27	---	---	32	66	8.9	10	2	31	24	94
Chloride (Cl)	1	2	26	27	---	---	4	17	11.6	14	2	8	1.6	9.6
Fluoride (F)	---	---	.2	.1	---	---	.1	.2	---	---	.1	.1	<.05	.21
Nitrate + Nitrite (as NO_3)	.04	3.3	.49	.53	<.04	.84	.04	1.7	<.04	.13	.04	.49	<.04	6.6
Dissolved solids	---	---	160	169	---	---	---	---	---	---	---	---	159	291
Discharge ($m^3\ sec^{-1}$)	---	---	8,383	7,080	---	---	---	---	2.66	4.70	---	---	---	---
Specific Conductance ($\mu S\ cm^{-1}$, 25°C)	50	90	270	370	162	273	240	477	73	73	160	254	293	594
pH	6.5	8.3	7.1	7.8	7.5	8.1	7.4	8.4	6.2	6.6	6.1	8.2	7.1	8.9

1. Churchill River at Muskrat Falls, Labrador, 53°15'N, 60°45'W. 18 samples 1979-81.
2. Columbia River below Trail B.C., 49°1'39"N, 117°36'11"W. 72 samples 1979-81.
3. Exploits River at Grand Falls, Newfoundland, 48°55'N, 55°40'W. 26 samples 1979-81.
4. Fraser River near Vancouver, B.C., 49°29'15"N, 121°27'0"W. 20 samples 1979-81.
5. Great Bear River at Norman Wells, N.W.T., 65°08'N, 123°31'W. 17 samples 1979-81.
6. Liard River above Fort Simpson, N.W.T., 61°44'22"N, 121°13'9"W. 32 samples 1979-81.
7. McKenzie River at Arctic Red River, N.W.T., 67°27'N, 133°42'W. 21 samples 1979-81.
8. Mill River, Prince Edward Island. 41 samples 1979-81.
9. Nelson River near mouth, Manitoba, 56°54'20"N, 93°21'50"W. 10 samples 1979-81.
10. Niagara River at Fort Niagara, N.Y.—Ontario, 43°9'N, 79°2'W. Analyses of samples of minimum and maximum conductance Oct. 1976 through Sept. 1977, Oct. 1979 through Sept. 1980. Drainage area 686,350 km². Long-term mean discharge at Buffalo, NY 5,777 $m^3\ sec^{-1}$. USGS "NASQAN" station represents outflow from Lake Erie into Lake Ontario.
11. Ottawa River at mouth at Ste. Anne de Bellevue, Quebec, 45°24'N, 73°57'W. 15 samples 1979-81.
12. Rainy River near Fort Francis, Ontario (at Manitou Rapids) 48°38'N, 93°54'W. Analyses for samples of minimum and maximum conductance during periods Oct. 1978 to Sept. 1979 and Oct. 1980 to Sept. 1982. Drainage area 50,200 km², long term mean discharge 362.2 $m^3\ sec^{-1}$. USGS "NASQAN" station.
13. Red River of the North at Emerson, Manitoba, 49°1'N, 97°10'W. Samples of minimum and maximum conductance during periods Oct. 1977 to Sept. 1978, Oct. 1979 to Sept. 1980, and Oct. 1982 to Sept. 1983. Drainage area 104,100 km². Long-term mean discharge 93.74 $m^3\ sec^{-1}$. USGS "NASQAN" station.
14. Saint John River near Frederickton, N.B., 47°14'55"N, 68°36'20"W. 17 samples 1979-81.
15. Saint Lawrence River at Cornwall, Ontario, 45°0'N, 78°48'W. Analyses for samples of minimum and maximum conductance Oct. 1976 to Sept 1977 and Oct. 1979 to Sept. 1980. Drainage area 773,890 km², mean discharge long term 6,873 $m^3\ sec^{-1}$. USGS "NASQAN" station, represents outflow from Lake Ontario.
16. St. Lawrence River at Levis, Quebec, filtration plant, 46°48'0"N, 71°11'0"W. 12 samples 1979-81.
17. Saskatchewan River at The Pas, Manitoba, 53°50'30"N, 101°20'6"W. 34 samples 1979-81.
18. Shubenacadie River at Elmsdale, Nova Scotia, Lat. 44°58'N. 63°30'W. 4 samples 1979-81.
19. Slave River at Fitzgerald, Alberta, 59°52'10"N, 111°35'10"W. 16 samples 1979-81.
20. South Saskatchewan River at Highway 41 bridge, Alberta, 50°44'15"N, 110°5'44"W. 32 samples 1979-81.

GEOCHEMICAL PROCESSES IN THE HYDROLOGIC CYCLE

Properties of water substance

Water, H_2O, has certain unique physical and chemical properties that govern its occurrence and behavior toward other components of the earth's surface and atmosphere. In the gas, or vapor state, the molecules lead separate existences. The distance between the centers of the oxygen and hydrogen ions in the molecule is 0.0958 nm and the two O–H bonds describe a fixed angle of almost 105° (104°27′). This arrangement gives the molecule a dipolar nature; that is, the side with the two hydrogens has a positive electrostatic charge compared to the opposite side of the molecule.

TABLE 2. MAJOR-ION CONCENTRATIONS AND RELATED DATA FOR SAMPLES HAVING MINIMUM AND MAXIMUM SPECIFIC CONDUCTANCE COLLECTED AT SELECTED U.S. GEOLOGICAL SURVEY SAMPLING STATIONS DURING TIME PERIODS INDICATED

(Analyses in mg L^{-1}, except for specific conductance and pH. Table 2 continues on following pages)

Description	1		2		3		4		5	
Date of Sample (day/mo/year)	10/02/82	20/10/81	27/03/79	13/02/80	06/03/79	08/09/81	03/01/68	02/10/68	17/05/82	18/08/81
Silica (SiO_2)	5.5	6.0	4.0	5.5	7.2	7.4	6.1	8.7	9.1	14
Calcium (Ca)	8.8	15	2.9	7.3	5.5	15	17	51	53	78
Magnesium (Mg)	1.7	4.2	.5	1.2	.8	.9	3.0	11	21	58
Sodium (Na)	2.8	16	3.3	24	3.0	7.7	13	190	47	890
Potassium (K)	1.6	1.7	.7	1.4	1.1	1.1	2.3	4.4	5.7	51
Bicarbonate (HCO_3)	26	57	7.3	30	23	59	51	146	244	378
Sulfate (SO_4)	6.5	13	5.1	18	5.9	7.2	16	51	34	110
Chloride (Cl)	3.0	11	3.6	24	3.3	6.1	18	292	65	1,500
Fluoride (F)	---	.1	.0	.1	.1	.1	.1	.2	.2	.4
Nitrate (NO_3)	.89	.53	.62	.89	1.5	.27	.1	.2	1.1	1.6
Dissolved solids	44	90	24	98	39	86	92	686	355	2,890
Discharge at time of sampling ($m^3\ sec^{-1}$)	3,682	177	1,328	79.6	2,351	272	920	374	133	2.8
Specific Conductance ($\mu S\ cm^{-1}$, 25°C)	72	185	44	190	63	138	184	1,200	600	4,900
pH	7.0	7.7	6.4	6.6	7.0	7.7	7.3	8.2	8.0	---
Drainage area (km^2)	57,000		8,832		44,500		409,300		18,205	
Long-term mean water discharge ($m^3\ sec^{-1}$)	1,028		173.8		634.1		1,099		48.4	

1. Alabama River at Claiborne, Alabama, 31°33'N, 87°31'W. Time period Oct. 1978-Sept. 1979, Oct. 1980-Sept. 1982.
2. Androscoggin River near Auburn, Maine, 44°4'N, 70°13'W. Time period Oct. 1978-Sept. 1981.
3. Apalachicola River at Chattahoochee, Florida, 30°42'N, 84°52'W. Time period Oct. 1978-Sept. 1981.
4. Arkansas River at Little Rock, Arkansas, 34°40'07", 92°9'18". Time period Oct. 1967-Sept. 1970.
5. Bear River at Corinne, Utah, 41°35'N, 112°6'W. Time period Oct. 1978-Sept. 1979, Oct. 1980-Sept. 1982.

In the condensed forms, liquid water and ice, the molecules are much closer together, and the dipolar forces interact to produce characteristic effects such as the crystal structure of ice. In the liquid state, water molecules form insulating sheaths around inorganic solute ions, and the molecules are attracted to inorganic surfaces where chemical reactions of dissolution or precipitation can occur. Water is an excellent solvent for inorganic substances and plays a key role in rock-weathering processes. It is also essential for the existence of living organisms.

The tendency for attraction between liquid-state molecules is indicated by the high surface tension of liquid water. The density of the liquid is at a maximum near the temperature of 4°C. Below this temperature the density decreases and ice therefore forms first at the surface of lakes. Above 4°C the density decreases with temperature to the boiling point, 100°C at normal atmospheric pressure. Water exposed to the atmosphere vaporizes to some extent from both liquid and solid forms. The volume percentage of H_2O in the gas phase at equilibrium is near 3.0 at 25°C. If the temperature of the air is 0°C, however, it can contain only about 0.6 percent H_2O. Thus, when warm air that has accumulated water vapor near the earth's surface is cooled, it becomes supersaturated with water vapor, and condensation to the liquid form occurs. The liquid droplets, or micro–ice crystals, that result commonly form around solid nucleii such as dust particles, airborne sea salt, or various kinds of atmospheric particulates and aerosols introduced by human activities. Dissolution of these solids and solution of gases occur in clouds, and, when moisture is precipitated as rain or snow, a significant solute content may be present.

Isotopic composition of water

Oxygen and hydrogen occur in several different isotopic forms. The two most abundant stable isotopes of oxygen are ^{16}O and ^{18}O. In ocean water their abundances are 99.763 percent and 0.1995 percent, respectively. Hydrogen has two stable isotopes—^{1}H and ^{2}H, or deuterium (D)—with abundances in ocean water of 99.984 percent and 0.0156 percent, respectively. Water may therefore contain a fraction of molecules that are abnormally heavy because they contain one or more of the heavier isotopes. Heavy molecules tend to be less readily vaporized, and there is also some isotopic fractionation in biological and chemical proc-

TABLE 2. (continued)

Description	6		7		8		9		10	
Date of Sample (day, mo, yr)	23/05/79	24/08/81	14/06/82	05/02/79	12/07/78	06/04/79	29/11/79	16/09/80	19/04/77	11/08/77
Silica (SiO_2)	8.0	11	9.6	12	7.5	12	4.6	2.6	4.3	1.5
Calcium (Ca)	38	160	87	110	13	18	8.5	20	14	29
Magnesium (Mg)	11	49	32	35	3.6	5.7	2.4	7.1	2.5	4.9
Sodium (Na)	21	150	130	190	5.0	8.5	3.2	10	3.3	11
Potassium (K)	2.2	5.7	4.9	6.1	1.1	1.1	1.2	1.9	.6	1.7
Bicarbonate (HCO_3)	110	207	174	216	48	80	27	56	40	86
Sulfate (SO_4)	68	520	340	390	7.7	16	12	26	8.3	22
Chloride (Cl)	15	180	120	185	3.1	5.4	5.1	14	5.6	17
Fluoride (F)	.2	.5	.5	.6	---	---	.1	.1	.1	.0
Nitrate (NO_3)	1.8	7.1	.80	.70	.00	2.9	2.6	5.3	2.8	1.6
Dissolved solids	219	1,190	810	1,040	66	110	54	138	58	130
Discharge at time of sampling (m^3 sec^{-1})	994	54.1	218	83.8	---	---	952	85.8	620	120
Specific Conductance (µS cm^{-1}, 25°C)	355	1,770	1,150	1,630	115	188	81	225	107	350
pH	7.9	7.8	8.0	7.9	8.1	8.1	7.1	8.3	7.4	7.6
Drainage area (km^2)	62,420		488,200		(see notes)		17,560		20,953	
Long-term mean water discharge (m^3 sec^{-1})	213.6		308.7		(see notes)		332.8		387.7	

6. Colorado River near Cisco, Utah, 38°49'N, 109°18'W. Time period Oct. 1978-Sept. 1979, Oct. 1980-Sept. 1982.
7. Colorado River above Imperial Dam, Arizona, 32°53'N, 114°28'W. Time period Oct. 1978-Sept. 1979, Oct. 1980-Sept. 1982.
8. Columbia River at Bradwood, Oregon (60 km upstream from mouth). Time period Oct. 1976-Sept. 1979. Flow not measured at this site. Estimated flow at mouth averaged 6,340 m^3 sec^{-1} for this three-year period. Drainage area at mouth 665,900 km^2.
9. Delaware River at Trenton, New Jersey, 40°13'N, 74°47'W. Time period Oct. 1978-Sept. 1980, Oct. 1981-Sept. 1982.
10. Hudson River at Green Island, New York. 42°45'8"N, 73°41'22"W. Time period Oct. 1976-Sept. 1977, Oct. 1979-Sept. 1980.

esses. The degree of enrichment or depletion of ^{18}O and D can be determined by mass spectrometry; such data can be useful in many kinds of hydrologic studies (Fritz and Fontes, 1980).

Isotopic enrichment or depletion as $\pm\delta$ (per mil) for a particular isotope (e.g., $\delta^{18}O$ or δD) is calculated by the formula

$$\delta = \left(\frac{R_a}{R_{std}} - 1\right) 1{,}000,$$

where R_a is the ratio of isotopes in sample A, and R_{std} is the ratio in a standard. The standard composition commonly used in hydrologic studies is that of "standard mean ocean water" (SMOW), whose composition was indicated above.

Because of the evaporative fractionation, water in rain and snow is relatively depleted in heavy isotopes.

Composition of rain and snow

Some of the solutes contributed to rainfall by the atmosphere can constitute an important fraction of those dissolved species appearing in terrestrial surface water. For example, it is commonly supposed that chloride in surface water draining from areas near the ocean comes mostly from airborne sea salt. Certain trace metals, notably lead, are present in the atmosphere because of human activities, and they enter surface waters as components of rainfall.

Carbon dioxide from the atmosphere dissolves in rain water and, if other solutes affecting pH are not present in excessive amounts, buffer the pH of rainfall at a value near 5.6. Sulfur and nitrogen compounds that enter the atmosphere and undergo oxidation may produce strong acids that lower rainfall pH to values near 4.0 or even lower. In recent decades the pH of rainfall in parts of the eastern United States and Canada has declined substantially, and lakes in certain geologic terranes have become acidified (Cowling, 1982).

Some data on composition of rainfall in the United States are given in Table 4. The amounts and kinds of dissolved material tend to vary greatly, but in areas remote from the ocean and human-produced atmospheric pollution, most of the solutes in rainfall and snow are not a large factor in the total quantities of dissolved matter in lake and stream water.

The results obtained from Canadian studies of atmospheric deposition summarized by Whelpdale and Barrie (1982) indicate that pH values near or below 5.0 may occur in winter snowfall in arctic regions.

TABLE 2. (continued)

Description	11		12		13		14		15	
Date of Sample (day/mo/yr)	13/05/69	04/11/69	14/03/77	23/12/81	28/04/82	02/10/79	25/01/83	12/01/81	28/03/79	06/01/82
Silica (SiO_2)	11	24	9.2	18	8.1	7.5	7.6	7.2	10	7.1
Calcium (Ca)	9.3	18	41	136	29	58	34	51	55	66
Magnesium (Mg)	4.0	9.4	13	55	10	22	11	18	18	27
Sodium (Na)	2.3	15	16	31	4.3	8.3	11	23	37	85
Potassium (K)	.6	2.2	7.4	4.8	3.1	2.9	2.7	3.4	8.5	5.3
Bicarbonate (HCO_3)	47	113	132	475	106	219	118	173	146	219
Sulfate (SO_4)	3.0	19	54	180	8.0	55	35	72	150	250
Chloride (Cl)	.8	4.8	22	48	9.6	12	12	26	9.0	12
Fluoride (F)	.1	.2	.4	.4	.2	.2	.1	.2	.3	.5
Nitrate (NO_3)	1.2	1.9	17	43	6.6	10	8.0	6.6	5.8	---
Dissolved solids	55	151	228	708	125	284	173	287	360	561
Discharge at time of sampling ($m^3 sec^{-1}$)	1,274	123	123	---	4,588	717	22,430	16,570	1,272	510
Specific Conductance ($\mu S cm^{-1}$, 25°C)	91	233	370	1,160	275	495	315	512	600	860
pH	7.7	8.0	7.7	8.1	7.7	8.5	7.3	7.5	7.6	8.0
Drainage area (km^2)	31,340		42,000		221,700		2,914,500		814,800	
Long-term mean water discharge ($m^3 sec^{-1}$)	484.6		93.4		1,336		15,210		905	

11. Klamath River near Klamath, California (near mouth), 41°31'N, 123°58'W. Time period Oct. 1967-Sept.1970.
12. Minnesota River near Jordan, Minnesota, 44°42'N, 93°38'W. Time period Oct. 1976-Sept. 1977, Oct. 1978-Sept. 1979, Oct. 1981-Sept. 1982.
13. Mississippi River at Clinton, Iowa, 41°47'N, 90°15'W. Time period Oct. 1978-Sept. 1980, Oct. 1981-Sept. 1982.
14. Mississippi River near Saint Francisville, Louisiana, 30°46'N, 91°24'W. Time period Oct. 1978-Sept. 1983. Drainage area and discharge are for gaging station at Tarbert Landing.
15. Missouri River at Sioux City, Iowa, 42°29'N, 96°25'W. Time period Oct. 1978-Sept. 1980, Oct. 1981-Sept. 1982.

Climate and runoff composition

After rainfall reaches the earth's surface, the water and solutes it contains participate in chemical reactions with solid minerals, gases, and organic matter in soils and rocks. These reactions are the major source of solutes that finally appear in river and lake water. A major role of rainfall and snowmelt throughout the continent is related to the transport of soluble and insoluble reaction products toward the oceans, or to the lower portions of hydrologically closed basins. Where the ratio of runoff to rainfall is relatively high, the solute concentrations to be expected in runoff will be low. The diluting effect of unusually intense rainfall and runoff events can be discerned in the records of water composition at any fixed stream-sampling site. Thus, the diluting effect of runoff can induce extensive variability of solute concentrations wtih time in water at a single stream sampling point. The effects of differing climatic regimes are evident in persistent seasonal patterns of surface-water chemical composition. For example, a large part of the annual runoff in some drainage basins occurs from melting snow. Usually this water is very low in total solute concentrations, as surface soil may be frozen and biological activity is minimal. Snowmelt may have a low pH in some regions, and spring runoff to lakes and streams may cause a sudden lowering of pH. Where base flow is maintained by ground-water discharge from aquifers that contain readily soluble minerals, solute concentrations may reach high levels at low flow stages, and such effects can be magnified if low flows occur during warm seasons when flow may be decreased by evaporation and transpiration losses. Streams affected by these processes tend to display a wide seasonal range in solute concentrations. An extreme example is the Pecos River near the New Mexico–Texas border (Station 17 in Table 2). The maximum dissolved solids concentration shown for a 3-year period is more than 58 times as great as the minimum.

In desert climates many streams are ephemeral. Generally, except where soluble minerals occur or where soluble weathering products accumulate at the surface, the runoff that occurs in response to heavy rains has low solute concentrations. However, substantial amounts of sediment may be transported in the flood waters.

In the arctic regions of the continent, all or most of the precipitation occurs as snow, and ground water is present as permafrost. Here, the local runoff during warm seasons represents atmospheric water that had been stored in solid form, and weath-

TABLE 2. (continued)

Description	16		17		18		19		20	
Date of Sample (day/mo/yr)	21/05/80	19/10/78	27/09/78	06/04/78	14/12/76	16/11/76	02/04/79	18/11/82	24/07/79	12/02/80
Silica (SiO_2)	5.1	1.5	8.9	2.5	11	17	7.7	6.1	25	20
Calcium (Ca)	26	45	77	690	71	140	5.1	6.9	6.8	12
Magnesium (Mg)	6.4	13	12	400	18	36	2.1	2.9	2.8	5.3
Sodium (Na)	8.4	19	24	7,100	90	220	3.6	9.7	4.3	4.5
Potassium (K)	1.3	3.2	4.0	280	3.8	6.0	2.0	2.1	1.3	.8
Bicarbonate (HCO_3)	84	113	122	150	108	263	21	46	37	65
Sulfate (SO_4)	38	66	130	2,800	180	310	8.0	8.3	1.1	1.2
Chloride (Cl)	11	26	35	11,000	100	300	4.0	11	2.0	3.2
Fluoride (F)	.1	---	.1	.9	.4	.2	.0	.1	.1	.1
Nitrate (NO_3)	3.0	4.3	1.8	2.1	1.6	2.8	1.3	.40	.40	1.4
Dissolved solids	146	253	382	22,400	527	1,160	43	70	61	80
Discharge at time of sampling ($m^3 sec^{-1}$)	12,600	2,832	137	.18	193	42.2	527	237	65.1	108
Specific Conductance ($\mu S\ cm^{-1}$, 25°C)	215	420	525	30,800	895	1,870	60	125	81	134
pH	---	8.4	7.2	8.1	8.0	7.9	7.2	7.1	7.6	7.4
Drainage area (km^2)	526,000		50,600		456,702		21,780		10,202	
Long-term mean water discharge ($m^3 sec^{-1}$)	7,624		5.01		---		230.5		174.6	

16. Ohio River at Lock and Dam 53 near Grand Chain, Illinois, 37°12'N, 89°2'W. time period Oct. 1978-Sept.1981.
17. Pecos River at Red Bluff, New Mexico, 32°4'N, 104°2'W. Time period Oct. 1977-Sept. 1979, Oct. 1981-Sept. 1982.
18. Rio Grande at Brownsville, Texas, 25°53'N, 97°27'W. Time period Oct. 1976-Sept. 1977.
19. Roanoke River at Roanoke Rapids, North Carolina, 36°27'N, 77°38'W. Time period Oct. 1978-Sept. 1980, Oct. 1981-Sept. 1982.
20. Rogue River near Agness, Oregon, 42°35'N, 124°4'W. Time period Oct. 1979-Sept. 1981.

ering of rocks by aqueous solutions is minimal. Where snow accumulates from year to year, glaciers form. The continental ice sheets of past geologic history spread by plastic deformation of the heavy ice accumulation until the margins reached warm enough climates to permit melting to balance the movement rate. Modern glaciers and ice fields occur on a smaller scale in mountainous regions in the northwestern part of the continent. Some of these, as in Alaska and northern British Columbia, discharge ice into the Pacific Ocean with only minor amounts of melting. Farther south, the alpine glaciers discharge meltwater to mountain streams where they terminate. According to Reynolds and Johnson (1972), the rate of solute erosion in the temperate climate of the Cascade Mountains of Washington where glaciers are present can be very high. In part, this may be due to the grinding action of moving ice, which provides fine-grained debris having large surface areas per unit weight. The rate of dissolution of many natural minerals is controlled by the extent of solid-liquid interfaces.

Large ice accumulations, such as the Greenland ice cap, represent a fossil record of the chemical composition of atmospheric precipitation that can cover thousands of years. Ice cores from Greenland studied by Herron and others (1977) showed increased concentrations of sulfate, Zn, Cd, and Pb in ice more recent than about A.D. 1900.

Rock weathering and alteration

Weathering processes include the physical and chemical reactions that rock minerals undergo when they are exposed at the earth's surface. The processes that are most significant as controls over the chemical composition of natural water are reactions between rock minerals and water or dissolved material derived from the atmosphere or other sources. In these reactions the original rock minerals may be totally dissolved or altered to give new solid structures and release solute products.

Weathering processes that consist simply of the dissolution of solid minerals such as halite represent surrounding of Na^+ and Cl^- ions at the surface of the solid with water molecules, whence the hydrated ions are carried off in the solution. A somewhat more complex interaction is that of crystalline silica, such as quartz

$$SiO_2 + 2H_2O = Si(OH)_{4aq}.$$

Here, water interacts with the oxygen-silicon structure to produce hydroxyl ions that are tetrahedrally coordinated with silicon.

TABLE 2. (continued)

Description	21		22		23		24		25	
Date of Sample (day/mo/yr)	04/02/70	04/09/68	02/12/76	14/09/77	21/04/80	18/12/79	07/05/80	30/10/78	06/06/78	10/12/81
Silica (SiO_2)	15	20	10	9.9	20	31	11	24	4.2	4.1
Calcium (Ca)	10	15	3.3	4.3	21	45	77	240	19	24
Magnesium (Mg)	4.2	10	1.3	1.2	8.9	19	26	60	2.8	5.1
Sodium (Na)	4.7	18	5.4	10	21	43	50	220	3.7	11
Potassium (K)	1.1	1.6	1.5	2.1	3.2	5.1	8.9	21	1.4	1.8
Bicarbonate (HCO_3)	47	108	22	27	105	232	146	268	82	79
Sulfate (SO_4)	7.0	14	4.7	5.0	32	68	250	970	10	19
Chloride (Cl)	2.6	11	4.1	5.0	12	26	26	100	3.9	10
Fluoride (F)	.1	.2	.1	.1	.4	.6	.1	.6	.1	.1
Nitrate (NO_3)	.9	1.2	.66	1.7	.97	5.8	5.3	9.3	2.1	1.2
Dissolved solids	69	144	41	51	171	358	529	1,770	78	114
Discharge at time of sampling ($m^3 sec^{-1}$)	2,096	323	---	---	649	300	248	1.87	2,464	1,425
Specific Conductance ($\mu S cm^{-1}$, 25°C)	98	230	64	71	275	562	860	2,300	145	210
pH	7.2	8.2	6.4	6.8	8.7	8.6	8.0	8.4	7.2	7.5
Drainage area (km^2)	60,890		25,510		179,230		59,927		85,830	
Long-term mean water discharge ($m^3 sec^{-1}$)	668.1		343		511		13.45		1,550	

21. Sacramento River at Freeport, California, 38°27'N, 121°30'W. Time period Oct. 1967- Sept. 1970.
22. Savannah River near Clyo, Georgia, 32°32'N, 81°16'W. Time period Oct. 1976- Sept. 1977, Oct. 1979-Sept. 1980.
23. Snake River at Weiser, Idaho, 44°14'N, 116°58'W. Time period Oct. 1978-Sept. 1981.
24. South Platte River at Julesburg, Colorado, 40°59'N, 102°15'W. Time period Oct. 1978-Sept. 1980, Oct. 1981-Sept. 1982.
25. Tennessee River at Pickwick Landing Dam, Tennessee, 35°4'N, 88°15'W. Time period Oct. 1977-Sept. 1978, Oct. 1980-Sept. 1982.

Quartz is not very reactive, however, and at saturation with respect to this mineral the dissolved silica content of water at 25°C is only about 6 mg/L (as SiO_2). Less well crystallized forms of silica, such as cristobalite or chalcedony, are more soluble (Hem, 1985).

The solvent power of water can be substantially increased by the presence of other solutes. For example, carbon dioxide dissolves to give carbonic acid, which then dissociates in two steps to give hydrogen ions:

$$CO_{2g} + H_2O = H_2CO_3$$
$$H_2CO_3 = HCO_3^- + H^+$$
$$HCO_3^- = CO_3^{2-} + H^+.$$

Feldspars and other aluminosilicate minerals commonly break down in weathering to yield clay minerals and solute cations. For anorthite decomposition, for example,

$$\underset{\text{(anorthite)}}{CaAl_2Si_2O_8} + H_2O + 2H^+ = \underset{\text{(kaolinite)}}{Al_2Si_2O_5(OH)_4} + Ca^{2+}.$$

The kaolinite produced in this reaction may be further altered:

$$\underset{\text{(kaolinite)}}{Al_2Si_2O_5(OH)_4} + 5H_2O = \underset{\text{(gibbsite)}}{2Al(OH)_3} + 2Si(OH)^{o}_{4aq}.$$

Reactions of aluminosilicate minerals, like the one shown for anorthite, entail extensive rearrangement of the original structure to produce the clay mineral.

Another important weathering process that affects the chemical composition of most natural water is the dissolution or precipitation of Ca and Mg carbonates. These minerals are present in sedimentary rocks, as limestone and dolomite, or as coatings or cement between sandstone grains, and as coatings or impurities in other materials in contact with the water. The reactions involved are chemical dissolution equilibria of the type

$$CaCO_3 + H^+ \rightleftharpoons HCO_3^- + Ca^{2+}.$$

Oxygen from the atmosphere and generated in water bodies by photosynthetic mechanisms participates in many redox processes as an electron acceptor. Sulfide minerals such as pyrite may be attacked, for example,

$$7/2\, O_2 + FeS_2 + H_2O = 2SO_4^{2-} + Fe^{2+} + 2H^+.$$

The ferrous iron may later be oxidized and precipitated as $Fe(OH)_3$.

The rates at which dissolution and alteration reactions involved in rock weathering occur are related to both chemical and physical factors. Chemical factors include the composition and

TABLE 2. (continued)

Description	26		27		28		29	
Date of Sample (day/mo/yr)	24/05/77	25/04/79	07/03/79	04/11/77	08/04/80	02/10/80	02/06/81	27/02/79
Silica (SiO_2)	6.7	4.9	6.1	7.4	4.8	1.3	5.0	12
Calcium (Ca)	35	86	29	80	12	41	22	47
Magnesium (Mg)	20	92	8.2	26	4.1	15	4.0	10
Sodium (Na)	18	220	7.7	17	2.4	11	2.0	3.0
Potassium (K)	2.8	8.7	2.7	3.2	1.0	2.1	1.1	1.1
Bicarbonate (HCO_3)	150	366	89	270	8.5	26	56	167
Sulfate (SO_4)	76	730	29	72	37	140	21	24
Chloride (Cl)	1.6	7.4	12	30	3.9	15	.7	1.1
Fluoride (F)	.2	.3	.1	.2	.1	.2	.1	.1
Nitrate (NO_3)	.75	.13	9.3	12	2.7	2.5	.09	.23
Dissolved solids	234	1 ,330	139	369	72	241	71	181
Discharge at time of sampling ($m^3\ sec^{-1}$)	38.2	13.7	4,336	371	677	35.7	13,910	1,558
Specific Conductance ($\mu S\ cm^{-1}$, 25°C)	390	1,500	240	655	120	406	144	297
pH	8.4	8.4	---	8.2	7.0	7.8	7.0	7.1
Drainage area (km^2)	13,932		75,716		17,734		831,000	
Long-term mean water discharge ($m^3\ sec^{-1}$)	12.86		690.5		307.8		6,128	

26. Tongue River at Miles City, Montana, 46°22'N, 105°48'W. Time period Oct. 1976-Sept. 1977, Oct. 1978-Sept. 1979, Oct. 1983-Sept. 1984.
27. Wabash River at New Harmony, Indiana, 38°8'N, 87°56'W. Time period Oct. 1977-Sept. 1979, Oct. 1983-Sept. 1984.
28. West Branch Susquehanna River at Lewisburg, Pennsylvania, 40°58'N, 76°53'W. Time period Oct. 1978-Sept. 1981.
29. Yukon River at Pilot Station, Alaska. 61°56'N, 162°53'W. Time period Oct. 1978-Sept. 1982.

crystal structure of the solid minerals, the concentrations of dissolved reactants and products in weathering solutions in contact with the mineral surfaces, and the degree to which chemical energy relationships favor specific interactions to occur. Physical factors include the temperature, the area of solid surface that is in contact with a unit volume of aqueous solution, and the rate at which solution or solute movement can bring reactants to the surface of the solid or remove soluble or insoluble reaction products.

A state of chemical equilibrium for a particular reaction is attained when the reactants and products are at concentrations where chemical energy relationships are balanced and no thermodynamic driving force for the reaction to go on is available. This condition is commonly not reached in river water, as the rates of reactions of the types cited above are too slow compared to rates of solute and solid movement.

A broadly generalized summary of chemical weathering rates in a specific region can be attained by determining the rate at which solutes are transported out of a drainage basin by a river corrected for input of solutes in rain and snowfall and for inputs by human activities or other effects not related to weathering. Estimates of rates of erosion of the land surface by chemical dissolution of rock and soil minerals have been made by many earth scientists beginning nearly 100 years ago. These early estimates were limited to areas where data were available. An estimate by Clarke (1924a, p. 118) of the "chemical denudation" rate for North America is 31 metric tons per square kilometer per year ($t\ km^{-2}\ yr^{-1}$). Later estimates, based on more complete data, were 33 $t\ km^{-2}\ yr^{-1}$ (Livingstone, 1963; Garrels and Mackenzie, 1971).

In small drainage basins the rate of chemical denudation can be more specifically related to rates of weathering of rock or mineral types. Drever (1982) cited studies of weathering rates on igenous terrane in the Absaroka Mountains in Wyoming, where the chemical denudation rate was only half as great as the North American average, and a study in an igneous terrane in the Washington Cascade Mountains, where the rate was double the continental average.

The average annual precipitation at the Cascade site is nearly ten times as great as that at the Wyoming site. Chemical denudation rates are strongly dependent on climatic factors and rock type. A limestone terrane in a temperate region with abundant vegetation can readily attain denudation rates approaching twice the continental average (Hem, 1985) owing to high fluxes of CO_2 and water for attacking the calcium carbonate of the rocks. In cold or dry climates, vegetation is much less abundant, water and CO_2 supplies are lower, and denudation rates correspondingly are low.

TABLE 3. MAJOR-ION CONCENTRATIONS AND RELATED DATA FOR SOME STREAMS IN THE CARIBBEAN-CENTRAL AMERICAN REGION
(Analytical data in mg L^{-1} except for specific conductance and pH.)

Description	1		2	3	4		5	
Date of Sample (day/mo/yr)	02/11/83	14/08/84	1973	1970-1973	Min.	Max.	Min.	Max.
Silica (SiO_2)	14	12	12.6	11.5	—	—	13.4	29.5
Calcium (Ca)	17	22	15	3.3	42	136	10	29.5
Magnesium (Mg)	5.6	6.5	3.3	1.0	21	34	2	31
Sodium (Na)	9.9	10	8.3	1.5	27	40	3	44
Potassium (K)	2.0	2.1	1.8	.65	4.0	7.3	.0	2.7
Bicarbonate (HCO_3)	72	102	49.3	11	118	271	24	74
Sulfate (SO_4)	12	15	14.4	3.4	31	340	5	34
Chloride (Cl)	9.4	12	13.4	2.9	9	38	3	31
Fluoride (F)	<.1	.1	—	—	—	—	—	—
Nitrate (NO_3)	2.7	—	—	—	4.1	24	.04	1.9
Dissolved solids	110	130	93.1	29.7	—	—	—	—
Discharge (m^3 sec^{-1})	5.07	.51	—	—	.1	2,854	1.8	203
Specific Conductance (μS cm^{-1}, 25°C)	178	235	—	—	351	1,131	81	301
pH	7.6	7.4	—	—	7.7	8.4	7.2	8.1
Drainage area (km^2)	436		240,000	950,000	—		—	
Long-term mean water discharge (m^3 sec^{-1})	—		7,500	30,000	—		—	

1. Rio Grande de Arecibo below Lago dos Bocas, near Florida, Puerto Rico, 18°20'50"N, 66°40'2W". Analyses for samples having minimum and maximum conductance collected during time period Oct. 1983-Sept. 1984.
2. Rio Magdalena at mouth, Columbia, South America. Analysis from Meybeck (1979).
3. Rio Orinoco at mouth, Venezuela, South America. Average of 15 analyses for samples taken in 1970-1973 (Meybeck, 1979).
4. Rio Panuco, near Tampico, Mexico, 23°3'N, 98°18'30"W. Minimum and maximum concentrations and discharge observed during period 24 Jan. 1979 to 7 Dec. 1981. Data from GEMS (1983).
5. Rio San Felix, Panama, 9°19'N, 82°50'W. Minimum and maximum concentrations and discharge observed during period 9 Mar. 1979 to 16 Dec. 1981. Data from GEMS (1983).

Studies of river water composition by Meybeck (1982, 1983) cite many specific controlling factors and illustrate them with data from around the world.

Effects of soil and vegetation

Photosynthesis, the basis of plant life processes, produces carbohydrates from CO_2 and water and releases oxygen, utilizing solar energy. Organisms capable of photosynthesis occur as free-floating aquatic species and as soil-rooted species. Both influence surface-water composition. Aquatic vegetation during daylight hours depletes dissolved CO_2 in the surrounding water and releases oxygen. Respiration by the organisms at night releases CO_2. Thus, the water of a lake in which there is a standing crop of algae, or other aqueous photosynthesizing organisms, may show diurnal pH and dissolved oxygen fluctuations. Maximum pH reached during the day may be a unit or more greater than the nighttime minimum. Oxygen in solution may exceed the equilibrium value expected for water in contact with the atmosphere.

Influence of plants on the carbon dioxide content of soil moisture also affects chemical reactions that might occur between the water and soil minerals. Respiration of CO_2 from plant roots enriches the air in the soil pores with CO_2. The decay of dead plant parts also furnishes CO_2 as well as soluble organic compounds. Plants and microorganisms can convert atmospheric nitrogen to water soluble nitrate ions. Many elements are taken up by plants during growth cycles and released to soil or runoff later. Biologic cycling also may be an important factor in the occurrence of Si, K, Fe, Mn, P, and many minor solutes in river and lake water. These minor solutes can be metal ions (Welsh and Denny, 1976) or nonmetals such as selenium (Hem, 1985, p. 145).

Hydrochemistry of lakes and reservoirs

Characteristically, the water in a surface stream is in continuous motion in response to gravitational forces, although the rate of movement varies greatly in time and space. A lake or impoundment in a surface water drainage system could be viewed simply as a holding and mixing basin in which movement be-

TABLE 4. COMPOSITION OF RAIN AND SNOW
(Analytical data in mg L^{-1} except for pH)

Constituent	1	2	3	4	5	6	7
SiO_2	0.0	—	1.2	0.3	—	0.1	—
Al	.01	—	—	—	—	—	—
Fe	.00	—	—	—	—	.015	—
Ca	.0	0.65	1.2	.8	1.41	.075	.86
Mg	.2	.14	.7	1.2	—	.027	.15
Na	.6	.56	.0	9.4	.42	.220	.23
K	.6	.11	.0	.0	—	.072	.12
NH_4	.0	—	—	—	—	—	.34
HCO_3	3	—	7	4	—	—	—
SO_4	1.6	2.18	.7	7.6	2.14	1.1	4.03
Cl	.2	.57	.8	17	.22	—	.36
NO_2	.02	—	.00	.02	—	—	—
NO_3	.1	.62	.2	.0	—	—	2.05
Total dissolved solids	4.8	—	8.2	38	—	—	8.2
pH	5.6	—	6.4	5.5	—	4.9	4.29

1. Snow, Spooner Summit, U.S. Highway 50, Nevada, east of Lake Tahoe (Feth and others, 1964).
2. Average composition of rain August 1962 to July 1963 at 27 points in North Carolina and Virginia (Gambell and Fisher, 1966).
3. Rain, Menlo Park, California, 7:00 pm Jan. 9 to 8:00 am Jan. 10, 1958 (Whitehead and Feth, 1964).
4. Rain, Menlo Park, California, 8:00 am to 2:00 pm Jan. 10, 1958 (Whitehead and Feth, 1964).
5. Average for inland sampling stations in the U.S. for 1 year. Data from Junge and Werby (1958) as reported by Whitehead and Feth (1964).
6. Average composition of bulk perecipitation Williamson Creek, Snohomish County, Washington, 1973 to 1975. Also reported: As = 0.045, Cu = 0.0025, Pb = 0.0033, An = 0.0036 (Dethier, 1977).
7. Mean composition of bulk precipitation at Hinckley, New York, 1965 to 1975. Also reported specific conductance 34 μS/cm, H^+ 0.051 meq/L (Peters and others, 1982).

tween inlet and outlet is relatively slow. In the lake, however, biochemical processes affecting water chemistry are likely to occur. In closed-basin lakes, the only water leaving does so as vapor, and solutes accumulate.

Among the factors more strongly influencing lake rather than river water chemistry are the increased importance of algal and other biological growths, the effects of thermal stratification, and longer periods of time for approach to chemical equilibrium. Lake surfaces are commonly near the regional water table. However, interchange of water between the lake and the ground-water body may be very slow.

Lakes that have no surface outlets are common in the Basin and Range physiographic province and adjunct areas in the United States and Mexico. Such lakes also occur in recently glaciated portions of the northern United States and Canada.

Closed-basin lakes may be ephemeral or permanent, but their water becomes saline as the solutes from evaporated water accumulate. Livingstone (1963) gave chemical analyses for many closed-basin lakes. An evolutionary path can be predicted as the solute concentrations increase. Calcium carbonate precipitates first, followed by calcium sulfate. Finally, the water becomes a concentrated brine whose composition may be dominated by sodium and sulfate, chloride, or carbonate (Eugster and Jones, 1979).

Most lakes having outlet streams do not have high solute concentrations. However, certain kinds of deposited material can occur in them. Iron and manganese oxides, calcium carbonate, calcium phosphate, and silica are common in lake bed sediment.

Deep lakes tend to become thermally stratified during warm seasons. As the sun warms the surface layer, its density decreases; this surface layer of low-density warm water prevents replenishment of oxygen from the atmosphere in the deeper water. Biota such as algae and diatoms may multiply in the upper layer where light intensity is greatest, if nutrient solutes are available.

Dead organic debris settles into the deeper water and is oxidized by reacting with dissolved oxygen. After a time, the deeper water may become anoxic, and iron or manganese oxide in the sediments at the lake bottom may be reduced and brought into solution.

When the weather becomes cooler in the autumn the warm-water layer radiates its heat to the atmosphere. As a result, the warm layer decreases in thickness and finally disappears, and the lake water returns to a vertically uniform temperature. As the surface water continues to cool, it becomes more dense and sinks toward the bottom, carrying oxygen to the full depth of the lake. When the water temperature drops below 4°C its density again decreases, and finally ice begins to form in the 0° layer at the water surface. When the ice melts in the spring, there will again be a tendency for density-induced mixing.

Saline lakes may develop a chemical stratification where the bottom water is higher in solute concentration than overlying water. This type of stratification may be induced by changes in the runoff patterns and can persist for long periods of time.

As noted above, aquatic biota tend to become established in lakes. The terms "oligotrophic" and "eutrophic" have been coined from Greek terms equivalent to "nutrient-poor" and "nutrient-rich" to describe lakes in which biota are likely to be rare or plentiful, respectively. Eutrophic lakes experience growth cycles, sometimes called "blooms", of algae and related floating species. The growth cycle stops when the nutrient in lowest available quantity is used up. At this point the organisms may die off rapidly, leaving a residue of dissolved and suspended organic debris. The effects on water chemistry are seen most strongly in fluctuations in concentrations of nutrients, such as nitrogen and phosphorus, dissolved oxygen, and silica. The latter is utilized by diatoms and some other species to form skeletal parts.

Surface water–ground water relations

Perennial streams generally are in contact with the regional ground water. Typically, the river at high runoff stages may recharge the ground water system. At low flow the discharge of the river is maintained by ground water that enters the river channel. The latter is commonly referred to as "base flow." In a simplified sense, the river flow may be viewed as having two components, base flow and direct runoff. As the ground water can be expected to contain higher solute concentrations than the runoff, an inverse relationship between solute concentration and discharge is to be expected.

In actual drainage systems, may complications affect this simple model. It is, for example, generally observed that the solute concentration in a river during the early part of a rising flow stage remains about the same as in the preceding lower-stage water As the flow continues to increase, the solute concentration declines to a minimum, which may not be reached until after the discharge peak has passed. On the falling stage, the increase in solute concentration is relatively slow and for a given discharge will generally be substantially lower than it was at that discharge during the rising stage. These effects can be observed for individual floods brought about by intense rainfall. They also can be seen in seasonal records for streams that reach high flow stages in response to spring snowmelt followed by lower flow during the summer. Figure 1 shows the seasonal pattern of ion concentration in relation to discharge for a Canadian stream. Cycles for individual flood events on Kentucky streams were described by Hendrickson and Krieger (1964).

These complications can be explained in part by consideration of hydraulic relationships in the river, and between river water and immediately adjacent ground water in the flood plain and channel. The first rising stages of a flood event involve base-flow water swept up in the more rapidly moving runoff front, or flood wave. This may involve a substantial volume of water. Some of the more dilute flood water can be expected to infiltrate into the banks and stream bed, and will be forced out into the stream again as the flow stage declines and the ground water–surface water head relationships change. This prevents the flow composition from returning to normal base-flow concentrations for some time after the flow stage has reached the normal level.

More detailed studies of the chemical changes occurring during rising and falling stages show tha the two-source (base flow and runoff) model is an oversimplification for many, or perhaps most, streams. Kennedy (1971) observed that some solute components, notably SiO_2, of the water of the Mattole River in northern coastal California did not change much during flood events. This suggests that much of the flood water, which appeared quickly in the river after rainfall began, had spent enough

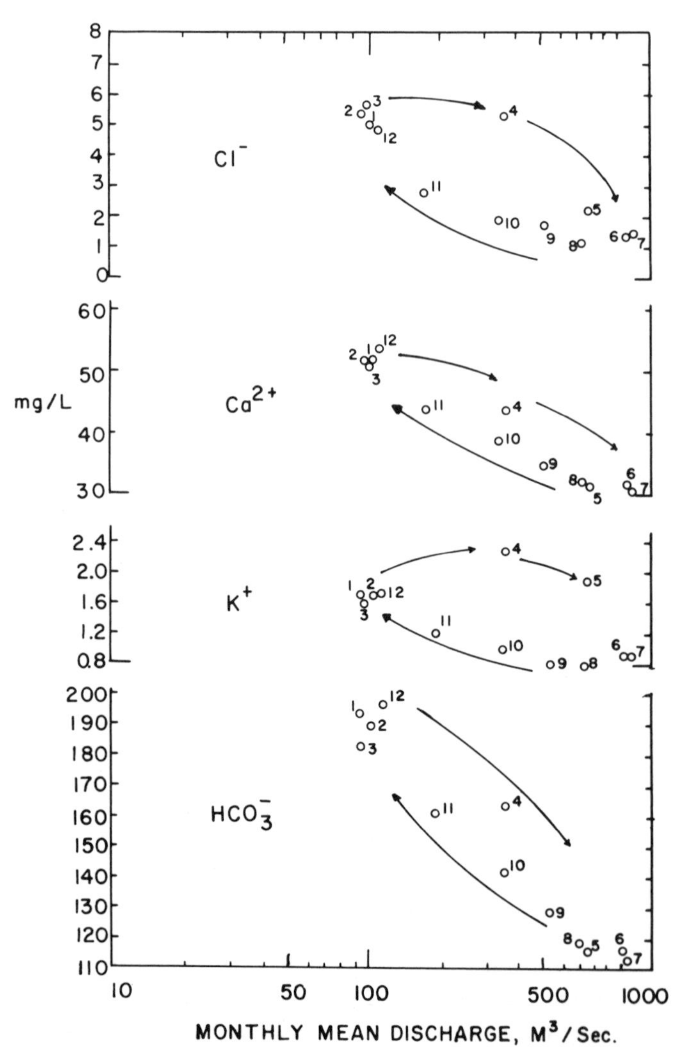

Figure 1. Graph showing seasonal variation of Cl, Ca, K, and HCO_3 concentrations, Athabasca River at Athabasca, Alberta (Mackenzie River Basin Committee, 1985).

time in contact with soil to take up significant amounts of these solutes. The effect of ground-water inflow was believed by Kennedy and others (1986) to be insignificant, and on the basis of stable isotopic measurments these investigators believed that a major fraction of the storm runoff in the Mattole was displaced soil moisture.

In regions of permanently frozen ground (permafrost) the exchange of water between surface streams and ground water is greatly inhibited.

Analyses and related data in Tables 1 through 3 show that although the observed minimum solute concentrations generally occur at high-flow stages, and maxima are observed at low flow, the relation is not easy to quantify. For example, the maximum concentration for the Mississippi River near Saint Francisville, Louisiana, during a 3-year period (Station 14, Table 2) occurred at a discharge that was a little above the long-term average, and for the Rogue River near Agness, Oregon, the minimum concentration observed during such a time period occurred at a flow stage that was substantially below average. Explanation of these relations may require consideration of both anthropogenic and natural hydrologic effects.

In statistical studies of water-quality data, several techniques have been used to compensate for and evaluate discharge versus concentration relationships. The subject is discussed further under the heading "Data organization and correlation" later in this chapter.

Episodic natural events

Certain types of catastrophic events can have major effects on river and lake water composition over large areas. Until the 1980 eruption of Mount St. Helens in the state of Washington, the effects of volcanic eruptions on stream-flow composition were poorly documented. A large amount of volatile material is ejected in explosive type eruptions, and quieter volcanic eruptions also release large amounts of gases. Water vapor is a major constituent of these volatile materials, and large quantities of CO_2 and sulfur dioxide or other sulfur-containing gases also are given off by some volcanoes. Lesser amounts of fluoride and chloride also are commonly present.

Estimates in the literature (Kaplan, 1972) suggest that natural sulfur sources such as rock weathering and volcanism may contribute 160 to 180 million tons of SO_4 per year to fresh waters. Other estimates of the sulfur cycle (Berner, 1971) suggest that additions of sulfur through anthropogenic sources are now at least equal to the natural sources.

Some volcanic eruptions produce very large quanitities of sulfur-containing gases that form aerosols in the upper stratosphere and have global effects on weather and optical properties of the atmosphere for long periods of time. A classical example is the 1815 eruption of Mount Tambora, located in Indonesia. Gases and dust emitted into the stratosphere by this eruption are commonly thought to have brought on unusually cold summer weather in 1816 in parts of North America and Europe. Effects of that event on surface-water quality are, of course, uncertain, although some evidence of sulfur fallout may have been preserved in ice accumulated on Greenland.

Chemical composition of stream water in the areas directly affected by Mount St. Helens was described by Klein and Taylor (1980). Some streams northeast of the mountain showed sulfate and chloride increases at the time of the eruption, lasting only a few days. Other streams, more directly affected by mud flows, remained relatively high in chloride and sulfate for a month or more. An interesting side effect was the introduction of organic compounds such as polynuclear aromatic hydrocarbons (PAH) produced when the hot, eruptive debris buried or destroyed forests on the slopes of the volcano. El Chichon, in Mexico, which erupted in 1982, contributed much larger amounts of sulfur to the atmosphere than did the Mount St. Helens' eruption.

Influence of humans

Human activities have very strong influences on water-quality regimes of rivers and lakes. Structures that impound or regulate rates of river flow, diversion of water from one drainage basin to another, irrigation of land adjacent to streams, or lowering of water tables by pumping are kinds of activities that bring about water-quality changes by altering flow characteristics. Waste disposal, directly or indirectly into streams and lakes, influences water quality in a more direct way by adding solutes and suspended matter.

Effects on stream-water quality can also be expected from land-use changes and related developments. Examples include urbanization, clearing of forests, and various agricultural practices such as use of fertilizers and pesticides, and return flow of drainage water from irrigated fields.

Urbanization and industrialization lead to various side effects. Mining for coal and metals generally impairs water quality during and after mining activity. Smelting ores to recover metals and burning coal to generate power release air pollutants that find their way into water supplies.

Metal ores, coal, and other organic fuels commonly contain, or are associated with, reduced sulfur. Oxidation of the sulfur by burning the fuel and smelting the ores, and oxidation in the mines or in waste dumps when sulfides are exposed to air constitute major sources of sulfate for river water. High metal concentrations are commonly found in surface waters draining metal-mining areas.

Common salt, NaCl, is widely dispersed in the environment by such activities as melting ice from highways by spreading salt and sand mixtures.

Disposal of untreated organic waste into streams was common practice in urban and rural areas of the continent until the early twentieth century and it is still being practiced in some parts of the continent now. Besides the pathogenic bacteria, large amounts of organic solute and particulate material depleted the dissolved oxygen of receiving waters and killed much of the aquatic biota in some streams.

Although more enlightened waste-disposal practices are now generally followed, wastes that are incompletely treated, or which reach surface streams through accidental spills or storm sewers, remain a problem in urban areas.

Many organic and inorganic substances reaching lakes and rivers cannot be broken down by aquatic and soil organisms or other aerobic life forms, or, even worse, they are converted to more toxic forms and thus pose a threat to aquatic and human life.

QUANTIFICATION OF GEOCHEMICAL PROCESSES

One of the aims of water quality data collection programs generally has been to provide a basis for understanding how geochemical and related processes work, and ultimately to attain the ability to predict the effects of actual or potential resource development or other kinds of stresses that might be imposed on the system.

In theory it should be possible to develop a predictive model based on long-term records of water quality and related data. This type of model, sometimes called the "black box" type, has been developed and used successfully for projecting behavior of dissolved oxygen and nutrients (Ditoro and others, 1971). Because of the long time period required to develop the data base and many complications in the systems involved, such models are not always practicable.

Chemical principles and processes

A more fundamental approach to water-quality modeling is to develop a conceptual process-oriented framework based on laws of physics and chemistry—the principles of conservation of mass and energy as matter is transported in the system. Chemical interactions in such systems can be described and predicted by applying principles of chemical equilibrium and reaction rates. These chemical principles are based in general on the Second Law of Thermodynamics, which states that energy transfers occur along favorable potential gradients (e.g., water flows downhill). In a chemical reaction, energy transfers occur, and for a chemical reaction to occur spontaneously there must be a chemical energy gradient favoring the process. Some important reactions in natural water bodies, however, are driven by energy from external sources. Photosynthesis, for example, uses light to convert carbon dioxide and water to carbohydrates.

Another consequence of the Second Law is that a chemical reaction that occurs in a closed system (where reactants and products of the reaction cannot leave and additional materials cannot enter) will tend toward a state of equilibrium. At equilibrium the reaction appears to have stopped as energy gradients for both the forward and the reverse reaction are near zero.

The chemical law of mass action states that the rate of a chemical reaction is proportional to the effective molar concentrations of the reacting substances. At equilibrium in the hypothetical reaction between reactants A and B to produce products C and D,

$$aA + bB \rightleftharpoons cC + dD.$$

This law leads to the mathematical statement,

$$\frac{[C]^c\,[D]^d}{[A]^a\,[B]^b} = K$$

where K is the equilibrium constant, the brackets represent active molar concentrations, and the exponents are the relative proportions of the substances in the balanced chemical reaction.

From equilibrium constants available in the literature, known chemical composition of a water, and known minerals, gases, or other substances in contact with the water, one can determine whether a postulated chemical equilibrium can exist and predict the direction in which possible reactions may go.

Chemical reactions in river and lake waters commonly do not occur under closed-system conditions, as defined above, and states of thermodynamic equilibrium may not be reached. However, mathematical models using mass-law calculations can be used to show which solid minerals might be attacked and which ones would be stable in the presence of water having a given chemical composition. Because chemical thermodynamic calculations can show which chemical processes are feasible in a given system and which ones are not, they have broad usefulness even where equilibrium is never reached.

Chemical reaction rates

Rates of many chemical reactions of interest in surface water systems are not well known, and rates determined in laboratory studies may not adequately reflect the dynamics of real systems. However, rates of such processes as reaeration (the restoration of dissolved oxygen after depletion by oxidation of wastes) and oxidation of metal ions added by mining wastes or other sources have been determined in the field (Langbein and Durum, 1967; Nordstrom, 1977). Models of stream behavior can be developed for particular stream systems by applying such data, or by developing non–process oriented models based on observed behavior in a river reach under various conditions (Ditoro and others, 1971).

Surface chemical processes

Surfaces of oxide, silica, and aluminosilicate minerals in contact with water commonly have a net negative electrostatic charge that tends to attract and retain a layer of cations from solution. Where there are broken chemical valence bonds at the mineral surface, the attraction for cations may be stronger, and the smaller, more electropositive cations are more strongly retained at such sites than are larger cations of lower valence.

The general phenomena involved in these interfacial reactions that do not involve dissolution or precipitation are commonly termed "adsorption." The processes at specific sites of surface charge are referred to as "ion exchange," or, because usually cations are the principal species involved, "cation exchange." Ion exchange processes can be evaluated by mass-law equilibria.

Suspended and bed sediments of streams can be expected to participate in adsorption and desorption or ion-exchange processes. These may sometimes be evaluated theoretically using chemical equilibrium or adsorption isotherm relationships (Hem, 1976).

Estimates of the load of ions carried to the ocean by rivers in dissolved and suspended forms were made by Martin and Meybeck (1979). Their figures suggest that, for most metal cations, more than 50 percent of the total was in the suspended load. These calculations, however, did not separate adsorbed forms from the much less available ions that were incorporated in mineral structures.

The picture is further complicated by the tendency for some solute species to form coatings of new solid phases, such as iron oxide, on the surfaces of suspended and bed sediments, and the adherence of hydrophobic organic material to solid surfaces. This effect was noted for certain streams by Gibbs (1973). Some organic compounds that have significant water solubilities also tend to be adsorbed at solid surfaces.

Redox processes

Oxidation and reduction reactions in natural surface waters tend ultimately to be dominated by processes that involve atmospheric oxygen. The atmosphere maintains the dissolved oxygen content of water that is in contact with air. However, the effectiveness of this control is limited by rates at which oxygen can cross the water-air phase boundary and physical factors that influence these rates. Oxygen in natural water oxidizes organic material. The oxidation of organic matter by oxygen is a complex process, generating various reaction products and intermediates. In the presence of sunlight, hydrogen peroxide may be formed (Cooper and Zika, 1983).

Depletion of oxygen occurs in water that carries large amounts of readily oxidizable material. One of the major indices of organic pollution is the degree of oxygen depletion below the saturation value for water in contact with air at ambient temperature. Other reactants also may deplete oxygen, notably wastes that contain oxidizable metal ions. Dissolved oxygen is required for survival of most aqueous life forms that humans consider desirable.

In sediments at the bottom of lakes, or sluggish or impounded streams, the oxygen content of pore water can be almost completely used up by bacterial reactions. A highly reduced environment may occur in such sediments, and an upward flux of reduced metal ions and other solutes may occur. As this flux reaches the sediment surface, if oxygenated water is available, a crust of metal oxides may be precipitated. In some sediments the oxygen may be low enough to permit growth of anaerobic biota such as the sulfate reducers or those that convert nitrate to nitrogen or ammonia.

Buffering effects

Although many of the possible chemical reactions that might dissolve mineral matter from soils and sediment or cause material to precipitate from solution are not at a state of chemical equilibrium, there is a tendency toward a steady state for many such reactions. This condition results from the availability of reserve unreacted supplies of many of the species that are involved in the chemical reactions, and the fact that rivers and lakes are not closed chemical systems. The stabilization of pH that occurs in many natural systems is an illustration of interactions among solutes.

The pH (hydrogen ion activity) of a simple aqueous solution is an indication of the degree or intensity of acidity or basicity. A $10^{-4.0}$ molar aqueous solution of hydrochloric acid (HCl) contains 0.0001 moles per liter of H^+ and Cl^- ions and would have a pH near 4.0. In terms of mg/L, the concentrations would be about 0.10 for H^+ and 3.5 for Cl^-. Adding $10^{-4.0}$ moles of a solid base, such as sodium hydroxide (NaOH) to a liter of this solution would bring about the reactions

$$NaOH = Na^+ + OH^-$$

$$Na^+ + OH^- + H^+ + Cl^- = H_2O + Na^+ + Cl^-.$$

The Na^+ and Cl^- ions do not participate in this second reaction, which may simply be written

$$H^+ + OH^- = H_2O.$$

The chemical equilibrium expression for the reverse of this reaction is

$$[H^+]\,[OH^-] = K_w$$

and at 25°C K_w, the dissociation constant for water, is $10^{-14.00}$. The reaction of H^+ and OH^- to produce H_2O would remove those ionic species and leave their concentrations at the levels seen in pure water. At 25°C the concentrations of both are near $10^{-7.0}$ molar (pH 7.0). The H^+ and OH^- reacting directly produce large and simply predictable changes in pH of the solution. In this example, the amount of base added returns the pH to 7.0.

In most natural water the pH is influenced or controlled by other reactions, especially those of dissolved carbon dioxide. Dissolved CO_2 forms carbonic acid, which dissociates in solution to form bicarbonate and carbonate ions. A large external supply of CO_2 is generally available in the atmosphere and in higher concentrations in air spaces in soil. Soil air is supplied with CO_2 by plant-root respiration and decay of organic matter.

A natural river water that contains $10^{-3.0}$ moles/L (61 mg/L) of HCO_3^- and has a pH of 7.0 will also contain some

H_2CO_3. It may be assumed that the measured pH is controlled by an equilibrium among these species and the amount of H_2CO_3 present can then be calculated from the mass-law relationship

$$\frac{[HCO_3^-][H^+]}{K_1} = [H_2CO_3].$$

If K_1 is $10^{-6.35}$ (Hem, 1985, p. 62), $[H_2CO_3] = 10^{-3.65}$.

Adding $10^{-4.0}$ moles of H^+ to a liter of this solution will decrease $[HCO_3^-]$ to $10^{-3.05}$ and raise $[H_2CO_3]$ to $10^{-3.49}$. If no reactants or products escape, the pH of the solution will change to 6.8. This behavior is in marked contrast with the previously cited example where $10^{-4.0}$ moles of H^+ or OH^- added to a liter of solution changed the pH by 3.0 units. Here the same amount of acid in the same volume of solution as before produces only 0.2 pH unit of change.

Effects of this kind are referred to as "buffering." They result from the involvement of H^+ or OH^- in chemical reactions with other solute ions. Similar effects involving other reactants could be expected for most of the other solutes that might be introduced into natural-water systems.

Figure 2 illustrates graphically the relationships among pH and the relative amounts of the three different species of dissolved carbon dioxide, in a system that does not contain a gas phase. The curved lines show how the proportions of these solutes relate to pH. This relation is rather highly sensitive to temperature, especially between 0°C and 25°C, which is a common seasonal range for surface water. At the lower temperatures the dissociation of dissolved CO_2 is substantially less than it is at 25°C.

The literature contains extensive discussions of buffering effects and the behavior of CO_2 species (Stumm and Morgan, 1981). These authors also discuss the nature of buffering in systems where the excess supply of reactants is in solid form.

Natural hydrogen ion buffering is most effective in water that contains at least moderate amounts of dissociated dissolved CO_2, and in regions where soils contain solid carbonates. Thus, "acid rain" (precipitation with a pH below about 4.5) will influence lakes and streams most strongly if soil and bedrock are low in carbonate or other acid-consuming matter. In terranes where carbonates are abundant, the amount of H^+ added by rainfall will have an insignificant effect unless the condition becomes very severe and prolonged.

Properties of this type in natural waters and in the drainage system through which the water passes contribute to stability of the chemical composition of the water toward deleterious pollution effects. Sensitivity toward acid rain for streams and lakes can be predicted approximately from such considerations. Small- and medium-sized lakes in ancient igneous or metamorphic terranes (Canadian Shield, for example) are much more likely to be affected by rainfall composition than are lakes in sedimentary terranes.

It should be noted that addition of a cation such as H^+ to a natural water also entails adding equivalent amounts of some anion such as SO_4^{2-}, Cl^-, or NO_3^-.

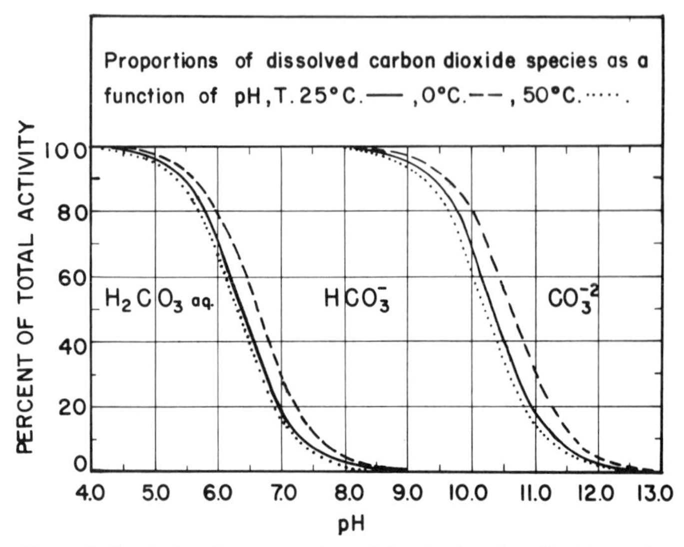

Figure 2. Graph showing proportions of dissolved carbon dioxide species as a function of pH for temperatures of 0°, 25°, and 50°C.

NATURE OF WATER-QUALITY DATA AND METHODS FOR DATA COLLECTION

In the introductory chapter of a well-known textbook on water chemistry (Stumm and Morgan, 1981, p. 3) the authors state, "Aquatic chemistry is of practical importance because water is a necessary resouce for humans. We are not concerned with the quantity—water as a substance is abundant—but with the quality of the water and its distribution."

To a hydrologist whose career has been devoted to measuring quantities and distribution of usable water supplies this statement may seem somewhat heretical. However, it is true that chemical, physical, and biological properties that define water quality are decisively important in evaluation of the usefulness of surface-water resources. Water-quality studies done in the past and most of those now going on have had the aim of resource evaluation, or to document the effects of human activities such as flow control, diversions, or waste disposal.

Water-quality studies of interest in this chapter are primarily related to solute concentrations. Some of the solutes are intimately related to biological processes, but strictly biological aspects are discussed in only a minimal fashion here. The importance of suspended sediment and organic material as a water-quality parameter is obvious, but aspects of the topic appropriate for discussion here are limited to those relating to the influence of these particulate solids on water chemistry.

To determine the suitability of a water for use as a public supply, for industry, or irrigation, a water-quality study has normally been concerned with concentrations of the major ionic solutes, along with such constituents and properties as dissolved silica, pH, hardness, conductivity, and certain dissolved elements—present in generally minor concentrations—such as fluoride, boron, chromium, and various others for which specific upper concentration limits have been established for the uses proposed. Physical properties such as sediment concentrations or temperature are sometimes included as well.

Water intended for human consumption has also been examined for "sanitary" quality, generally by laboratories specializing in the health-science field. The sanitary analyses of "raw" (untreated) waters include determinations of oxygen demand, various forms of nitrogen, and microbiological examination, as well as certain other biological and chemical tests. The microbiological tests also are required for monitoring the quality of the finished water as delivered to consumers.

Although some agencies involved in sanitary analyses also did some chemical quality work, and some in the chemical-quality field had active interests in biological evaluations, the two disciplines generally were not in close communication. Beginning in about the 1960s, with an increase in public awareness of environmental quality and programs to control and decrease water pollution, the concept of water quality was seen as involving both "chemical" and "sanitary" aspects. At the same time, analytical instruments began to become available that could determine very small concentrations of many chemical elements and organic compounds, and public attention was focused on some of those thought to be threats to public health or aquatic life forms.

Because of the differing background and experience of practitioners in the chemical and sanitary disciplines, there have been some differences in approach to the collection of water-quality data. Perhaps, because microbiological tests ordinarily need to be run on unfiltered samples, the need for phase separation for other kinds of testing has not been fully recognized by some.

Aquatic environmental systems

A lake or river may be viewed as part of an open 3-phase physicochemical system. The three phases are the liquid water, the solids in the bed or banks that confine or at least restrict or restrain the movement of the water, and the overlying atmosphere. In alluvial streams a fraction of the bed material is in motion along with the water, although generally at a much slower and more irregular rate. Solutes come into the aqueous phase, for example, by dissolution of atmospheric gases or particulates, are dissolved from the solid phases already present, or may be added as solutes in aqueous inflow or as pollutant spills.

Solutes that are present as dissolved material tend to interact in various ways with solid particulates. For some types of dissolved material, such interactions strongly favor some type of immobilization of solute ions or molecules at charged sites on mineral particles or favorable points of attachment on organic solids. The strength of the attractive forces can vary from relatively weak electrostatic adsorption to total incorporation of the solute in a layer of precipitate.

Surficial reactions tend to be important controls on the behavior of solutes that are present in small amounts compared to the availability of surface sites to which they might be attracted, and for organic material of low aqueous solubility.

Biota that are present in aquatic systems add to the complications in behavior of solutes. Organisms extract nutrients from the solution and control the behavior of many solutes. They also may be attached to solid surfaces.

The complexity of chemical, physical, and biological interrelationships in natural aqueous systems obviously makes it very difficult to attain a complete understanding of all the processes that govern water quality in surface waters. Water-quality data collection programs can provide useful information only if these complexities are recognized and sampling and analysis procedures are properly designed. For major dissolved species, including SiO_2, Ca, Mg, Na, HCO_3, SO_4, and Cl, the amounts present in dissolved form are normally much larger than amounts associated with suspended particulates, and methods of separating particulate from dissolved forms do not have a major effect on final results. For minor metals and some other constituents, however, the separation technique is a critical factor.

PREANALYSIS SAMPLE TREATMENT

Some surface-water-quality measurements are made and data gathered, as noted above, to help provide answers to specific questions about usability of the resource or the effects of certain known activities that are influencing a particular stream or lake. Other programs involving such measurements have more diffuse aims—commonly to provide a set of basic data that can be used for many different purposes. The methods by which water-quality data are obtained need to be appropriate for the problems that are to be addressed. Geochemical studies that stress concentration and variability of the major components of dissolved material, for example, require separation of nonsolute material from the aqueous phase. This has generally been done by filtration. Other kinds of studies may require analysis of various properties of the separated solids. Major fractions of trace constitutents, inorganic and organic, may be adsorbed or otherwise immobilized on surfaces of suspended particles. Studies concerned with biochemical or sanitary conditions may be centered mostly on identification and evaluation of suspended organics or biota.

Phase separation

Separation of nonsolute material and the analysis or evaluation of this fraction represent an asepct of surface-water quality investigation that has been extensively studied. Some representive papers on these topics will be cited in this section to provide some background and to indicate the many complexities that are involved.

Many investigators have studied the effects of filtration and other types of sample pretreatment on minor metals and organic substances in water, and results are in general conflicting and inconclusive. In a series of experiments by Davies (1976), it was found that neither cell dialysis (mean pore diameters of 4.8 nanometers [nm] and 1.6 nm) nor filtration (0.45 μm and 0.025 μm) provided suitable means of separating the truly dissolved species of lead from the colloidal species. The dialysis tubing permitted more than the dissolved species to pass through, whereas the filtration gave erratic results by removing some of the dissolved lead and permitting some of the colloidal lead to pass.

Specifying that a set of analytical data represent correctly only the material that is present in solution implies that a precise definition of the dissolved state can be made and that simple practical methods exist for removing nonsolute components from water samples prior to analysis. By general informal agreement among institutions and individuals active in the field of water-quality data collection, an operational definition of dissolved material is in use: that which remains in the water sample after it has been filtered through a filter membrane having pores 0.45 micrometer (μm) in diameter, is dissolved. This filter porosity will allow some particulate matter in the colloidal size range to pass through, although in practice a layer of coarser particulates that partially blocks the pores and retains some material considerably smaller than the nominal 0.45 μm limit forms on the membrane during filtration.

A precise definition of the dissolved state cannot be based solely on particle diameter. The size range for individual particles in colloidal suspensions is commonly given as 1.0 nm to 1.0 μm. Such material will only be partly removed by a 0.45 μm porosity filter.

Other methods for removing particulate material include centrifugation and dialysis. Although a more rigorous separation of particulate from dissolved material is possible by some of these procedures, they have not been widely used in surface-water quality studies.

Dissolved organic matter of natural origin is present in small amounts in all river and lake waters. In some regions, especially where there are many swamps and bogs and in areas of tundra, there may be enough dissolved organic matter to give the water a brown or amber color. This colored material has commonly been referred to in the literature of water chemistry as humic, or fulvic, matter, a nomenclature derived from soil chemistry. Organic matter that can be extracted from soil samples by treatment with a strong solution of sodium hydroxide is divided arbitrarily into humic and fulvic fractions by acidifying and filtering the extract. The organic matter that precipitates is termed "humic acid" and that remaining in solution is termed "fulvic acid."

An operational definition has been proposed by Thurman (1985, p. 274) for aquatic humic substances. They are stated to be polyelectrolytic colored organic acids, nonvolatile, with molecular weights from 500 to 5,000. Both carboxyl and phenolic H^+ sites give acidic character to these materials, and the larger molecules may be in the colloidal size range. Both kinds of H^+ sites may become points of attachment for metal ions adsorbed from solution. The humic materials may also adsorb other kinds of organic material from water, such as organic pollutants from industrial waste, or pesticide residues.

Colloidal organic matter, if present, complicates the separation of dissolved from suspended matter in water samples. This generally is a matter of concern only for samples to be analyzed for trace metals and organics.

Guy and Chakrabarti (1976) reported that ultrafiltration of metal (Cu, Pb, Cd, and Zn)–humic acid solutions indicated that at pH 6 the metal ions were present almost exclusively in the fraction larger than 3.2 nm in size. At pH 2, however, 100 percent of the metal ions were in the fraction less than 1.9 nm in size. It was also determined that at pH $>$ 6, 50 percent of the copper was present in the particulate fraction and 50 percent of the copper was complexed. At pH 2, 100 percent of the copper was present as free ion (Guy and others, 1976). Copper in natural waters was found excluisvely in two fractions, $>$ 5.1 nm and $<$ 1.9 nm, in various proportions depending on the origin of the water sample (Guy and Chakrabarti, 1976).

Marvin and others (1970) have tested five different types of filters with respect to the concentration of copper in salt water and in laboratory-prepared solutions in distilled water. The results obtained were erratic, due both to contamination from the filter and to sorption of copper by the filter. Andelman (1973) quoted data from a Missouri creek that showed no consistent difference between 0.45 μm and 0.1 μm filters. Kennedy and others (1974), however, reported that iron concentrations in river water passed through a 0.45 μm filter could be 10 times as great as the amount present after filtration with a 0.10 μm filter. Iron in aerated water can form colloidal ferric hydroxide particles less than 0.10 μm in diameter (Lee, 1973). In the presence of natural organic matter, a dominant size fraction occurred in the range from 0.10 to 0.45 μm (Lee, 1975). Obviously, some colloidal iron hydroxide can be present in filtered river water.

Acidification of an unfiltered alkaline natural water sample will dissolve metal oxides and release metals bound to clay suspensions present in the sample. The final dissolved concentration of a given ion after such treatment depends on the final pH and the concentration of other species, including any adsorbents present (Farrah and Pickering, 1977). Obviously, the original state of the metals in unfiltered acidified samples cannot be determined.

In summary, it appears that: (1) Filtration through 0.45 μm porosity membranes does not remove all particulate material from water samples. In some instances it may remove some dissolved material. (2) Treatment of water samples by filtration or other phase-separation techniques is essential for meaningful chemical analyses for most inorganic constituents of surface water systems, if it is concluded that it is important to know the mode of occurrence of that constituent in the system being sampled. (3) No practical methodology of phase separation can quantitatively separate fine colloidal suspended trace metal oxides and hydroxides from dissolved forms of trace metals. The behavior of iron is an obvious example. However, this problem does not significantly affect analytical results for major solutes.

FORMS OF DISSOLVED MATERIAL

Most analytical procedures used in determining dissolved constituents yield results in terms of the total amount of the ion or element that is present. For some purposes it is important to know more exactly the forms of the dissolved species. Metal cations, for example, may form complex ions with inorganic ligands such as hydroxide, carbonate, or fluoride, or with organic ligands such as humic derivatives or oxalate. The complex ions can be expected to behave differently from uncomplexed species.

Various studies have indicated that the free ions in dissolved state are the most important when health of the aquatic environment is the issue of concern. For example, laboratory experiments on the toxicity of lead to rainbow trout showed that the results could be interpreted only in relation to the concentration of the dissolved lead measured by pulsed polarography. The results for the "total" (dissolved plus colloidal) measured by atomic absorption became less indicative as the increased alkalinity reduced the availability of the lead by forming complex ionic precipitates and colloidal particles (Davies, 1976).

Andrew (1976) showed that the toxicity of copper to fish and Daphnia correlated well with the free cupric ion activity measured using an ion specific electrode. Neither the dissolved copper (defined as that passing through a 0.45 μm filter) nor total copper, both measured by atomic absorption, explained the results of the toxicity experiments. Most of the dissolved copper was in the form of dissolved but undissociated $CuCO_3$. The survival time of rainbow trout in hard water was also related to the cupric ion activity measured by specific ion electrode (Shaw and Brown, 1974).

The complexation of metals with organic compounds affects the concentration of metals in dissolved phase, as well as their availability and toxicity to aquatic life. Both the nature of the metal and the organic group are important. In general, organic complexation reduced the biological availability of trace metals in solution compared with availability of inorganic forms (Jenne and Luoma, 1977). The toxicity of copper and zinc to juvenile fish may be reduced by the presence of humic acids or other organics. In experiments conducted with guppies, the toxicity of copper was reduced in the presence of NTA, EDTA, humic acid, and glycine (Chynoweth and others, 1976). Lewis and others (1972), in studying the Cu(II) toxicity to *Euchaeta japonica* (copepoda, calanoida, a zooplankter) in seawater, found that toxicity decreased when water-soluble compounds of unknown composition but with soillike characteristics (humic acids and soil extracts) were added to the water-copper solutions.

The type of complexing ligand plays a major role in the biological availability of metals. For example, the iron-EDTA (ethylenediaminetetraacetic acid) chelate has sufficient stability to prevent the removal of iron by precipitation and yet permits the iron to be incorporated into the cell growth. The iron-DTPA (diethylenetriaminepentaacetic acid) complex also prevents the precipitation of iron as ferric hydroxide, but in this case iron is no longer available for growth because of the higher stability of the complex (Lee, 1973). Similarly, copper citrate is toxic to aquatic organisms, whereas the copper-EDTA complex is not (Fitzgerald and Faust, 1965).

In surface waters, the most common complexes are those between metal ions and fulvic and humic acids with molecular weights ranging from 300 to more than 200,000. For example, in river-water samples from the Ottawa area (Rideau Canal and Brewery Creek) ions of the form M^{2+} (e.g., Pb^{2+} and Cd^{2+}) at concentrations of 1 mg/L or less were completely bound to compounds with molecular weights of less than 1,400. Most of the copper, 82.6 percent was bound to compounds with molecular weights greater than 45,000. At higher concentrations, some of the metal ions could still be found as "free ions" (dissolved). In waters with total inorganic carbon (TIC) of approximately 20 mg/L or more, some of the binding capacity could be attributed to HCO_3^- and CO_3^{2-} ions (Ramamoorthy and Kushner, 1975).

Model calculations have shown that in fresh waters the complexation of metals with humic materials depends on (a) the pH, which controls the competition from carbonate and hydroxy ligands; (b) the concentration of major ions (e.g., calcium and magnesium), which are competing with the trace metals for humic materials, chloro-ligands, and sulphato-ligands; and (c) the stability constant of the metal-humic substances interaction (Mantoura and others, 1978).

ANALYSIS AND CHARACTERIZATION OF SOLID-PHASE COMPONENTS

Particulate material of suspended or bottom sediment that is associated with river or lake water may interact in various ways with solutes in the water, and as indicated earlier, adsorption at the solid-liquid interface is a major factor in controlling the behavior of many of the minor inorganic ions, and of most organics, especially those having hydrophobic character. The mineral surfaces also may have coatings of precipitated metal oxides or organic material.

It is possible to determine the bulk chemical composition of suspended or bed material by techniques of rock analysis. These amounts, expressed as concentrations in the water sample plus the amounts in solution, represent "total" concentrations. Determination of characteristics such as surface area or cation exchange capacity per unit weight may be more useful than particle size distribution for explaining and predicting adsorption effects.

Many investigators have sought to develop analytical techniques that could determine the kinds and amounts of ions on solid surfaces that might be available for exchange with solutes or that could be available for biota. Most of these techniques involve treatment of the sediment with extractants of various types. These have included strong and weak acids and bases, complexing agents, reductants such as hydroxylamine, and neutral salt solutions to displace exchangeable cations. There is no agreement on the form of treatment that gives best results, nor on the interpretation of data obtained by extraction techniques.

The protocol used in trace element analyses in surface waters by the U.S. Geological Survey (Brown and others, 1970) has been to take duplicate aliquots of the sample at the time of collection, to filter one and acidify it with ultra pure HCl or HNO_3 to a pH near 2.0 (to prevent loss of minor constituents by precipitation or adsorption on the container surface) and to leave the other unfiltered. When analyzed in the laboratory, the unacidified aliquot is made approximately 0.3 molar with HCl and heated to 90°C for 30 minutes. The mixture is then filtered, and the filtrate is analyzed. Concentrations are reported as "total recoverable" referred back to the original sample volume. The concentration in the original filtered and acidified aliquot is also determined and reported as "dissolved."

Analyses of metal or other solutes of interest associated with stream-bed sediments can also be done by a similar procedure. The results of such analyses may be useful in certain types of geochemical investigations or studies of stream pollution.

Comparisons of various extraction techniques have been made by Luoma and Jenne (1976), Malo (1977), and Agemian and Chau (1976). Agemian and Chau (1976) evaluated a series of extracting agents for metals from the <80-mesh portion of sediments. As expected, the amount extracted depended on the extracting agent, the metal extracted and the type of sample. Treatment with a mixture of HF-HNO_3-$HClO_4$ was necessary to measure the "total metals" (including those in the rock matrix). To measure the "weakly held" metals the authors recommended extraction with 0.5 N HCL. Jenne and Luoma (1977) reviewed the use of a series of reagents (oxalate, pyrophosphate, and dithionite-citrate) in the extraction of a few particular elements from certain fractions of the sediment. The same authors (Luoma and Jenne, 1976, p. 351) have concluded that "it is unlikely that any single chemical extractant will adequately describe the biologically available fraction of numerous trace metals."

Soil chemists, in trying to determine which fraction of the metals present in the soils is available to plants, have used a variety of extraction agents ranging from cold and hot water to mixtures such as acetic acid and ammonium acetate or EDTA solutions (Welch and Cary, 1975; Vanselow, 1966; Page, 1974, p. 27). Amounts of metals released by these techniques are commonly referred to as "extractable." Stream sediments can include hydrous metal oxides, such as those of Fe(III) and Mn(IV), clay particles, and large-size organic molecules. Organic compounds and metals can adsorb on the metal oxides and on clay particles, or can be associated with large-size organic molecules. The bonding can be by electrostatic attraction or chemical coordination. At the normal pH of natural waters (6.5 to 8.5), a variety of inorganic solids, prevalent in natural waters, such as Fe_2O_3, $Al(OH)_3$, MnO_2, SiO_2, and kaolinite are negatively charged, thus exhibiting an electrostatic attraction for metal cations (Stumm and Morgan, 1981, p. 631).

In a study of the Yukon and Amazon rivers, previously mentioned (Gibbs, 1973), 71 percent to 93 percent of the total Fe, Ni, Co, Cr, and Cu was carried in the crystalline structure or the metallic coating of the suspended materials. In another study, it was found that 50 percent to 80 percent of the total copper present in river waters was in the suspended solids phase, possibly as oxide, sulphide, and carbonate precipitates, adsorbed on clay particles or complexed by organic molecules (Montgomery and Stiff, 1971).

Part of the assimilative capacity of natural-water systems for metals, pesticides, and herbicides is the ability of the sediment to bind these substances, thus removing them from the water. On the other hand, many toxic substances stored in the sediment can be released to the surrounding waters by a variety of chemical and biochemical reactions, thus making them available to the organisms living in these waters. Stokes and Szokalo (1977) have pointed out that lake and stream sediments often reflect recent additions of heavy metals before elevations of such elements are detectable in the overlying water. Although tests of water quality, may, therefore, indicate no elevated metal levels in the soluble phase, a water body may still be heavily polluted with organic and inorganic material in sediments. The possibility always exists of reentry of sedimented material into the water due to physical, chemical, or biological processes in natural situations. With bottom-feeding fish, sediments may be more important than water as a source of heavy metals (Florence and Batley, 1977).

The transfer of metals between bottom materials and the water above them is a complex process depending on a multitude of factors such as the oxygen concentration (aerobic or anaerobic conditions), the nature and the concentration of heavy metals and the organic and inorganic ligands present, the pH, and the particle size of the bottom sediments (Lee and Mariani, 1977; Lu and Chen, 1977). Some of the individual processes that may be involved in this transfer are adsorption-desorption, diffusion into or out of the sediment layer, precipitation, coprecipitation, complexation, and dissolution (Jenne and Luoma, 1977). Another transfer mechanism from bottom sediments to water is by absorption through the roots of aquatic plants and subsequent release into the water from the decaying leaves (Welsh and Denny, 1976).

Lee (1975) reviewed the role of manganese and iron oxides, which are nearly ubiquitous in suspended solids and sediments, as partial coatings on other minerals and as discrete oxide particles. He supported Jenne's (1968) proposal that these oxides act as a major sink for trace metals (Co, Ni, Cu, and Zn).

McDuffie and others (1976) have shown that the total metal concentrations in bed sediment increase as the particle size decreases. For example, the average ratio of metal concentration in the silt plus clay to that in the sand fraction ranged from 4.3 for manganese to 23.1 for silver. In the overall transport, 10 percent to 24 percent of the metals were carried in the silt fraction (0.062 to 0.004 mm), 21 percent to 35 percent in the clay fraction (<0.004), and only 1 percent to 2 percent in the sand fraction (2.0 to 0.062 mm) of suspended sediment. The rest (38 percent to 63 percent) was carried in the dissolved phase (<0.45 μm). In studying the distribution of aerially deposited trace elements in an aquatic system, Peyton and others (1974) found the highest metal concentration in the silt aggregate, that is, particle size of <0.050 mm. A study of the Kuskokwim River (Nelson and others, 1977) showed that the mercury concentration in the finest size fraction (<0.062 mm) of bottom materials was much higher than in the coarse fraction (>0.25 mm). The respective concentrations ranged from approximately 1 mg/kg to 25 mg/kg and from 0.1 mg/kg to 2 mg/kg. The dispersal distance of the finer sized sediments also was greater. The general trend of increased metal concentration with decreasing grain size was also confirmed by Delisle and others (1975) and Perhac (1974). Oliver (1973) suggested that the metal concentration in the sediments was related to the surface area of the variously sized particles. The metal concentrations (Cu, Pb, Ni, Co, Fe, Mn, Cr, but not Hg) in the sediments from the Ottawa and Rideau rivers significantly correlated with the surface area of the sediment fraction.

Rickert and others (1977) were able to distinguish between "natural" background concentrations and polluted conditions for Cr, Cu, Pb, and Zn in the bottom sediments of the Willamette River in Oregon by analyzing the <20 μm fraction of the sediments. Wittman and Forstner (1976) suggested that the sediment fraction of <2 μm is a good pollution indicator.

Ramamoorthy and Rust (1976) have shown that mercury is rapidly and efficiently transferred from water to the sediment—almost 100 percent in 5 minutes—and is little affected by variation in pH. No significant differences were observed in the uptake rates in aerobic and anaerobic conditions. The uptake phenomena were surface dependent, and the depth of the bed sediment affected was less than 1 mm (Kudo and others, 1975). The uptake rate of methylmercury was only about 10 percent higher than the value for inorganic mercury (Mortimer and Kudo, 1975).

The mobility of certain elements, notably mercury, in the presence of organic material is greatly increased. Brinckman and Iverson (1975) found that the distribution of mercury in the surface sediments of Chesapeake Bay reflected the use of its waters and the corresponding metal inputs. In heavily contaminated areas the total mercury concentration increased with depth. The reverse was true for unpolluted sites. In an aquatic ecosystem, most of the total mercury, 90 to 99 percent, is found in the sediment, 1 to 10 percent in the water, and less than 1 percent in the biota. The distribution of methylmercury is completely different: 90 to 99 percent in the biota, 1 to 10 percent in the sediment, and less than 1 percent in the water (Jernelov and Lann, 1973).

Lead is also rapidly removed from surface waters by bottom materials. In the Saline Branch Watershed Study of Getz and others (1977) the concentration of lead in the top 5 cm of the bottom sediments ranged from 2,330 to 6,300 mg/kg in urban areas and from 133 to 342 mg/kg in rural areas. In the water, dissolved lead concentrations ranged between nondetectable and 15 μg/L.

Kemp and others (1976) concluded that the elements in the sediments of Lake Erie could be divided into six groups: conservative, enriched, nutrient, carbonate, mobile, and miscellaneous elements. The conservative elements (Al, K, Na, Mg, and Si), indicative of major mineralogical species, were the most abundant in the sediments, and their concentration has remained constant with time. Both the enriched metals (Hg, Pb, Zn, Cd, and Cu) and nutrient elements (organic-C, N, and P) had higher concentrations than those expected from the soil composition in the basin. This enrichment was attributed to anthropogenic sources. The mobile elements (Mn, Fe, and S) had chemical profiles related to the chemical conditions in the sediments. For example, in a zone of permanent aerobic conditions, manganese was concentrated in the surface oxidized zone. Finally, the miscellaneous elements, Be and V, showed no clear trend in their concentrations or enrichment factors over base-line concentrations.

Organic pollutants

Many organic compounds of anthropogenic origin occur in small amounts in rivers and lakes of populated regions. Many of these are hydrophobic—that is, they are nearly insoluble in water and tend to form separate films on water surfaces or be adsorbed on sediment surfaces or to be present as particulate organic material. Other compounds may have hydrophylic properties—that is, they are relatively soluble in water and are molecularly dispersed. Relatively hydrophobic chlorinated aliphatic hydrocarbons of various types and polychlorinated phenolics such as PCB (polychlorinated bi-phenyl) are widely distributed in the environment.

McCrea and others (1985) studied the occurrence of 17 organchlorine pesticide residues and PCBs in suspended sediment and extracts of the aqueous phase collected at eleven locations in the lower Great Lakes region. In terms of total pesticides 78 to 100 percent were found in the aqueous phase. The partition of individual compounds between aqueous and suspended sediment phase varied greatly between compounds and between water bodies. For example, alpha-BKC β-BMC dieldrin, endrin, and transchlordane were found almost exclusively in the aqueous phase. Mirex, when found, was almost 100 percent present in the suspended sediment. Most of the pesticides, however, although they were found to be most abundant in the water phase, were partitioned in different proportions between the aqueous and suspended sediment phases in different water bodies. The same applied to PCBs; the percent of PCBs in the aqueous phase ranged from 22 in the Niagara River, to 100 in Lake Erie samples.

From the foregoing discussion and the results of the research referred to, some general statements can be made regarding techniques for characterizing solids and for extracting material from them. (1) Procedures for analysis and characterization of solids associated with river and lake waters are in a rather primitive state of development. (2) Total dissolution of particulate material and other less drastic treatments to give "total extractable" concentrations of minor elements or organics give data that can be used in a qualitative fashion to indicate existence of environmental pollution. Such data have little usefulness in evaluating probable biological availability or geochemical behavior of the substances determined. (3) Determination of surface chemical properties of solids associated with surface water, such as surface areas per unit weight, and ion exchange capacity, should help in understanding and predicting the behavior of these solids toward solute species. Almost no data of this kind have been obtained for natural stream sediments.

DESIGN OF MONITORING PROGRAMS

The quality of surface water of North America has been extensively studied in Canada and the United States since the early years of this century. Systematic evaluation of the water quality in a river requires a sampling program extensive enough to cover variations in water composition both in space and time, and an ongoing program of chemical and biological analyses. Some of the governmental agencies active in this field have also conducted streamflow measurement programs so that both quantity and quality data are available for assessment of surface-water resources. Sources of published data are noted later in this chapter, and the major-ion composition of North American surface waters is summarized by selected analytical data in Tables 1 through 3 and in Plate 1.

A major monitoring program currently operating in the United States is the National Stream Quality Accounting Network, NASQAN (Ficke and Hawkinson, 1975). This program was begun in 1973 by the U.S. Geological Survey and included more than 500 stream sampling points by the end of the decade. It supplements water-quality monitoring by the U.S. Geological Survey at a large number of additional sites, providing data for many diverse purposes. Surface-water quality monitoring networks also are operated in Canada, Dominion-wide thrugh Environment Canada, and in the provinces.

A worldwide monitoring effort is sponsored by World Health Organization and some other agencies of the United Nations. This monitoring program for water supplies is the Global Environmental Monitoring System–Water (GEMS/WATER), which began in 1978, has been described by Meybeck (1985). It is planned ultimately to include about 1,200 sampling stations for rivers, lakes, and ground waters. The computerized data bank for the system is operated by Environment Canada, and a preliminary water data summary for this program for the period 1979 to 1981 (GEMS, 1983) provided some of the surface-water quality data given in this chapter for the Caribbean region and the more remote locations in Canada.

Some of the considerations in planning and operating water-quality monitoring systems are summarized briefly here. Some aspects are covered more fully in the references given. The summary is based to a considerable extent on the procedures followed in planning and establishing the international GEMS/WATER network. Determining and then monitoring changes in river quality can be accomplished through operation of a network of strategically located long-term periodic sampling stations. Local environmental influences also can be evaluated by repeated short-term surveys, each providing limited spatial coverage, but with many samples taken nearly simultaneously. For some systems it may be necessary to use a combination of these approaches. The exact approach to be used in a given situation will depend on the objectives set out for the program.

Objectives for monitoring programs

The establishment of objectives for a water-quality program is a combination of a scientific task and often of a political process which requires a critical assessment of national and regional priorities (UNESCO-WHO, 1978, p. 33). While the details and emphases in the selection of objectives vary with circumstances, the fundamental long-range objectives of any water-quality-monitoring system are one or more of the following:

1. Increase knowledge of existing water-quality conditions and understanding of the aquatic environment, both under natural and man-affected conditions.
2. Determine the amount and the quality of water available to meet future needs (inventory of water resources).
3. Provide information for predicting the effects of significant future natural and man-made changes on the aquatic environment, including water projects such as dams, diversions, stream enlargements, massive irrigation projects, and aquifer recharge.
4. Assess the effectiveness of pollution control measures.
5. Assess the effects of water quality on aquatic and terrestrial life and on human health and well-being.
6. Monitor the effects of pollution sources, such as industrial complexes and urgan areas, on water quality.
7. Detect trends in water quality and provide an early warning system.
8. Monitor compliance with water-quality objectives or standards.

Some of the more specific short-term objectives to implement the above long-range plans are:

1. Collection of information on current water quality and its trends.
2. Identification of areas that need improvement and an assessment of the urgency.
3. Monitoring trends to warn of changes that may be damaging to the aquatic environment.
4. Establishment of early warning systems for sudden changes.
5. Identification of sources of pollutants, their loads, and movement through the aquatic ecosystem.
6. Identification of possible precautions to be taken and cleanup procedures in the event of accidental spills.
7. Determination of the suitability of natural waters for an intended use.
8. Determination of compliance with guidelines, regulations, and standards.
9. Determination of water quality where international frontiers are crossed or where the water is shared by two or more jurisdictions.
10. Identification of gaps in knowledge requiring research or special studies.

In addition to defining its objectives, the major elements in planning a water-quality-monitoring system are: choosing the location of sampling stations; deciding the frequency of sampling; selecting the water quality parameters to be measured and in which medium (i.e., water, bottom or suspended sediment, and/or biota); selecting the sampling methods; selecting the field and laboratory analytical methods; selecting the methods for data storage and retrieval; establishing the appropriate quality assurance program; and choosing the methods for reporting and interpreting the results.

Location of sampling stations

For either long-term monitoring or for a short-term survey, the location of sampling points is decided by considering: monitoring objectives; existing problems and conditions; potential growth centers (industrial and municipal); population trends; climate, geography, and geology; accessibility of locations; available manpower, funding, and laboratory facilities; degree of inter-jurisdictional cooperation; travel time to the laboratory to minimize changes that might occur in samples during shipping and storage; and safety of the site.

The goal of network planning and design is to accomplish the objectives of the water-quality program in a given area with the minimum of stations, effort, and expense.

Frequency of sampling

The processes, natural and societal, that affect water quality have random features superimposed on the hydrologic, the climatic, and possibly other cyclic factors. This requires a sampling schedule that can adequately evaluate these variable effects. The common approach in establishing the frequency of sampling is statistical. Frequency of sampling is established based on the variability of the data, the concentrations to be measured, and the changes to be detected (UNESCO-WHO, 1978, p. 27). In the absence of sufficient background data, some arbitrary frequency is chosen, based on some knowledge of local conditions. After sufficient data have been collected, the variability can be calculated and the frequency adjusted accordingly. The frequency is also influenced by the relative importance of the station and whether the variability can include values that approach critical levels of some contaminant. Unfortunately, the cost of ideal sampling frequency may be high, and budget constraints may become a factor.

Water-quality parameters

The parameters selected for evaluation at a station will be determined largely by the objectives of the monitoring program (UNESCO-WHO, 1978, p. 28). To minimize cost of operation, the selection must also consider known characteristics of the water resource and of the polluting sources. Samples may not always be analyzed for all selected parameters. Choices can be based on their relative importance, cost of determination, and correlation between parameters.

The parameters that characterize water quality may be classified in several ways:

1. By kind of measurements: physical, for example, temperature, electrical conductivity, color, turbidity; chemical, for example, inorganic and organic constituents and respective subclassifications; biological, for example, macro and microbiological components and the respective subclassifications.

A more detailed list of parameters according to this classification (except for biological parameters) is given in Tables 5 through 7. The biological parameters are not of direct concern in this chapter. They are given in the reference cited above.

2. According to the importance of the parameters with respect to the objectives of the monitoring program.

For example, the Global Environmental Monitoring System (GEMS)/Water (GEMS, 1978) classifies the water-quality variables as either basic, to be measured at all locations, or use-related, such as drinking water supplies, irrigation, and general (e.g., aquatic life).

Tables 8 and 9 show the grouping of the parameters according to this classification.

Some of the properties measured in water-quality evaluation, for example, water temperature, can be measured only at the sampling site. Other parameters, such as dissolved oxygen, must be measured at the time of sample collection or very soon thereafter. Automated monitoring equipment that can be installed at the sampling site is now available for measuring specific conduc-

TABLE 5. WATER-QUALITY PARAMETERS

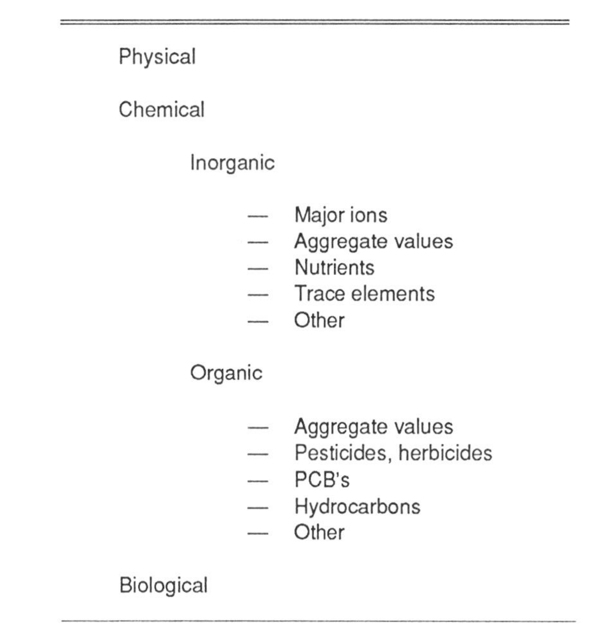

- Physical
- Chemical
  - Inorganic
    - — Major ions
    - — Aggregate values
    - — Nutrients
    - — Trace elements
    - — Other
  - Organic
    - — Aggregate values
    - — Pesticides, herbicides
    - — PCB's
    - — Hydrocarbons
    - — Other
- Biological

TABLE 6. PHYSICAL PARAMETERS

Discharge (for rivers)
Water temperature
Turbidity
Light penetration (Secchi disc)
Suspended sediment
Color
Odor
Taste
Specific conductance
Particle-size distribution (suspended and bottom sediments)

TABLE 7. CHEMICAL PARAMETERS (INORGANIC)

Major Ions		
Anions	Bicarbonate	(mg HCO_3^-/L)
	Carbonate	(mg CO_3^{2-}/L)
	Sulphate	(mg SO_4^{2-}/L)
	Chloride	(mg Cl^-/L)
Cations	Sodium	(mg Na^+/L)
	Potassium	(mg K^+/L)
	Calcium	(mg Ca^{2+}/L)
	Magnesium	(mg Mg^{2+}/L)
Others	pH, Silica	(mg SiO_2/L)
Aggregate Values		
	Total dissolved solids (TDS, mg/L)	
	Alkalinity (HCO_3^- + CO_3^{2-}, mg $CaCO_3$/L)	
	Total Hardness (Ca^{2+} + Mg^{2+}, mg Ca CO_3/L)	
	Sodium absorption ratio (SAR)	
	Inorganic carbon (total and dissolved)	
	Radioactivity (gross alpha, beta + gamma, pCi/L)	
Nutrients		
	Nitrogen, nitrate + nitrate (NO_3^- + NO_2^-, mg N/L)	
	Phosphorus, orthophosphate	(PO_4^{3-}, mg P/L)
	Inorganic	(mg P/L)
	Total	(mg P/L)
Trace Elements: (Dissolved and total; other elements may be determined as well.)		
	Aluminum	(mg Al/L or ug Al/L)
	Antimony	(mg Sb/L or ug Sb/L)
	Arsenic	(mg As/L or ug As/L)
	Cadmium	(mg Cd/L or ug Cd/L)
	Chromium	(mg Cr/L or ug Cr/L)
	Cobalt	(mg Co/L or ug Co/L)
	Copper	(mg Cu/L or ug Cu/L)
	Iron	(mg Fe/L or ug Fe/L)
	Lead	(mg Pb/L or ug Pb/L)
	Manganese	(mg Mn/L or ug Mn/L)
	Mercury	(mg Hg/L or ug Hg/L)
	Nickel	(mg Ni/L or ug Ni/L)
	Selenium	(mg Se/L or ug Se/L)
	Silver	(mg Ag/L or ug Ag/L)
	Tin	(mg Sn/L or ug Sn/L)
	Vanadium	(mg V/L or ug V/L)
	Zinc	(mg Zn/L or ug Zn/L)
Other		
	Radioactive Nuclides	(pCi/L)
	Carbon, organic	(Total and dissolved, mg C/L)
	Nitrogen, organic	(Total and dissolved, mg N/L)
	Oxygen, chemical demand	(mg O/L)
Pesticides		
	Chlorinated	(ug/L)
	Carbamate	(Nitrogen containing, ug/L)
	Phosphorus	(ug/L)
Herbicides		
	Phenoxy acids	(ug/L)
	MCPA, MCPB	(ug/L)
Other		
	Phenols	(ug/L)
	Pentachlorophenols	(ug/L)
	Oil and grease	(ug/L)
	Nitrilotriacetic acid	(NTA, ug/L)
	Polychlorinated biphenyls	(PCB's, ug/L)
	Chloroform	
	Carbon tetrachloride	
	Polyaromatic hydrocarbons	(PAH's)
Others		
	Sugar, sterols, halogenated aliphatic compounds	

TABLE 8. GEMS/WATER BASIC VARIABLES

	Rivers	Lakes and Reservoirs
Temperature	X	X
pH	X	X
Electrical conductivity	X	X
Dissolved oxygen	X	X
Nitrate	X	X
Ammonia	X	X
Calcium	X	X
Magnesium	X	X
Sodium	X	X
Potassium	X	X
Chloride	X	X
Sulfate	X	X
Alkalinity	X	X
BOD	X	—
Total suspended solids	X	—
Chlorophyll a	—	X
Transparency	—	X

tance, pH, DO, and some other unstable water-quality parameters, and for transmitting results of such measurements to a central point, should this be desirable.

FIELD AND LABORATORY METHODOLOGY

Time and space variability

Sampling is the process of collection of a representative portion of a body of material. The significance of reported analytical data on the sample is limited by the confidence that can be placed in the representativeness of the sample.

As mentioned before, water-quality parameters exhibit both time and space variability. The time variability can be detected and measured by an approrpriate sampling frequency. It has been recommended that to define a cyclic variation properly, the water body must be sampled over a period 10 times as long as the period of the longest cycle and at intervals no longer than one-third of the shortest cycle (Sanders and others, 1983, p. 195).

Rivers are usually sufficiently shallow and turbulent that vertical homogeneity is quickly attained below an influx of new material. Lateral homogeneity is much more slowly attained. Thus wide, swift-flowing rivers may not be completely mixed for many kilometers below the outfall point (Sanders and others, 1983, p. 119–27). In one study, mixing in the Columbia River was incomplete 5 km below an outfall at Trail, British Columbia, but was complete after 16 km. Calculations predicted complete mixing by 60 km for a source located on shore, and 15 km for a midstream source. Large inflows from tributary streams may not be completely mixed with the receiving stream for even greater distances (Hem, 1985, p. 43).

Various protocols are recomended for testing whether a sample at any point will be representative: (1) six samples analyzed in duplicate, at three positions across the river and two depths (BSI, 1983, section 6.1.4), and (2) middepth samples at third, quarter, or various decile points depending on the river flow (GEMS, 1978, ch. 1, p. 12). If a conveniently obtainable sample is not representative because of lack of complete mixing, it is preferable to sample farther downstream, since the use of flow-weighted average samples across a set of vertical cross sections is time consuming.

Longitudinal mixing, that is, the smoothing out downstream of irregular or cyclic discharges into a river, will have a secondary influence on the location of sampling stations, insofar as distance from such a discharge can be important in deciding frequency of sampling. Its effect must be taken into account when interpreting data (BSI, 1983, p. 11; UNESCO-WHO, 1978, p. 261–2).

Because of the important relationships between water quality and stream discharge and the need for facilities such as sampling cableways, river-water sampling points should generally be located at or near stream gaging stations.

The degree of mixing in lakes will depend on the configuration of the lake and the location of inlets and outlets. Where the lake is elongated or dentritic, with many branches, or consists of a number of basins, mixing will be poor. Many lakes also exhibit the phenomenon of seasonal thermal stratification. Mechanisms involved were described earlier in this chapter.

Saline closed-basin lakes may exhibit salinity stratification, with the dilute inflows floating on the more dense saline water at greater depth. Such stratification may persist for long periods of time because no seasonal overturn can occur and anaerobic bacterial activity can bring about extensive reduction of sulfate to H_2S (Hem, 1985, p. 117).

Artificial lakes created by dams have characteristics similar to those of natural lakes in that temporary stratification caused by seasonal water-temperature changes is commonly observed. Reservoirs having storage capacities that are large compared to stream flow rates may have major effects on downstream water quality, as high and low flows entering the reservoir are stored and mixed, and released water shows only minor seasonal variation in composition.

TYPES OF SURFACE-WATER SAMPLES

"Grab" samples

A discrete "grab" (or spot) sample is one that is taken at a selected location, depth, and time, and then analyzed for the constituents of interest. A "depth-integrated" sample is collected over a predetermined part or the entire depth of the water column, at a selected location and time in a given body of water, and then analyzed for the constituents of interest.

The collection of grab samples is appropriate (Huibregtse and Mosher, 1976) when it is desired to: (a) characterize water quality at a particular time, (b) provide information about min-

TABLE 9. GEMS/WATER-USE RELATED VARIABLES

	Rivers	Lakes and Reservoirs
(a) Drinking water supply[1]		
Total coliforms	X	X
Fecal coliforms	X	X
Arsenic	X	X
Cadmium	X	X
Chromium	X	X
Lead	X	X
Mercury	X	X
Selenium	X	X
Cyanide	X	X
Fluoride	X	X
Nitrate[2]	X	X
Dieldrin	X	X
Alidrin[3]	X	X
DDT[3]	X	X
Copper[4]	X	X
Iron[4]	X	X
Manganese[4]	X	X
Zinc[4]	X	X
(b) Irrigation:		
Sodium[2]	X	X
Calcium[2]	X	X
Chloride[2]	X	X
Boron	X	X
(c) General water quality (e.g., aquatic life):		
Silica, reactive	X	X
Kjeldahl nitrogen	X	X
COD	X	—
TOC	X	—
Chlorophyll a[2]	X	X
Hydrogen sulfide	—	X
Iron	—	X
Manganese	—	X
PCB's	X	X
Aluminum	X	X
Sulfate[2]	X	X
pH[2]	X	X

[1] As in WHO Guidelines for Drinking Water Quality (World Health Organization, 1982).

[2] Already mentioned in basic variables.

[3] Total organochlorine compounds (TOCl), dieldrin, aldrin, and DDT are considered as representative of the major categories of organic pollutants listed in WHO Guidelines.

[4] Selected aesthetic quality variables listed in WHO Guidelines but not covered under basic variables.

imum and maximum concentrations, (c) allow collection of variable sample volumes, (d) deal with a stream that does not flow continuously, (e) analyze for parameters likely to change, or to (d) establish the history of water quality based on relatively short time intervals.

Composite samples

There are two main types of composite samples (WQB, 1983). Sequential, or time, composites are made up by: (a) continuous, constant sample pumping, or by (b) mixing equal water volumes from samples collected at regular time intervals. Flow proportional composites are obtained by: (a) continuous pumping at a rate proportional to the flow, (b) mixing equal volumes of water collected at time intervals inversely proportional to the volume of flow, or by (c) mixing volumes of water proportional to the flow collected during or at regular time intervals.

A composite sample provides an estimate of average water-quality condition over the period of sampling. Composite sampling reduces the number of samples to be analyzed. On the other hand, composite samples cannot detect changes in parameters occurring during the period of time represented by the composite samples.

Collecting a representative water sample

For water-quality sampling sites located on a homogeneous reach of a river or stream, the collection of depth-integrated samples in a single vertical may be adequate. For small streams a grab sample taken at the centroid of flow is usually adequate (USGS, 1981, section 5.B.4.b.1).

For sampling sites located on a nonhomogeneous reach of a river or stream, it is necessary to sample the channel cross section at the location at a specified number of points and depths. The number and type of samples taken will depend on the width, depth, discharge, the amount of suspended sediment being transported, and aquatic life present. Generally, the more points that are sampled along the cross section, the more representative the composite sample will be. Three to five verticals are usually sufficient, and fewer are necessary for narrow and shallow streams.

One common method is the EQI (equal-width increment) method, in which verticals are spaced at equal intervals across the stream. The EDI or equal-discharge increment method requires detailed knowledge of the streamflow distribution in the cross section in order to subdivide the cross section into verticals proportional to incremental discharges (USGS, 1981, section 5.B.4.b.1).

FIELD AND LABORATORY ANALYTICAL METHODS

The analysis of a water sample can be carried out in situ, in the field, or in a regular analytical laboratory.

A number of parameters, including pH, conductivity, dissolved oxygen, temperature, turbidity, color, and transparency, are likely to change on storage of a sample, and should, therefore, be measured "in situ" or in the field as soon as possible after sample collection. Equipment is also available for continuous on-site measurements of these parameters, as noted earlier in this discussion. Research is underway on in situ and field methods and instrumentation for measuring other parameters such as heavy metals and organic compounds.

To aid in obtaining comparably accurate results, standardized methodologies for water analyses have been recommended (APHA, 1985; BSI 1983). Methods in use by government-operated or contractor laboratories generally follow these standards (Skougstad and others, 1979). Modern laboratory instruments and procedures can detect much lower concentrations of minor elements and trace organics than those of only a decade or two ago, and further improvements can be expected in the future. Data aquisition by computers and microcomputers directly from the instrument and data handling systems are also common features in today's analytical laboratory.

Analyses for minor constituents in water samples require special attention to prevent loss of solutes during sample collection and storage and to prevent contamination from sampling equipment or other sources that may add material not present in the source sampled.

The following elements are considered when choosing the most appropriate method for a water-quality determination: (1) parameter to be measured; (2) parameter form (e.g., dissolved, total); (3) medium (e.g., water sediment, biota); (4) stability of sample and preservation; (5) range of concentration that can be measured by the method; (6) precision; (7) limit of detection and limit of quantitation; (8) number of determinations that can be made per unit of time; and (9) availability of instruments, chemicals, and trained personnel.

QUALITY ASSURANCE

Quality assurance (QA) is a systematic process that ensures a specified degree of confidence in the data collected for a water-quality-measurement program. It is made up of a series of procedures and practices covering the planning, field, laboratory, and data-handling phases of the data-collection program.

The QA of the planning process must ensure that the objectives of the monitoring program are well defined, that the progress toward accomplishing them can be quantitatively measured, and that the resources available are adequate for the objectives. It also must ensure that the criteria for the selection of sampling locations, and the methodologies for sampling, field, and laboratory analysis, for data storage and retrieval, and for data analysis exist and are adequate given the objectives and the resources available.

The field QA must ensure that the actual sampling locations chosen are adequate for the purposes of the program, that the samples collected are representative of that location, that sample contamination and deterioration are not occurring, and that all the necessary field data and information are properly recorded. Replicate, spiked, and blank samples may be prepared in the field and submitted to the laboratory as a part of the field QA process.

Similarly, the laboratory quality assurance program ensures the reliability of the analytical data. The laboratory QA has two main components. The first one is the internal laboratory quality control (QC), which ensures the reproducibility of data within a single laboratory. The accuracy of the results can be determined only if suitable certified reference materials are available. If the accuracy is not known, then one cannot establish the degree of confidence of the analytical results. The interlaboratory QC, the second component of the QA program, can also assist the laboratory personnel in establishing the accuracy of their results. This QC, best conducted by an independent agency, assesses the overall performance of different laboratories, including the effectiveness of their internal laboratory QC, and ensures the compatibility of data generated by these various laboratories. This assessment can be based on analytical performance on standard "blind" samples submitted to participating laboratories.

Data storage and retrieval

Data obtained in water-quality monitoring should be filed in an orderly and systematic manner to ensure proper access to it. It is important that the data filing system be organized in such a manner that data are easily assembled for quick scanning and review of results. The data should also be arranged so that statistical analyses and report preparation may be readily carried out. The importance of developing and maintaining an efficient datas processing and storage system cannot be overemphasized. The purposes of such a system may be summarized as follows: (1) To provide a rapid warning system of environmental problems on a national or regional basis; (2) to provide rapid information to water resource managers on matters pertaining to noncompliance with standards or objectives; (3) to provide data on short notice for international purposes; (4) to provide data input to statistical computations such as time series analysis; water-quality trends and forecasting; hydrological, physical, chemical, and biological correlations; and (5) to provide data input to determinations of material balance and loadings in lakes, rivers, and streams.

Water-quality data may be stored manually on laboratory cards or report forms and filed in standard filing cabinets, or the data may be stored on computer-readable forms. Although manually filed field and laboratory cards might still be in use in some agencies, the development of computers has provided a capability for handling large volumes of data quickly and efficiently. Some countries have developed central data banks into which data may be fed and retrieved from different locations through the use of special transmitting and receiving equipment, commercial telephone lines, or on other communication channels. This method is convenient and economical when large volumes of data are being produced and analyzed. It is, however, essential that all laboratory data, before they are filed, are carefully reviewed for accuracy and reasonableness by informed professional staff members who have direct familiarity with sampling sites and past records obtained there.

Considerable flexibility and versatility exist in electronic data processing. Computers will produce print-out graphs, or maps showing water-quality data and related information in a very short time period. This technique is particularly useful in developing water-quality trend analyses, forecasts, and other types of statistical summaries and indicators, which are of vital importance to water resource managers.

To ensure maximum confidence in data collected in water-quality survey programs, a data evaluation mechanism should be established as part of the survey program. The evaluation should be carried out concurrently with the data collection to make certain that the data truly represent the water-quality conditions being investigated and that the data are meeting the requirements of the study.

Evaluation of data should be carried out continuously during the data collection program to ensure the following. (1) Information on each sample should be properly recorded. (2) Analytical results should be properly recorded. Particular attention should be paid to detection limits and the number of significant figures to make sure that they remain consistent throughout the program. (3) Analytical results at each location should be critically evaluated to ensure that the detection limits meet the requirements of the study. (4) Analytical results should be compared with previous results. This will help catch any abnormal values immediately and alert the analyst to reexamine the sample. (5) Check analytical results to ensure correct values. Laboratories should enter into quality-control arrangements with other laboratories to compare results on standard or identical samples in order to ensure that standard methodologies are being followed, that proper analytical techniques are being adhered to, and that correct analytical results are being obtained.

Reporting and interpreting water-quality data

Once water-quality data have been collected and assembled in data storage systems, the next step is to interpret the data with respect to specific questions, environmental problems, and requirements of water resource management. Water-quality concerns are wide and varied. Some of the most commonly asked questions about water are: (1) What is the water quality of any specific location or area? (2) What are the water-quality trends in the country or region; is the quality improving or getting worse? (3) How do certain parameters relate to one another at given sites; that is, how does specific conductance relate to total dissolved solids and how do these relate to stream discharge, how does turbidity relate to chlorophyll, or how does dissolved oxygen relate to time or temperature? (4) Are sampling frequencies adequate and are sampling stations suitably located to represent water-quality conditions in an area? (5) What are the total mass loadings of materials moving in and out of water systems, and from what sources and in what quantities do these originate? (6) Is it possible to develop a water-quality model based on the available data that will predict water-quality characteristics under varying conditions?

Although general water-quality conditions may be ascertained by scanning long columns of tabulated data, for more reliable, scientifically correct conclusions the statistical approach is the most commonly used. A statistical analysis of the data may provide a deeper understanding of the data and enables the water resource manager to arrive at better decisions in water-quality management.

DATA ORGANIZATION AND CORRELATION

Water composition in relation to stream discharge

As noted earlier in this chapter, the concentrations of water-quality constituents are frequently correlated with stream discharge. The physical basis for a relation between concentration and flow stems from the interaction of two opposing processes. First, additions to streamflow dilute dissolved and suspended materials in the water column and tend to decrease concentrations. Simultaneously, increases in runoff contribute new material to the channel and tend to increase concentrations. Thus, the degree and, indeed, the direction of the correlation between streamflow and concentration varies with the substance in question and the location of sampling. As a rule, the correlation is positive for particulate material and negative for dissolved substances or substances of flow-independent origin such as pollutants from point-source discharges. A summary of flow-concentration relations for a variety of common constituents in U.S. rivers is given in Table 10. The data are based on 6- to 10-year records of monthly samples from 504 stations in the United States National Stream Quality Accounting Network (NASQAN). The concentrations of certain nutrients, such as total phosphorus and dissolved nitrate, may exhibit either a positive or a negative relation to flow depending on circumstances in the basin. Figures 3 and 4 provide examples of these relations. The Klamath River of California (Fig. 3) is a stream with relatively high sediment concentrations, and soil erosion is the primary contributor to both sediment and total phosphorus concentration. In contrast, the Black River in South Carolina (Fig. 4) has much lower sediment concentrations and exhibits a pattern typical of the dilution of point sources of phosphorus.

TABLE 10. AVERAGE r^2 VALUES FOR THE REGRESSION OF DISSOLVED CONCENTRATION ON FLOW FOR 504 STATIONS IN THE NATIONAL STREAM QUALITY ACCOUNTING NETWORK (U.S. GEOLOGICAL SURVEY)

Constituent	Average r^2	Fraction Positive	Fraction Negative
Dissolved solids	0.46	0.04	0.96
Sodium	0.44	0.09	0.91
Potassium	0.23	0.11	0.89
Calcium	0.40	0.02	0.98
Magnesium	0.43	0.03	0.97
Chloride	0.34	0.03	0.97
Sulfate	0.29	0.24	0.76
Nitrate	0.31	0.44	0.56
Suspended sediment	0.34	0.95	0.05
Total phosphorus	0.26	0.72	0.28

Relations between concentration and streamflow have been described by a variety of mathematical functions. For a majority of common constituents, these are of the general form:

$$C = a + b \cdot f(Q) \quad (1)$$

where C is the estimated concentration, Q is the instantaneous discharge, and f(Q) may have one of the following forms (Smith and others, 1982):

*Functional Form**	*Description*	
$f(Q) = Q$	linear	(2a)
$f(Q) = \ln Q$	log	(2b)
$f(Q) = \frac{1}{1 + \beta Q}$	hyperbolic	(2c)
$f(Q) = \frac{1}{Q}$	inverse	(2d)

*β is a positive constant.

The function (2c) was introduced by Johnson and others (1969) to describe the behavior of the major dissolved species under varying flow. It is included here, along with other more common model forms 2a, 2b, and 2d, because of its considerable flexibility and demonstrated usefulness with many constituents.

A second general form of equation relating concentration and flow is the exponential

$$\ln C = b_0 + b_1 \ln Q + b_2(\ln Q)^2.$$

This equation has been found (Smith and Alexander, 1983a) to be most useful for nondissolved or sediment-related constituents (e.g., suspended sediment, total phosphorus, bacterial counts) because of the tendency for particulate material to increase nonlinearly with increasing flow.

In addition to estimating concentrations at specific values of flow (i.e., between samples or at flow extremes) regression equations relating flow and concentration are also useful in a variety of statistical applications where an adjustment is required for the effect of flow variability on water-quality records. Flow adjustment is sometimes necessary, for example, in making estimates of long-term mean concentrations and transport rates, as well as in analyzing for time trends. Estimation of mean values and trend analysis are discussed in the following section.

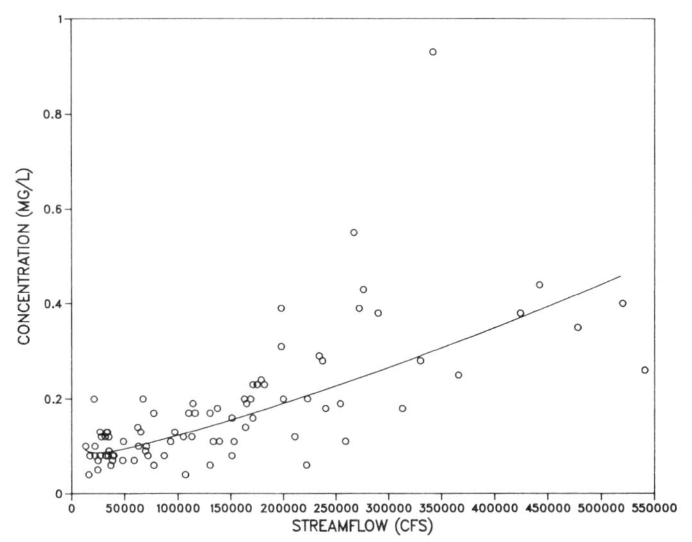

Figure 3. Total phosphorus concentration in relation to stream discharge for the Ohio River at Cannelton Dam, Kentucky. The data were collected approximately bimonthly between 1974 and 1985.

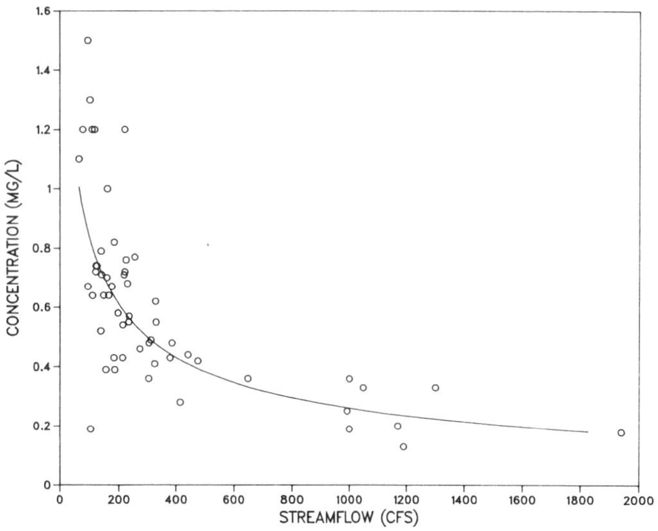

Figure 4. Total phosphorus concentration in relation to stream discharge for the Patuxent River near Bowie, Maryland. The data were collected approximately bimonthly between 1978 and 1985.

Statistical description

If water-quality sampling occurs at approximately regular intervals or randomly in time, and if all samples contain measurable concentrations of the constituents of interest, then simple arithmetic averages of the data provide unbiased estimates of mean concentrations. If, however, sampling has occurred at other than regular or random intervals (e.g., more frequently at some seasons than others), then the samples are likely to reflect a bias toward either high or low flows, and some form of discharge weighting may be required to correct for the sampling bias. Regression of concentration on flow (see above) may be used in conjunction with the flow-frequency distribution to achieve the correction. To date, however, this approach has been used far more extensively in estimating mean transport rates (Johnson, 1979), for which the problem of sampling bias is considerably more severe, than it has in estimating mean concentrations.

Because of the large temporal variability contained in most surface-water-quality records, it is generally of interest to provide some information on the distribution of sample values that surround means and medians. Much of the variability in water-quality data is periodic in nature, occurring seasonally in response to flow and temperature cycles, and diurnally in the case of constituents which are influenced by plant photosynthesis. The latter may include dissolved oxygen, pH, dissolved CO_2, bicarbonate, phosphorus, and various forms of nitrogen. As a result of the variety of influences, water-quality data are frequently non-normally distributed and are, thus, somewhat ill-suited to classical statistical techniques. There is a growing tendency, instead, toward the use of techniques that require no a priori assumptions concerning the mathematical form of the underlying distribution.

A simple and increasingly popular method of graphic presentation of water-quality records is by means of "box" (or "box and whisker") plots (Tukey, 1977). Some examples appear in Figure 5. Box plots provide a quick impression of the general features of the distribution: its central tendency (median), the spread of the data (interquartile range), and the length of the whiskers (extreme values). The location of the median within the interquartile range indicates the degree of asymmetry of the distribution, and the "outside values" show the location of extreme values. By far the most effective use of box plots comes in comparing distributions (as in Fig. 5) to show differences in medians, data spread, or extremes.

A significant and recurring problem encountered in estimating means and other distributional parameters for trace metals and organic compounds is the occurrence of "censored" or "less-than" values in the record. Censored values are those recorded as falling below the laboratory detection limit for the constituent. Despite technological progress in lower detection limits, the problem has become more serious in recent years due to increased interest in the possible health effects of trace contaminants in water. Several approaches have been taken to working with censored data. The most convenient—but least satisfactory statistically—of these has been to simply discard the censored values. Slightly better (because the resulting bias is more recognizable) is the "conservative" approach of replacing all censored values in the record with either zero or the value of the detection limit, depending on the circumstances or application. Average values based on such edited data sets represent minimum and maximum estimates of the mean. Thus, the reporting of both estimates has the advantage of showing the range of uncertainty inherent in the data set. As the fraction of censored values in the record approaches zero, the difference between maximum and minimum estimates of the mean also approaches zero.

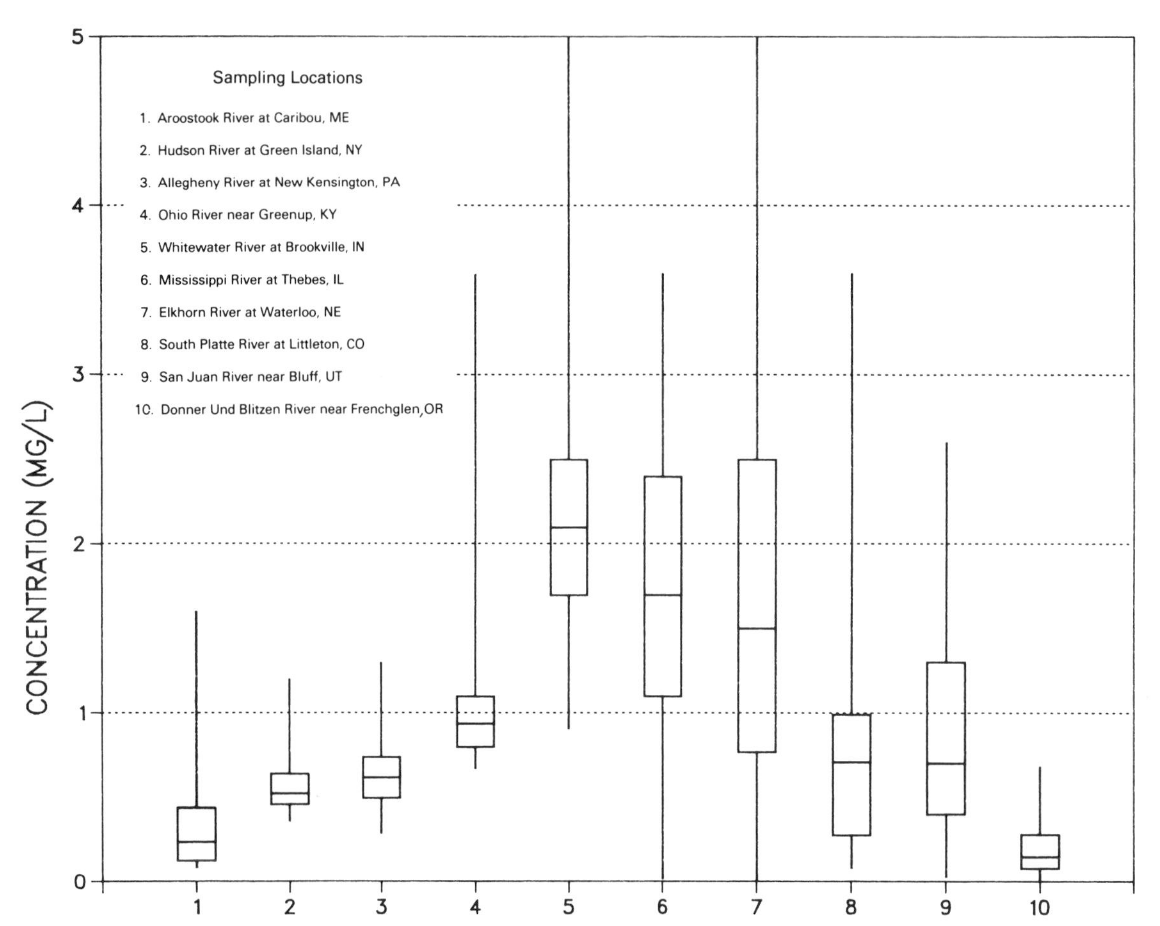

Figure 5. Nitrate concentration records for ten United States rivers summarized in the form of "box and whisker" plots. The median concentration for each station is shown by the horizontal line inside the box; the top and bottom of each box show the 75th and 25th percentiles respectively; and the vertical lines extend to the observed values that lie closest to 1.5 times the interquartile range above and below the 75th and 25th percentiles respectively. Nitrate concentrations are seen to be somewhat higher and more variable at midcontinent stations than at those closer to either the eastern or western coasts.

One recent and very promising approach to working with water-quality records containing censored data has been to fit a mathematical distribution to the uncensored data and then extrapolate this function to the censored portion of the record when estimating means and other parameters. Based on a comparison of eight methods for working with censored data, using both actual data sets (Helsel and Gilliom, 1986) and Monte-Carlo simulations (Gilliom and Helsel, 1986), the most accurate estimates of means and other distributional parameters are obtained with this method assuming a lognormal distribution for the censored values in the data set.

Trend analysis

There has been much interest in recent years in the analysis of trends in water-quality data. This interest can be traced in part to a growing public concern over various sources of water-quality degradation and to growing public curiosity concerning the effectiveness of expensive pollution control programs. An additional reason for recent interest in water-quality trend analysis may simply be that certain large data-collection programs established in the early 1970s are now providing an abundance of records greater than 10 years in length.

Although there is no general agreement on terminology, it is of some value to distinguish here between short-term and long-term analysis. Owing to the current abundance of 10 to 20 year records, short-term trend analysis has become increasingly a statistical matter, whereas, the near absence of records longer than 40 years has required the development of less direct methods for the detection of long-term water-quality trends. Some of these, in fact, do not involve water-quality data at all, and are only briefly described here.

Many of the common statistical characteristics of water-quality records increase the difficulty of detecting short-term trends. These characteristics include nonnormal distributions, seasonality, flow-relatedness, frequent missing values, censored data, and serial correlation. As a consequence of these characteristics, certain of the assumptions inherent in classical techniques of trend testing are violated. In statistical terms, trend is defined as monotonic (though not necessarily linear) correlation with time, and the classical approach to trend testing is to perform regressions against time. A great deal of research has been devoted in recent years to accommodating water-quality records in regression tests (Harned and others, 1981) as well as to developing nonparametric alternatives to regression (Hirsch and others, 1982). Perhaps the most widely used example of the latter is the seasonal Kendall test (Hirsch and others, 1982) based on the nonparametric correlation statistic known as Kendall's tau (Kendall, 1975). The seasonal Kendall test adjusts for seasonality in water-quality records by applying Kendall's tau separately to 12 time series consisting of the data for each month of the year. An example of the application of the seasonal Kendall test to the dissolved solids record for the Bear River in Utah is given in Figure 6.

The characteristic of water-quality data that generally has the most confounding influence on trend testing is flow dependence. Due to the large influence flow changes may have on concentrations, trends in flow may easily induce trends in concentrations, or may obscure trends stemming from other causes. Because flow trends generally occur independently of the factors of interest in water-quality investigations (e.g., anthropogenic effects), adjustment for flow effects is often desired. Flow adjustment is accomplished by regressing concentration on flow (see above) and applying the trend test to the regression residuals. An example is given in Figure 7. It appears that the pronounced trend in dissolved solids in the Bear River from 1974 to 1979 is removed after flow adjustment. The results of flow-adjusted trend tests applied to dissolved chloride records for sampling stations on major United States rivers is given in Figure 8. As noted by Smith and Alexander (1985, p. 66), there has been a large increase in the use of salt for clearing snow and ice from highways since the 1950s, but most of the increase occurred prior to 1970.

Because of the existence of many transitory influences on water quality, short-term records cannot, in general, be extrapolated to indicate long-term trends. One alternative approach to interpreting the long-term changes in water quality that have accompanied European colonization and development in North America has been the study of remote basins largely isolated from the effects of development. A nationwide sampling program with this objective (the Hydrologic Benchmark Program of the U.S. Geological Survey) was established in 1958. Since that time only a few comparative studies of water quality in these and more developed basins have been published (Biesecker and Leifeste, 1975; Janzer and Saindon, 1972). One difficulty has been the fact that few of the basins are entirely free of the effects of human habitation, and essentially all of the basins are subject to anthropogenic influences via the atmosphere (see Smith and Alexander, 1983b). Nevertheless through careful study and perhaps adjustment for atmospheric and other effects, benchmark data would seem to have the potential for providing a general impression of the effects of industrial and agricultural development on North American water quality.

A second, and perhaps more promising approach to identifying certain kinds of long-term water-quality trends has been the chemical analysis of sediment cores. Recent interpretations have appeared of trends in weathering and erosion rates (MacKereth, 1966), chlorophyll and productivity (Brush, 1984), and several trace metals (Nriagu, 1983). Time scales range from tens to thousands of years, but many recent studies have emphasized the past

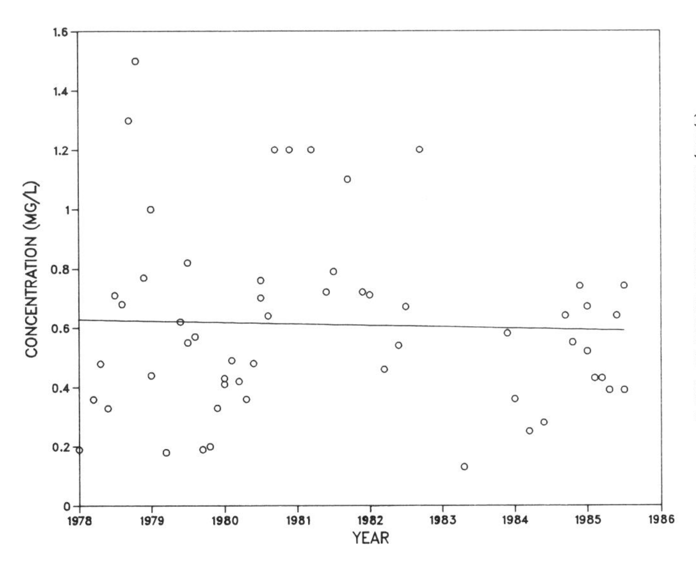

Figure 6. Total phosphorus concentrations plotted against time for the Patuxent River near Bowie, Maryland. The data show some tendency towards declining values with time, but application of the seasonal Kendall test indicates the trend is not statistically significant (p = 0.947).

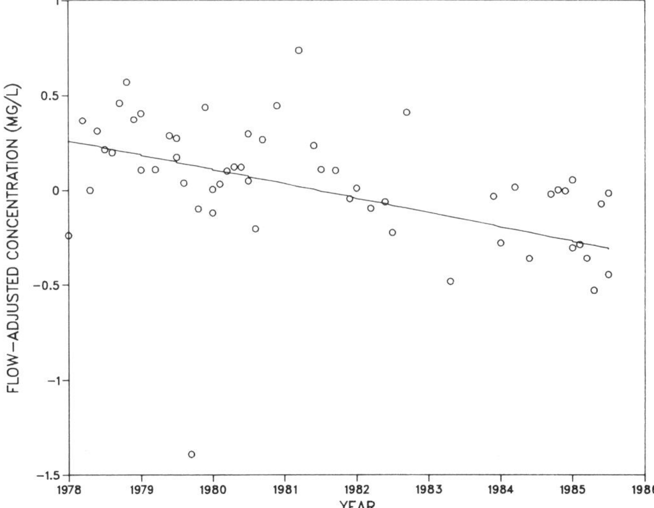

Figure 7. Flow-adjusted concentrations of total phosphorus plotted against time for the Patuxent River near Bowie, Maryland. Flow-adjusted values are the residuals (differences between estimated and observed values) from a regression of concentration on streamflow. Application of the Seasonal Kendall test indicates a highly significant ($p < 0.001$) decrease in flow-independent concentrations occurred during the 1978 to 1985 period.

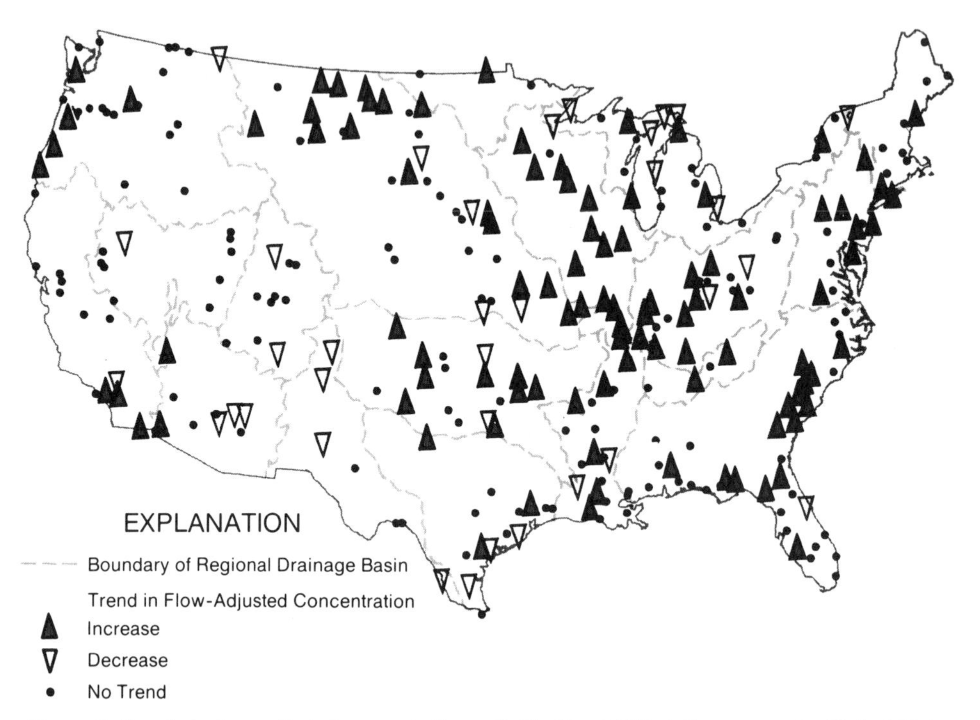

Figure 8. Trends in flow-adjusted concentrations of chloride at 289 stations on major United States rivers. Trends were identified by the Seasonal Kendall test and were considered to be significant at the $p < 0.1$ level.

150 years. The ability to determine sedimentation rates is a fundamental prerequisite to assessing trends in these studies. To date, water-quality records have not been directly compared to time series based on sediment cores because there are few cases where a common period of record exists. In coming years, however, this will likely be fertile ground for statistical analysis.

Material fluxes

A broad-scale view of the geochemistry of rock-weathering processes can be obtained by computing the rates at which the various chemical elements are transported in dissolved or suspended form from the continents into the ocean. Most of this transport is accomplished by rivers. According to Martin and Meybeck (1979), the average total amount of dissolved matter carried to the ocean by streams each year is 4.1×10^9 metric tons. These authors also estimated the total annual load of particulate material to be almost four times greater. These fluxes can be calculated in a straightforward fashion from concurrent records of flow and water quality, if they are available, but for some streams very little of this kind of data exists.

In order to develop estimates of fluxes of dissolved material in metric tons per square kilometer per year (t km^{-2} yr^{-1}) where records are sparse, Meybeck (1979) assigned river drainage areas of the continents to 15 different "morphoclimatic types of dissolved solids transport." Enough records for streams draining each type area were available to permit calculation of solute flux rates for each type. The global flux was then computed from the total continental areas assigned to the different morphoclimatic types.

The bases for classification of morphoclimatic type areas used by Meybeck (1979) included approximate areal runoff rate, average temperature, and physical relief, but rock type was not considered. The 15 classes included glaciers, tundra, taiga (subdivided into three groups based on rate of runoff), temperate (subdivided into four groups based on rate of runoff), deserts, tropical (subdivided into four classes based on runoff characteristics and relief), and arid regions from which little runoff was expected to occur. The highest expected total solute flux was computed to be 120 t km^{-2} yr^{-1} from the most humid of the temperate morphoclimatic type regions. Because lithologic factors were omitted from these estimates, they may not give reliable fluxes for specific terranes.

Measured data for North America are more complete than for most of the other continents. Livingstone (1963) computed a dissolved solids flux of 33 t km^{-2} yr^{-1} for the continent. Meybeck (1979) gives essentially the same flux based on measured values for streams draining about half of the continent and on estimated values for the remainder.

Fluxes for individual U.S. streams discharging to the oceans were given by Leifeste (1974). These ranged from 80.9 to 9.1 t km^{-2} yr^{-1}. The value for the Mississippi was

48.4 t km^{-2} yr^{-1}. The average flux for the area considered, which excluded the St. Lawrence–Great Lakes basin and areas draining to Hudson Bay, was given as "about" 35 t km^{-2} yr^{-1}.

All these estimates refer to actual stream loads of solutes. Meybeck (1979) made some calculations that might be used to reduce the fluxes to actual rock weathering rates, by subtracting anthropogenic and atmospheric sources. As noted elsewhere in this chapter, the sulfate and chloride concentrations have apparently increased greatly in the Mississippi, St. Lawrence, and other rivers in highly developed areas of the continent in the past 80 to 100 years.

The relative importance of some of the various factors that may influence solute fluxes has been studied by Peters (1984). In his study the records of major ionic concentrations and loads at 56 sampling sites on U.S. rivers were tested using multiple-linear-regression equations for correlation with rock type, annual precipitation, basin population density, and average water temperature. Best correlations with annual flux were obtained with annual precipitation especially in basins where the rock type was predominantly crystalline (igneous and metamorphic).

Maximum dissolved solids fluxes observed in the streams studied by Peters reached or slightly exceed 100 t km^{-2} yr^{-1}. In basins where the predominant rock type was limestone this loading was attained in several streams where annual precipitation was a little less than 100 cm. A higher level of annual precipitation was needed to reach this flux rate in crystalline rock- and sandstone-dominated basins. For most individual solute species, fluxes did not correlate well with population density or water temperature.

A general correlation between solute flux and annual precipitation is perhaps to be expected in basins where exposed rocks are resistant to chemical attack. Weathering reactions of igneous rock minerals in natural systems probably can be expected to proceed at rates proportional to the surface area of solids exposed per unit volume of water. Where the water supply is plentiful the surfaces will tend to be kept wet almost continuously, and the chemical weathering processes will thus be able to operate most of the time. In drier or colder surroundings the reactions may occur intermittently, and the rate of solute tranport out of the area owing to rock weathering will be lower, although solute concentrations in the smaller amount of water may be higher. The direct contribution of solutes from precipitation also becomes more important in regions with high rainfall.

In carbonate rocks the rate of weathering is strongly influenced by the available CO_2 supply, which would be a function of biological activity in most areas. The relationship of solute fluxes to precipitation and runoff in such terranes is more complicated, and a simple correlation with runoff rate is less likely.

An extensive review of global CO_2 fluxes and factors affecting them has been made by Youngquist (1985).

A summary by Dethier (1985) gives solute transport rates ranging from 10 to 100 t km^{-2} yr^{-1} in the streams draining westward from the Cascade and Klamath mountain ranges in California, Oregon, and Washington. This study noted some correlation between dissolved solids flux and annual precipitation. However, Dethier's study also took rock type into account, and he observed substantial differences in weathering rates attributable to differences in mineralogy and land utilization or development in the region studied.

GEOCHEMICAL CHARACTERISTICS OF THE RIVER WATERS OF NORTH AMERICA

During the past several decades there has been an enormous increase in the public interest in, and attention to, the quality of water supplies, and the adverse effects of many human activities on water quality is a matter of continuing concern. One result of this interest has been a great increase in water-quality monitoring of rivers by governmental and other agencies in North America, as well as elsewhere in the world.

Although the usual pollution indicators, such as dissolved oxygen, biochemical oxygen demand (BOD), and bacterial counts, are not of major geochemical interest, the monitoring programs commonly include many chemical determinations. The accumulated records for rivers of Canada and the United States form a great mass of geochemical data. Some interpretive studies of the data have been made, as noted elsewhere in this chapter. Generally these studies are limited to specific geographical areas, and nationwide evaluations have not been made. Data available for Mexico and other Central American countries are much more limited in quantity.

A detailed inventory of the chemical composition of river waters of North America is beyond the scope of this chapter. However, a generalized view of the major features of present-day North American river-water composition can be gained from selected recent records, presented in Tables 1, 2, and 3. River sampling-station records were selected to represent the major streams both with respect to water discharge and drainage areas, and to provide a wide geographical distribution. Selections were made with the aim of showing some cultural effects as well as the influences of geology and climate on river-water major-ion chemistry.

The complete records for stations in the U.S. since 1971 have been published in the annual series "U.S. Geological Survey Water Resources Data for (state and year)." Earlier data were published annually in a U.S. Geological Survey Water-Supply Paper series entitled "Quality of Surface Water of the United States (year)."

Most of the U.S. records in Table 2 were obtained at stations in the National Streamflow Quality Accounting Network (NASQAN). Records for the Caribbean region in part were obtained from the GEMS network (GEMS, 1983), which has been mentioned earlier in this chapter, and a few records were obtained from other published sources. Canadian data were obtained from water-quality studies of Environment Canada, and for stations near the International boundary, from NASQAN data.

Publications related to these studies are listed in annual indices (e.g., Environment Canada, 1985). The complete records of water-quality data for Canadian streams are published in the series "Detailed surface water quality data (province or territory and time period)," by Environment Canada, Ottawa. The time periods covered in a single volume have generally ranged from a few years to more than a decade. A general summary of surface-water quality in Canada was published by Thomas (1964) based on data then available.

Sampling and analysis protocols differ among these stations and have changed with time. Commonly, for a fixed number of "grab" samples obtained each year, a chemical analysis for all major inorganic solutes is made. The number of such samples and their distributions in time is variable; however, most records used here provided about 12 analyses a year. Some of the sampling programs paid special attention to timing samples with runoff events, but for the most part the samples were about evenly spaced in time through the year.

In order to summarize this kind of data, we have chosen to present only the major dissolved constituent concentrations in the tables along with the specific electrical conductance of the water sample and a field-measured pH value. Other parameters are of less direct geochemical interest or, as in the case of organic and minor inorganic constituents, involve greater uncertainties in reliability of data. All the analyses directly quoted in the tables were checked for cation-anion balance. Specific conductance values are in microsiemens per centimeter ($\mu S/cm^{-1}$). One μS is equal to one μmho.

For most sites, a 3-year period was selected for evaluation. Generally this period was centered around the years 1979 to 1981, but for some of the stations it was necessary to use earlier or later years. From the data for the period chosen, the analysis having the highest total ionic concentration and that with the lowest were selected (based on reported conductivities). In addition to the analytical data for each site, the drainage area in square kilometers (km^2) and long-term mean water discharge in cubic meters per second (m^3/sec^{-1}) when available, are given in the tables. This mean discharge represents all the flow data available for that sampling site. For most analyses in Table 2, the instantaneous discharge value also is given for the stream at the time the samples were collected.

For data from the GEMS network in Tables 1 and 3, the tabulated parameters represent the minimum and maximum values observed for each parameter during the period of record (1979 to 1981). These concentrations generally do not represent a single water sample. The GEMS data also give median values for each parameter, and medians were used in the tables for a few sampling sites where the extremes did not give an approximate anion-cation balance. For the GEMS stations in Tables 1 and 3, the maximum observed discharge appears in the same column as the maximum solute concentrations.

The data in the tables give an indication of the commonly encountered annual or seasonal variability of water composition and give a basis for computing a median concentration for the record period used. Obviously a 3-year period is not sufficient to give highly reliable values for medians or extremes to be expected over longer periods of time.

Some of the constituents in the original analytical data were recalculated to other units. Specifically, alkalinity values have all been converted to equivalent concentrations of bicarbonate ions, and the dissolved nitrate + nitrite nitrogen values have been converted to equivalent quantities of nitrate and are reported as NO_3.

Most of the chemical analyses in Tables 1, 2, and 3 are shown on a map (Plate 1) by means of bar-graph symbols. Besides indicating the total solute concentration, by overall height of the bar, the relative importance of seven dissolved components is indicated by subdivision of the bar graph, identified by different colors. The units used are milliequivalents per liter (millimoles per liter in the case of silica, SiO_2) computed from mg/L data in Tables 1 to 3. Details of this method of graphic representation of water analysis data are given elsewhere (Hem, 1985, p. 173–175).

In general, for most large rivers the maximum dissolved solids concentration occurs at minimum flow rates and is from two to four or five times as high as the minimum expected at high discharge rates. Exceptions may represent streams with large upstream lakes or reservoirs, such as the St. Lawrence flowing from Lake Ontario (Station 7, Table 1) and the Colorado River above Imperial Dam, Arizona (Station 7, Table 2). Such streams may show long-term variability in solute concentrations, but seasonal or year-to-year changes are generally minor. At the opposite extreme, a smaller river affected by large point or regional solute sources, and having a wide range in discharge may be extremely variable in solute concentration. The Pecos River at Red Bluff, New Mexico, near the New Mexico–Texas boundary (Station 17, Table 2), had a maximum solute concentration more than 58 times as great as the minimum during the three years of data. This stream drains an area in which evaporite rocks are common and is also influenced by irrigation return-flows. Analyses for this station are not shown on the map because the maximum concentration is too large to be represented conveniently at the scale used for the other sampling stations. For the same reason the analyses for the Bear River at Corinne, Utah (no. 5 in Table 2) are not shown on the map.

Climatic effects are evident in the low solute concentrations of many Arctic streams compared with rivers of similar size in temperate regions. Rock weathering by dissolution is strongly inhibited in cold climates.

Geologic effects also are evident in data in Tables 1 through 3. For example, the influence of igneous rock weathering can be seen in the water of the Snake and Rogue rivers in Idaho and Oregon (Stations 20 and 23, Table 2). Both streams have high proportions of dissolved silica, compared to concentrations of other solutes.

Combinations of geologic and climatic effects increase the contrasts in solute concentration shown for the continent in Plate 1. In a temperate climate, where precipitation is heavy and evenly distributed, and where exposed rock and soil minerals are relatively resistant to dissolution, a low solute concentration is continuously maintained in runoff, as shown for the Alabama, Appalachicola, and Savannah rivers (Stations 1, 3, and 22, Table 2). In cold climates, erosion by solution occurs much more slowly; where rocks are resistant, as in the Canadian Shield, the solute concentrations observed are extremely low. Several streams in northeastern Canada show these effects. Water of the Churchill River at Muskrat Falls, Labrador (Station 1, Table 1), ranged in conductance from 12 to 23 $\mu S\ cm^{-1}$ and is the most dilute of the river waters represented in Tables 1 through 3. The solutes contributed by atmospheric fallout and rainout are a major part of the dissolved material in such waters, and because they are poorly buffered, their pH can easily be lowered by acidic precipitation.

Fewer data for tropical streams of the continent are available. Two of the data points in Table 3 are on large streams entering the Caribbean Sea from northern South America. The analyses for the Orinoco are stated by Meybeck (1979) to be a mean of samples taken over a 4-year period. There is only a single analysis for the Magdalena, and the discharge represented is unknown. For the Orinoco the average dissolved solids content given is below the value given in the same reference (Meybeck, 1979) for the Amazon. A point of interest for the streams in Table 3, however, is that the silica concentrations, although low in an absolute sense, are consistently a rather high proportion of the dissolved solids. This tendency may be explainable as related to the intense weathering of silicates that can occur in tropical climates.

Anthropogenic effects of various types can be discerned in the tabulated data. The influence of irrigated agriculture is shown strongly by the South Platte River at Julesburg, Colorado (Station 24, Table 2) near the Colorado-Nebraska boundary. Intensive use and reuse of the river water for irrigation in eastern Colorado has greatly increased the solute concentrations at all flow stages over what might be expected under pristine conditions. The effects of irrigation return flows can be seen in other streams of the western United States, notably the Colorado River, which now rarely discharges any significant amount of water to the Gulf of California.

There are noticeable effects of human activities on even the largest rivers of the intensively developed parts of the continent. Some of these have been noted elsewhere in this chapter. Such effects, however, cannot totally obscure the stronger geochemical influences that are present in some parts of the continent. Extensive areas of the interior of North America are underlain by sedimentary rocks of the Mesozoic and Cenozoic eras in which relatively soluble, or readily oxidizable sulfur-bearing minerals occur. Weathering of these rocks produces rather large amounts of sulfate and, in some places, chloride that can appear in runoff. Examples in Table 2 include the Tongue River at Miles City, Montana (Station 26), and the Pecos River at Red Bluff, New Mexico (Station 17). The low runoff and high evaporation rates characteristic of these drainage areas intensify the effects so that very high solute concentrations may occur during low-flow periods. The Missouri River at Sioux City, Iowa (Station 15, Table 2), shows the effect of inflows from Great Plains streams in its relatively high sulfate concentration compared with other streams having comparable mean discharge rates. Streams in the High Plains region of Canada also may have relatively high sulfate concentrations at times as shown by the data for the South Saskatchewan River (Station 20, Table 1).

Impact of human activities on surface-water quality

In industrialized and thickly populated river basins such as those in western Europe and parts of North America, many stresses are placed on surface-water resources. Many of these stresses tend to degrade water quality, and some of their effects are now plainly evident. Drever (1982), for example, cites studies by European investigators who concluded that more than 90 percent of the chloride and two-thirds of the sulfate in the water of the lower Rhine is of anthropogenic origin. Berner (1971) suggested that from 40 to 60 percent of the sulfate load in many rivers can be attributed to human activities.

Statistical analyses, described elsewhere in this chapter, can readily detect relatively small consistent increases or decreases of dissolved concentrations with time. Records to which such techniques can be rigorously applied, however, cover only the past few decades for most of the rivers of North America. Earlier records are available for some streams, but because of changing sampling and analysis protocols, and changes in analytical methods and instruments, as well as the large gaps in time for which data are not available, a rigorous statistical approach does not seem viable for comparing these data with more recent ones. However, if there have been major changes in sulfate and chloride concentrations, rather simple comparisons of old with new records should reveal them.

In an earlier section of this chapter, some chemical reactions between water and rock minerals were cited. It was pointed out that much of the bicarbonate present in natural water comes from carbon dioxide produced in biochemical processes. Such processes also strongly influence nitrate concentrations. Sulfate is a product of oxidation of reduced-sulfur mineral species, a process that is promoted by human activities such as burning of fossil fuels, smelting of ores, and mining. Sulfur is also involved in biochemical reactions, as it is an essential nutritional element for plants and animals. Chloride is extensively utilized in many ways in industrial societies and tends to remain in solution, as almost all chloride compounds are readily water soluble. Industrial uses of various metals may introduce much larger quantities of certain elements to surface waters than would normally be expected from rock weathering, and synthetic organic material produced by industry includes compounds that do not occur naturally at all. Accordingly, one may expect that the most obvious anthropogenic effects would be observed in the concentrations of industrially utilized metals, organics, and their by-products. However, these substances are affected by many chemical interactions with water and sediment. The complications that are thus introduced into their chemical behavior have been pointed out. Of the remaining potential indices, the bicarbonate and nitrate nitrogen concentrations tend to be controlled by cyclic biochemical processes. Thus, the sulfate and chloride anions constitute the most feasible indices of long-term anthropogenic influences.

Long-term trends in river- and lake-water composition

The first comprehensive study of the chemical composition of United States surface waters was made in the first decade of the twentieth century by the U.S. Geological Survey. Samples of water were generally obtained daily for a period of a year at the sites included in this program. About 100 sites in the eastern states and 55 in the western states were included, and results were published by Dole (1909) and Stabler (1911) and in several other U.S. Geological Survey reports for individual states. These data and some additional material obtained from other sources later were used by Clarke (1924b) in his summary of the composition of river and lake waters of the U.S. These publications form a major data base that can be used for comparison with current and recent observations at the same or nearby sampling points.

Studies of the chemistry of surface waters of Canada on a Dominion-wide scale began in 1936. Results were summarized by Leverin (1942, 1947) and by Thomas (1953, 1964, 1965). Sampling and analysis techniques used were similar to those used by the U.S. Geological Survey. Data for the Central American and Caribbean region are too spare to be useful in trend studies.

Descriptions of sampling and analytical procedures for the early USGS data suggest that, in general, the results for major constituents should be comparable in precision and accuracy with those given in later records. However, in some respects the procedures used in the early work may have introduced inaccuracy and bias. The daily samples were collected in 120-ml-capacity bottles, giving a limited volume of water for analysis. The rather low sensitivity of some of the procedures available at that time probably affected the analytical accuracy for solutes present at low concentrations. Analyses reported by Stabler (1911) for stations in the western U.S. give only partial anionic and total dissolved solids concentrations for the individual samples and weekly composites. Methods used to prepare composite samples for more complete analysis for these stations were inconsistent and introduced substantial bias, hence cationic compositions given for these waters probably are not geochemically useful. Analyses given by Dole (1909) represent composites of equal volumes of daily samples. Generally there were three composites for each month. The complete analyses for these stations are much more reliable as indicators of the actual water composition.

Both Dole (1909) and Stabler (1911) reported difficulties in removing suspended sediment before the analyses were started. It is evident from the reported concentrations that the various filtration and pretreatment methods used sometimes left particulate material in the samples and caused improbably high iron concentrations. Incomplete filtration possibly may have also affected the reported silica concentrations.

Unfortunately, after about 1910, the U.S. Geological Survey's quality of surface water data collection program was greatly curtailed, and for many sites there is little intervening information until about 1950 and later. The older data did not include any determinations of trace metals except for iron and aluminum in some samples. The values, as noted above, obtained for these metals were affected by incomplete filtration and are unreliable.

Because river-water quality varies greatly in response to water discharge, a single year of record may not be representative. In any exacting comparison of data, this factor needs to be taken into account. Up to about 1970, many of the USGS river-water-quality records continued to be based on daily sampling with complete analyses of composited samples. Composites contained from 10 to 30 daily samples. Annual averages of these analyses, weighted by time or discharge, were used to summarize the records. After 1970, complete analyses were done on single samples collected at various time intervals ranging from semi-monthly to quarterly, and analyses of composite samples were no longer made.

Data for the Great Lakes–St. Lawrence River system give the best evidence that important changes in anionic composition have occurred at sampling points in the eastern U.S. and Canada since the period 1905 to 1907. Although bicarbonate concentrations in the Great Lakes remained essentially constant, the sulfate and chloride concentrations in the St. Lawrnce at Ogdensburg, New York, and in all the lakes except Lake Superior, increased by at least a factor of two between 1906 and 1980. Beeton (1965) showed that major sulfate increases occurred in the lower four lakes from about 1850 to 1960. The amount of water stored in the lakes is so great that annual stream discharge fluctuations have little influence on the composition of water discharged from Lake Ontario. Figure 9, adapted from Dobson (1967), demonstrates the increase in SO_4 from 1906 to 1980. The trend line fitted by Dobson (1967) fits the additional points for 1970 and

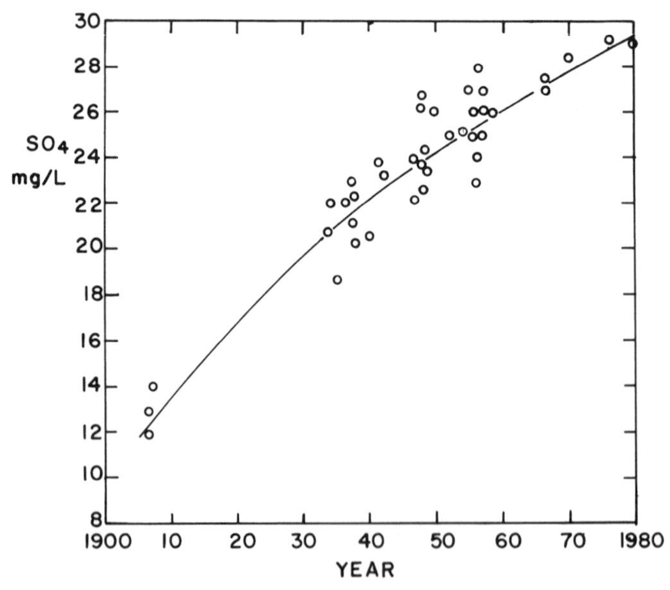

Figure 9. Graph showing sulfate concentrations in Lake Ontario, 1907 to 1980. (After Dobson, 1967.)

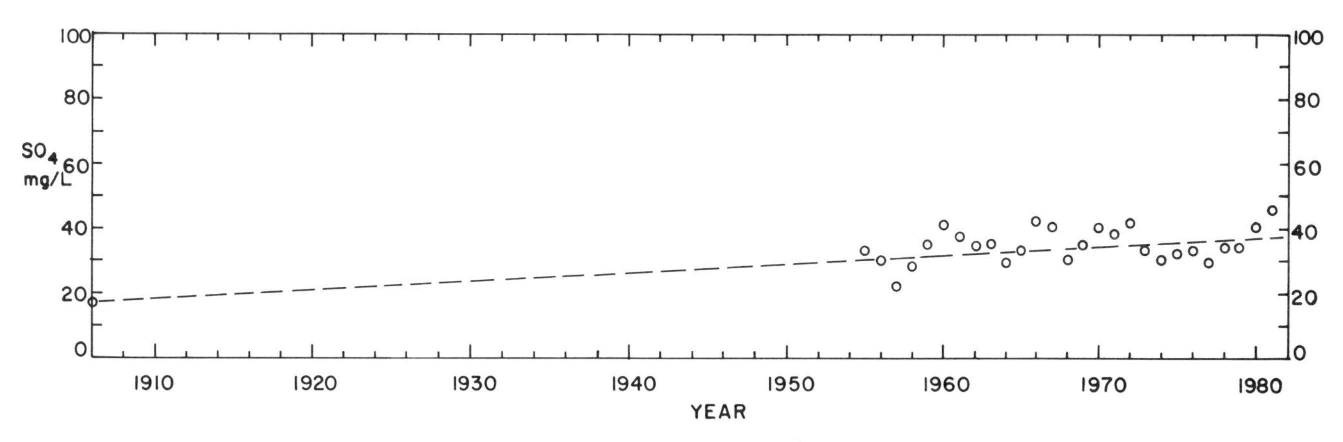

Figure 10. Graph showing sulfate concentrations at times of annual minimum dissolved solids, Mississippi River near St. Francisville, Louisiana, 1955 to 1981 and at New Orleans, Louisiana, 1905 to 1906.

1980, which are from more recent analytical data for the St. Lawrence River near the lake outlet.

Increases in sulfate and chloride concentrations in U.S. river waters comparable to those documented for the St. Lawrence appear to have occurred at many of the sampling sites between 1907, and the 1950s and 1960s. The reality of some of these apparent changes may be questioned in some instances, as differences in flow and various other hydraulic factors—such as reservoir construction, diversion, and augmentation of flow by man-made structures—can affect the comparability of past and present records. However, the increases are too large and widespread to be readily explained by differences in runoff rate.

Changes over the past 80 years in chemical composition of the Mississippi River are of particular interest because this is the largest North American river in terms of discharge and drainage area. The U.S. Geological Survey has collected water-quality data continuously at a sampling station near St. Francisville, Louisiana, since 1954 (Station 14, Table 2). This station is about 48 km upstream from Baton Rouge, Louisiana. In the publication by Dole (1909), mentioned earlier, there is a record of water composition for the Mississippi River at New Orleans during a 12-month period in 1905 and 1906.

Although New Orleans is more than 160 km downstream from Baton Rouge, the composition of the water in the river is evidently not significantly different at New Orleans from what is observed at the St. Francisville station, based on concurrent records obtained at the two sites during several years between 1955 and 1985. Factors that did not exist in 1906, such as the controlled Atchafalaya bypass, may have some effect on the more recent data. However, as a first approximation of water-quality changes, it is of interest to examine and compare these records.

It is immediately evident that sulfate and chloride concentrations at New Orleans in 1905 and 1906 were substantially lower than those observed at the St. Francisville site in the 1955 to 1984 period. The time-weighted average concentrations of sulfate and chloride at New Orleans were 24 and 9.7 mg/L, respectively, in 1905 and 1906. In almost all the years from 1955 to 1967, the time-weighted average concentrations of these ions at St. Francisville were more than twice as great. Changes in sampling and analysis protocols after 1967 make this kind of comparison for more recent records more difficult, but there is an indication that sulfate concentrations were continuing to increase slowly through all or most of the 1955 to 1984 period.

In order to compensate, at least in part, for effects of water discharge and utilize all the available data, we selected, from the entire array of records, the complete analysis having the lowest total dissolved solids concentration for each year. The 1905–1906 data for New Orleans represent weekly composite samples. For most years, from 1955 to 1967, at the St. Francisville station a total of 36 complete analyses were made on 10-day composites of daily samples. For the remainder of the period of record the complete analyses are for single samples taken at monthly intervals. In some years these samples were taken at semimonthly or bimonthly intervals. Obviously, these differences bias the data to some extent. The selection of the most dilute samples for the correlation compensates, in part, for effects of water discharge, as these samples should all represent high, and possible similar flow rates.

The sulfate concentrations in mg/L for these analyses are plotted against time in Figure 10. The dashed line in this graph is fitted by a least-squares linear regression ($r = 0.60$) and passes through the minimum observed in 1905 to 1906. Although the points for the 1955–1981 period are scattered considerably, the fit is not unreasonable and probably is meaningful. If only the 1955 to 1981 data are used for the regression, they give a considerably smaller value for r, but a linear extrapolation of the fitted straight line back to 1906 gives a sulfate value of 21 mg/L, compared to the measured minimum of 17 mg/L. This suggests that the simple linear regression is not greatly in error.

Assuming a mean discharge rate of 18,400 cubic meters per second ($m^3 sec^{-1}$) for the Mississippi and Atchafalaya to the Gulf of Mexico (Leifeste, 1974), at the 1905–1906 mean concentration of sulfate of 24 mg/L at New Orleans (Dole, 1909), the rate of discharge of sulfate was 14 million metric tons per year in

1906. From the recent data for the Mississippi near St. Francisville, Louisiana (Table 2), it seems likely that the present sulfate discharge rate is about 30 million metric tons per year.

At least a part of the 1905–1906 sulfate load must have been anthropogenic in origin also. One of the major sources of sulfate for the Mississippi system is drainage of water from coal mines and mine waste piles in the upper reaches of the Ohio River and its tributaries, especially in Pennsylvania, West Virginia, Ohio, and Kentucky. The pyrite and marcasite that occur in association with the coal are oxidized by aerated water to yield sulfate, ferrous iron, and H^+. Barton (1978) reported that the total sulfuric acid load contributed to streams in the Appalachian region exceeds 4,000 tons per day. However, some of this drainage enters streams draining to the east, and does not reach the Ohio.

The Youghiogheny River, sampled near its confluence with the Monongahela above Pittsburgh, Pennsylvania, in 1906–1907, was substantially affected by mine drainage (Dole, 1909), and its water was acidic most of the time during that period, with sulfate concentrations as high as 300 mg/L. Sulfate originating from mine drainage therefore must have been affecting the quality of the Ohio and Mississippi waters in 1906–1907. Amounts of sulfate originating from this source cannot account for all the apparent increase in loading of the Mississippi mentioned above, but may be an important factor.

The average sulfate concentration of the Mississippi at its mouth in the seventeenth century can only be guessed at from these calculations, but it probably was not much more than 15 mg/L. Chloride and sulfate enter the western tributaries of the Mississippi naturally in substantial quantity. The evaporite facies of Permian and related rocks occur near the surface in parts of the Arkansas River and Red river basins, and their effect on water composition in these and some other tributaries is substantial. The low-flow analysis for the Pecos River at Red Bluff, New Mexico (Station 17, Table 2), demonstrates the effect of these formations on water quality, although the Pecos is a tributary of the Rio Grande. Relatively high sulfate concentrations also occur in streams farther north, as shown by data for the Tongue River at Miles City, Montana (Station 26, Table 2), a tributary of the Missouri.

Chloride concentrations in the Mississippi at New Orleans ranged from 5.0 to 16 mg/L^{-1} in 1905–1906, compared with the range of 12 to 26 mg/L^{-1} at St. Francisville shown in Table 2. This implies an approximate doubling, an increase of about the same magnitude as that given for sulfate in Figure 10.

Ackermann and others (1970) summarized historical data for Lake Michigan, the Ohio River at Cairo, Illinois, and the Illinois and Mississippi rivers at Peoria and Alton, Illinois, respectively. They showed upward trends similar to those pointed out above for sulfate and chloride. The data used by Ackermann and others cover a time period generally from about 1900 to 1965, but there are many gaps in the records.

A review of records covering the period 1960 to 1975 for some major Canadian streams and a comparison with conditions a decade or so earlier as shown by Thomas (1964) were prepared by Environment Canada (1977). This paper noted that small increases in specific conductance appears to have occurred in some streams, most notably the Ottawa River at Sainte-Anne de-Bellevue, Quebec, and in the St. Lawrence River at Levis, Quebec.

The large and increasing amounts of nitrogen fertilizers used in modern agriculture would be expected to cause upward trends in nitrate concentrations of streams draining farmed land. Nitrate from this source probably explains high NO_3 concentrations observed in streams and ground water in some irrigated areas of the western United States (Hem, 1985, p. 125). In many streams in the corn belt of the U.S. there appear to have been substantial increases in average NO_3 concentrations. The average NO_3 concentration for the Iowa River at Iowa City, Iowa, for 1906 and 1907, given by Dole (1909), is 2.8 mg/L. In 1951, the measured discharge weighted NO_3 concentration at this site was 14 mg/L (Love, 1956, p. 24). Other examples of high NO_3 concentrations from this region include the Minnesota River near Jordan, Minnesota (Station 12, Table 2), the Mississippi at Clinton, Iowa (Station 13, Table 2), and the Wabash at New Harmony, Indiana (Station 27, Table 2).

Because surface water nitrate is utilized by biota, such as algae and diatoms, the concentration in most streams is seasonally variable. Another constituent of river and lake water that is involved in biological uptake is the dissolved silica, which is utilized by diatoms. As a result of this assimilation the surface layers of seawater, to the depth to which light penetrates readily, are depleted in silica, with dissolved concentrations commonly well below 1.0 mg SiO_2/L (Hem, 1985, p. 73). Similar effects occur in eutrophic lakes. Data given by Dobson (1984) for Lake Ontario show seasonal variations with near-surface water containing less than 100 μg/L SiO_2 (0.10 mg/L) at times in the summers from 1969 to 1980. During this period the concentrations in water near the bottom were higher, also showing seasonal fluctuations, typically ranging between 500 μg/L and 1.2 mg/L.

Silica concentration in the water of the St. Lawrence river and of the Great Lakes system, in general, appears to have decreased during the past several decades. Livingstone (1963) quoted an average of eight analyses for the St. Lawrence at Kingston, Ontario, covering the period from 1934 to 1942, in which the silica concentration was 3.9 mg/L. The average silica concentration in Lake Ontario, sampled at Toronto during 1934 and 1938, was 3.7 mg/L. The silica values reported by Dobson (1984) for the 1969 to 1981 period for Lake Ontario are generally well below 1.0 mg/L, and the highest concentration measured in water near the lake surface in this period was only about 0.6 mg/L. Analyses for the St. Lawrence (Station 7, Table 1) show silica concentrations ranging from 0.5 to 0.2 mg/L^{-1}.

Average silica concentrations in Lake Erie, given by Livingstone (1963, p. G12), were 6.0 mg/L in 1934 and 1938, and 2.1 mg/L in 1950 and 1951. Analyses for Lake Michigan and Lake Huron, reported by Dole (1909), represent monthly samples taken in 1906 and 1907. The average SiO_2 concentrations were 10 mg/L for Lake Michigan and 12 mg/L for Lake Huron. More recent data (Livingstone, 1963) indicate silica concentrations were between 5 and 6 mg/L by 1940. Analyses for raw water from Lake Michigan at the Chicago municipal intakes in February and August of 1961 (Durfor and Becker, 1962) give silica concentrations of 1.8 and 1.1 mg/L, respectively.

The removal of silica from lake water by diatoms seems a plausible explanation for these data. It seems likely that increasing nutrient availability in the lake water has increased the quantity of these organisms during the period covered by the analytical records. Seasonal and vertical nonhomogeniety of dissolved SiO_2 in lakes may add some uncertainty to the analyses quoted here, but can not explain the large changes observed.

Some loss of silica through uptake by diatoms is likely to occur in most reservoirs created by damming streams. The clear, stored water permits more sunlight for photosynthesis, and the rate of movement of water is greatly reduced, giving more time for the organisms to assimilate dissolved silica. Data summarized by Irelan (1971) for the Colorado River, above and below Lake Mead, Nevada and Arizona, show a decrease in average silica concentration between inflow and outflow water of 3 mg/L, for the period 1935 to 1965. This suggests that about 20 percent of the dissolved silica in the river water entering the reservoir was precipitated, most likely assimilated by diatoms.

The tendency for silica to be removed from solution in impounded water is of considerable scientific interest but does not appear to have been investigated closely by hydrologists. The effect noted for the lower Colorado River suggests that other streams that are now controlled by reservoirs might show downward trends in silica concentrations.

Some of the stream waters most strongly affected by human activities are those of rivers intensively used for crop irrigation. Data for the Gila river system in Arizona, and the Rio Grande in New Mexico, Texas, and adjacent Mexico, show that depletion of the water by evapotranspiration leads to saline solutions with little resemblance to the initial waters (Hem, 1966). In some of the intensively developed basins the return flow from upstream areas may be successively recycled downstream until it is no longer usable. Analyses for the South Platte River at Julesburg, Colorado (Station 24, Table 2), show the effects of intensive irrigation use along the river and its tributaries in Colorado.

Effects of flow control on water quality

The utilization of rivers for the benefit of mankind generally has entailed building dams to control flow and to retain water in the reservoirs thus created for future use. In some areas a primary purpose of reservoir construction has been to provide water to augment low river flows for navigation, and to improve the quality of downstream water that may be impaired by waste disposal. In other areas the primary goal may be to generate power or to provide a dependable water supply that can be diverted and used for irrigation, or by public water-supply utilities. The low-flow quality improvement that commonly results may be considered in part a fringe benefit in these areas.

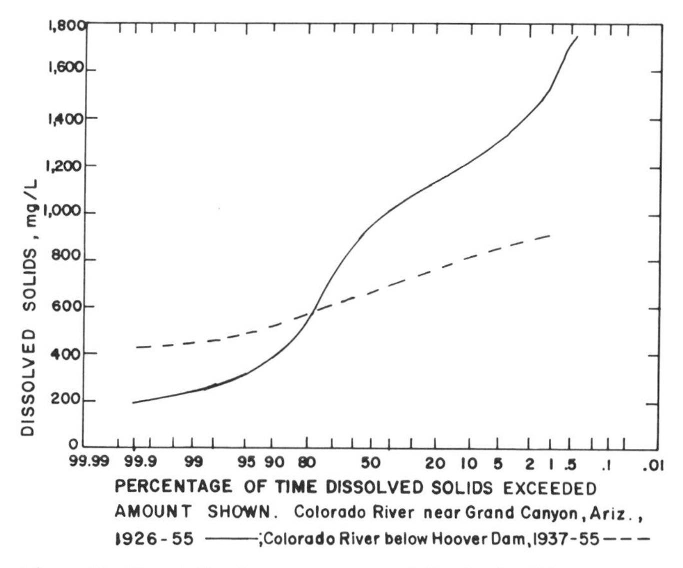

Figure 11. Cumulative frequency curve of dissolved solids concentrations for Colorado River above and below Hoover Dam, Arizona and Nevada.

Many of the streams in water-deficient areas of North America are now partly or totally controlled in flow for long reaches by large storage reservoirs. Conspicuous examples are the lower Colorado River (below Lees Ferry, Arizona); its major tributary, the Gila, which drains much of Arizona; and the Rio Grande, which rises in the southern Colorado mountains and forms the border between Texas and adjoining states of Mexico. The Columbia and Missouri rivers also have large flow-control and storage structures.

Prior to the construction of Hoover Dam the dissolved solids concentration of the river water at Yuma, Arizona, ranged from a minimum near or below 300 mg/L to a maximum near 1,500 mg/L in most years, and in some years it had an even wider variation. After Hoover Dam was completed and storage began in 1935, much of this quality fluctuation disappeared. Lake Mead has a storage capacity equal to about two years of average discharge of the river, but it has been operated so as to maintain some reserve storage capacity.

Figure 11 (Hem, 1985) shows the frequency of occurrence of dissolved-solids values for the river at the Grand Canyon sampling point upstream from the dam and at the reservoir outlet downstream from it. For the periods shown, the median values of dissolved solids are about 950 mg/L in the uncontrolled stream and about 650 in the stored water. This represents a highly significant improvement in average water quality, considered simply on a time base. In practical terms, however, the most important advantages of controlled flow in the lower Colorado are that a larger dependable supply of water becomes continuously available.

Reservoir storage alters water composition in various ways. Early in the history of Lake Mead a substantial amount of Ca^{2+} and SO_4^{2-} was added to the stored water when the rising lake contacted exposures of evaporite rocks. Similar effects were observed by Bolke (1979) in Flaming Gorge Reservoir, a more recently completed impoundment on the Green River, which is a major tributory of the upper Colorado. The concentrations of sulfate in stored water and total load in the river below the dam are increased by the dissolution process, and there is also a decrease in bicarbonate owing to precipitation of $CaCO_3$ as the water becomes supersaturated with respect to calcite. These effects are particularly noticeable in the water leaving the Flaming Gorge Reservoir, where the annual dissolved solids load of outflow exceeds that of inflow by 100,000 to 200,000 tons. Bolke (1979) noted annual average dissolved solids concentration increases of 100 mg/L or more as a result of these effects.

The decreased variability of solute concentrations caused by flow impoundments is also demonstrated by the data for Stations 6 and 7 in Table 2. The Colorado River near Cisco, Utah, is upstream from all major storage dams, and the maximum dissolved solids concentration shown is more than 5 times as great as the minimum. The Imperial Dam sampling site is downstream from all the storage facilities, and the maximum is only 28 percent above the minimum over a similar 3-year period.

Reservoirs in dry climates may lose substantial amounts of water by evaporation, which tends to raise concentrations of all the solutes. Development of anaerobic zones in water at and near the bottom of reservoirs may bring about reduction of metal oxides in the bed sediments. Documentation of this effect in Sam Rayburn Reservoir in eastern Texas was given by Rawson and Lansford (1971). These investigators reported that as much as 14 mg/L Fe and 6.2 mg/L Mn were present in the bottom 5 m of water at times. Some of these reduced metal ions can be released downstream if intake structures intercept the deep water at the dam.

The depletion of oxygen observed in stratified lakes and reservoirs also is related to growth cycles of aquatic vegetation and is described in more detail elsewhere in this volume.

SUMMARY AND CONCLUSIONS

Existing water-quality data bases are sufficient to define major element content and seasonal or annual variability in these components rather well for most of the regionally important streams of the United States and the developed parts of Canada. Controls of these characteristics of surface waters can be identified with some confidence as related to climatic influences and geologic factors, with a substantial human influence in many areas. There have been substantial increases in chloride and sulfate concentrations in the Great Lakes–St. Lawrence River system, for example, since the early twentieth century, and a similar effect is observable in the lower Mississippi. The data base does not permit close quantification of these trends.

More refined statistical analyses can be made using the more complete data base now available for the period 1970 to date. Although some of these analyses clearly show short-term trends for some regions for certain constituents, much work remains to be done, and a clear definition of North American surface-water composition free from human effects cannot be made at this time.

At present, surface-water quality in most of Mexico and the Caribbean region is still rather poorly known. Programs such as GEMS/Water should improve this part of the data base if they are continued.

The operational definition of "dissolved" as the material still in the aqueous phase after filtration through a 0.45 μm porosity filter membrane is a satisfactory approximation for the principal dissolved components of river and lake waters. The nine solute ions and aqueous silica for which concentration data are given in Tables 1 through 3 in this chapter are in this category. Although the concentrations of minor inorganic solutes and many specific organic compounds can be determined in the laboratory, problems in phase separation and analysis of solid-phase material in water samples make it very difficult to predict the distribution and fate of small amounts of these materials that may enter river or lake water.

It may be justifiably concluded that many aspects of the geochemistry of surface water require further research. Although some progress in understanding the subject can be made through study of data produced by past and ongoing water quality data collection programs, many of the questions that need answering will require collection of specialized field data by staff and procedures that are under the direct control of the researcher.

REFERENCES CITED

Ackermann, W. C., Harmeson, R. H., and Sinclair, R. A., 1970, Some long-term trends in water quality of rivers and lakes: EOS American Geophysical Union Transactions, v. 51, p. 516–522.

Agemian, H., and Chau, A.S.Y., 1976, Evaluation of extraction techniques for the determination of metals in aquatic sediments: Analyst, v. 101, no. 1207, p. 761–767.

American Public Health Association, American Water Works Association and Water Pollution Control Federation, 1985, Standard methods for the examination of water and wastewater (sixteenth edition): Washington, D.C., American Public Health Association, 1268 p.

Andelman, J. B., 1973, Incidence, variability, and controlling factors for trace elements in natural, fresh waters, *in* Singer, P. C., ed., Trace metals and metal-organic interactions in natural waters: Ann Arbor, Michigan, Ann Arbor Science Publishers, p. 57–88.

Andrew, R. W., 1976, Toxicity relationships to copper forms in natural waters, *in* Andrew, R. W., Hodson, P. V., and Konasewich, D. E., eds., Proceedings of the Workshop on Toxicity to Biota of Metal Forms in Natural Water, Duluth, Minnesota, October 7–8, 1975: Great Lakes Research Advisory Board, International Joint Commission, p. 127–143.

Barton, P., 1978, The acid mine drainage, *in* Nriagu, J. O., ed., Sulfur in the environment; Part 2, Ecological impacts: New York, Wiley Interscience, p. 313–358.

Beeton, A. M., 1965, Eutrophication of the Saint Lawrence Great Lakes: Limnology and Oceanography, v. 10, p. 240–254.

Berner, R. A., 1971, Worldwide sulfur pollution of rivers: Journal of Geophysical Research, v. 76, p. 6597–6600.

Biesecker, J. E., and Leifeste, D. K., 1975, Water quality of hydrologic benchmarks; An indicator of water quality in the natural environment: U.S. Geological Survey Circular 460-E, 21 p.

Bolke, E. L., 1979, Dissolved-oxygen depletion and other effects of storing water in Flaming Gorge Reservoir, Wyoming and Utah: U.S. Geological Survey Water-Supply Paper 2058, 41 p.

Brinckman, F. E., and Iverson, W. P., 1975, Chemical and bacterial cycling of heavy metals in the estuarine system, *in* Church, T. M., ed., Marine chemistry in the coastal environment: American Chemical Society Symposium Series 18, p. 319–342.

Brown, E., Skougstad, M. W., and Fishman, M. J., 1970, Methods for collection and analysis of water samples for dissolved minerals and gases: U.S. Geological Survey Techniques of Water Resources Investigations, book 5, Ch. A1, p. 16.

Brush, G. S., 1984, Stratigraphic evidence of eutrophication in an estuary: Water Resources Research, v. 20, p. 531–541.

BSI, 1983, Water quality; Sampling of rivers and streams: British Standards Institution, ISO/TC 147/SC 6/WG 4 N8 (working draft), 19 p.

Chynoweth, D. P., Black, J. A., and Mancy, K. H., 1976, Effects of organic pollutants on copper toxicity to fish, *in* Andres, R. W., Hodson, P. V., and Konasewich, D. E., eds., Proceedings of the Workshop on Toxicity to Biota of Metal Forms in Natural Water, Duluth, Minnesota, October 7–8, 1975: Great Lakes Research Advisory Board, International Joint Commission, p. 144–157.

Clark, F. W., 1924a, The data of geochemistry (fifth edition): U.S. Geological Survey Bulletin 770, 841 p.

——, 1924b, The composition of the river and lake waters of the United States: U.S. Geological Survey Professional Paper 135, 199 p.

Cooper, W. J., and Zika, R. G., 1983, Photochemical formation of hydrogen peroxide in surface and ground waters exposed to sunlight: Science, v. 220, p. 711–712.

Cowling, E. B., 1982, Acid precipitation in historical perspective: Environmental Science and Technology, v. 16, p. 110A–121A.

Davies, P. H., 1976, The need to establish heavy-metal standards on the basis of dissolved metals, *in* Andres, R. W., Hodson, P. V., and Konasewich, D. E., eds., Proceedings of the Workshop on Toxicity to Biota of Metal Forms in Natural Water, Duluth, Minnesota, October 7–8, 1975: Great Lakes Research Advisory Board, International Joint Commission, p. 93–126.

Delisle, C. E., Hummel, B., and Wheeland, K. G., 1975, Uptake of heavy metals from sediments by fish: Symposium Proceeding, International Conference on Heavy Metals in the Environment, Toronto, v. 2, pt. 2, p. 821–827.

Dethier, D. P., 1977, Geochemistry of Williamson Creek, Snohomish County, Washington [Ph.D. thesis]: Seattle, University of Washington, 315 p.

——, 1986, Weathering rates and the chemical flux from catchments in the Pacific Northwest, U.S.A., *in* Colman, S. M. and Dethier, D. P., eds., Rates of chemical weathering of rocks and minerals: Orlando, Florida, Academic Press, p. 503–530.

Ditoro, D. M., Thomann, R. V., and O'Connor, D. J., 1971, A dynamic model of phytoplankton production in the Sacramento–San Joaquin delta, *in* Nonequilibrium systems in natural water chemistry: Washington, D.C., American Chemical Society Advances in Chemistry Series 106, p. 131–180.

Dobson, H.F.H., 1984, Lake Ontario water chemistry atlas: Environment Canada Scientific Series no. 139, 59 p.

Dobson, H. H., 1967, Principal ions and dissolved oxygen in Lake Ontario: Proceedings, Tenth Conference on Great Lakes Research 1967, p. 337–356.

Dole, R. B., 1909, The quality of surface waters in the United States; Part 1, Analyses of waters east of the one hundredth meridian: U.S. Geological Survey Water-Supply Paper 236, 123 p.

Drever, J. I., 1982, The geochemistry of natural waters: Englewood Cliffs, New Jersey, Prentice-Hall, 388 p.

Durfor, C. N., and Becker, E., 1962, Public water supplies of the 100 largest cities in the United States, 1962: U.S. Geological Survey Water-Supply Paper 1812, 364 p.

Environment Canada, 1977, Surface water quality in Canada; An overview: Ottawa, Inland Waters Directorate, Water Quality Branch, 45 p.

——, 1985, Water Quality Branch publications, 1984: Ottawa, Inland Waters Directorate, Water Quality Branch, 37 p.

Eugster, H. P., and Jones, B. F., 1979, Behavior of major solutes during closed basin brine evolution: American Journal of Science, v. 279, p. 609–631.

Farrah, H., and Pickering, W. F., 1977, Influence of clay-solute interactions on aqueous heavy metal ion levels: Water Air Soil Pollution, v. 8, p. 189–197.

Feth, J. H., Rogers, S. M., and Roberson, C. E., 1964, Chemical composition of snow in the northern Sierra Nevada and other areas: U.S. Geological Survey Water-Supply Paper 1535-J, 39 p.

Ficke, J. F., and Hawkinson, R. O., 1975, The national stream quality accounting network (NASQAN); Some questions and answers: U.S. Geological Survey Circular 719, 23 p.

Fitzgerald, G. P., and Faust, S. L., 1965, Effect of bacteria on the solubility of copper algicides: Water and Sewage Works, v. 112, p. 271–275.

Florence, T. M., and Batley, G. E., 1977, Determination of the chemical forms of trace metals in natural waters, with special reference to copper, lead, cadmium, and zinc: Talanta, v. 24, p. 151–158.

Fritz, P., and Fontes, J. Ch., eds., 1980, Handbook of environmental isotope geochemistry, v. 1: Amsterdam, Elsevier, 545 p.

Gambell, A. W., and Fisher, D. W., 1966, Chemical composition of rainfall, eastern North Carolina and southeastern Virginia: U.S. Geological Survey Water-Supply Paper 1535-K, 41 p.

Garrels, R. M., and Mackenzie, F. T., 1971, Evolution of sedimentary rocks: New York, W. W. Norton, 397 p.

GEMS, 1978, GEMS/WATER operational guide: Geneva, World Health Organization, 386 p.

——, 1983, GEMS/WATER data summary 1979–1981: Ottawa, Environment Canada, 453 p.

Getz, L. L., and 9 others, 1977, Transport and distribution in a watershed ecosystem, *in* Boggess, W. R., and Wixson, B. G., eds., Lead in the environment: Report prepared for the National Science Foundation, Washington, D. C., NSF/RAO770214, p. 105–134.

Gibbs, R. J., 1973, Mechanisms of trace metal transport in rivers: Science, v. 180, p. 71–73.

Gilliom, R. J., and Helsel, D. R., 1986, Estimation of distributional parameters for censored trace-level water-quality data: 1. Estimation techniques: Water Resources Research, v. 22, p. 135–146.

Goldschmidt, V. M., 1954, Geochemistry: Oxford, England, Clarendon Press, 730 p.

Guy, R. D., and Chakrabarti, C. L., 1976, Studies of metal-organic interactions in model systems pertaining to natural waters: Canadian Journal of Chemistry, v. 54, p. 2600–2611.

Guy, R. D., Chakrabarti, C. L., and Schramm, L. L., 1974, The application of a simple chemical model of natural waters to metal fixation in particulate matter: Canadian Journal of Chemistry, v. 55, p. 611–669.

Harned, D. A., Daniel, C. C., III, and Crawford, J. K., 1981, Methods of discharge compensation as an aid to the evaluation of water quality trends: Water Resources Research, v. 17, p. 1389–1400.

Helsel, D. R. and Gilliom, R. J., 1986, Estimation of distributional parameters for censored trace-level water-quality data: 2. Verification and applications: Water Resources Research, v. 22, p. 147–155.

Hem, J. D., 1966, Chemical controls of irrigation drainage water composition: Proceedings American Water Resources Conference, 2nd, Chicago, 1966, p. 64–77.

——, 1976, Geochemical controls on lead concentrations in stream water and sediments: Geochimica et Cosmochimica Acta, v. 40, p. 599–609.

——, 1985, Study and interpretation of the chemical characteristics of natural water (third edition): U.S. Geological Survey Water-Supply Paper 2254, 263 p.

Hendrickson, G. E., and Krieger, R. A., 1964, Geochemistry of natural waters in the Blue Grass Region, Kentucky: U.S. Geological Survey Water-Supply Paper 1700, 135 p.

Herron, M. M., Langway, C. C., Jr., Weiss, H. V., and Cragin, J. H., 1977, Atmospheric trace metals and sulfate in the Greenland ice sheet: Geochimica et Cosmochimica Acta, v. 41, p. 915–920.

Hirsch, R. M., Slack, J. R., and Smith, R. A. 1982, Techniques of trend analysis for monthly water quality data: Water Resources Research, v. 18, no. 1,

p. 107–121.
Huibregtse, K. R., and Mosher, J. H., 1976, Handbook for sampling and sample preservation of water and wastewater: Cincinnati, Ohio, U.S. Environmental Protection Agency Environmental Monitoring and Support Laboratory, 257 p.
Irelan, B., 1971, Salinity of surface water in the lower Colorado River–Salton Sea area: U.S. Geological Survey Professional Paper 486-E, E1–E40.
Janzer, V. J., and Saindon, L. G., 1972, Radiochemical analyses of surface water from U.S. Geological Survey hydrologic bench-mark stations: U.S. Geological Survey Open-File Report, 41 p.
Jenne, E. A., 1968, Controls on Mn, Fe, Co, Ni, Cu, and Zn concentrations in soils and water; The significant roles of hydrous Mn and Fe oxides, *in* Baker, R. A., ed., Trace inorganics in water: American Chemical Society Advances in Chemistry Series, v. 73, p. 337–387.
Jenne, E. A., and Luoma, S. N., 1977, Forms of trace elements in soils, sediments, and associated waters; An overview of their determination and biological availability, *in* Proceedings of the Symposium on Biological Implications of Metals in the Environment: Energy Resources Development Administration Symposium Series, no. 42, p. 110–143.
Jernelov, A., and Lann, H., 1973, Studies in Sweden on feasibility of some methods for restoration of mercury-contaminated bodies of water: Environmental Science and Technology, v 7, p. 712–718.
Johnson, A. H., 1979, Estimating solute transport in streams from grab samples: Water Resources Research, v. 15, p. 1224–1228.
Johnson, N. M., Likens, G. E., Bormann, F. H., Fisher, D. W., and Pierce, R. S., 1969, A working model for the variation in stream water chemistry at the Hubbard Brook Experimental Forest, New Hampshire: Water Resources Research, v. 5, p. 1353–1363.
Jung, C. E., and Werby, R. T., 1958, The concentration of chloride, sodium, potassium, calcium, and sulfate in rain water over the United States: Journal of Meteorology, v. 15, p. 417–425.
Kaplan, I. R., 1972, Sulfur cycle, *in* Fairbidge, R. W., ed., The encyclopedia of geochemistry and environmental sciences: New York, Van Nostrand Reinhold, p. 1148–1152.
Kemp, A.L.W., Thomas, R. L., Dell, C. I., and Jaquet, J. M., 1976, Cultural impact on the geochemistry of sediments in Lake Erie: Journal of Fisheries Research Board Canada, v. 33, p. 440–462.
Kendall, M. G., 1975, Rank correlation methods: London, Charles Griffin and Company, Limited, 202 p.
Kennedy, V. C., 1971, Silica variation in stream water with time and discharge, *in* Hem, J. D., ed., Nonequilibrium systems in natural water chemistry: American Chemical Society Advances in Chemistry Series, v. 106, p. 94–130.
Kennedy, V. C., Zellweger, G. W., and Jones, B. F., 1974, Filter pore-size effects on the analysis of Al, Fe, Mn, and Ti in water: Water Resources Research, v. 10, p. 785–790.
Kennedy, V. C., Kendall, C., Zellweger, G. W., Wyerman, T. A., and Avanzino, R. W., 1986, Determination of the components of storm flow using water chemistry and environmental isotopes, Mattole River basin, California: Journal of Hydrology, v. 84, p. 107–140.
Klein, J. M., and Taylor, H. E., 1980, Mount St. Helens, Washington, volcanic eruption; Part 2, Chemical variations in surface waters affected by volcanic activity [abs.]: EOS American Geophysical Union Transactions, v. 61, p. 956.
Kudo, A., Mortimer, D. C., and Hart, J. S., 1975, Factors influencing desorption of mercury from bed sediments: Canadian Journal of Earth Sciences, v. 12, p. 1036–1040.
Langbein, W. B., and Durum, W. H., 1967, The re-aeration capacity of streams: U.S. Geological Survey Circular 542, 6 p.
Lee, G. F., 1973, Chemical aspects of bioassay techniques for establishing water quality criteria: Water Research, v. 7, p. 1525–1546.
——, 1975, Role of hydrous metal oxides in the transport of heavy metals in the environment: Progress in Water Technology, v. 17, p. 137–147.
Lee, G. F. and Mariani, G. M., 1977, Evaluation of the significance of waterway sediment-associated contaminants on water quality at the dredged material disposal site, *in* Mayer, F. L., and Hamelink, J. L., eds., Aquatic toxicology and hazard evaluation: American Society for Testing and materials, ASTM STP 634, p. 196–213.
Leifeste, D. K., 1974, Dissolved-solids discharge to the oceans from the conterminous United States: U.S. Geological Survey Circular 685, 8 p.
Leverin, H. A., 1942, Industrial waters of Canada: Candian Department of Mines and Resources, Mines Geology Branch Reports, v. 807, 112 p.
——, 1947, Industrial waters of Canada: Canadian Department of Mines and Resources, Mines Geology Branch Reports, v. 819, 109 p.
Lewis, A. G., Whitfield, P. H., and Ramnarine, A., 1972, Some particulate and soluble agents affecting the relationship between metal toxicity and organism survival in Calanoid Copepod *Euchateta japonica*: Marine Biology, v. 17, p. 215–221.
Livingstone, D. A., 1963, Data of geochemistry; Chemical composition of rivers and lakes (sixth edition): U.S. Geological Survey Professional Paper 440-G, 64 p.
Love, S. K., 1956, Quality of surface waters of the United States 1951: U.S. Geological Survey Water-Supply Paper 1198, parts 5 and 6, 586 p.
Lu, J.C.S., and Chen, K. Y., 1977, Migration of trace metals in interfaces of seawater and polluted surfical sediments: Environmental Science and Technology, v. 11, p. 174–182.
Luoma, S. N., and Jenne, E. A., 1976, Estimating bioavailability of sediment-bound trace metals with chemical extractants, *in* Hemphill, D. D., ed., Proceedings of University of Missouri 10th Annual Conference on Trace Substances in Environmental Health: Columbia, Missouri, University of Missouri, p. 343–351.
Mackenzie River Basin Committee, 1985, Water quality Mackenzie River Basin: Ottawa, Environment Canada Study Report Supplement 9, 201 p.
MacKereth, F.J.H., 1966, Some chemical observations on post-glacial lake sediments: Philosophical Transactions of the Royal Society, v. B250, p. 165–213.
Malo, B. A., 1977, Partial extraction of metals from aquatic sediments: Environmental Science and Technology, v. 11, p. 277–282.
Mantoura, R.F.C., Dickson, A., and Riley, J. P., 1978, The complexation of metals with humic materials in natural waters: Estuarine Coastal Marine Science, v. 6, p. 387–408.
Martin, J.-M., and Maybeck, M., 1979, Elemental mass-balance of material carried by major world rivers: Marine Chemistry, v. 7, p. 173–206.
Marvin, K. T., Proctor, R. R., Jr., and Neal, R. A., 1970, Some effects of filtration on the determination of copper in freshwater and saltwater: Limnology and Oceanography, v. 15, p. 320–325.
McCrea, R. C., Fischer, J. D., and Kuntz, K. W., 1985, Distribution of organochlorine pesticides and PCBs between aqueous and suspended sediment phases in the lower Great Lakes region: Water Pollution Research Journal Canada, v. 20, p. 67–78.
McDuffie, B., El-Barbary, I., Hollod, G. J., and Tiberio, R.C., 1976, Trace metals in rivers; Speciation, transport, and role of sediments, *in* Hemphill, D. D., ed., Proceedings of the University of Missouri 10th Annual Conference on Trace Substances in Environmental Health: Columbia, Missouri, University of Missouri, p. 85–95.
Meybeck, M., 1979, Concentrations des eaux fluviales en éléments majeurs et apports en solution aux oceans: Revue de Géologie Dynamique et de Géographie Physique, 21, fasc. 3, p. 213–246.
——, 1982, Carbon, nitrogen, and phosphorus transport by world rivers: American Journal of Science, v. 270, p. 109–135.
——, 1983, Atmospheric inputs and river transport of dissolved substances, *in* Proceedings, Symposium on Dissolved Loads of Rivers and Surface Water Quantity/Quality Relationships, Hamburg, August 1983: International Association of Hydrological Sciences Bulletin 141, p. 173–192.
——, 1985, The GEMS/WATER program, 1978–1983: World Health Organization Water Quality Bulletin, v. 10, p. 167–174.
Montgomery, H.A.C., and Stiff, M. J., 1971, Differentiation of chemical states of toxic species, especially cyanide and copper, in water, *in* Proceedings of the International Symposium on the Identification and Measurement of

Environmental Pollutants: Ottawa, National Research Council of Canada, p. 375–379.

Mortimer, D. C., and Kudo, A., 1975, Interaction between aquatic plants and bed sediments in mercury uptake from flowing water: Journal of Environmental Quality, v. 4, p. 491–495.

Nelson, H., Larsen, B. R., Jenne, E. A., and Sorg, D. H., 1977, Mercury dispersal from lode sources in the Kuskokwim River drainage, Alaska: Science, v. 198, p. 820–824.

Nordstrom, D. K., 1977, Hydrogeochemical and microbiological factors affecting the heavy metal chemistry of an acid mine drainage system [Ph.D. thesis]: Stanford, California, Stanford University, 210 p.

Nriagu, J. O., 1983, Arsenic enrichment in lakes near the smelters at Sudbury, Ontario: Geochimica et Cosmochimica Acta, v. 47, p. 1523–1526.

Oliver, B. T., 1973, Heavy metal levels of Ottawa and Rideau River sediments: Environmental Science and Technology, v. 7, p. 135–137.

Page, A. L., 1974, Fate and effects of trace elements in sewage sludge when applied to agricultural lands; A literature review study: University of California at Riverside, prepared for the Environmental Protection Agency, National Technical Information Service Report PB, p. 231–271.

Perhac, R. M., 1974, Water transport of heavy metals in solution and by different sizes of particulate solids: University of Tennessee Water Resources Center Research Report no. 32, National Technical Information Service Report PB, p. 232–427.

Peters, N. E., 1984, Evaluation of environmental factors affecting yields of major dissolved ions of streams in the United States: U.S. Geological Survey Water-Supply Paper 2228, 39 p.

Peters, N. E., Schroeder, R. A., and Troutman, D. E., 1982, Temporal trends in the acidity of precipitation and surface waters of New York: U.S. Geological Survey Water-Supply Paper 2188, 35 p.

Peyton, T., McIntosh, A., and Anderson, V., 1974, A field study on the distribution of aerially deposited trace elements in an aquatic ecosystem; 1. Sediment component, *in* Hemphill, D. D., ed., Proceedings of the University of Missouri 8th Annual Conference on Trace Susbtances in Environmental Health: Columbia, Missouri, University of Missouri, p. 193–201.

Ramamoorthy, S., and Kushner, D. J., 1975, Heavy-metal binding components of river water: Research Board of Canada Journal of Fisheries, v. 32, p. 1755–1766.

Ramamoorthy, S., and Rust, B. R., 1976, Mercury sorption and desorption characteristics of some Ottawa River sediments, *in* Distribution and transport of pollutants in flowing water ecosystems: University of Ottawa and National Research Council of Canada Ottawa River Project Report no. 3, Ch. 7, 20 p.

Rawson, J., and Lansford, M. W., 1971, The water quality of Sam Rayburn Reservoir, eastern Texas: U.S. Geological Survey Water-Supply Paper 1999-J, 67 p.

Reynolds, R. C., Jr., and Johnson, N. M., 1972, Chemical weathering in the temperate glacial environment of the northern Cascade Mountains: Geochimica et Cosmochimica Acta, v. 36, p. 537–554.

Rickert, D. A., Kennedy, V. C., McKenzie, S. W., and Hines, W. G., 1977, A synoptic survey of trace metals in bottom sediments of the Willamette River, Oregon: U.S. Geological Survey Circular 715-F, 27 p.

Sanders, T. G., Ward, R. C., Loftis, J. C., Steele, T. D., Adrian, D. D., and Yevjevich, V., 1983, Design of networks for monitoring water quality: Littleton, Colorado, Water Resources Publications, 328 p.

Shaw, T. L., and Brown, V. M., 1974, The toxicity of some forms of copper to rainbow trout: Water Research, v. 8, p. 377–382.

Skougstad, M. W., Fishman, M. J., Friedman, L. D., Erdmann, D. E., and Duncan, S. S., eds., 1979, Methods for determination of inorganic substances in water and fluvial sediments: U.S. Geological Survey Techniques of Water Resources Investigation, book 5, Ch. A1, 626 p.

Smith, R. A., and Alexander, R. B., 1983a, A statistical summary of data from the U.S. Geological Survey's national water quality networks: U.S. Geological Survey Open-File Report 83-533, 30 p.

——, 1983b, Evidence for acid-precipitation-induced trends in stream chemistry at hydrologic bench-mark stations: U.S. Geological Survey Circular 910, 12 p.

——, 1985, Trends in concentrations of dissolved solids, suspended sediments, phosphorus, and inorganic nitrogen at U.S. Geological Survey National Stream Quality Accounting Network Stations, *in* U.S. Geological Survey National Water Summary 1984: U.S. Geological Survey Water-Supply Paper 2275, p. 66–73.

Smith, R. A., Hirsch, R. M., and Slack, J. R., 1982, A study of trends in total phosphorus measurements at NASQAN stations: U.S. Geological Survey Water-Supply Paper 2190, 34 p.

Stabler, H., 1911, Some stream waters of the western United States: U.S. Geological Survey Water-Supply Paper 274, 188 p.

Stokes, P. M., and Szokalo, A. M., 1977, Sediment-water interchange of copper and nickel in experimental aquaria, *in* Proceedings of the 12th Canadian Symposium on Water Pollution Research in Canada: p. 157–177.

Stumm, W., and Morgan, J. J., 1981, Aquatic Chemistry (second edition): New York, Wiley Interscience, 780 p.

Thomas, J.F.J., 1953, Industrial water resources of Canada, scope, procedure, and interpretation of survey studies: Canadian Department of Mines, Technological Surveys, Industrial Minerals Division Water Survey Report 1, 69 p.

——, 1964, Surface water quality in major drainage basins and northern areas of Canada: Journal of the American Water Works Association, v. 56, p. 1173–1193.

——, 1965, Industrial water resources of Canada; The upper Great Lakes Basin in Canada 1957–63: Mines Branch Monograph 870, 137 p. (Similar papers by Thomas cover other drainage basins.)

Thurman, E. M., 1985, Organic geochemistry of natural waters: Dordrecht, The Netherlands, Martinius Nijhoff/Dr. W. Junk, 497 p.

Tukey, J. W., 1977, Exploratory data analysis: Reading, Massachusetts, Addison Wesley, 688 p.

UNESCO-WHO, 1978, Water quality studies: Studies and Reports in Hydrology no. 23, 350 p.

USGS, 1981, National handbook of recommended methods for water-data acquisition: Reston, Virginia, U.S. Geological Survey, 10 p.

Vanselow, A. P., 1966, Nickel, *in* Chapman, H. D., ed., Diagnostic criteria for plants and soils: Davis, University of California Division of Agricultural Sciences, p. 302–309.

Welch, R., and Carey, E. E., 1975, Concentration of chromium, nickel and vanadium in plant materials: Journal of Agricultural and Food Chemistry, v. 23, p. 479–482.

Welsh, P., and Denny, P., 1976, Waterplants and the recycling of heavy metals in an English lake, *in* Hemphill, D. D., ed., Proceedings of the University of Missouri 10th Annual Conference on Trace Substances in Environmental Health: Columbia, Missouri, University of Missouri, p. 217–223.

Whelpdale, D. M., and Barrie, L. A., 1982, Atmospheric monitoring network operations and results in Canada: Water, Air, and Soil Pollution, v. 18, p. 7–23.

Whitehead, H. C., and Feth, J. H., 1964, Chemical composition of rain, dry fallout, and bulk precipitation at Menlo Park, California, 1957–59: Journal of Geophysical Research, v. 69, p. 3319–3333.

Wittmann, G.T.W., and Forstner, U., 1976, Heavy metal enrichment in mine drainage; 1. The Rustenburg planinum mining area: South Africa Journal of Science, v. 72, p. 242–246.

World Health Organization, 1982, WHO guidelines for Drinking Water Quality: Geneva, World Health Organization, 130 p.

WQB, 1983, Sampling for water quality: Ottawa, Environment Canada, Water Quality Branch, 55 p.

Youngquist, E. T., 1985, Geological perspectives on carbon dioxide and the carbon cycle, *in* Sundquist, E. T., and Broecker, W. S., eds., The carbon cycle and atmospheric CO_2; Natural variations, Archean to present: American Geophysical Union Geophysical Monograph Series, v. 32, p. 5–59.

Manuscript Accepted by the Society May 1, 1987

Printed in U.S.A.

The Geology of North America
Vol. O-1, Surface Water Hydrology
The Geological Society of America, 1990

Chapter 10

Aquatic biota in North America

R. Patrick
Academy of Natural Sciences, 19th and The Parkway, Logan Square, Philadelphia, Pennsylvania 19103
D. D. Williams
Division of Life Sciences, Scarborough Campus, University of Toronto, 1265 Military Trail, Scarborough, Ontario M1C 1A4, Canada

INTRODUCTION

This chapter consists of two sections that represent complementary approaches to the description of the distribution of aquatic biota on the North American continent. The first section, by Patrick, deals with aquatic life in rivers in the United States and emphasizes the geologic, hydrologic, and hydraulic characteristics that affect the spatial distribution of the biota. These principles, as indicated in the section by Williams, also relate to distribution throughout the North American continent. The geographic distribution itself is elucidated in the text and is organized around the factors influencing that distribution.

The second section, on the distribution of biota in Canadian and other North American waters, describes the spatial or geographic distribution of major aquatic biota in the river systems of Canada. It includes maps depicting the geographic distribution throughout the North American continent of selected species of fish, indicative of the wide range of spatial characteristics of these distributions. Together the two sections emphasize both the geographic distribution throughout the North American continent of selected species of fish, indicative of the wide range of spatial characteristics of these distributions, and the factors influencing those distributions throughout the continent.

FACTORS INFLUENCING AQUATIC LIFE IN RIVERS IN THE UNITED STATES

R. Patrick

INTRODUCTION

In order to discuss aquatic communities in rivers, community must first be defined. Basically, a community is a group of interacting species. In most natural aquatic ecosystems, it consists of a great many species belonging to many major groups of organisms that interact in their functioning and also in their selection of habitats. Currents and substrates are probably the main characteristics of streams that influence the distribution of organisms.

STRUCTURE OF AQUATIC COMMUNITIES

In natural communities one typically finds many species with relatively small populations. A few species will be much more common than others, while some will be exceedingly rare (Patrick, 1948, 1961).

Algae form an important base of the food web in streams. The importance of this group of organisms is dependent upon the type of stream. In heavily shaded areas, which typically occur in the eastern part of the United States, detritus may be more important than the algae, whereas in open prairie streams, algae are a more important food source than detritus. Heavy shading is more common in headwater streams than in downstream areas. Typically a stream has a great many species of diatoms and green algae. Blue-green algae are represented by relatively small populations in most natural streams, although there are some exceptions, such as in Florida, where considerable amounts of phosphates are present as well as nitrates. In these instances, blue-green algae may be very common. Even in streams where the nitrogen content is low, if there is sufficient phosphate, nitrogen-fixing blue-green algae such as the Nostocaceae may be fairly prevalent.

Vasculate plants and cryptogams such as mosses, equisetums, and in some cases, ferns (*Azolla* in southern streams) may be fairly common in streams. The vascular plants are more important as a source of detritus than as a direct source of food. Likewise, the mosses and the ferns and the Equisteales are not nearly as important a food source as are the algae, particularly the diatoms and the unicellular green algae. If these higher plants are completely submerged, they may be an important source of oxygen in the water. They often furnish excellent habitats for many species.

Invertebrates in streams represent a great many phyla and may be herbivores, omnivores, or carnivores. The types of species

Patrick, R., and Williams, D. D., 1990, Aquatic biota in North America, *in* Wolman, M. G., and Riggs, H. C., ed., Surface water hydrology: Boulder, Colorado, Geological Society of America, The Geology of North America, v. O-1.

that predominate vary greatly with the chemistry of the water and the characteristics of the channel. For example, in high-gradient, cool-water streams many species of stoneflies (Plecoptera) are found; in lesser gradient streams, often the mayflies (Ephemeroptera) may be represented by many more species than are the Plecoptera. In Ohio, in the headwaters of the Ottawa River, for another example, Crustacea, particularly crayfish, are extremely common and dominate the invertebrate community.

In streams where large amounts of detritus accumulate in pools, chironomids (Diptera) are often prevalent as are tubificid worms. These organisms typically occur in beds of organic materials where the oxygen level may be fairly low. However, many chironomids that are typically herbivores are found in well-oxygenated water.

In moderately hard or hard-water streams, snails (Gastropoda) and the bivalves (Unionidae) are often very common. This is particularly true in streams in the Midwest as well as those Eastern streams where the water is typically moderately hard or hard.

Typically, in natural streams, the fish community also consists of many species. The number of species is usually larger in large rivers than in headwater streams and is greatly influenced by the type and amount of food present and the types of available habitats.

FUNCTIONAL INTERRELATIONSHIPS

Species interaction

Interactions among species are mainly those of predator-prey relationships, although there is a fair amount of interaction between species for available habitats (e.g., some species of caddisflies may be very pugnacious if another species of caddisfly tries to invade their habitat; Glass and Bovbjerg, 1969). Chemicals excreted by some algae may greatly influence the presence of other species of algae (Keating, 1976; Proctor, 1957; Rice, 1954).

Food web

The flow of energy through the food web is complex. It has been estimated by many workers that about one-third of the energy stored in protoplasm is transferred to the next higher stage in the food web (e.g., about one-third the nutrient and energy value of algae is transferred to herbivores, and about one-third of the nutrients in herbivore tissue is transferred to omnivores and, in turn, to the carnivore). These are only very rough estimates but they do express the great loss of nutrients that occurs during progression up the food web.

The various algae and other organisms have different nutritive values, and for this reason the predator pressure is much higher on certain species than on others. As a result, productivity may be equal in two species, but the standing crop may be very different because of predator pressure.

Algae. Algae perform two important functions in streams; they are an important food source as the base of the food web, and they produce oxygen, thus regenerating the oxygen supply in a stream. The main primary production in a natural stream is accomplished by benthic algae that live close to the bed of the stream or attached to vegetation or to the substrate. Sometimes they are scuffed up by feeding organisms or freshets and become floating species. The kinds of species of algae vary greatly in their importance in the food web. Diatoms are recognized as the most important algal food source. The size and shape of these diatoms cause them to be selected by various types of invertebrates and fish. In general, single-celled diatoms and green algae are preferred over filamentous algae. Blue-green algae, although a food source for some species, are a less common food source.

Planktonic algae (algae that spend most of their life floating) are not produced in free-flowing streams except, for example, in the West where large, low gradient, free-flowing streams are relatively deep. Where such algae are found in association with free-flowing streams they are typically produced in flood-plain ponds, reservoirs, or lakes that are made by man, but are part of the riverine system.

Detritus. Detritus is also a main food source at the base of the food web. It may consist of relatively large pieces such as leaves that fall into the stream from trees, or small particulate matter such as algae. It is, therefore, produced in the watershed or within the stream proper. Detritus in the form of leaf packs is a desirable habitat for many invertebrates during the fall and winter months (Vannote and Sweeney, 1985). The leaves form a direct food source. It is well established that the microflora growing on these leaves, as well as the nutrients of the leaf tissue, make them a desirable food source. Vascular plants living within the stream also produce detritus that may be a valuable food source.

Effects of drift

The effects of drift carried by the current on the structure and functioning of aquatic communities is quite significant. Drift, when it consists of organic matter such as bits of leaves or dead algal cells or other plant tissues, transfers inanimate food from one section of the stream to a downstream section. Observations made at Stroud Water Research Laboratory indicate that leaves in the upstream reach, for example, do not travel very far downstream but rather are retained in pools and behind riffles a short distance downstream; leaves produced in the downstream reach are transferred to close lower reaches (Vannote and others, 1980).

Another effect of drift is the transfer of organisms and, hence, a control of the invasion rate of species from upstream (Waters, 1966). The value of drift in the establishment of species in a given area is variable and depends upon the type of species and various environmental factors.

TYPES OF COMMUNITIES

Influence of river channel characteristics on types of communities

The characteristics of the river channel are greatly influenced by the volume of flow, the sediment load, the slope, the vegetation of the landscape, the surface or near-surface geology of the area, and associated soils (Leopold and others, 1964; Dunne and Leopold, 1978). Other important factors are the amount and pattern of precipitation, the seasonal presence of ice, and freezing and thawing of the bank materials. These various factors affect the shape and density of the riverine system and the stability and shape of the banks of the channel. The kinds of species and the distributions of species that form the aquatic communities are greatly influenced by these factors as is the functioning of the ecosystem.

Although many classifications of communities are possible, three types, roughly defined, are used in this discussion. The first type is closely integrated communities consisting of species that live attached or move only very short distances in a given area. They spend their entire lifespan in that area and are dependent for all of their functions upon the material produced or brought into the area. The second type of community includes the elements of type one plus nekton that move variable distances. The nekton may be small fish or crayfish that do not move very far or nekton that have a much greater range. A third community includes anadromous species that move long distances. For example, fish such as shad will move from the open ocean to the headwaters of a stream in order to spawn. Likewise, crabs (as in the Guadalupe River, Texas) may move from the open ocean up into the freshwater areas of the river during their lifespan. These species become predators upon many different communities, and their energy and nutrients are derived from the several areas that they transgress during their lifespan. The populations of these anadromous species, and nekton that move small distances, may have a great effect on the structure of a given indigenous community because of predator pressure. For example, the structure of the indigenous community may be very different depending on the number of individuals or the anadromous species that happens to spend part of their lifespan, as breeding fish and juveniles, in a given area.

CHARACTERISTICS OF COMMUNITIES IN VARIOUS STREAM SECTIONS

The classification of stream sections used in this part of the chapter is based on channel characteristics and includes headwater streams, mainstem or trunk streams, and estuaries.

Headwater streams

The headwater streams are first- to third- or fourth-order streams. They are characterized by small volume of flow and are typically shallow. Those that flow over consolidated substrates have a very different character from those that flow over unconsolidated substrates. Headwater streams may be further divided into high- and intermediate-gradient streams with consolidated or unconsolidated bed material, and low-gradient streams usually with unconsolidated bed material.

High-gradient headwater streams. One type of headwater stream is the very high-elevation stream with a steep gradient, which often carries meltwater from glaciers. Streams of this type, because of their very heavy and heterogeneous sediment load, are often braided. A high concentration of very fine sediment often causes the water to be turbid, which in turn restricts algal growth. Coarser sediments that settle out of the flow often move as bedload to these very shallow streams and, hence, greatly reduce the stable habitats available for species to occupy. The species of organisms found in such streams are very small crustaceans and some algae, which often coat the sand or fine particulates.

High-gradient streams not influenced by glaciers may consist of a series of pools; the water may be either hard or soft. Hunting Creek, Maryland, is an example of this type of stream. In these streams, large boulders form the breast of a succession of pools. The beds of the pools are mainly sand and gravel and under normal conditions most of the flow filters through the sand and gravel into the pool downstream. Of course, some of it may flow in and around the boulders that form the breast of the pool. The water temperature of these pools is less than 18°C in the summer. This may restrict the number of species of invertebrates, particularly insects, that may be present. However, a few species may have fairly large populations. The numbers of species and the sizes of their populations are also limited by the low nutrient content of the water in these streams draining small areas. The algae such as *Achnanthes* and *Cocconeis* typically grow attached to the rock surfaces. The pools are often favorite habitats for trout. They lay their eggs in the sand or gravel near the breast of the pool. The cool, well-oxygenated water flows through these gravels and provides a protected and desirable habitat for the eggs and young larvae.

The kinds of insects in the pools of these high-gradient streams are typically the larvae of stoneflies that cling to the surfaces of rocks and a few species of mayflies, chironomids (midges), and simuliids (blackflies). The algae in such pools are typically diatoms.

Intermediate-gradient headwater streams with unsorted beds. *Materials.* The riffle-pool sequence, typical of stream channels of intermediate gradient, is found over much of North America. Such channels are present in areas underlain by a great range of bedrock types. As a result, the hardness of the river water varies from soft (less than 60 ppm hardness) to very hard (greater than 250 ppm hardness).

Dissipation of energy is accomplished in this type of stream by the riffle, slack water, or pool regime (which covers relatively short reaches of the stream) rather than by the plunging water as in very high gradient streams. Riffles occur typically in undisturbed natural headwater streams at intervals of about five to seven times the breadth of the stream.

The bed material consists of rock and rubble in the riffles and smaller materials and sand in the pools; mud, silt, and organic debris may accumulate in the slow currents at the edges of the stream and in the pools (Leopold and others, 1964).

Because of rapid mixing of the water in the riffles, and the pattern of flow, the water is well oxygenated except at or just below the surface of muds in pools, where very little current is present except during floods. This low oxygen condition is also greatly influenced by the accumulated organic matter.

The great variety of substrates, coupled with the diversity of flow rates, provide many habitats for aquatic life, and hence, the communities consist of many species. If the nutrient content of the water is low and the temperature in the summer rarely rises above 12° to 13°C, the populations are often smaller and the numbers of species less than in streams with warmer summer temperatures and more nutrients.

In headwater streams, the species that make up the community may be quite different in various parts of the United States. However, for streams of the same type, the total number of species in a Utah stream would be about the same as in a Pennsylvania stream, as would be the relative sizes of the populations of the species, although different species would be present. In this type of stream in the summer, in Utah, many species of stoneflies occur in and among the rocks that form the riffles, whereas in Pennsylvania at the same time, more species of mayflies than of stoneflies may be present. In sunny Pennsylvania streams, purse caddisflies (*Hydrophilidae*), are often found closely pressed to the rocks in the riffles. Stone and sand-case caddisflies are found on rocks in various current velocities in Utah as well as in Pennsylvania. In such streams, a rock, if it is rough, will have different current patterns (e.g., the current pattern on the upstream end of the rock is different from the current on the sides of the rock or on the downstream end where the current is much reduced). Thus, various species of caddisflies are oriented to the different types of current that may be present.

Many insect larvae that move freely about, such as those of mayflies and stoneflies, will retreat to the undersides of rocks during the daytime, and come out to the surface at night to feed. This pattern of movement contributes to a drift of algae that is higher at night than in the daytime (Waters, 1966). As these insects scurry over the surfaces of rocks, they often loosen the algae. Records at White Clay Creek, Pennsylvania, clearly show that the drift of diatoms is two to four times higher at night than in the daytime (Patrick, unpublished data).

Planaria (free-living flatworms) are often found on rocks in areas where the current is slower, particularly the under-surfaces of rocks, although they may be found on the surfaces of rocks in slower water in the riffle.

Algae, particularly diatoms, commonly grow on the surfaces of rocks in riffles. The diatoms that grow fairly close to the surface of the rock are much more common than filamentous algae, although some filamentous green algae with basal attachments may be found on these rocks.

The primary productivity of a typical riffle-pool stream depends on the amount of shading as well as on the dissolved nutrients present. For example, in the eastern part of the United States where most headwater streams are in forested areas, the primary production by algae is not nearly as great as in the open streams of the Midwest and far West. The algae of the forested stream are quickly cropped by the invertebrate fauna, and the standing crop of algae is very low.

In shallow streams where erosion is low, the photosynthetic zone extends across the stream. In contrast, in the areas where erosion is continually taking place, the photosynthetic zone is limited by the turbidity and is often confined to the edges of the stream where the current is slack, and to pool areas.

In slack water, gravel and sand are typically present, with silt and finer sediments near the edge of the stream. Rocks that were deposited during previous high water may also be found. The most common groups of organisms are those found in the riffles, but represented by species that prefer slower water. Several species of mayflies are typically found in this habitat as are some of the caddisflies. Water pennies (Coleoptera) are often found attached to the rocks. Planaria are more common in the slower current. Snails and limpets (*Ferrissia*) are also found. Hellgrammites (Megaloptera) are often associated with rocks in this habitat, as are crayfish. In the gravel and sand mixed with some mud are members of the Unionidae. These clams filter their food from the flowing water. They prefer water with considerable current, but not as fast as that found in riffles. Also, the sand/gravel habitat is one into which they can burrow and become attached. In the finer gravel, fingernail clams (Sphaeriidae) are often found, and in the mud on the edges of the slack water are worms (Oligochaetes).

Filamentous algae attached to the rocks in slack-water areas are fairly common, particularly species of the genus *Melosira* and green algae that have hold-fasts. Blue-green algae are sometimes found in small patches. Diatoms, which form a brown covering on the sediments, are common in such slow-flowing areas and form favorite habitats for chironomids (midges). Some snails may also be found grazing on diatoms in this type of habitat.

Along the sides of the slack-water channel, roots of various plants living on the banks of the stream extend into the water. In these trailing roots, members of the Odonata (damselflies and dragonflies) are often fairly common. In some streams, dragonflies live in the benthos, especially where there is a fair amount of organic matter.

Pools support organisms that like very slow water and gain their nourishment from the detritus in the bottom of the pool or from the algae coating the fine sediments in the pool.

Protozoans are common in this habitat. Some small crustaceans, Odonata, whirlygig beetles (*Gyrinus* sp.), and water striders (Gerridae) are also found associated with pool conditions. Living in the fine sediments at the bottom of the pool, particularly if it is rich in organic matter, are members of the Oligochaeta and often some Chironomidae. If oxygen levels are low, red chironomids may be present. Several species of fish may be present, as many prefer pool habitat.

Intermediate-gradient headwater streams with clayey bed material, usually unconsolidated. In areas where the soil is very deep, as in certain parts of Texas, Missouri, and Kansas, the bed of the stream may often be mainly clay and silt, and rocks are absent or extremely rare.

In these areas, the streams tend to form large meanders; their energy is also dissipated by this flow pattern. In areas of more rapid flows, clumps of clay, almost as hard as rocks, occur. On these clay clumps are species of insects, algae, and flatworms that are commonly associated with rocks in the riffle pool sequence described above. Where the channel-bottom silt is stable but loose, such as below the meander, there are areas where oligochaetes and chironomids accumulate.

The turbidity of these intermediate-gradient headwater streams greatly decreases the amount of algae present, and it is only in the slow-flowing areas with pool-like conditions, often on the edges of the stream, that algae develop. In this type of stream, productivity is very low. Detritus plays a primary part in the base of the food web. This detrital material is produced both within the stream and also from the debris from grass and shrubs and trees that live near the banks.

Low-gradient headwater streams. Low-gradient headwater streams are typically found in coastal plains and in some of the plains areas of the midwest. In many coastal streams, the bed is typically sand, and water hardness is extremely low. This means that the water is not well buffered and the pH may be quite variable. If humates (acid organic materials) are present, the stream may be quite acid. Such natural acid streams with pH from 3.5 to 4 are common in areas such as the Pine Barrens of New Jersey, and in the Coastal Plains of the southeastern states.

A diversity of current patterns is often produced in such streams by fallen trees, which act as dams, producing more or less pool-like conditions behind them. Riffle-like conditions are produced as the water flows over the surface of the fallen tree. Under these conditions, the burrowing mayfly *Hexagenia* is often found in the sediments (e.g., in the Upper Three Runs in South Carolina). The number of those species that live on hard substrates, such as mayflies or caddisflies, is restricted unless there is a large amount of fairly stable hard surface debris.

A great variety of algae is found in this kind of stream. Diatoms are usually common and an excellent source of food. They are species of very different genera from those found commonly in the Piedmont or medium-gradient headwaters. Populations of red algae such as *Batracospermum* and *Compsopogen* are often fairly common, particularly in the spring of the year. Unicellular green algae (desmids) are often common as are green filamentous algae, *Zygnemia* and *Mougeotia*.

Mainstem streams or rivers

Mainstem here refers to streams usually fifth order or greater. They are characterized by much larger flow and greater sediment loads than headwater streams. In general, there are two types of channel form. One type of channel, usually found in foothills or in the Piedmont, has a consolidated bed composed of poorly sorted material including boulders, rubble, gravel, sand of various sizes, and silt. A second mainstem type has unconsolidated bed materials that are fairly well sorted.

Consolidated-bed rivers. Examples of mainstem streams with consolidated beds and unsorted materials are the Potomac (Maryland), the upper Guadalupe (Texas), and the Susquehanna (Pennsylvania) Rivers. The difference between the Susquehanna River and the Potomac and Guadalupe Rivers is that much of the Susquehanna River has a braided channel. Where this is the case, the flow pattern is more varied, resulting in poor mixing of the water and nonuniform chemical characteristics. Pools that develop in such a braided river often have longer storage times than pools in a nonbraided river. Thus, the dissolved solids, particularly of minor elements, released from the sediments may differ among such pools and may be different from those found in nonbraided rivers.

In the Potomac and Guadalupe Rivers, the meanders are typically larger for similar flows than in a braided river. Energy loss in the flowing water is reduced by meandering rather than by braiding and meanders. More uniform mixing of the water enhances the predictability of the chemical characteristics of the water in any given area.

The photosynthetic zones in these relatively shallow rivers extend across almost the entire channel. Thus, the basic productivity from algae is high, particularly if the nutrient level is balanced and fairly high. The great variety of physical habitats and current patterns produces habitats suitable for a large number of species.

As in headwater streams, the orders of organisms to be found can be predicted, although the particular species will vary depending on the chemical and physical characteristics of the water. For example, a great assortment of mayflies and stoneflies will be found in the Potomac River on the rocks in fairly rapid current, both in the Piedmont area and in the headwaters. Stoneflies are particularly prevalent during the winter months, except in those streams having cooler summer temperatures, in which certain species may be present all year. Caddisflies, particularly those that build stone or sand cases, are very common in these habitats. Fresh-water sponges and bryozoans are often found pressed to rock surfaces. Hellgrammites (Megaloptera) are often found on the undersides of rocks in faster water. The main algae are diatoms and those green algae that have a hold fast.

On hard substrates in somewhat slower water, snails such as *Goniobasis virginica* and *Nitocris carinatus* and other gastropods belonging to the genera *Physa, Limnaea,* and *Ferrissia* will be found. As in headwaters, flatworms are often found on rocks, particularly where there is considerable organic matter and some algal growth.

In bed material rich in organic matter, dragonfly larvae belonging to the Gomphidae are often found. Other Odonata, particularly damselflies, are found, as in the headwaters, on trailing branches of trees and among the roots of bankside vegetation that extend into the water. The root habitat is a particularly favorable one for the damselflies.

Also fairly common in these broad, shallow-water streams are various decapods, particularly crayfish of various genera and species and, among algae and floating higher plant aquatics, isopods and amphipods. In this latter habitat there are also chironomids and some gastropods.

The diversity of fish in rivers of this type is similar, although there tend to be fewer species in very hard-water rivers such as the Guadalupe, as compared to the Potomac; the same families seem to be represented, although the species are different.

In very slow waters, particularly in pools, there are often a number of whirlygig beetles and water striders. In the mud, as in the head-water streams, a number of chironomids are typically found, and if the organic material in the mud is fairly high, tubificids (Oligochaeta) will be present. On the surface of this mud, rather dense growths of diatoms occur, and in some cases, blue-green algae may be found. Associated with these algae are often various species of Cladocera and copepods.

A diatom flora typically dominates the algae of this type of main stream, although green algae may be quite prevalent, particularly *Cladophora glomerata* and *Stigeoclonium lubricum* if the pH is near neutral and the nutrient level is fairly high.

In these broad, shallow-water channels, with sand/gravel substrates and medium-hard to hard waters, there are typically larger populations of molluscs, particularly Unionidae and snails, than in softer water rivers. However, in the Savannah River, which is a soft-water river, there is a fairly large number of species of Unionidae, but the sizes of the populations are small.

Unconsolidated-bed rivers. Many large rivers or main trunks of rivers in the United States have unconsolidated beds. This type of river is found in the western states—the Missouri and Kansas Rivers are examples—where the unconsolidated materials may be sand or mud or a mixture of both. Such rivers are also found in the coastal plain from Maryland through the Gulf states. The flow through unconsolidated bed material produces a deeper channel with steeper banks than in rivers with consolidated beds. The photosynthetic zone is much narrower because of the high sediment load and is confined largely to point bars of the meanders and to floating material, to oxbows, and to sloughs. This restriction of the photosynthetic zone in the main channel greatly reduces the productivity. The stream system is much more dependent on the sloughs and the oxbows to enhance its productivity than is the case with broad, shallow-water rivers with consolidated substrates. Rivers with unconsolidated substrates have a bed load of sediments that moves quite frequently and thus forms unsuitable habitats for most invertebrates, though some burrowing mayflies (*Hexagenia*) can live in these areas, as can a few worms.

Most of the worms are found in the areas of net deposition on the downstream side of a meander where fine sediments settle out. It is here that chironomids typically live in bed material. The upstream side of the meander, where the current is faster and the sediments are eroding, is often a habitat for caddisflies, if there is a substrate to which they can attach. Clams are also often found in areas, associated with point bars, that have moderate current speed and very little erosion.

The floating materials in these streams form the equivalent of the solid substrate habitats of the Piedmont area. On floating logjams, mayflies may be very common, stoneflies are sometimes present, and caddisflies, beetles, and various small Diptera may be found; algae are also quite common. If a log is dead and covered with diatoms, there will probably be many insects present, particularly mayflies.

The caddisflies that are grazers may also be found in this habitat. Those that are filter feeders usually are present on the smaller twigs in the faster current. Another favorable habitat in this type of riverine channel for caddisflies and some damselflies is composed of the branches of riparian vegetation that trail in the water. In more southern streams, Spanish moss is often caught in these trailing branches. Roots that extend out from the bankside vegetation are an excellent habitat for damselflies and some of the dragonflies.

Many favorable habitats for organisms are found in oxbows. There is little current, sediments settle out, and a diatom film is often found on the sediment. As a result, there are excellent feeding grounds for fish that either feed directly on the algae, or on the insects that are abundant in these habitats. A great assortment of insects, depending upon the kinds of substrates, is found here. Sloughs also form very productive areas for algal growth, for chironomids, various kinds of worms, burrowing mayflies, crayfish, and smaller crustaceans.

In some large, unconsolidated channels in the coastal zone, marine species are present. For example, during the summer months in the Guadalupe River, crabs come far upstream into fresh water, and mating takes place. Usually the males penetrate farther upstream than the females. Migration of crabs upstream from an estuary is characteristic of many coastal-plain rivers.

The fish fauna in coastal-plain rivers is very diverse. Certain species are anadromous and traverse the river in the spring and again in later summer when they return to sea. Some species live their complete lifespan in the river, while for others it is only the habitat for juveniles.

In rivers with unconsolidated beds in the Midwest and far West, the heavy suspended-solid loads greatly limit productivity. For example, in the Kansas River below Topeka, the only place where algae were found to be common was in pools on the edge of the river, and in shallow protected areas. In the main channel, few algae were present. Because pools are not common and oxbows infrequent, the general productivity of this system is much less than in the Savannah River where there are many large oxbows.

Rivers altered by dams. In large rivers that have been altered by dams and are no longer free-flowing systems, an altered fauna and flora are present. Lake-like conditions are produced in the reservoirs, and sediments are deposited. The fauna and flora consist of limnetic species typically not found in the riverine system. The channel downstream from the dam is likely to change in character for many years as a result of the new flow and sediment regime.

Estuaries

An estuary is that part of a river that enters the sea and is under tidal influence; the water is characterized by a salinity gradient. The shape of an estuary is largely controlled by the geology of the area, prehistoric climatic conditions, freshwater flow, sediment load, and tidal currents. Three general types of estuaries occur in the United States: open-mouth estuaries; barrier-island estuaries, in which the mouth of the estuary is fringed by islands; and fjord-type estuaries.

The open-mouth estuary. This estuary is usually characterized by the great influence of the flow of a large river. In prehistoric times, the mouths of such rivers were farther seaward. With the melting of the glaciers, the increase in sea level has drowned the mouths of these former rivers. The Chesapeake and Delaware Bays are such estuaries.

The barrier-island estuary. This type of estuary is characteristic of many of the smaller estuaries on the New England and New Jersey coasts, but the best development is along the coast of southeastern United States and in the Gulf of Mexico area. The flow of the rivers entering this type of estuary is not nearly as large as that in the open-mouth estuary. The formation of barrier islands is a result of lesser flow and the deposition of a fairly large sediment load, together with the current patterns of the tides, which deposit sediments across the mouths of the estuaries. This type of estuary does not as freely exchange water with the sea, and thus, river water often has a greater hold-up time within the estuary than is characteristic of the open-mouth estuary.

The fjord-type estuary. Puget Sound and some of the small estuaries in Maine are typical fjord-type estuaries. These estuaries have steep banks and deep areas behind a sill at the mouth, which permits little exchange of water.

Biology of estuaries

The open-mouth estuary and the barrier-island estuary are characterized by large grass marshes into which the waters of the main channel merge. These grass marshlands are common on all coasts, but are particularly prevalent on the northeastern Atlantic coast, southeastern coast, and on the Gulf coasts. They are present on the Pacific coast but not as extensively as in the eastern United States. Tides have a great influence on these marshlands. Some parts of the marshland are continually under water during high tide, whereas other parts are dry except during periods of very high tides. They are usually dominated by one or two grasses, particularly *Spartina,* that typically form a significant source of food after they have decayed and become detritus; not many organisms eat the growing *Spartina.*

On the surface muds are large mats of diatoms and, if the estuary is polluted by water rich in organics, blue-green and green algae may also be very common. These algae grow not only on the mud surface, but also on the bases of the grass stems. Marshlands have some of the highest primary productivity in the world.

The open grasslands are an excellent breeding and nursery ground for many invertebrates and fish. On the eastern coast of the United States, various species of crabs are common inhabitants of marsh areas and their drainage canals. It is the extensive marshlands along the Gulf Coast and some parts of Florida and southern South Carolina that are the nursery grounds for the shrimp that form an important industry in the southern waters.

Worms, particularly polychaetes, inhabit the muds of marshlands. They play a significant role in mixing the detritus from a few centimeters down to several centimeters depth.

Grass marshlands are not only productive habitats for aquatic life, but also serve to purify polluted river water before it enters the ocean. Studies by Grant and Patrick (1970) in Tinicum Marsh (Delaware Estuary), show that large amounts of oxygen are generated each day as waters flow over the marshlands and return to the estuary proper, and in a similar way, large amounts of nitrogen and phosphorus are removed from the water either by being sorbed onto the sediments or taken up directly by the plants. Later, some of these nutrients are returned to the estuary in detritus, often at seasons when other food sources are scarce. Detritus is very important in maintaining the aquatic life of the estuary and coastal areas. The complex forms of nitrogen and phosphorus that are returned are less soluble in water than are the nitrates and phosphates in the riverine water that enters the marshland.

In southern Florida, marshlands of open-mouth estuaries and barrier-island estuaries may be mainly dominated by mangroves. These mangrove forests form a unique habitat for many kinds of invertebrates and have their own characteristic fauna. They are particularly rich in crabs, molluscs, and barnacles (Odum and others, 1982).

The "open waters" of the open-mouth and barrier-island estuaries are characterized by large populations of phytoplankton and zooplankton, which, depending on the time of year, are dominated by marine or brackish-water species. The nekton consist of fish, crabs, and shrimp, many of which are anadromous (i.e., they spend part of their life cycle in the sea and part in the estuary and even in the tributary fresh-water streams). Therefore, the fish fauna is a mixture of marine and brackish-water species. Because of the salinity gradient in estuaries, mainly fresh-water fish and some insects are found only in the fresh-water parts. In the brackish-water parts, there are organisms that have a wide salt tolerance. Usually the only insects in this part of the estuary are the Chironomidae and some of the other Diptera. Other arthropods occupy the habitats that would be filled by aquatic insects in the fresh-water sections. In higher salinity areas, a well-developed molluscan fauna is typical; oysters are very common (particularly in the open-mouth estuaries), as are mussels. Polychaetes, barnacles, bryozoans, and sponges are often common. Benthic algae are also very common in the open-water estuaries and grade from fresh-water diatoms, green algae, and blue-green algae in the upper parts of the estuary, to typically marine species such as the brown algae, *Ascophyllum,* and *Fucus,* and some of the red algae such as *Agardhiella, Ceramium,* and *Polysiphonia,* in the lower parts.

The Fjord-type estuary has a different type of biological community. Because of the steep banks that form the sides of the estuary, the benthic fauna and flora are much more restricted than they are in estuaries with broad marshlands bordering their edges.

The main species are plankton and nekton. In the mouths of

small streams that enter large fjords such as Puget Sound, a typical assortment of benthic species is found, but this type of fauna is not nearly as rich as it is in the grassland estuaries. The species of fish and other nekton are a mixture of brackish- and fresh-water species. In the mouths of small streams that enter the fjord, fresh-water, or at least euryhaline species, can be found.

CONCLUSIONS

The results of many studies clearly indicate that if the geology and topography of an area and the gradient and flow of a stream are known, the type of channel and bed materials that will be present can be predicted. These characteristics, which apply not only to the headwaters but also to the main trunk and estuaries, greatly influence the kinds of communities of aquatic life that will exist.

As pointed out by Vannote and others (1980), there is a continuum and a transition in faunas and floras from the headwaters, to the main trunk, to the estuary. The drift and the active migration of species are very important in developing biological communities. However, the flow pattern, chemistry of the water, and characteristics of the substrates largely determine what species will be present in each area and what the productivity of a system will be.

DISTRIBUTION OF THE BIOTA IN CANADIAN AND NORTH AMERICAN WATERS

D. D. Williams

INTRODUCTION

The first section of this chapter has dealt not only with the overall distribution of lotic biotas (biotas in flowing waters) in the United States but also with those characteristics that control the distribution of aquatic communities within individual river systems. Much of the latter are applicable to Canadian waters. Thus, for example, streams in deciduous forested watersheds in southern Ontario and Quebec have habitat characteristics and faunas similar (though the latter may be less diverse) to those found in the eastern and midwestern United States. Coastal river faunas of British Columbia are similar to those of Washington and Oregon, while the lowland faunas of rivers of the Canadian prairies resemble those of the northern U.S. plains. Canada lacks species characteristic of streams of the southwestern deserts, but it has an extensive arctic element. In this section, therefore, I propose to deal primarily with the macrodistribution of the Canadian lotic biota.

At the outset, however, it will be useful to consider the different running-water habitats that are to be found in Canada. Although these will be present in virtually all parts of the country, local climate, geology, vegetation, and postglacial colonization routes will have affected the exact nature of the biota in each.

The precise distinction between a river and a stream is rather vague. Macan (1974) suggested that streams can be forded by someone wearing hip-waders, but rivers are deeper. Ricker (1934), in a classification of southern Ontario streams, defined rivers as having a discharge of greater than 0.28 m^3/s on 1 June, with a width greater than 3 m. Classifying running waters has always been problematic, and various schemes have been proposed, as seen in the first section of this chapter, involving local geology, sources of water, size, current speed, gradient, discharge, substrate type, temperature, dissolved oxygen and carbon dioxide contents, fauna and flora, and productivity. Two noteworthy systems with universal application are Horton's (1945) stream-order classification and the classification by Illies and Botosaneanu (1963), based on faunal zones. The latter divides rivers, longitudinally, into four regions: (1) the *Eucrenon* is the site of emergence of ground water at a spring; (2) the *Hypocrenon* is the short part of the stream flowing directly from the spring (springbrook); (3) the *Rithron* is the region extending from the hypocrenon to the point where the monthly mean temperature rises to 20°C, where the water flow is fast and turbulent and the substrate is composed of large particles with some sand, silt, and mud in sheltered patches, and where dissolved oxygen is near saturation; (4) the *Potamon* is the region below the rithron extending to the mouth where the monthly mean temperature rises above 20°C, where flow is slow and nearly laminar over substrates of mud, sand, and silt, and where oxygen deficiencies may develop. The biota of springs is mostly stenothermic and is characteristically dominated by just a few species; that of the rithron is largely cold stenothermic and characteristic of flowing water and thus benthic rather than planktonic; while that of the potamon is eurythermic or warm stenothermic and contains forms that reach their maximum diversity in lentic (slow-moving or ponded stream) habitats. In the Canadian Arctic, there is a tendency for rivers to be short and typically rithronic throughout their entire length, whereas in the more temperate regions of Canada there is a balance between rithron and potamon. It is also possible to further subdivide both the rithron and potamon on a faunal basis.

Apart from macrodistributional differences, there are microdistributions of stream faunas to be considered. For example, streams and rivers with deep gravel beds, in areas of glacial deposition, support extensive interstitial faunas. This fauna, the hyporheos, consists of a wide variety of taxa, many of them insects or related arthropods such as mites or crustaceans (including copepods, ostracods, cladocerans, and amphipods). Much work on interstitial faunas in general has concentrated on mites and crustaceans, as some species seem to be particularly well adapted to this kind of environment (Williams, 1984).

The extent of the hyporheos is largely controlled by physical and chemical conditions in the interstices. In terms of vertical depth, Williams and Hynes (1974) found insect larvae regularly to a depth of 70 cm beneath the surface of the bed of the Speed River, Ontario. Numbers were greatest around 10 cm and declined gradually with increasing depth until at 80 cm few individuals were present, although interstitial water clearly was present beyond this. In the Matamek River, Quebec, however, these same authors (Williams and Hynes, 1974) found animals down to the maximum depth capable of being sampled (100 cm), although at reduced densities. Below about 40 cm, the numbers of insect larvae decreased, but two forms, the oribatid mites and the elongate harpacticoid copepod *Parastenocaris starretti,* became dominant. This copepod genus is typical of true subterranean habitats (Schwoerbel, 1961), and both these and the oribatid mites may possibly be indicative of the boundary between hyporheic water and ground water. Schwoerbel (1961) and Husmann (1966) have suggested that the stream interstitial habitat could be subdivided into two zones on the basis of the distribution of mites: an upper layer characterized by the presence of mites in families belonging to the Hydrachnellae, and a lower zone characterized by mites of the subfamily Limnohalacarinae. The point where the limnohalacarids disappear marks the upper boundary of the true ground-water zone (Husmann, 1966).

The horizontal extent of the hyporheos has not been well studied. The distribution of hyporheos at a depth of 10 cm on a transect from midstream to 2 m into the bank of the Speed River was examined by Williams (1981). Typical stream insect taxa (e.g., mayflies, caddisflies, stoneflies) were entirely restricted to the stream interstitial environment, whereas some of the Chironominae (midges) and Elmidae (riffle beetles) were taken right up to the stream margin, and other midges (Orthocladiinae and Tanypodinae) were found up to 2 m into the interstitial water under the bank. Among the noninsect arthropods, the Ostracoda, Cyclopoida, Harpacticoida (microcrustaceans), and Acari (mites) all showed fairly continuous distributions from midstream to at least 2 m into the bank. This supports Schwoerbel's (1961) idea that the hyporheic zone extends several meters beyond the margin of the stream. Here again, the hyporheic zone presumably merges with the true ground water and its associated fauna. The lateral extent also may vary according to local conditions. For example, hyporheic insects have been found as much as 50 m laterally from a river channel (Stanford and Gaufin, 1974).

Other microdistribution patterns of stream faunas have recently been reviewed by Williams (1979, 1981).

In addition to permanent streams and rivers, temporary streams are common across the country, but most are associated with the upper regions of drainage basins. Control of flow in these streams is due to a combination of factors, foremost among which is the balance between precipitation and the infiltration capacity of the stream bed and surrounding soil (Williams and Hynes, 1977). The rate of evaporation may also be important, particularly during low-flow conditions. There have been relatively few studies on temporary streams in Canada.

In general, temporary water bodies exhibit much greater amplitudes in both physical and chemical parameters than most permanent aquatic habitats. Animals that live there have, therefore, to be extremely tolerant of these conditions if they are to survive. In temporary streams, the period and range of flow are significant factors, as the current may vary from torrential proportions during spring floods to zero in summer pools. The degree and rate of descent of the ground-water table and the permeability of the bed substrate are also important insofar as they control the formation and duration of the pool stage, which inevitably follows the cessation of flow. These pools support certain species of insect that would not normally be able to live in a lotic environment, and at the same time they invite colonization from many purely lentic forms that use these habitats as temporary breeding sites (e.g., Coleoptera and Hemiptera; Fernando and Galbraith, 1973). In Canada, the water temperature of these habitats varies from near 0°C under the winter ice, to near 30°C in the summer pools.

The faunas of temporary water bodies show seasonal succession. In a study of temporary streams in southern Ontario, Williams and Hynes (1977) showed the following succession of insects from the fall-winter stream stage, through the spring pool stage, to the summer terrestrial stage.

Stream stage. Plecoptera, Trichoptera, and Diptera (some Tipulidae, Simuliidae, and Chironomidae)

Pool stage. Ephemeroptera, Odonata, Hemiptera, Coleoptera, and Diptera (some Tipulidae, Culicidae, Ephydridae, Syrphidae, and Psychodidae)

Terrestrial stage. Coleoptera (Heteroceridae, Staphylinidae, and Scarabaeidae), Hymenoptera (Formicidae), and Diptera (Sepsidae, Sphaeroceridae, and some Ceratopogonidae).

On the basis of the species found in 16 temporary streams in this area, these authors postulated that the fauna belonged to three main groups. The first consists of permanent stream insects that are not particularly well adapted to life in temporary streams, but because of wide tolerance are able to survive in streams dry for a short period. The second group contains facultative species occurring in both lentic and lotic waters. The third group consists of species highly adapted, and often restricted, to temporary waters.

THE CANADIAN DRAINAGE BASINS

In Canada there are four major drainage basins (each having a mean annual discharge in excess of 16,000 m^3/s) and one minor drainage basin (mean annual discharge 25 m^3/s; Fig. 1; Hydrological Atlas of Canada, 1978).

Arctic Ocean drainage basin (mean discharge 16,400 m^3/s)

This primarily consists of the Mackenzie River, Northwest Territories, and its feeder rivers the Peace, Liard, Slave, Peel, and Smoky Rivers, which flow north into the Beaufort Sea at Mackenzie Bay. The system also receives water from Great Bear Lake, Great Slave Lake, and Lake Athabasca. In the eastern Northwest Territories the Back River flows north into McClintock Channel. Total dissolved solids are generally in the 100 to 200 ppm range but they are lower in the mountains where rainfall is greater, and much lower (50 ppm) in areas where the bedrock resists weathering (Livingstone, 1963).

Large rivers such as the Mackenzie originate in more temperate zones and they are therefore not wholly arctic in character;

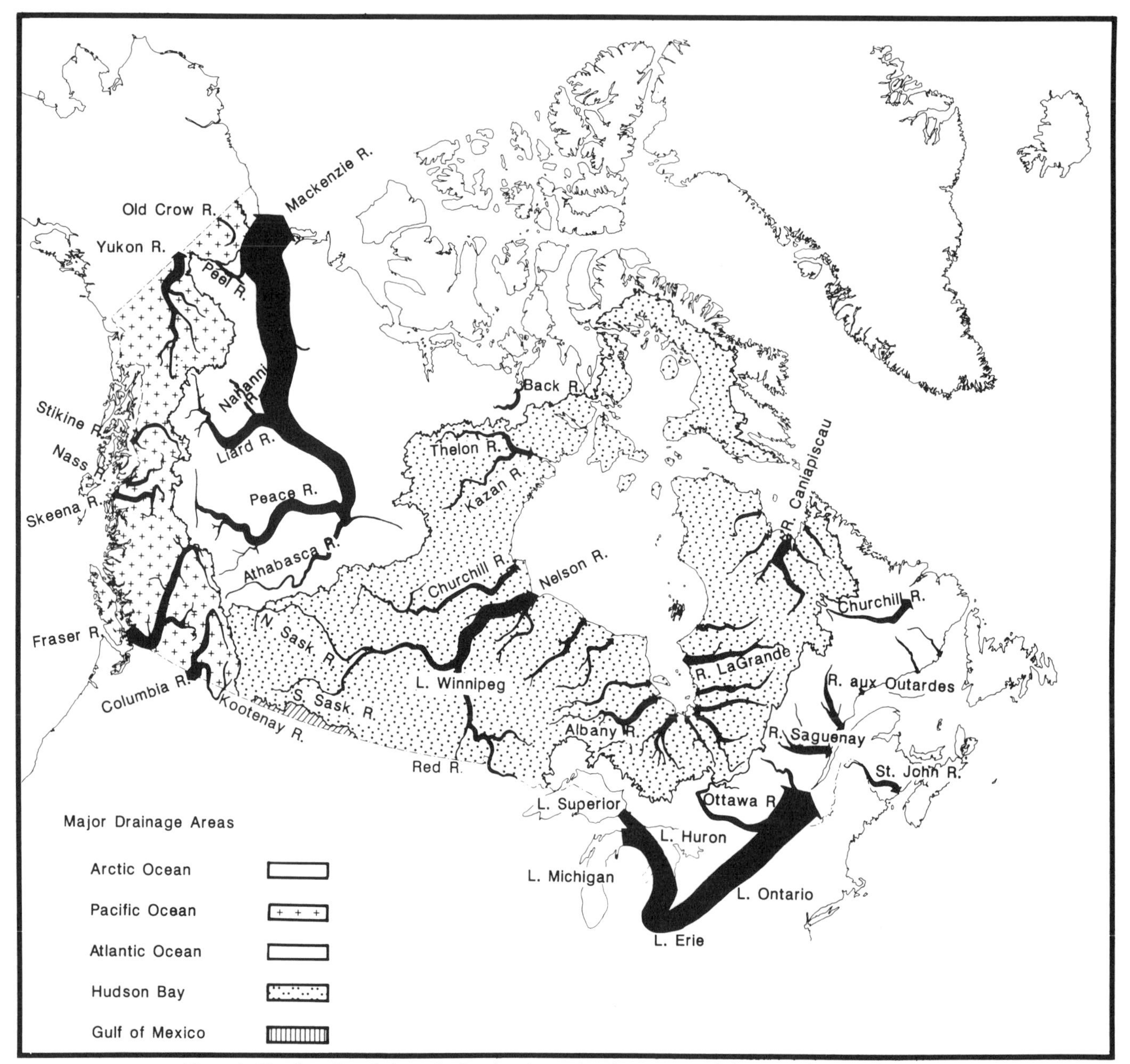

Figure 1. The Canadian drainage basins showing major river systems. (The mean discharge is proportional to the width of the arrows and only rivers with a mean discharge of greater than 1,000 m/s are shown. (Redrawn from the Hydrological Atlas of Canada, 1978.)

they are sometimes seen as protrusions of the boreal forest biome into the north. Peak discharge is in early summer, ice forms in the fall, and flow generally continues throughout the winter.

Mountain streams above the tree line are short lived; often 80 percent of their annual discharge occurs in a two-week period at spring snowmelt; maximum temperatures of between 5° and 15°C are achieved by late July, and by late summer the streams refreeze. Water chemistry varies over the summer, the water being low in total dissolved solids at first (meltwater), but subsequently increasing with contributions from deeper in the substrate.

Coastal plain, lowland streams are somewhat warmer than mountain streams, and although flow is still restricted to summer, discharge is somewhat more uniform. Nutrient levels are generally low, as is pH, and leachates from the tundra may color the water brown.

The benthos of lowland streams is dominated by chironomid midges, but stoneflies, mayflies, and oligochaete worms may be abundant also. Mountain streams, too, are dominated by chironomids (subfamilies Diamesinae and Orthocladiinae) together with blackflies (*Prosimulium*), although densities are low. Large rivers again are dominated by chironomids, biting midges

(Ceratopogonidae), and oligochaetes (Harper, 1981). High densities of a few specialized chironomid species occur on the extensive, unstable sand substrates of large rivers such as the Athabasca (Barton and Wallace, 1980).

Pacific Ocean drainage basin (mean discharge 24,100 m^3/s)

This consists of the Yukon River and its major tributary the Porcupine River, which flows from the Yukon Territories west through Alaska; the Frazer River in southern British Columbia, and its major tributary the Thompson River, which flow out into the Straits of Georgia; the Columbia River in southern British Columbia, and its major tributary the Kootenay River flow south into Washington State; and in northern British Columbia, three rivers flow more or less directly west from the Rocky Mountains into the Pacific: they are the Skeena, Nass, and Stikine Rivers. Maximum stream discharge along the coast is generally in late fall and winter, corresponding to the rainfall pattern. Most streams are very clear, apart from times of spate, and maximum water temperatures of 10° to 15°C are reached in late summer. They typically have very low levels of dissolved solids (50 ppm). Streams in the lower Frazer Valley are small, slow flowing, and meandering, and may be highly colored due to leachates from agriculture and peatlands. Small, clean, softwater woodland streams, such as those on Vancouver Island, are typically dominated by mayflies, chironomids, and oligochaete worms, with stoneflies, caddisflies, and blackflies represented also (Mundie and others, 1983). Small tributary streams in the insular lowland may become dry in the summer (Northcote and Larkin, 1963). Rivers of the Canadian interior receive much of their water from snow and glacial melt in summer, and total dissolved solids are frequently about 100 ppm (chiefly calcium carbonate). To the northeast, the Hay River is stained brown as are many of the meandering streams that drain this bogland area (Northcote and Larkin, 1963).

Atlantic Ocean drainage basin (mean discharge 33,400 m^3/s)

The St. Lawrence River drains the St. Lawrence Great Lakes (Superior, Michigan, Huron, Erie, and Ontario) and flows northeast. Its major feeders are mostly rivers that flow from the north, the Ottawa, St. Maurice, and Saguenay Rivers, Riviere aux Outardes, and Moisie River. Two other major rivers flow south into the Gulf of St. Lawrence, the Riviere Natashquan and Riviere du Peut-Mecatina. The St. John River flows south through New Brunswick into the Bay of Fundy. The Churchill River flows east through Newfoundland into the Labrador Sea. In the southern part of Ontario and southwestern Quebec, sedimentary rocks overlain by mixed or deciduous forest result in hard, highly alkaline waters that support a rich fauna of insects (most aquatic orders), crustaceans, molluscs, mites, and worms (Hallam, 1959). To the north the rocks of the Precambrian Shield, overlain by thin soils, produce soft, acidic waters that support a high diversity of aquatic insects, particularly mayflies, stoneflies, caddisflies, blackflies, and chironomids (Sprules, 1947). Precipitation, for the most part, is fairly evenly distributed throughout the year, but streams are subject to spates at snowmelt and after thunderstorms. In the Maritime region, the majority of streams are short and flow over igneous rock overlain by thin soils, resulting in a low pH. Flow maintenance is chiefly from surface runoff, and smaller streams tend to become dry in the summer (Smith, 1963). Precipitation is maximal in winter, though plentiful in summer, and the climate is continental rather than maritime (Danks, 1979). In New Brunswick and Nova Scotia streams, common elements in the fauna are mayflies, stoneflies, caddisflies, chironomids, and blackflies, while in Newfoundland, although the important Nearctic families of most lotic insect orders are represented, diversity is somewhat reduced (e.g., as in some of the mayfly families; Larson and Colbo, 1983).

Hudson Bay drainage basin (mean discharge 30,900 m^3/s)

The Nelson River, Manitoba, flows northeast into Hudson Bay. It is fed by the North and South Saskatchewan Rivers, the Red River, the Assiniboine River, Hayes River, and by Lake Winnipeg and Lake of the Woods. The Churchill River, Manitoba, flows parallel to the Nelson River, as do the Severn and Winisk Rivers of northwestern Ontario. In the Northwest Territories, the Thelon and Kazan Rivers flow into Baker Lake and thence into Hudson Bay. A number of large rivers flow into James Bay: the Attawapiskat and Albany Rivers flow from the west; the Moose River, Riviere Harricana, and Riviere Nottaway flow from the south; Riviere de Rupert, R. Eastmain, La Grande Riviere, and Grande Riviere de la Baleine flow from the east. Rivieres Arnaud, aux Feuilles, Canaipiscan, a la Baleine, and George drain into Ungava Bay. (Most of the abovementioned tributaries have a mean discharge greater than 85 m^3/s.) Much of this area is characterized by either the exposed bedrock of the Canadian Shield or by thin (often 10 cm deep), rocky soils. Where the land is poorly drained, such as in the Hudson Bay Lowlands, organic soils occur that support black spruce and muskeg. Although surface waters are generally soft and acidic, outcrops of Ordovician-Devonian rock, together with widespread Pleistocene marine deposits, provide some regions that are rich in calcium compounds. Climate within this vast area tends to parallel phytogeographic regions, which range from Tundra-Grassland to Boreal Forest (Clarke, 1973).

Gulf of Mexico drainage basin (mean discharge 25 m^3/s)

A small portion of the northern part of this drainage area occurs in southern Alberta and southern Saskatchewan. There are no major river systems in the Canadian sector of the basin, but there are a number of smaller rivers. These may have been an important route for recolonization of the southwestern part of the Hudson Bay drainage area by the biota in postglacial times, for it is in this region that the upper Missouri River system comes in close proximity to the South Saskatchewan River.

HISTORICAL PERSPECTIVE

Discussion of the present diversity and distribution of river biotas in Canada must include consideration of climatic changes that took place in this country during the Quaternary Period (approximately the last 1.6 m.y.). This span has been seen as the most influential in the development of modern floras and faunas (Ross, 1965). Matthews (1977) points out that for northern species of insects, at least, the rate of speciation was probably considerably slower than the frequency of climatic oscillation in the Quaternary. Despite this, Quaternary events are still fundamental to any discussion of the existing biota of Canada; even though many species may have been derived from Tertiary times, their present distribution is primarily the outcome of the Quaternary climate (Matthews, 1979). There has been a relatively short time, geologically speaking, since the last (Pleistocene) glaciation. Most of Canada was ice covered as little as 15,000 years ago and a substantial area remained glaciated up to 8,000 years ago, until the rapid melting of the Laurentide ice sheet. The present-day biota is therefore essentially an interglacial one, and one that can be expected to change in the near geological future as the climate of Canada continues to change, responding to changes in the general circulation of the atmosphere or perhaps to the re-expansion of the glaciers.

During Pleistocene glaciation, certain areas around the ice margin remained free from ice. Some or all of these areas may have acted as refugia for a limited number of tolerant species. The four largest refugia were: (1) unglaciated parts of Alaska and the Yukon, connected intermittently by a land bridge to unglaciated parts of eastern Siberia, and collectively known as Beringia (Hulten, 1937); (2) Banks Island (Vincent and Gauthier, 1976) and, possibly, eastern parts of Baffin Island (England and Bradley, 1978) in the Arctic Archipelago during the Late Wisconsinan; (3) small regions of western Newfoundland and perhaps the other Maritime Provinces, also during Late Wisconsinan times (Grant, 1977); (4) a transcontinental refugium to the south of the ice margin—essentially most of the continental United States. Some Holarctic running water species may have recolonized through Beringia, as there appears to have existed a major river that flowed north over the land bridge, to the east of what is now Wrangel Island (USSR), into the Arctic Ocean. This river had tributaries arising both on the Chukchi Peninsula and in northwestern Alaska (possibly the Kobuk River, Alaska, was one of these) (Ager, 1982). Thus, a continuous waterway was available for the transfer of aquatic species between the old world and the new. In addition, there was a large fresh-water lake, some 100 km in length, located to the southwest of the present-day Bering Straits (Hopkins, 1972). Banks Island, eastern Baffin Island, and western Newfoundland all have been shown to have sheltered some plants and animals (Danks, 1981); however, none appear to have been fresh-water lotic forms (but see Larson and Colbo, 1983). The extensive refugium of the southern United States was undoubtedly the most important so far as repatriation of the Canadian fresh-water fauna is concerned.

The rate of colonization can be expected to vary greatly between taxa, depending to a large extent on their physiological tolerance to the cold environments left in the wake of the retreating ice, and their relative mobilities. For example, fishes have to rely almost entirely on routes created by drainage patterns, and either may be able, consequently, to invade new river systems quite quickly where they are confluent, or be prevented for long periods from spreading between adjacent drainage systems where there is lack of a suitable physical connection. Freshwater invertebrates other than insects have similar colonization characteristics that may be compounded by a lesser degree of mobility. However, some species occasionally may be carried passively between water bodies by agents such as wind, fishes, aquatic birds, amphibia, and mammals (Maguire, 1963; Proctor, 1964). Most aquatic insects have an aerial adult stage in their life cycle that can bridge the gaps between watersheds, but many (e.g., mayflies and stoneflies) are not strong fliers and are thus somewhat limited in the distance they can cover. Aquatic macrophytes and algae are perhaps the least restricted in terms of distribution, as many can be spread through seeds or drought-resistant disseminules that may be carried by the wind.

THE BIOTA

Fishes

Despite Canada's vast fresh-water resources (7.5 percent of the country's total surface area is fresh water) the diversity of fresh-water fishes is low. This is primarily a consequence of the relatively short time since the Pleistocene glaciation. The total complement of freshwater fish species is 181 (including four introduced species), and these belong to 24 families. However, 70 percent of the species belong to just five families, the Salmonidae, Cyprinidae, Catastomidae, Percidae, and Cottidae. Scott and Crossman (1973) emphasize this low diversity by pointing out that the state of Ohio alone (which was minimally affected by the Wisconsinan ice) has 170 species of fish. Although fish diversity in Canada is low, many species, particularly the salmonids, are extremely abundant. Due to space limitations, we will discuss only the five common families.

Salmonidae. Three genera occur in the family Salmonidae. The genus *Oncorhynchus* contains five species of anadromous fishes: *O. gorbuscha,* the pink salmon; *O. keta,* the chum salmon; *O. kitsuch,* the coho salmon; *O. nerka,* the sockeye salmon; and *O. tshanytscha,* the chinook salmon. All are common in coastal rivers and streams of British Columbia from the Washington to Alaska borders. In addition, the chum and pink salmon occur in the Mackenzie River, Northwest Territories, where the chum ranges as far upstream as the Hay and Slave Rivers.

The genus *Salmo* contains four species: the cutthroat trout (*S. clarki*) is primarily western and coastal, but there is an inland form common in western Alberta. The two forms are separated by an area in southern British Columbia in which rainbow trout (*S. gairdneri*) predominate. Rainbow trout are native to the eastern Pacific Ocean, in rivers west of the Rocky Mountains and the Peace and Athabasca river systems to the east; it has been widely introduced to eastern and central Canada and is firmly established in the St. Lawrence drainage system. The Atlantic salmon (*S. salar*) is native to both sides of the Atlantic Ocean and, in Canada, ranges from the coasts of Labrador to New Brunswick and to Quebec and Ontario along the St. Lawrence River. *S. trutta,* the brown trout, is native to Europe and western Asia but has been introduced into streams of the Maritime Provinces, the St. Lawrence system, and parts of Manitoba and Alberta (Fig. 2; Scott and Crossman, 1973).

The genus *Salvelinus* contains four species: (1) *S. alpinus,* the arctic char, has the most northerly distribution of any freshwater fish, ranging from the Mackenzie through the Arctic islands to Hudson Bay, the coasts of Labrador and Newfoundland to southeastern Quebec (Fig. 2). (2) *S. fontinalis,* the brook trout, is endemic to streams in eastern North America and, in Canada, is widespread in the Atlantic Ocean drainage basin and in the eastern half of the Hudson Bay system; it has been introduced into western Canada (Fig. 4). (3) *S. malma,* the dolly varden, is a western species that ranges from the headwaters of the South Saskatchewan River, north through British Columbia to the headwaters of the Liard, Yukon, Peace, Athabasca, and Nahani Rivers. (4) *S. namaycush* the lake trout, is widely distributed in Canadian lakes but occurs in some rivers in the Northwest Territories (Scott and Crossman, 1973).

The subfamily, Coregoninae, is generally widely distributed in Canada; however, most of its 18 species inhabit lakes, although some are common in large rivers. For example, *Coregonus autumnalis,* the Arctic cisco, occurs in rivers of the Arctic Ocean drainage system; *C. artedii,* the cisco or lake herring, has been recorded from the Mackenzie River system north to Great Bear Lake; *C. canadensis,* the Atlantic whitefish, is found only in the Tusket River system, Nova Scotia; the mountain whitefish (*Prosopium williamsoni*) occurs in rivers in western Alberta and British Columbia.

The subfamily Thymallinae is represented in Canada by *Thymallus arcticus,* the Arctic grayling. It occurs in northern drainage areas including most rivers of the western portion of Hudson Bay, Alaska, and the northern part of the Pacific Ocean drainage basin. Populations extend as far south as the Flathead River in British Columbia and the Churchill River in Saskatchewan (Fig. 2; Scott and Crossman, 1973).

Cyprinidae. This family includes more than 1,500 species of minnows and carps from most parts of the world. 68 species are found in Canadian fresh waters. A few species are strictly western, such as the Redside shiner (*Richardsonius balteatus*), the Northern Squawfish (*Ptychocheilus oregonensis*), and the Chiselmouth (*Acrocheilus alutaceus*); many occur in southern parts of the Hudson Bay drainage area and the Great Lakes region (e.g., Hornyhead chub, *Nocomis biguttatus*; Rosyface shiner, *Notropis rubellus*; Bluntnose minnow, *Pimephales notatus*); while others are limited to southern Ontario (e.g., River chub, *Nocomis micropogon*; Redfin shiner, *Notropis umbratilis*). Few are found in the Maritimes and they are absent from the island of Newfoundland. A few species, such as the Northern redbelly dace (*Chrosomus eos*), the Lake chub (*Couesius plumbeus*), the Emerald shiner (*Notropis atherinodes*), and the Flathead chub (*Platygobio gracilis*; Fig. 3) occur in southern or western regions of the Arctic Ocean drainage area. However, the family is generally dominant in the more southerly parts of Canada, which they have recolonized in postglacial times from the large river systems of the United States, particularly the Mississippi–Missouri River system in central and eastern areas, and the Columbia River system in the west (Scott and Crossman, 1973).

Catostomidae. This large family occurs in most of North America (Fig. 2), eastern Siberia, and China. In Canada, two species are widespread, *Catostomus catostomus,* the longnose sucker, and *C commersoni,* the white sucker. Two species are restricted to British Columbia, the remaining 11 species occur in the Atlantic Ocean and/or Hudson Bay drainage basins. Many species of *Moxostoma* (redhorses) have very restricted distributions in Ontario and Quebec (Scott and Crossman, 1973). In postglacial times, eastern and central species probably re-invaded from the Mississippi-Missouri River system while western species recolonized from the Columbia River system. Widespread species may have survived glaciation in suitable habitats all along the southern limit of the glaciers.

Percidae. This family is circumpolar in distribution and consists of two subfamilies, the perches and the darters, many species of which are restricted to North America. The yellow perch (*Perca flavescens*) and walleye (*Stizostedion vitreum*) have the widest and most northerly distribution of the perches (the former extending as far north as Great Slave Lake and the latter as far as the Mackenzie River Delta), while the Iowa darter (*Etheostoma exile*) is the most widespread of the darters (reaching as far north as the Alberta–Northwest Territories border). The remainder of the darters and the sauger (*Stizostedion canadense*) have a more southerly distribution in the Hudson Bay drainage area and in the upper St. Lawrence River–Great Lakes region. Some darters such as the greenside darter (*E. blennoides*), the rainbow darter (*E. caeruleum*), and the channel darter (*Percina copelandi*) have very restricted distributions in southeastern Canada, which probably represent the northern limits of their primary range, the Mississippi River system.

Cottidae. Sculpin are primarily marine fishes, but eight species occur in Canadian fresh waters. Four are confined to the Pacific region (*Cottus aleuticus, C. asper, C. confusus,* and *C. rhotheus*; Fig. 3); one has a disjunct western and eastern distribution (*C. bairdi*); two, *C. cognatus,* the slimy sculpin and *C. ricei,* the spoonhead sculpin, have northern distributions; and *Myoxocephalus quadricornis,* the deep-water sculpin, is a glacial relict with a very patchy present-day distribution in the cold bottomwater of deep lakes (Scott and Crossman, 1973).

Distribution patterns of other families worthy of note include the catfishes (Ictaluridae) and sunfishes (Centrarchidae), both of which are "warm-water" forms and have a distinctly southern (south of 50° N latitude) distribution in Canada, and the

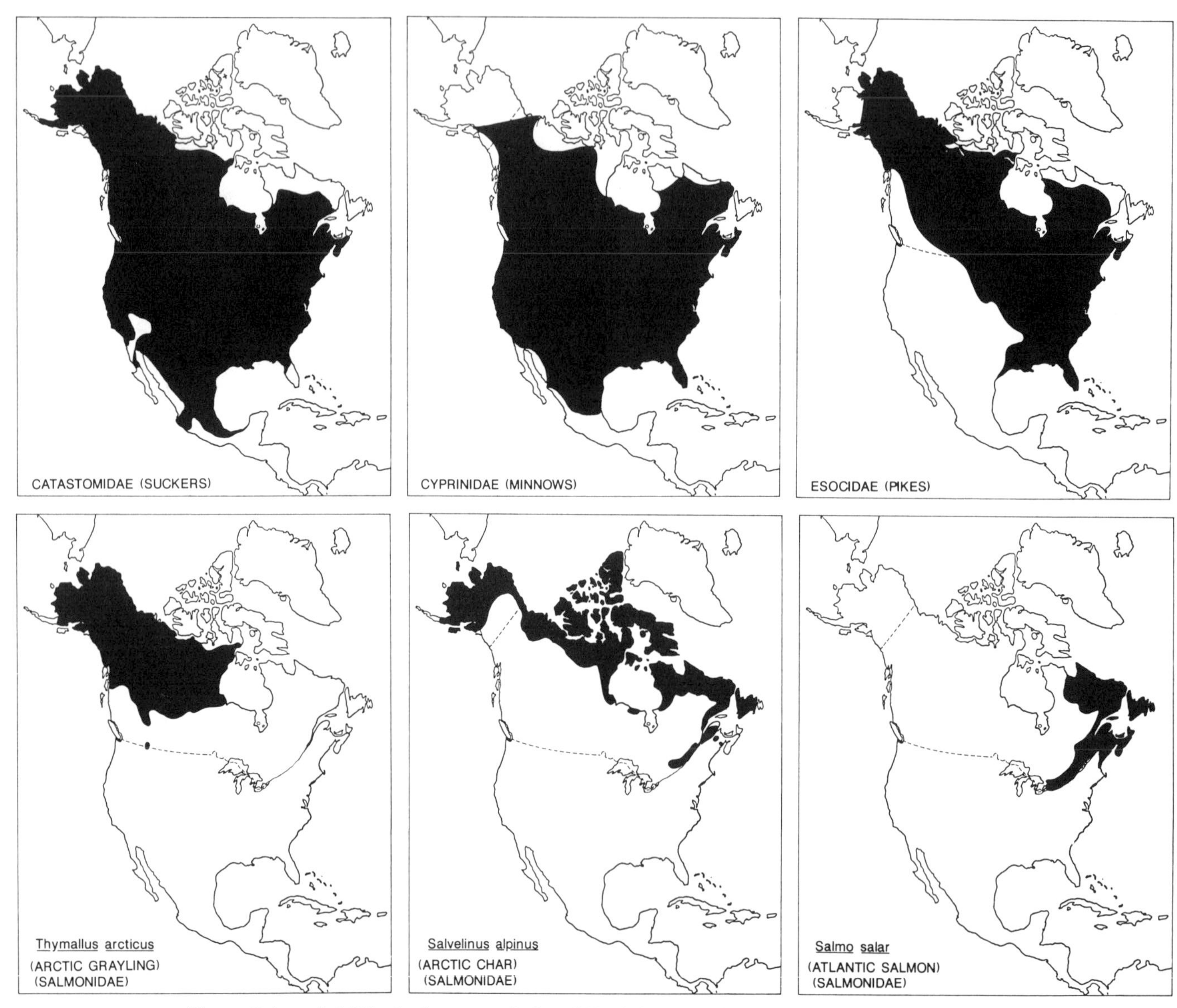

Figures 2 through 4. Distribution maps of selected North American fishes. (Redrawn from Kuehne and Barbour, 1983; Miller, 1981; Scott and Crossman, 1973.)

eel *Anquilla rostrata* (Anguillidae)—this single species is abundant in all rivers in Quebec and New Brunswick that flow into the St. Lawrence basin and also in tributaries to the Lake Ontario–St. Lawrence River system together with salt and fresh waters of the other Maritime Provinces. It has gained a foothold in the upper Great Lakes via access through the Welland Canal (Scott and Crossman, 1973).

Figures 2 to 4 show the distribution patterns of some of the above-mentioned Canadian lotic fish species, together with some that occur in the United States, in the larger context of their distribution on the entire North American continent. The purpose of this exercise is to show the remarkable differences in range between certain families and species. The factors determining these various continent-wide patterns include features of the habitat (see river channel characteristics in the first section of this chapter), physiological requirements of the fish species, and colonization abilities and opportunities. Some types of fish, such as the Catostomidae (suckers) and Cyprinidae (minnows) cover virtually the entire continent, while the pikes (Esocidae) are absent only from the western United States (Fig. 2). The arctic grayling (*Thymallus articus*) has, as we have noted, a northwestern distribution that contrasts with the northeastern distribution of another salmonid, the Atlantic salmon (*Salmon salar*). The arctic char (*Salvelinus alpinus*) is distributed along the Arctic coast (Fig. 2).

Cottus aleuticus, a sculpin, occurs only in west-coast streams from southern California to the Aleutian Islands of Alaska (Fig. 3). The sturgeon, *Acipenser oxyrhynchus,* has a similarly restricted coastal distribution but on the east coast, from Florida to Labrador. *Platygobio gracilis,* the flathead chub, and *Hiodon alosoides,* the goldeye, have similar north-south, midcontinent

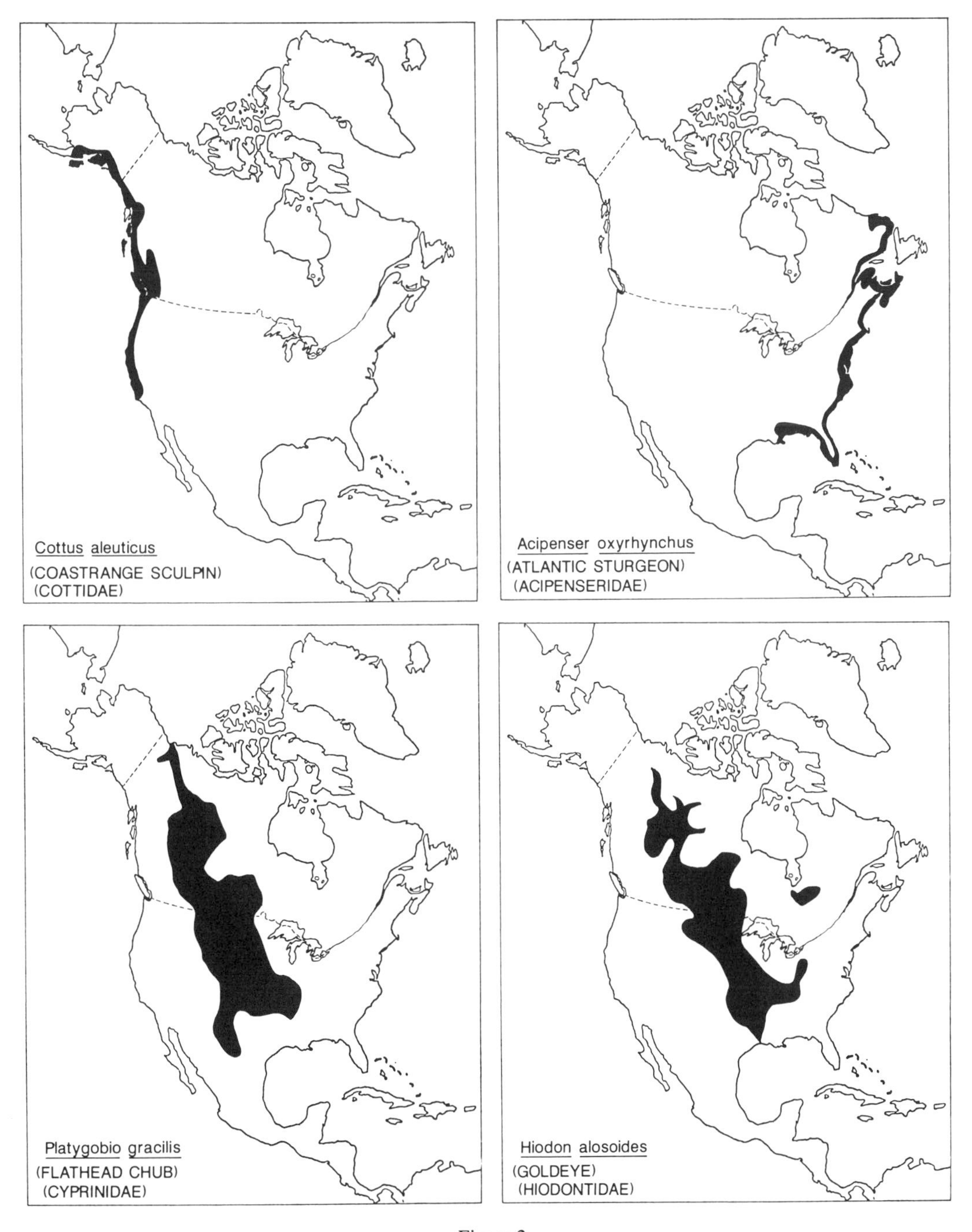

Figure 3.

distributions stretching from the lower Mississippi River to the Mackenzie River (Fig. 3).

Hybognathus hankinsoni, the brassy minnow, has a midcontinent distribution that runs east-west from the upper St. Lawrence River to the rivers of Colorado, Wyoming, and Montana (Fig. 4). The garfishes (Lepisosteidae) have a broadly southeastern distribution, mostly within the Mississippi River basin, although it extends out as far as Montana and southern Quebec (Fig. 4). In contrast, although a few species of pupfishes (*Cyprinodon*) occur in brackish and salt waters of eastern North America, the greatest concentration of species is in desert waters of the southwest (Fig. 4; Miller, 1981). *Salvelinus fontinalis,* the brook trout, has an apparent disjunct east and west distribution in North America (Fig. 4); however, as we saw, it is endemic only to the northeast and has been introduced extensively by man in the west. Two highly endemic distributions are seen in the darters. *Etheostoma australe* is found only in the upper Rio Conchos basin in the states of Chihuahua and Durango, Mexico, while *E. neopterum* is restricted to the tributaries of the lower Tennessee River from Shoal Creek in Alabama and Tennessee northward to

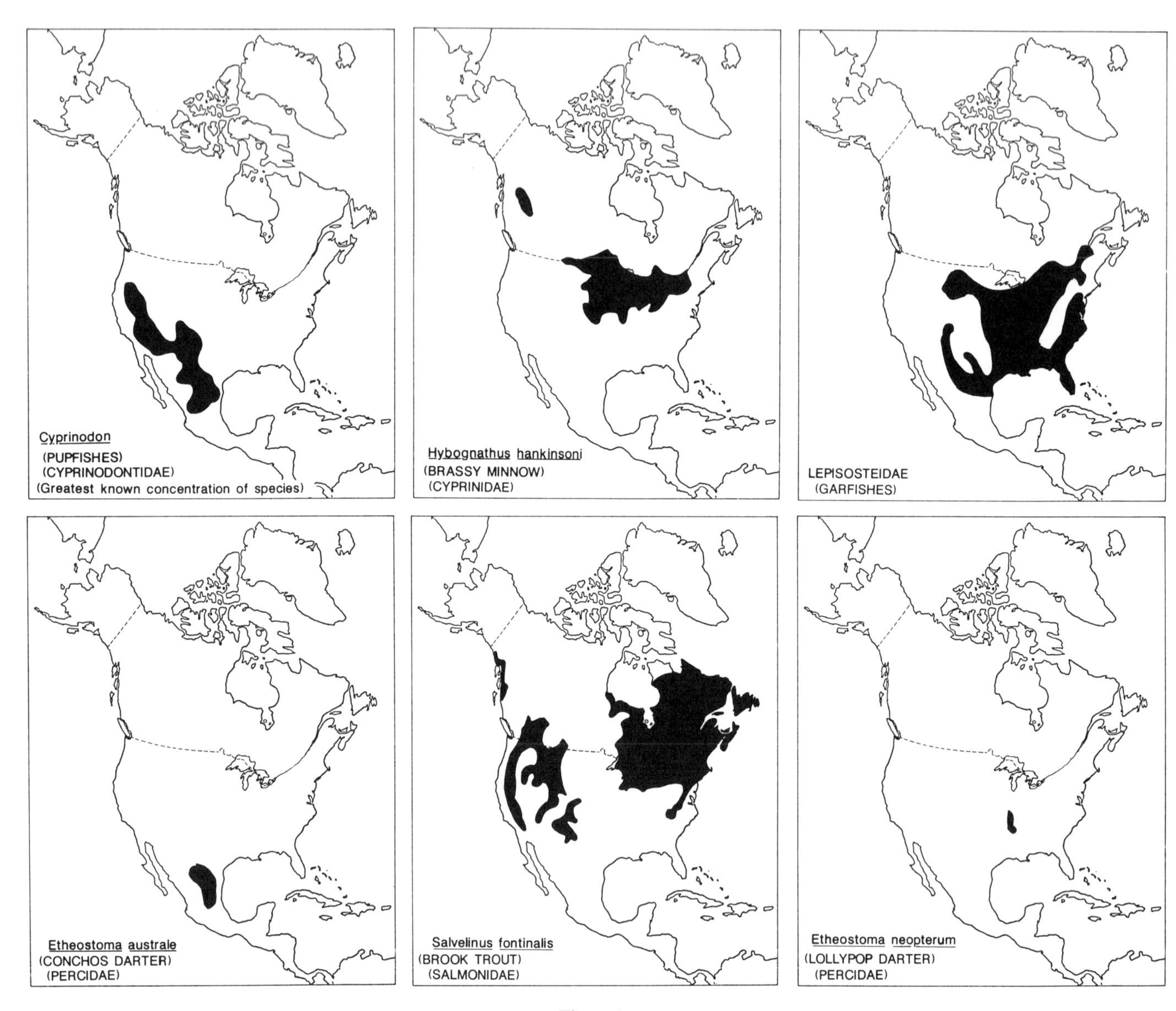

Figure 4.

the mouth of the river in Kentucky (Fig. 4; Kuehne and Barbour, 1983).

Invertebrates

Insects. Canadian running-water insects belong to 11 of the 22 orders of insects. The Plecoptera are mostly restricted to running waters, where together with the Ephemeroptera, Megaloptera, and Trichoptera, they reach their maximum development and diversity (Hynes, 1970). In addition, dipteran families, such as the Simuliidae, Deuterophlebiidae, and Blephariceridae, are confined largely to running waters, as are a few specialized families of the Odonata (e.g., Cordulegasteridae and Gomphidae) and Coleoptera (Elmidae and Psephenidae).

Ephemeroptera. The greatest diversity of mayfly species occurs in unpolluted rivers and streams. Most genera are transcontinental in distribution, but some are typically western (e.g., *Cinygma*) while others are eastern (e.g., *Pentagenia, Potamanthus,* and *Neoephemera*). Many genera occur as far north as the northern edge of the boreal forest, but only a few species of *Baetis* are found in the Arctic tundra (Lehmkuhl, 1979). A particularly rich and somewhat unique mayfly fauna occurs in the Saskatchewan River. Of the approximately 30 genera, many are known to have migrated from the southwestern United States as the Pleistocene glaciers retreated (Lehmkuhl, 1976a). Specific examples include the invasion of *Baetisca,* through connections between the Saskatchewan River and the Missouri-Mississippi system; and the entry of *Analatris eximia* Edmunds to the Saskatchewan system, possibly from the Colorado system via tributaries of the Missouri system (Lehmkuhl, 1972, 1976b). In Manitoba, *Pentagenia vittigera* Walsh and *Tortopus primus* (McD.) now run in the Red River; Ide (1955) thought that they had entered from the Mississippi when the Red River was its southerly draining tributary, its normal Hudson Bay outlet being blocked by receding Wisconsinan ice.

Odonata. Although more diverse in lentic habitats, dragonflies can be abundant in lotic habitats, particularly in the lower reaches of slow streams where there are beds of macrophytes and the summer water temperature is reasonably warm. In Canada, dragonflies are more common in the south, as low northern water temperatures generally do not suit their physiology. In the east, forest streams with pool and riffle regions support a number of species representing different families. Among the damselflies (Zygoptera), *Agrion* is common, with *A. maculatum* predominating in the south but being replaced by *A. aequabile* farther north to the limit of Hudson Bay. Another species, *A. moesta* is found in large, rapid streams, while *A vivida* is predominantly found in springs. Other primarily eastern genera, include *Enallagma* and *Chromagrion,* the latter being common in springs. Amongst the dragonflies proper, (Anisoptera) the family Aeshnidae is represented in running water by *Boyeria, Basiaeschna,* and *Aeshna*; the Gomphidae by *Ophiogomphus, Gomphus, Hagenius,* and *Lanthus* in the east and by *Ophiogomphus, Gomphus,* and *Octogomphus* in the west. The genera *Macromia, Didymops,* and *Stomatochlora* represent some of the Macromiidae and Corduliidae. The large anisopteran family Libellulidae seldom occurs in small forested streams, but some species are abundant in calm, marshy backwater regions of other streams (Walker, 1953).

Plecoptera. The nymphs of stoneflies are mostly restricted to cool, unpolluted fast-flowing streams. In Canada they are most diverse in mountain regions such as the Rocky Mountains in the west, the Appalachians in the east, and the Laurentian Shield; their most northerly distribution is Point Barrow, Alaska, and Victoria, Southampton, and Baffin Islands within the Canadian Arctic. Approximately 124 species have been recorded in the east, and a similar number—135—in the west. Only nine species are common to both regions, and the distribution of these tends to be northern transcontinental; four species have a Holarctic distribution (Harper, 1979). The great plains area of North America, including the Canadian Prairies, has a poor stonefly fauna, and *Isoperla longiseta* seems to be the only typical prairie species (Ricker, 1964).

Some hardy western species are thought to have survived Pleistocene glaciation in three unglaciated areas: near the Yukon River valley, along the Arctic coastal plain from the Bering Strait east past the Mackenzie River, and in the western Arctic islands. Species that may have colonized from this northwestern refugium are likely to be present-day tundra species such as *Nemoura arctica, Capnia nearctica,* and *Diura bicaudata.* A very limited number of species may have survived on isolated areas of high relief above the level of ice (nunataks). Most species, though, survived south of the glacier and, according to Ricker (1964), the primary recolonization routes were northward from the region east of the Mississippi and from southern Washington, Idaho, and Montana.

Postglacial dispersion in Canada for this and other cold-water–adapted invertebrate groups may have been via a process of "island-hopping between adjacent cold springs and springbrooks. Ross and Ricker (1971) proposed, for example, that the winter stonefly genus *Allocapnia* dispersed from the Appalachians to the Ozarks along a corridor of deep valleys lined with spring-fed streams that formed after coastal uplift in the Pliocene. This uplift increased local gradient and thus erosion, which subsequently produced the steep-sided valleys. Breaches in previous water tables along the valley walls produced a series of springs that formed a dispersal corridor for cold-adapted species.

Hemiptera. Aquatic members of this order are confined to the subgroup Heteroptera, and only six families are wholly aquatic. Although found in slow-flowing streams and at the edges of larger rivers, the group is more typical of standing-water habitats. The Corixidae (waterboatmen) and Saldidae (shore bugs) are cold-adapted (Polhemus, 1984), but in general, few species of Heteroptera occur in the Arctic; those that do are mostly restricted to the low Arctic (Scudder, 1979). Most families are widely distributed across southern Canada.

Megaloptera. This is a small, ancient group of insects belonging to two families. In Canada, the Dobson-flies, Corydalidae, occur in a wide variety of habitats including streams, rivers, spring seeps, and temporary streams, as well as in ponds, lakes, and swamps. Four species occur in the east and two in the west. Of the 10 species of alder-flies (Sialidae), five are eastern, four are western, and one, *Sialis velata,* appears transcontinental (Kevan, 1979). Sialids are typically found in areas of streams and rivers that have a soft substrate rich in detritus (Evans, 1984).

Trichoptera. Many of the 18 families of caddisflies found in Canada occur in running water; in fact, this may be their ancestral habitat (Ross, 1956). Canadian species largely represent components of broad Holarctic distributions. Some, however, are endemic to the Nearctic where they may be primarily either eastern, western, or transcontinental. A few species in the Yukon Territories have a restricted Beringian distribution, while some found in southern Ontario and Quebec have been derived from ancestors to the south. More than 30 percent of Canadian species belong to one family, the Limnephilidae (Wiggins, 1979). Corbet (1966) recorded the most northern distribution of the group from Ellesmere Island. Some widespread species occur in a wide range of habitats, including both lotic and lentic ones (e.g., *Ptilostomis ocellifera*). Others are restricted; for example, *Chyranda centralis* occurs primarily in cold-water springs and spring-fed brooks (Williams and Williams, 1987). Many of the genera found in springs are either endemic to the west or exhibit their greatest diversity of species there (e.g., *Psychoglypha, Parapsyche,* and *Lepidostoma*).

Wiggins and Parker (1984) have reported a distinct Beringian component to the caddisfly fauna. One species set of this component is endemic to Beringia, another is widespread in Canada but does not occur in the Palearctic, and a third set is amphi-Beringian, being widespread in both the Nearctic and Palearctic. *Ylodes kaszabi,* for example, occurs both in rivers in Mongolia and in the Yukon River basin (Manuel and Nimmo, 1984).

Coleoptera. Although beetles compose the largest order of insects, with more than 5,000 aquatic species, they do not dominate running-water communities. The lotic families are: the Elmidae, typically found in fast-water sections of both rivers and streams but with only three widely distributed genera—*Dubiraphia, Stenelmis,* and *Optioservus*—the majority of which are restricted to mountain regions; the Psephenidae, also inhabitants of fast-water, rocky substrate areas; the Gyrinidae, restricted to slow streams and backwater regions of rapid streams; the Hydraenidae, typically found around the waterline; some of the Haliplidae, found predominantly in weedy, slow-water areas; and the Dytiscidae, which occur in slow water, but occasionally also in riffle areas or in springs (Pennak, 1953; White and others, 1984). In Canada, the Dytiscidae are widely distributed, including Arctic and alpine regions. The Gyrinidae and Elmidae occur in all faunal regions but only to the limit of the tree line. The Haliplidae again are widely distributed, but apart from a few species that occur in the Northwest Territories and in northern Manitoba, most are restricted to the more southerly rivers. The distributions of the Psephenidae and Hydraenidae in Canada are poorly known (Campbell and others, 1979).

Diptera. The running-water Diptera in Canada belong to the following families: Chironomidae (particularly the Diamesinae, Orthocladiinae, and Tanytarsini), Simuliidae, Blephariceridae, Deuterophlebiidae, Tipulidae (particularly the Limoniinae and Pediciinae), Rhagionidae, and Muscidae (Hynes, 1970).

Canadian chironomids, in general, extend as far north as the limit of land, but the distribution of species is determined primarily by the environmental requirements of the larvae. The Tanypodinae and Chironominae are characteristic of relatively warmer waters, although both subfamilies contain cold, lotic species, and hence are more abundant in southern Canada (Oliver, 1981). The Orthocladiinae, on the other hand, are dominant in the Arctic but become less diverse, proportional to the other subfamilies, in the south (Oliver, 1968). The Diamesinae, too, are cold-water adapted and live mostly in streams and rivers in montane and circumpolar regions. Change in relative proportions of the various subfamilies may sometimes be seen along the course of a single river as, for example, Ward and Williams (1986) found that in the Rouge River in southern Ontario, species of Orthocladiinae predominated in the colder headwaters while species of Chironomini dominated the comparatively warmer water some 36 km downstream at the estuary.

Simuliidae are restricted to running water. In Canada, *Prosimulium* is one of the major genera of blackflies, and its species show a variety of distribution patterns. For example, *Prosimulium pleurale* has a distinct distribution occurring only in the east and west, usually in medium to large streams and rivers. *P. onychodactylum* is a western endemic occurring in a wide variety of small streams and large rivers, though generally on sand or gravel substrates. *P. fontanum* is an eastern endemic and is confined to small, cold stenothermal bog or spring-fed streams usually in shady forests. *P. ursinum* has a northern distribution from the Yukon to Baffin Island and the Labrador coast and is typically found in small, cold trickles. In contrast, *P. magnum* is a southern Nearctic species, restricted in Canada to streams along the Niagara Escarpment of southern Ontario (Paterson, 1970).

Canadian species of Blephariceridae live mostly in swift, cool, western montane streams, but one species occurs in the low-lying Saskatchewan River. The Deuterophlebiidae similarly live in fast mountain streams in the west, north to the Yukon (Downes, 1979).

The craneflies, Tipulidae, occupy a number of different types of running-water habitats in Canada, including rapid streams (*Antocha, Hesperoconopa, Cryptolabis*), stream margins (many genera), submerged logs, and vertical rock faces kept moist by percolating or slow-flowing water (*Limonia, Orimarga, Elliptera*; Alexander and Byers, 1981).

Oligochaeta. In running water, aquatic oligochaetes (worms) are generally found in areas of low flow where silt and mud accumulate—the amount and quality of the organic component frequently is an important factor in determining species distribution (Brinkhurst and Cook, 1974). However, there are some species, particularly among the Naididae, that are found on fast-water riffles in sand and gravel substrates. In Canada, many species are widespread, for example *Nais variabilis, N. elinquis,* and *Dero digitata* among the Naididae; *Tubifex tubifex, Limnodrilus hoffmeisteri,* and *Aulodrilus americanus,* among the Tubificidae; and *Stylodrilus heringianus* and *Lumbrilicus variegatus* among the Lumbriculidae. Western endemics include *Rhynchelmis elrodi* (Lumbriculidae), *Specaria fraseri* (Naididae), and *Ilyodrilus mastix* (Tubificidae), while solely eastern species are rare but include *Tubifex nerthus.* A number of the more widespread species extend into the Yukon and Northwest Territories, but *Pristina idrensis* (Naididae) appears to be restricted to northern Canada.

Mollusca. Running-water clams and snails are generally more common in areas of shallow depth, slow current, and relatively warm temperature, particularly where there is abundant macrophyte growth. However, in Canada, some species are cold-water adapted, for example, *Sphaerium nitidum,* which occurs in many transarctic localities to the south (Clarke, 1973). In contrast, many species appear restricted to individual drainage areas. The Pacific Coastal region contains a distinct molluscan fauna characterized by, for example, the snail *Physa columbiana* and the clams *Anodonta nuttaliana* and *Gonidae angulata.* The Yukon has a few endemic species thought to have been derived from the Beringian Refugium. Over 100 species occur in the Hudson Bay and Arctic drainage areas; species such as *Stagnicola arctica* and several in the genus *Pisidium* occur as far north as the southern parts of Baffin and Victoria Islands. In the subarctic, distributional patterns of some species are correlated with regions of different phytogeography. In southern Manitoba, the Red River system has, comparably, a high species diversity thought to be the consequence of its proximity to, and intermittent connection with, the upper Mississippi–Missouri river systems. The St. Lawrence–Great Lakes region of the Atlantic Ocean drainage area has a rich molluscan fauna, and again, this is linked to past connections with the Mississippi-Ohio system. Farther east, the Atlantic Coastal Plain contains characteristic species such as *Anodonta cataracta, Margaretifera margaretifera,* and *Lampsilis ochracea*; the first two species are common in sand and mud substrates in small to medium-sized streams, while the latter is found on similar substrates in the slow-moving parts of rivers (Clarke, 1981).

Amphipoda. Canadian freshwater amphipods are primarily cold-stenotherms. The common surface-water genera are *Crangonyx* and *Gammarus.* These live mostly in waters that are cool in summer, especially springs and spring-fed streams(Bousfield, 1958). *Stygobromus* is a common genus in subterranean waters. During the Mesozoic and much of the Cenozoic, many species in the *Crangonyx* group are thought to have been distributed widely in Canada. During Pleistocene glaciation, they were pushed south, but some of the hardier ones moved north again as the glaciers retreated. Their northern limit, today, seems to be that of the hard-water drainage basins, high water tables and/or extensive winter freezing; a few species have colonized soft-water streams in the boreal forest and on the Canadian Shield (Bousfield, 1958). The genus *Gammarus* is Holarctic in distribution. In Canada, the common species are: *G. lacustris,* which occurs primarily in cold lentic habitats but also in their outflows from Baffin Island and the Hudson Bay drainage area west to the Yukon and as far south as 40°N; *G. fasciatus,* which occurs in large, slow-flowing, somewhat turbid, summer-warm rivers throughout the Great Lakes and St. Lawrence drainage systems; and *G. pseudolimnaeus,* which in Canada, occurs in Ontario and western Quebec in large rivers but also in tributary streams and springs where it breeds (Bousfield, 1958; Barnard and Barnard, 1983). Although primarily a neotropical genus, one species of *Hyalella* has spread transcontinentally throughout a wide variety of Canadian lotic and lentic habitats as far north as the tree line.

Decapoda. Two of the four subfamilies of the Astacidae (crayfishes) are represented in Canada. On the west coast, the genus *Pacifasticus* (Cambaroidinae) is found in lakes and rivers. It is thought to have been derived from an ancestor in east Asia that migrated along the shores of the North Pacific Ocean (Ortmann, 1902). In eastern Canada, the subfamily Cambarinae is represented by the genera *Cambarus* and *Orconectes,* which are most abundant and diverse in southern Ontario (4 and 5 species, respectively). Ortman (1902) proposed that the Cambarinae were derived from another part of the ancestral stock of *Pacifasticus* that had managed to cross the Continental Divide into Mexico and from there had spread in a northeasterly direction. Speciation took place during this dispersion, giving rise to *Procambarus* and *Cambarellus* in the south and *Cambarus* and *Orconectes* in the north. The first two genera do not occur in Canada, and the diversities of the latter two are low compared with the United States, which has about 43 and 59 species in these genera, respectively (Chace and others, 1959). *Orconectes* is thought to have evolved at the confluence of the Mississippi, Missouri, and Ohio Rivers, while *Cambarus* may have arisen in the Ozarks and the southern Appalachian Mountains (Pennak, 1953). The Ontario crayfishes are all recent immigrants from glacial refugia in the central and eastern United States (Crocker and Barr, 1968). The present distribution of the species in the province is thought to be linked to their individual abilities to cope with water currents. Superior streamlining abilities of *C. bartoni* and *O. rusticus* correlate with their occurrence in fast-flowing rivers, while those species that fare poorly in water currents, such as *O. immunis, C. fodiens, O. virilis,* and *O. obscurus,* are all found predominantly in slow- or still-water habitats. The ability to move effectively through fast water may account for the rapid and widespread expansion of *O. rusticus* in Ontario (Maude and Williams, 1983).

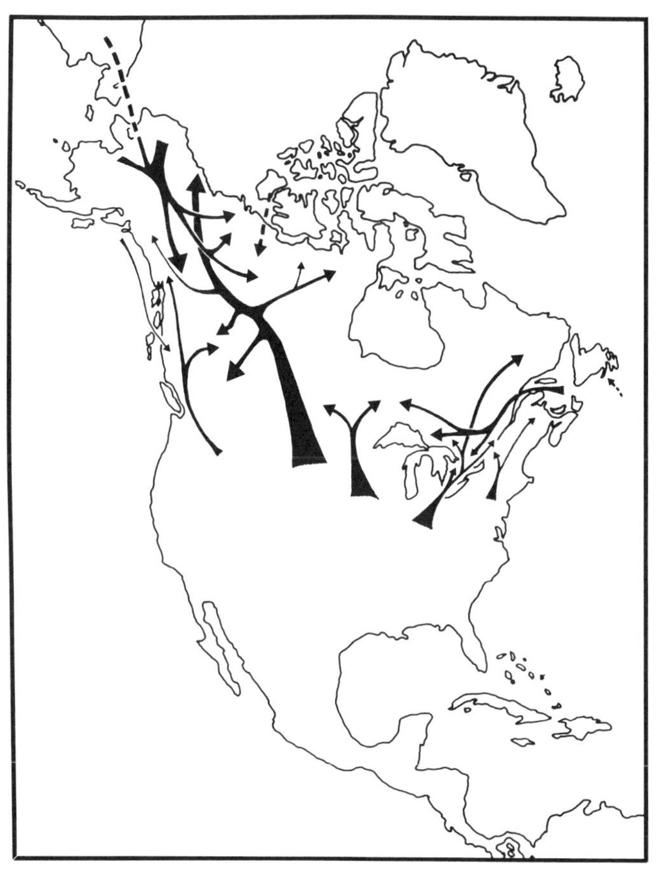

Figure 5. Summary of colonization routes of the Canadian running water biota after the retreat of the Wisconsinan Ice Sheet. Width of the arrows gives an approximation of the importance of the various routes. Broken lines indicate more speculative routes. Based on McPhail and Lindsey, 1970; Scott and Crossman, 1973; plus a variety of other sources as indicated in the text.

CONCLUDING REMARKS

As exemplified by the fishes and many of the invertebrates, the diversity of Canadian running-water faunas is low compared with unglaciated parts of North America. This is due mainly to the relatively short time that has elapsed since the last glaciation. Nevertheless, Canadian rivers contain species that represent most of the taxa found worldwide in running water.

Figure 5 summarizes the postglacial colonization routes and emphasizes the importance of major United States river systems (such as the Columbia, Mississippi, Missouri, and Ohio) as sources of the Canadian lotic biota. Other elements in the biota have been derived from more northerly refugia, particularly Beringia.

It is clear that different colonization rates have, in part, been responsible for individual species distribution patterns, but physiography, temperature, and phytogeography have had pronounced influences also. There are, for example, many lotic species that are endemic in the west, especially in streams on the Pacific side of the Rocky Mountains. Again, some cold-adapted species thrive in Arctic rivers and are transcontinental at these high latitudes, while other, warm-adapted forms have barely gained a foothold in the most southerly parts of the country. Some areas, such as the island of Newfoundland, have a depauperate biota even though the climate is mild. Such anomalies may simply reflect the short time since glacial retreat coupled with a saltwater barrier, although Larson and Colbo (1983) have cautioned against these being regarded as the only reasons and have suggested that reduced habitat diversity in this region may limit faunal diversity.

REFERENCES

Ager, T. A., 1982, Vegetational history of western Alaska during the Wisconsin glacial interval and the Holocene, *in* Hopkins, D. M., Matthews, J. V., Schweger, C. E., and Young, S. G., eds., Paleoecology of Beringia: New York, Academic Press, p. 75–93.

Alexander, C. P., and Byers, G. W., 1981, Tipulidae, *in* Manual of Nearctic Diptera, v. 1: Agriculture Canada Monograph Research Branch 27, p. 153–190.

Barnard, J. L., and Barnard, C. M., 1983, Freshwater Amphiboda of the World; v. 1, Evolutionary patterns: Mt. Vernon, Virginia, Hayfield Associates, 259 p.

Barton, D. R., and Wallace, R. R., 1980, Ecological studies of the aquatic invertebrates of the Alberta Oil Sands Environmental Research Programme Study Area of northeastern Alberta: Freshwater Institute, Fisheries and Environment Canada Report 88, 216 p.

Bousfield, E. L., 1958, Freshwater amphipod crustaceans of glaciated North America: Canadian Field Naturalist, v. 72, p. 55–113.

Brinkhurst, R. O., and Cook, D. G., 1974, Aquatic earthworms (Annelida, Oligochaeta) *in* Hart, C. W., and Fuller, S.L.H., eds., Pollution Ecology of Freshwater Invertebrates: New York, Academic Press, p. 143–156.

Campbell, J. M., and 8 others, 1979, Coleoptera, *in* Danks, H. V., ed., Canada and its Insect Fauna: Entomological Society of Canada Memoir 108, p. 357–388.

Chace, F. A., Mackin, J. G., Hubricht, L., Banner, A. H., and Hobbs, H .H., 1959, Malacoastraca, *in* Edmonson, W. T., ed., Freshwater Biology: New York, John Wiley and Sons, p. 869–901.

Clarke, A. H., 1973, The Freshwater Molluscs of the Canadian Interior Basin: Malacologia, v. 13, p. 1–509.

—— , 1981, The Freshwater Molluscs of Canada: Ottawa, National Museum of Natural Sciences, National Museums of Canada, 446 p.

Corbet, P. S., 1966, Parthenogenesis in caddisflies (Trichoptera): Canadian Journal of Zoology, v. 44, p. 981–982.

Crocker, D. W., and Barr, D. W., 1968, Handbook of the crayfishes of Ontario: Toronto, University of Toronto Press, 158 p.

Danks, H. V., 1979, Canada and its Insect Fauna: Entomological Society of Canada Memoir 108, 573 p.

—— , 1981, Arctic Arthropods: Ottawa, Entomological Society of Canada, 608 p.

Downes, J. A., 1979, Infraorder Psychodomorpha, *in* Danks, H. V., ed., Canada and its Insect Fauna: Entomological Society of Canada Memoir 108, p. 397.

Dunne, T., and Leopold, L. B., 1978, Water in environmental planning: San Francisco, W. H. Freeman and Company, 818 p.

England, J., and Bradley, R. S., 1978, Past glacial activity in the Canadian high arctic: Science, v. 200, p. 265–270.

Evans, E. D., 1984, Megaloptera and aquatic Neuroptera, *in* Merritt, R. W., and Cummins, K. W., eds., An Introduction to Aquatic insects: Dubuque, Iowa, Kendal/Hunt Publishing Company, 722 p.

Fernando, C. H. and Galbraith, D. F., 1973, Seasonality and dynamics of aquatic insects colonizing small habitats: Verhandlungen Internationale Vereinigung fur Theoretische und Angewandte Limnologie, v. 18, p. 1564–1575.

Glass, L. W., and Bovbjerg, R. V., 1969, Density and diversity in laboratory populations of caddisfly larvae (*Cheumatopsyche,* Hydropsychidae): Ecology, v. 50, p. 1082–1084.

Grant, D. R., 1977, Glacial style and ice limits, the Quaternary stratigraphic record, and changes of land and ocean level in the Atlantic provinces, Canada: Geographie et Physique Quaternaire, v. 31, p. 247–260.

Grant, R. G., and Patrick, R., 1970, Tinicum Marsh as a water purifier; Two studies of Tinicum Marsh: Conservation Foundation, p. 105–123.

Hallam, J. C., 1959, Habitat and associated fauna of four species of fish in Ontario streams: Journal of the Fisheries Research Board of Canada, v. 16, p. 147–173.

Harper, P. O., 1979, Plecoptera, *in* Danks, H. V., ed., Canada and its Insect Fauna: Entomological Society of Canada Memoir 108, p. 311–313.

—— , 1981, Ecology of streams at high latitudes, *in* Lock, M. A., and Williams, D. D., eds., Perspectives in Running Water Ecology: New York, Plenum Publishers, p. 313–337.

Hopkins, D. M., 1972, The paleogeography and climatic history of Beringia during late Cenozoic time: Internord, v. 12, p. 121–150.

Horton, R. A., 1945, Erosional development of streams and their drainage basins—hydrophysical approach to quantitative morphology: Geological Society of America Bulletin, v. 56, p. 275–370.

Hulten, E., 1937, Outline of the history of arctic and boreal biota during the Quaternary Period: Stockholm, Bokforlags Aktiebolaget Thule, 168 p.

Husmann, S., 1966, Versuch einer okologischen Gliederung des Interstitiellen Grundwassers in Lebensbereiche eigener Pragung: Archiv fur Hydrobiologie, v. 62, p. 231–268.

Hydrological Atlas of Canada, 1978: Ottawa, Fisheries and Environment of Canada Publication EN 37–26/1978, Map 22, scale.

Hynes, H.B.N., 1970, The Ecology of Running Waters; Toronto, University of Toronto Press, 555 p.

Ide, F. P., 1955, Two species of mayflies representing southern groups occurring at Winnipeg, Manitoba (Ephemeroptera): Annals of the Entomological Society of Canada, v. 48, p. 15–16.

Illies, J., and Botosaneanu, L., 1963, Problemes et methodes de la classification et de la zonation ecologique des eaux courantes, considerees surtout du point de vue faunistique: Mitteilungen Internationale Vereinigung fur Theoretische and Angewandte Limnologie, v. 12, p. 1–57.

Keating, K. I., 1976, Algal metobolite influence on bloom sequence in eutrophied fresh water ponds: Washington, D.C., Office of Research and Development Corvallis Environmental Research Laboratory, EPA 600/3–76–081, July 1976, 109 p.

Kevan, D. K., McE., 1979, Megaloptera, *in* Danks, H. V., ed., Canada and Its Insect Fauna: Entomological Society of Canada Memoir 108, p. 351–352.

Kuehne, R. A., and Barbour, R. W., 1983, The American Darters: Lexington, University of Kentucky Press, 177 p.

Larson, D. J., and Colbo, M. H., 1983, The aquatic insects, biogeographical considerations, *in* South, G. R., ed., Biogeography and Ecology of the Island of Newfoundland: The Hague, W. Junk, p. 593–677.

Lehmkuhl, D. M., 1972, *Baetisca* (Ephemeroptera, Baetiscidae) from the western interior of Canada with notes on the life cycle: Canadian Journal of Zoology, v. 50, p. 1015–1017.

—— , 1976a, Mayflies: Blue Jay, v. 34, p. 70–81.

—— , 1976b, Additions to the taxonomy, zoogeography, and biology of *Analetris eximia* (Acanthametropodinae; Siphloneuridae; Ephemeroptera): Canadian Entomologist, v. 108, p. 199–207.

—— , 1979, Ephemeroptera, *in* Danks, H. V., ed., Canada and its insect fauna: Entomological Society of Canada Memoir 108, p. 305–308.

Leopold, L. B., Wolman, M. G., and Miller, J. P., 1964, Fluvial processes in Geomorphology: San Francisco, W. H. Freeman and Company, 522 p.

Livingstone, D. A., 1963, Alaska, Yukon, Northwest Territories, and Greenland, *in* Frey, D. G., ed., Limnology in North America: Madison, The University of Wisconsin Press, p. 559–579.

Macan, T. T., 1974, Running water: Mitteilungen Internationale Vereinigung fur Theoretische und Angewandte Limnologie, v. 20, p. 301–321.

Maguire, B., 1963, The passive dispersal of small aquatic organisms and their colonization of isolated bodies of water: Ecological Monographs, v. 33, p. 161–185.

Manuel, K. L., and Nimmo, A. P., 1984, The caddisfly genus *Ylodes* in North America (Trichoptera; Leptoceridae), *in* Morse, J. C., ed., Proceedings of the Fourth International Symposium on Trichoptera: The Hague, W. Junk, Series Entomologica, v. 30, p. 219–224.

Matthews, J. V., 1977, Tertiary Coleoptera fossils from the North American arctic: Coleopterist's Bulletin, v. 1, p. 297–298.

—— , 1979, Tertiary and Quaternary environments; historical background for an analysis of the Canadian insect fauna, *in* Danks, H. V., ed., Canada and its insect fauna: Entomological Society of Canada Memoir 108, p. 31–86.

Maude, S. H., and Williams, D. D., 1983, Behavior of crayfish in water currents—hydrodynamics of eight species with reference to their distribution patterns in southern Ontario: Canadian Journal of Fisheries and Aquatic Sciences, v. 40, p. 68–77.

McPhail, J. D., and Lindsay, C. C., 1970, Freshwater fishes of Northwestern Canada and Alaska: Fisheries Research Board of Canada Bulletin 173, 81 p.

Miller, R. R., 1981, Coevolution of deserts and pupfishes (Genus *Cyprinodon*) in the American Southwest, *in* Naiman, R. J., and Soltz, D. C., eds., Fishes in North American Deserts: New York, Wiley and Sons, p. 39–94.

Mundie, J. H., McKinnell, S. M., and Traber, R. E., 1983, Responses of stream zoobenthos to enrichment of gravel substrates with cereal grain and soybean: Canadian Journal of Fisheries and Aquatic Sciences, v. 40, p. 1702–1712.

Northcote, T. G., and Larkin, P. A., 1963, Western Canada, *in* Frey, D. G., ed., Limnology in North America: Madison, University of Wisconsin Press, p. 451–485.

Odum, E. P., McIvor, C. C., and Smith, T. J., III, 1982, The ecology of the mangroves of Southern Florida—A community profile: Bureau of Land Management Fish and Wildlife Service Biology Series Program, 130 p.

Oliver, D. R., 1968, Adaptations of Arctic Chronomidae: Annales Zoologici Fennica, v. 5, p. 111–118.

—— , 1981, Chironomidae, Manual of Nearctic Diptera, v. 1: Research Branch, Agriculture Canada Monograph 27, p. 423–458.

Ortmann, A. E., 1902, The geographical distribution of the freshwater decapods and its bearing on ancient geography: Proceedings of the American Philosophical Society, v. 41, p. 267–400.

Paterson, B. V., 1970, The *Prosimulium* of Canada and Alaska: Entomological Society of Canada Memoir 69, 216 p.

Patrick, R., 1948, A proposed biological measure of stream conditions based on a survey of Conestoga Basin, Lancaster County, Pennsylvania: Philadelphia, Proceedings of the Academy of Natural Sciences, p. 277–341.

—— , 1961, A study of the numbers and kinds of species found in rivers in the United States: Philadelphia, Pennsylvania, Proceedings of the Academy of Natural Sciences, v. 112, p. 129–293.

Pennak, R. W., 1953, Freshwater Invertebrates of the United States: New York, Ronald Press, 369 p.

Polhemus, J. T., 1984, Aquatic and semiaquatic Hemiptera, *in* Merritt, R. W., and Cummins, K. W., eds., An Introduction to the Aquatic Insects: Dubuque, Iowa, Kendal/Hunt Publishing Company, p. 231–260.

Proctor, V. W., 1957, Studies of algal antibiosis using *Haematococeus* and *Chlamydomonas*: Limnology and Oceanography, v. 2, p. 125–139.

—— , 1964, Viability of crustacean eggs recovered from ducks: Ecology, v. 45, p. 656–658.

Rice, T. R., 1954, Biotic influences affecting population growth of planktonic algae: U.S. Fish and Wildlife Service Fisheries Bulletin 87, p. 227–245.

Ricker, W. E., 1934, An ecological classification of certain Ontario streams: Toronto, University of Toronto Study Series in Biology, v. 37, p. 1–114.

—— , 1964, Distribution of Canadian stoneflies: Gewasser und Abwasser, v. 34–35, p. 50–71.

Ross, H. H., 1956, Evolution and classification of the mountain caddisflies: Urbana, University of Illinois Press, 21 p.

—— , 1965, Pleistocene events and insects, *in* Wright, H. E., and Frey, D. G., eds., The Quaternary of the United States: Princeton, New Jersey, Princeton University Press, p. 583–596.

Ross, H. H., and Ricker, W. E., 1971, The classification, evolution, and dispersal of the winter stonefly genus *Allocapnia:* Illinois Biological Monograph 45, 166 p.

Schwoerbel, J., 1961, Uber die Lebensbedingungen und die Besiedlung des hyporheischen Lebensraumes: Archiv fur Hydrobiologie Supplement, v. 25, p. 182–214.

Scott, W. B., and Crossman, E. J., 1973, Freshwater Fishes of Canada: Fisheries Research Board of Canada Bulletin 184, 966 p.

Scudder, G.G.E., 1979, Hemiptera, *in* Danks, H. V., ed., Canada and Its Insect Fauna: Entomological Society of Canada Memoir 108, p. 329–348.

Smith, M. W., 1963, The Atlantic Provinces of Canada, *in* Frey, D. G., ed., Limnology in North America: Madison, University of Wisconsin Press, p. 521–534.

Sprules, W. M., 1947, An ecological investigation of stream insects in Algonquin Park, Ontario: Toronto, University of Toronto Studies in Biology Series 56, 81 p.

Stanford, J. A., and Gaufin, A. R., 1974, Hyporheic communities in two Montana rivers: Science, v. 185, p. 700–702.

Vannote, R. L., and Sweeney, B. W., 1985, Larval feeding and growth rate of the stream cranefly, *Tipula abdominalis,* in gradient of temperature and nutrition: Philadelphia, Pennsylvania, Academy of Natural Sciences, v. 137, p. 118–128.

Vannote, R. L., Marshall, G. W., Cummins, K. W., Sedall, J. R., and Cushing, C. F., 1980, The river continuum concept: Canadian Journal of Fisheries and Aquatic Science, v. 37, p. 130–137.

Vincent, J. S., and Gauthier, R. C., 1976, Inventaire des depots de surface de L'Isle Banks, District de Franklin: Geological Survey of Canada Paper 76–1A, 27 p.

Walker, E. M., 1953, The Odonata of Canada and Alaska, v. 1: Toronto, University of Toronto Press, 292 p.

Ward, A. F., and Williams, D. D., 1986, Longitudinal zonation and food of larval chironomids (Insecta, Diptera) along the course of a river in temperate Canada: Holarctic Ecology, v. 9, p. 48–57.

Waters, T. F., 1966, Production rate, population density, and drift of a stream invertebrate: Ecology, v. 2, p. 595–604.

White, D. S., Brigham, W. U., and Doyer, J. T., 1984, Aquatic Coleoptera, *in* Merritt, R. W., and Cummins, K. W., eds., An Introduction to the Aquatic Insects: Dubuque, Iowa, Kendal/Hunt Publishing Company, p. 361–437.

Wiggins, G. B., 1979, Trichoptera, *in* Danks, H. V., ed., Canada and its Insect Fauna: Entomological Society of Canada Memoir 108, p. 482–484.

Wiggins, G. B., and Parker, C. R., 1984, Beringian Trichoptera, a preliminary report, *in* Morse, J. C., ed., Proceedings of the Fourth International Symposium on Trichoptera: The Hague, W. Junk Series Entomologica, v. 30, p. 445–446.

Williams, D. D., 1979, Aquatic Habitats of Canada and their Insects, *in* Danks, H. V., ed., Canada and Its Insect Fauna: Entomological Society of Canada Memoir 108, p. 211–214.

——, 1981, Migrations and Distributions of Stream Benthos, *in* Lock, M. A., and Williams, D. D., eds., Perspectives in Running Water Ecology: New York, Plenum Press, p. 155–207.

——, 1984, The Hyporpheic Zone as a Habitat for Aquatic Insects and Associated Arthorpods, *in* Resh, V. H., and Rosenberg, D. M., eds., The Ecology of Aquatic Insects: New York, Praeger Scientific Press, p. 430–455.

Williams, D. D., and Hynes, H.B.N., 1974, The occurrence of benthos deep in the substratum of a stream: Freshwater Biology, v. 4, p. 233–256.

——, 1977, The ecology of temporary streams: Internationale Revue der gesamten Hydrobiologie, v. 62, p. 53–61.

Williams, D. D., and Williams, N. E., 1987, Trichoptera from cold freshwater springs in Canada; records and comments: Proceedings of the Entomological Society of Ontario, v. 118, p. 13–23.

Manuscript Accepted by the Society August 17, 1988

ACKNOWLEDGMENTS

I thank the following people for information helpful in compiling this review: N. E. Williams, N. Eyles, C. H. Eyles, J. C. Ritchie, J. A. Westgate, and J. V. Matthews, G. K. Dunn provided technical assistance.

Printed in U.S.A.

The Geology of North America
Vol. O-1, Surface Water Hydrology
The Geological Society of America, 1990

Chapter 11

Movement and storage of sediment in rivers of the United States and Canada

Robert H. Meade
U.S. Geological Survey, M.S. 413, Box 25046, Denver Federal Center, Denver, Colorado 80225
Ted R. Yuzyk and Terry J. Day
Sediment Survey Section, Water Survey of Canada, Water Resources Branch, Environment Canada, Ottawa, Ontario K1A 0H3, Canada

INTRODUCTION

Sediment in river systems is of interest to earth and water scientists working at problems that span several time scales. On the longest time scale (10^8 to 10^9 years), sediment is the major form in which material is transferred from continents to oceans, and the rates at which river sediment has been produced in the geologic past are major concerns of those who study long-term geochemical cycling. At somewhat shorter time scales (10^6 to 10^8 years), the properties of the existing sedimentary rocks and deposits are the major clues to past hydrologic and geologic conditions, and, to the extent that the present really is the key to the past, it behooves us to understand how today's observable conditions influence today's river sediments. On a more secular time scale, sediment in rivers is of immediate concern as a reflection of soil erosion, as a major design consideration for reservoir sedimentation, river navigation, and other engineering works, as a transporter of various materials (toxic and otherwise) that are adsorbed onto sediment particles in river systems, and as an influence on the habitat of aquatic wildlife.

These secular-scale considerations have prompted research and monitoring activities during the past half century that have provided much of our basic knowledge of sediment in the river systems of North America. Studies of soil erosion and valley sedimentation became especially intensive and extensive during the 1930s; these studies continue today, mostly under the aegis of the national agricultural agencies of the United States and Canada. The construction of large dams and reservoirs for hydropower and irrigation, beginning in the 1920s and 1930s and peaking during the 1950s and 1960s, has led to the accumulation of long records of river-sediment discharge from which design and maintenance criteria could be derived (for example, predicting the number of years of useful life before a reservoir becomes filled with sediments). The most recent impetus to sediment research and monitoring has been the growing awareness that many persistent contaminants (heavy metals, pesticides, radionuclides) are adsorbed readily onto fine sediment particles, and that the fate of these contaminants in river systems cannot be predicted without a clear knowledge of the movement and storage of river sediments.

In this chapter, we consider the following aspects of sediment in the river systems of North America: how sediment is moved by rivers, the natural and anthropogenic factors that influence sediment yields, the storage of sediment in river systems at different time scales, the quantities of sediment transported by rivers of the continent, and the principal sites on the continental edges at which river sediment is deposited.

SUSPENDED LOAD AND BED LOAD

Sediment in rivers is usually divided into bed load and suspended load. Bed load is the sediment that is moved along a riverbed by rolling, sliding, or skipping within a few grain diameters of the bed. Suspended load is the sediment that is supported by the fluid flow of a river and maintained in suspension by the net upward component of turbulence. Bed-load transport is difficult to measure accurately but, because many of its relations to the hydraulics of water flow have been quantified by experimental studies, it is frequently predicted from known properties of the flow and the sediment in a channel. Suspended-load transport, on the other hand, is easier to measure but more difficult to predict because it depends to a significant degree on processes outside the river channel that are not always susceptible to accurate quantification.

The proportion of bed load to suspended load varies from river to river but, in general, the larger the river the smaller the proportion that is attributable to bed load. In glacier-fed streams draining small watersheds on Baffin Island (drainage areas of 200 to 400 km^2, partly covered with glacier ice), bed load accounts for 80 to 90 percent of the total sediment transport (Church,

Meade, R. H., Yuzyk, T. R., and Day, T. J., 1990, Movement and storage of sediment in rivers of the United States and Canada, *in* Wolman, M. G., and Riggs, H. C., eds., Surface water hydrology: Boulder, Colorado, Geological Society of America, The Geology of North America, v. O-1.

1972, p. 61). In a small stream that drains 500 km^2 of the Wind River Range of Wyoming, bed-load transport accounts for about 50 percent of the total sediment transport (Nordin, 1985, p. 192). In the larger Trinity River, which drains 7,700 km^2 of the northern California Coast Range, bed load accounts for about 20 percent of the total sediment load (Knott, 1974). In the Snake and Clearwater Rivers of Idaho, the Fraser River of British Columbia, and the Tanana and Susitna Rivers of Alaska, the largest North American rivers in which comprehensive series of bed-load samples have been collected, bed load accounts for 1 to 10 percent of the total sediment load (Burrows and Harrold, 1983; Emmett, 1981, p. 14; Knott and Lipscomb, 1985; McLean and Church, 1986). In the Mississippi River, in which bed load has been computed rather than sampled, bed load is estimated to account for about 5 percent of the total sediment load (Jordan, 1965, p. 86).

Although suspended load is much more significant on a continental scale, bed load continues to command a disproportionately large share of the recent literature of geomorphology and sedimentology. The present-day fascination with bed load is partly due to its role in determining many of the properties and dimensions of river channels that are important to river engineering and aquatic ecology. In the context of longer term geochemical cycling on a global scale, however, bed loads may be insignificant. Bed load accounts for less than 10 percent of the present-day transfer of sediment from the continental uplands to the continental margins, which places it within the limits of error of measurement of the suspended load of the world's large rivers. Nevertheless, a significant proportion of the geologic column consists of sandstones, and to the extent that these sandstones are alluvial in origin and contain particles coarser than about half a millimeter (the maximum sand size transported in suspension by today's large rivers), fluvial bed-load transport must have been important in older geochemical cycles.

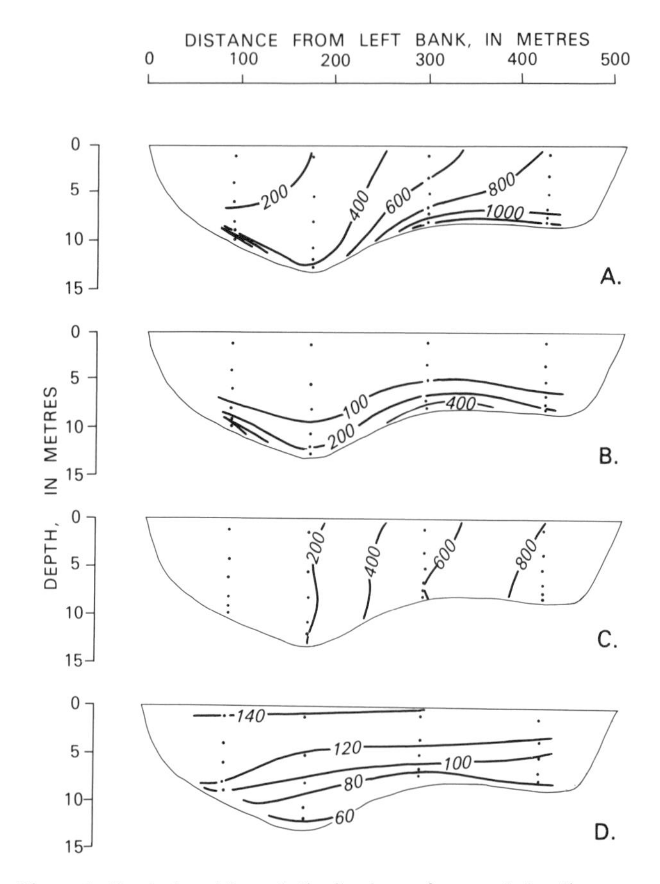

Figure 1. Vertical and lateral distributions of suspended-sediment concentration and water velocity in a cross section of the Mississippi River at St. Louis, Missouri, May 9, 1956; based on point samples and point measurements (Jordan, 1965, p. 50, 61–62). Viewer is facing downstream. The sediment-laden Missouri River enters the right side at a point 25 km upstream of St. Louis. A: Total suspended-sediment concentration, in milligrams per liter. B: Concentration of suspended sand (material coarser than 0.062 mm), in milligrams per liter. C: Concentration of suspended silt and clay (material finer than 0.062 mm), in milligrams per liter. D: Water velocity, in centimeters per second.

Measurement, computation, and estimation of suspended load

The central problem of measuring the instantaneous suspended-sediment discharge is the heterogeneity of the distributions of water velocity and suspended-sediment concentration in rivers. Water velocity is usually greatest near the water surface in midriver, and it diminishes toward the bed and banks of the channel. Suspended sediment, on the other hand, is often more concentrated near the river bed, and it diminishes upward toward the river surface. An accurate measurement of suspended-sediment discharge, which is the product of the cross-sectional area and the discharge-weighted average of the velocity and the suspended-sediment concentration, requires that these heterogeneities be integrated in some way (Nordin, 1981).

Data from a cross section of the Mississippi River at St. Louis (Fig. 1) demonstrate these heterogeneities, and also show the different distributions of the two components into which suspended load is usually subdivided: bed-material load and wash load. Bed-material load comprises the suspended particles large enough to be found in appreciable quantities on the surface of the stream bed, conventionally taken to be those suspended particles coarser than 0.063 mm. Concentrations of suspended bed-material load (Fig. 1B) are greatest near the river bed and diminish upward, a reflection of the exchange of particles between the river bed and the river suspension. Wash load (as usually defined) comprises the suspended particles so fine grained that they make up only an inappreciable fraction of the sediment on the stream bed, conventionally taken to be those particles finer than 0.063 mm. Concentrations of wash load (Fig. 1C) are often uniform from the river bed to the water surface, but they can be markedly inhomogeneous in the cross-channel direction. The cross-channel inhomogeneity shown in Figure 1C is due to the incomplete mixing of the waters of the sediment-laden Missouri River, which enters the right side of the Mississippi 25 km upriver from St. Louis.

The principal techniques and equipment that have been

developed during the past 50 years to resolve these inhomogeneities involve two different approaches: point sampling and depth-integrated sampling. In point sampling, velocity and suspended-sediment concentration are measured at numerous individual points in a river cross section (as shown in Fig. 1), and the total suspended-sediment discharge is computed from the measurements. The process of depth-integrated sampling, on the other hand, provides a mechanical integration of the distributions of velocity and suspended-sediment concentration, and it usually yields a single determination of discharge-weighted suspended-sediment concentration that can be combined with the measured water discharge to compute the suspended-sediment discharge. Most measurements of the sediment loads of the rivers of Canada and the United States are based on depth-integrated sampling. Equipment and techniques for sampling and computing suspended-sediment loads are described in the manuals prepared by Guy (1969), Guy and Norman (1970), and Porterfield (1972), in the summary report by Stichling (1974), and in the articles by Feltz and Culbertson (1972) and Nordin and others (1983). In practice, even the most careful measurements of instantaneous suspended-sediment discharge should be considered accurate only within ±10 percent. Considerably less accuracy can be expected where the suspended sediment contains large proportions of sand (Burkham, 1985; Guy and Norman, 1970, p. 40–41).

Once a program of accurate measurements of suspended sediment has been established, the next step involves estimating the long-term average sediment discharge. The most accurate data for such an estimate are those collected daily (or more frequently when water discharge is changing rapidly) for a period of decades. If the period of record has been one of stable conditions in the watershed (that is, if the record has the property the stochastic hydrologists call "stationarity"; see Nordin, 1985, p. 197–201), the record should yield the average sediment discharge of that river under those conditions, and the accuracy of the average would be related directly to the length of the period of record. In one example, a record of daily sediment discharge that was continuous for 15 consecutive years yielded an estimate of mean annual sediment discharge whose standard error was 20 to 25 percent (Day and Spitzer, 1985, p. 48). If the period has been one of changing conditions in the watershed ("non-stationarity"), the data may not be useful for computing long-term average sediment discharges. Both types of record are worth collecting, however, because they both have predictive value: the first type (stationary) as a guide to the magnitude and frequency of sediment-discharge events under the prevailing hydrologic conditions, and the second type (nonstationary) as a guide to the effects of changing watershed conditions (land use, vegetative cover) or storage conditions (dams) on sediment yields.

Collecting daily records of sediment discharge for periods of decades, however, is expensive and in many cases impractical. The establishment and maintenance of long-term daily sediment stations requires an uncommon blend of foresight and persistence. The more usual practice is to collect sediment data over as full a range of water discharges as possible within the time and budget available for a study, and to use these data to construct a relation between sediment discharge (or concentration) and water discharge. Such a relation is called a sediment rating curve (Colby, 1956) or sediment-transport curve (Porterfield, 1972, p. 21), and it usually takes the approximate form of a simple power function (Fig. 2A); the power function is specific to the site at which the data are collected and is not readily transferable to another site. The sediment rating curve is then combined with the available record of water discharge, which almost always is longer and more detailed than the measured sediment record, to synthesize a long-term record of sediment discharge (Colby, 1956; Miller, 1951). This method can sometimes entail errors as great as 50 percent in estimating long-term sediment discharges (Walling and Webb, 1981; Yorke and Ward, 1986). However, if it is applied with care and discrimination, especially to rivers in which the relations between sediment concentration and water discharge are demonstrably simple (Fig. 2A), the method can produce estimates whose accuracies are comparable to those based on many years of daily measurement (Abrahams and Kellerhals, 1973; Church and others, 1985, p. 141–193; Dickinson, 1981; Kellerhals and others, 1974).

Two types of departures from the simple power function in the relations between sediment concentration and water discharge have been observed in some rivers. First, the relations between sediment concentration and water discharge may show marked seasonal differences. The record shown in Figure 2B, for example, is typical of many rivers that drain the Piedmont region of the eastern United States (Guy, 1964; Meade, 1982, p. 236–237), and the seasonal differences have been attributed to differences in the sediment loads transported by runoff following snowmelt or winter rains versus those transported in response to the more intense convective rainfalls of summer. Seasonal differences in sediment rating curves in midwestern and western parts of the United States have been discussed by Colby (1956), Miller (1951), and Strand (1975). Second, the relation between sediment concentration and water discharge in many rivers, when graphed on a log-log scale with water discharge as the abscissa, forms a clockwise loop (Fig. 2C). These looped relations are usually explained as showing the "exhaustion" or "depletion" effect (Walling, 1977; Wood, 1977): fine-grained sediment, which is stored on the beds or along the banks of river channels during the recession of floods or during low-water periods, is in plentiful supply as the river begins to rise, but the stored material is soon resuspended, and it eventually becomes depleted before the river reaches its maximum water discharge. Such clockwise looped relations are typical of some of North America's largest rivers, among them the Mississippi, Fraser, and Rio Grande (Robbins, 1977, Figs. 39–41; Milliman, 1980, Fig. 4; Whitfield and Schreier, 1981; Nordin and Beverage, 1965, Figs. 5–6), and of some smaller rivers as well. Looped or seasonally variable relations between sediment concentration and water discharge can be used to compute long-term averages if one takes their loopedness, seasonality, or other discernible variability into account (Dickinson, 1981; Nolan and others, 1986).

Temporal distribution of suspended-sediment transport

The power-function relation between sediment discharge and water discharge implies that most of the river sediment is transported at the highest water discharges. Because these water discharges occur during only a few days each year, most of the sediment carried by rivers is transported during these few days (Wolman and Miller, 1960). The long-term sediment records represented in Figure 3 show that, over periods of 2 to 3 decades, moderate-sized rivers in different parts of the United States and Canada have transported about 50 percent of their annual sediment loads during 3 to 4 days each year (1 percent of the time) and 80 to 90 percent of their annual sediment loads during 10 percent of the time. Table 1 shows numerical statistics for these and other rivers of North America for which long records of daily sediment discharge are available during periods of fairly stable watershed conditions.

In most of the rivers listed in Table 1, 40 to 60 percent of the total sediment load during the period of record was discharged during 1 percent of the time, and 85 to 95 percent was discharged during 10 percent of the time. However, data from a few other rivers are listed in the table to demonstrate some other types of load-frequency distributions. In the Nenana River of Alaska, the transport of suspended sediment is closely related to the melting of glacial ice, a process which is spread over several summer months rather than being confined to a few days per year. The Tar River of North Carolina mostly drains an area of the Coastal Plain where the concentration of suspended sediment does not increase sharply with increasing water discharge (the relation is similar to that shown in Fig. 7B) and, consequently, the transport of suspended sediment was distributed over more days of the period of record: only 10 percent of the sediment load was discharged during 1 percent of the period. At the other extreme, 93 percent of the suspended sediment recorded during an 18-year period in the Santa Clara River of southern California was transported during 1 percent of the time; this distribution of sediment loads is related to the infrequency of intensive rainstorms in southern California.

A further confounding factor in estimating long-term sediment discharge is the infrequent large event that transports enormous sediment loads. Judging from the few available sediment records, Atlantic hurricanes such as Hurricane Agnes in 1972 seem to generate sediment discharges within a few days that are equivalent to about 3 years of average sediment discharge (Fig. 3A). In prairie regions, intense convective rainstorms can produce similarly large sediment loads; one such rainstorm in the basin of Wilson Creek in Manitoba moved, in only two days, a quantity of sediment equivalent to four years of average sediment discharge. In the most spectacular single sediment-discharge event preserved in the daily sediment records of North America, the Eel River of California carried more sediment in 3 days following an intense Pacific storm in December 1964 (1965 water year) than it

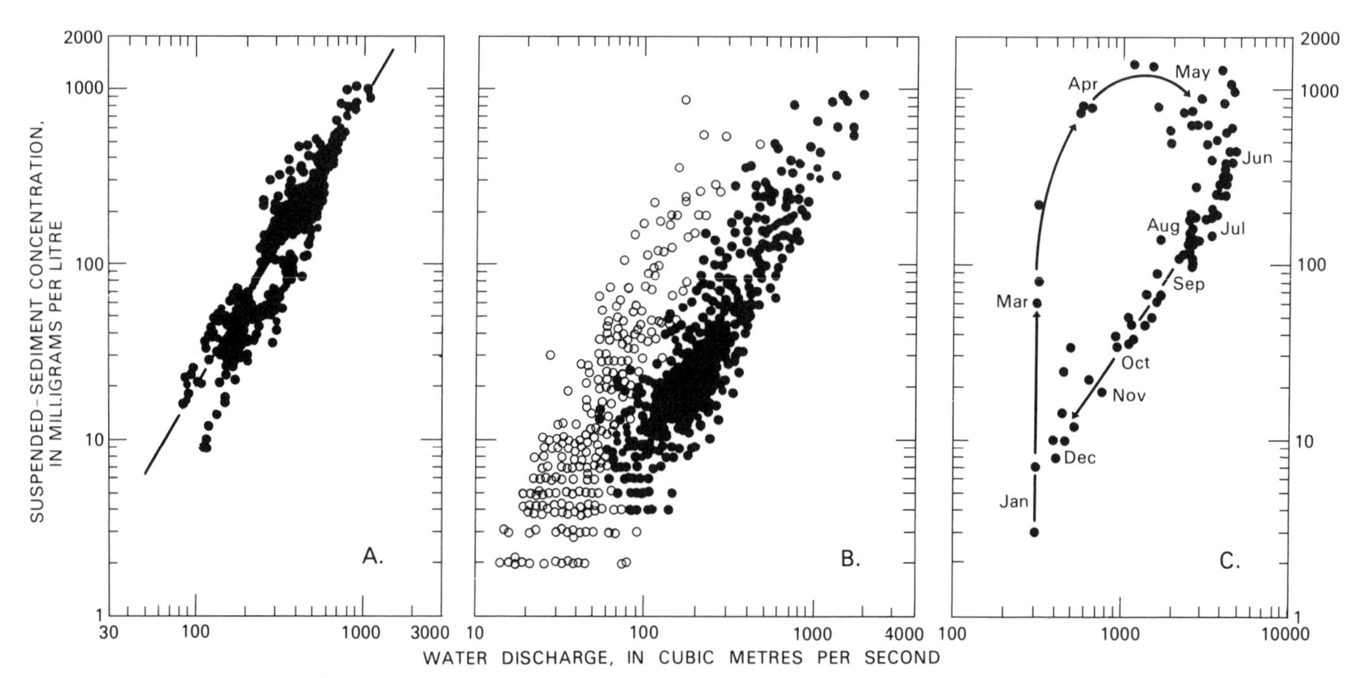

Figure 2. Relations between suspended-sediment concentrations and water discharges. A: North Saskatchewan River at Prince Albert, Saskatchewan, 1966–1968. Drainage area 131,000 km^2. (Abrahams and Kellerhals, 1973, p. 109.) B: James River at Scottsville, Virginia, water years 1951–1953, 1955. Drainage area 11,900 km^2. Data from U.S. Geological Survey Water-Supply Papers 1197, 1250, 1290, and 1400. Open circles, data from July through first heavy late-autumn storm. Dark circles, data from first late-autumn storm through May. Data from June and from hurricane-induced runoffs omitted to emphasize seasonal contrast. C: Fraser River near Marguerite, British Columbia, 1982. Drainage area 114,000 km^2. Data from Water Survey of Canada (1984, p. 87).

had carried during the previous 7 years (Fig. 3C); in 10 days, it carried an estimated quantity of sediment equivalent to that transported during 10 average years (Brown and Ritter, 1971). The influences of Hurricane Agnes and the December 1964 storm on the calculated long-term frequencies of suspended-sediment discharge are shown in the comparison of the dual entries for the Juniata and Eel rivers in Table 1. Such rare events seem to belong to different populations than the more normal year-to-year peak flows, and therefore they are not predictable by simple extrapolations based on magnitudes and frequencies of the more normal events (Nordin, 1985).

Measurement and computation of bed load

The two basic problems that complicate the direct measurement of bed-load transport in rivers are the inherent variability of bed-load movement and distortions of the natural flow fields caused by bed-load samplers. The transport of bed load by rivers is often more variable, in both time and space, than that of suspended load. At any given instant in a river cross section, bed load may be moving at rates that vary spatially and temporally by several orders of magnitude. Only rarely is the entire bed of a river cross section in a state of uniform transport at any single time (Hubbell, 1987; Hubbell and Stevens, 1986).

TABLE 1. FREQUENCY OF SUSPENDED-SEDIMENT DISCHARGE IN SELECTED RIVERS OF NORTH AMERICA
(Compiled from records of daily measurement of sediment, U.S. Geological Survey and Water Survey of Canada)

River and Gauging Station	Drainage Area (10^2 km^2)	Years of Daily Record	Period of Daily Record (water years)	Mean Annual Sediment Discharge (10^3 tonnes)	Percent of Suspended Sediment Discharged in			
					1% of time	2% of time	5% of time	10% of time
Appalachian Piedmont								
Juniata River at Newport, PA	87	33	1952-84	248	44	57	74	85
Juniata River at Newport, PA	87	32*	1952-71, 1973-84	225	41	54	72	84
Rappahannock River at Remington, VA	16	33	1952-84	90	48	64	80	89
Atlantic Coastal Plain								
Tar River at Tarboro, NC	57	10	1958-67†	112	10	17	34	52
Rocky Mountains-Central Plains								
Fifteen Mile Creek near Worland, WY	13	21	1952-72	547	60	79	94	98§
Oldman River near Brocket, Alberta	44	17	1967-83	275	59	72	87	94
Pacific Coast Ranges								
Eel River at Scotia, CA	81	23	1958-80	20,550	60	73	88	95
Eel River at Scotia, CA	81	22**	1958-64, 1966-80	14,600	48	65	85	95
Santa Clara River at Montalvo, CA	42	18	1968-85	6,020	93††	—	—	—
Glacier-fed Stream								
Nenana River near Healy, Alaska	49	14	1953-66	2,880	21	32	53	73

*Minus 1972, the year of Hurricane Agnes.
†Calendar years.
§Fifteen Mile Creek flowed only 24% of days during period 1952-72.
**Minus water year 1965, which included the record storm of December 1964.
††Santa Clara River flowed 64% of days during period 1968-85, mostly at very low discharges (irrigation return flow).

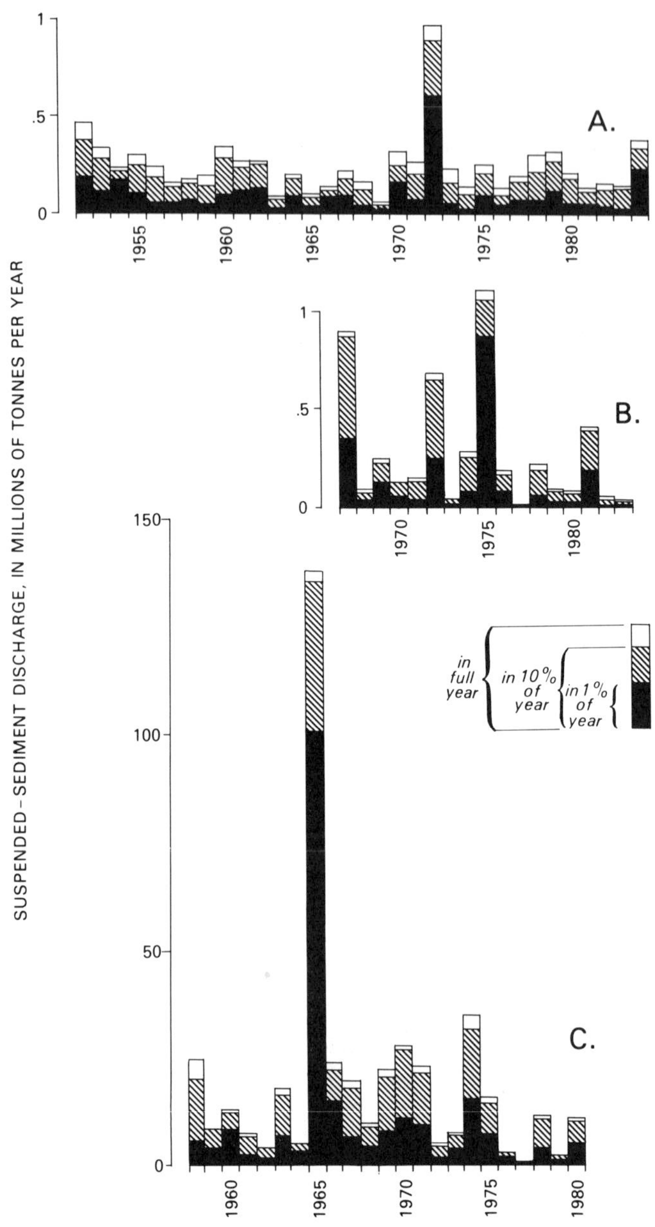

Figure 3. Annual suspended-sediment discharge of three rivers showing variations between years and frequencies within individual years. 1 tonne = 1 megagram = 10^3 kilograms = 10^6 grams. A: Juniata River at Newport, Pennsylvania. Drainage area 8,700 km^2. Compiled from data reported in U.S. Geological Survey Water-Supply Paper series, "Quality of Surface Waters of the United States. Part 1" (water years 1952–1970); and Water-Data Report series, "Water Resources Data for Pennsylvania" (water years 1971–1984). B: Oldman River near Brocket, Alberta. Drainage area 4,400 km^2. Compiled from data of Water Survey of Canada, reported by Day and Spitzer, 1985, appendix D1). C: Eel River at Scotia, California. Drainage area 8,100 km^2. Compiled from data reported in U.S. Geological Survey Water-Supply Paper series, "Quality of Surface Waters of the United States. Part 11" (water years 1958–1970), and Water-Data Report series, "Water Resources Data for California" (water years 1971–1980).

The most nearly ideal solution to the problem of sampling bed load is to construct some sort of trap that is large enough to collect the entire bed load of a river for periods of time long enough to average out the spatial and temporal variations. The largest, and perhaps the most successful, of the bed-load traps in North America was built in 1973 on the East Fork River of Wyoming and operated there every runoff season until it was dismantled in 1981. This facility was able to measure channel-wide rates of bed-load transport on the order of 3 to 5 kg/s (Emmett, 1980, p. 4–8; Nordin, 1985, p. 190–193). Such facilities are limited by necessity to small rivers.

A number of more portable bed-load samplers have been developed in Europe and North America (Hubbell, 1964). The most widely used in North America is the Helley-Smith sampler (Emmett, 1980, 1981), which shares with all other bed-load samplers the problem of changing the hydraulics of the flow field and thereby changing the very transport of bed load it is designed to sample. The comparative study by Emmett (1980) shows a scatter of plus or minus 50 to 100 percent, depending on the particle sizes being collected, between bed-load transport rates measured with the Helley-Smith sampler and those measured concurrently with the East Fork River bed-load trap.

Data from moderately large rivers of western North America show that bed-load discharge, like suspended-load discharge, can sometimes be described as a simple power function of water discharge. Examples in Figures 4A and 4B show bed-load discharge amounts to 1 to 5 percent of the suspended-load discharge. In rivers where the supply of material for bed-load transport is limited, the graphed relations between bed-load discharge and water discharge can be looped, and, depending on the location of the measuring point relative to the river reaches where the material is stored, the loops may be either clockwise or counterclockwise (Meade and others, 1981). Sediment rating curves for bed-load transport are therefore subject to many of the same types of errors as those discussed above for suspended-load transport (Church, 1985).

Longer term rates of bed-load transport can be estimated from repeated surveys of deltas that form where rivers flow into lakes or reservoirs. If the reservoir-survey data are combined with measurements of water discharge and suspended-sediment concentration made upriver and downriver of the reservoir during the time periods between reservoir surveys, an accurate picture can be inferred of the yields and discharges of bed load.

In rivers where bed-load transport rates cannot be measured readily, they are frequently computed. Among the computational methods most widely used in North America is that devised originally by Einstein (1950) and subsequently simplified (Colby and Hubbell, 1961). Other bed-load computational methods and their histories are summarized in the extensive reviews by Vanoni (1975, p. 190–230; 1984).

FACTORS INFLUENCING SEDIMENT YIELD

Sediment yield is usually expressed as the mass of suspended sediment per unit drainage area per unit time—for example, as

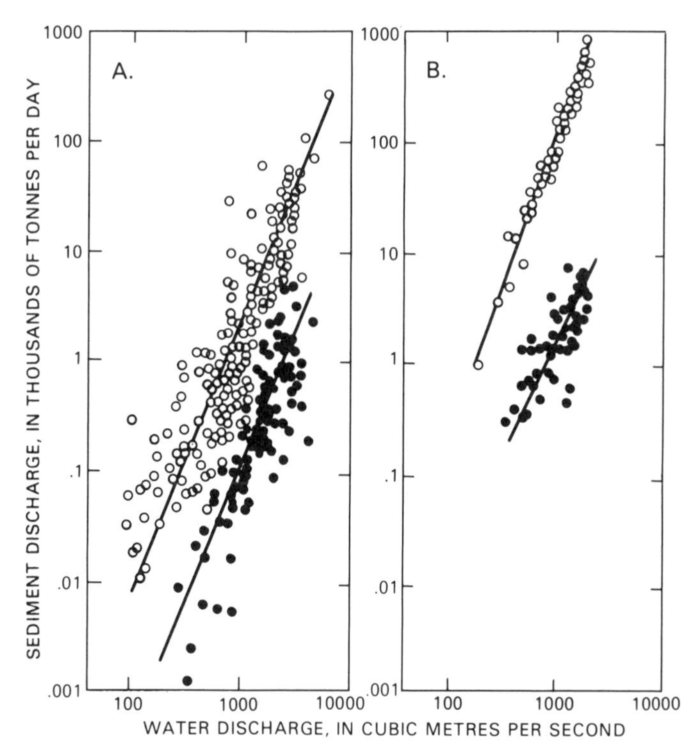

Figure 4. Comparison of measured discharges of suspended load (open circles) and bed load (dark circles) in two rivers. A: Clearwater and Snake Rivers at Lewiston, Idaho, 1972–1979 (Emmett, 1981, p. 14). Drainage area approximately 266,000 km^2. B: Tanana River near Fairbanks, Alaska, 1977–1982 (Burrows and Harrold, 1983). Drainage area approximately 65,000 km^2.

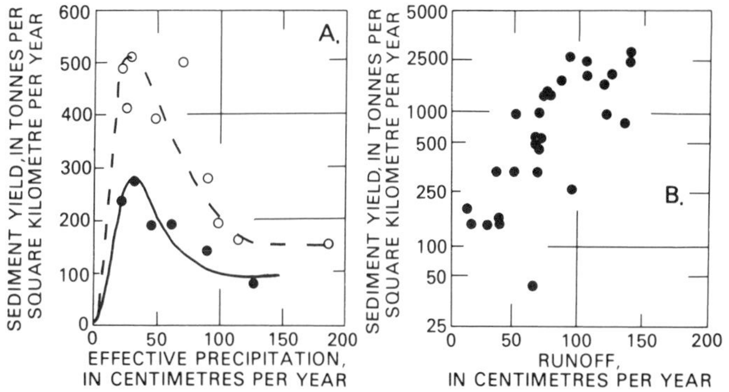

Figure 5. Relations of sediment yield to effective precipitation and runoff. A: In midcontinent and Atlantic drainages of the United States (Langbein and Schumm, 1958). "Effective precipitation" was employed by Langbein and Schumm to normalize the effects of temperature on the relation between rainfall and runoff. All other factors being equal, a given amount of rainfall will produce less runoff in warmer climates than in colder climates (Langbein and others, 1949). To lessen this source of variability in the effects of precipitation on runoff and sediment yield, Langbein and Schumm used the known values of runoff to convert the actual precipitation to "effective" precipitation, which they defined as the quantity of precipitation that would have produced the known amount of annual runoff in an area whose mean annual temperature was 50°F. Open circles and dashed line, data from reservoir surveys. Dark circles and solid line, data measured at gauging stations. The difference between the curves for sediment yields measured at gauging stations versus those from reservoir surveys can be attributed to differences in drainage-basin areas. Sediment yield in these regions, expressed in tonnes per km^2 per year, is inversely related to drainage area (see Figs. 9C and 16). Drainage areas of the gauged watersheds averaged 4000 km^2, whereas those of the reservoir watersheds averaged only 80 km^2. B: Northwestern California (Janda and Nolan, 1979). Sediment yield is plotted on a logarithmic scale.

tonnes per square kilometer per year. Present-day sediment yields are influenced by natural factors, anthropogenic factors, or (perhaps most often under today's conditions) combinations of natural and anthropogenic factors. These influences can be assigned to four major categories: climate, geology (including soil type), relief, and land use (including vegetation). In practical usage, these four categories provide the basic inputs for predictive models of soil erosion such as the universal soil-loss equation (Foster, 1977; Wischmeier and Smith, 1965), but in most areas the influences of different factors on sediment yield are difficult to discriminate from each other (Dickinson and Wall, 1977; Jordan, 1979; Osterkamp, 1976, Slaymaker and McPherson, 1977). In extrapolating modern sediment loads into the geologic past, a central concern is that of separating the natural factors from the anthropogenic. In the discussion that follows, we shall consider first the natural factors, and then the anthropogenic.

Climatic factors

The principal climatic factor influencing river sediment on a continental scale is precipitation. Sediment yields have been considered in relation to two main aspects of precipitation: the mean annual precipitation, and the frequency and magnitude of rainfall events.

Langbein and Schumm (1958) proposed the relation that has been used most widely in North America to describe and explain the effects of mean annual precipitation on sediment yield. The "Langbein-Schumm rule," as it has come to be called by many (Hadley and others, 1985, p. 57; Walling and Webb, 1983, p. 77), states that sediment yields are greatest where effective precipitation (defined in the caption to Fig. 5A) is about 300 mm/yr, and that it declines as effective precipitation becomes progressively smaller or greater than that amount. Where effective precipitation is less than 300 mm/yr, runoff is insufficient to move whatever sediment may be available. As effective precipitation increases between 300 mm/yr and about 1,000 mm/yr, sediment yields decrease because the increase in precipitation encourages progressively more vegetation, which, in turn, impedes rainfall erosion and runoff. Data to develop this relation were taken from stream-gauging stations and from reservoir surveys of small watersheds (Fig. 5A). Although the sediment yields of the gauged watersheds averaged about half those of the reservoir watersheds (Fig. 5A), the basic relation was the same in both data sets. The general aspect of this relation is supported by a subsequent summary of sediment accumulation in about 500 reservoirs in the United States made by Dendy and Bolton

(1976); their analysis, however, showed that sediment yields are greatest where runoff is 50 to 70 mm/yr, or significantly less than the 300 mm/yr suggested by Langbein and Schumm.

The Langbein-Schumm rule, as interpreted from their data, is subject to several lines of criticism (see especially Hadley and others, 1985, p. 57–62, or Walling and Webb, 1983, p. 77–84). The data from the 94 gauged watersheds and the 163 reservoirs were combined into 6 and 9 groups, respectively, according to ranges of effective precipitation, and the average sediment yield for each group was used to construct the graph. Sediment yields within the groups, however, showed a great deal of scatter; they differed by factors as great as 10, and their standard deviations were about 30 percent. Furthermore, the sediment-gaging data were selected from the station records that were available in the mid 1950s, and most of these were collected in the central part of the United States. Virtually none were available at that time from New England, the Gulf of Mexico region, or the Pacific coastal states. Similarly, the reservoir data summarized by Dendy and Bolton (1976) are strongly biased toward sites in the central states. The suspicion that the Langbein-Schumm rule might apply only to the midcontinent and Atlantic region is reinforced by data from the northern California Coast Range that show a quite different relation (Fig. 5B): a steady increase of sediment yield with mean annual runoff whose form approaches that of a power relation. Although it shows little similarity to the Langbein-Schumm relation, the relation in Figure 5B is similar to that in streams of South Island, New Zealand, which, like northwestern California, is an area of tectonically oversteepened slopes underlain by rock suites typical of subduction regions (the California and New Zealand data are presented side by side in the summary papers by Hadley and others, 1985, p. 61, and Walling and Webb, 1983, p. 85). And as Schumm (1977, p. 29–30) stated, "the Langbein-Schumm relationship tends to be valid for continental types of climates, but where there is marked seasonal variability of precipitation, sediment yields can be much greater than indicated by mean annual precipitation alone." The relations between sediment yield and annual precipitation are in need of a comprehensive reassessment, using the more recent and more complete sediment-load data that have accumulated during the past 30 years. The likely outcome of such a reassessment may well be a number of different relations, each of which characterizes a different geologic, physiographic, or climatic region of the continent.

The effect of climate on sediment yields also can be considered in the context of the intensity and frequency of floods. Knox (1983, 1984, 1985) has shown that the magnitude-frequency relations of large floods are not fixed for long periods of time. For example, in the upper Mississippi Valley between the years 1865 and 1975, the magnitude of the 2-percent-probability flood was only about 2,000 m^3/s for one 29-year period (1920–49), whereas it was greater than 3,000 m^3/s during other periods of approximately the same length (1867–1895 and 1950–1980). Knox (1983, p. 29–30) related these differences in flood magnitude and frequency to differences in the prevailing pattern of atmospheric circulation. The circulation was mainly meridional before 1895 and after 1950, during which periods polar air masses extended far south and tropical air masses extended far north, resulting in intense storms in the middle latitudes that engendered larger floods. Circulation was zonal between 1895 and 1950; strong westerly flow prevailed and intense storms were less likely. We have seen that most of the sediment in natural rivers is transported in only a few days each year (Table 1), and that large floods can play a disproportionately large role in the long-term transport of sediments (Fig. 3). Therefore, changes in the distribution of large floods in time and space should be expected to cause large differences in sediment yields.

Another manifestation of climate that affects sediment yields in North America is the presence of glaciers in the wetter, colder regions of the continent. Relations of suspended-sediment concentration to water discharge in two large drainage basins of Alaska are shown in Figure 6. The Copper River drains the south slope of the Alaska Range and the northern slopes of other ranges, all of which contain a fair number of large glaciers. The Kuskokwim River drains a larger area of southwestern Alaska that contains only a few glaciers of small to moderate size in some of its headwaters. Although the discharges of water from these two basins are approximately equal (within about 10 percent), the discharges of sediment differ by a factor near 10. Although the Copper River drains an area of comparatively modest size (about 70,000 km^2 overall), it discharges the third-largest sediment load to the ocean among the rivers of North America. This unusually large sediment yield is attributable to the erosive activity of the glaciers in the Copper River basin.

Even after glaciers have receded or have disappeared altogether, their effects on sediment yield may continue to be large.

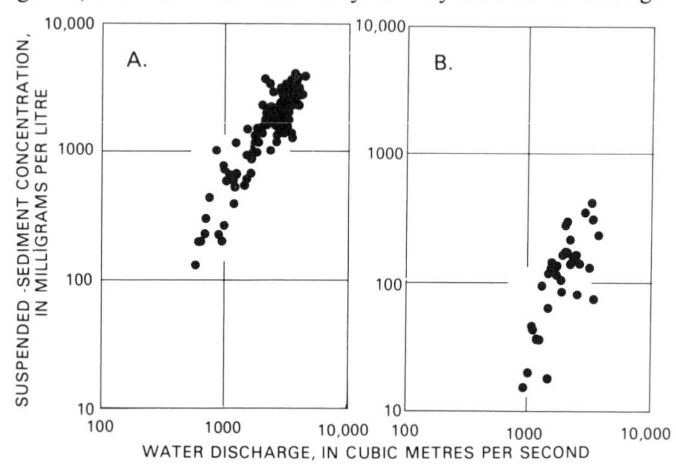

Figure 6. Relations between suspended-sediment concentrations and water discharge in two rivers of Alaska, showing the effects of glaciers. Compiled from data reported in U.S. Geological Survey Water-Supply Papers 1500, 1959, 1966, and 1996, and in Water-Data Report Series, "Water Resources Data for Alaska." A: Copper River at Chitina, May–October 1957, 1964–1965, 1978–1984. Drainage area 53,400 km^2; 29-year (1955–1984) mean water discharge 1,070 m^3/s. B: Kuskokwim River at Crooked Creek, May–October, 1966, 1975–1984. Drainage area 80,500 km^2; 33-year (1951–1984) mean water discharge 1,160 m^3/s.

Glacial drift left behind on a landscape as glaciers recede can be eroded and transported away at rates considerably greater than material supplied by normal subaerial weathering processes. Church and Ryder (1972) presented convincing evidence from parts of Canada, particularly Baffin Island and British Columbia, to show that sediment yields may be abnormally large for some time after the recession of glacier ice. However, given enough time for the removal of most of the loose drift material, sediment yields will eventually decrease.

Information on the effects of present-day precipitation, vegetation, flood intensities, and glaciers has been used to infer patterns of erosion and sedimentation during earlier periods of the Earth's history. Schumm (1968) extrapolated the Langbein-Schumm relation to infer rainfall-runoff relations and their possible effects on alluvial sedimentation during four periods in the history of terrestrial plant life: (1) before the appearance of land plants, and after the appearances of (2) primitive land plants, (3) flowering plants and conifers, and (4) grasses. "During prevegetation time, bed-load channels moved coarse sediments from their sources and spread them as sheets on piedmont areas. With increased plant cover, alluvial deposits were stabilized, but large floods caused periodic flushing of sediment from the system, thereby creating cyclic sedimentary deposits. The influence of climate change on the volume and type of sediment moved from an erosional system became more pronounced as the effect of vegetation on the hydrologic cycle increased. Finally, with the appearance of grasses during the Cenozoic Era, the relations between climate, vegetation, erosion, and runoff became much as today except for the influence of man" (Schumm, 1968, p. 1573).

Inferences concerning climatic effects during more recent epochs of Earth history are less speculative; the data base for the more recent epochs is more complete, and the inferences drawn from it presumably carry a greater probability of being correct. Baker (1983) summarized the influences of the climatic transition from glacial to interglacial between 25,000 and 10,000 years ago, with emphasis on the effects of massive discharges of glacial meltwaters through the Mississippi drainage and catastrophic flooding by outbreaks of glacially dammed lakes in the Columbia River drainage. Knox (1983) used a combination of evidence and inference concerning flood patterns and vegetation cover to construct the alluvial chronologies in different parts of the United States during the past 10,000 years. In the formerly glaciated northern parts of the continent, however, the effects of loose glacial drift should also be taken into account in estimates of post-glacial sediment yields (Church and Ryder, 1972).

Geologic Factors

Rock types and soil types strongly influence sediment yields. In Georgia and adjacent parts of the southeastern United States, the deeply weathered soils developed on the crystalline rocks of the Piedmont produce much larger sediment yields than the sandier soils developed on the poorly consolidated sedimentary rocks of the Coastal Plain (Kennedy, 1964). The data in Figure 7 show that a Piedmont Stream (Broad River) may be expected to carry about ten times the concentration of suspended sediment that a Coastal-Plain stream (Ogeechee River) would carry at the same water discharge. Although the Broad River (Fig. 7A) carried less water past the gaging station near Bell than the Ogeechee River (Fig. 7B) carried past the station near Eden during 1964, and although it drains a smaller area, it carried ten times as much sediment, as the following figures for the 1964 water year show.

Station	Drainage Area (km^2)	Water Discharge (10^9 m^3/yr)	Suspended-Sediment Discharge (tonnes/yr)
Broad River near Bell, GA	3,720	2.91	566,000
Ogeechee River near Eden, GA	6,890	4.14	56,200

Because the climate is virtually the same (considered on a continental scale) in both drainage basins, the apparent influence of rock and soil types on these sediment yields is not complicated by climatic differences. The influence of man, however, is an important factor here. Although the extent of the conversion of the original forests to croplands was similar in the Piedmont and Coastal Plain, the Piedmont soils were much more susceptible to accelerated erosion than were the Coastal-Plain soils. Among the streams of the southeastern United States, which reliable early observers reported originally to be mostly clear of suspended sediment, the streams that drained areas of Piedmont soils became most laden with anthropogenic sediment (Trimble, 1974).

The effect of underlying rock type on sediment yield is strongly evident in three river basins of moderate size (20,000–40,000 km^2) on the flanks of the Great Plains of Alberta and

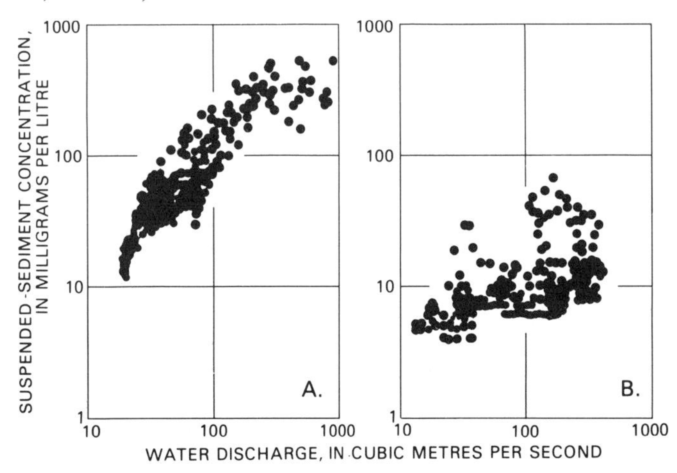

Figure 7. Relations between suspended-sediment concentrations and water discharge in two rivers of Georgia during the 1964 water year (Meade, 1976). A: Piedmont stream, the Broad River near Bell. Drainage area 3,720 km^2. B: Coastal-Plain stream, the Ogeechee River near Eden. Drainage area 6,890 km^2.

Wyoming. In these river basins, most of the runoff is derived from mountainous areas underlain by limestone and other crystalline rocks, whereas most of the suspended sediment is derived from areas of lower relief underlain by shales and siltstones. Underlying rock type is a more important factor than relief in controlling sediment yield from these basins. In the Red Deer River basin of Alberta, a small area of badlands that covers only 2 percent of the total drainage area is capable of producing 80 percent of the total sediment discharge of the river near its mouth (Campbell, 1977). In the Wind River basin of Wyoming, more than half the water but less than 1 percent of the sediment is derived from the 14 percent of the area that is underlain by crystalline Precambrian rocks, whereas 63 percent of the sediment but less than 3 percent of the water is derived from the 16 percent of the drainage area that is underlain by poorly consolidated Quaternary deposits (Colby and others, 1956, p. 85). In the Powder River basin of Wyoming, most of the water but very little sediment is derived from the mountainous areas underlain by Precambrian crystalline rocks and Paleozoic limestones and dolomite, whereas most of the sediment but very little water is derived from areas of less relief underlain by Mesozoic and Cenozoic siltstones and shales (Hembree and others, 1952, p. 47–48).

Tectonic framework is also among the geologic factors that can be said to influence sediment yield, but it is difficult to separate this factor from others, particularly rock type and relief. Among the greatest sediment yields in North America are those in rivers that drain some of the western coastal regions. Where the picture is not complicated by the presence of active glaciers (for example, California), these large sediment yields are clearly associated with tectonically oversteepened slopes developed on sheared bedrock, including the clay-rich subduction mélanges and other rocks of the Jurassic and Cretaceous Franciscan assemblage (Janda, 1979; Scott and Williams, 1978). The steepness of the relief as well as the susceptibility of the rocks to erosion are functions of the tectonic setting. This pattern of very large sediment yields connected with intense tectonism continues around the Pacific rim and includes the high Andes of South America, the coast ranges of North America, and the islands that rim the western Pacific Ocean: Japan, Taiwan, Philippines, New Guinea, and New Zealand (Milliman and Meade, 1983; Walling and Webb, 1983).

Relief

Schumm and Hadley (1961) assessed the influence of relief on sediment yield from measurements of sediment deposited in stock-water reservoirs in 59 small drainage basins (0.3 to 47 km^2) in Wyoming, Colorado, Utah, New Mexico, and Arizona. The sediment yields were averaged into 12 groups by rock type, and the averages were plotted against relief ratio, which is the ratio of vertical relief to straight-line horizontal length of the basin. The resulting plot (Fig. 8) shows an exponential relation of sediment yield to relief ratio. Sediment yields of drainage areas underlain by sandstone and conglomerate plot closer to the lower left corner of the graph, because both relative relief and sediment yields are low. Shale areas tend to have greater relief and greater sediment yields, partly because of gullying and badland development.

One might question whether relief is really more important than rock type in this relation, or whether such a relation between sediment yield and relief can be extrapolated to larger drainage basins and larger scale relief features (mountain ranges, for example). Even in these small basins, the effect of relief as such is obscured by a correlation between relief and rock type: How much of the larger sediment yields in the shale areas portrayed in Figure 8 is due to relief alone and how much is related to the greater erodibility of shale relative to sandstone? In the larger drainage basins of Wyoming discussed above, the areas of low elevation underlain by shale and siltstone contribute much more sediment than the elevated mountain ranges underlain by less-erodible rocks. Is relief, by itself, an independent influence on sediment yield?

Anthropogenic factors that increase sediment loads

Crop-land farming and timber harvesting. Crop farming is the most widespread human activity that has increased sediment loads in the rivers of North America. The settlement of the eastern parts of Canada and the United States during the 17th, 18th, and 19th centuries was marked by the replacement of native forests by croplands. During the 19th century, and especially after the general adoption of the moldboard plow, prairie grass-

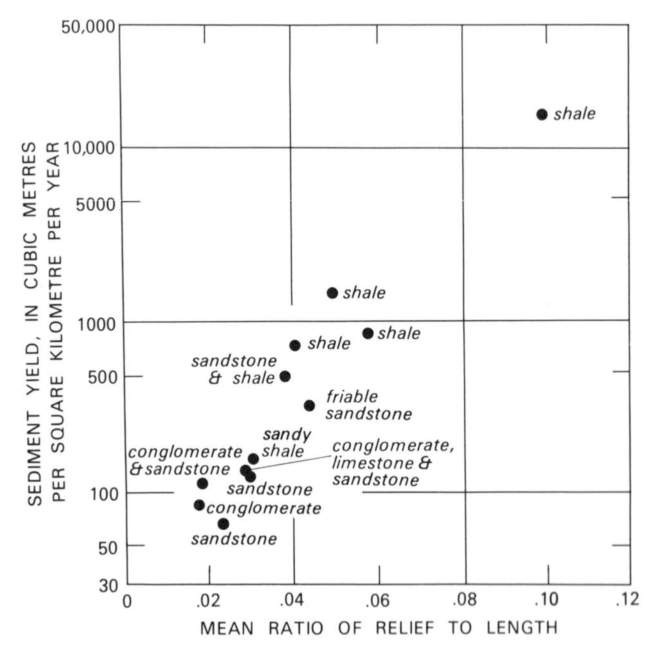

Figure 8. Relation of sediment yield to mean ratio of relief to length of small drainage basins underlain by similar basement materials. Data were from a total of 59 basins, all unglaciated, in Wyoming, Colorado, Utah, New Mexico, and Arizona (Schumm and Hadley, 1961; Schumm, 1977, p. 21). Each point is an average value for several basins underlain by the same rock type in the same general area.

lands were converted to deeply plowed fields. These conversions caused orders-of-magnitude increases in soil erosion and corresponding increases in the sediment yields of rivers. Although there are qualitative descriptions by early observers, we have no numerical data from the rivers to describe sediment yields during the pre-settlement times. However, we can compare presettlement and postsettlement rates of the deposition of sediments in natural lakes (Davis, 1976; Kemp and others, 1974; Last, 1984; Murchie, 1985; Warwick, 1980). We also can compare modern sediment yields from drainage basins that are comparable in all respects except their patterns of land use (for example, forest versus cropland), and infer the increases in sediment yields that must have accompanied the replacement of the natural vegetation by crops.

The relation of sediment yield to cropland is shown in Figure 9A, in which the quantity of sediment transported by 32 tributaries of the Potomac River is plotted against the percent of cropland in each tributary basin. Sediment yield increases with the increase in percent of the tributary basin that is devoted to cropland. Graphs of sediment yield versus forested land in tributary basins of the Potomac, as well as in other river valleys of the eastern United States, show a complementary inverse relation (Wark and Keller, 1963; Patric and others, 1984; Simmons, 1976; Williams and George, 1968). The scatter in the graph (Fig. 9A) is caused by other influences on sediment yield, such as differences in drainage areas of the tributaries and the influences of past patterns of cropland distribution whose sedimentary inputs may still be moving slowly through the tributary basins. In general, despite the scatter in Figure 9A, cropland in this part of North America can be expected to yield about 10 times as much sediment as land that is forested or in pasture.

Fairly extreme differences between sediment yields from croplands and those from forested lands have been observed in the loess region of northern Mississippi. Figure 9B shows that sediment yields from very small watersheds (0.04 to 0.12 km^2) decrease according to land use, in the order cultivated (cornfields) > pasture > abandoned fields > woodlands, and that sediment yields from cornfields can be 100 times those from woodlands. Similar differences in sediment yield have been observed in small drainage basins in the loess regions of Iowa, where native grassland has been converted to cropland (Piest and others, 1975). Loessial soils are extremely susceptible to erosion, and apparently the destruction of original vegetation and disruption by plowing can entail hundred-fold increases in sediment yields from small drainage basins. Although the order of the increase of sediment yield should be somewhat smaller in larger drainage basins, it is well to remember that the largest sediment yields on Earth have been recorded in the loess region of China (Walling and Webb, 1983).

Figure 9C shows sediment-yield data compiled by Brune (1948) from watersheds ranging in area from 1 to 350,000 km^2 in the upper Mississippi River basin. Of the 58 points plotted in the graph, 39 represent suspended-sediment measurements in streams and 16 represent reservoir-sediment surveys. Because most of the suspended-sediment measurements represented short periods of record (some as short as one year), Brune estimated the long-term sediment yields according to the relation between the runoff during the periods of record and the long-term mean runoff. Because most of the surveyed reservoirs had trap efficiencies of less than 100 percent, Brune estimated the trap efficiency of each reservoir and adjusted the sediment yield accordingly. The estimated sediment yields are plotted against drainage area in Figure 9C to reflect the observation that sediment yields commonly decrease with increasing drainage area. Despite the approximate nature of some of the estimates, the greater sediment yields are clearly associated with the greater proportions of cultivated and idle cropland (versus pasture and woodland) in the drainage basins. Similar relations of sediment yield to drainage area and cropland have been shown to exist in the Ohio River basin and Great Lakes drainage (Brune, 1948) and in the Columbia River basin (Flaxman and Hobba, 1955).

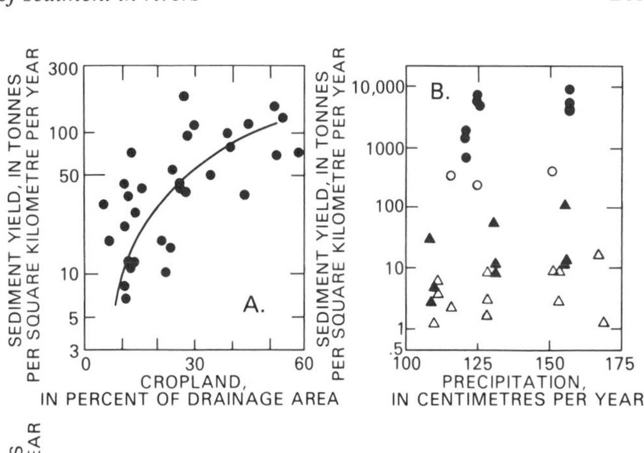

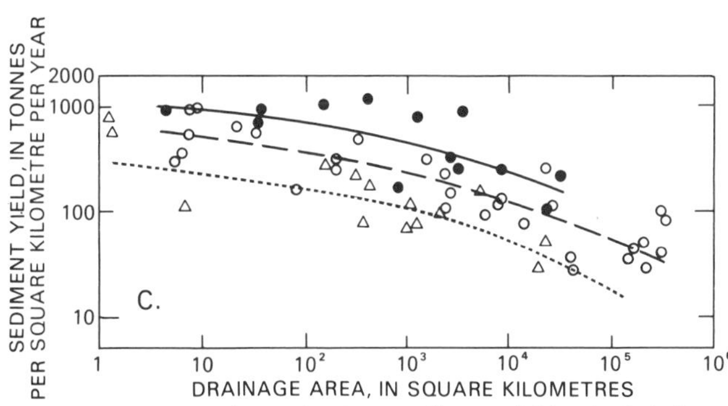

Figure 9. Influence of crop lands on sediment yields. A: Relation of sediment yield to cropland area in tributary basins of the Potomac River, 1959–1962 (Wark and Keller, 1963, p. 25). Drainage areas, 10–1300 km^2. B: Relations of sediment yield to different land uses in small drainage basins (0.04–0.12 km^2), in mostly loessial soils of northern Mississippi (Ursic and Dendy, 1965). Dark circles, cultivated crop lands. Open circles, pasture. Dark triangles, abandoned fields. Open triangles, woodlands. C: Relations of sediment yield to drainage area and crop land in the upper Mississippi River basin (data from Brune, 1948, Table 4). Dark circles and solid line, more than two-thirds of drainage area is crop land. Open circles and dashed line, one-third to two-thirds cropland. Open triangles and dotted line, less than one-third of drainage area is cropland.

Timber harvesting in the western mountain regions of North America also has strong effects on sediment yields. The many studies of the effects of logging have shown complex relations of sediment yield to road construction and timber-harvesting practices (Megahan, 1975; Swanston and Swanson, 1976). When these logging activities take place—as they do in many parts of the Pacific Cordillera—in tectonically active areas where slopes are steep and sheared rocks are especially susceptible to erosion, the various influences on sediment yield are difficult to sort out (Nolan and Janda, 1981). In general, however, the apparent effect of decades of logging in these mountainous areas (which has been intermittent in both space and time) has been an increase by a factor of two or three in the sediment yields from drainage areas of moderate size.

Mining activities. Mining activities have added substantial sediment loads to some rivers. Perhaps the most spectacular case in North America is that of the impact of hydraulic mining for gold in the Sacramento River valley of California, which is well known to geologists because it was the subject of a classic study by Gilbert (1917). Gold-bearing terrace deposits in the Sierra Nevada foothills were mined intensively with water cannons for nearly 30 years before the activity was stopped by a court decision. Gilbert estimated that hydraulic mining activities between 1849 and 1909 contributed 1.2×10^9 m^3 of sediment to the Sacramento River system. "The volume of earth thus moved was nearly eight times as great as the volume moved in making the Panama Canal" (Gilbert, 1917, p. 43). The estimated sediment from sources other than mining in the Sacramento River basin during approximately the same period (1850–1914) included about 0.21×10^9 m^3 from agricultural activities (grazing and crop farming) and about 0.11×10^9 m^3 from natural sources. The sediment added by hydraulic mining in only part of the Sacramento River basin therefore caused a ten-fold increase above the natural sediment yield of the entire basin.

On a smaller scale, the effects of strip mining for coal were studied in a pair of small drainage basins (1.7 and 2.2 km^2) in eastern Kentucky by Collier and Musser (1964). Strip mining began in the smaller of the two drainage basins, Cane Branch, in 1955, and the sediment yields in the basin increased markedly. Figure 10 shows daily sediment yields from Cane Branch compared with those from Helton Branch, a nearby unmined drainage basin, during a two-year period. Sediment yields from the strip-mined basin were at least 10 times those from the unmined basin. A similar increase in sediment yield has been shown to be related to surface mining in a small drainage basin in Pennsylvania (Stump and Mastrilli, 1985).

Urbanization. Urbanization is one of the more recent human influences that contributes large sediment loads to streams, and its effects have been studied most intensively and extensively in the areas surrounding the cities of Washington and Baltimore (Roberts and Pierce, 1974; Vice and others, 1969; Yorke and Herb, 1978). Wolman and Schick (1967, p. 454) reported annual sediment yields from construction sites in the Washington-Baltimore area that ranged from 2,000 to 50,000 tonnes/km^2. Guy (1965, p. 37) estimated that the extra sediment produced by urbanization near Washington was sufficient to double the discharge of suspended sediment by the Potomac River to its estuary.

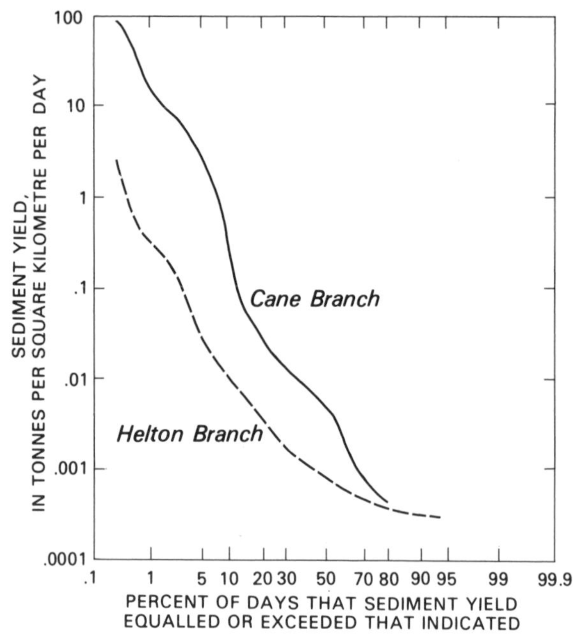

Figure 10. Duration curves of daily suspended-sediment yield for Cane Branch (basin strip-mined) and Helton Branch (basin not mined), eastern Kentucky, October 1956–September 1958 (Collier and Musser, 1964, p. 56). Drainage areas: Cane Branch, 1.7 km^2; Helton Branch, 2.2 km^2.

Wolman's (1967) schematic summary of changes in sediment yield with changing land use in a typical area between Washington and Baltimore is shown in Figure 11. Until the end of the 18th century, the area remained in its original forested condition, and sediment yields were low. The area was converted to crop farms in the 19th century, and the sediment yields increased accordingly. During the first half of the 20th century, soil-conservation measures were introduced and some lands reverted to woods and pasture while awaiting their conversion to suburbs and cities; both these effects caused the sediment yields to decrease. During the period when the lands are converted to urban use, the sediment yields are extraordinarily large, but this period is relatively short. After the area becomes a city with paved streets and planted lawns, the sediment yields become small again.

Anthropogenic influences that decrease sediment loads

One of the most pervasive influences on sediment loads is exerted by the dams and reservoirs that have been built in large numbers across the rivers of North America (Hathaway, 1948; Williams and Wolman, 1984, 1986). Dams are built to impound water for various purposes, and the reservoirs they form interrupt the down-river flow of sediment. Although the river water that enters a reservoir is released eventually (through a power plant, into a diversion canal, or over a spillway), much of the sediment is trapped permanently in the reservoir. Nearly all reservoirs on major rivers of North America trap at least half of the river sediment that flows into them. Some of the largest reservoirs on the continent, like Lake Powell and Lake Mead on the Colorado River, trap virtually all the sediment that flows into them (Brune, 1953). The effects of reservoirs on sediment loads are apparent in rivers in all parts of the continent; however, they are most obvious in the large western rivers where the original sediment loads were naturally large and where the construction of dams has been especially intense.

Perhaps the classic example in North America of the interruption of a large discharge of river sediment to the oceans is that of the Colorado River. Before about 1930, the Colorado River delivered an average of 125 to 150 million tonnes of suspended sediment per year to its delta at the head of the Gulf of California. Since the closure of Hoover Dam, which began in 1935, this rate of sediment delivery has declined, first precipitously and then more gradually, to an average annual amount today of about 100,000 tonnes. Figure 12 graphically shows this decline in sediment and also the more gradual decline of water flow in the lowermost Colorado River since the turn of the century. Except for a period between 1934 and 1938, when 30×10^9 m^3 of the river water was appropriated for the initial filling of Lake Mead behind Hoover Dam, the quantity of water carried by the Colorado River past Yuma, Arizona, has declined more or less progressively. This decline has been in response to the increasing diversion of water from the Colorado River for irrigation of crop lands and for municipal water supplies. The more abrupt decline in sediment discharge at Yuma clearly was related to a single event, the closing of Hoover Dam.

Another large river system whose sediment loads are strongly influenced by reservoirs is the Missouri-Mississippi. Annual discharges of suspended sediment measured at six gaging stations on the Missouri River and two stations on the Mississippi River over a period of about four decades are shown in Figure 13. The Missouri River has always been the principal supplier of sediment to the lower Mississippi River; the other two large components of the Mississippi River system, the upper Mississippi River and the Ohio River, supply large quantities of water but

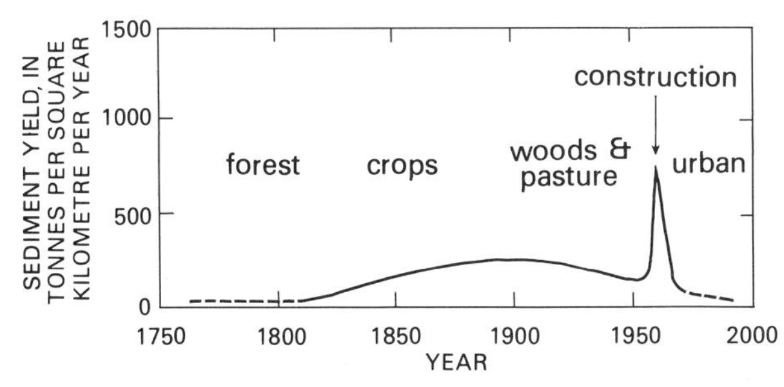

Figure 11. Schematic representation of changes in sediment yield related to changes in land use between the years 1800 and 2000 in a fixed area of the Maryland Piedmont (modified after Wolman, 1967, p. 386).

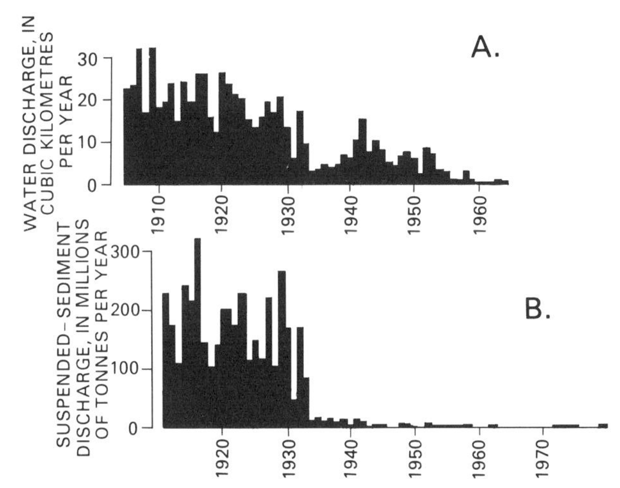

Figure 12. Annual discharges of water (1905–1964) and suspended sediment (1911–1979) in the Colorado River at Yuma, Arizona (Curtis and others, 1973, p. 9; Meade and Parker, 1985, p. 55). A: Water discharges. Compiled from data of U.S. Geological Survey. 1 km^3 = 10^9 m^3. B: Suspended-sediment discharges. Compiled from data of U.S. Bureau of Reclamation. The abrupt decrease in suspended-sediment discharge in the middle 1930s coincided with the closure of Hoover Dam, 500 km upriver of Yuma.

comparatively small amounts of sediment. When five large dams were completed for irrigation and hydroelectric power above Yankton, South Dakota, between 1953 and 1963, the flow of sediment from the upper Missouri basin virtually stopped. Following the closure of Fort Randall Dam and Gavins Point Dam in 1953, downstream sediment loads were diminished immediately, and the effect could be observed all the way down to the mouth of the Mississippi River. Partly because of the construction of reservoirs, and partly as a result of other causes (Keown and others, 1981, 1986), sediment discharges to the Gulf of Mexico by the Mississippi River are now less than half of what they were before 1953.

Reservoirs in other parts of North America have caused similar decreases in river-sediment loads downstream, but the data are not so extensive as those portrayed in Figures 12 and 13. Perhaps the outstanding Canadian example is that of Lake Diefenbaker, which has decreased the sediment loads of the South Saskatchewan River downstream of the dam by 90 percent. Sediment discharges measured in the river above the reservoir average 6×10^6 tonnes per year; those measured at Saskatoon, 115 km down river, during the post-dam period averaged only 0.7×10^6 tonnes per year (Rasid, 1979). Even though 6×10^6 tonnes of sediment are deposited annually in Lake Diefenbaker, this amounts to less than 0.1 percent loss in storage per year (Yuzyk, 1983). Reservoir sedimentation problems are mostly negligible in Canada; the most dramatic example is that of the Bassano Reservoir, which has been losing capacity at a rate of 1.0 percent per year (Yuzyk, 1984).

To avoid giving the impression that interruption of the seaward transport of sediment by dams and reservoirs is a phenomenon confined to the western parts of the continent, Figure 14 is presented to show the effects of reservoirs on the sediment loads of rivers in the southeastern United States. Although continuous records of sediment discharge, such as those shown in Figures 12 and 13, are not available for these rivers, enough data were available to compare measurements made about 1910 with data collected about 1980. During the years between the two World Wars, many dams were built across these rivers, mostly for hydroelectric power and flood control. A comparison of the sediment loads before (about 1910) and after (about 1980) shows the large influence of these reservoirs in trapping sediment. As shown in Figure 14, five major rivers, which previously carried a total of 10×10^6 tonnes per year of sediment to the coastal zone, now carry only about one-third of that amount. However, the 10 ×

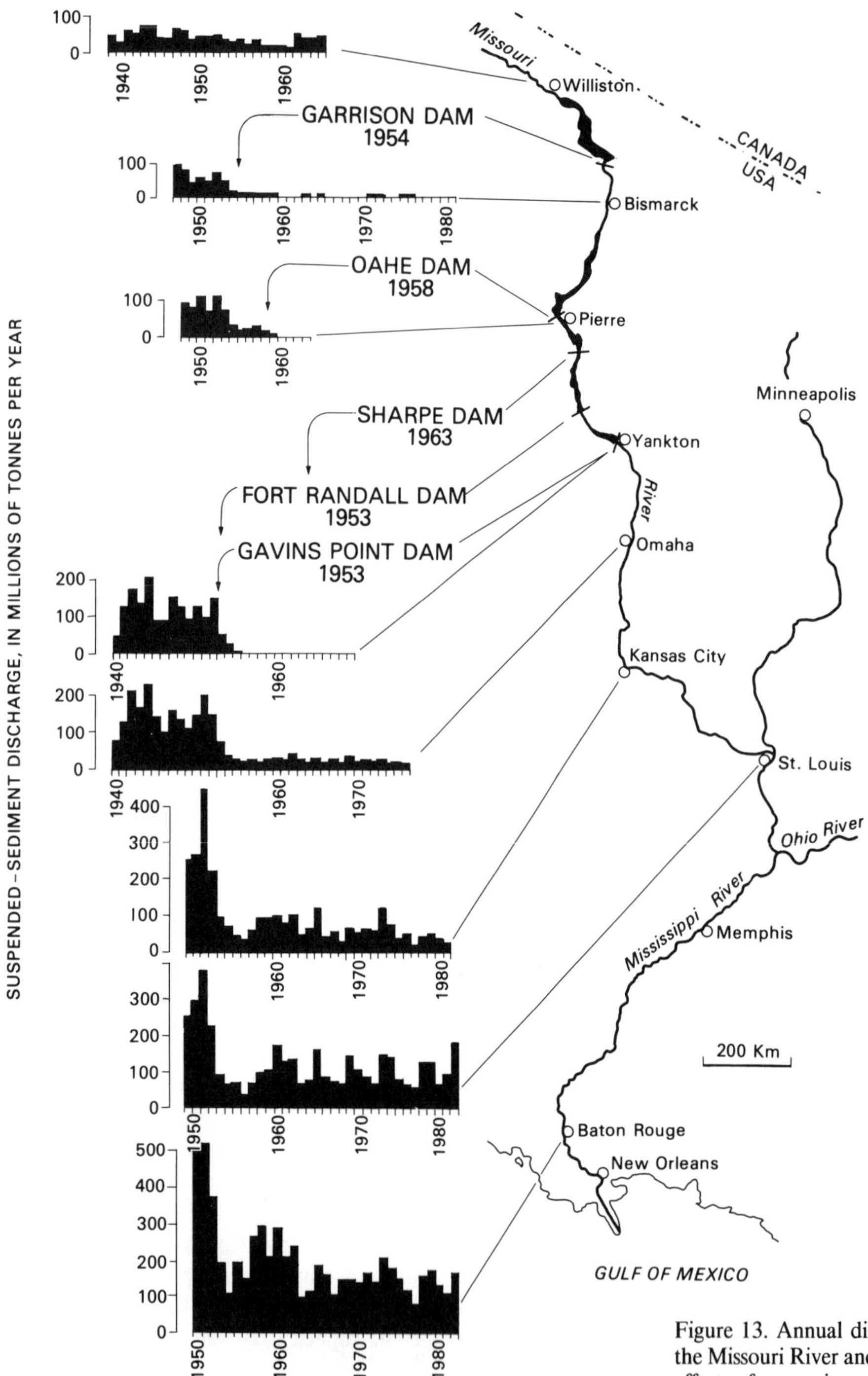

Figure 13. Annual discharges of suspended sediment at six stations on the Missouri River and two stations on the Mississippi River showing the effects of reservoirs on downstream sediment loads, 1939–1982 (Meade and Parker, 1985, p. 52; compiled from data of U.S. Army Corps of Engineers and U.S. Geological Survey).

10^6 tonnes per year that were measured about 1910 represented artificially accelerated sediment loads, and today's total of about 3×10^6 tonnes per year represents more sediment than these rivers transported before the time of European settlement (Trimble, 1974).

ROUTING AND STORAGE OF SEDIMENT IN RIVER SYSTEMS

Although we know that, in general, sediment moves off uplands and hillslopes into stream channels, and that streams convey sediment toward the sea, the precise routing of sediment particles in river systems is obscure. At present, we are unable to predict, within any acceptable limits of accuracy or with any reasonable degree of certainty, the pathways and sinks of sediment particles in river systems. As much as we would like to be able to consider sediment transport in river systems as a simple linear continuum, this is rendered impossible by the overwhelming importance of sediment storage.

Sediment storage is important at many scales, both spatial and temporal. On a semicontinental spatial scale, upland erosion in the continental United States (exclusive of Alaska) has been estimated at about 5×10^9 tonnes per year (Holeman, 1981), whereas the discharge of river sediment from the same region to the coastal zones is only one-tenth that amount, or about 0.5×10^9 tonnes per year (Curtis and others, 1973). That is, ninety percent of the sediment presently being eroded off the land surface of the conterminous United States is being stored somewhere in the river systems between the uplands and the sea. Storage in artificial reservoirs may account for some of this—we have seen that $0.3–0.4 \times 10^9$ tonnes per year are stored in reservoirs on the Colorado and Missouri Rivers (Figs. 12 and 13), and another $0.1–0.2 \times 10^9$ tonnes per year are stored in reservoirs on the Rio Grande, Columbia River, and elsewhere (Meade and Parker, 1985). But this leaves 70 to 80 percent of the sediment eroded off the uplands to be accounted for by storage in other compartments of the river systems, probably hillslopes and floodplains. When temporal factors are taken into account, we find that storage occurs over a large spectrum of time scales and, in many cases, the time lags between storage and remobilization are such that the sediments being carried by rivers today may represent episodes of erosion that occurred decades or centuries ago. Examples of patterns and problems of sediment storage in North America are discussed below in the context of different time scales. As Schumm (1977, p. 99) stated, "Confusion can only be avoided if discussion is restricted to a consideration of one time span at a time."

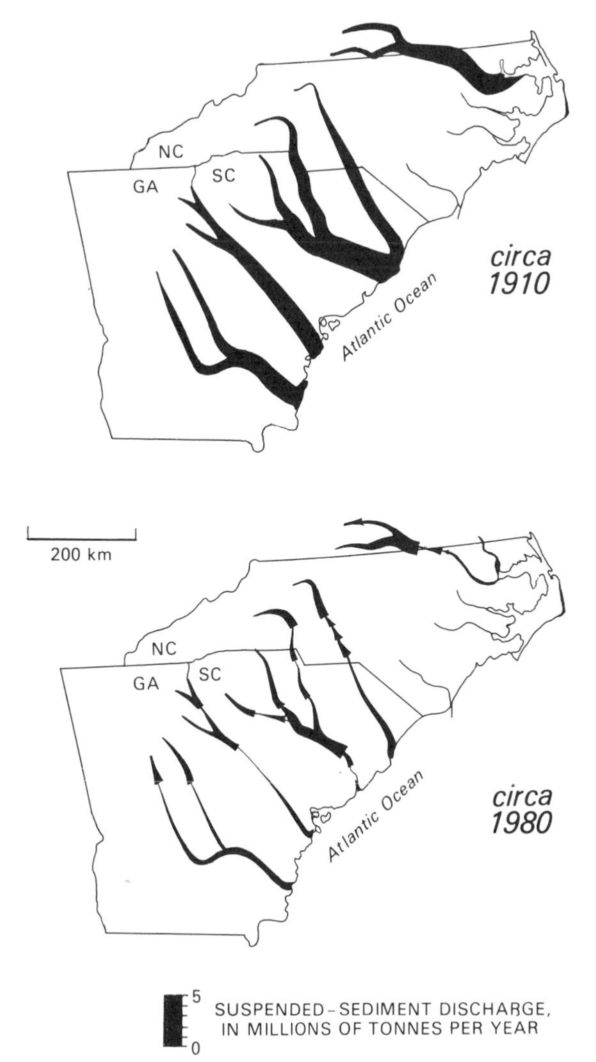

Figure 14. Average suspended-sediment discharges of major rivers of Georgia (GA), South Carolina (SC), and North Carolina (NC) during two periods, circa 1910 and circa 1980, that show the decrease in sediment loads caused by reservoirs constructed during the intervening years (Meade and Parker, 1985, p. 56; compiled from Dole, 1909, Dole and Stabler, 1909, and from later data of U.S. Geological Survey). Width of river represents mean annual suspended-sediment discharge.

Storage at seasonal time scales

Seasonal storage of sediment, both suspended load and bed load, is probably typical to some extent of all rivers. Seasonal fluctuations in water discharge cause seasonal differences in the movement and storage of sediment, and seasonal storage of sediment in river channels is probably easier to understand and predict than long-term storage in river systems. Some of the short-term changes in storage of suspended sediment in the lower reaches of the Mississippi River in Louisiana are shown in Figure 15. The pattern of storage is shown particularly well in Figure 15 because the data portrayed are based on multiple samples collected repeatedly, in 8 to 10 cross sections of the river spaced 20 to 50 km apart, in downstream sequence over periods of only a few days. No dams obstruct these reaches of the Mississippi, no tributaries bring in sediment, and no outlets drain sediment away until it reaches the mouth of the river. Any down-river changes that are observed in the discharges of suspended sediment represent deposition of material onto the riverbed or resuspension of

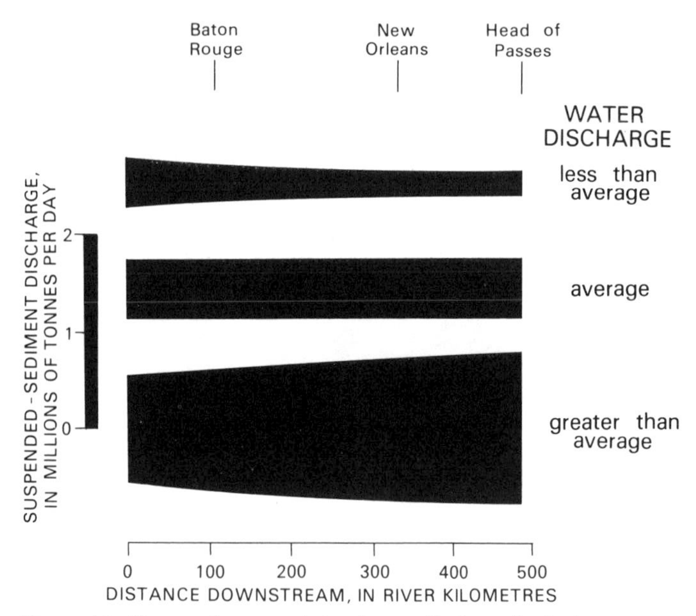

Figure 15. Changes in suspended-sediment discharge in the lowermost 500 km of the Mississippi River at three different stages of river flow—less than average, average, and greater than average. (Meade and Parker, 1985, p. 57; compiled from data reported by Everett, 1971, p. 14, and Wells, 1980, p. 13.)

material from the riverbed. At average water discharge, the sediment load remains the same through the entire 500-km reach of the lower Mississippi; on a net basis, sediment is neither stored nor resuspended at average water discharge. At less-than-average water discharge, the suspended load decreases down the reach; sediment is being dropped by the flowing river and stored on the riverbed. At greater-than-average water discharge, the sediment load increases down the reach; at least part of the previously stored sediment is being resuspended from the riverbed. The short-term pattern, therefore, shows sediment being deposited and stored on the riverbed at lower flows and being resuspended and flushed out to the Gulf of Mexico on higher flows (Curwick, 1986). Analogous patterns of seasonal storage and remobilization of sediment have been observed and described from rivers that range in size from small (Emmett and others, 1983; Meade, 1985) to the largest in the world (Meade and others, 1983, p. 1139–1140; 1985).

Storage at decade-to-century time scales

Most studies of sediment storage in river systems fall into the secular time scale of 10 to 100 years, the scale of the human life span and consequently the scale at which we most naturally perceive and approach environmental problems. For example, the rates at which sediment went into storage during the years of intense, artificially accelerated soil erosion in the Piedmont region of the southeastern United States are suggested by the graph in Figure 16, which is based on data from periods of several decades that ended in the 1930s and 1940s. The sediment delivery ratio (defined in the caption to Fig. 16) is a measure of the disparity between the amount of sediment delivered by a stream and the amount that has been eroded upstream. The disparity represented in Figure 16 is so great that streams draining areas on the order of 100 km^2 were transporting only about 10 percent of the soil eroded off the uplands. A significant portion of the other 90 percent of the eroded material is stored on the floodplains of the southern Piedmont. Happ (1945) estimated that floodplains in Piedmont valleys of South Carolina are covered with an average of 1.2 m of upland soil that was deposited since the onset of European settlement. Similar thicknesses of upland soil have been deposited on floodplains in Piedmont valleys of Maryland, the Driftless area of southwestern Wisconsin, and the loess region of northern Mississippi (Costa, 1975; Happ and others, 1940; Knox, 1977; Magilligan, 1985; Trimble, 1977).

Natural processes can also engender periods of storage and remobilization of river sediment at time scales of decades to centuries. In the Bella Coola River of British Columbia, coarse sediments derived from Neoglacial (18th and 19th centuries) moraines of alpine glaciers have been stored and slowly moved through the 60-km reach of river between the glaciers and the sea-level delta (Church, 1983). Maps and aerial photographs of the lowermost 25 km of the river show a steady increase in channel stability during the past 90 years (and especially during the past 35 years) that can be interpreted as marking the downstream passage of the locus of maximum sediment storage.

The classic case study of the movement and storage of sediment in a river system at time scales of 10 to 100 years is that of the hydraulic-mining debris in the Sacramento River valley of California (Gilbert, 1917; Kelley, 1959). Between 1855 and

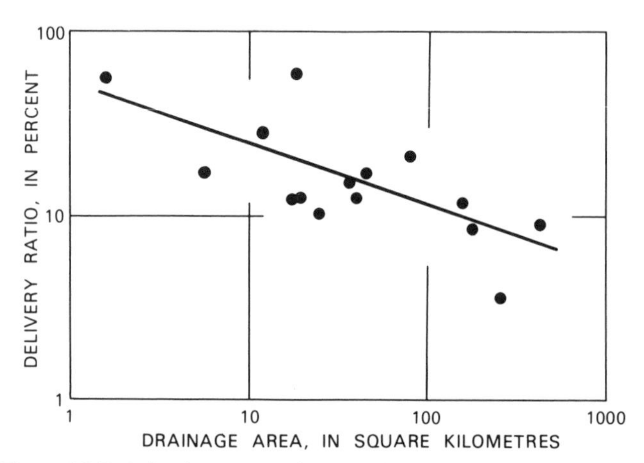

Figure 16. Relation between sediment delivery ratio and drainage-basin area in the Piedmont of Georgia, South Carolina, and North Carolina (Roehl, 1962). Delivery ratio is defined as the ratio between the amount of sediment being carried by a stream draining an area and the estimated amount of sediment eroded off the soils of the same area. The amount of sediment being carried by the stream usually is measured, whereas the amount of soil being eroded usually is estimated by an empirical equation. Similar relations between delivery ratio and drainage area have been described from several parts of the central United States (Piest and Miller, 1975, p. 458–464; Renfro, 1975).

1885, enormous quantities of coarse sediment were washed into some of the tributaries of the Sacramento River during hydraulic mining for gold. The resulting problems that developed downstream (flooding, filling of navigation channels, destruction of flood-plain farms) became so serious that hydraulic mining was curtailed by a court decision in 1884. By that time, however, the large mass of sediment, characterized as a "wave" by Gilbert (1917), was already into the stream channels and was moving slowly down the tributaries and into the Sacramento River. As the mass of sediment advanced, it raised the levels of the channel beds, much as an ocean swell raises the level of the sea as it passes through (Fig. 17). Bed levels rose 5 m in the tributary Yuba River at Marysville and nearly 3 m in the Sacramento River at Sacramento. The riverbeds at these towns reached their greatest elevations 10 to 20 years after the mining was stopped; then they declined steadily to their previous elevations during the next 30 to 40 years. All in all, the great wave of hydraulic-mining debris took nearly a century to pass through the channels of the Sacramento River system and finally reach San Francisco Bay.

A similar example, but on a smaller scale, has been described by Trimble (1969) from a mill site in the upper Altamaha River basin of Georgia that has gone through a cycle of burial and excavation since 1865 (Fig. 18). A mill dam was built here in 1865 on a bedrock channel. By 1930, not only had the reservoir behind the dam been filled with sediment, but the dam itself and the adjacent floodplain were also buried. Subsequently, with a decrease in the supply of sediment from the uplands, the river is now eroding its bed and the remnants of the dam are exposed again.

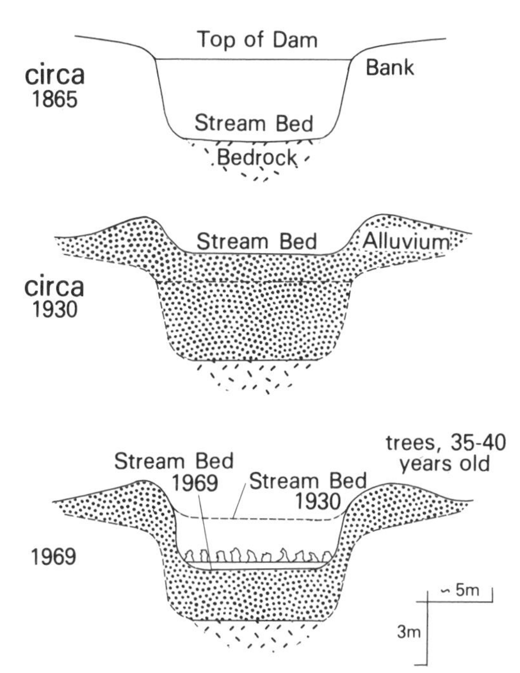

Figure 18. Cross sections showing the aggradation and subsequent degradation of a mill site on a small tributary of the Oconee River in the Piedmont of Georgia between 1865 and 1969. Modified after Trimble (1969).

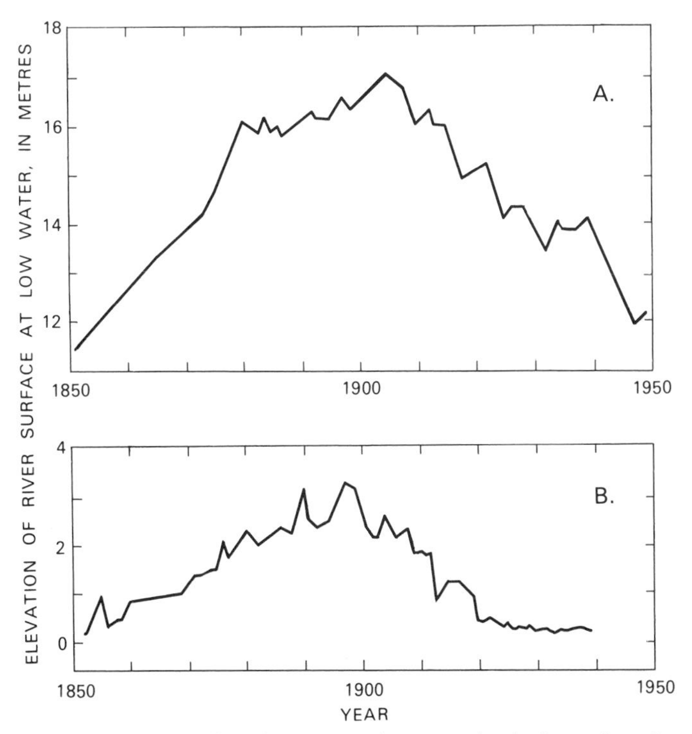

Figure 17. Rise and fall of the annual low-water level of two rivers in California between 1850 and 1950, due mainly to the deposition, storage, and subsequent erosion of hydraulic mining debris in their channels (Gilbert, 1917; Graves and Eliab, 1977). A: Yuba River at Marysville. Drainage area 3,500 km^2. B: Sacramento River at Sacramento. Drainage area 55,000 km^2.

Storage at century-to-millennium time scales

The pattern of storage and remobilization described in the previous two paragraphs applies only to the sediment in and near the river channels. It does not apply to the debris that overflowed onto the floodplains. The hydraulic-mining debris that was carried out of the Sacramento River channel during floods and deposited on the floodplains was sufficient in many places to cover entire houses and orchards (Kelley, 1959, p. 134–135, 203–204). Most of that debris still remains where it was deposited a century ago. The time required to remove sediment from storage on the floodplain is much longer than the century that was required to remove the debris from the main river channels. Floodplain deposits are removed mainly by erosion of channel banks as streams slowly migrate laterally, a process that proceeds at a substantially slower pace than the vertical removal of material stored in the bottom of the river channel. The complete remobilization of floodplain deposits may require time periods on the order of a millennium or more (Leopold and others, 1964, p. 328). However, as the studies cited by Schumm (1977, p. 131–132) indicate, there is considerable variation in the length and time scales at which floodplains are remobilized.

A. 1853-1938

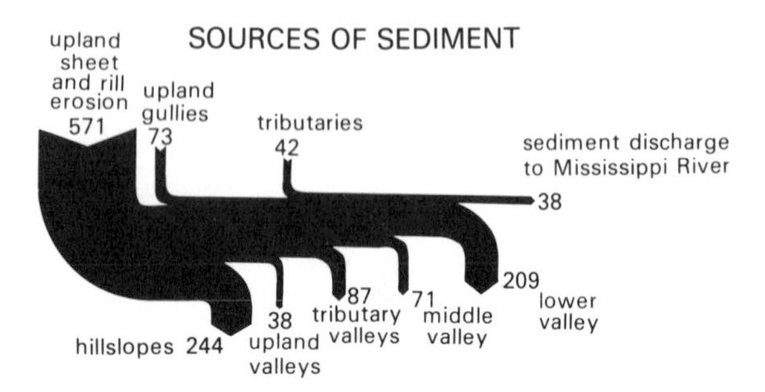

SINKS AND STORAGE OF SEDIMENT

B. 1938-1975

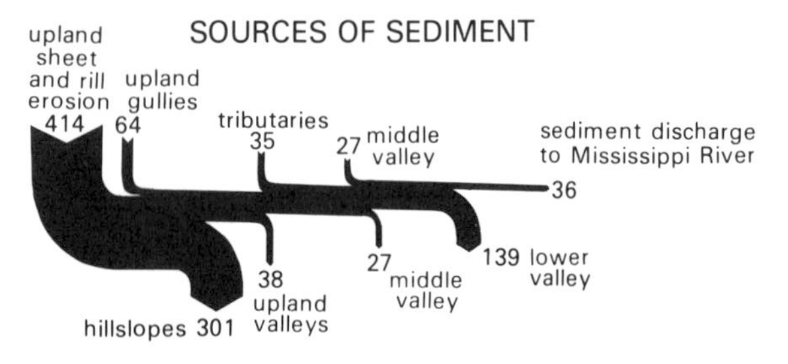

SINKS AND STORAGE OF SEDIMENT

Figure 19. Sources, sinks, and storage of sediment in the Coon Creek basin, Wisconsin, during two periods, 1853–1938 (A) and 1938–1975 (B). Numbers on the diagrams are estimated annual averages for the period in thousands of tonnes per year. Of the total amount of sediment contributed by uplands and tributaries during the 122-year period from 1853 to 1975, less than 7 percent has left the Coon Creek basin. Modified after Trimble (1983).

Coon Creek, Wisconsin. Many of the problems associated with the prediction of sediment storage at time scales of 100 to 1,000 years are demonstrated in a study carried out on Coon Creek, a small stream that drains 360 km^2 of southwestern Wisconsin (Trimble and Lund, 1982). Originally covered by forests, Coon Creek basin was settled by European immigrants and cleared for farming about 1850. As the forests were cleared and the land was plowed, a cycle of erosion and sedimentation began, the consequences of which are still strongly in effect today. In 1933, after about 80 years of land-management practices that resulted in excessive erosion, soil-conservation efforts were begun in earnest. These efforts continue. Two time periods (1853–1938 and 1938–1975) are described below. The year 1938 was selected as the transitional date because of an extensive sedimentation study that was carried out that year.

Figure 19A shows the accelerated erosion of sediment from the uplands and tributary areas and the transfer of sediment to the lower hillslopes and valleys of the Coon Creek basin between 1853 and 1938. Much less than 10 percent of the sediment eroded off the uplands during this period was exported out of the basin by the creek. More than 90 percent of the sediment was deposited along the way, on hillslopes and floodplains, where most of it still remains in storage.

From 1938 to 1975, improved soil conservation and land management reduced the rates of upland erosion. However, the quantity of sediment that passes out the mouth of Coon Creek is still less than 10 percent of the total upland erosion (Fig. 19B). The other 90 percent or more of the eroded sediment still is being stored within the creek basin. The only important difference in recent years is that some of the sediment formerly stored in the middle valley is now being remobilized and transported out of the basin.

Redwood Creek, California. Temporal as well as spatial aspects of sediment storage have been demonstrated in a study carried out in Redwood Creek, which drains 720 km^2 of the northern California Coast Range. Drawing heavily on concepts that had been applied earlier by Dietrich and Dunne (1978), Madej (1984, 1989) developed a sediment budget for the Redwood Creek basin. She used aerial photographs and the ages of trees and shrubs growing on various deposits in the channel and on the floodplain and terraces of Redwood Creek to characterize four different categories of alluvial deposits: active, semiactive, inactive, and stable. Madej also estimated the storage times of each category of deposit in each of three reaches (33–36 km long) of the channel: an upper reach, where the channel is steep (slope = 0.012) and the valley is narrow; a middle reach, where the channel is braided (slope = 0.0045) and the valley is wide; and a lower reach, where the gradient becomes more gentle (0.001–0.002) and the valley even wider. In general, as shown in Figure 20 and Table 2, the deposits that are progressively farther from the active channel are stored for progressively longer times than the material that is closer to the active channel. The storage times of the sediments in the different compartments range over more than two orders of magnitude (Table 2).

Storage at multimillennial time scales

At time scales of 1,000 to 10,000 years, the most significant factors in the storage of sediment in large river systems seem to be associated with the melting of the late Pleistocene ice sheets. The most likely of these factors are the eustatic rise of sea level and the changes in river regimes related directly to changes in the quantities and sediment loads of glacial meltwaters.

About 15,000 years ago, at the time of the last major ice advance (Wisconsin, Würm), the world's large rivers were graded to a sea level 100 m lower than today's, and the lowest reaches of their valleys were 100 m deeper than they are today. During the past 15,000 years, as the great ice sheets melted and sea level rose (the Flandrian transgression), the lower valleys of the large rivers have accumulated and stored large quantities of sediment. A convincing interpretation of the sequence of storage of alluvial sediment in the lower Mississippi Valley, for example, was constructed by Fisk (1944) from an extensive collection of

TABLE 2. STORAGE TIMES OF SEDIMENTS IN DIFFERENT ALLUVIAL COMPARTMENTS OF REDWOOD CREEK, CALIFORNIA*

Category of Deposit	Vegetation	Recurrence interval of Flow that Mobilizes Deposit	Mean Storage Time (yr)		
			Upper Reach	Middle Reach	Lower Reach
Active	Absent or sparse	1-5 yr	26	11	9
Semi-active	Shrubs and young trees	5-20 yr	50	19	15
Inactive	Young to mature trees	20-100 yr	104	99	106
Stable	Old-growth forest	>200 yr	700	3100	7200

*Data from Madej (1984).

data from core holes drilled into the floodplain. This interpretation was expressed in a compelling series of block diagrams (Fisk, 1947, Plate 5, reproduced by Schumm, 1977, p. 166–167) that showed the sequence of filling of the previously incised lower Mississippi Valley during the past 15,000 years. At the lowest stands of sea level, the times of maximum glacial advance, the floor of the river valley was eroded and dissected by a network of dendritic drainage. As the sea level rose, coarse sands were deposited first, probably by a braided stream. Fine sand, silt, and clay were deposited later, on top of the coarse sands, as sea level continued to rise and the valley slope diminished. When the rate of sea-level rise slowed and large quantities of sediment were no longer supplied by glacial meltwaters, the river changed from a braided stream to a meandering stream.

Studies made during the decades since 1947 have contradicted some of Fisk's interpretations (Saucier, 1974, 1981; Saucier and Lindfors-Kearns, 1989). More detailed coring has suggested that many parts of the pre-Holocene valley floor were not dissected by dendritic drainage. Furthermore, the effects of the lowered base level of 15,000 years ago are clearly discernible in the river valley only below the present location of Baton Rouge, Louisiana, or perhaps a bit farther north at Natchez, Mississippi. The most recent interpretations (Saucier and Lindfors-Kearns, 1989) correlate episodes of aggradation of the middle Mississippi Valley with climatic episodes (changes in water discharge and sediment load, and the deposition of large quantities of glacial outwash) rather than with interglacial stands of higher sea level. Regardless of their principal causes, major phases of aggradation and degradation of river valleys have taken place at multimillennial time scales.

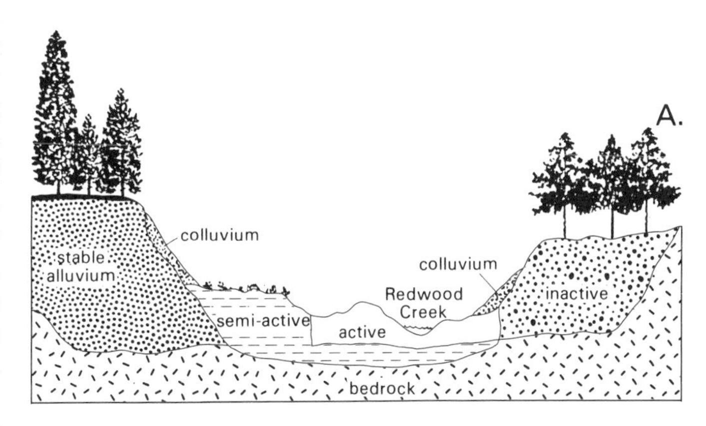

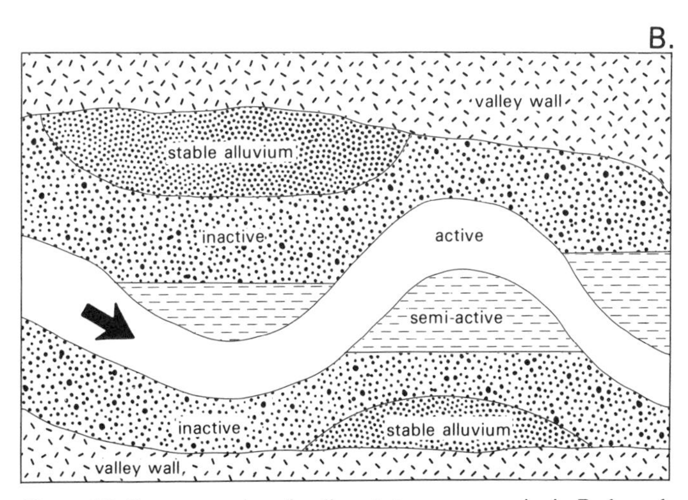

Figure 20. Four categories of sediment-storage reservoirs in Redwood Creek, California (Madej, 1984, 1989). A: Schematic cross section. B: Schematic plan view.

Storage: some unresolved questions

Several questions confound the present understanding of the storage and delivery of sediment in the river systems of North America. For example: To what extent are low delivery ratios (large proportions of sediment going into storage in sites, such as hillslopes, between the eroding uplands and the stream channels) due to excessive inputs of anthropogenic sediment? That is, how much of the sediment going into storage is in response to a man-caused overloading of the sediment-delivery systems? If all were in natural equilibrium (assuming that such an equilibrium really exists), would so much sediment be going into storage? To what extent are the low delivery ratios that have been observed during the past half-century only part of a longer term (one to several centuries) cycle of aggradation and degradation? A century ago, the southwestern United States underwent an episode of intense gullying. Was this the degradational phase of a cycle that is now in its aggradational phase? A third question involves an even longer time scale: To what extent are the sediment-storage phenomena we observe today related to the world-wide rise in sea level that has continued for the past 15,000 years? If base level has risen and continues to rise steadily, should we not expect the delivery system to back up and to store progressively more sediment? As outlined in a recent summary by Walling (1983), the problems of sediment delivery and long-term storage in river valleys are among the principal challenges for future studies of sediment.

DELIVERY OF RIVER SEDIMENT TO COASTAL ZONES

Figure 21 shows the quantities of sediment being delivered to the coastal zones by the principal rivers of North America during a year of average runoff before the continent was settled extensively by Europeans. Not portrayed in the figure are sediment discharges smaller than about 10×10^6 tonnes per year. The figure shows clearly that the large sediment discharges are associated only with the rivers that drain the western half (mostly the western third) of the continent. Sediment discharges of rivers in the eastern and north-central portions of the continent are smaller because of the effects of the erosion-resistant Canadian Shield, the generally smaller sizes of drainage basins, the low grassed topography of the central Canadian plains, and the storage of sediment in large lakes. Table 3 lists the pre-settlement sediment discharges of major rivers of the continent and compares them with today's sediment discharges. The principal sites of deposition of river sediment around the fringes of North America are deltas, estuaries, and continental shelves. Although these deposition sites can be considered permanent on a secular time scale, they only can be considered temporary storage sites on a multimillennial time scale.

Deltas receive most of the sediment discharged by rivers to the coastal zones of North America. Nearly all the large rivers whose sediment discharges are portrayed in Figure 21 have formed deltas that have either filled the upper ends of large embayments or extended well onto the continental shelves. The largest in North America is the delta of the Mississippi River, which has been the subject of many intensive and excellent studies over the years (Bates, 1953; Coleman and Gagliano, 1964; Fisk and others, 1954; Fisk and McFarlan, 1955; Frazier, 1967; Gould, 1970; Gould and McFarlan, 1959; Kolb and Van Lopik, 1966; Scruton, 1960). Perhaps the second-best-studied modern delta in North America is that of the Fraser River (Clague and others, 1983; Luternauer and Murray, 1973; Mathews and Shepard, 1962). Other extensive studies have been made in the deltas of the Colorado River (Thompson, 1968), MacKenzie River (Mackay, 1963), and Copper River (Reimnitz, 1966).

Estuaries are the sink for a small but significant portion of the sediments discharged at the mouths of the rivers of North America. Along the Atlantic seaboard, where sediment yields of rivers are relatively small and where the lower valleys of the principal rivers have been drowned by rising sea level, virtually all the river sediment is trapped in the estuaries (Meade, 1969, 1982; Nichols, 1986). Estuarine entrapment of sediment also occurs where large embayments have been formed by tectonic activity or glacial erosion; for example, San Francisco Bay (Smith, 1965), Puget Sound (Baker, 1984), and James Bay (Kranck and Ruffman, 1982). In some coastal areas of North America, the embayments have been completely filled by river sediments. The Brazos and Colorado Rivers of Texas, for example, supplied sediment at rates that kept pace with the Holocene transgression (Wilkinson and Basse, 1978). Some coastal embayments of Alaska also have become filled with sediment within the past 200 years (Molnia, 1979).

The Great Lakes are important sinks for the river sediment discharged into them. According to measurements of the quantities of sediment deposited above the *Ambrosia* (ragweed) pollen horizon, an average of $25–30 \times 10^6$ tonnes of sediment per year have been deposited in the Great Lakes since 1850 (Kemp and Harper, 1976, 1977; Kemp and others, 1974, 1977, 1978). Of this quantity, less than half ($10–15 \times 10^6$ tonnes per year) seems

TABLE 3. DISCHARGES OF SEDIMENT TO THE COASTAL ZONES OF NORTH AMERICA BY MAJOR RIVERS

River	Mean Sediment Discharge (10^6 tonnes/year) ca. 1700	ca. 1980
Mississippi	400	210
MacKenzie	100	100
Colorado	100	0.1
Copper	70	70
Yukon	60	60
Susitna	25	25
Fraser	20	20
Grande	30	0.8
Eel		15
Brazos		11
Columbia		
before Mount St. Helens	15	10
after Mount St. Helens	—	40
St. Lawrence	1	1.5

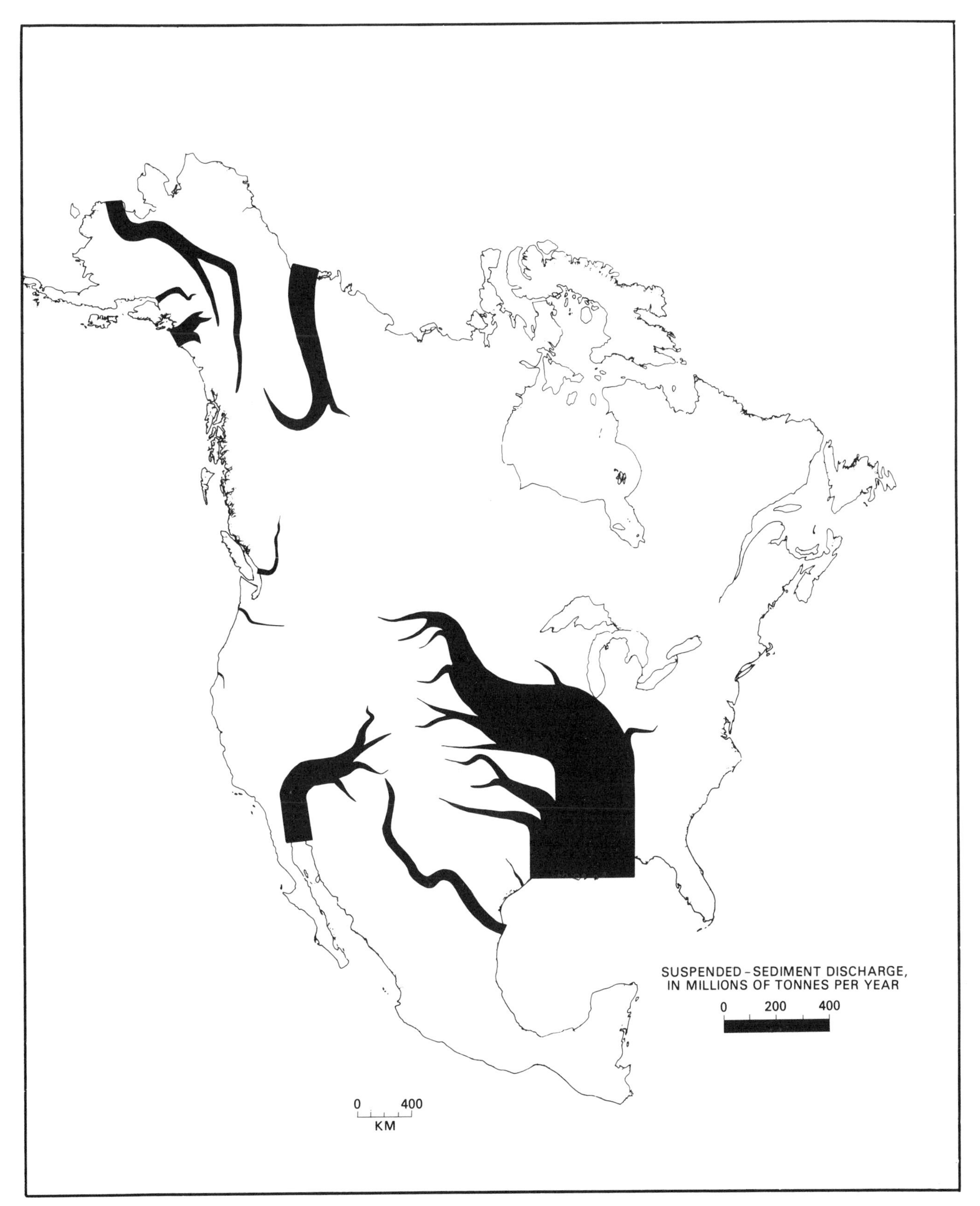

Figure 21. Average discharges of suspended sediment by major rivers of North America, estimated as of circa 1700, before the advent of any significant human impact. Width of river represents suspended-sediment discharge.

to be derived from rivers; most of the remainder of the deposited sediment is derived from erosion of the bluffs on the shores of the lakes. The export of sediment from the Great Lakes is relatively small. Suspended-sediment data collected monthly during a period of 8 years (1975–1982) from the St. Lawrence River at Massena, New York, below Lake Ontario (the lowest of the Great Lakes) suggests a mean suspended-sediment concentration of 6 mg/L (range 0 to 20 mg/L) which, when combined with the nearly constant water discharge that has averaged 6860 m^3/s during the past 120 years (1860–1980), gives a sediment discharge of about 1.3×10^6 tonnes per year.

The continental shelf is a significant sink for river sediment, especially along the west coast of the continent, where sediment yields are large and the river mouths are not separated from the Pacific Ocean by large estuaries. Especially prominent are the accumulations of glacio-fluvial sediments on the continental shelf in the Gulf of Alaska (Molnia and Carlson, 1978) and the river sediments that have accumulated off the coasts of Oregon and California (Griggs and Hein, 1980; Kulm and others, 1975; Schwalbach and Gorsline, 1985). Muds derived from modern rivers are present, but to a lesser degree, on the continental shelf of Texas (Shideler, 1977, 1978). On other continental shelves of North America, modern river sediment is an insignificant fraction of the existing sedimentary cover (Dinter, 1985; Milliman and others, 1972).

Only a small fraction, less than 5 percent, of the river sediment delivered to the coastal zones of North America under present conditions ever reaches the floor of the deep sea beyond the continental shelves. The few likely places for this to happen are in the southern California borderland (Schwalbach and Gorsline, 1985), the Middle America Trench that lies off the west coast of Mexico and Central America (Baltuck and others, 1985), or the outer Mississippi delta, which is draped over the continental slope (Davies, 1972).

REFERENCES CITED

Abrahams, A. D., and Kellerhals, R., 1973, Correlations between water discharge and concentration of suspended solids for four large prairie rivers, *in* Fluvial processes and sedimentation: Canada National Research Council, Subcommittee on Hydrology, Hydrology Symposium 9, p. 96–113.

Baker, E. T., 1984, Patterns of suspended particle distribution and transport in a large fjordlike estuary: Journal of Geophysical Research, v. 89, p. 6553–6566.

Baker, V. R., 1983, Late-Pleistocene fluvial systems, *in* Wright, H. E., Jr., ed., Late-Quaternary environments of the United States; Volume 1, The late Pleistocene, Porter, S. C., ed.: Minneapolis, University of Minnesota Press, p. 115–129.

Baltuck, M., von Huene, R., and Arnott, R. J., 1985, Sedimentology of the western continental slope of Central America, *in* von Huene, R., and Aubouin, J., Initial reports of the Deep Sea Drilling Project: Washington, D.C., U.S. Government Printing Office, v. 84, p. 921–937.

Bates, C. C., 1953, Rational theory of delta formation: American Association of Petroleum Geologists Bulletin, v. 37, p. 2119–2162.

Brown, W. M., III, and Ritter, J. R., 1971, Sediment transport and turbidity in the Eel River basin, California: U.S. Geological Survey Water-Supply Paper 1986, 70 p.

Brune, G. M., 1948, Rates of sediment production in midwestern United States: U.S. Soil Conservation Service SCS–TP–65, 40 p., 8 figs.

——, 1953, Trap efficiency of reservoirs: American Geophysical Union Transactions, v. 34, p. 407–418.

Burkham, D. E., 1985, An approach for appraising the accuracy of suspended-sediment data: U.S. Geological Survey Professional Paper 1333, 18 p.

Burrows, R. L., and Harrold, P. E., 1983, Sediment transport in the Tanana River near Fairbanks, Alaska: U.S. Geological Survey Water-Resources Investigations Report 83–4064, 116 p.

Campbell, I. A., 1977, Stream discharge, suspended sediment, and erosion rates in the Red Deer River basin, Alberta, Canada, *in* Erosion and solid matter transport in inland waters: International Association of Hydrological Sciences Publication 122, p. 244–259.

Church, M., 1972, Baffin Island sandurs; A study of arctic fluvial processes: Geological Survey of Canada Bulletin 216, 208 p.

——, 1983, Pattern of instability in a wandering gravel bed channel, *in* Collinson, J. D., and Lewin, J., eds., Modern and ancient fluvial systems: International Association of Sedimentologists Special Publication 6, p. 169–180.

——, 1985, Bed load in gravel-bed rivers; Observed phenomena and implications for computation: Canadian Society for Civil Engineering, Annual Conference, Saskatoon, p. 17–37.

Church, M., and Ryder, J. M., 1972, Paraglacial sedimentation; A consideration of fluvial processes conditioned by glaciation: Geological Society of America Bulletin, v. 83, p. 3059–3071.

Church, M., Kellerhals, R., and Ward, P.R.B., 1985, Sediment in the Pacific and Yukon region; Review and assessment: Environment Canada, Sediment Survey Section Report Series, 140 p.

Clague, J. J., Luternauer, J. L., and Hebda, R. J., 1983, Sedimentary environments and postglacial history of the Fraser Delta and lower Fraser valley, British Columbia: Canadian Journal of Earth Sciences, v. 20, p. 1314–1326.

Colby, B. R., 1956, Relationship of sediment discharge to streamflow: U.S. Geological Survey Open-File Report, 170 p., 7 pl.

Colby, B. R., and Hubbell, D. W., 1961, Simplified methods for computing total sediment discharge with the modified Einstein procedure: U.S. Geological Survey Water-Supply Paper 1593, 17 p., 8 pl.

Colby, B. R., Hembree, C. H., and Rainwater, F. H., 1956, Sedimentation and chemical quality of the surface waters of the Wind River basin, Wyoming: U.S. Geological Survey Water-Supply Paper 1373, 336 p.

Coleman, J. M., and Gagliano, S. M., 1964, Cyclic sedimentation in the Mississippi River deltaic plain: Gulf Coast Association of Geological Societies Transactions, v. 14, p. 67–80.

Collier, C. R., and Musser, J. J., 1964, Sedimentation, *in* Influences of strip mining on the hydrologic environment of parts of Beaver Creek basin, Kentucky, 1955–59: U.S. Geological Survey Professional Paper 427–B, p. B48–B64.

Costa, J. E., 1975, Effects of agriculture on erosion and sedimentation in the Piedmont Province, Maryland: Geological Society of America Bulletin, v. 86, p. 1281–1286.

Curtis, W. F., Culbertson, J. K., and Chase, E. B., 1973, Fluvial-sediment discharge to the oceans from the conterminous United States: U.S. Geological Survey Circular 670, 17 p.

Curwick, P. B., 1986, Model of diffusion and settling for suspended-sediment transport, *in* Wang, S. Y., Shen, H. W., and Ding, L. Z., eds., Proceedings, International Symposium on River Sedimentation, 3rd, Jackson, Mississippi: University of Mississippi School of Engineering, p. 1690–1700.

Davies, D. K., 1972, Deep sea sediments and their sedimentation, Gulf of Mexico: American Association of Petroleum Geologists Bulletin, v. 56, p. 2212–2239.

Davis, M. B., 1976, Erosion rates and land-use history in southern Michigan: Environmental Conservation, v. 3, p. 139–148.

Day, T. J., and Spitzer, M. O., 1985, Sediment station analysis, Oldman River near Brocket 05AA024: Environment Canada, Sediment Survey Section Report Series, 54 p.

Dendy, F. E., and Bolton, G. C., 1976, Sediment yield-runoff-drainage area relationships in the United States: Journal of Soil and Water Conservation, v. 31, p. 264–266.

Dickinson, W. T., 1981, Accuracy and precision of suspended sediment loads, *in* Erosion and sediment transport measurement: International Association of Hydrological Sciences Publication 133, p. 195–202.

Dickinson, W. T., and Wall, G. J., 1977, Temporal and spatial patterns in erosion and fluvial processes, *in* Research in fluvial geomorphology; Proceedings, Guelph Symposium on Geomorphology, 5th: Norwich, England, Geo-Abstracts Ltd., p. 133–148.

Dietrich, W. E., and Dunne, T., 1978, Sediment budget for a small catchment in mountainous terrain: Zeitschrift für Geomorphologie, neue folge, supplementary v. 29, p. 191–206.

Dinter, D. A., 1985, Quaternary sedimentation of the Alaskan Beaufort shelf; Influence of regional tectonics, fluctuating sea levels, and glacial sediment sources: Tectonophysics, v. 114, p. 133–161.

Dole, R. B., 1909, The quality of surface waters in the United States; Part I, Analyses of waters east of the one hundredth meridian: U.S. Geological Survey Water-Supply Paper 236, 123 p.

Dole, R. B., and Stabler, H., 1909, Denudation, *in* Papers on the conservation of water resources: U.S. Geological Survey Water-Supply Paper 234, p. 78–93.

Einstein, H. A., 1950, The bed-load function for sediment transportation in open channel flows: U.S. Department of Agriculture Technical Bulletin 1026, 71 p.

Emmett, W. W., 1980, A field calibration of the sediment-trapping characteristics of the Helley-Smith bedload sampler: U.S. Geological Survey Professional Paper 1139, 44 p.

——, 1981, Measurement of bed load in rivers, *in* Erosion and sediment transport measurement: International Association of Hydrological Sciences Publication 133, p. 3–15.

Emmett, W. W., Leopold, L. B., and Myrick, R. M., 1983, Some characteristics of fluvial processes in rivers, *in* Proceedings, International Symposium on River Sedimentation, 2nd, Nanjing: Beijing, Water Resources and Electric Power Press, p. 730–754.

Everett, D. E., 1971, Hydrologic and quality characteristics of the lower Mississippi River: Louisiana Department of Public Works Technical Report 5, 48 p.

Feltz, H. R., and Culbertson, J. K., 1972, Sampling procedures and problems in determining pesticide residues in the hydrologic environment: Pesticides Monitoring Journal, v. 6, p. 171–178.

Fisk, H. N., 1944, Geological investigation of the alluvial valley of the lower Mississippi River: Vicksburg, Mississippi River Commission, 78 p.

Fisk, H. N., 1947, Fine-grained alluvial deposits and their effects on Mississippi River activity: Vicksburg, U.S. Corps of Engineers Mississippi River Commission Waterways Experiment Station, v. 1, 82 p.; v. 2, 74 pl.

Fisk, H. N., and McFarlan, E., Jr., 1955, Late Quaternary deltaic deposits of the Mississippi River (local sedimentation and basin tectonics), *in* Poldervaart, A., ed., Crust of the earth; A symposium: Geological Society of America Special Paper 62, p. 279–302.

Fisk, H. N., McFarlan, E., Jr., Kolb, C. R., and Wilbert, L. J., Jr., 1954, Sedimentary framework of the modern Mississippi delta: Journal of Sedimentary Petrology, v. 24, p. 76–99.

Flaxman, E. M., and Hobba, R. L., 1955, Some factors affecting rates of sedimentation in the Columbia River basin: American Geophysical Union Transactions, v. 36, p. 293–303.

Foster, G. R., editor, 1977, Soil erosion; Prediction and control: Soil Conservation Society of America Special Publication 21, 393 p.

Frazier, D. E., 1967, Recent deltaic deposits of the Mississippi River; Their development and chronology: Gulf Coast Association of Geological Societies Transactions, v. 17, p. 287–315.

Gilbert, G. K., 1917, Hydraulic-mining débris in the Sierra Nevada: U.S. Geological Survey Professional Paper 105, 154 p.

Gould, H. R., 1970, The Mississippi delta complex, *in* Morgan, J. P., ed., Deltaic sedimentation, modern and ancient: Society of Economic Paleontologists and Mineralogists Special Publication 15, p. 3–30.

Gould, H. R., and McFarlan, E., Jr., 1959, Geologic history of the Chenier plain, southwestern Louisiana: Gulf Coast Association of Geological Societies Transactions, v. 9, p. 261–270.

Graves, W. P., and Eliab, P. L., 1977, Sediment study, alternative delta water facilities, peripheral canal plan: California Department of Water Resources, Central District, Sacramento, 117 p.

Griggs, G. B., and Hein, J. R., 1980, Sources, dispersal, and clay mineral composition of fine-grained sediment off central and northern California: Journal of Geology, v. 88, p. 541–566.

Guy, H. P., 1964, An analysis of some storm-period variables affecting stream sediment transport: U.S. Geological Survey Professional Paper 462–E, p. E1–E46.

——, 1965, Residential construction and sedimentation at Kensington, Md., *in* Proceedings, Federal Inter-Agency Sedimentation Conference, 1963: U.S. Department of Agriculture Miscellaneous Publication 970, p. 30–37.

——, 1969, Laboratory theory and methods for sediment analysis: U.S. Geological Survey Techniques of Water-Resources Investigations, book 5, chapter C1, 58 p.

Guy, H. P., and Norman, V. W., 1970, Field methods for measurement of fluvial sediment: U.S. Geological Survey Techniques of Water-Resources Investigations, book 3, chapter C2, 59 p.

Hadley, R. F., Lal, R., Onstad, C. A., Walling, D. E., and Yair, A., 1985, Recent developments in erosion and sediment yield studies: United Nations Educational Scientific and Cultural Organization, International Hydrological Programme, Technical Document in Hydrology, 127 p.

Happ, S. C., 1945, Sedimentation in South Carolina Piedmont valleys: American Journal of Science, v. 243, p. 113–126.

Happ, S. C., Rittenhouse, G., and Dobson, G. C., 1940, Some principles of accelerated stream and valley sedimentation: U.S. Department of Agriculture Technical Bulletin 695, 133 p.

Hathaway, G. A., 1948, Observations on channel changes, degradation, and scour below dams: International Association for Hydraulic Structures Research meeting, 2nd, Stockholm, Appendix 16, p. 287–307.

Hembree, C. H., Colby, B. R., Swenson, H. A., and Davis, J. R., 1952, Sedimentation and chemical quality of water in the Powder River drainage basin, Wyoming and Montana: U.S. Geological Survey Circular 170, 92 p.

Holeman, J. N., 1981, The national erosion inventory of the Soil Conservation Service, U.S. Department of Agriculture, 1977–79, *in* Erosion and sediment transport measurement: International Association of Hydrological Sciences Publication 133 p. 315–319.

Hubbell, D. W., 1964, Apparatus and techniques for measuring bedload: U.S. Geological Survey Water-Supply Paper 1748, 74 p.

——, 1987, Bedload sampling and analysis, *in* Thorne, C. R., Bathurst, J. C., and Hey, R. D., eds., Sediment transport in gravel-bed rivers: New York, John Wiley & Sons, p. 89–118.

Hubbell, D. W., and Stevens, H. H., Jr., 1986, Factors affecting accuracy of bedload sampling, *in* Proceedings, Federal Interagency Sedimentation Conference, 4th, Las Vegas, Volume 1: Interagency Advisory Committee on Water Data, Subcommittee on Sedimentation, p. 4-20–4-29.

Janda, R. J., 1979, Summary of regional geology in relation to geomorphic form and process, *in* Guidebook for a field trip to observe natural and management-related erosion in Franciscan terrane of northern California; Geological Society of America Cordilleran Section: Menlo Park, California, U.S. Geological Survey, p. II-1–II-17.

Janda, R. J., and Nolan, K. M., 1979, Stream sediment discharge in northwestern California, *in* Guidebook for a field trip to observe natural and management-related erosion in Franciscan terrane of northern California; Geological Society of America Cordilleran Section: Menlo Park, California, U.S.

Geological Survey, p. IV-1–IV-27.

Jordan, P. R., 1965, Fluvial sediment of the Mississippi River at St. Louis, Missouri: U.S. Geological Survey Water-Supply Paper 1802, 89 p.

—— , 1979, Relation of sediment yield to climatic and physical characteristics in the Missouri River basin: U.S. Geological Survey Water Resources Investigations 79–49, 26 p.

Kellerhals, R., Abrahams, A. D., and von Gaza, H., 1974, Possibilities for using suspended sediment rating curves in the Canadian Sediment Survey Program: Environment Canada, Sediment Survey Section Report Series, 86 p.

Kelley, R. L., 1959, Gold vs. grain; The hydraulic mining controversy in California's Sacramento Valley: Glendale, California, Arthur H. Clark, 327 p.

Kemp, A.L.W., and Harper, N. S., 1976, Sedimentation rates and a sediment budget for Lake Ontario: Journal of Great Lakes Research, v. 2, p. 324–339.

—— , 1977, Sedimentation rates in Lake Huron and Georgian Bay: Journal of Great Lakes Research, v. 3, p. 215–220.

Kemp, A.L.W., Anderson, T. W., Thomas, R. L., and Mudrochova, A., 1974, Sedimentation rates and recent sediment history of Lakes Ontario, Erie, and Huron: Journal of Sedimentary Petrology, v. 44, p. 207–218.

Kemp, A.L.W., MacInnis, G. A., and Harper, N. S., 1977, Sedimentation rates and a revised sediment budget for Lake Erie: Journal of Great Lakes Research, v. 3, p. 221–233.

Kemp, A.L.W., Dell, C. I., and Harper, N. S., 1978, Sedimentation rates and a sediment budget for Lake Superior: Journal of Great Lakes Research, v. 4, p. 276–287.

Kennedy, V. C., 1964, Sediment transported by Georgia streams: U.S. Geological Survey Water-Supply Paper 1668, 101 p.

Keown, M. P., Dardeau, E. A., Jr., and Causey, E. M., 1981, Characterization of the suspended-sediment regime and bed-material gradation of the Mississippi River Basin: U.S. Army Corps of Engineers, Lower Mississippi Valley Division, Potamology Program (P-1) Report 1, v. 1, 493 p.; v. 2, 379 p.

—— , 1986, Historic trends in the sediment flow regime of the Mississippi River: Water Resources Research, v. 22, p. 1555–1564.

Knott, J. M., 1974, Sediment discharge in the Trinity River basin, California: U.S. Geological Survey Water-Resources Investigation 49–73, 56 p.

Knott, J. M., and Lipscomb, S. W., 1985, Sediment discharge data for selected sites in the Susitna River basin, Alaska, October 1982 to February 1984: U.S. Geological Survey Open-File Report 85–157, 68 p.

Knox, J. C., 1977, Human impacts on Wisconsin stream channels: Association of American Geographers Annals, v. 67, p. 323–342.

—— , 1983, Responses of river systems to Holocene climates, *in* Wright, H. E., Jr., ed., Late-Quaternary environments of the United States; Volume 2, The Holocene: Minneapolis, University of Minnesota Press, p. 26–41.

—— , 1984, Fluvial response to small scale climate changes, *in* Costa, J. E., and Fleisher, P. J., eds., Developments and applications of geomorphology: Berlin, Springer-Verlag, p. 318–342.

—— , 1985, Responses of floods to Holocene climatic change in the upper Mississippi valley: Quaternary Research, v. 23, p. 287–300.

Kolb, C. R., and Van Lopik, J. R., 1966, Depositional environments of the Mississippi deltaic plain; Southeastern Louisiana, *in* Shirley, M. L., and Ragsdale, J. A., eds., Deltas in their geologic framework: Houston, Texas, Houston Geological Society, p. 17–61.

Kranck, K., and Ruffman, A., 1982, Sedimentation in James Bay: Naturaliste Canadienne, v. 109, p. 353–361.

Kulm, L. D., Roush, R. C., Harlett, J. C., Neudeck, R. H., Chambers, D. M., and Runge, E. J., 1975, Oregon continental shelf sedimentation; Interrelationships of facies distribution and sedimentary processes: Journal of Geology, v. 83, p. 145–175.

Langbein, W. B., and Schumm, S. A., 1958, Yield of sediment in relation to mean annual precipitation: American Geophysical Union Transactions, v. 39, p. 1076–1084.

Langbein, W. B., and others, 1949, Annual runoff in the United States: U.S. Geological Survey Circular 52, 14 p.

Last, W. M., 1984, Modern sedimentology and hydrology of Lake Manitoba, Canada: Environmental Geology, v. 5, p. 177–190.

Leopold, L. B., Wolman, M. G., and Miller, J. P., 1964, Fluvial processes in geomorphology: San Francisco, W. H. Freeman and Company, 522 p.

Luternauer, J. L., and Murray, J. W., 1973, Sedimentation of the western delta front on the Fraser River, British Columbia: Canadian Journal of Earth Sciences, v. 10, p. 1642–1663.

Magilligan, F. J., 1985, Historical floodplain sedimentation in the Galena River basin, Wisconsin and Illinois: Association of American Geographers Annals, v. 75, p. 583–594.

Madej, M. A., 1984, Recent changes in channel-stored sediment, Redwood Creek, California: U.S. National Park Service, Redwood National Park Technical Report 11, 54 p.

—— , 1989, Recent changes in channel-stored sediment, Redwood Creek, California, *in* Nolan, K. M., Kelsey, H. M., and Marron, D. C., eds., Geomorphic processes and aquatic habitat in the Redwood Creek basin, northwestern California: U.S. Geological Survey Professional Paper 1454 (in press).

Mackay, J. R., 1963, The MacKenzie delta area, N.W.T., Canada: Geological Branch Memoir 8, 202 p.

Mathews, W. H., and Shepard, F. P., 1962, Sedimentation of Fraser River delta, British Columbia: American Association of Petroleum Geologists Bulletin, v. 46, p. 1416–1443.

McLean, D. G., and Church, M. A., 1986, A re-examination of sediment transport observations in the lower Fraser River: Environment Canada, Sediment Survey Section Report Series (IWD–HQ–WRB–SS–86–6), 56 p.

Meade, R. H., 1969, Landward transport of bottom sediments in estuaries of the Atlantic Coastal Plain: Journal of Sedimentary Petrology, v. 39, p. 222–234.

—— , 1976, Sediment problems in the Savannah River basin, *in* Dillman, B. L., and Stepp, J. M., eds., The future of the Savannah River: Clemson University Water Resources Research Institute, p. 105–129.

—— , 1982, Sources, sinks, and storage of river sediment in the Atlantic drainage of the United States: Journal of Geology, v. 90, p. 235–252.

—— , 1985, Wavelike movement of bedload sediment, East Fork River, Wyoming: Environmental Geology and Water Sciences, v. 7, p. 215–225.

Meade, R. H., and Parker, R. S., 1985, Sediment in rivers of the United States, *in* National water summary 1984: U.S. Geological Survey Water-Supply Paper 2275, p. 49–60.

Meade, R. H., Emmett, W. W., and Myrick, R. M., 1981, Movement and storage of bed material during 1979 in East Fork River, Wyoming, USA, *in* Davies, T.R.H., and Pearce, A. J., eds., Erosion and sediment transport in Pacific rim steeplands: International Association of Hydrological Sciences Publication 132, p. 225–235.

Meade, R. H., Nordin, C. F., Jr., Pérez-Hernández, D., Mejía-B., A., and Pérez-Godoy, J. M., 1983, Sediment and water discharge in Rio Orinoco, Venezuela and Colombia, *in* Proceedings, International Symposium on River Sedimentation, 2nd, Nanjing: Beijing, Water Resources and Electric Power Press, p. 1134–1144.

Meade, R. H., Dunne, T., Richey, J. E., Santos, U. de M., and Salati, E., 1985, Storage and remobilization of suspended sediment in the lower Amazon River of Brazil: Science, v. 228, p. 488–490.

Megahan, W. F., 1975, Sedimentation in relation to logging activities in the mountains of central Idaho, *in* Present and prospective technology for predicting sediment yields and sources: U.S. Agricultural Research Service ARS–S–40, p. 74–82.

Miller, C. R., 1951, Analysis of flow-duration, sediment-rating curve method of computing sediment yield: Denver, U.S. Bureau of Reclamation, 55 p.

Milliman, J. D., 1980, Sedimentation in the Fraser River and its estuary, southwestern British Columbia (Canada): Estuarine and Coastal Marine Science, v. 10, p. 609–633.

Milliman, J. D., and Meade, R. H., 1983, World-wide delivery of river sediment to the oceans: Journal of Geology, v. 91, p. 1–21.

Milliman, J. D., Pilkey, O. H., and Ross, D. A., 1972, Sediments of the continental margin off the eastern United States: Geological Society of America Bulletin, v. 83, p. 1315–1333.

Molnia, B. F., 1979, Sedimentation in coastal embayments, northeastern Gulf of Alaska: Houston, Texas, Offshore Technology Conference, 11th, Proceedings, p. 665–670.

Molnia, B. F., and Carlson, P. R., 1978, Surface sedimentary units of northern Gulf of Alaska continental shelf: American Association of Petroleum Geologists Bulletin, v. 62, p. 633–643.

Murchie, S. L., 1985, ^{210}Pb dating and the recent geologic history of Crystal Bay, Lake Minnetonka, Minnesota: Limnology and Oceanography, v. 30, p. 1154–1170.

Nichols, M. M., 1986, Storage efficiency of estuaries, *in* Wang, S. Y., Shen, H. W., and Ding, L. Z., eds., Proceedings, International Symposium on River Sedimentation, 3rd, Jackson, Mississippi: University of Mississippi School of Engineering, p. 273–289.

Nolan, K. M., and Janda, R. J., 1981, Use of short-term water and suspended-sediment discharge observations to assess impacts of logging on stream-sediment discharge in the Redwood Creek basin, northwestern California, U.S.A., *in* Davies, T.R.H., and Pearce, A. J., eds., Erosion and sediment transport in Pacific rim steeplands: International Association of Hydrological Sciences Publication 132, p. 415–437.

Nolan, K. M., Janda, R. J., and Galton, J. H., 1986, Sediment sources and sediment-transport curves, *in* Proceedings, Federal Interagency Sedimentation Conference, 4th, Las Vegas, Volume 1: Interagency Advisory Committee on Water Data, Subcommittee on Sedimentation, p. 4-70–4-79.

Nordin, C. F., Jr., 1981, The sediment discharge of rivers; A review, *in* Erosion and sediment transport measurement: International Association of Hydrological Sciences Publication 133, v. 2, p. 3–47.

—— , 1985, The sediment loads of rivers, *in* Rodda, J. C., ed., Facets of hydrology II: New York, John Wiley and Sons, p. 184–203.

Nordin, C. F., Jr., and Beverage, J. P., 1965, Sediment transport in the Rio Grande, New Mexico: U.S. Geological Survey Professional Paper 462–F, p. F1–F35.

Nordin, C. F., Jr., Cranston, C. C., and Mejía-B., A., 1983, New technology for measuring water and suspended-sediment discharge in large rivers, *in* Proceedings, International Symposium on River Sedimentation, 2nd, Nanjing: Beijing, Water Resources and Electric Power Press, p. 1145–1158.

Osterkamp, W. R., 1976, Variation and causative factors of sediment yields in the Arkansas River basin, Kansas, *in* Proceedings, Federal Interagency Sedimentation Conference, 3rd, Denver: Water Resources Council Sedimentation Committee, p. 1-59–1-70.

Patric, J. H., 1976, Soil erosion in the eastern forest: Journal of Forestry, v. 74, p. 671–677.

Patric, J. H., Evans, J. O., and Helvey, J. D., 1984, Summary of sediment yield data from forested land in the United States: Journal of Forestry, v. 82, p. 101–104.

Piest, R. F., and Miller, C. R., 1975, Sediment sources and sediment yields, *in* Vanoni, V. A., ed., Sedimentation engineering: American Society of Civil Engineers Manual and Report on Engineering Practice 54, p. 437–493.

Piest, R. F., Kramer, L. A., and Heinemann, H. G., 1975, Sediment movement from loessial watersheds, *in* Present and prospective technology for predicting sediment yields and sources: U.S. Agricultural Research Service ARS–S–40, p. 130–141.

Porterfield, G., 1972, Computation of fluvial-sediment discharge: U.S. Geological Survey Techniques of Water-Resources Investigations, book 3, chapter C3, 66 p.

Rasid, H., 1979, The effects of regime regulation by the Gardiner Dam on downstream geomorphic processes in the South Saskatchewan River: Canadian Geographer, v. 23, p. 140–158.

Reimnitz, E., 1966, Late Quaternary history and sedimentation of the Copper River delta and vicinity, Alaska [Ph.D. thesis]: San Diego, University of California, 160 p.

Renfro, G. W., 1975, Use of erosion equations and sediment-delivery ratios for predicting sediment yield, *in* Present and prospective technology for predicting sediment yields and sources: U.S. Agricultural Research Service ARS–S–40, p. 33–45.

Robbins, L. G., 1977, Suspended sediment and bed material studies on the lower Mississippi River: U.S. Army Engineer District, Vicksburg, Potamology Investigations Report 300–1, 29 p.

Roberts, W. P., and Pierce, J. W., 1974, Sediment yield in the Patuxent River (Md.) undergoing urbanization, 1968–1969: Sedimentary Geology, v. 12, p. 179–197.

Roehl, J. W., 1962, Sediment source areas, delivery ratios, and influencing morphological factors: International Association of Scientific Hydrology Publication 59, p. 202–213.

Saucier, R. T., 1974, Quaternary geology of the lower Mississippi Valley: Arkansas Archeological Survey Research Series 6, 26 p.

—— , 1981, Current thinking on riverine processes and geologic history as related to human settlement in the Southeast: Geoscience and Man, v. 22, p. 7–18.

Saucier, R. T., and Lindfors-Kearns, F. E., 1989, Geomorphology, stratigraphy, and chronology, *in* Autin, W. J., Saucier, R. T., and Lindfors-Kearns, F. E., Lower Mississippi Valley, *in* Morrison, R. B., ed., Quaternary non-glacial geology; conterminous U.S.: Boulder, Colorado, Geological Society of America, The Geology of North America, v. K–2 (in press).

Schumm, S. A., 1968, Speculations concerning paleohydrologic controls of terrestrial sedimentation: Geological Society of America Bulletin, v. 79, p. 1573–1588.

—— , 1977, The fluvial system: New York, John Wiley and Sons, 338 p.

Schumm, S. A., and Hadley, R. F., 1961, Progress in the application of landform analysis in studies of semiarid erosion: U.S. Geological Survey Circular 437, 14 p.

Schwalbach, J. R., and Gorsline, D. S., 1985, Holocene sediment budgets for the basins of the California continental borderland: Journal of Sedimentary Petrology, v. 55, p. 829–842.

Scott, K. M., and Williams, R. P., 1978, Erosion and sediment yields in the Transverse Ranges, southern California: U.S. Geological Survey Professional Paper 1030, 38 p.

Scruton, P. C., 1960, Delta building and the deltaic sequence, *in* Shepard, F. P., Phleger, F. B., and van Andel, T. H., eds., Recent sediments, northwest Gulf of Mexico: Tulsa, American Association of Petroleum Geologists, p. 82–102.

Shideler, G. L., 1977, Late Holocene sedimentary provinces, south Texas outer continental shelf: American Association of Petroleum Geologists Bulletin, v. 61, p. 708–722.

—— , 1978, A sediment-dispersal model for the south Texas continental shelf, northwest Gulf of Mexico: Marine Geology, v. 26, p. 289–313.

Simmons, C. E., 1976, Sediment characteristics of streams in the eastern Piedmont and western Coastal Plain regions of North Carolina: U.S. Geological Survey Water-Supply Paper 1798–O, p. O1–O32.

Slaymaker, O., and McPherson, H. J., 1977, An overview of geomorphic processes in the Canadian Cordillera: Zeitschrift für Geomorphologie, v. 21, p. 169–186.

Smith, B. J., 1965, Sedimentation in the San Francisco Bay system, *in* Proceedings, Federal Interagency Sedimentation Conference, 1963: U.S. Department of Agriculture Miscellaneous Publication 970, p. 675–708.

Stichling, W., 1974, Sediment loads in Canadian rivers: Canada Inland Waters Directorate, Water Resources Branch Technical Bulletin 74, 27 p.

Strand, R. I., 1975, Bureau of Reclamation procedures for predicting sediment yield, *in* Present and prospective technology for predicting sediment yields and sources: U.S. Agricultural Research Service ARS–S–40, p. 10–15.

Stump, D. E., Jr., and Mastrilli, T. M., 1985, Effects of surface mining on streamflow, suspended-sediment, and water quality in the Stony Fork drainage basin, Fayette County, Pennsylvania: U.S. Geological Survey Water-Resources Investigations Report 84–4362, 28 p.

Swanston, D. N., and Swanson, F. J., 1976, Timber harvesting, mass erosion, and steepland forest geomorphology in the Pacific Northwest, *in* Coates, D. R., ed., Geomorphology and engineering: Stroudsburg, Pennsylvania, Dowden, Hutchinson and Ross, p. 199–221.

Thompson, R. W., 1968, Tidal flat sedimentation on the Colorado River delta, northwestern Gulf of California: Geological Society of America Memoir 107, 133 p.

Trimble, S. W., 1969, Culturally accelerated sedimentation in the middle Georgia

Piedmont [M.S. thesis]: Athens, University of Georgia, 110 p. (Reissued in 1971 by U.S. Soil Conservation Service, Fort Worth, Texas.)

——, 1974, Man-induced soil erosion in the southern Piedmont, 1700–1970: Ankeny, Iowa, Soil Conservation Society of America, 180 p.

——, 1977, The fallacy of stream equilibrium in contemporary denudation studies: American Journal of Science, v. 277, p. 876–887.

——, 1983, A sediment budget for Coon Creek basin in the Driftless area, Wisconsin, 1853–1977: American Journal of Science, v. 283, p. 454–474.

Trimble, S. W., and Lund, S. W., 1982, Soil conservation and the reduction of sedimentation and erosion in the Coon Creek basin, Wisconsin: U.S. Geological Survey Professional Paper 1234, 35 p.

Ursic, S. J., and Dendy, F. E., 1965, Sediment yields from small watersheds under various land uses and forest covers, *in* Proceedings, Federal Interagency Sedimentation Conference, 1963: U.S. Department of Agriculture Miscellaneous Publication 970, p. 47–52.

Vanoni, V. A., ed., 1975, Sedimentation engineering: American Society of Civil Engineers Manual and Report on Engineering Practice 54, 745 p.

——, 1984, Fifty years of sedimentation: American Society of Civil Engineers Journal of Hydraulic Engineering, v. 110, p. 1022–1057.

Vice, R. B., Guy, H. P., and Ferguson, G. E., 1969, Sediment movement in an area of suburban highway construction, Scott Run basin, Fairfax County, Virginia, 1961–64: U.S. Geological Survey Water-Supply Paper 1591–E, p. E1–E41.

Walling, D. E., 1977, Suspended sediment and solute response characteristics of the River Exe, Devon, England, *in* Research in fluvial geomorphology; Proceedings, Guelph Symposium on Geomorphology, 5th: Norwich, England, GeoAbstracts, Ltd., p. 169–197.

——, 1983, The sediment delivery problem, *in* Rodríguez-Iturbe, I., and Gupta, V. K., eds., Scale problems in hydrology: Journal of Hydrology, v. 65, p. 209–237.

Walling, D. E., and Webb, B. W., 1981, The reliability of suspended sediment load data, *in* Erosion and sediment transport measurement: International Association of Hydrological Sciences Publication 133, p. 177–194.

——, 1983, Patterns of sediment yield, *in* Gregory, K. J., ed., Background to palaeohydrology: New York, John Wiley and Sons, p. 69–100.

Wark, J. W., and Keller, F. J., 1963, Preliminary study of sediment sources and transport in the Potomac River basin: Interstate Commission on the Potomac River Basin Technical Bulletin 1963–11, 28 p.

Warwick, W. F., 1980, Palaeolimnology of the Bay of Quinte, Lake Ontario; 2,800 years of cultural influence: Canadian Bulletin of Fisheries and Aquatic Sciences 206, 117 p.

Water Survey of Canada, 1984, Sediment data, Canadian rivers, 1982: Water Survey of Canada, 268 p.

Wells, F. C., 1980, Hydrology and water quality of the lower Mississippi River: Louisiana Office of Public Works Technical Report 21, 83 p.

Whitfield, P. H., and Schreier, H., 1981, Hysteresis in relationships between discharge and water chemistry in the Fraser River basin, British Columbia: Limnology and Oceanography, v. 26, p. 1179–1182.

Wilkinson, B. H., and Basse, R. A., 1978, Late Holocene history of the central Texas coast from Galveston Island to Pass Cavallo: Geological Society of America Bulletin, v. 89, p. 1592–1600.

Williams, G. P., and Wolman, M. G., 1984, Downstream effects of dams on alluvial rivers: U.S. Geological Survey Professional Paper 1286, 83 p.

——, 1986, Effects of dams and reservoirs on surface-water hydrology; Changes in rivers downstream from dams, *in* National water summary 1985: U.S. Geological Survey Water-Supply Paper 2300, p. 83–88.

Williams, K. F., and George, J. R., 1968, Preliminary appraisal of stream sedimentation in the Susquehanna River basin: U.S. Geological Survey Open-File Report, 49 p.

Wischmeier, W. H., and Smith, D. D., 1965, Predicting rainfall-erosion losses from cropland east of the Rocky Mountains: U.S. Department of Agriculture Agricultural Handbook 282, 47 p.

Wolman, M. G., 1967, A cycle of sedimentation and erosion in urban river channels: Geografiska Annaler, v. 49A, p. 385–395.

Wolman, M. G., and Miller, J. P., 1960, Magnitude and frequency of forces in geomorphic processes: Journal of Geology, v. 68, p. 54–74.

Wolman, M. G., and Schick, A. P., 1967, Effects of construction on fluvial sediment, urban and suburban areas of Maryland: Water Resources Research, v. 3, p. 451–464.

Wood, P. A., 1977, Controls of variation in suspended sediment concentration in the River Rother, West Sussex, England: Sedimentology, v. 24, p. 437–445.

Yorke, T. H., and Herb, W. J., 1978, Effects of urbanization on streamflow and sediment transport in the Rock Creek and Anacostia River basins, Montgomery County, Maryland, 1962–74: U.S. Geological Survey Professional Paper 1003, 71 p.

Yorke, T. H., and Ward, J. R., 1986, Accuracy of sediment discharge estimates, *in* Proceedings, Federal Interagency Sedimentation Conference, 4th, Las Vegas, Volume 1: Interagency Advisory Committee on Water Data, Subcommittee on Sedimentation, p. 4-49–4-59.

Yuzyk, T. R., 1983, Lake Diefenbaker, Saskatchewan: A case study of reservoir sedimentation: Environment Canada, Sediment Survey Section Report Series, 125 p.

——, 1984, Reservoir sedimentation studies in Canada: Environment Canada, Sediment Survey Section Report Series, 35 p.

MANUSCRIPT ACCEPTED BY THE SOCIETY AUGUST 28, 1987

ACKNOWLEDGMENTS

We thank our predecessors and colleagues in the U.S. Geological Survey and the Water Survey of Canada for collecting the samples, analyzing the data, and originating many of the ideas contained in our review. We also thank the people of the U.S. Bureau of Reclamation, U.S. Soil Conservation Service, U.S. Agricultural Research Service, U.S. Army Corps of Engineers, and the Prairie Farm Rehabilitation Administration of Canada, who have measured and studied sediment in North American rivers. Without their decades of dedication, we would not have had much to write about. We thank Reds Wolman, who not only invited us to prepare this chapter, but who, over the years, has been an inexhaustible source of ideas, enthusiasm, and encouragement to our generation of students of river sediment. We thank Dale Bray (University of New Brunswick) and Waite Osterkamp (U.S. Geological Survey) for their thoughtful reviews of the manuscript.

Printed in U.S.A.

The Geology of North America
Vol. O-1, Surface Water Hydrology
The Geological Society of America, 1990

Chapter 12

The riverscape

M. Gordon Wolman
Department of Geography and Environmental Engineering, The Johns Hopkins University, Baltimore, Maryland 21218
Michael Church
Department of Geography, University of British Columbia, 217, 1984 West Mall, Vancouver, British Columbia V6T 1W5, Canada
Robert Newbury
Box 1173, Gibsons, British Columbia V0N 1VO, Canada
Michel Lapointe
Department of Geography, Burnside Hall, McGill University, 805 rue Sherbrooke Ouest, Montreal Quebec H3A 2K6, Canada
Marcel Frenette
Faculte des Sciences et de Genie, Departement de genie civil, Universite Laval, Cite Universitaire, Quebec, G1K 7P4, Canada
E. D. Andrews
U.S. Geological Survey, MS 413, Box 25046, Denver Federal Center, Denver, Colorado 80225-0046
Thomas E. Lisle
Redwood Service Laboratory, U.S. Department of Agriculture, 1700 Bayview Drive, Arcata, California 95521
John P. Buchanan
Department of Geology, Eastern Washington University, Cheney, Washington 99004
Stanley A. Schumm
Department of Earth Resources, Colorado State University, Fort Collins, Colorado 80523
Brien R. Winkley
P.O. Box 543, Vicksburg, Mississippi 39108

INTRODUCTION

M. G. Wolman

A tenet of geomorphology and hydrology holds that rivers are ultimately the products of processes occurring on the landscape. Geology, topography, and climate reflected in rainfall, runoff, and vegetation determine the characteristics of the flow of water in the river channel as well as the quantities and characteristics of clastic and dissolved materials transported by rivers. Where river channels are unimpeded by bedrock and are truly alluvial—capable of adjusting to changes in flow and sediment input, and flowing within materials transported by the river itself—it has been demonstrated that rivers adjust to the flows of water and sediment derived from the watershed. At the same time, geologists recognize that many river reaches reflect history and local influences as well.

These essays on the riverscape attempt to capture these diverse hydrologic and other influences determining the look of a river at a given place. The riverscape here means the river itself and its immediate surroundings. The riverscapes described are ones for which good information is available. They demonstrate a variety of influences and conditions under which particular ones dominate the scene. Each, of course, could warrant a book, and these vignettes are not intended to be comprehensive, but rather illustrative. (The editors, not the authors, are responsible for the constrained space.) The rivers chosen have distinct qualities. The Nelson River, for example, is one of the large rivers of the world; it flows northward from proximate sources in Lake Winnipeg, but its sources are also in the Rocky Mountains, draining to the Saskatchewan River. Its behavior locally is related both to the bedrock of the Canadian Shield and to freeze-up and thaw of the river in draining northward at high latitudes. By contrast, the vast alluvial plain of the Mackenzie Delta is a system dominated not only by the hydrology of a high-latitude climate, but one in which the alluvial pattern changes rapidly within short periods of time.

The setting of the lower St. Lawrence reflects hydrologic control by the Great Lakes but within a geologic path determined by the contact of crystalline in the north and sedimentary rocks to the south. In contrast, the Fraser is markedly influenced by the complex history of aggradation and incision associated with the Pleistocene, which determines the form of the valley and its relief. Each of the rivers in the drier climate of the west and southwest United States—the Colorado, and Eel and the Niobrara—represents a different facet of both the hydrologic regime and the geography of the drainage basin. The Eel River in northern California carries one of the highest sediment loads of any river in the

Wolman, M. G., Church, M., Newbury, R., Lapointe, M., Frenette, M., Andrews, E. D., Lisle, T. E., Buchanan, J. P., Schumm, S. A., and Winkley, B. R., 1990, The riverscape, *in* Wolman, M. G., and Riggs, H. C., eds., Surface water hydrology: Boulder, Colorado, Geological Society of America, The Geology of North America, v. O-1.

world. The high concentration of sediment is determined by the climate and the erosive bedrock topography of the watershed. Hillslope processes combine with high seasonal rainfall to generate the sediment that influences the behavior of the river itself. The Eel is relatively unaffected by human activities.

Although rapids in the northern rivers have been dammed, and the Eel has recently been influenced by land use, none reflects the same magnitude of human influence as the other riverscapes. The present pattern and form of the Niobrara on the Great Plains, for example, is dominated by the effects of human activities. Both flow and flow variability in the Niobrara have been reduced by human interference, and the channel has metamorphosed from a highly variable braided river to a smaller, more stable one.

The Colorado River is much larger than the Eel and the Niobrara, and the behavior of water and sediment are not only affected by major changes in climate but by the construction of large reservoirs. The river in the Grand Canyon is clearly constrained by the canyon itself, but nevertheless the riverscape displays the deposition of sediment in bars and terraces as well as recurring cycles of cut and fill. Marked changes in the sediment load in the Colorado River antedate reservoir construction and are related to the complex response of rivers and gully systems to climatic change. Thus, the Colorado illustrates the complexity of climatic responses in arid and semiarid regions as well as the influence of human modification.

The riverscape of the lower Mississippi, the largest river on the North American continent, demonstrates the interplay of natural and man-made forces on a large alluvial system. The lower Mississippi is often characterized as a textbook example of a true alluvial river in which channel form and valley topography are adjusted to the interaction of water and sediment delivered from upstream. Its classic meandering pattern has been much described. At the same time, as the essay here shows, this huge and classic river has, in fact, been markedly altered by levees, dams, and other control works.

While the principle that water and sediment from the landscape control the characteristics of the riverscape is a valuable generalization, the brief descriptions of the riverscapes given here indicate a pervasive effect of both history and man on the behavior and configuration of a river at a particular location. Geologic history may be reflected in the local topography as well as in the nature and distribution of sediment and flow contributed to the river. Human activities, as the chapter by Hirsch and others (this volume) also documents, may markedly alter the riverscape both by altering the hydrology of the river drainage basin and the riverscape itself.

FRASER RIVER IN CENTRAL BRITISH COLUMBIA

Michael Church

INTRODUCTION

The Fraser River drains the heartland of British Columbia. The source of the river traditionally is placed near Mount Robson, the highest peak in the Canadian Rockies (in fact, it rises 90 km to the southeast), and the river then flows 1,360 km to the sea near Vancouver, Canada's principal west coast city. It drains 230,000 km^2 of the country between the Rocky Mountains and Columbia Mountains to the east, and the Coast Mountains to the west (see Fig. 1). Most of it is rolling upland on the Chilcotin, Cariboo, Fraser, and Thompson Plateaus, a region of ranching, logging, mining, and agricultural activities that form the traditional economic base of interior British Columbia life.

The Fraser remains unimpounded and has the greatest run of Pacific salmon (*Onchorhyncus* spp.) remaining. The Indian cultures of the region have depended on the salmon for the more than 10,000 years they have been here, as does a major commercial fishery today.

Although the first overland European explorer, Alexander Mackenzie, travelled the upper river on his journey to the Pacific in 1793, the North West Company's second exploring party into the region made the first descent of the awesome canyons of the lower Fraser. Simon Fraser—for whom the river is named—and John Stuart reached the sea in 1808. Within 50 years, placer gold was discovered in the river. Many of the California gold seekers moved north for this second western gold rush. Ultimately, the richest strikes were made along the Quesnel and Cariboo Rivers—left-bank tributaries in the Interior Plateau—and in the hills to the north. Sternwheel riverboats began to ply the lower river, the Cariboo wagon road was pushed northward, and eventually a road was forced through the canyons. In the 1880s the Canadian Pacific Railway came to British Columbia: laying the track through the Fraser Canyon was an engineering triumph of its day. The feat was achieved largely by indentured Chinese "navvies," whose descendents have been a part of British Columbian society ever since. The river has been a dominating factor in many of the major events of British Columbia's history, and directs the patterns of commerce today.

HISTORY OF THE FRASER RIVER

The hydrography of the Fraser River is remarkable. It holds a course to the sea—as do several of the largest west coast rivers—across a major mountain range. Inland, the tributaries north of 52° latitude, and eventually the Fraser itself above Prince George, flow with a northerly set.

Elements of the history of the region were first worked out during studies of the Cariboo placers (cf., Lay, 1940). The ancestral drainage of the region goes back more than 10 m.y. The Tertiary history of the Interior Plateau region has been punctuated by episodes of volcanism, the dates of which indicate the antiquity of intercalated fluvial and lacustrine deposits, the latter perhaps resulting from interference with normal drainage by the lavas. Middle Eocene lavas (Mathews and Rouse, 1984: the following descriptions are based on their paper) are overlain by cobble conglomerates indicating north to northeasterly drainage near 51.5°N. Palynomorphs indicate a warm, probably subtropical, climate. Above an unconformity, fining-upward Miocene

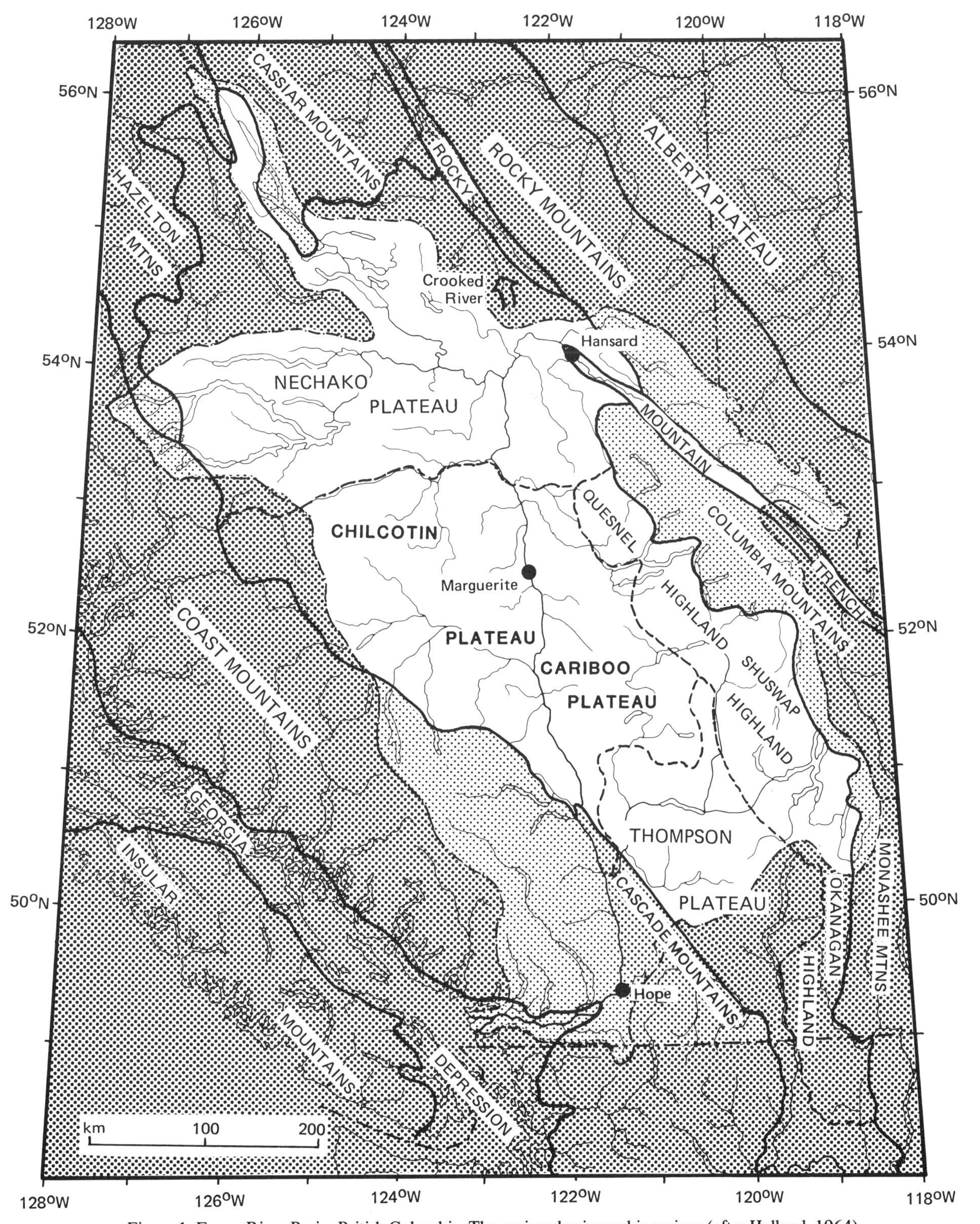

Figure 1. Fraser River Basin, British Columbia. The major physiographic regions (after Holland, 1964) are shown and the Interior Plateau region is highlighted.

Figure 2. Fraser River near Pavilion, British Columbia (view upstream), near mean flow. The Quaternary fill is seen within the older valley. Bedrock outcrops are visible at river level and eroding bluffs are evident along the recent trench.

conglomerates still indicate northerly streamflow, and warm temperature conditions. Finally, Pliocene lavas (ca. 2 Ma) occupy a linear zone along Fraser Valley—as if they originally flooded the ancestral valley. The most recent uplift of the Coast Mountains occurred during the Plio-Pleistocene period. There is no record of the southern Fraser River through all this time.

The history of Quaternary glaciation in the region is known to be long, but no details are known of events more than 50,000 years ago. The final glaciers spread into the plateaus from the Coast Mountains to the west and from the Columbia and Rocky Mountains to the east (cf., Tipper, 1971), and eventually coalesced in the vicinity of the Fraser River. Subsequent ice flow was both northward and southward from near 52°N. This pattern presumably repeated what had occurred many times previously during the Pleistocene Epoch. At the end of at least the final glacial period, the breakup and downwasting of plateau ice left the main valleys blocked so that large proglacial lakes developed. Major spillways were associated with these lakes.

The reversal of the northern drainage and the development of the modern hydrography remain enigmatic. There are at least three candidate mechanisms: (1) reversal as a result of drainage interference by lava flows (possibly the Pliocene lavas); (2) reversal as a result of drainage interference by glacier ice in middle or early Pleistocene time (the scale of late-glacial spillways indicates the feasibility of this); and (3) stream piracy by a southern proto–Fraser River as the result of regional uplift and steepening of that river.

In the latter case, the pirate might have been a former tributary of Chilcotin River working its way north near Williams Lake. Insofar as Fraser River, from Chilcotin River south, follows a major fault zone all the way through the canyons, this development could have been relatively late and rapid. It is supposed that before capture the northern proto–Fraser River drained to the ancient Peace River via Crooked River, north of Prince George (Fig. 1).

In the Fraser Valley of the central interior, the Pleistocene valley fill constitutes a complex sequence of lacustrine sediments, tills, and outwashes, capped in many places by Holocene alluvial fans and slopewash deposits. The river has, in the last 10,000 years, degraded through these deposits, so that today it flows on or near bedrock confined between bluffs of Quaternary and older materials (Fig. 2). The river flows, then, in a trench within the Pleistocene valley, itself in places cut within the late Tertiary lavas and the valley-fill gravels that they cap (Fig. 3). North of 52°, the broad, mid-Tertiary valley still exists. The river flows 150 to 500 m below the plateau surface of the surrounding country. This setting determines the hydrologic regime of the river today.

PRESENT-DAY HYDROLOGY AND SEDIMENT TRANSPORT

The interior plateaus of British Columbia lie within the rain shadow of the Coast Mountains and are subhumid. Nonetheless, a substantial accumulation of snow occurs on the plateau in winter, and in the surrounding mountains, accumulation is heavy. The nival runoff in spring is the major hydrologic event of the Fraser River. Indeed, so large is the drainage area that the spring snowmelt is the only annual event to substantially influence the entire basin within a relatively short time period—the event dominates flow to the point that the annual hydrograph resembles a single-event hydrograph (see Fig. 4).

Of course, there are additional details. The pattern of spring warming influences the speed of freshet rise and the timing of peak flow, which may fall any time between the last days of May and late June. The flood of record on the river, in 1948, was produced by a very cold spring and large snow accumulation, followed by abrupt warming and very wet conditions in May. In this case, a prolonged wet period influenced the peak flows. More usually, autumn rainstorms produce noticeable, though relatively minor rises of the river. A consequence of the entrenchment of the river is that, despite the characteristic summer drought, most irrigation water is obtained not from the river but from its small upland tributaries.

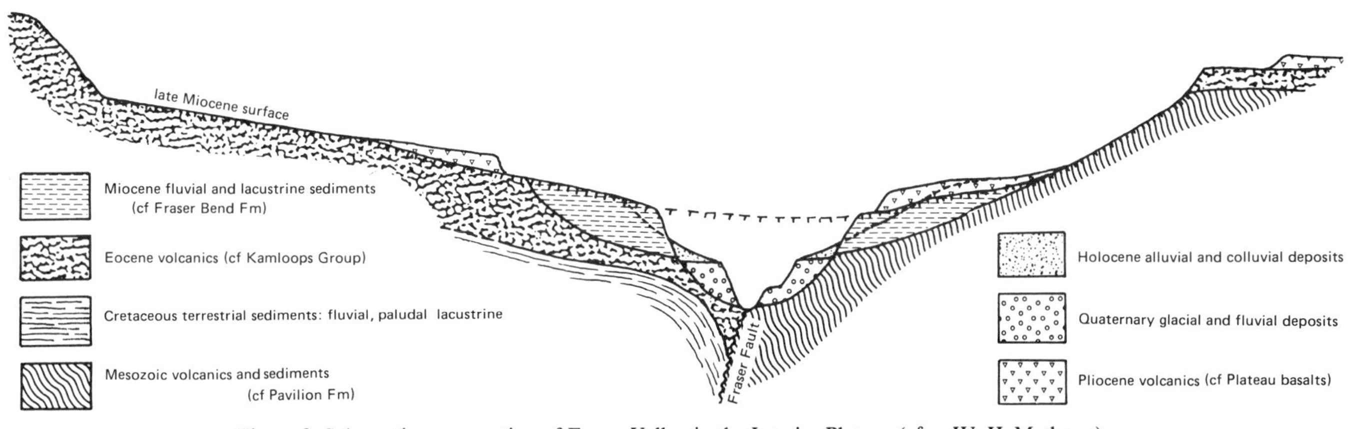

Figure 3. Schematic cross section of Fraser Valley in the Interior Plateau (after W. H. Mathews).

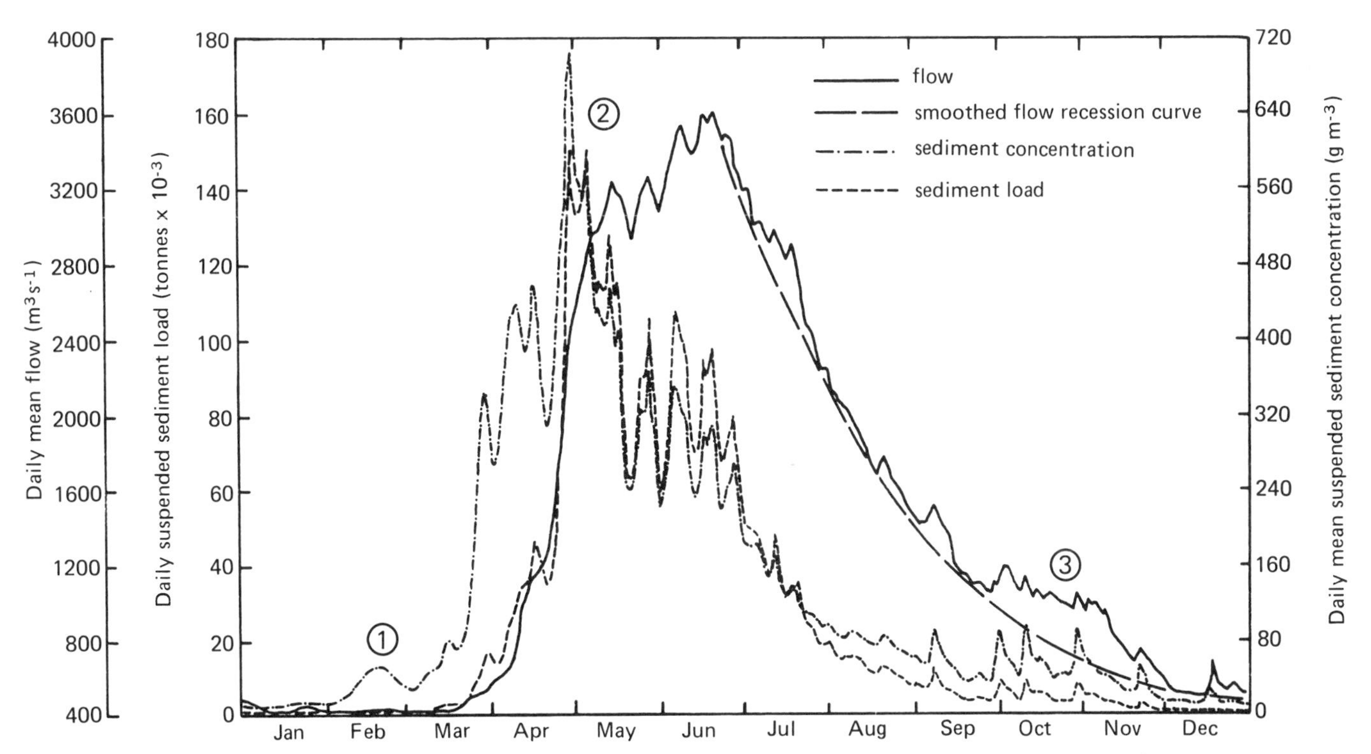

Figure 4. Hydrograph and suspended-sediment graph for Fraser River at Marguerite (Water Survey of Canada Station 08MC018), daily means of 1971 to 1981 observations. 1, Early spring pulse of high suspended sediment concentration; 2, Peak of suspended sediment load leads the peak flow; 3, Autumn rainstorm runoff.

The Quaternary valley fills in the Fraser River basin are unconsolidated and highly erodible. Underlying Tertiary and Cretaceous sediments along the valleysides—particularly lacustrine silts and clays and volcanic clays—give rise to massive, slow-moving earthflows, many of which are kilometers in length and descend to the river (cf., Bovis, 1985). The high local relief, developed by incision of the river into the sedimentary fill of its ancient valley, has been an important factor in mobilizing these flows. Most of the sediment carried by the river originates in the banks from these Cretaceous, Tertiary, and Quaternary materials. Hence the pattern of sediment yield is strongly associated with the annual flow regime.

Through autumn and winter, dry material ravels from noncohesive Quaternary material to river level, while winter frost prepares a surface layer of cohesive sediments for erosion. Snowmelt and thaw near river level introduces the first pulse of

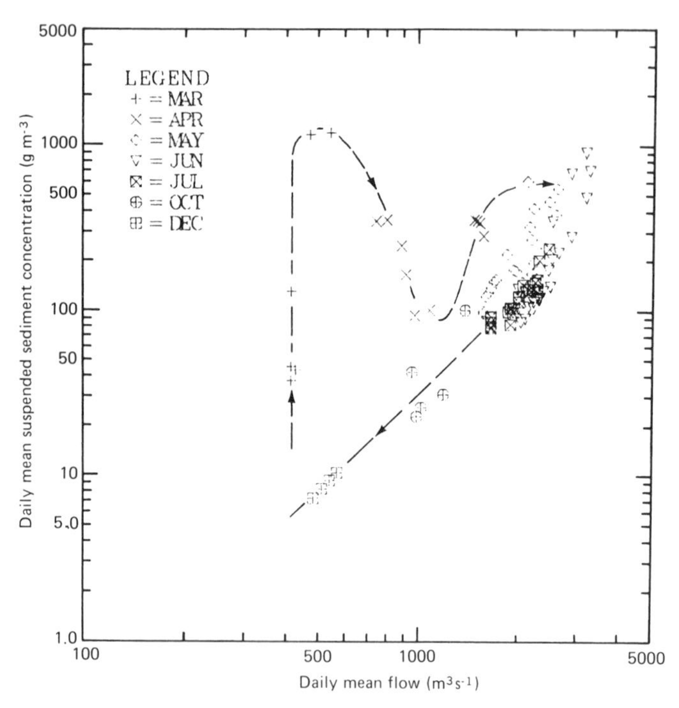

Figure 5. Suspended sediment rating curve for Marguerite, 1978. Note the counterclockwise hysteresis. This is typical in all years at all stations along the river.

fine material to the river in late winter: a pulse that is evident mainly in the increased concentration of sediment in the water column (see Fig. 4). The rising river entrains much material in early spring, with the result that the peak of suspended sediment concentration and of sediment load leads the runoff peak by a month or more. By the time of peak runoff from snowmelt in the mountain headwaters, much of the available material along the river—including sand deposited by the river in the previous year—has been entrained and moved on. River banks under direct attack by the current, particularly Pleistocene glaciolacustrine deposits susceptible to retrogressive failure and flow, continue to yield material throughout the flood. A remarkable individual sediment source of this type is the Big Slide area at Quesnel (see Evans, 1982, for a description of the contemporary slide). Sediment delivered from the land surface and minor tributaries is most important during autumn rains. The net result of this annual pattern of sediment yield is a markedly hysteretic relation between flow and sediment transport (Fig. 5).

A further result is that the mean sediment concentration increases downstream from the mountain headwaters. Sediment yield per unit area of the drainage basin declines downstream; however, yield per unit distance along the main channel consistently increases (Table 1). Further, all the data obtained for Fraser River mainstem stations indicate sediment parameters an order of magnitude higher than those obtained from mountain tributaries.

The landscape develops as a result of erosion and sediment transport over a long period of time. It is interesting to ask what has been the pattern of sediment yield in postglacial time (roughly the last 12,000 years). It is apparent that, following deglaciation, sediment yield from the exposed uplands into the major valleys was high. Alluvial fans and colluvial aprons are widespread over the glacial valley-fill sediments. Much of this material moved into the valley as mudflows. Such events still happen in the region after summer and early autumn storms, but their frequency seems to be much less than in early postglacial time: we estimate that the yield has declined by an order of magnitude (cf., Church and Ryder, 1972). We have called this early postglacial sediment pulse "paraglacial sedimentation." Presumably, it ended in the uplands after the growth of forests and the exhaustion of relatively erodible glacial sediments reduced areal sediment yields.

With the reduction in upland sediment yield, incision of the valley fills began, and proceeds today. The Fraser River, and its major tributaries, have degraded to bedrock in many places. This outcome probably required some millennia to achieve. As the main sediment sources today are Pleistocene sediments along the entrenched rivers and sediment yielded from massive earth movements that occur along the steep inner valley sides, one could argue that extended periglacial effects still dominate the central Fraser basin (see Fig. 6).

PERSPECTIVE

So we return to the viewpoint from which we began this description of the central Fraser basin. The history of the river is a million-year-old alternation of excavation and fill of the major valleys, cut well below the plateau of central British Columbia. The summary effect has been the development of valley elements on several scales, reflecting the time scale of various stages of development. Quaternary events have affected a smaller valley within the major Tertiary valley north of about 51° latitude. Spectacular disturbances—episodes of volcanism and major gla-

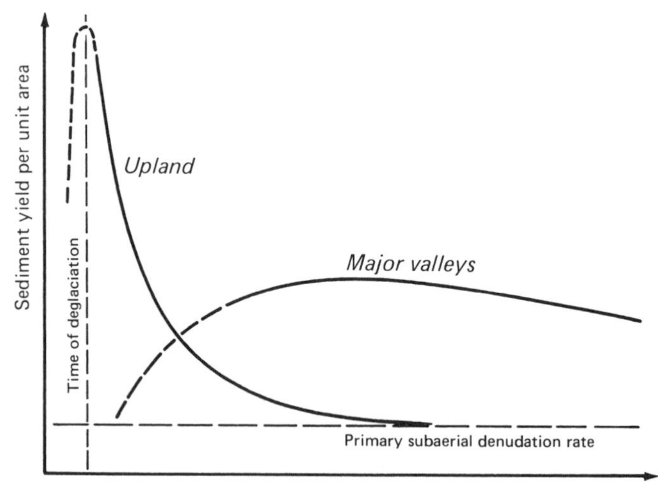

Figure 6. Schematic trend of "paraglacial" sediment yield from upland tributaries (very shortly after deglaciation) and from the main rivers (persistent).

TABLE 1. SEDIMENT IN FRASER RIVER*

Station	Drainage area (km^2)	Mean flow (m^3s^{-1})	Mean sediment transport ($Mg\ a^{-1}$)	Mean suspended sediment concentration ($mg\ 1^{-1} = g\ m^{-3}$)	Sediment yield ($Mg\ km^{-2}a^{-1}$)	Linear sediment yield† ($Mg\ km^{-1}a^{-1}$)
Fraser/ Hansard	18,000	480	2,990,000	197	166	22,300
Fraser/ Marguerite	114,000	1,460	10,800,000	235	95	32,100
Fraser/ Hope	217,000	2,730	18,300,000	212	84	39,300
Silverhope Creek	350	14.4	4,120	9.1	11.8	220

*Based upon suspended sediment data of the Water Survey of Canada; the bedload is known to be on the order of 1 percent of suspended load. The number of years of record varies from station to station. See Figure 1 for station locations.
†The linear measure is $\sqrt{\text{drainage area}}$. It underestimates length of channeled flow, which is approximately proportional to drainage area; it is, however, a reasonable estimate of the distance along the mainstem river from the headwaters.

ciation, and a major reversal of drainage—have interrupted normal fluvial development. The hydrology and sediment yields today reflect the recent incision of the major rivers through the sediments of the last major glaciation and the continuation of an upland nival environment.

Man has had relatively little impact on this landscape—neither the Indians who have lived on the terraces and fished the salmon for 10,000 years, nor modern, engineering man. The Fraser is entirely a river that reflects its geomorphologic setting and geologic history. Man's will has, thus far, been bent to that of the river.

THE NELSON RIVER

Robert Newbury

In 1611, Henry Hudson was abandoned on an unknown subarctic sea in the New World. A year later, after Hudson's mutinous crew returned to England, Sir Thomas Button was sent to rescue him. Button became locked in the ice and was forced to overwinter in 1612 to 1613. He built a land-based camp at the mouth of a large river, which he named "Nelson" after his sailing master. Hudson was never found, and the northern sea became known as "Hudson Bay" (Fig. 7).

In the early 1960s, eight hydroelectric dam sites were identified on the lower Nelson within 330 km of Hudson Bay. Between 1966 and 1986, two of the dams were completed, one at Kettle Rapids and the other at Long Spruce Rapids. Construction has started for a third dam at Upper Limestone Rapids.

The Nelson is an ideal power river. The 1.1 million km^2 drainage basin lies south and west of Hudson Bay, collecting water from the interior plains of western Canada. The principal tributaries of the upper basin, the Saskatchewan, Red, and Winnipeg Rivers, flow into a natural 26,000-km^2 headwater reservoir, Lake Winnipeg, the remnant of glacial Lake Agassiz. Mean monthly flows from the upper basin vary from 1,400 m^3/sec in February to 3,100 m^3/sec in June (unregulated).

From the outlet of Lake Winnipeg, the Nelson River valley follows a topographic break or trough in the raised arc of the Precambrian Shield that surrounds Hudson Bay. The trough follows a heavily mineralized zone of gneissic and metamorphic rocks that lie between the Churchill and Superior provinces of the Shield. In the first 350 km below Lake Winnipeg, the river flows through a series of short cascading reaches between bedrock-controlled lake basins, ending in Split Lake. The glacial till overburden is less than a few meters in thickness, and in many places, rounded bedrock knolls up to 30 m high protrude through the smoothly glaciated surface.

Below Split Lake, the lower Nelson flows in a single channel over a series of steps terminated by steep rapids formed on the irregular bedrock surface. In the last 150 km before entering Hudson Bay, the rapids are extended in long chutes over horizontally bedded limestones that overlie the Precambrian surface in the southwest corner of Hudson Bay. The channel is deeply incised in dense glacial tills, forming a steep-walled canyon that increases in height to more than 60 m at Hudson Bay (Fig. 8).

The lower 150 km of the valley lie within the marine intrusion zone that has emerged above sea level since deglaciation of

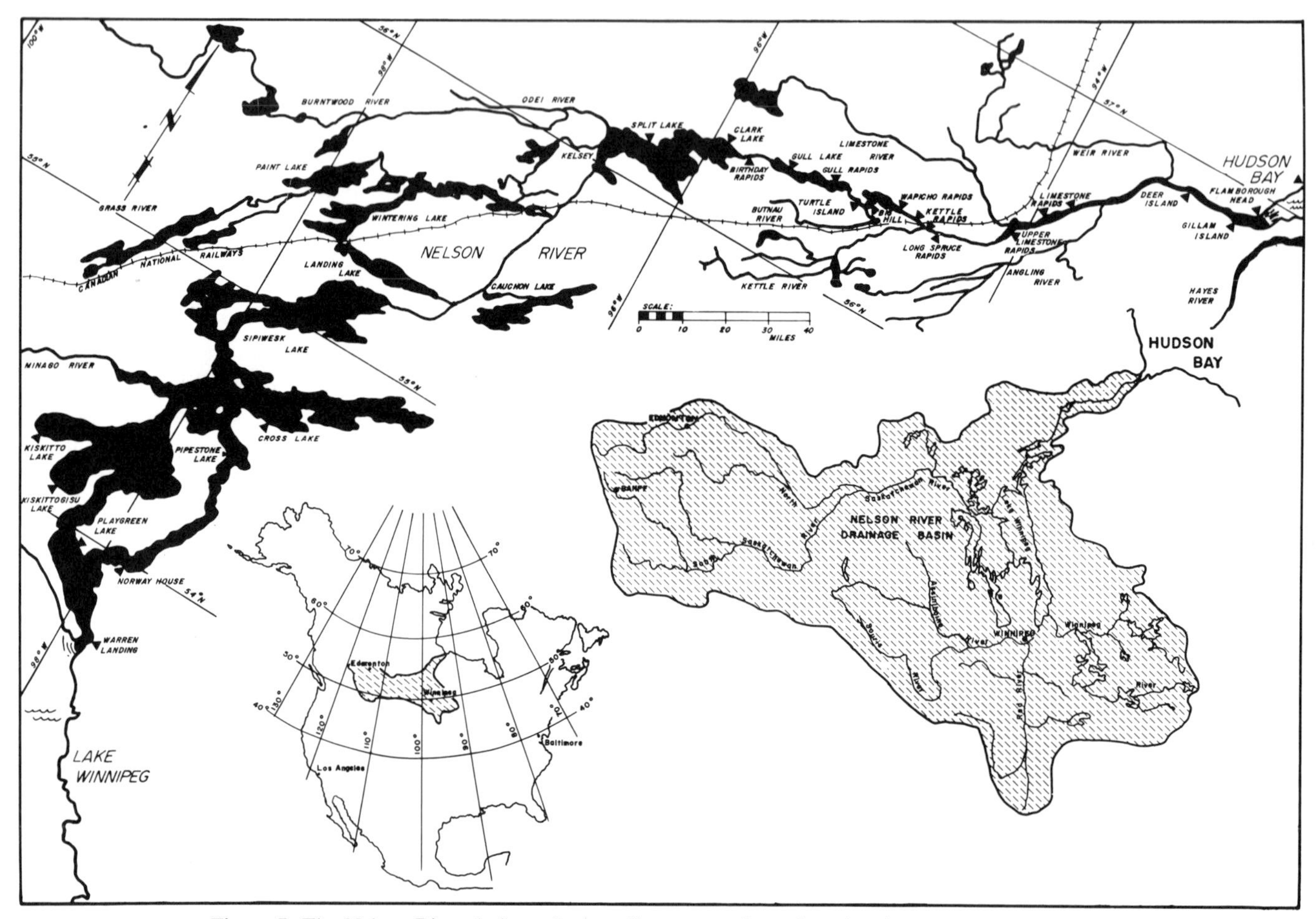

Figure 7. The Nelson River drainage basin collects water from three interior headwater rivers; the Saskatchewan, the Red-Assiniboine, and the Winnipeg. The upper Nelson flows from Lake Winnipeg to Split Lake through Playgreen, Cross, and Sipiwesk Lakes. The lower Nelson flows from Split Lake to Hudson Bay in a single channel.

the region 7,000 to 9,000 yr BP (Simpson, 1972). The flat uplands on either side of the valley are poorly drained, forming an endless monotony of bogs, fens, peat plateaus, palsas, and other microrelief permafrost features. In many places on the north bank of the river, the frozen valley walls are nearly vertical, taking the form of a series of bare, sharply pointed bluffs or "flatirons" set on their heels. In contrast, the sloping (20° to 50°) southern banks are vegetated with stunted subarctic species of black spruce (*Picea mariana*), jackpine (*Pinus banksiana*), aspen (*Populus tremuloides*), and willow (*Salix* spp.).

In several reaches that end in steeply sloping rapids and chutes, the larger vegetation is neatly "trimmed" well above the maximum open-water stage in a long, tapered zone that grows in height to as much as 20 or 30 m at the end of the reach. Only grasses and a few stunted willows grow within the trimmed zone. The distinct trimline gives islands in the channel a "top-hat" appearance with a central core of trees and a surrounding "brim" of grasses (Fig. 8). In some rapids zones, neatly trimmed terraces of grass and willows extend into the valley walls for several hundred meters at a level slightly below the trimline (Fig. 8, left bank at Middle Limestone Rapids). There are also reaches and rapids where the trimline does not exist, and the trees extend down the valley walls to the maximum open water level.

In the 1960s, for the energy consumers of the Nelson basin above Lake Winnipeg, the Nelson was a remote northern river waiting to be rediscovered, after a 350-year pause, as a source of hydroelectric power. In contrast, for the people of the valley below Lake Winnipeg, the Swampy Cree, the Nelson was their traditional home. Their discovery of Europeans began when the Hudson's Bay Company established a central sea-level fur-trading port at York Factory on Hudson Bay in 1715 (Newbury, 1981).

York Factory is located on the mouth of the Hayes River a few kilometers south of the Nelson estuary. The smaller Hayes River runs parallel to the Nelson along the southern boundary of the gneissic zone. The northern headwaters of the Hayes are shared with a small Nelson River tributary near the outlet of Lake Winnipeg (McLeod, 1976). The small connecting channel is called the Echimamish in Swampy Cree, roughly translated as the "river that flows in two directions" (into the Nelson and the Hayes). For more than 200 years of the fur trade and early

Figure 8. The lower Nelson River looking upstream in the reach below middle Limestone Rapids 80 km from Hudson Bay in 1928. The same reach is shown looking downstream in the inset photograph taken in 1968, 40 years later. The hanging ice dam developed in this ice accumulation reach raises the midwinter stage by 11 m trimming the larger vegetation from the channel banks and edges of "top-hat" islands. High-level terraces are formed as the flow is diverted under and around the ice dam, creating and maintaining an enlarged channel geometry. In the 1968 view, steep "flatirons" of permanently frozen glacial till form the north wall of the deeply incised river channel.

settlement, the Hayes-Echimamish route provided the main navigable trnasportation link between Lake Winnipeg, and consequently western Canada, and sea level at Hudson Bay.

In 1913, a rail line from southern Canada was built across the upper Nelson above Split Lake and down the lower valley as far as Kettle Rapids. Although a grade was built to the Nelson estuary, the harbor works at Port Nelson were abandoned because of the shallow approach to the river from Hudson Bay. For 15 years, the rail line ended on the south bank of the Nelson at Kettle Rapids. In 1928, port facilities were completed on the deeper estuary of the Churchill River, 250 km farther north. A bridge was built across the river on the upper edge of Kettle Rapids (Fig. 9, background), and the railway was completed to the Port of Churchill.

As the northern arm of the railway was completed, the traditional Hayes-Echimamish route to York Factory was abandoned, and navigation routes were shortened to join with the rail line rather than Lake Winnipeg. Travelers to York Factory left the railway at the Kettle Rapids bridge and followed the lower Nelson channel into Hudson Bay and back up the Hayes estuary. The traditional river guides of the upper Nelson, the Swampy Cree, discovered new routes through the four sets of immense rapids that lay between Kettle Rapids and the Bay. When the construction of the first dam began in 1966, Duncan McLeod of

Figure 9. The ice accumulation reach above the Kettle Rapids railway bridge looking downstream on February 1, 1966, after the leading edge of the accumulated ice cover has moved upstream in the central channel between the border ice boundaries. The abruptly elevated stages of the river coincide with the trimmed line of vegetation that surrounds islands in the rapids zone and extends upstream through the reach along the channel boundaries. A short section of open water has been eroded from the cover at the head of the hanging ice dam under the Kettle Rapids bridge. In the foreground, shear stress on the bottom of the central ice cover has driven a segment of the partially consolidated cover further downstream.

Cross Lake, an 80-year-old veteran of Nelson River travel, provided the first insight into the origins of the peculiarly trimmed valley walls and terraces of the lower Nelson channel.

The first power dam on the lower Nelson was constructed a few kilometers above the railway bridge at Kettle Rapids. The channel was dewatered behind temporary cofferdams that stretched in a ring, first from the south bank and then from the north bank of the river. The height of the cofferdams was determined by estimating the maximum probable summer flood that would occur from the upper basin during the period of construction. There were reports from railway engineers of winter "ice jams" on Kettle Rapids that rose 15 m to the bottom of the bridge girders.

When freeze-up began in October 1966, the river was several meters below the cofferdam crest. Initially, particles of ice were observed on the surface of the river that moved with the flow. A fixed ice boundary began to grow across the surface from every edge of the river, surrounding islands and growing inward from the sides of the channel to constrict the open surface of the flow between "border ice" boundaries. The concentration of ice moving on and just below the surface of the narrowing central channel increased dramatically as the temperature dropped.

Loose aggregations of particles moved as slush ice pans in wider sections and as a semisolid sludge in narrower sections of the open central channel. The ice moved smoothly under the railway bridge and through Kettle Rapids, causing no significant rise in water level. By mid-November, when the mean daily temperature reached −20°C (−4°F), the slush ice was passing the cofferdam at a rate of 40 m^3/sec. To the relief of the design engineers, there were no "ice jams" in Kettle Rapids. Duncan McLeod explained that there were no jams because "the ice would not arrive from Hudson Bay until Christmas." A flight over the lower channel made his observation abundantly clear.

Beginning in October, the slush ice had moved through the lower Nelson to Hudson Bay. As the slowly flowing estuary froze over, the slush ice was piled against the upstream edge of the solid cover, accumulating a thick conglomerate of ice pans and particles that built slowly upstream, filling the open central channel between the border-ice boundaries. When the leading edge of the accumulating cover approached the foot of the first set of rapids, the incoming slush ice was carried down under the existing cover to form a thick hanging-ice dam that grew until the stage of the river was raised to the elevation of the top of the rapids. Before the ice could continue to build upstream through the rapids, sufficient impoundment had to occur to decrease the approaching velocity of the flow to a rate that would allow the accumulating cover to remain on the surface butted against the leading edge. In the lower Nelson rapids, this required stage increases of up to 20 m before the ice cover stabilized. The maximum observed Froude number of the flow at a stable leading edge as the rapids were surmounted was 0.09, one-tenth of the ice-free critical flow value.

The raised winter stages of the river occurred sequentially, starting at the last rapids and moving upstream to form a series of nearly horizontal steps in the river profile as the winter progressed. The elevations of the winter stages coincided with the vegetation trimlines. As each rapids was surmounted by ice, the accumulating dam forced the flow to the edges of the channel, freeing the border ice boundaries and scouring the river banks. At some locations, elevated side channels were cut into the valley walls adjacent to the rapids (Fig. 8).

At the Kettle Rapids cofferdam, an additional few meters of fill were added hurriedly to the crest until the top of the dam coincided with the elevation of the trimlines on either side of the river. In the last week of December, the water level rose 8 m (26 ft) as Kettle Rapids was surmounted by the accumulating ice cover (Fig. 9). When the leading edge of the cover moved upstream past the cofferdam, there was no freeboard. Duncan McLeod's Christmas ice had arrived.

Following the Kettle Rapids experience, ice observations were extended along the entire lower Nelson channel. The reason for trimlines, terraces, and "top-hat" islands occurring in some reaches but not in others became apparent. Early in the ice season, riverine lakes, the Nelson estuary, and several other wide, slowly flowing segments of the channel form fixed ice covers that extend from shore to shore, dividing the river into a sequence of ice generation and ice accumulation reaches. As the river flow emerges from under the fixed cover at the head of an ice generation reach into the subzero winter environment, the surface layers of the flow are cooled until crystals of "frazil ice" form. As the flow proceeds downstream in the central channel, the concentration of crystals increases until slush-like mats of ice are formed. The ice mats are broken and rebroken as each set of turbulent rapids is encountered. The ice continues to move downstream until the first fixed cover is encountered. The ice cover then fills the central channel and the leading edge of the accumulated cover moves upstream, pausing to build hanging ice dams below each rapids. However, as the ice accumulation reach extends upstream, the reach of open water in which the slush ice is generated grows shorter. The upstream movement of the leading edge of the accumulated cover slows as the supply of ice diminishes and eventually stops when the mean daily temperatures rise above freezing in May. Consequently, rapids at the upper end of an ice-generation reach flow under open-water conditions in all seasons. In these reaches, forest vegetation grows undisturbed down to the midsummer high-water level. In contrast, rapids in ice-accumulation reaches are impounded every winter, neatly maintaining a tapering trimmed strip in the backwater zone that coincides with the stage required for the ice to surmount each particular configuration of rapids. The side channels and trimlines at Limestone Rapids shown in Figure 8, for example, have changed little in the 40-year period between the first photograph and the last.

On the lower Nelson, there were no ice "jams" in the sense of having the river surface bridged over from bank to bank by drifting ice pans, as has been observed in other rivers during the spring breakup period (Ashton, 1979a). With perhaps too much imagination, the processes and forms of the winter ice regime viewed from the bottom of the river looking upward may be thought of as being similar to the transport and depositional processes that govern the more familiar mineral sediment regime of a river (Newbury, 1968). The ice "sediment," being lighter than water, is deposited on the river surface to infill and braid the top of the river with hanging ice dams or ice "deltas." Terraces, and the scoured edges of the river channel observed after the ice has melted, are the channel geometry of the ice-filled condition rather than open-water flood stages.

The natural ice regime of the lower Nelson River changed as water was impounded behind the Kettle Rapids dam. The reservoir divided the river into two ice-generation and accumulation reaches, and Kettle Rapids, at the head of the shortened lower reach, was never surmounted by ice again. As more dams were completed in the 1970s, there were other changes in the Nelson valley as well. The Churchill River was diverted into the Nelson River to augment power generation, causing irreversible losses to the environment and people of the upper lakes. The environmental effects of the Churchill River diversion and the impoundment of Southern Indian Lake in the permafrost zone of northern Manitoba are presented in a summary form in a special issue of the Canadian Journal of Fisheries and Aquatic Sciences (v. 41, no. 4, 1984). As roads and airports follow the construction boom towns, the need to navigate or to find navigators for the fur-trade rivers has disappeared, to the loss of river hydrologists.

THE MACKENZIE DELTA, NORTHWEST TERRITORIES

Michel Lapointe

The Mackenzie River presently captures all of the flow that runs off the eastern slope of the Rocky Mountains between the latitude of Edmonton, Alberta, and the Arctic Ocean; in the terminology of Audley-Charles and others (1977), it is an example of grain-parallel drainage between a fold belt and a craton. Where the Mackenzie reaches the Beaufort Sea, it has created the second largest delta in North America (after the Mississippi Delta); as an Arctic delta, it is second only to that of the Lena River in Siberia. Major terminus for both Pacific and Central flyways of American migrating waterfowl, rich habitat for fish and both marine and land mammals, it is a crucial component of Canada's Arctic ecosystem. Because of the potential wealth of oil and gas reserves buried under the Beaufort Sea's sediments, it has also become in recent decades a new frontier for petroleum exploration in Canada, and an increasing focus of studies aimed at managing the environment in the face of impending development.

The Mackenzie River drains one of the largest river basins in Canada (area 2×10^6 km^2), most of which is underlain by discontinuous permafrost. Just upstream from the Mackenzie Delta, it carries an average annual discharge of 9,100 m^3/s, with peak yearly flows of the order of 30,000 m^3/s during breakup in late May or early June. An average yearly sediment load estimated at 126 million tons (C. P. Lewis, personal communication, 1986) feeds its immense delta (Fig. 10). At Point Separation, the Mackenzie River is joined by the Peel River draining part of the Richardson Mountains to the southwest (average discharge 760 m^3/s) and splits into numerous smaller channels that intricately cover the 13,000 km^2 subaerial delta plain.

The Mackenzie is an estuary delta, constructed, since the retreat of Late Wisconsinan ice from the area 12 to 13,000 years ago, within an erosional/structural trough bounded on the west by the Richardson Mountains and on the east by the Caribou Hills and fragments of a higher level Pleistocene Delta (Mackay, 1974). The present delta plain has a long axis of 210 km and a mean breadth (area/long axis) of 62 km; possibly because of its large aspect ratio, it does not display any major abandoned lobes (contrary to some other modern deltas such as the Mississippi and the Yukon). While the two primary input rivers have a combined width of some 2.25 km upstream of Point Separation, 57 active output distributaries (sensu Coleman and Wright, 1971) with almost 86 km of total width discharge into the Beaufort Sea (Lewis, 1986, op. cit.). Nonetheless, Middle Channel is the main distributary, and in the coastal part of the delta plain, its flow is split essentially three ways, westward into Shallow Bay, northwestward through Richard's Island, and northeastward into Kugmallit Bay. From the distributary mouths, an extremely shallow delta front (area 6,930 km^2, average slope 1 minute of arc) extends some 50 km to a break in slope near the 12-m water depth. From there a shallow continental shelf, dissected by two main troughs off Shallow and Kugmallit Bays, carries on some 74 km to the upper edge of the shelf break at about the 70 m isobath. The Mackenzie Delta front is thus notably shallower than even the Mississippi, already recognized among other large modern deltas for its gentle prodelta slope (Lewis, 1986, op. cit.).

It is instructive to apply to the Mackenzie Delta Wright and Coleman's (1972, 1973) model of delta-front morphologic response to fluvial versus marine processes. The microtidal Beaufort Sea regime (tidal range ~0.2 m) and relatively weak wave climate should result in a strongly river-dominated environment, of which the Mississippi Delta itself is the classic example (Lewis, 1986). However, funnel-shaped channel mouths with large "middle ground bars" within a lobate front clearly distinguish the Mackenzie from the birdfoot prototype. In addition, both offshore and onshore evidence suggest that the delta may be undergoing continued submergence. As a consequence, the delta front itself does not appear to have undergone net progradation during recent decades, and indeed, important segments of the coast have been retreating at rates up to 5 to 10 m/year. Lewis (1986, op. cit.) suggests that the shallow slope of the delta front impedes salt water intrusion into distributary mouths: the effluent jets thus sustain much bed friction and diverge rapidly, in contrast to the salt-buoyed Mississippi outlet jets that diverge less, and tend to deposit levees rather than middle ground bars.

Small-scale maps such as Figure 10 do not do justice to the intricacy of the anastamosing channel network and the abundance of lake storage on the delta plain, which is permafrost-bound but riddled with taliks under all large water bodies. Figure 11 gives a more accurate portrait of conditions in an area equivalent to one-seventieth of the total delta plain surface, centered near 67°55′N and 135°07′W. Over large parts of the delta, there is as much as 50 percent coverage by shallow lakes (1 to 4 m summer depths), partly as a result of thermokarst subsidence of the ice-rich topmost 10 to 15 m of sandy-silt delta plain sediments. These lakes are flooded, and to various degrees infilled by breakup flood sediment inputs, at recurrence intervals that depend on the nature of the connection and the distance to major distributaries. Except for localized ice-jam releases during breakup, channel-energy gradients are of course fairly low: Middle Channel surface slopes average some 5×10^{-5} at flood and only 2×10^5 in summer, when thalweg depths beyond 68°N are in the 20- to 45-m range. While much of the main thoroughfare, Middle Channel, and upstream reaches of the Peel are floored with medium to coarse sands (and gravels near the apex of the delta) the height of the threshold to smaller distributaries (e.g., East Channel, Fig. 10, with summer thalweg depths averaging 4 to 6 m) ensures that only a limited amount of sands find their way into these channels, mostly during high-flow stages. As a rule, distributaries over large parts of the Mackenzie Delta thus have easily

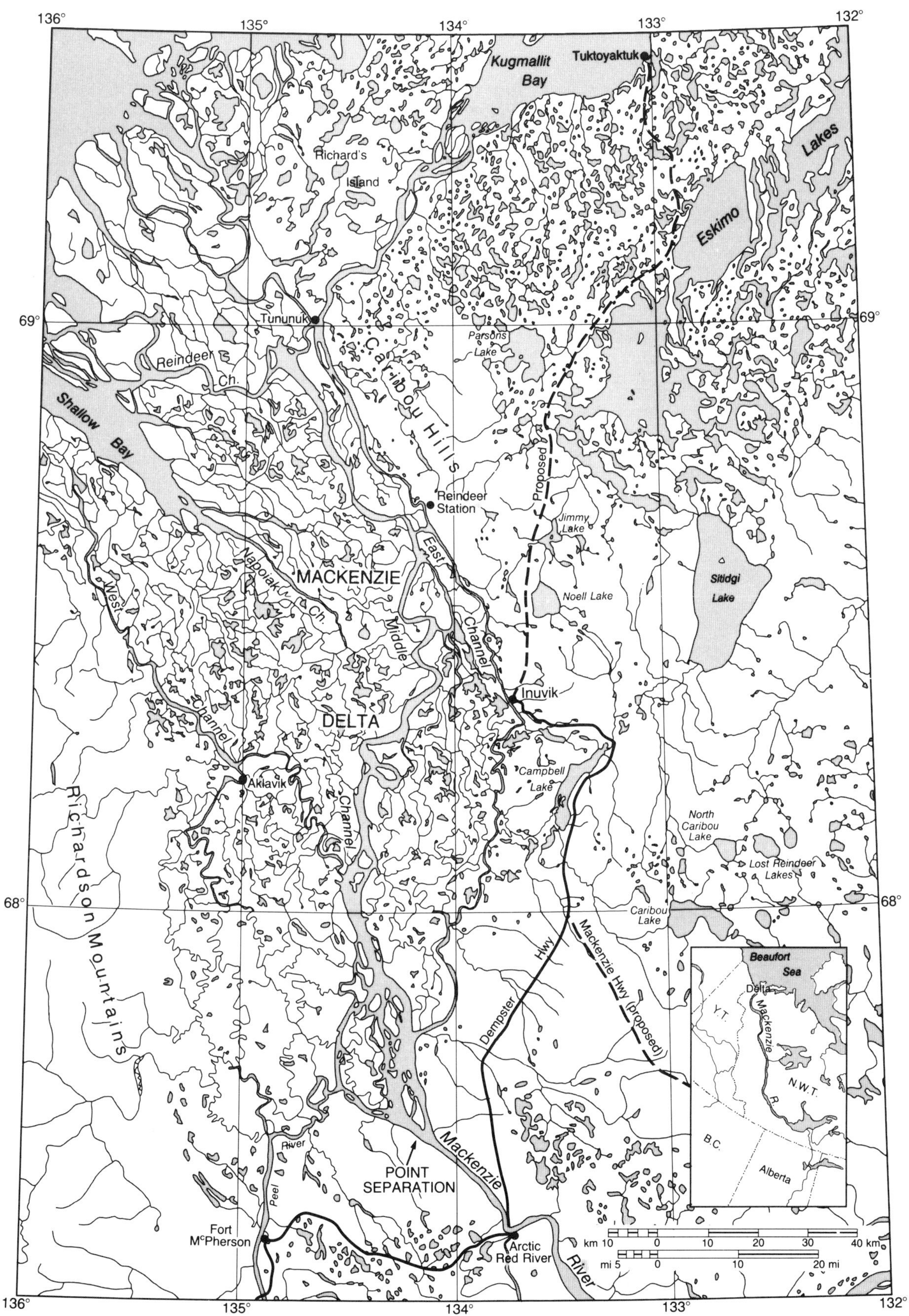

Figure 10. Map of the Mackenzie Delta, Northwest Territories, Canada.

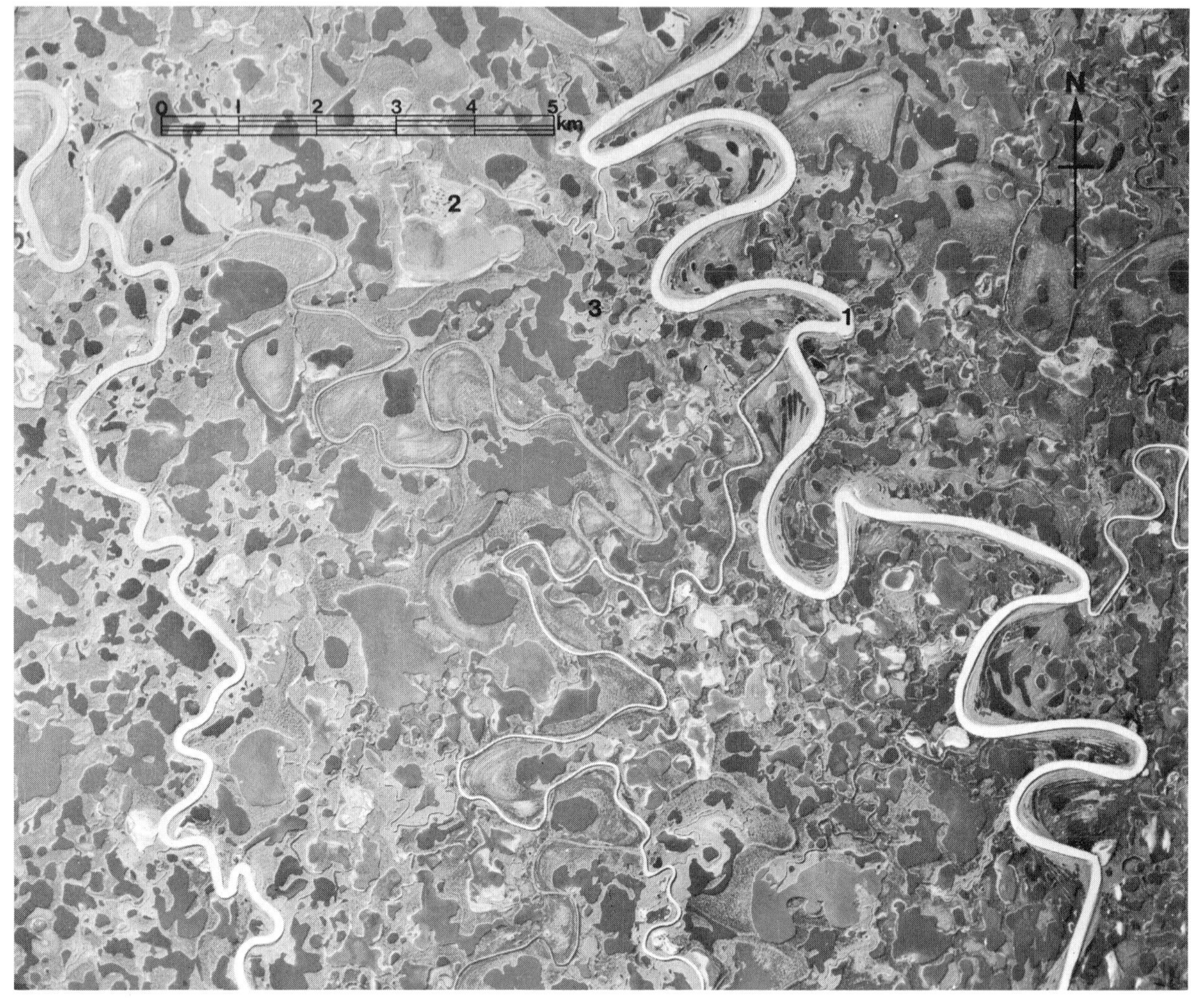

Figure 11. Detail from Government of Canada air-photo. A13489–3, August 1952. Site 1, large embayment associated with deep scour hole; Site 2, small avulsion channel through lake system; Site 3, "reverse delta."

suspended fine-sandy muds and, locally, more consolidated clayey silts as boundary materials.

Although many channel reaches are irregularly sinuous, more or less regular asymmetric meanders are also common, and most channels display subdued shifting activity adjusted to their planform and limited flow power. Peak bend migration rates average 10 to 20 (locally 30) m/yr on the 1.4-km-wide Middle Channel, while East, West, Napoiak Channels, and others mostly undergo 1 (locally 2) m/yr peak local shifting in their bends. It appears that channel size (as a surrogate for flow power) and local curvature largely control bend shifting rates, with scatter possibly related to subtle variations in bank conditions (Lapointe, 1984, 1985). Mackenzie Delta evidence does not, however, shed much new light on the vexed question of the effects of permafrost on bank erosion (Scott, 1978; Lawson, 1983). Although thermo-erosional niching of Middle Channel cutbanks is prevalent, it is quite uncommon in lower power streams and wherever shifting is less than some 2 m per year. Meaningful comparisons are extremely difficult, but it is interesting to note that, with similar channel dimensions, sinuosities, and slopes, "thermally eroded" Middle Channel banks do not seem to retreat significantly faster than those of the Mississippi near Baton Rouge, Louisiana (some 320 to 400 km above Head of Passes), where the latter also flows through sands (Kolb, 1963; Brunsden and Kesel, 1973; Carey, 1969). Differential shifting rates are one mechanism explaining the abundance of small abandoned channels that litter the delta plain: scroll evidence suggests that often, as main channels gradually shift depositional zones over the mouth of some of the smaller, more stable distributaries, these are plugged and abandoned. Conversely, small avulsions regularly occur when overbank splays extend into lake systems that can maintain throughflow by linking to "downdelta" distributaries; as in the Mississippi-Atchafalaya diversion (Cunningham, 1981) where "reverse deltas" extending through the lake system eventually create a new channel (e.g., Fig. 11, site 2).

Some of the most intriguing aspects of channel dynamics on the Mackenzie Delta have been uncovered during bathymetric reconnaissance tied to the possibility of pipeline crossings under the beds of small channels. Complex trough, shelf, and mound features, reminiscent of erosional beds formed through coherent materials, have been identified in a few channel reaches. Even more widespread are unusually deep scour holes, attaining up to 3 to 4 times average thalweg depths at high stages; these most often occur at or slightly *upstream* of bend apices and are often associated with prominent local bankline embayments (e.g., Fig. 12 and Fig. 11, site 1). These scour holes do not appear to be entirely hydraulically controlled, however, and they are not restricted to areas of tight channel curvature. Indeed, most tight bends display more moderate scour depths (less than twice the thalweg average), while marked holes occasionally occur even in very gently curved reaches. Further research into the stratigraphic and geocryological variability of sediments, over different areas and depositional environments of the delta plain, may reveal that some of these features reflect inhomogeneities of channel boundary materials (Lapointe, 1985, 1986).

Bend hydraulic processes, however, do underlie one notable aspect of the bathymetry of the mud-bedded distributaries. Even in the absence of an exceptional scour hole, they systematically lack the point-bar "platform" near bend apex, that typically forces the thalweg to the outer bank in sand or gravel-bedded meanders (Bluck, 1971; Dietrich and Smith, 1983). Figure 13a illustrates the apex bathymetry of a typical delta distributary: transverse slopes off the accreting inner bank are remarkably steep from the bank-top vegetation boundary down to the thalweg, situated near the center of the channel. The inner-bank slope flattens only slightly farther down the bend. Jackson (1981) also noted the steepness of the inner bank in mud-bed meanders of the American Midwest. This dichotomy of form exists within the Mackenzie Delta itself: in contrast to conditions in mud-bedded distributaries such as in Figure 13, coarser sand-bedded middle and upper Peel Channels do have prominent unvegetated point bar platforms that constrict water surface widths at low-flow stages.

The availability on meander beds of sufficient coarse traction load seems to largely control the extent of the apex platform. Studies of bend-flow structure suggest that near-bed inward-directed flow, coupled with the progressive crossover of high velocities to the outer bank, force much of the coarse sand or gravel traction load in the thalweg to stall near the inner bank at the entrance to an abrupt bend. In fine sandy mud bedded distributaries, in contrast, easily entrained bed materials are held in suspension by the strong upwelling, endemic to the inner bank zone at a sharp apex, and fine sands only settle along the inner bank downstream from the apex (Fig. 13B; Lapointe, 1986).

Sediment-flow interactions and morphologic responses alluded to in these brief paragraphs indicate, it is hoped, the geomorphic wealth of large deltas such as the Mackenzie. Conditions on the delta provide a variety of end members in the spectrum of alluvial landforms, here determined by the interaction of low-gradient, diverse particle sizes, a broad range of discharges and channel sizes, and a cryogeologic environment.

THE ST. LAWRENCE

Marcel Frenette

INTRODUCTION

Before explaining the St. Lawrence, it is interesting to recall remarks of two famous writers who describe the "grandeur and the misery" of this extraordinary seaway from the Atlantic into the heart of North America.

> . . . Immense channel cut into the country rock, an artery ramified into practically an entire continent, witness of the first settlement, cradle of civilization, the St. Lawrence.

This description by Jacques Rousseau poetically expresses the "grandeur" of the river.

> Singular country, this French Canada, of which only the southernmost extremity could be settled, the gateway to the land remaining subject to a nearly continual and implacable winter. Settlement was forced southward nearly one thousand miles from the sea (a reference to Quebec City) to begin its course. Every winter the gateway was blocked by ice, so that the passage to the settlement became a trap. The isolation was complete: there was no communication, no supplies for five months, no way out.

Here is the opposite side of the coin, the "misery," expressed by the French geographer Pierre Deffontaines in his book *Winter in Canada.*

These first descriptions of the river delineate a subject of surpassingly wide interest. Similarly broad must be the attempt to describe in only a few paragraphs this major "personality" of the continent, which carries a name redolent of history yet still hides many secrets within its depths. Nonetheless, in the following paragraphs we will attempt to describe the general character of this river, which carries in its "veins" a great deal of the life of the North American people.

SOME HISTORY

What more can be said about a river that has been explored, analyzed, described, and exploited since the first days of the French colony, and even during the prehistoric period? In fact, it is only within the last 7,000 years that the formerly vast Laurentide ice sheet permitted the intrusion of humans onto the newly exposed land. Archaeological investigations—scattered and incomplete—show that the Iroquois Indians were the first inhabitants of the river, centuries before the arrival of European explorers.

The French succeeded the Amerindians—or cohabited with them—and integrated themselves into what became a comfortable and prosperous landscape within the St. Lawrence Valley. Settlement spread slowly, first French then Anglo-Saxon. The

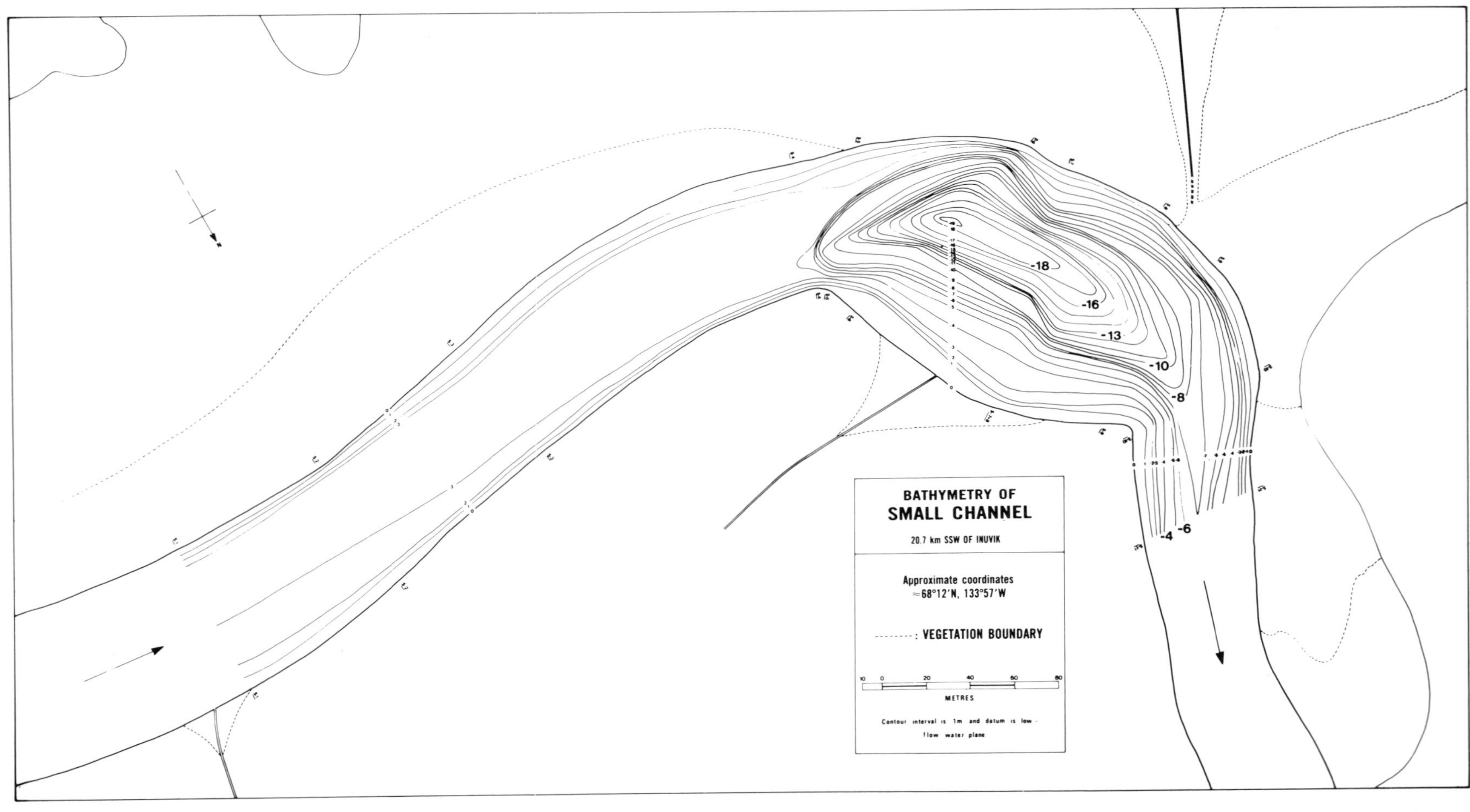

Figure 12. Bathymetric map of a deep scour hole at the bend apex of a small channel. Contour interval in 1 m and datum is low-flow water surface.

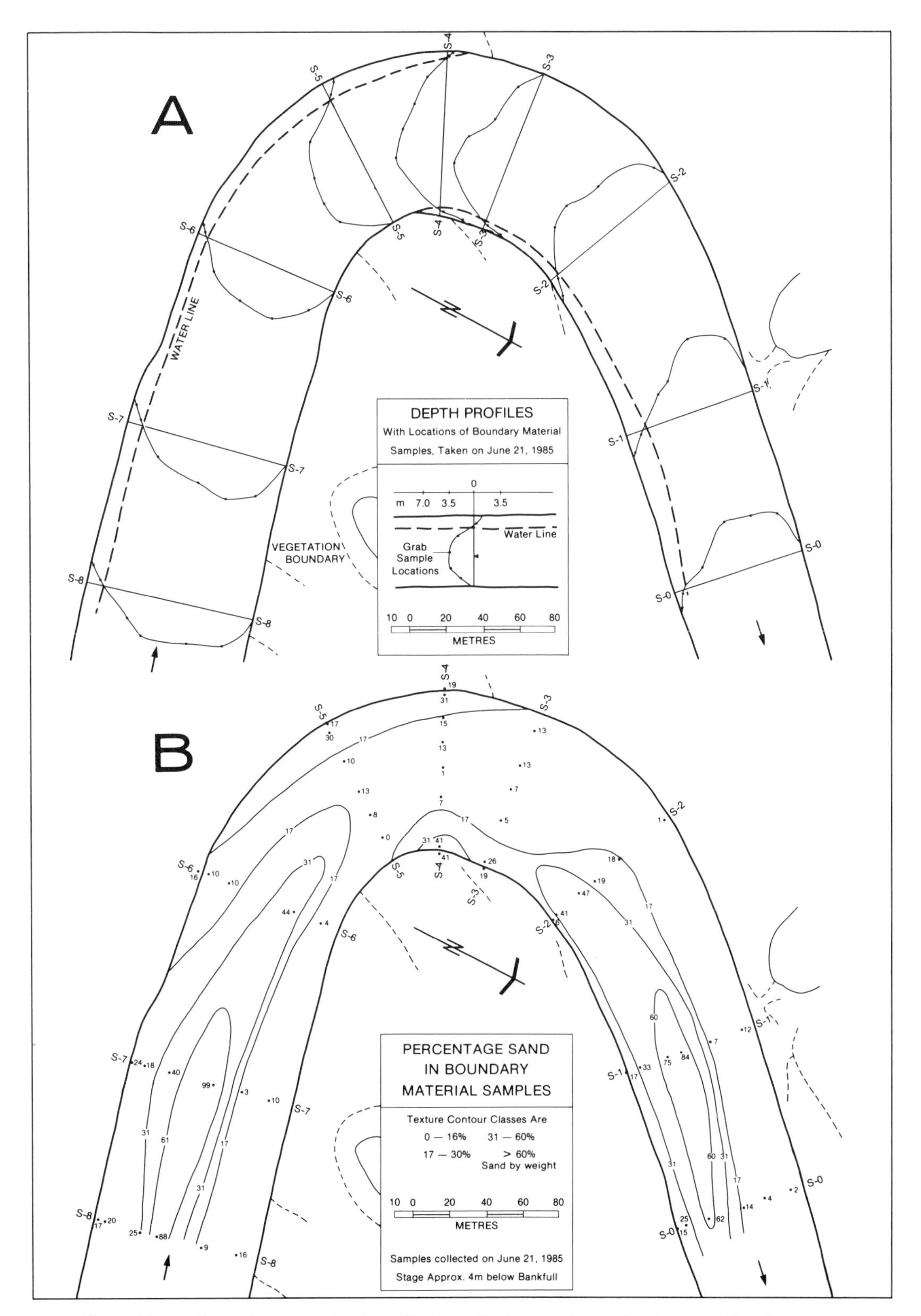

Figure 13. A. (Top) Transverse depth profiles in a distributary channel bend at a medium flow stage. Location: 68° 13′ N, 133°51′W. B. (bottom) Percentage fine-sand in boundary material samples. Non–fine-sand fraction averages 60 percent silt and 20 percent clay. Clay percentages reach 45 to 55 percent in thalweg, sections 3 to 5.

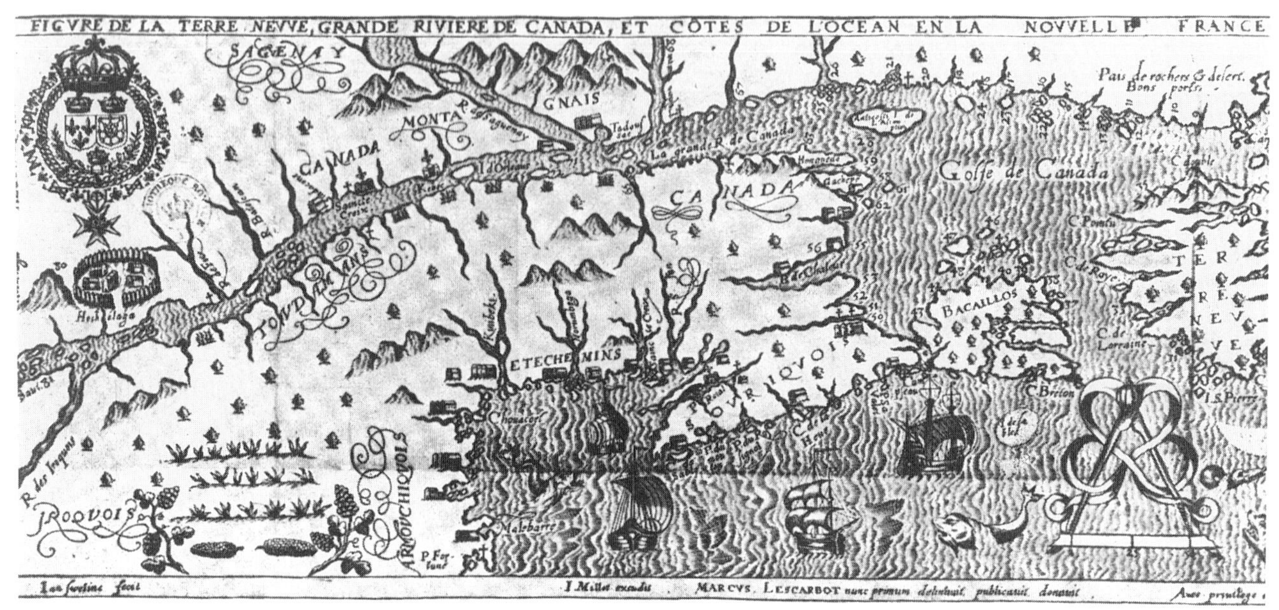

Figure 14. First map of "La Nouvelle France," including "La Grande Riviere de Canada," by Marcus Lescarbot (1609).

Franco-Quebecois have always remained the principal settlers along the river, occupying a narrow band between the shore and the forest. The Amerindians, on the other hand, hunters and fishermen by nature, have withdrawn into the forests. Today, nearly five million people live along the St. Lawrence, principally in the metropolitan centers: Montreal (the second largest French city in the world), Quebec, and Trois-Rivieres.

We owe the name "Saint-Laurent" to Samuel de Champlain, founder of Quebec in 1608, who also characterized the river as the "jewel of the land." Before that, in 1535 to 1536, the French navigator Jacques Cartier ascended the river to Stadacona (Quebec) and Hochelaga (Montreal), having planted a cross at Gaspe in 1534 in the name of the King of France. It was on this voyage into the interior that Cartier named the majestic waterway "La Grande Riviere du Canada," as is shown on the celebrated map of Marcus Lescarbot, dated 1609 (Fig. 14). The change of name, after only a short time, was due to an error in identifying the strait between the Ile d'Anticosti and the North Shore, which Jacques Cartier had named "Baye Saint-Laurens" on 10 August 1535, the saint day of Saint-Laurent (martyred in AD 258). In translations of his account, first into Spanish and then Italian, the name of this "arm of the sea" was extended to the entire gulf, and then to the whole river by Champlain in his book of 1632. Cartographers were more reluctant to adopt the name "Saint-Laurent" for the river. Hence, the maps of Sanson d'Abbeville (1656) and of Deshyes (1695) still indicated "Riviere du Canada ou (or) St.-Laurens."

HOW THE RIVER WAS FORMED

The physiography of a river always reflects its geologic history. The Saint Laurent (Fig. 15) is no exception, for an examination of the geological map of Quebec reveals a remarkable difference between the north and south shores of the river. The plain to the south is underlain by sediments of Cambrian and Ordovician ages, while the steep north side is composed of igneous and metamorphic rocks of Precambrian age—they are in fact very old. Why this difference? To explain it, one must review the origins and history of the river.

At the beginning of the Ordovician Period, about 500 Ma, sediments derived from the erosion of the Appalachians to the southeast accumulated in the Saint Lawrence Valley, especially following mountain bulding toward the end of this period. The sediments were directed toward the northwest and came into contact with the igneous rocks of the Canadian Shield. This line of control has been named the "Logan fault" in honor of the celebrated Canadian geologist who first described it. This was the "origin" of the Saint Lawrence, which today occupies this contact zone, one of the oldest depressions in the world.

In the course of its history, the Saint Lawrence has experienced many glaciations. The last invasion of ice, the Wisconsinan, dates to about 70,000 yr BP. Ice blocked the river drainage and prompted the development of glacial lakes. Glacier advance and retreat has left behind many features in the Canadian and Quebec landscape; the Great Lakes, glaciofluvial valleys, the Saguenay fjord, the rounded hills of the Laurentides, lake-studded plateaux, gravelly moraines, erratic boulders, and so on. It was not until 50,000 years later that the glaciers began their final melt and retreat to the north.

With the retreat of the last ice lobe, between 14,000 and 11,000 yr BP, at Isle Verte (Fig. 16), fresh and salt waters mixed to form the Champlain Sea, which flooded the valley. Marine clays deposited in the sea underlie most of the Saint Lawrence Valley lowland today.

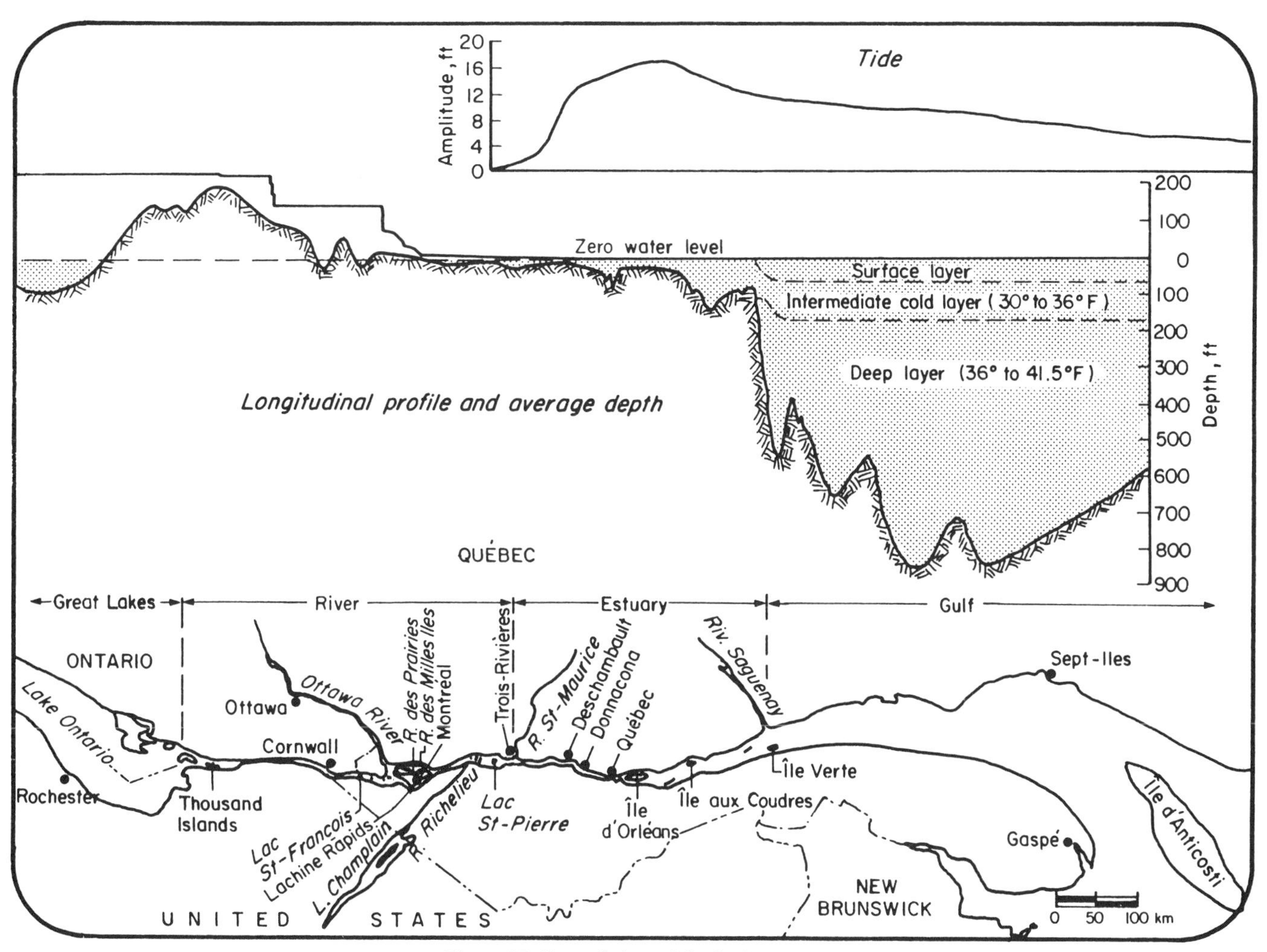

Figure 15. General view of the St. Lawrence.

The Champlain Sea gradually retreated toward the Atlantic Ocean as the remaining ice-sheet melted and the Earth's crust rebounded from its depression under the ice load. With each lowering of water level, bed and bank erosion occurred in the estuary and tributary rivers, producing the staircase flights of terraces that are found today at many places in the valley. In this way, millions of tonnes of sand and finer sediments were transported by the tributaries and deposited in the main river, notably in the estuary. The river attained its present form only about 5,000 years ago, after having formed its channel in the clays of the Champlain Sea. Hence, from the geological point of view, the Saint Lawrence is a young river, even though the valley through which it runs is very old.

A LANDSCAPE OF GRANDEUR

The Saint Lawrence is part of the Laurentian drainage system, delimited by two chains of mountains: the Laurentides of the Canadian Shield to the north, and the Appalachians on the south shore. Thus, an impressive and mature landscape parallels the river all along its course to the sea. The ice age has also left its mark, so that the Saint Lawrence is notable for having its source in the most important system of lakes in the world, while its mouth is lost in a veritable inland sea. Its riverine corridor is, on counterpart, particularly shallow.

From Lake Superior to the limits of the gulf, the Saint Lawrence extends over nearly 4,020 km. The size and diversity of the river are such that one must divide its hydrographic system into four major regions: the Great Lakes, the riverine corridor, the estuary, and the gulf.

The Great Lakes

The Great Lakes drain an area of nearly 722,400 km^2 nearly 70 percent of the entire drainage area of the Saint Lawrence, and yield a strongly regulated flow of about 7,000 m^3/sec. The total area of the lakes covers nearly 233,100 km^2; that is, nearly a third of the contributary basin, and the 42,400 km^3 of water in the lakes constitutes the largest fresh-water reservoir in the world (one-tenth of all fresh water).

As has been implied, the Great Lakes are the product of a gigantic excavation by the most powerful engine of erosion ever seen on earth. From glaciation to glaciation, the shales and limestones of their beds have been so deeply eroded that the bottoms of all the lakes but Lake Erie are today below sea level.

The river

The Saint Lawrence River consists of two major reaches: the fluvial reach from the Great Lakes to Lac Saint-Pierre, and the estuarine reach from Lac Saint-Pierre to the Saguenay.

The river leaves the Great Lakes basin still in the guise of a lake through the well-known Thousand Islands. After that it becomes wild and rapid, with a fall of nearly 61 m in less than 113 km. This section is known particularly for the international rapids that separate Canada and the United States. After becoming tranquil again in Lac Saint-Francois, the Saint Lawrence gains its most important tributary, the 1,126-km-long Ottawa River. This tributary has four outlets; Vaudreuil and Sainte-Anne channels upstream flowing into Lac Saint-Louis and the Riviere des Prairies and Riviere des Milles Iles that form the archipelago of Hochelaga (Montreal) and join the Saint Lawrence 48 km further down (see Fig. 17).

The Lachine rapids form the lower end of this steep reach. There are important hydroelectric power generating stations in the reach at the Beauharnois, Moses Saunders, and Des Cedres dams, and the St. Lawrence Seaway, which have profoundly altered the natural form of the river. In effect, in the rapid reach, the appearance of the river today owes as much to man as it does to nature.

Between Montreal and Lac Saint-Pierre, the river is transformed into a quiet yet imposing force. In the port of Montreal, the river is only 6.8 m above sea level, and about 3 m above Lac Saint-Pierre, 121 km downstream. The discharge in Montreal is about 7,700 m^3/sec. The Richelieu River, draining Lake Champlain (located in the United States) reaches the Saint-Lawrence upstream of Lac Saint-Pierre. This reach is also known for numerous islands: Iles de Vercheras, Iles de Contrecoeur, but especially the islands of Sorel that form an immense delta at the entrance to Lac Saint-Pierre. This lake is wide but very shallow, less than 8 m deep except in the navigation channel dredged the entire length of the lake.

The estuary

Because of the very small drop that remains between Lac Saint-Pierre and the sea—about 4 m—the tide influences the lake, but only over a range up to 15 cm.

At the eastern extremity of the lake on the north shore, the Riviere Saint-Maurice meets the Saint Lawrence from the Canadian Shield. In the estuarine reach downstream, characterized by fresh water only, the Saint Lawrence slowly widens, from about 1.6 ka at Trois-Rivieres to 3 km at Donnacona only, to contract again at Quebec, where two imposing cliffs only 1 km apart squeeze the river, notably at the site of the Quebec Bridge. On the other hand, the river reaches a depth of nearly 55 m between the cliffs, even though upstream soundings are in general less than 9 m, except in the seaway where dredging maintains the navigation channel at a minimum of 13 m at low-level tide, up to Montreal. At Deschambault, about 48 km upstream, Quebec Richelieu rapids are exposed at low tide (3 to 4 m amplitude on average).

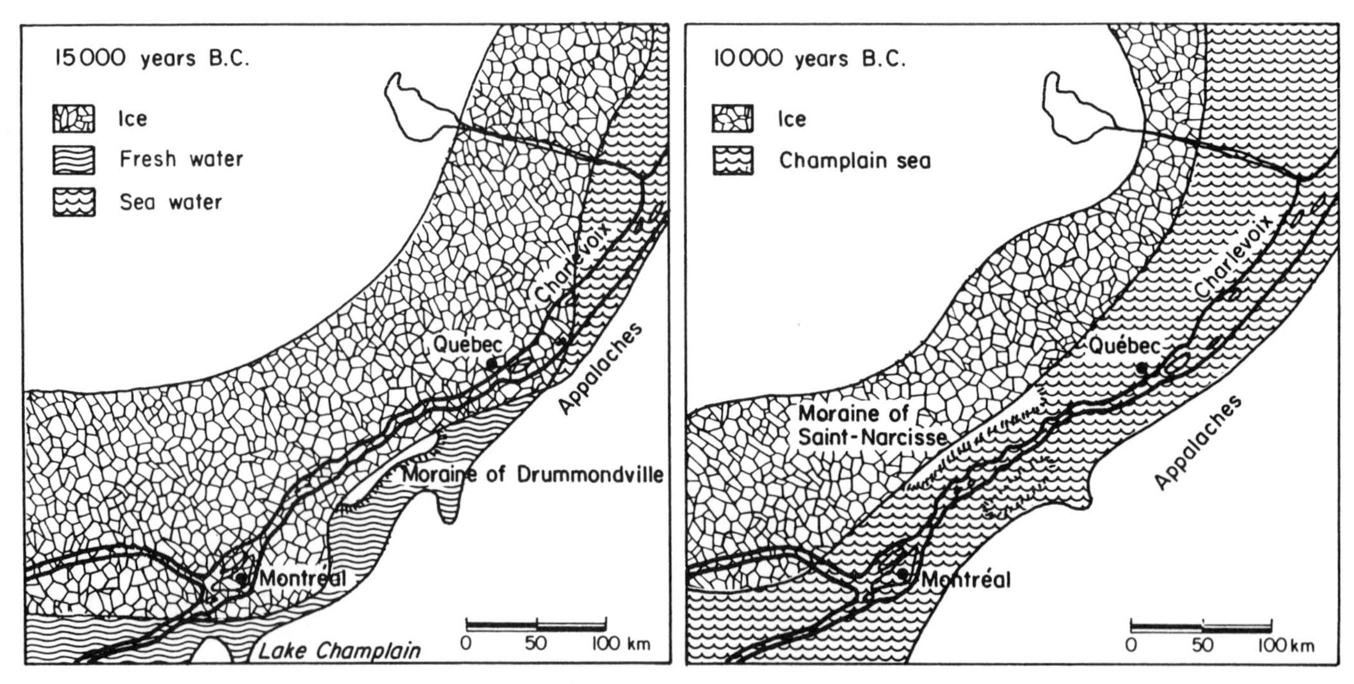

Figure 16. The Champlain sea through the ice age.

Figure 17. Satellite view of the St. Lawrence at Montreal, (water flowing from left to right): Lac St-Louis (left), Lachine Rapids (middle), La Prairie basin (right) and fluvial reach (upper right corner); Old Lachine canal build in 1824 (north shore) and new navigation Seaway with two locks built in 1959 (south shore); Riviere-des-Prairies (upper left corner), one of the four effluents of the Ottawa River; the limited lateral mixing of the Ottawa River (light) and the St. Lawrence (dark); and the Montreal Island, between the two rivers, the heart of the city.

Upstream from this rapids, the influence of the tide is felt in a reduction of current and consequent rise of water level. Downstream from the rapids, it is sufficient to cause a complete reversal of current on the rising tide. This increase in discharge between the Great Lakes and Quebec City is on the order of 30 percent, for a total of about 9,300 m^3/sec at Quebec. Tidal flows are, however, five times greater, to give a total flow of more than 56,000 m^3/s at the peak underneath the Quebec Bridge.

At the Ile d'Orleans just below Quebec, spectacular changes occur, with a rapidly increasing cross section: width changes from about a kilometer to more than 19 km, and depth from 15 m to more than 107 m between Ile d'Orleans and Ile aux Coudres (Fig. 18). The tidal range here is 4 to 6 m, but it then declines progressively downstream toward the gulf. The salinity of the water gradually passes from zero at the eastern end of Ile d'Orleans to its maximum value at Ile aux Coudres. This part of the estuary is therefore characterized by the mixing of fresh and saltwater, after which the water remains salt to the limit of the estuary (i.e., to the well-known Saguenay fjord). Here the relief of the shore is very high. This is the impressive Charlesvoix region, the massive plateau of the Laurentides that rises to 1,070 m. On the south shore, in contrast, low plains continue.

The gulf

Downstream from the Saguenay, the river attains more than 500 m depth and widens progressively, so that at Pointe des Monts more than 96 km separates Sept-Isles (North Shore) from the Gaspe coast. With such dimensions, the river waters mingle so subtly and slowly with the water of the gulf that it is difficult to fix the limits of the river.

Tributary basins to the lower estuary and the gulf are mainly located on the North Shore, draining a rugged terrain of about 154,800 km^2 area. The streams that drain the North Shore and the Gaspe are known as the finest habitat in the world for spawning and rearing Atlantic salmon (*Salmo salar*). Finally, at the limit of the gulf, east of Anticosti Island, the continental shelf is reached.

FLOW AND SEDIMENTATION

The diversity of reaches—lacustrine, rapid, estuarine, or island studded—imposes great variation on the nature of the flow. The rapids above Montreal and the portion of the fresh-water estuary with reversing flow, from Deschambault to Quebec, are

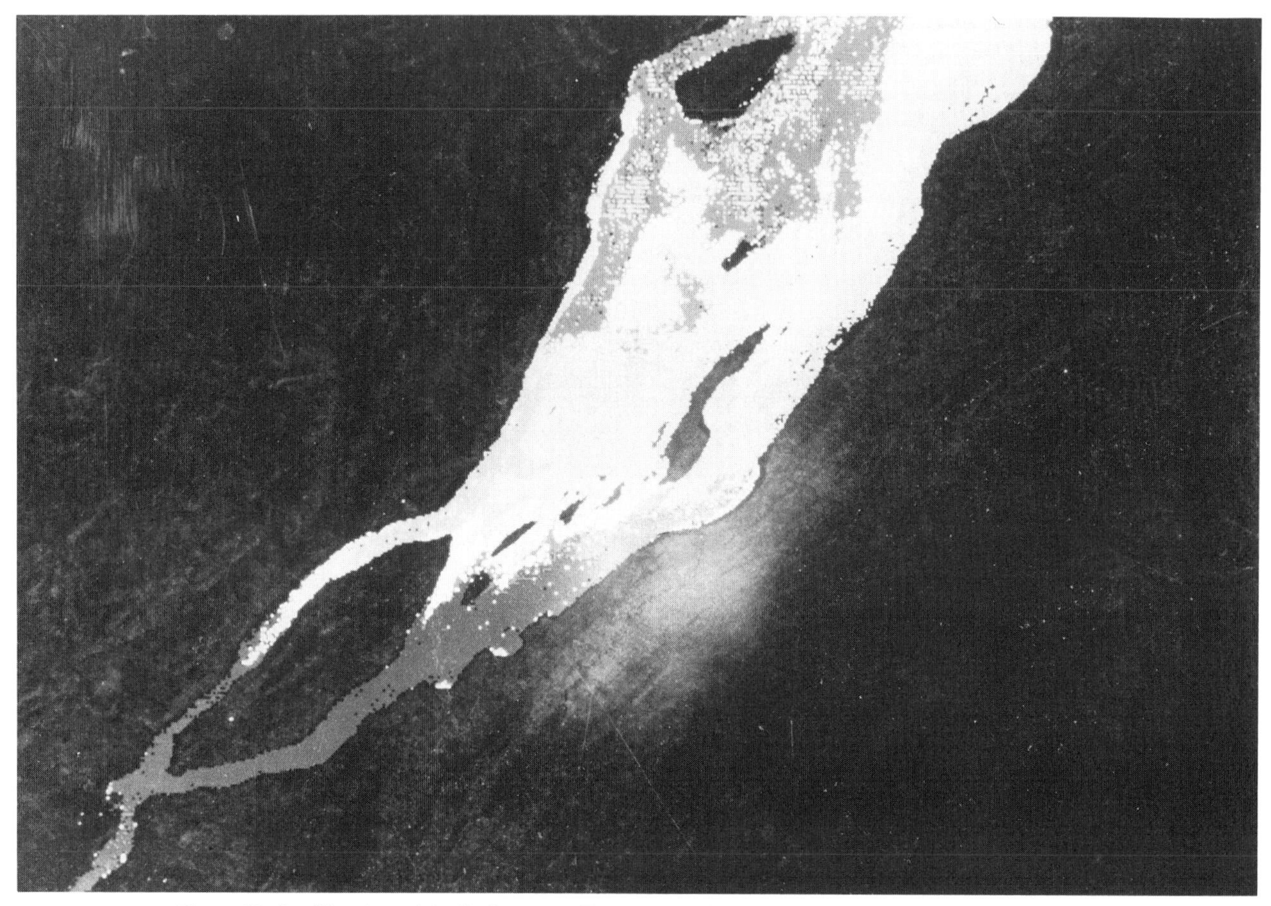

Figure 18. Satellite view of the St. Lawrence Estuary near Quebec: Narrow channel with Ile d'Orleans (lower left corner), wide cross section going to Ile-aux-Coudres (upper right corner); fresh water (gray, lower left), mixing fresh and salt water (white), and pure salt water (gray, upper center). The heart of Quebec City is located along the bay, upstream Ile de'Orleans.

the zones of greatest velocity. Aside from the lacustrine section, the current varies between 0.5 and 1.5 m/s^{-1}. It requires two or three days, then, for Great Lakes water to reach the brackish-water estuary.

One of the remarkable hydrodynamic characteristics of the river is the very limited lateral mixing. After their arrival in the Saint Lawrence, tributary waters are forced to remain close to the shores, to form various identifiable streams and mixing zones that run side by side for dozens of kilometers, leaving the center of the river to pure Great Lakes water, which can be identified essentially all the way to Quebec.

Unlike other major rivers of its type, the Saint Lawrence transports very little sediment, less than 5 million tonnes per year compared, for example, with 300 million tonnes for the Mississippi. Bedload transport is also very small, since the bed is composed principally of clay in the fluvial reach. The sediments are also very fine and clayey. The Great Lakes, which traps nearly all of the upstream sediment supply, and the small population in Quebec combine to explain this phenomenon.

The mixing of fresh and brackish waters downstream from Quebec produces, however, a considerable increase in suspended matter. Between Ile d'Orleans and Ile aux Coudres are the most turbulent waters in the entire length of the river, producing a palpable "plug of turbidity."

Nonetheless, the greatest sediment-related problems in the Saint Lawrence are concerned with water quality, given that the very fine particles have a large affinity for adsorbing dissolved contaminants. Thus, zones of sediment deposition (lakes and bays—both harbors and recreation areas) are subject to sedimentary contamination.

ICE AND FLOODS

In a northern country like Canada, the winter ice regime and, particularly, spring breakup play an important role in fluvial processes. With the exception of the brackish and salt waters of the estuary, the water temperature hovers at the freezing point throughout the four winter months—from December to April—and reaches an extreme value, above 20°C at the end of August. It is possible to discern three "winter reaches" according to the behavior of the ice. A first zone, with continuous ice cover about Montreal, is closed to navigation; a second zone called the "zone

Figure 19. Ice breaker destroying an ice jam formed at the Montreal Harbor.

of jamming" between Montreal and Quebec, is travelled by ice breakers to permit winter navigation to Montreal; and finally the drift-ice zone occurs in the estuary, from Ile d'Orleans to the gulf.

From Montreal to Quebec, the ice consists essentially of frazil formed in the Lachine rapids. The frazil agglomerates into slush downstream and forms an ice cover over the entire freshwater reach. Before the intensive use of ice breakers, most of the river became ice covered from one bank to the other during the winter, a phenomenon that we will see often in those places favorable for the development of ice jams: Montreal harbor (Fig. 19), the entrance to Lac Saint-Pierre, Portneuf, and under the Quebec Bridge. Once a jam begins to develop, ice accumulates upstream very rapidly, often for many miles.

Erosion of the banks and sediment transport are greatly affected by the action of shore ice, which is capable of displacing large erratic stones as well as grooving the bed.

Moreover, on the Saint Lawrence and on many of its tributaries, ice effects control flooding. The worst jam on the river occurred in 1886 and inundated a large part of the city of Montreal. The struggle against ice jams, hence against floods, was joined in 1906 with the placement into service of the first two ice breakers—the *Lady Gray* and the *Montcalm*—which then prompted the desire for a fleet of ice breakers that would eventually permit winter navigation by preventing ice-bridge formation. Even into the 1950s, this winter navigation seemed to be a physical and economic impossibility, but it has been a reality since 1960.

AN EXTRAORDINARY SEAWAY

Joining the Atlantic with the center of the North American continent, the Saint Lawrence provides an extraordinary maritime incursion. It leads directly to the Great Lakes and follows a practically linear course, which aids navigation. Furthermore, being situated in the same latitude as the English Channel—at the 50th parallel—in the Atlantic, it is, in comparison with the Mississippi or with Hudson Bay, incontestably the shortest route between Europe and the center of North America.

The Great Lakes, with the exception of Lake Michigan, form the border between Canada and the United States. In Canada, the lakes are situated entirely within the province of Ontario. From Lake Ontario to Cornwall, the Saint Lawrence also serves as the border. Supervision of the international portion of the river is exercised by a Canada–United States Joint Commission. When it enters Quebec, the Saint Lawrence loses its status as an international river. Still, since 1955, the flow at the outlet of Lake Ontario has been regulated by order of the commission in order to ensure that all downstream interests are respected.

Ships of all shapes and sizes, mostly ocean going, travel the Saint Lawrence and the Great Lakes. More than 125 million tonnes of merchandise are transported each year (1976 evaluation) by more than 10,000 ships (20,000 passages of the seaway). The port of Montreal receives ships with capacities up to 30,000 or 40,000 tonnes, while the port of Quebec can accommodate ships up to 150,000 tonnes.

Because of the rapids, seven locks have been constructed between Montreal and Lake Ontario. Eight others have been necessary to bypass Niagara Falls, between Lakes Erie and Ontario, and finally one more series (the Sault St. Marie locks) between Lake Superior and Lake Huron. The maximum dimensions of ships that may enter the Saint Lawrence Seaway are 222.5 m length and 22.8 m beam.

The Saint Lawrence Seaway has been officially open since April 1959. The Saint Lawrence Seaway Development Corporation has the responsibility for managing locks in the United States, while in Canada this responsibility falls to the Saint Lawrence Seaway Authority.

Thus, thanks to the Saint Lawrence and the Great Lakes, the heart of the continent is accessible from ports the world over. Ships flying the flags of nearly 40 countries—coming from ports of call on all continents—ply the Saint Lawrence and so confirm its rank as one of the great rivers of the world. Clearly, the river plays a major role in international relations and in the economic progress of many nations, beginning with Canada and the United States.

THE COLORADO RIVER; A PERSPECTIVE FROM LEES FERRY, ARIZONA

E. D. Andrews

INTRODUCTION

"At night we stop at the mouth of a creek coming in from the right, and suppose it to be the Paria. . . .

Here the canyon terminates abruptly in a line of cliffs, which stretches from either side across the river.

August 5.—With some feeling of anxiety, we enter a new cañon this morning. We have learned to closely observe the texture of the rock. In the softer strata, we have a quiet river; in harder, we find rapids and falls. Below us are the limestones and hard sandstones which we found in Cataract Cañon. This bodes toil and danger" (John Wesley Powell, 1875, p. 73).

Powell's camp at the confluence of the Colorado and Paria Rivers is known today as Lees Ferry. The break in the canyons is short. Between the end of Glen Canyon upstream and the beginning of Marble Canyon downstream is a reach of less than 3 km. It is, however, one of only a few places for a distance of 1,300 km where the Colorado River can be crossed. For centuries, Native Americans used this river crossing. The year after Powell's expedition in 1869, John D. Lee began farming along the banks of the Paria River just upstream of its mouth. In 1874, Lee built a ferry within the backwater created by the debris fan of the Paria River. The ferry remained in use until 1929, when Marble Canyon Bridge was constructed 7.2 km downstream.

The Colorado River receives runoff from seven states: Wyoming, Colorado, Utah, New Mexico, Arizona, Nevada, and California. A 1922 compact among these seven states apportioned the flow of the Colorado River between the upper and lower parts of the basin. A point 1.6 km downstream from the confluence of the Paria and Colorado Rivers was chosen as the Compact Point or division between the Upper and Lower Colorado River Basins. The actual Compact Point, however, is rather inaccessible because Marble Canyon is already 50 m deep. Consequently, gaging stations were established on the Colorado River at Lees Ferry and on the Paria River just upstream from the mouth. Suspended-sediment concentration was sampled daily in the Colorado River at Lees Ferry from October 1947 until September 1965 and in the Paria River at Lees Ferry from October 1948 until September 1976. These long-term records of streamflow and sediment transport, together with historical accounts and photographs, provide a rich insight into how the Colorado River and its tributaries have changed during the past 100 years.

SOURCE AREAS OF WATER AND SEDIMENT

Water and sediment are not contributed uniformly to the channel network throughout the Upper Colorado River Basin (Iorns and others, 1965). Furthermore, the principal source areas of water and sediment differ greatly. Most of the annual water discharge comes from the headwater areas near the crest of the Rocky Mountains. Conversely, most of the annual sediment load is contributed by the semiarid parts of the basin at lower elevations. The mean annual water discharge and sediment load measured at ten gaging stations in the Upper Colorado River Basin are compared in Figure 20. The information summarized in Figure 20 was computed from various periods of record selected to represent natural conditions. After 1958, the water discharge and sediment load of the Colorado River and many of its principal tributaries were altered substantially due to construction of several reservoirs. For the most part, the information summarized in Figure 20 was collected prior to 1958. Where it is appropriate, however, gaging station records collected after 1957 have been included.

Mean annual water discharge of the Colorado River at Lees Ferry during the period 1922 to 1957 was about 480 m^3/s or 15.1×10^9 m^3 from a drainage area of 279,200 km^2. The combined mean annual water discharge at the six farthest upstream gaging stations shown in Figure 20 was 287 m^3/s or 60 percent of the water discharge at Lees Ferry. The drainage area contributing to the six upstream gaging stations is 59,980 km^2 or 21 percent of the total drainage basin area. The estimated combined mean annual sediment load at these gaging stations was only 5.7 $\times 10^6$ tons or 8.6 percent of the total sediment load. Thus, the headwater tributaries contribute about 60 percent of the water

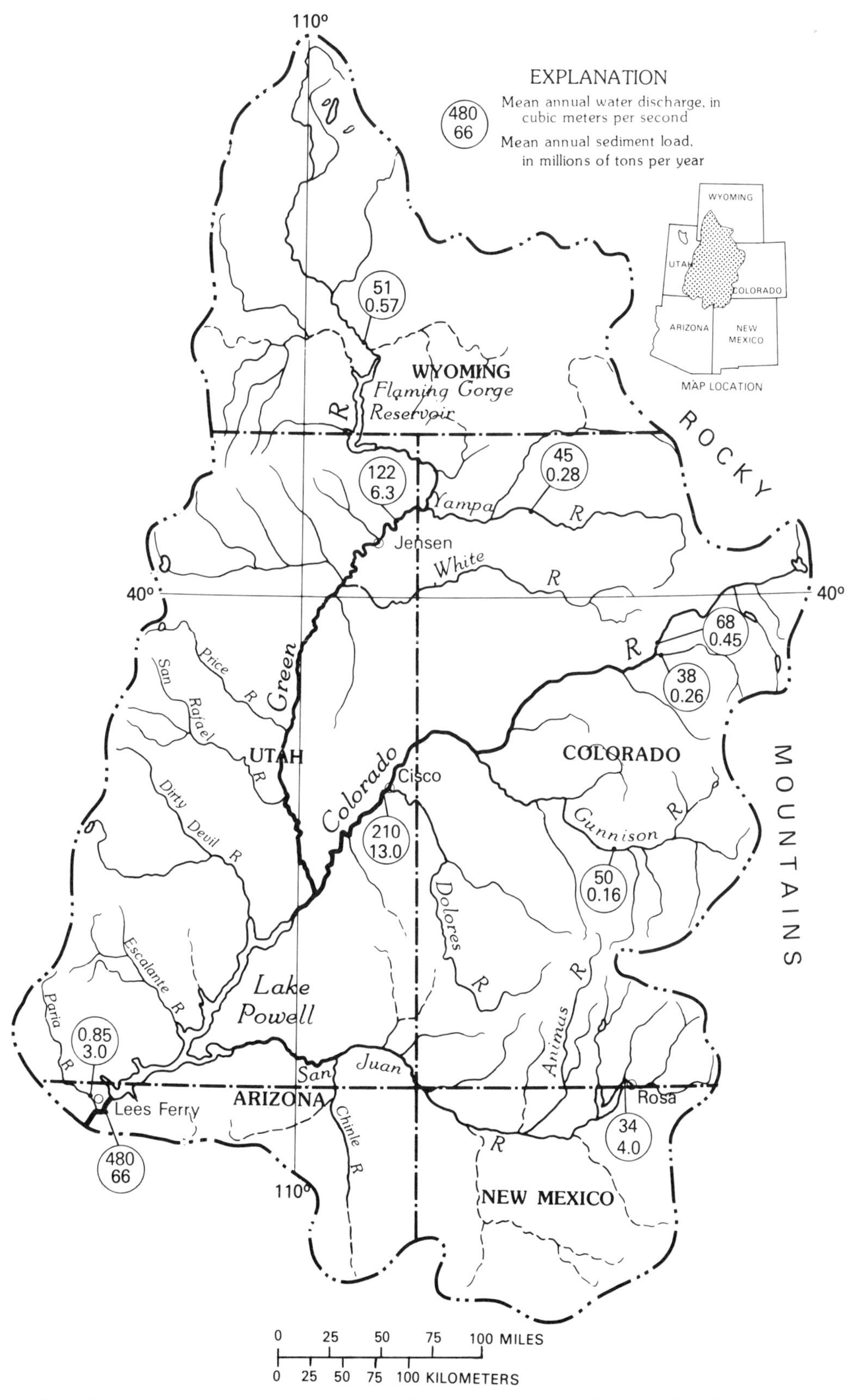

Figure 20. Map showing mean annual water discharge and sediment load at selected gaging stations in the Upper Colorado River Basin prior to extensive regulation and diversions.

discharge, but less than 10 percent of the sediment load in the Colorado River at Lees Ferry.

Most of the sediment discharged from the Upper Colorado River Basin is contributed by tributaries draining relatively low, semiarid parts of the basin. The largest of these tributaries, notably the Price, San Rafael, Dirty Devil, Escalante, Paria, Chaco, and Chinle Rivers, are shown in Figure 20. The part of the Upper Colorado River Basin that lies downstream from Jensen, Utah, on the Green River; from Cisco, Utah, on the Colorado River; and from Rosa, New Mexico, on the San Juan River contributed about 42.7×10^6 tons/yr on an average, or about 65 percent of the basinwide sediment load, but supplied only 26 percent of the basinwide water discharge. The large sediment-contributing part of the Upper Colorado River Basin is 135,200 km^2, or about 51 percent of the total basin area. The most conspicuous sediment sources are areas of badland topography that developed on Cenozoic and Mesozoic mudstone and shale, principally the Wasatch, Mancos, Morrison, Chinle, and Moenkopi Formations.

The characteristics of streamflow and sediment transport in those tributaries that supply most of the sediment to the Colorado River are not well known. Few gaging stations have been operated on these streams and then only for a short time. The long-term contributions of water and sediment from these streams were determined by comparing the quantities measured at adjacent mainstem gaging stations. The only long records of water discharge and sediment load for any of these tributaries have been collected on the Paria River to meet the requirements of the Colorado River Compact.

The mean annual water discharge of the Paria River was only 0.16 percent that of the Colorado River at Lees Ferry. Mean annual sediment load, however, was 3.02×10^6 tons or 4.6 percent of that transported by the Colorado River at Lees Ferry. Compared to the Colorado River Basin, the Paria River basin yields one-tenth as much water and three times as much sediment per unit area. Mean monthly water discharge and sediment load for the Colorado and Paria Rivers at Lees Ferry before the construction of Glen Canyon Dam are compared in Figure 21. Maximum monthly water discharges in the Paria River occur during August through October. During the period of record from 1923 to 1982, 41 of the 61 maximum instantaneous water discharges occurred in August, September, and October. These floods were a result of intense, but short-duration thunderstorms that increased the discharges from a base flow of less than a cubic meter per second to several tens of cubic meters per second for a few days. These floods have transported most of the sediment outflow from the Paria River basin. Between 1947 and 1976, 50 percent of the sediment contributed by the Paria River to the Colorado River was transported on just 74 days, or just 0.7 percent of the time.

The largest suspended-sediment concentrations ever measured in the United States prior to the eruption of Mount St. Helens in Washington were collected from the Paria River (Beverage and Culbertson, 1964; Pierson and Scott, 1985). Mean daily concentrations more than 400,000 mg/L have occurred on

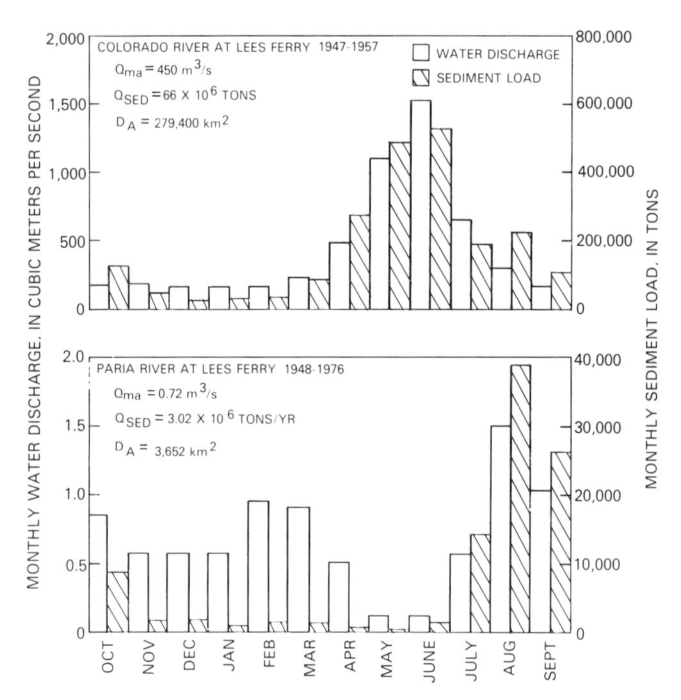

Figure 21. Graph showing the mean monthly water discharge and sediment load of the Colorado River and Paria River at Lees Ferry, Arizona.

an average of twice a year. The maximum mean daily concentration was 780,000 mg/L. River flows that have suspended-sediment concentrations greater than 400,000 mg/L are classified as hyperconcentrated. The physical properties of hyperconcentrated flow differ from typical river flows with suspended-sediment concentrations of a few tens to thousands of milligrams per liter (Nordin, 1963) in that the fluid density and viscosity are substantially larger. The fall velocity of sand-sized particles also is decreased. Furthermore, flows that have very large concentrations of silt and clay (>200,000 mg/L) have a greatly enhanced capacity to transport sand compared to hydraulically similar flows with small concentrations of silt and clay. Nordin (1963) found that the transport rate of sand increased by as much as an order of magnitude for a given water discharge when the concentration of silt and clay exceeded 200,000 mg/L. Greater fluid density and viscosity alone probably are not sufficient to account for the increased transport of sand. Grain-to-grain interactions and the suppression of momentum exchange have been suggested by several investigators as possible mechanisms responsible for enhancing sand transport (Nordin, 1963; Beverage and Culbertson, 1964; Pierson and Scott, 1985).

Instantaneous and monthly peak flows of the Colorado River at Lees Ferry, prior to 1958, occurred during May and June and were the result of spring snowmelt in the Rocky Mountains. These floods spanned two to three months, with a gradual rise to the peak and recession. The spring floods also transported most of the mean annual sediment load. In general, the magnitude of monthly water discharge and sediment load were well correlated (Fig. 21). During the spring floods, suspended-sediment concentration tended to increase with water discharge in the Colorado River. The annual peak suspended-sediment concentration, however, did not coincide with the peak water discharge; rather it occurred during August, September, and October when flash floods on upstream tributaries similar to the Paria River contributed large quantities of sediment, but relatively little water.

The maximum suspended-sediment concentration measured in the Colorado River at Lees Ferry between 1947 and 1957 was 32,400 mg/L, or less than 5 percent of the maximum concentration measured in the Paria River. Suspended-sediment concentrations of the magnitude that occurred in the Colorado River at Lees Ferry do not appreciably alter fluid density or viscosity. Similarly, the grain-to-grain interactions, an important characteristic of hyperconcentrated flows, are not significant. The observed concentrations, however, are sufficient to create a vertical density gradient that will suppress the exchange of mass and momentum.

EFFECTS OF RESERVOIR STORAGE AND FLOW REGULATIONS ON THE COLORADO RIVER

The Colorado River has one of the most extensively regulated drainage basins in the world. Total storage provided by two large reservoirs, Lake Powell and Lake Mead, each with a capacity greater than 32×10^9 m^3, and numerous smaller reservoirs exceeds 4.5 times the mean annual runoff of about 18.5×10^9 m^3. Upstream from Lees Ferry, total reservoir storage is slightly less than 41×10^9 m^3, or more than 2.6 times the mean annual flow. Construction and operation of several reservoirs in the Upper Colorado River Basin has radically altered both the duration of particular water discharges and sediment concentrations during the past 30 years. As a result, river channels throughout the Colorado River basin are adjusting to new conditions of water and sediment supply.

Due to regulation of flow by the Glen Canyon Dam, the range of water discharges in the Colorado River at Lees Ferry has been greatly reduced. Mean annual peak water discharge has decreased from more than 2,800 m^3/sec to about 850 m^3/sec. Furthermore, the seasonal variation in water discharge of the Colorado River at Lees Ferry (see Fig. 21), has been replaced by a nearly uniform flow regime. Between 1965 and 1982, the mean monthly water discharge varied from 290 m^3/sec in March to 437 m^3/sec in August. The daily variation in water discharge, however, has increased greatly. Releases from Glen Canyon Dam are scheduled to maximize the generation of electric power given a maximum turbine capacity of about 890 m^3/s, and the Compact commitment to supply a mean annual volume of 10.15×10^9 m^3 to the Lower Colorado River Basin. Thus, releases from the dam fluctuate within a broad daily range in response to changes in the immediate demand for electrical power.

Lake Powell is located downstream from major sediment source areas, and a large volume of sediment is deposited annually in the reservoir. The mean annual sediment load of the Colorado River at Lees Ferry decreased from about 66×10^6 tons prior to construction of Glen Canyon dam to only about 91×10^3 tons during 1980 to 1985. The large concentrations of suspended sediment, which caused the Colorado River to be reddish brown (hence, its name), have been greatly decreased. Today, the Colorado River at Lees Ferry is pale green.

Suspended-sediment concentrations also were sampled during the pre- and post-reservoir periods at the Colorado River near a Grand Canyon gaging station located 140 km downstream from Lees Ferry. Between 1947 and 1957, the mean annual sediment load of the Colorado River near the Grand Canyon was about 86×10^6 tons, or an increase of about 20×10^6 tons compared to the sediment load at Lees Ferry during the same period. This additional sediment was contributed primarily by tributaries to the reach between the two gaging stations. The Paria and Little Colorado Rivers are the two largest tributaries along this reach and contributed about 70 percent of the increase. The mean annual sediment load at the Grand Canyon gaging station during the period 1975 to 1985 is estimated to have been 9 to 11×10^6 tons. Uncertainty of this estimate is large, because relatively few suspended-sediment samples were collected between 1975 and 1985 compared to the pre-1957 period. Despite these uncertainties, the mean annual sediment load of the Colorado River at the Grand Canyon seems to be appreciably less than the amount of sediment contributed to the upstream reach by tributaries. Based on a similar analysis, Howard and Dolan (1981) concluded that about 1.5 m of sediment, on an average, had accumulated on the river bed in the reach from Lees Ferry to the Grand Canyon gaging station during the period from the completion of Glen Canyon Dam until 1980. A three-fold decrease in the mean annual peak water discharge, plus the large contribution of sediment by tributaries, results in a surplus rather than a deficit of sediment.

RECENT CHANNEL CHANGES NEAR LEES FERRY

During 1872, Powell led a second expedition down the Colorado River by boat. Among the crew was photographer J. K. Hillers. At about the same time, an expedition led by G. M. Wheeler traveled upstream from Fort Mohave along the Colorado River and included photographers T. H. O'Sullivan and William Bell. These men took several thousand photographs, which are our best source of information concerning river-channel change in the Colorado River during the past 115 years. Most of the original sites have been rephotographed since 1969 (Stephens and Shoemaker, 1987; Turner and Karpiscak, 1980). A photograph of Lees Ferry taken by T. H. O'Sullivan in 1873 is shown in Figure 22A (Turner and Karpiscak, 1980, Fig. 33A). A photograph taken from the same location in 1972 by R. M. Turner (Turner and Karpiscak, 1980, Fig. 33B) is shown in Figure 22B. The view is from the right bank, looking downstream, southwest, across the confluence of the Colorado and Paria Rivers. The large alluvial deposit shown near the center of the photograph is the debris fan formed by the Paria River, which enters from the right.

Figure 22. Matched photographs of the Colorado River 400 m downstream from Lees Ferry crossing taken 1873 (A) and 1972 (B) showing changes in vegetation and channel configuration.

Both the changes and stability evident in this pair of photographs are typical of those found at a majority of the rephotographed sites. The most prominent change is the establishment of a dense stand of riparian vegetation, mainly saltcedar, along the previously bare river bank. Saltcedar, a non-native species, was introduced into the southwestern United States sometime before 1900. Saltcedar trees were established in the Grand Canyon by 1936, but did not form dense thickets along the river banks until after flow regulation by Glen Canyon Dam greatly decreased the range of water discharge (Turner and Karpiscak, 1980).

Certain channel adjustments, such as bank erosion, change in width, and longitudinally continuous aggradation or degradation of the channel bed, would be detected most easily through a comparison of historical photographs. No such channel adjustments are apparent from a comparison of Figures 22A and 22B. An additional 10 pairs of similar photographs compiled by Turner and Karpiscak (1980), likewise, reveal no major channel changes along a reach of 40 km approximately centered about Lees Ferry. General stability of the Colorado River channel in the vicinity of Lees Ferry is caused by the abundance of large particles in the river bed and banks. Cobbles and boulders are abundant in the surface of the debris fan at the mouth of the Paria River. This material is most readily discernible in the 1872 photograph, Figure 22A. When I inspected the debris fan at low flow during March 1986, coarse material was present on the surface. The discharge required to move the largest particles can be estimated from the hydraulic measurements made at the Lees Ferry gaging station. The largest particles having a median diameter of 1 to 1.5 m probably were entrained only by the largest recorded floods, $>5,500\ m^3/sec$. Thus, the debris fan was rather stable even before flood discharges were decreased. A large fraction of the material in the fan surface is immobile under the flow regime that has existed since construction of the Glen Canyon Dam.

Stability of the tributary debris fans, such as the one formed by the Paria River, has limited general degradation of the river bed. As a result of sediment-free releases from Glen Canyon Dam, however, relatively fine-grained material, mainly sand, has been scoured from the pool reaches between debris fans. Pemberton (1976) concluded that about $10 \times 10^6\ m^3$ of bed material were scoured from the channel in the reach from Glen Canyon dam to the mouth of the Paria River between 1956 and 1975. Most of the material was removed within a few years after the completion of the dam and channel scour, even in the pool reaches, ceased by 1965.

In the vicinity of Lees Ferry, banks of the Colorado River are composed primarily of talus shed from the canyon walls. Large blocks of material as much as several meters in diameter commonly occur and make the banks quite stable. Fine-grained alluvium is important locally and usually is associated with zones of flow separation (Schmidt, 1986). Some of these deposits have been completely eroded since 1957, whereas others are little changed.

The notable stability of the Colorado River channel contrasts markedly with the major channel changes that have occurred in the Paria River and other similar tributaries during the past 100 years. Beginning about 1880, channels of the Paria and Little Colorado Rivers, as well as other tributaries to the Colorado River, were entrenched, and large arroyos were formed (Graf, 1983). This widespread geomorphic change has been studied for more than 90 years and continues to be a topic of interest. Webb (1985) compiled a list of 116 journal articles and books that described the chronology of arroyo development on the Colorado Plateau and evaluated the importance of various causes.

Hereford (1986) reconstructed the sequence of channel changes in the Paria River basin using stratigraphic, botanical, and photographic evidence. An arroyo began to form in the Paria River channel in 1883, and attained its maximum size by 1890 (Gregory and Moore, 1931). The arroyo remained deep and wide until the early 1940s when channels throughout the basin began to aggrade. By 1950, the Paria River channel had narrowed appreciably and a vegetated flood plain had developed. A comparison of the stage-discharge relations measured at the Paria River at Lees Ferry gaging station prior to 1939 and after 1964 indicates that the channel aggraded by as much as 2 m. Since 1980, the Paria River channel has begun to degrade and widen (Hereford, 1986).

Many investigators have concluded that large floods were the immediate cause of arroyos throughout the Colorado Plateau (e.g., Gregory and Moore, 1931; Thornthwaite and others, 1942; Graf, 1983; Hereford, 1984, 1986; Webb, 1985). In the case of Kanab Creek and the Fremont River, historical accounts identify the dates of floods that enlarged the arroyos (Graf, 1983). For most streams, the period of arroyo formation is known only to within a few years. Although arroyos developed in all major tributaries draining the semiarid part of the Colorado Plateau during the period from 1880 to 1910, the period of intense arroyo formation varied from basin to basin. For example, entrenchment and arroyo development appears to have occurred somewhat later in the Escalante River, which drains the area to the east of the Paria River. Using historical accounts and slackwater deposits, Webb (1985) determined that the Escalante River arroyo was initiated by a large summer flood that occurred on August 29, 1909. The arroyo was subsequently enlarged by recorded floods during 1910, 1911, 1914, 1916, and 1921.

Slack-water deposits along the Escalante River studied by Webb (1985) indicate that the frequency of floods on the Escalante River has varied during the past 2,000 years. Large floods occurred most frequently between 1200 and 900 yr BP, 600 and 400 yr BP, and during the past 100 yr. The earlier periods of frequent floods appear to be correlative with ancestral arroyos along the Escalante River.

A change in the frequency of large floods is indicated by the time series of annual peak flows recorded at the Paria River at Lees Ferry gaging station, as shown in Figure 23. Large floods occurred more commonly prior to 1942 than since. Seven of the eight largest floods of the Paria River since 1924 occurred before 1942. The mean annual peak flood of the Paria River since 1942

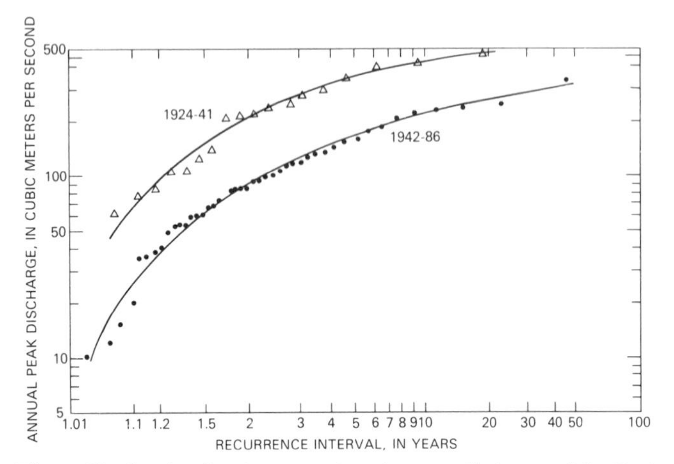

Figure 23. Graphs showing annual peak water discharge of the Paria River at Lees Ferry, Arizona, from 1924 to 1986.

is 106 m^3/s or 55 percent of mean annual peak flood, 220 m^3/sec, recorded prior to 1942. Hereford (1986) noted that the period of frequent larger floods corresponded to channel degradation and arroyo development. In contrast, the period of relatively small floods corresponded to channel aggradation and partial filling of arroyos.

Mean annual water discharge of the Paria River at Lees Ferry also changed during the period from 1924 to 1986, but the relative magnitude was less than occurred in the mean annual peak water discharge. The mean annual water discharge prior to 1942 was 1.03 m^3/sec, compared to 0.77 m^3/sec since 1942. The change is confirmed at the 1-percent level of significance.

One of the most important and interesting mysteries of the Colorado River is an apparent decrease in sediment transport during the period 1925 to 1957. Sampling of suspended-sediment concentrations in the Colorado River began in 1925 at the Grand Canyon gaging station located 140 km downstream from Lees Ferry. By 1929, suspended-sediment concentrations were being sampled at three additional gaging stations upstream from Lees Ferry. A similar pattern of change in annual sediment loads was determined at each of the four gaging stations (Hadley, 1974). During the period of record through 1942, annual sediment loads were appreciably larger than those measured after 1942. For example, the mean annual sediment loads at the Grand Canyon gaging station were 180×10^6 tons during 1926 to 1941, compared to 92×10^6 tons during 1942 to 1957.

One factor contributing to the large annual sediment loads in the decades prior to 1942 was a much larger than normal runoff. Stockton (1975) estimated long-term annual runoff for the Colorado River using tree-ring information. On the basis of this analysis, the mean annual runoff that occurred during 1905 to 1940 was the largest of any time in the past 400 yr.

A second factor contributing to the relatively large sediment loads determined for the years prior to 1942 was the generally greater suspended-sediment concentrations at any water discharge. Differences in the suspended-sediment concentrations before and after 1942, however, have not been firmly established or quantified. Unfortunately, a new sampler and new sampling procedure were introduced at all four gaging stations, and a new laboratory procedure was developed at the same time the change in suspended sediment loads and concentrations appears to have occurred: 1941 to 1943. A thorough evaluation of the old and new samplers and methods was never conducted. It seems likely that the suspended-sediment records collected before and after 1942 are not directly comparable. Evidence from three sources, however, indicates that most of the difference is real. First, the volume of sediment deposited during 1935 to 1948 in Lake Mead (Smith and others, 1960), located 238 km downstream of the Grand Canyon gaging station, agrees fairly well with the estimated quantity transported past the gaging station during the same period. Secondly, a comparison of water-discharge measurements made during 1922 to 1962 indicates that the river-bed elevation at the surveyed cross sections began to degrade about 1940 at both the Lees Ferry and Grand Canyon gaging stations (Burkham, personal communications, 1986). Thirdly, the apparent change in the quantity of sediment transported by the Colorado River corresponds in time with the shift from channel degradation to channel aggradation that occurred in the tributaries that contribute most of the sediment to the Colorado River, as described above. Thus, it appears probable that suspended-sediment concentrations in the Colorado River were greater during 1926 to 1941 than during 1942 to 1957.

CONCLUSIONS

Since John Wesley Powell's expedition camped at the mouth of the Paria River nearly 120 years ago, significant and unlikely hydrologic and geomorphic events have occurred at Lees Ferry. Mean annual runoff of the Colorado River during the first few decades of the twentieth century probably was the largest in more than 500 years. Similarly, the development of arroyo channels in Colorado River tributaries, such as the Paria River, ended 600 years of relative channel stability. These natural events alone would have altered the fluvial processes and river landscape Powell observed.

During the past 40 years, the effects of human activities upon the Colorado River have accelerated greatly, especially the introduction of salt cedar and the construction of Glen Canyon Dam. These activities have appreciably changed the channel of the Colorado River. Upstream from the mouth of the Paria River, adjustment of the Colorado River to the new conditions is relatively complete. Beginning at the mouth of the Paria River, however, and to an increasing degree with distance downstream, the Colorado River is still adjusting to a radically altered regime of water discharge and sediment transport. Evidence indicates that channel adjustment will continue for decades and perhaps centuries.

THE EEL RIVER, NORTHWESTERN CALIFORNIA; HIGH SEDIMENT YIELDS FROM A DYNAMIC LANDSCAPE

Thomas E. Lisle

The Eel River draining the Coast Range of northwestern California has the highest recorded average suspended sediment yield per drainage area of any river of its size or larger unaffected by volcanic eruptions or active glaciers in the conterminous United States (1,720 t/km^{2}yr from 9,390 km^2; Brown and Ritter, 1971). These high rates of erosion and sediment transport result from a combination of widespread tectonic deformation of the underlying rocks, recent rapid uplift of the landscape, high seasonal rainfall, and widespread disruption of the ground surface by man in the last century. Not surprisingly, the basin has some unusual geomorphologic characteristics. Sediment-transporting processes on hillslopes and in channels are closely linked, and as a result, high-magnitude, low-frequency climatic events are more responsible for forming channels than in most other areas.

BASIN CHARACTERISTICS

Geology

The Eel River basin is underlain almost entirely by the Franciscan assemblage of complexly deformed, continental margin deposits of Late Jurassic to mid-Tertiary age (Bailey and others, 1964; Jones and others, 1978). The area has undergone uplift since mid-Miocene time (Bailey and others, 1964). Franciscan rocks are predominantly sandstone and shale, but also include tectonically emplaced blocks of volcanics and low-grade metamorphic rock. Bedrock has been pervasively sheared to various intensities over the basin. Zones of weakness trending generally north-northwest have created a trellis network of drainages. Narrow, deeply cut canyons incised below moderately dipping upper slopes, on which older soils are developed, attest to recent or ongoing uplift of the area, although local downwarping has formed isolated depositional basins in the Eel valley (Kelsey, 1982).

Hydrology

The Mediterranean climate of the area is conducive to the production of high sediment yields. Annual precipitation is heavy (averaging 1,500 mm basinwide and 2,800 mm at high elevations) and seasonal, with 90 percent falling between October and April. During winter, northern California has the highest latitudinal temperature gradients of any area in the Pacific Northwest (Janda and Nolan, 1979). This produces intense storms that commonly travel perpendicular to the trend of the Coast Range, which are as high as 2,000 m in the Eel basin. As a result, large cyclonic storms lasting several days have produced widespread rainfall totaling more than 250 mm on several occasions in the last 40 years (Harden and others, 1978).

Runoff from the basin, averaging 890 mm annually, is highly variable because of seasonality of rainfall, infrequent large storms, and poor retention of water in the basin. At Scotia (Fig. 24), the discharge equaled or exceeded 99 percent and 1 percent of the time equals 0.0004 m^3 sec^{-1}km^{-2} and 0.8 m^{3}sec^{-1}km^{-2}, respectively (Rantz, 1972). Most importantly from a geomorphic standpoint, large flood flows are generated by moderately intense rain falling over the entire basin for a number of days and, in some cases, by snowmelt during warm winter storms (Harden and others, 1978). Little flood runoff is stored in the basin due to the steep slopes and constricted valley bottoms.

Sediment yield

High suspended-sediment discharges from this area result from a combination of high sediment concentrations (averaging 3,000 ppm over discharge at Scotia; Holeman, 1968) and, particularly, high rates of runoff (Janda and Nolan, 1979). Gullying and mass movement accelerated by human disturbance of the erodible terrain provide inexhaustable supplies of fine sediment that can be carried quickly to stream channels (Nolan and Janda, 1982). With increasing precipitation, there is greater surface erosion of broken ground in active earthflows and on soil bared by grazing, timber harvesting, and road building. Also, increasing soil moisture and erosion of toes of streamside slides and earthflows can accelerate mass movement directly into channels. Finally, high annual precipitation in the basin does not promote a denser protective cover of vegetation than in areas with less precipitation. Little of the precipitation falling in winter can be utilized for plant growth, and under natural conditions the basin is already well vegetated except on steep hillslopes along downcutting channels. As a result, sediment discharge increases with annual precipitation in the Coast Range (Janda and Nolan, 1979), unlike most other areas (Langbein and Schumm, 1958; Wilson, 1973).

Also unlike most areas, suspended sediment discharge per unit area in the Eel River increases with basin size (Brown and Ritter, 1971; Janda and Nolan, 1979). Because of ongoing uplift, main channels are commonly more deeply incised than their tributaries, and so streamside landslides, which are major sources of sediment, are particularly abundant along main channels. Parent material is generally soft and friable, and thus, bed particles rapidly break down into smaller sizes (Knott, 1971). Consequently, suspended-sediment load increases downstream at the expense of bedload (Brown and Ritter, 1971).

VARIATIONS IN GEOMORPHIC FORMS AND PROCESSES

The geologic complexity and youthfulness of the landscape are reflected in the variety of hillslopes and channels. Lithology and the degree of fracturing of the bedrock control local erosion rates, erosional landforms, and channel morphology (Janda, 1979).

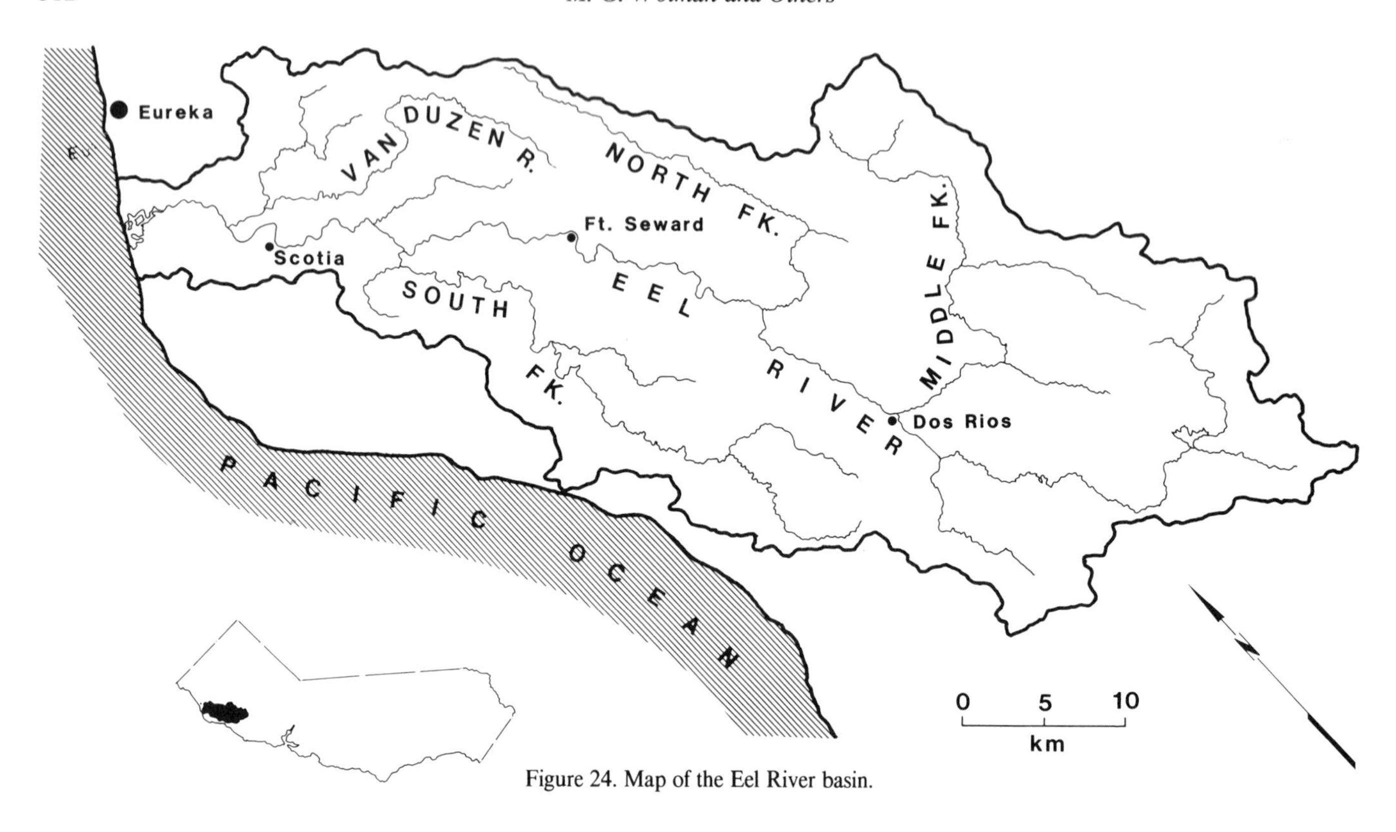

Figure 24. Map of the Eel River basin.

Mélange terrain

Highly fractured mélange units in the middle reaches of the Eel and Van Duzen basins contain abundant streamside slumps and earthflows that directly contribute large volumes of sediment to channels (Brown and Ritter, 1971; Kelsey, 1980). Estimated average annual sediment yield from a stream draining an earthflow is 24,000 t/km^2 (Kelsey, 1980)—about ten times that for the Eel basin as a whole. Sixty-eight percent of the suspended sediment discharge of the Eel River upstream of Scotia comes from 36 percent of the basin—the reach between Dos Rios and the junction with the South Fork (Fig. 24)—which contains the greatest areas of mélange, earthflows, and streamside slides (Brown and Ritter, 1971).

Most of the sediment from mélange terrain is sand or finer material eroded from toes of earthflows (Nolan and Janda, 1989) and from gullies cut on steep and disrupted hillslopes (Kelsey, 1980). However, earthflows that impinge on channels can contribute blocks of exotic material as large as 10 m and more in diameter and create extremely narrow, steep, coarse channels. These constrictions have led to the formation of depositional reaches upstream that have wide, alluvial channels and gentler streamside slopes. The alternation of these contrasting reaches produces large-scale steps in longitudinal channel profiles (Kelsey, 1980).

Competent terrain

Areas of more competent, graywacke sandstone are generally forested, have lower mass transport rates than mélange terrain, and contain "V"-shaped valleys with steep straight hillslopes. Debris slides and avalanches are the predominant sediment sources. These contribute abundant coarse material to channels, but maximum particle size is smaller than that from earthflows. Stream gradients are not unusually steep, and most coarse material entering from hillslopes can be transported downstream during annual floods. Average annual sediment yield from stable forested basins is estimated at 300 t/km^2 (Janda and Nolan, 1979; Kelsey, 1980)—only about one-tenth of the average for the Eel basin.

Effect of land use

Although soils are generally permeable and stable on slopes less than 30° (Brown and Ritter, 1971), disturbance of the ground cover can greatly accelerate surface and mass erosion in both stable and unstable areas. Despite the low population density, large areas of the basin are affected by grazing, timber harvesting, or associated road construction. Loss of tree-root strength in uncohesive soils (Ziemer, 1981) has probably helped to destabilize clearcut hillslopes; grazing and the replacement of native perennial grasses by European annuals with shallower roots has probably increased gullying of grasslands (Kelsey, 1980). Anderson (1970) estimated that intensive timber harvesting and associated road building from about 1950 to 1975 increased sediment yields several fold. Nolan and Janda (1981) measured a 10-fold increase in suspended-sediment discharge from tractor-yarded clearcuts in tributaries of Redwood Creek. The coincidence of concentrated timber harvesting and a series of large floods, however, makes it difficult to separate the effects of these two impacts on erosion and sediment yield (Harden and others, 1978; Kelsey, 1980).

EFFECTIVENESS OF LARGE FLOODS IN SHAPING THE LANDSCAPE

Several authors have concluded that high-magnitude, infrequent floods have a greater impact on the landscape relative to smaller floods in northwestern California than in other areas (Janda and Nolan, 1979; Kelsey, 1980; Lisle, 1981; Nolan and Marron, 1985). During the flood of December 1964, rainfall recorded at more than 550 mm during 48 hr in some locations produced stages in the Eel River 2 to 5 m above previous records (Waananen and others, 1971; Brown and Ritter, 1971). Peak flood discharge of the Eel River near its mouth was 26,500 m^3sec^{-1}, corresponding to runoff rates of 2.82 $m^3sec^{-1}km^{-2}$. This flood ranks among some of the world's great recorded floods for a basin of this size (Wolman and Gerson, 1978). Kelsey (1980) estimated the recurrence interval of the 1964 flood in the Van Duzen River, a major tributary, at approximately 100 yr. The flood caused profound changes in sediment transport rates and long-lasting changes in hillslopes and channels. Some morphologic changes persist today.

Sediment transport by large floods

Large, infrequent flows transport a relatively large proportion of sediment in the Eel River. At three gaging stations in the basin, discharges below which 90 percent of the suspended sediment load is carried have recurrence intervals between 3 and 16 years (Nolan and others, 1987). At these stations, the proportion of sediment carried by discharges of given frequencies increases with decreasing frequency of discharge and reaches a node at moderate frequencies (recurrence interval of 1.2 to 1.6 yr), as observed in other regions. The proportion remains high for infrequent discharges at the Van Duzen station, however, and increases again with further decrease in discharge frequency at the Fort Seward and Black Butte River stations. At Black Butte River, a major tributary upstream of Dos Rios, the greatest proportion of load has been transported by the most infrequent discharges.

During the 1964 flood, 105 million tonnes of suspended sediment were transported past Scotia during a 3-day period, compared to 85 million tonnes transported during the previous 8 years (Brown and Ritter, 1971). The flood accounted for 7 percent of the total sediment discharge of the Van Duzen River during a 35-yr period, and mobilized as much bed load as moves out of the basin in a century (Kelsey, 1980). Suspended-sediment concentrations at a given discharge increased several-fold and remained high for 2 to 5 years after the flood (Anderson, 1970; Knott, 1971).

Effects on channels and hillslopes

One reason why large floods are so important in shaping stream channels in the Coast Range is that material mobilized from landslides during large storms is commonly carried directly to stream channels instead of to lower hillslope sites or valley flats. Air photos of the basin taken before and after the 1964 flood (Fig. 25) show increased incidence of new landslides and long reaches of greatly widened channels (Brown and Ritter, 1971; Kelsey, 1977). For instance, the length of stream banks affected by debris avalanches increased 423 percent in the upper portion of the Van Duzen basin and 119 percent in the lower portion (Kelsey, 1977). Voluminous coarse debris from debris avalanches and torrents led to widespread channel braiding, channel widening commonly more than 100 percent, and aggradation more than several meters in some reaches (Hickey, 1969; Brown and Ritter, 1971; Knott, 1971; Kelsey, 1977). In areas where landslides were voluminous, aggradation and channel-widening downstream caused additional streamside failures by erosion of supporting material at the base of hillslopes (Kelsey, 1977; Janda and Nolan, 1979).

In addition to widening, channels adjusted to the increased sediment load by reducing bar-pool bed topography and thereby reducing hydraulic friction (Lisle, 1982). As a result, velocity increased and depth decreased at a given discharge, signifying an increase in bed-load transport capacity (Knott, 1971; Lisle, 1982). These adjustments may have accelerated the flushing of excess material from the channel networks. Associated changes in aquatic habitat may have contributed substantially to the decline in populations of anadromous salmonids in the basin (California Department of Water Resources, 1974).

Channel recovery

The 1964 flood appears to have been effective in shaping stream channels of the Eel basin, according to Wolman and Gerson's (1978) criteria, because the changes have persisted in some reaches up to the present (Lisle, 1981; Kelsey and Savina, 1985). In some reaches, channel patterns and flood deposits along the higher margins of channels will be altered little until a flood of equal or greater magnitude recurs (Kelsey, 1977).

Channels have recovered in overlapping stages dependent on a sequence of processes. First, suspended-sediment concentrations declined to pre-flood levels within about 5 years. Second, as excess bed material has been transported downstream, channel beds have degraded to stable levels at or above pre-flood elevations over periods of a few years or longer, and some reaches may remain aggraded into the next century (Kelsey, 1980; Kelsey and Savina, 1985; Lisle, 1981). These periods depend apparently on the volume and coarseness of aggraded material, channel gradient, and distance from sediment source. During channel-bed degradation, hydraulic geometries have recovered to some degree

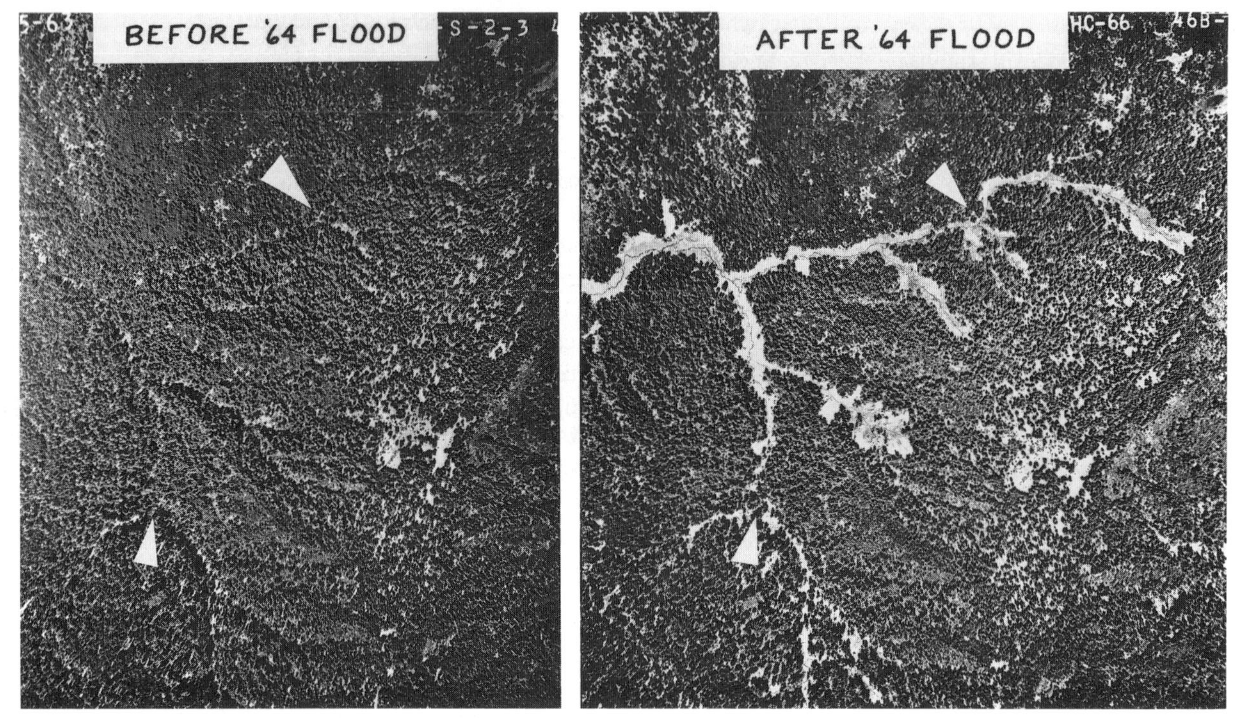

Figure 25. Aerial photographs taken in summers of 1963 and 1966 of the headwaters of the South Fork Van Duzen River, showing changes due to the 1964 flood. (From Kelsey, 1977, with permission). The white arrows identify the same channel reaches on both photos. Lighter areas in the 1966 photo were devegetated by debris avalanches, debris torrents, and widened, aggraded stream channels.

to pre-flood relations. The degree of recovery apparently depends on reestablishment of pre-flood channel widths (Lisle, 1982)—the third phase of channel recovery. Channels in alluvial reaches have incised into flood deposits, leaving a narrower channel bounded by sparsely vegetated flood deposits. Many tributary channels that are bounded on at least one bank by bedrock or colluvium have remained wide, however. Soil creep and dry ravel can be slow in replacing eroded banks, and new bank material is frequently scoured by high flows contained in narrow valley bottoms (Lisle, 1981). Riparian vegetation (primarily red alder and willow), which aids bank accretion along low-flow channel margins, is also subject to scour during high flows. Riparian trees are now well established along many reaches, however, due to the absence of large floods since 1975.

CONCLUSIONS

Erosive bedrock, rapid uplift, high seasonal rainfall, and recent disturbance by man have produced exceptionally high sediment yields from the Eel River basin. Because channels are commonly bounded by hillslopes in narrow valleys, channel morphology and sedimentology are strongly influenced by adjacent hillslope processes, which vary with the lithology and degree of shearing of bedrock. Because of the close linkage between channel and hillslope processes and the occurrence of high runoff events, large floods produce and transport a large proportion of fluvial sediment and cause widespread, persistent changes in channels. Subsequent remolding of channels by smaller discharges proceeds with the transport of excess sediment out of channels and the reconstruction of streambanks. These sequences of channel recovery can require as long as several decades.

NIOBRARA RIVER

John P. Buchanan and Stanley A. Schumm

The Great Plains physiographic province extends from the Mexican border in Texas to the Canadian border in Montana and North Dakota, and from the eastern edge of the Rocky Mountains to eastern Nebraska and Kansas. Therefore, as geomorphic, geologic, and climatic conditions vary substantially, it is difficult to select a typical Great Plains stream. In addition, the major rivers that have received the most attention, such as the South Platte, North Platte, and Arkansas, rise in the Rocky Mountains, and consequently they are strongly influenced by snow-melt runoff. Other major rivers such as the Canadian, Red, and Cheyenne, for example, have been significantly affected by flood-control reservoirs and irrigation projects. However there is at least one exception to these statements. The Niobrara River (Fig 26A) in northern Nebraska is one of the last relatively undeveloped Great Plains rivers that retains the braided pattern so characteristic of pre-settlement Great Plains rivers (Williams, 1978; Nadler and Schumm, 1981).

In order to determine the effects of construction of a dam near Norden, Nebraska, a detailed study of a 100-km reach of

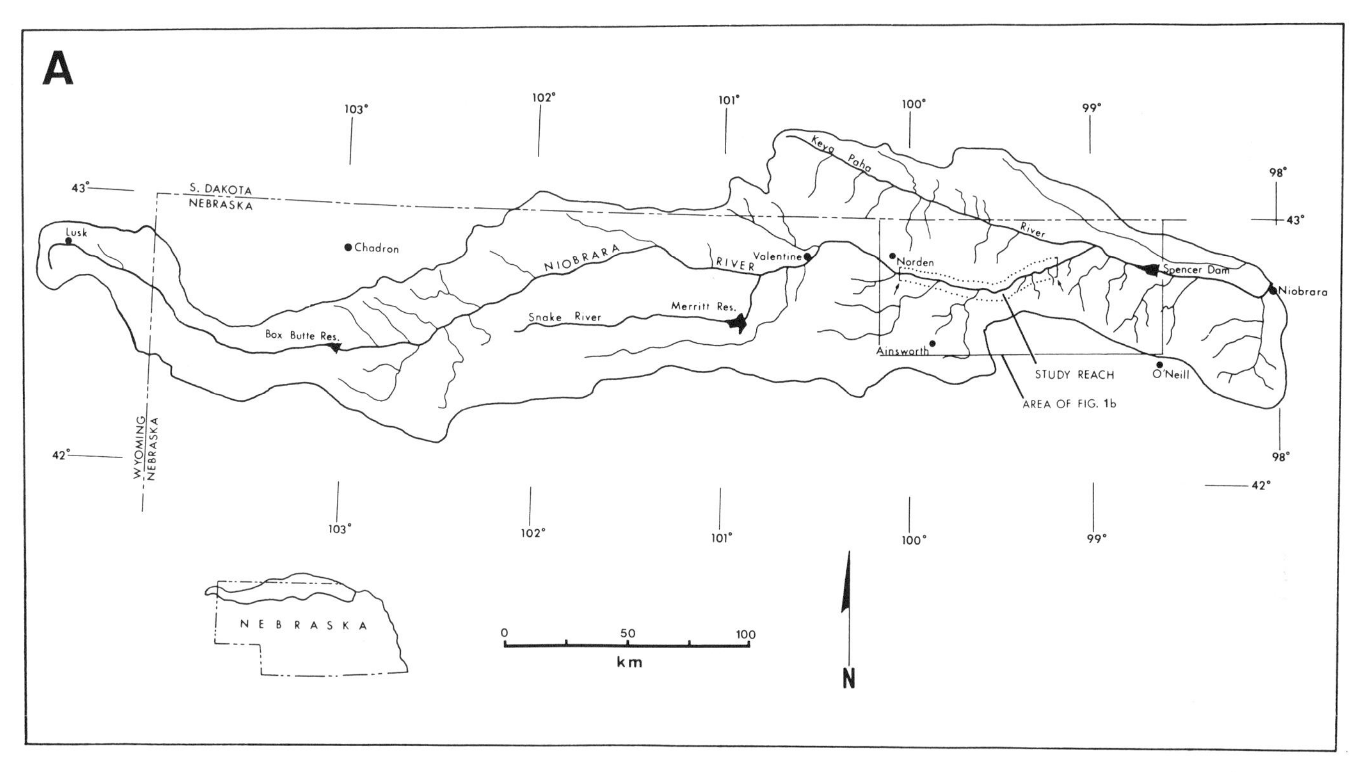

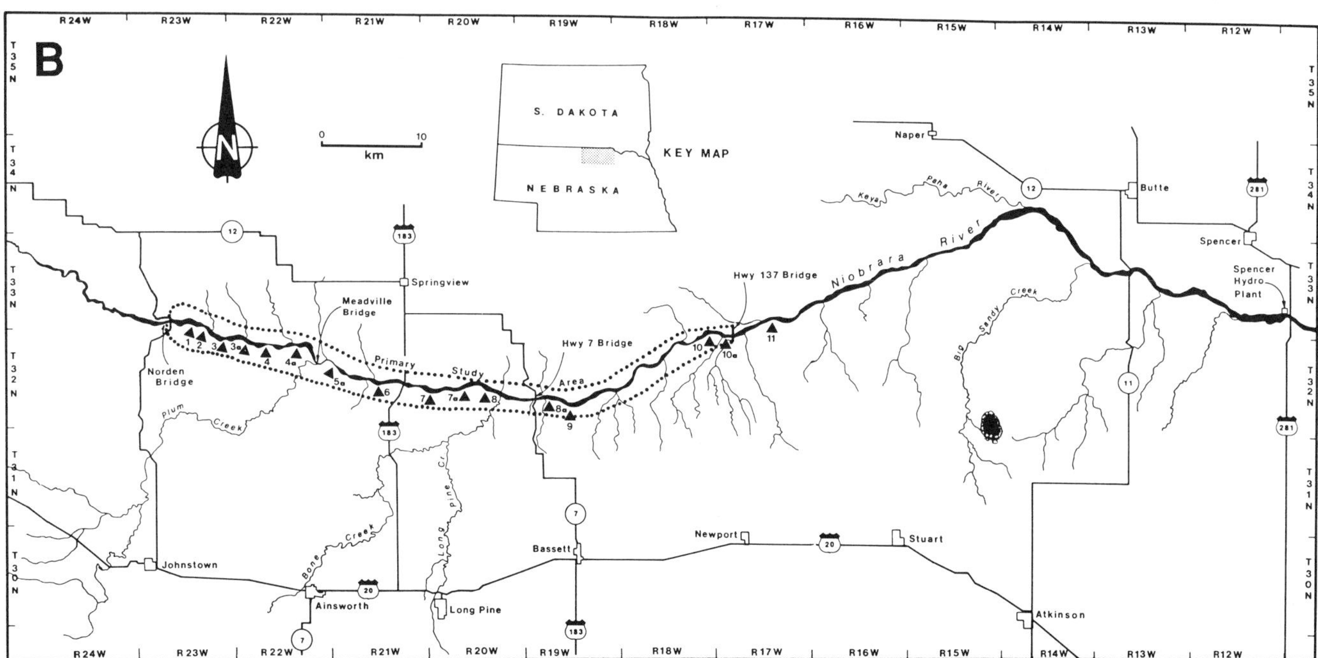

Figure 26. A. Map of the Niobrara River drainage basin showing major tributaries and water resource development projects. B. Study area of Buchanan (1981) showing locations of river ranges (dark triangles) surveyed in 1979 and 1980 by the Bureau of Reclamation.

Figure 27. Aerial photographs of Niobrara River near River Range 6. A. High-flow stage in April 1980 when discharge at Norden gage was about 28.4 m^3/sec. B. Low-flow stage in September 1980 discharge at Norden gage was about 13.2 m^3/sec. Channel is about 440 m wide; flow is from top to bottom of photograph.

Niobrara River downstream from Norden was undertaken (Fig. 26B). Seventeen permanent survey ranges were surveyed in 1979 and 1980 by the Bureau of Reclamation. They provide a base for studies of channel morphology and channel change, and a description of a significant reach of a Great Plains river (Buchanan, 1981).

Niobrara River (Fig. 26A) lies within the High Plains Section of the Great Plains physiographic province (Fenneman, 1931). It has a drainage area of 33,150 km^2 of gently rolling terrain, and it flows through the Sandhills area of the Nebraska panhandle. At the breakwaters near Lusk, Wyoming, the elevation is about 1,590 m above sea level. The river falls about 1,220 m in the 715 km of its course, and it enters the Missouri River at Niobrara, Nebraska, at an elevation of 366 m. In the lower one-third of its course the river is characterized by perennial flow, and it provides a classic example of braiding (Fig. 27).

Precipitation in the basin increases toward the east, ranging from about 41 cm at Lusk, Wyoming, to about 61 cm per year at Niobrara, Nebraska. Summers are usually very warm and winters extremely cold. Great variations in temperature and precipitation occur from year to year. Average annual temperature in the central part of the basin is 8.3°C, with 151 frost-free days in an average growing season. About 65 percent of the average annual rainfall occurs from May to September in the form of thunderstorms and cloudbursts. Drought conditions prevailed in the basin from 1931 to 1940, and there was an extended period of below-average precipitation from 1954 to 1977.

Developments on the Niobrara and its tributaries for the purpose of water storage, irrigation diversion, and hydroelectric power generation are few considering the potential of the basin. As a result, the Niobrara is relatively unaffected by discharge regulation. There are several projects, however, that have modified flows to some degree; these include Box Butte Reservoir, Merritt Reservoir, and Spencer Dam (Fig. 26A).

CHANNEL CHARACTERISTICS

The Niobrara River upstream from the study reach is typically a narrow meandering channel, about 30 m wide, which flows through a narrow valley at a gradient of about 0.0011 m/m. At a point 10 km upstream from Norden Bridge, the valley widens, and the stream gradient steepens to about 0.0017 m/m. Here the river changes to a braided pattern that persists to its mouth (Fig. 27). The braided channel ranges from 60 m to 550 m wide within a broad valley between bedrock bluffs. Water depths vary from a few centimeters to no more than 1.5 m at high flows. Discontinuous flood-plain remnants, which consist of wet meadow and recent terrace alluvium, are present in the study reach, and they suggest at least some local variation in stream-bed elevation through time.

At several points there are significant changes in the channel morphology along the river. The braided channel narrows drastically where flood-plain remnants are more extensive, and constricted reaches are common where the flow impinges against a bedrock outcrop on the otuside of a river bend. Narrowing is also related to the effect of tributaries, where their contribution of sediment to the main channel may cause formation of a local flood-plain downstream of their junction.

Although the Niobrara is a braided channel, it exhibits very different characteristics over a range of discharges. At high discharge, which is common in the spring during ice breakup and runoff, a well-defined sinuous thalweg is present in the braided channel (Fig. 27A). However, at lower flows the thalweg is absent, and the river assumes a classic braided pattern of divided flows around linguoid bars (Fig. 27B).

Linguoid bars, which are typical of sandy-braided rivers (Miall, 1977), are the most dynamic features of the Niobrara River channel. They range in height from 0.3 m to 1.2 m, and they may be hundreds of meters long. The bars are typically lobate or rhombic in planform; the downstream face is usually straight to sinuous with an avalanche slope, and the upstream surface dips gently beneath the succeeding bar. The linguoid bars are most active when low discharges cover the entire channel. The sandy bed of the Niobrara River displays a complex assemblage of various bed forms at a scale smaller than linguoid bars; these include ripples, dunes, sand waves, plane beds, and antidunes. Islands are covered with vegetation, and they are always emergent at the highest flows. They are elongate in the direction of flow, and they vary in size from tens to hundreds of meters long.

Bed sediments are dominantly sand size or finer, with only a very small amount coarser than sand. Mean grain diameter is between 0.2 and 0.3 mm, and 98 percent of the samples are sand size or finer. There is no significant trend of grain-size variation with channel distance downstream through the study area. Fifty-five surface bed-material samples were collected from 1950 to 1967 at the Norden gage. Inspection of these data reveals that there has been no change of grain size during that period. Drilling into the bed at River Range 1 (Fig. 26B) shows that the alluvium is mostly poorly sorted sand underlain by the Pierre Shale at a depth of about 18 m. There is no coarsening of the sediment with depth.

Fifty percent of the total sediment discharge of the Niobrara is carried in suspension (Colby and Hembree, 1955). Sediment loads range from about 365 metric tons per day at a water discharge of 14 m^3/sec to 10,000 metric tons per day at a water discharge of 65 m^3/sec.

HYDROLOGIC CHANGES

The Niobrara River and its tributaries are fed by ground water, which provides steady base flow throughout the year. Occasional high flows are due to spring ice breakup, snowmelt, and summer cloudbursts. The flow of the Niobrara River has been reduced only slightly by upstream water-resource developments. Ground-water irrigation in the Niobrara drainage basin began in 1938 and has increased steadily.

Diversions at Box Butte Reservoir began in 1946, as part of the Mirage Flats Irrigation Project south of Chadron, Nebraska (Fig. 26A). The effect of this project in the study area is generally accepted as being negligible, as the project is far upstream. More important, however, is the effect of the construction of Merritt Reservoir on the Snake River, southwest of Valentine (Fig. 26A). This project began diverting water in February 1964, as part of the Ainsworth Irrigation Unit. The Snake River is a major tributary of the Niobrara, and the total discharge of the Niobrara is doubled at their confluence. Therefore, the depletion of flow from the Snake River has had a significant effect on the discharge within the study area.

From 1950 to 1964, before construction of Merritt dam and subsequent regulation of flows, peak discharges were irregular and had high values (Fig. 28). However, mean annual discharge varied little from year to year during this period due to the steady ground-water inflow from the Sandhills. Since 1964, however, the regulation of outflow from the reservoir has reduced the peak discharges significantly. Mean annual discharge decreased, and it became more variable. Nevertheless, 1967, which was a year of unusually high precipitation, was the year of maximum discharge, 58.6 m^3/sec (U.S. Geological Survey, 1984).

For the period 1964 to 1979, mean monthly discharge reached a maximum in March of 28.5 m^3/sec and 22.4 m^3/sec in August. The mean annual discharge of the Niobrara River at the Norden gage has been reduced about 16 percent from 27 m^3/sec to 22.7 m^3/sec since the operation of Merritt dam (Fig. 28).

The Spencer dam, located 122 km downstream from the western end of the study area (Fig. 26A and B), is the largest project on the Niobrara. Constructed in 1910, this is now a run-of-the-river plant because the storage capacity of the reservoir has been lost owing to sediment deposition. This structure has caused width reduction, pattern change, and degradation of the channel downstream.

Climatic fluctuations in the basin have also affected streamflows to a degree. For example, the 1930s drought had a profound impact on the discharge of the river for that period. There was also an extended period of below-average precipitation in the 1960s, which resulted in below-average discharges as well.

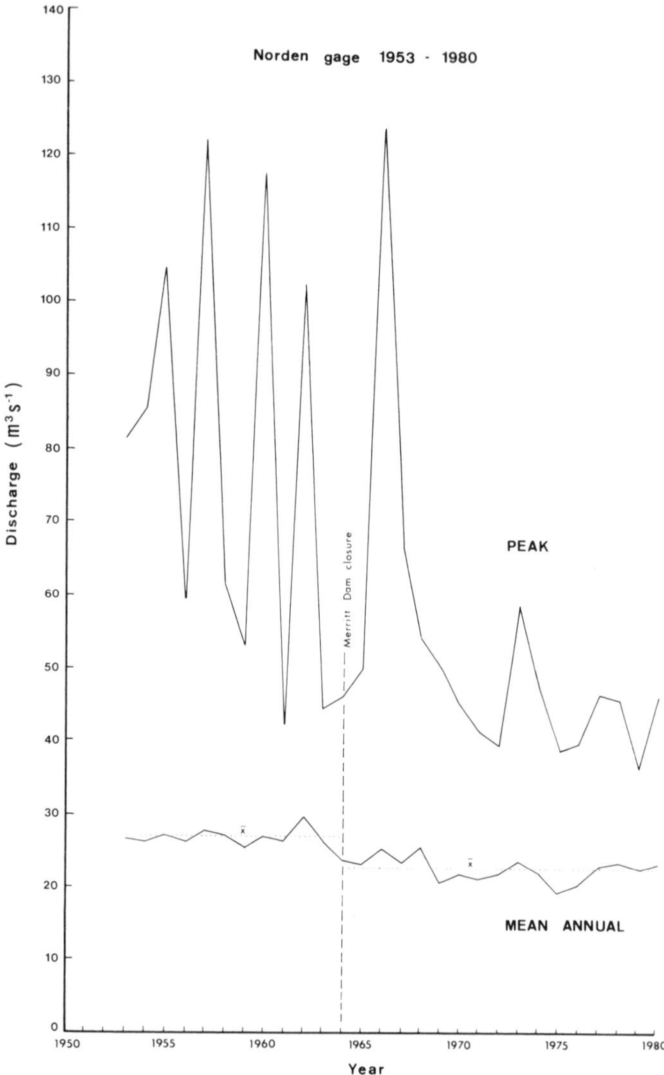

Figure 28. Peak and mean annual discharges of Niobrara River recorded at Norden gage, from 1953 to 1980. Note reduction of flows after closure of Merritt Dam upstream on the Snake River.

MORPHOLOGIC CHANGES

The earliest reliable maps of the river are the General Land Office surveys of about 1870. More recent information was obtained from aerial photographs taken in 1939, 1954, 1968, 1974, 1977, and 1980.

In 1939, The Niobrara River had a wide braided channel with divided flow around sand bars. Comparisons with the 1954 to 1980 aerial photos reveal significant changes in the bank line and vegetated areas adjacent to the channel. Channel width generally increased from 1939 to 1954 and decreased between 1954 and 1980 (Table 2). Nevertheless, the channel remained braided during the period of record. These data show the general variability of braided-channel dimensions, as islands and flood-plain areas form and are destroyed.

The Niobrara River in 1870, as shown by the General Land Office maps, was broad, with the widest channel of record and few islands. By 1939, its width had decreased, and several islands had formed as vegetation colonized portions of the channel. The drought period of the 1930s apparently allowed the stabilization of sand bars by vegetation, as flood peaks, which had previously removed vegetation from the sandy channel annually decreased. This nine-year drought caused a general narrowing of the channel and the formation of several large islands between River Ranges 4 and 5.

From 1939 to 1954 there was an increase in channel width. The removal of vegetation and scouring of the channel banks increased during this period of high precipitation and associated high flows.

After 1954 there was an extended period of below-average precipitation, and the river narrowed by island formation and island attachment to the flood plain. Closure of Merritt dam on the Snake River in 1964 also reduced flood peaks and mean annual flows. This may have accelerated the rate of vegetation encroachment into the sandy channel, and thereby increased the rate of channel-width reduction.

In summary, the changes in channel morphology of the Niobrara River have undoubtedly been the product of altered river regime. Episodes of channel widening can be associated with higher precipitation and discharge, and width reduction is linked with droughts and the closure of Merritt Reservoir.

ANNUAL MORPHOLOGIC CHANGE

In the study reach, the Niobrara River is a braided channel. However, in the course of one year, the river displays very different channel patterns over a range of discharges. A comparison of aerial photographs taken in April and September 1980 shows that a narrow, sinuous thalweg had developed in the channel in April, but it almost completely disappeared in September (Fig. 27A and B). The discharges at the Norden gaging station when the aerial photographs were taken were 28.4 m^3/sec in April and 13.2 m^3/sec in September. The photos, therefore, are representative of the channel morphology in the Niobrara River under high- and low-flow situations. The study reach was photographed again in December 1980, and the thalweg had reappeared during high flows, when the discharge at Norden was 24.4 m^3/sec.

Basically, the river is less braided at high discharge, as a result of concentration of discharge within a relatively narrow channel at high flow (Fig. 27A). At lower flows the channel assumes a more braided pattern, and the flow is more evenly distributed across the entire channel width, which produces more linguoid bars (Fig. 27B). Also, the linguoid bars tend to shorten at the low flows, thereby increasing bed roughness.

Impressive annual changes in characteristics of cross sections occur in the Niobrara River. Cross sections were periodically surveyed by the Bureau of Reclamation in 1979 and 1980. Generally, these surveys indicate shifting of the bed elevation and location of bars, as well as the presence or absence of the thalweg.

TABLE 2. WIDTH OF THE SAND-BED CHANNEL OF THE NIOBRARA RIVER AT THE RIVER RANGES AS MEASURED FROM AERIAL PHOTOGRAPHS

River Range	Channel width 1939 (m)	1954 (m)	1968 (m)	1980 (m)
1	423	512	293	216
2	302	341	244	137
3	161	244	256	226
3a	463	439	390	255
4	60	97	146	137
4a	261	354	390	275
5	201	232	244	206
5a	302	366	439	304
6	242	378	414	236
7	161	366	390	343
7a	362	463	365	334
8	463	512	561	393
8a	422	488	512	383
9	321	341	317	294
10	161	195	244	216
10a	121	73	98	108

Width did not change significantly between surveys. An example of this cycle of thalweg behavior is displayed at River Range 10. Figure 29 shows the thalweg during the June 1979 survey; however, by August it had disappeared completely at low flow, and the bed of the channel assumed a more uniform profile. In April 1980, the thalweg reformed at high discharge, but it had shifted laterally. An additional survey of this cross section in November 1980, after several months of low flow, reveals that the thalweg had aggraded. Presumably the thalweg re-forms at another position during the high discharge of the next spring.

Not only does the channel morphology change, which is most easily observable, but water-surface elevation and Manning's roughness coefficient (n) also change in varying amounts over a range of discharges. These variables were determined for several representative cross sections in order to contrast the high-and low-flow behavior of the Niobrara River channel (Table 3). Surprisingly, there is only a relatively small change of the water-surface elevation, or stage, for wide variations of discharge (Table 3).

In order for the water-surface elevation to remain relatively constant, and for the flow to spread more evenly across the entire channel at low flows, some adjustment of the bed of the channel is necessary. A simple mechanism, which can provide such an adjustment, is a change in the flow resistance or roughness of the channel. The linguoid bars shortened and became more numerous at low flow (Fig. 29B). This, in effect, increased bed roughness, as reflected by high values of Manning's roughness coefficient (n) at low discharge (Table 3). When discharge increases, channel roughness is decreased by concentration of flow in the thalweg (decreasing braiding indices) and by lengthening of the linguoid bars. The velocity, therefore, is increased, and the water-surface elevation, or stage, remains remarkably constant.

Because of the severely cold winters in Nebraska, the Niobrara River channel is frequently covered with ice. Ice usually begins to develop in December. It forms on the sand bars first and then spreads into the channels. The period of major ice breakup is usually in March, at the same time spring runoff is at its peak. During the breakup the channel is often choked by ice jams.

The effect of ice on river channels is a poorly understood phenomenon; nevertheless, Collinson (1971), Ashton (1979b) and Smith (1979) all comment that the presence of ice in a river channel usually results in scouring of the bed. Ice cover also reduces the cross-sectional area of a channel and causes higher flow stages. Flow velocities under the ice are lowered, and bed roughness increases.

In order to investigate the effects of ice cover on the Niobrara River channel, several flights were made over the river during the winter. In mid-December 1979 there was a sinuous open channel in the ice cover. Because scouring of the bed is

TABLE 3. HYDRAULIC PARAMETERS DETERMINED FOR SELECTED RIVER RANGES AT HIGH AND LOW DISCHARGE IN THE NIOBRARA RIVER

River Range	August 1979 Q (m^3s^{-1})	n	Water surface elevation (m MSL)	April 1980 Q (m^3s^{-1})	n	Water surface elevation (m MSL)	Rate of change of water surface elevation (m/m^3s^{-1})	Percent change n
2	15.3	0.0222	636.88	37.0	0.0145	636.97	0.00415	53
5	16.6	0.0220	619.05	36.2	0.0147	619.20	0.00765	50
6	19.0	0.0259	637.79	37.1	0.0153	637.87	0.00442	69
9	22.5	0.0280	578.66	37.4	0.0152	578.66	0.0	84
10	24.9	0.0218	550.59	42.1	0.0185	550.74	0.00872	18
11	26.7	0.0185	543.61	39.3	0.0151	543.91	0.0238	23

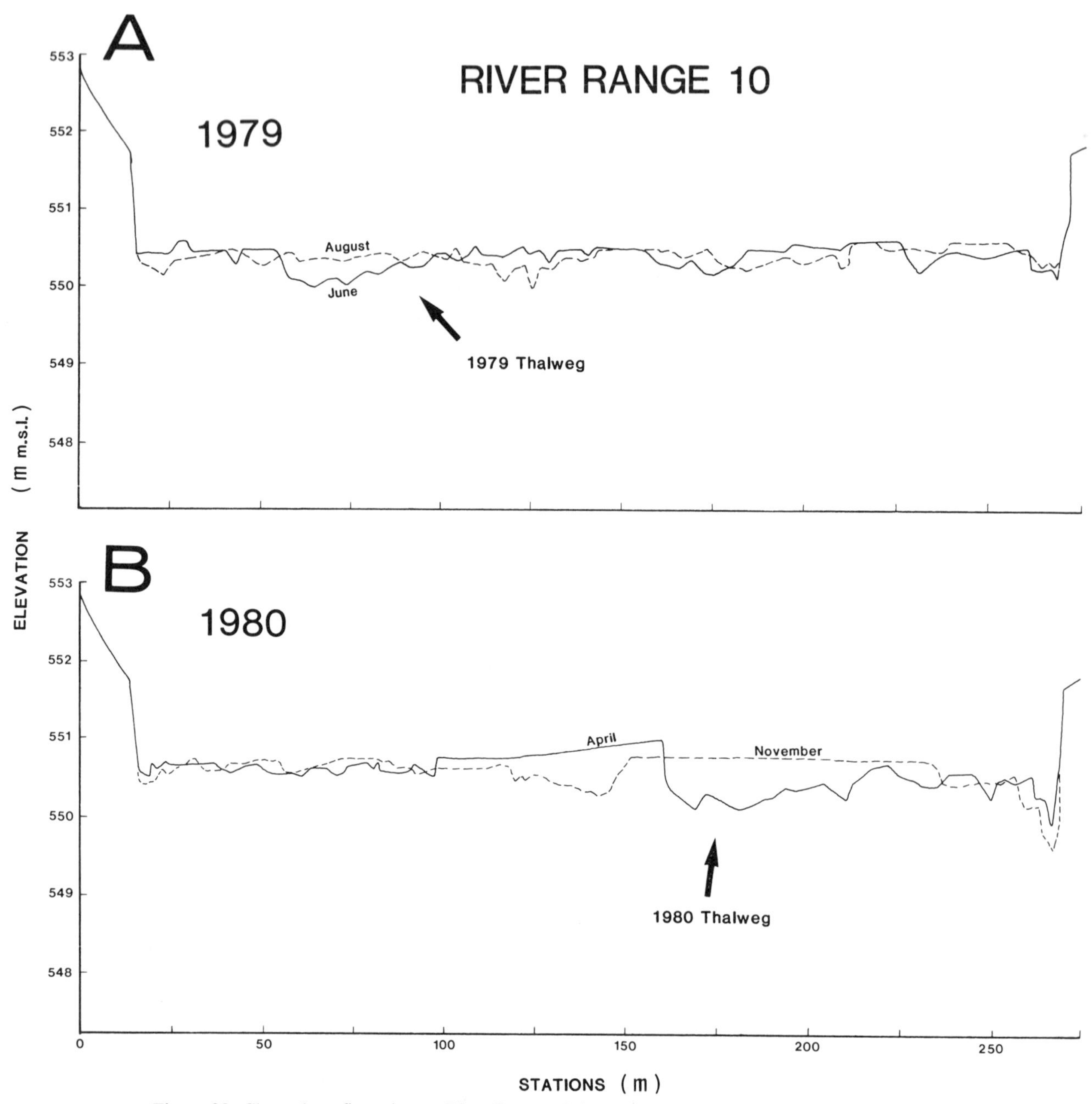

Figure 29. Channel configuration at River Range 10 in 1979 (A) and 1980 (B). Survey data from Bureau of Reclamation.

enhanced when there is ice cover, the formation of ice during the winter months must control the development of the Niobrara River thalweg at that time. It has been demonstrated that the greatest lateral shift of thalweg position occurs from one year to the next, rather than during the high-flow season. The position of the thalweg in the spring changes significantly from its position during the preceding fall. Therefore, the most dynamic movement of the thalweg may be controlled by the position of the open channel in the ice cover and the location of ice jams during spring breakup. A hypothetical thalweg-shift cycle is presented in Figure 30.

In summer (Fig. 30A), after ice breakup and spring runoff, a sinuous thalweg is well defined, and most of the flow is concentrated in it (Fig. 27A). A braided pattern occupies the point-bar-like areas on both sides of the thalweg meanders. During lower flows later in the year, the fine, sandy bed material is transported into the thalweg, where it is deposited (Fig. 30B). The flow then becomes more evenly spread over the entire channel, and the thalweg disappears (Fig. 27B). In winter, ice forms on the river and a narrow open channel develops randomly in the ice cover. Since the sandbars are frozen and essentially stabilized, most of the discharge of the river flows in the open ice channel, and it scours the bed below it (Fig. 30C). This process creates a new thalweg in the spring (Fig. 30D), and the cycle is repeated.

SUMMARY

Niobrara River in many ways appears to be typical of numerous Great Plains rivers. It is a sand-bed braided river that responds quickly to discharge fluctuations resulting either from climatic fluctuations or flow regulation. However, Niobrara River also has some unusual characteristics. The seasonal change of bed roughness maintains a remarkably constant water-surface elevation during substantial annual discharge changes. This may be a function of the unique effects of ice cover and fine-sand bed material, but, nevertheless, other Great Plains rivers may undergo similar annual changes to a lesser degree.

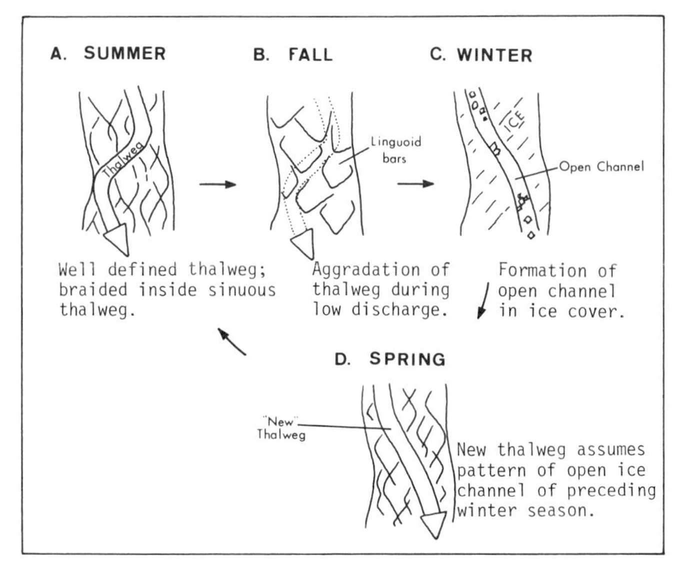

Figure 30. Hypothetical cycle of thalweg behavior attributed to ice development on the Niobrara River. See text for explanation.

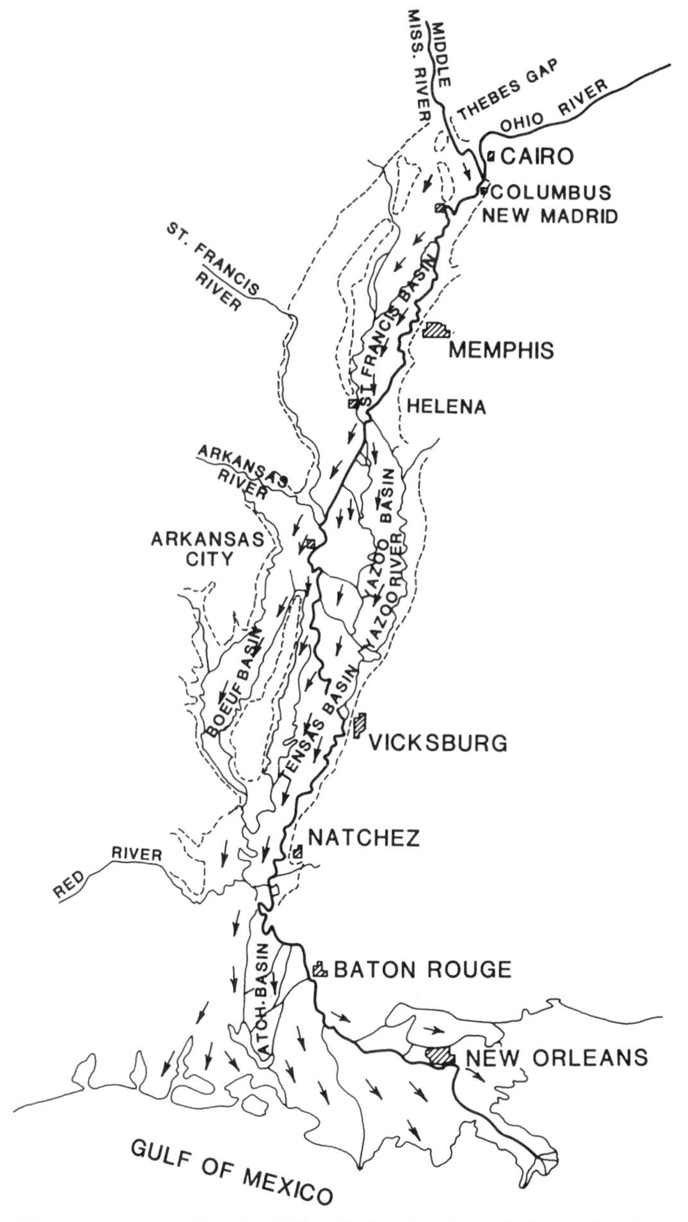

Figure 31. Lower Mississippi River Valley showing principal cities. Arrows show flow prior to construction of levees (Winkley, 1977b).

THE RIVERSCAPE OF THE LOWER MISSISSIPPI RIVER

Brien R. Winkley

The hydrologic and the geometric characteristics of the present Lower Mississippi River are a result of its geologic history and of man's endeavors to utilize its drainage basin. The Lower Mississippi River Valley was virtually unfinished when the European immigrants began settling there (Fig. 31). Each of the many braided and meandering belts are visible on the valley surface with distinct lateral elevation changes between them. The suballuvial deposits are mostly of Tertiary marine origin. In some reaches the thalweg has eroded into these silts, sands, and clays, which, if cohesive, tend to act as a bedrock. Figure 33 shows several profiles of a 64 km reach of the river: (1) mean valley surface slope; (2) average low-water plain, (3) graveliferous alluvium; (4) 1882, 1943, and 1968–1973 thalweg profiles with regression lines between obvious slope and shape changes in these profiles; (5) probable location of fault zones from geologic maps; and (6) Tertiary deposits. Quaternary sediments, ranging from consolidated clays to coarse gravels, were deposited during the waxing and waning of the North American glaciers. The most recent sediments are a stratified conglomeration of all of the above plus eroded soils from 41 percent of the continental United States.

Today's meander belt, number 5 on Figure 32, varies in age from 1,000 years old, near the Gulf of Mexico, to as much as 2,000 to 6,000 years old farther upstream. This belt indicates

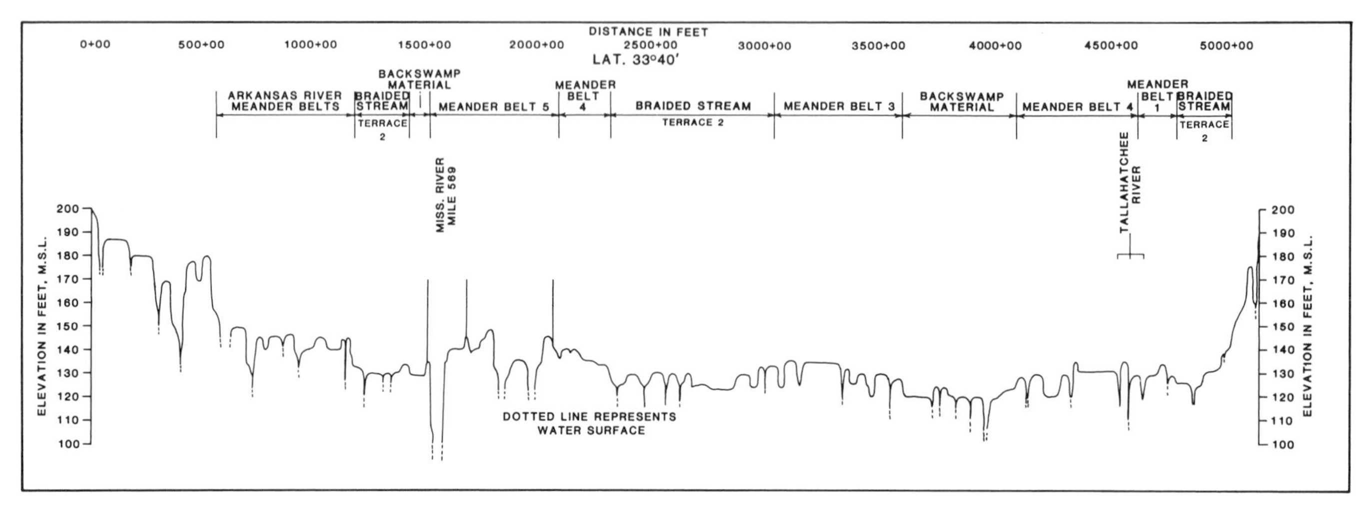

Figure 32. Valley cross section, Mississippi River, Lat. 33°41′ (Winkley, 1977a).

aggradation throughout the entire valley. Specific gage records (Fig. 34) also indicate a continual aggradation at all stations, during the period of record, until the man-made cutoffs of the 1930s.

Gradients varied as the relative sea level changed. Discharges, probably several times greater during glacier melt, deposited terraces of coarse sands and gravels within present thalweg depths (Fig. 33). Tributaries flowing through these varied deposits altered the sediment load of the Mississippi River, changing its geometry, at their confluence and often for several bendways downstream.

The valley slope undulates from as little as 5 cm/km to as much as 1.96 m/km as a result of tectonics, sediment surges, and tributary deltas. This energy-gradient variation resulted in sinuosity diversity as well as other channel geometric changes. The surge of sediment from the New Madrid Earthquake of 1811–12 could be influencing the river's current sediment activities. Remnants of seven delta outlets are still evident on the Gulf coast of Louisiana. Limestone formations at valley bluff-lines as well as in the suballuvium and clay-filled abandoned channels also influenced channel alignment. As a result of these varying sediments, gradients, and other geologic controls, the natural river had a wide assortment of channel geometric and hydraulic characteristics from reach to reach.

Abandoned meander belts indicate that the hydrograph could have been relatively constant over the past 4,000 to 6,000 years and also that the river had gained a dynamic equilibrium during the 500 years prior to the 1700s.

The Lower Mississippi River Valley is rich in nutrients—its topsoil is often many meters deep. There is an abundance of both surface and ground water for all of mankind's needs. However, as man began to convert the river and its valley to satisfy these needs, the significance of geologic history, and the residual sediments, and the sediments involved in hydraulic alterations were not adequately considered. Varying characteristics of individual reaches were not analyzed as local projects altered hydraulics and geometry. The flood plain, which is an integral part of the river, was occupied and the river stiff-armed away from it. In an endeavor to analyze the river's response, consideration must be given to emergencies and man's needs as well as to available knowledge and project funding levels. Moreover, politics often dictated where and when work was to be performed.

The water and sediment loads were constantly changed as a wide variety of flood control and navigation projects were initiated. Levees, drainage systems, dredging, reservoirs, cutoffs, floodways, and stabilization each had an impact on discharges and on hydraulics, and thus on sediment movement and channel geometry.

Originally, levees were built by local land owners. Levee districts were soon formed to coordinate construction efforts and techniques. The federal government began developing an integrated levee system with the creation of the Mississippi River Commission in 1879. Figure 35 is the maximum annual high water at Natchez, Mississippi, for each year of record. Levee heights have increased as more water was confined to the main channel and as the channel reacted to the many changes in the river's hydrology. Prior to levee construction, up to one-half of all flood discharges were overbank and a portion of all flows were released through the numerous distributary channels. Today's levees average 11 m in height above top bank and are intended to contain the maximum calculated flood plus 1 m of freeboard. Prior to 1928 there were numerous crevasses (Fig. 35); thus, the Natchez gage indicates limited overbank flooding. Originally, the entire valley below the confluence of the Ohio River was a delta with a distributary and collection system of channels throughout its entire length. Between Baton Rouge and New Orleans there were 144 outlets carrying a wide range of discharges. Before levees were built, the top-bank capacity of the entire 1,770 km of river was only 28,000 m^3 per second. The river between Memphis, Tennessee, and Vicksburg, Mississippi, had numerous side channels on the east bank that discharged 2,800 to 14,000 m^3/sec into the Yazoo Basin at all stages. As these distributary

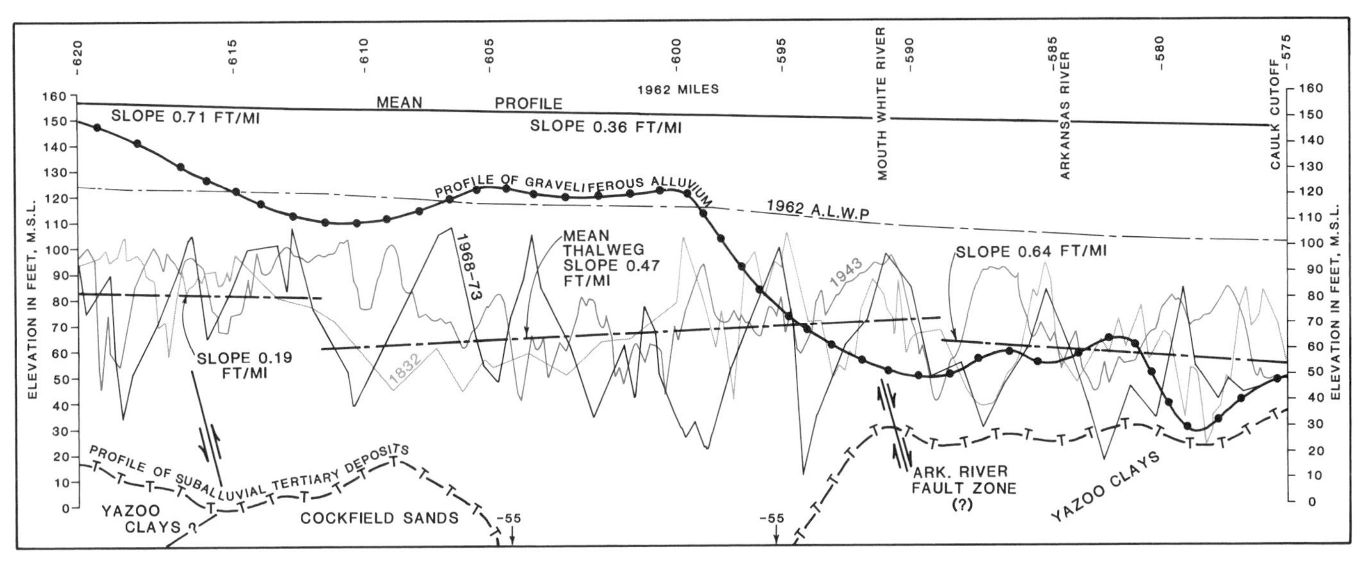

Figure 33. Sediment and thalweg profiles, Mississippi River, miles 575 to 615 (Winkley, 1977b). Thalweg profiles; black, 1968 to 1972; dark blue, 1943; light blue, 1832.

channels were closed, the levee heights were increased. Most of the levees are built on the natural levees of either the present river or abandoned channels; therefore, the distance between left- and right-bank levees varies from 3 to 30 km between Cairo, Illinois, and Baton Rouge. From Baton Rouge to the Gulf, levees are on top of the river bank, and so the distance between them only varies from 0.8 to 4 km.

As the in-channel flows increased, the river began adjusting its geometry to accommodate the extra discharge. The initial reaction was an increase in bank caving and channel widening. This created less flow efficiency and the need to increase levee heights. Also, banks were stabilized wherever levees or other man-made facilities were endangered. With these piecemeal stabilization efforts, malalignments began to occur and with them concomitant loss of sediment management.

As land clearing and drainage continued, a valley that originally contained more than 90 percent forests and swamps now has less than 10 percent forests and swamps. The need to move the increasing flood heights out of the valley more quickly lead to a series of man-made cutoffs during the 1930s and early 1940s (Figure 35). An 800-km reach in the central portion of the river was shortened 235 km. In addition, another 90 km of river length was eliminated by realignment through point-bar chutes. Some of these were a natural reaction to increased slopes, but many were constructed to align the river with the new cutoff channels and to develop the highwater channel; however, the highwater channel will not maintain itself because of sediment-activated changes in geometry during long periods of relatively lower flows. During the pre-1900 period, the Lower Mississippi River made an average of 16 cutoffs every 100 to 200 years over its entire 1,770-km length. The natural river's ability to adjust its hydraulic geometry to a cutoff took 30 to 80 years, depending on local gradients and

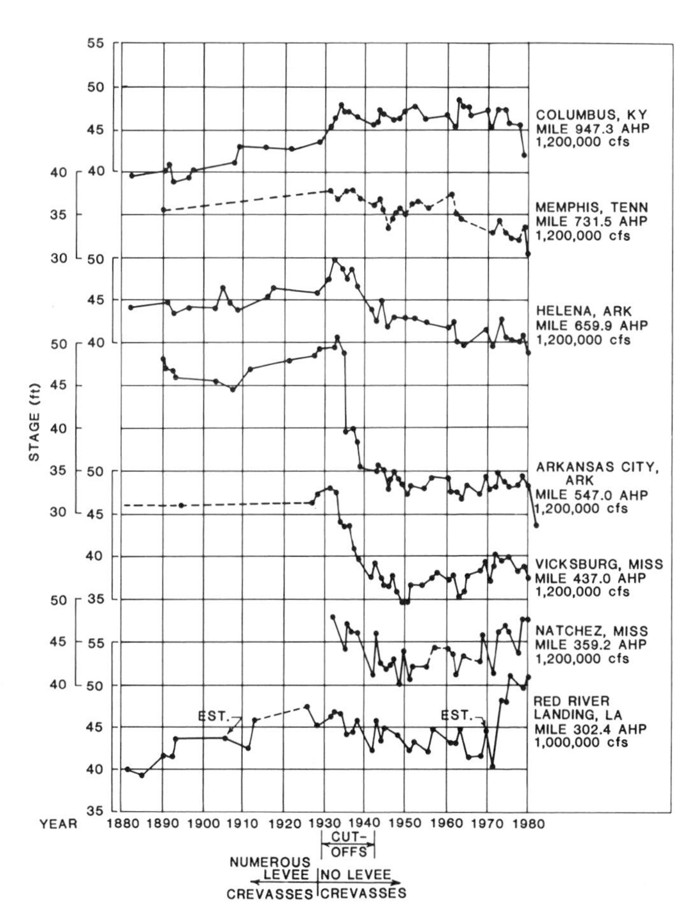

Figure 34. Specific-gage records, Lower Mississippi River showing changes in stage for selected constant flows over time. (Winkley, 1977b, updated 1982.)

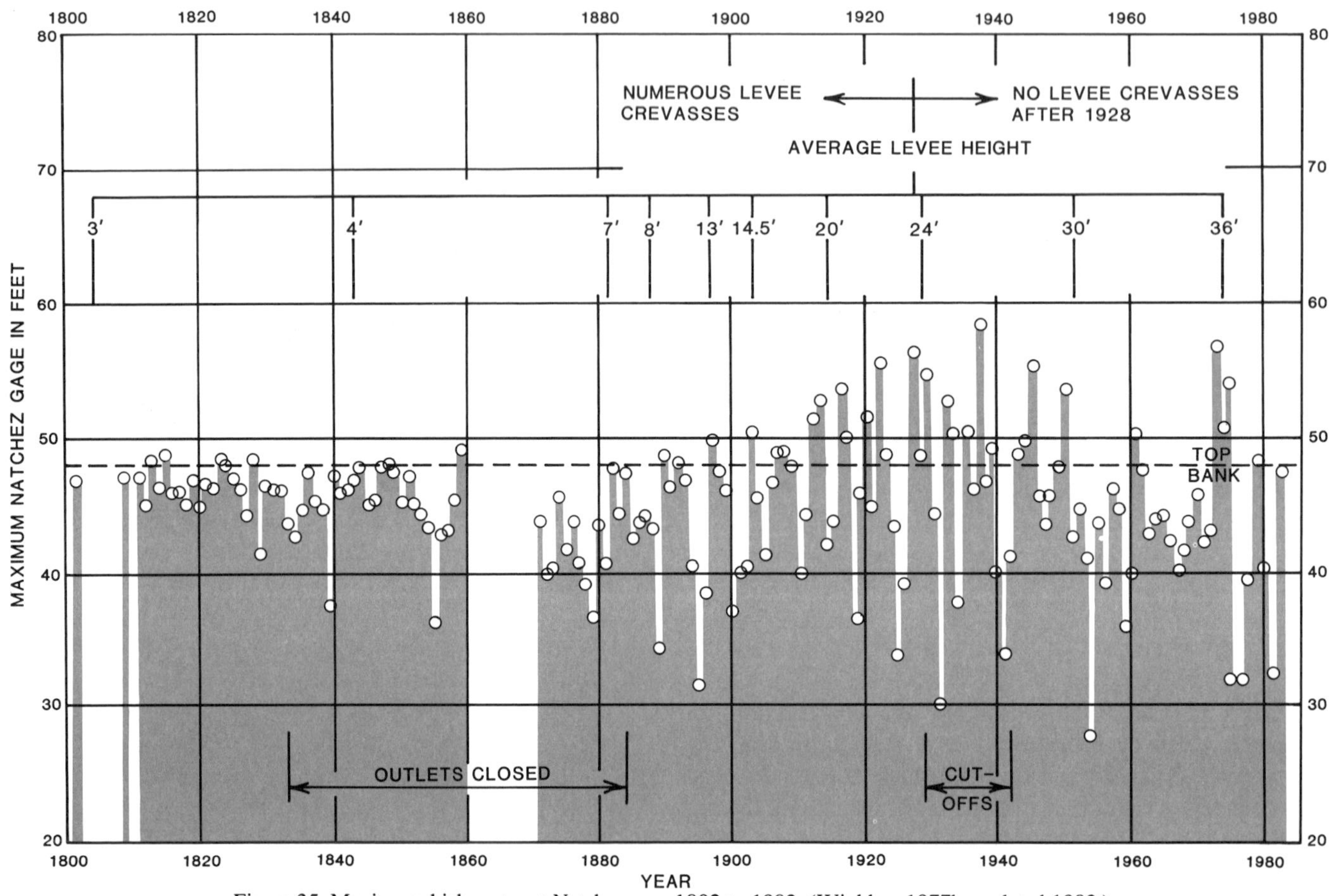

Figure 35. Maximum high water at Natchez gage 1802 to 1982. (Winkley, 1977b, updated 1982.)

sediments. Sixteen cutoffs were constructed in a 10-year period in the 1930's.

Dredging to improve navigation began in 1895, with 5 to 25 crossings per year requiring some improvement. During the next four decades, after the 1930 cutoffs, 26 to 69 crossings per year required navigation dredging. Since the late 1960s, revetments, land-use, reservoirs, better channel alignment, and river-training structures have decreased dredging needs 50 percent or more. Navigation has doubled every 10 years since 1940 so some of the increased dredging was due to the need to insure dependable channel depths.

The mining of gravel from the river bed for construction purposes has had some impact on channel stability; however, because this is a very economical method of obtaining gravel, it will probably be continued. Very little new coarse material is being introduced to today's locked-in channel, and some operators are already noting the increased difficulty of finding gravel sources in the river bed; however, there are several large gravel deposits within the levee system that could be developed at a slightly higher cost per unit of gravel.

Reservoirs of all sizes have changed the river's hydrograph. As an uncontrolled river, the average yearly high discharge was 42,812 m^3/sec at Vicksburg, Mississippi, with a flood every 1.3 years and an average low discharge of 4,054 m^3/sec. Now the average high discharge is 37,548 m^3/sec with only seven overbank flows in the past 35 years and no floods from 1961 to 1973, during which time most of the realignment, bank stabilization, and river-training structures were built. The average low flow is now 6,244 m^3/sec. In many reaches this work locked in the low-water navigation channel and allowed the high-water channel to deteriorate.

Several floodways have been built on the Lower Mississippi River. The Atchafalaya diversion and floodway at the old Red River confluence diverts 25 to 30 percent of all flows. Bonnet Carre floodway, just upstream from New Orleans, has been used seven times. Other floodways have been used only once or not at all. Some floodways only divert water from a length of channel, holding crests at a lower level and then allow the diverted flows to reenter at a slower rate. Others are designed to divert a large percent of flow from the channel and finally to the Gulf of Mexico. Not enough data are available to analyze the potential changes in sediment activity and alterations of channel geometry due to use of these flood diversions. Sediment movement is related to a power function of the discharge, so any diversion of flows would tend to accrete sediment in one or all of the channels from which diversion occurs.

TABLE 4. AVERAGE TOPBANK WIDTH*
MISSISSIPPI RIVER FROM 1821 TO 1975

Reach	Date of Survey					
	1821	1874	1911= 1915	1948- 1952	1973 1952	Percent Increase 1821-1975
Ohio River–St. Francis River	3,800	5.400	6.400	6.500	6.340	67
St. Francis–Arkansas River	3,000	4,900	5.200	5,000	7.220	141
Arkansas River–Yazoo River	2,400	4.700	5,500	6,000	6,660	177
Yazoo River–Red River	2,300	4,400	4.500	4.900	5,810	152
Average Ohio to Red River	3,100	5,000	5.800	5,900	6,420	107

*In feet.

More than 90 percent of the concave banks on the bends of the Lower Mississippi River are at least partially stabilized. The balance are mostly protected by a limestone bluff or a cohesive clay formation. Also, many convex banks have been stabilized, and there are 140 dike fields holding alignment in straight reaches, preventing diversion of flows, anchoring bars, and improving navigation depths. This has locked in the plan view of the river, preventing any lateral movement; therefore, the vertical axis is the only plain in which the river can attempt to balance its energies.

Two hundred years ago the Lower Mississippi River was in a state of dynamic equilibrium, balancing the wide variations of gradients and sediment sizes and distribution, with a discharge that was consistently repetitive if several hydrographs are considered. The channel from the Ohio River to the Gulf of Mexico held a flow of about 28,000 m^3/sec within its top banks. Overbank flows of 30,800 to 61,600 m^3/sec occurred every 4 out of 5 years. Yet, because of the perched elevation of the present meander belt occupied by the river with its 965 valley kilometers of distributory channels, the in-channel capacity seemed to have remained constant.

In an effort to improve the valley for his own use, man has constructed many types of projects, each of which had some effect on the river's water load and sediment load. Fortunately, the responses, along with the geologic controls, had compensating results. Since the 1811 to 1812 New Madrid earthquake, the river has been in a constant state of transition, never able to regain a form of equilibrium before another change in energy, sediment, or discharge was imposed.

Table 4 and Figure 36 show the average increases in top bank width of the river between successive major tributaries. The bank widening from 1821 to 1847 was the most pronounced and can be attributed primarily to the New Madrid earthquake and to a much lesser extent to levee construction and land clearing. The 1874 to 1915 changes could be equally weighed between the sediment surge from the 1811–12 quake, the added in-channel flow due to levee improvement, and elimination of the distributaries. The period from 1915 to 1952 was influenced mostly by

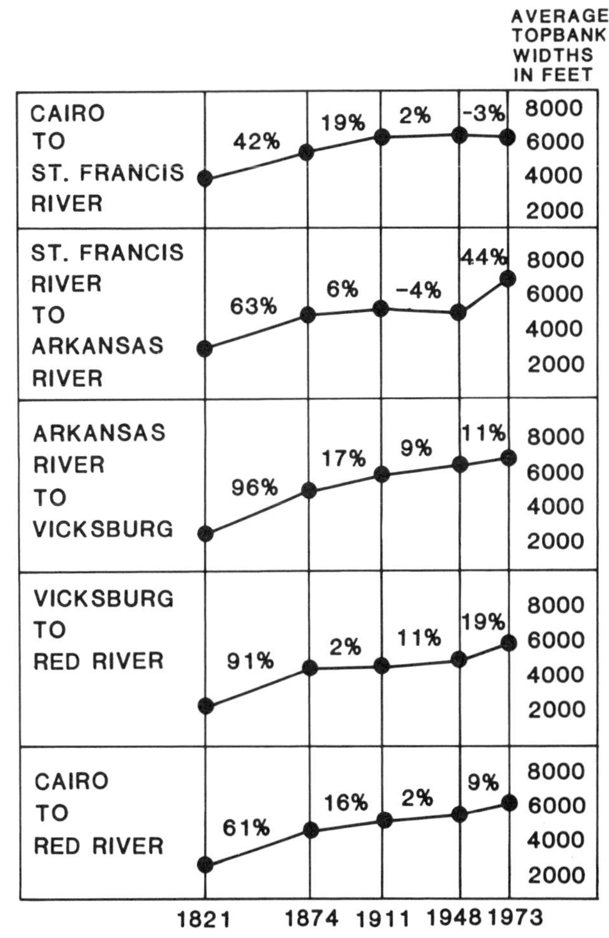

Figure 36. Changes in the topbank width of the Mississippi River from 1821 to 1973. Causes of successive changes are discussed in the text.

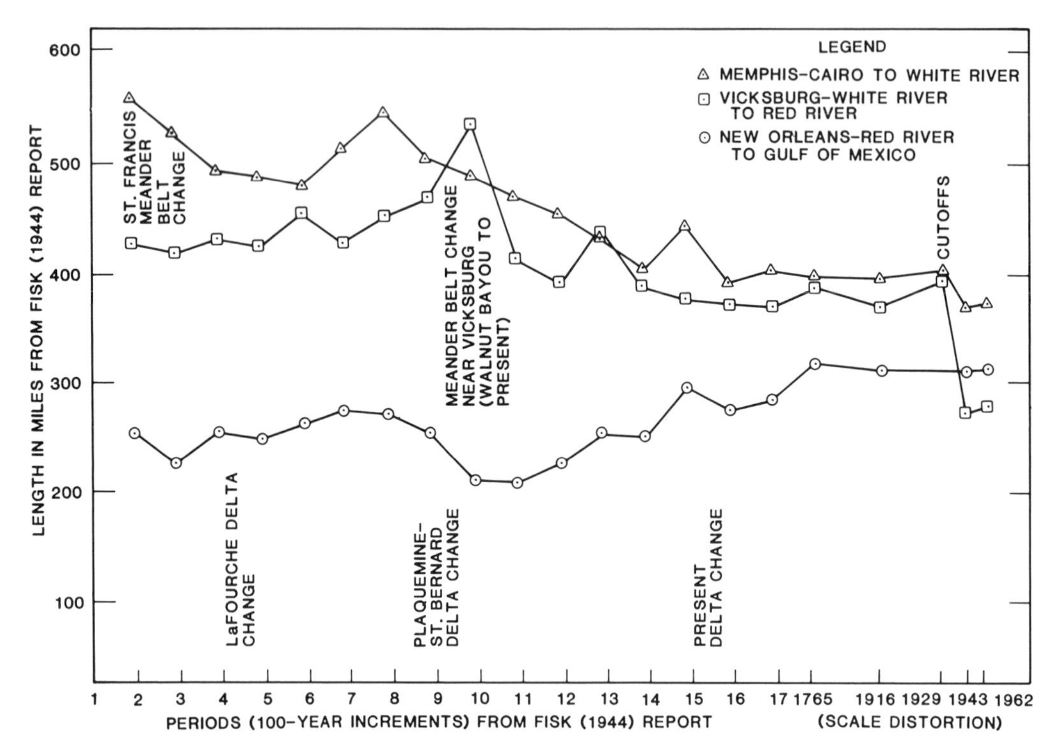

Figure 37. Two thousand years of length variation of the Lower Mississippi (Winkley, 1977a).

cutoffs, and secondarily by land-use, reservoirs, and some in-channel stabilization work. The period 1952 to 1975 included most of the dike and revetment work, chute development, reservoir building, and the continued response of the channel to all changes since 1811-12. The combined result was a decrease in the suspended sediment load of 60 percent, but increased gradients and channel efficiencies could at the same time be increasing the bed sediment movement, which remains unmeasured.

Figure 37 shows the length and thus, the slope of the river over three reaches during the past 2,000 years (data from Fisk, 1944). The Memphis reach continually shortened as it redistributed the coarse glacial sediments and built less-bulbous meander loops. During the past 600 years, lengths were relatively constant until the cutoffs of 50 years ago. The Vicksburg reach maintained a consistent length, except for a meander belt change and a slight shortening due to redistribution of coarse sediments, until the 1930 to 1940 cutoffs. The New Orleans reach initially remained constant in length by changing delta outlet locations, but has steadily increased its length with the present locked-in delta location and changes in energy, discharges, and sediment distribution.

Throughout the entire river length, and at different periods in various reaches, energies and sediment movements have varied. An increase in energy of 32 percent accumulated in the Memphis reach while the Vicksburg reach gained 35 percent in energy, 30 percent of which occurred recently as a result of the man-made cutoffs. The New Orleans reach experienced a 25 percent loss in energy. This suggests that the present and future river may have the ability to transport more coarse sediment in the upper two-thirds of its length becoming a sediment sink in the lower one-third of its length.

Figure 34, a specific-gage record of near bank-full discharges over the period of record, supports this analysis. All stations show aggradation until 1930 resulting from the almost yearly overflow of its sediments. The cutoffs in the middle one-third of the river's length caused extreme changes. However, as time progressed the river upstream of the cutoffs began to degrade while the downstream reach aggraded. During the past decade the effects of stabilization, slope changes, and a decrease in sediment input seem to be further aggravating these responses.

Because sediments will continue to move, man may have a busy future maintaining flood control and navigation on the continent's largest waterway.

REFERENCES CITED

Anderson, H. W., 1970, Principal components analysis of watershed variables affecting suspended sediment discharge after a major flood, *in* Proceedings, International Symposium on Results on Representative and Experimental BAsins: New Zealand, International Scientific Hydrology Publication 96, p. 404–416.

Ashton, G. D., 1979a, River ice: American Scientist, v. 67, p. 38–45.

——, 1979b, Modeling of ice in rivers, *in* Shen, ed., Modeling of rivers: New York, John Wiley and Sons, p. 14-1 to 14-26.

Audley-Charles, M. G., Curray, J. R., and Evans, G., 1977, Location of major deltas: Geology, v. 5, p. 341–344.

Bailey, E. H., Irwin, W. P., and Jones, D. L., 1964, Franciscan and related rocks, and their significance in the geology of western California: California Division of Mines and Geology Bulletin 183, 177 p.

Beverage, J. P., and Culbertson, J. K., 1964, Hyperconcentrations of suspended sediment: American Society of Civil Engineers Journal of Hydraulics, v. 90, p. 117–128.
Bluck, B. J., 1971, Sedimentation in the meandering river Endrick: Scottish Journal of Geology, no. 7, part 2, p. 93–140.
Bovis, M. J., 1985, Earthflows in the Interior Plateau, southwest British Columbia: Canadian Geotechnical Journal, v. 22, p. 313–334.
Brown, W. M., and Ritter, J. R., 1971, Sediment transport and turbidity in the Eel River basin, California: U.S. Geological Survey Water-Supply Paper 1986, 70 p.
Brundsden, D., and Kesel, R. H., 1973, Slope development of a Mississippi River bluff in historic time: Journal of Geology, v. 81, p. 576–598.
Buchanan, J. P., 1981, Channel morphology and sedimentary facies of the Niobrara River, north-central Nebraska [M.S. thesis]: Fort Collins, Colorado State University, 126 p.
California Department of Water Resources, 1974, Water management for fishery enhancement on north coastal streams: California Department of Water Resources, 68 p.
Canadian Journal of Fisheries and Aquatic Sciences, 1984, The Churchill River diversion and southern Indian Lake impoundment: Canadian Journal of Fisheries and Aquatic Sciences, v. 41, no. 4, p. 548–732.
Carey, W. C., 1969, Formation of flood plain lands: American Society of Civil Engineers Journal of the Hydrogeology Division, no. 95, p. 981–994.
Colby, B. R., and Hembree, C. H., 1955, Computations of total sediment discharge Niobrara River near Cody, Nebraska: U.S. Geological Survey Water-Supply Paper 1357, 187 p.
Coleman, J. M., and Wright, L. D., 1971, Analysis of major river systems and their deltas; Procedures and rationale, with two examples: Baton Rouge, Louisiana State University Technical Report 95, 125 p.
Collinson, J. D., 1971, Some effects of ice on a river bed: Journal of Sedimentary Petrology, v. 41, p. 557–564.
Church, M., and Ryder, J. M., 1972, Paraglacial sedimentation; A consideration of fluvial sedimentation influenced by glaciation: Geological Society of America Bulletin, v. 83, p. 3059–3072.
Cunningham, R., 1981, Atchafalaya Delta, subaerial development, *in* Cross, R. D., and Williams, D. L., eds., Proceedings, National Symposium on Freshwater Inflow to Estuaries: U.S. Department of the Interior, Fish and Wildlife Service, v. 1, p. 349–365.
Dietrich, W. E., and Smith, J. D., 1983, Influence of the point bar on flow through curved channels: Water Resources Research, v. 19, p. 1173–1192.
Evans, S. G., 1982, The development of Big Slide near Quesnel, British Columbia, between 1953 and 1982: Geoscience Canada, v. 9, p. 220–222.
Fenneman, N. M., 1931, Physiography of western United States: New York, McGraw-Hill Book Company, Inc., 534 p.
Fisk, H. N., 1944, Geological investigation of the alluvial valley of the lower Mississippi River: Vicksburg, Mississippi River Commission, 78 p.
Graf, W. L., 1983, The arroyo problem; Paleohydrology and paleohydraulics in the short term, *in* Gregory, K. J., ed., Background to paleohydrology: New York, John Wiley and Sons, p. 279–302.
Gregory, H. E., and Moore, R. C., 1931, The Kaiparowits region; A geographic and geologic reconnaissance of parts of Utah and Arizona: U.S. Geological Survey Professional Paper 164, 161 p.
Hadley, R. F., 1974, Sediment yield and land use in southwest United States: International Association of Scientific Hydrology Publication 113, p. 96–98.
Harden, D. R., Janda, R. J., and Nolan, K. M., 1978, Mass movement and storms in the drainage basin of Redwood Creek, Humboldt County, California; A progress report: U.S. Geological Survey Open-File Report 78–486, 161 p.
Hereford, R., 1984, Climate and ephemeral-stream processes; Twentieth century geomorphology and alluvial stratigraphy of the Little Colorado River, Arizona: Geological Society of America Bulletin, v. 95, p. 654–668.
——, 1986, Modern alluvial history of the Paria River drainage basin, southern Utah: Quaternary Research, v. 25, p. 293–311.
Hickey, J. J., 1969, Variations in low-water streambed elevations at selected stream gaging stations in northwestern California: U.S. Geological Survey Water-Supply Paper 1879–E, 33 p.
Holeman, 1968, The sediment yield of major rivers of the world: Water Resources Research, v. 4, p. 737–747.
Holland, S. S., 1964, Landforms of British Columbia; A physiographic outline: British Columbia Department of Mines and Petroleum Resources Bulletin 48, 138 p.
Howard, A., and Dolan, R., 1981, Geomorphology of the Colorado River in the Grand Canyon: Journal of Geology, v. 89, p. 269–298.
Iorns, W. V., Hembree, C. H., and Oakland, G. L., 1965, Water resources of the Upper Colorado River Basin: U.S. Geological Survey Professional Paper 441, 370 p.
Jackson, R. G., 1981, Sedimentology of muddy fine-grained channel deposits in meandering streams of the American Middle West: Journal of Sedimentary Petrology, v. 51, p. 1169–1192.
Janda, R. J., 1979, Summary of regional geology in relation to geomorphic form and process, *in* Field trip to observe natural and management-related erosion in Franciscan Terranes of northern California; Geological Society of America Cordilleran Section April 9-11, 1979: Menlo Park, California, U.S. Geological Survey, p. II-1 to II-17.
Janda, R. J. and Nolan, K. M., 1979, Stream sediment discharge in northwestern California, *in* Guidebook for a field trip to observe natural and management-related erosion in the Franciscan terrane of northern California; Geological Society of America Cordilleran Section, April 9–11, 1979: Menlo Park, California, U.S. Geological Survey, p. IV-1 to IV-27.
Jones, D. L., Blake, M. C., Jr., Bailey, E. H., and McLaughlin, R. J., 1978, Distribution and character of upper Mesozoic subduction complexes along the West Coast of North America: Tectonophysics, v. 47, p. 207–222.
Kelsey, H. M., 1977, Landsliding, channel changes, sediment yield, and land use in the Van Duzen River basin, north coastal California, 1941–1975 [Ph.D. thesis]: Santa Cruz, University of California, 370 p.
——, 1980, A sediment budget and an analysis of geomorphic process in the Van Duzen basin, north coastal California, 1941–1975: Geological Society of America Bulletin, v. 91, p. 1119–1216.
——, 1982, Hillslope evolution and sediment movement in a forested headwater basin, Van Duzen River, north coastal California, *in* Swanson, F. J., and Dunne, T., eds., Sediment budgets and routing in forested drainage basins: U.S. Department of Agriculture Forest Service General Technical Report PNW–141, p. 86–96.
Kelsey, H. M., and Savina, M. E., 1985, Van Duzen River basin, *in* Savina, M. E., ed., American Geomorphological Field Group Field Trip Guidebook, 1985 Conference, Northwestern California: Arcata, California, Redwood National Park, p. 3–36.
Knott, J. M., 1971, Sedimentation in the Middle Fork Eel River basin, California: U.S. Geological Survey Open-File Report 2001-06, 60 p.
Kolb, C. R., 1963, Sediments forming the bed and banks of the lower Mississippi River and their effect on river migration: Sedimentology, v. 2, p. 227–234.
Langbein, W. B., and Schumm, S. A., 1958, Yield of sediment in relation to mean annual precipitation: American Geophysical Union Transactions, v. 39, p. 1076–1084.
Lapointe, M. F., 1984, Patterns and processes of channel change, Mackenzie Delta, Northwest Territories: Ottawa, Ontario, Surface Water Division, National Hydrology Research Institute, Environment Canada, 11 p.
——, 1985, Aspects of channel bathymetry and migration patterns in the Mackenzie Delta, Northwest Territories: Ottawa, Ontario, Surface Water Division, Environment Canada, 22 p.
——, 1986, Deep scour holes on the bed of Mackenzie Delta channels, Northwest Territories: Ottawa, Ontario, Surface Water Division, Environment Canada, 36 p.
Lawson, D. E., 1983, Erosion of perennially frozen streambanks: U.S. Army Engineer District, Alaska, CRREL Report 83–29, 22 p.
Lay, D., 1940, Fraser River Tertiary drainage history in relation to placer gold deposits: British Columbia Department of Mines Bulletin 3, 30 p.
Lisle, T. E., 1981, Recovery of aggraded steam channels at gauging stations in northern California and southern Oregon, *in* Davies, T.R.H., and Pearce,

A. J., eds., Erosion and sediment transport in Pacific Rim Steeplands, Symposium: International Association of Hydrological Sciences Publication 132, p. 189–211.

——, 1982, Effects of aggradation and degradation on rifflepool morphology in natural gravel channels, northwestern California: Water Resources Research, v. 18, no. 6, p. 1643–1651.

Mackay, J. R., 1974, The Mackenzie Delta area, N.W.T.: Geological Survey of Canada Miscellaneous Report 23, 202 p.

Mathews, W. E., and Rouse, G. E., 1984, The Gang Ranch–Big Bar area, south-central British Columbia; Stratigraphy, geochronology, and palynology of the Tertiary beds and their relationship to the Fraser fault: Canadian Journal of Earth Sciences, v. 21, p. 1132–1144.

McLeod, S. L., 1976, The evolution of the Echimamish River in northern Manitoba [M.S. thesis]: Winnipeg, University of Manitoba, 158 p.

Miall, A. D., 1977, A review of the braided river depositional environment: Earth Science Review, v. 13, p. 1–62.

Nadler, C. T., and Schumm, S. A., 1981, Metamorphosis of South Platte and Arkansas Rivers, eastern Colorado: Physical Geography, v. 2, p. 95–115.

Newbury, R. W., 1968, The Nelson River; A study of sub-Arctic river processes [Ph.D. thesis]: Baltimore, Maryland, Johns Hopkins University, 318 p.

——, 1981, The river of spirits: Manitoba Nature, v. 22, no. 2, p. 32–39.

Nolan, K. M., and Marron, D. C., 1985, Contrast in steam channel response to major storms in two mountainous areas of California: Geology, v. 13, p. 135–138.

Nolan, K. M., and Janda, R. J., 1981, Use of short-term water and suspended-sediment discharge observations to assess impacts of logging on steam-sediment discharge in the Redwood Creek basin, northwestern California, U.S.A., *in* Davies, T.R.H., and Pearce, A. J., eds., Erosion and sediment transport in Pacific Rim Steeplands, Symposium: International Association of Hydrological Sciences Publication 132, p. 416–437.

——, 1982, The role of hillslope processes in controlling sediment yield in steep terrance [abs.], *in* Leopold, L. B., ed., American Geomorphological Field Group Field Trip Guide, Pinedale, Wyoming: Denver, Colorado, U.S. Geological Survey, p. 138.

——, 1989, Movement and sediment yield of two earthflows, northwestern California, *in* Nolan, K. M., Kelsey, H. M., and Marron, D. C., eds., Geomorphic processes and aquatic habitat in the Redwood Creek basin: U.S. Geological Survey Professional Paper (in press).

Nolan, K. M., Lisle, T. E., and Kelsey, H. M., 1987, Bankfull discharge and sediment transport in northwestern California, *in* Beschta, R. L., Blinn, T., Grant, G. E., Swanson, F. J., and Ice, G. G., eds., Erosion and sedimentation in the Pacific Rim: International Association of Hydrological Sciences Publication 165, p. 439–450.

Nordin, C. F., 1963, A preliminary study of sediment transport parameters Rio Puerco near Bernardo, New Mexico: U.S. Geological Survey Professional Paper 462–C, 21 p.

Pemberton, E. L., 1976, Channel changes in the Colorado River below Glen Canyon dam, *in* Proceedings, Third Federal Inter-Agency Sedimentation conference, March 22–25, 1976, Denver, Colorado: Washington, D.C., Federal Inter-Agency Sedimentation Committee of the Water Resources Council, p. 5-61 to 5–73.

Pierson, T. C., and Scott, K. M., 1985, Downstream dilution of a Lahar's transition from debris flow to hyperconcentrated steamflow: Water Resources Research, v. 21, p. 1511–1524.

Powell, J. W., 1875, Exploration of the Colorado River of the west and its tributaries: Washington, D.C., U.S. Government Printing Office, 291 p.

Rantz, S. E., 1972, Runoff characteristics of California streams: U.S. Geological Survey Water-Supply Paper 2009–A, 38 p.

Schmidt, J. C., 1986, Changes in alluvial deposits, Upper Grand Canyon, *in* Proceedings, Fourth Federal Inter-Agency Sedimentation Conference, Las Vegas, Nevada: Washington, D.C., Federal Inter-Agency Sedimentation Committee of the Water Resources Council, p. 2-48 to 2-57.

Scott, K. M., 1978, Effects of permafrost on stream channel behavior in Arctic Alaska: U.S. Geological Survey Professional Paper 1068, 19 p.

Simpson, S. J., 1972, The York Factory area, Hudson Bay [Ph.D. thesis]: Winnipeg, University of Manitoba, 293 p.

Smith, D. G., 1979, Effects of channel enlargement by river ice processes on bankfull discharge in Alberta, Canada: Water Resources Research, v. 15, p. 469–475.

Smith, W. O., Vetter, C. P., and Cummings, G. B., 1960, Comprehensive survey of sedimentation in Lake Mead, 1948–1949: U.S. Geological Survey Professional Paper 295, 254 p.

Stephens, H. G., and Shoemaker, E. M., 1987, In the footsteps of John Wesley Powell: Denver, Colorado, The Powell Society, Lts., 286 p.

Stockton, C. W., 1975, Long-term streamflow records reconstructed from tree rings: Tucson, University of Arizona Press, 111 p.

Thornthwaite, C. W., Sharpe, C.F.S., and Dosch, E. F., 1942, Climate and accelerated erosion in the arid and semiarid southwest with special reference to the Polacca Wash drainage basin, Arizona: U.S. Department of Agriculture Technical Bulletin 808, 134 p.

Tipper, H. W., 1971, Glacial geomorphology and Pleistocene history of central British Columbia: Geological Survey of Canada Bulletin 196, 89 p.

Turner, R. M., and Karpiscah, M. M., 1980, Recent vegetation changes along the Colorado River between Glen Canyon Dam and Lake Mead, Arizona: U.S. Geological Survey Professional Paper 1132, 125 p.

U.S. Geological Survey, 1984, Water resources data, Nebraska, Water Year 1984: U.S. Geological Survey Water-data Report NE–84–1, 501 p.

Webb, R. H., 1985, Late Holocene flooding on the Escalante River, south-central Utah [Ph.D. thesis]: Tucson, University of Arizona, 204 p.

Waananen, A. O., Harris, P. P., and Williams, R. C., 1971, Floods of December 1964 and January 1965 in the far western states: U.S. Geological Survey Water Supply Paper 1866–A, 265 p.

Williams, G. P., 1978, The case of the shrinking channels; The North Platte and Platte Rivers in Nebraska: U.S. Geological Survey Circular 781, 48 p.

Wilson, L., 1973, Variations in mean annual sediment yield as a function of mean annual precipitation: American Journal of Science, v. 273, p. 335–349.

Winkley, B. R., 1977a, Manmade Cutoffs on the Lower Mississippi River; Conception, construction, and river response: Vicksburg, Mississippi, U.S. Army Corps of Engineers, Report P.I. 300-2, 209 p.

——, 1977b, A Geomorphic Study of the Lower Mississippi River: San Francisco, California, American Society of Civil Engineers Preprint 2990, 31 p.

Wolman, M. G., and Gerson, R., 1978, Relative scales of time and effectiveness of climate in watershed geomorphology: Earth Surface Processes, v. 3, p. 189–208.

Wolman, M. G., and Miller, J. P., 1960, Magnitude and frequency of forces in geomorphic processes: Journal of Geology, v. 68, p. 54–74.

Wright, L. D., and Coleman, J. M., 1972, River delta morphology; Wave, climate, and the role of the subaqueous profile: Science, no. 176, p. 282–284.

——, 1973, Variations in morphology of major river deltas as functions of ocean wave and river discharge regimes: American Association of Petroleum Geology Bulletin, v. 57, no. 2, p. 370–398.

Ziemer, R. R., 1981, Roots and the stability of forested slopes, *in* Davies, T.R.H., and Pearce, A. J., eds., Erosion and sediment transport in Pacific Rim Steeplands: International Association of Hydrological Sciences Publication 132, p. 343–361.

Manuscript Accepted by the Society May 24, 1989

ACKNOWLEDGMENTS

W. H. Mathews generously made available unpublished information, and reviewed the section on the Fraser River. H. Kelsey and M. Nolan provided critical reviews of the section on the Eel River.

Printed in U.S.A.

The Geology of North America
Vol. O-1, Surface Water Hydrology
The Geological Society of America, 1990

Chapter 13

The influence of man on hydrologic systems

Robert M. Hirsch
U.S. Geological Survey, 436 National Center, Reston, Virginia 22092
John F. Walker
U.S. Geological Survey, 6417 Normandy Lane, Madison, Wisconsin 53719
J. C. Day
Department of Geography, Simon-Fraser University, Burnaby, British Columbia V5A 1S6, Canada
Raimo Kallio
Inland Waters Directorate, Environment Canada, Ottawa, Ontario K1A 0E7, Canada

INTRODUCTION

No discussion of the surface-water resources of North America would be complete without explicit consideration of the role that man plays in determining the flow of water through river systems. Human activities that may affect flows include diversion of water from one river basin to another, creation of artificial reservoir storage, destruction of natural wetland storage, and land changes that alter rates of erosion, infiltration, overland flow, or evapotranspiration. The effects of these human actions influence not only long-term average flows, but the magnitude and frequency of droughts and floods and year-to-year and season-to-season flow variations. These effects, in turn, have a variety of direct effects on man, related to availability of reliable water supplies for in-stream and withdrawal uses and to magnitude and frequency of flood damages. They also affect geomorphic features resulting from changes in the temporal distribution of erosive or transporting forces. The result of these are changes in the geometry and sedimentary characteristics of river channels, flood plains, and deltas.

Types of activities that affect river flows are illustrated in this chapter by using a number of examples. The extent and nature of the origins of the change are described where possible, and examples of their effects are given. Flood flows are emphasized because of their particularly great geomorphic significance in determining channel geometry and sediment characteristics. In addition to consideration of flood flows, the flow-duration curve (frequency distribution), patterns of seasonality, and extent of year-to-year variability are discussed. The chapter concludes with an appraisal of the effect of human activities on water budgets of various large North American regions by comparing present averaege streamflow with estimated natural flow before water development began.

There are four types of situations where flow changes are fairly clear and documented in several cases: dam construction, urbanization, interbasin transfers, and consumption of water by industry or agriculture. Other, less clear situations are artificial land drainage, surface mining, man-induced vegetation changes, and physical channel alterations. Because it is essential to an understanding of the effect of human activities on surface-water systems, the natural variability of streamflow is discussed first.

NATURAL STREAMFLOW VARIABILITY

Analysis of the extent of man's impact on streamflow is difficult and, in many cases, subject to controversy. The reason for the difficulty is the natural, climatically driven, year-to-year variability of flows, especially floods. It is not uncommon for annual floods to have a coefficient of variation (ratio of standard deviation to mean) of one or even more. Suppose the coefficient of variation of annual floods was one and the frequency distribution changed abruptly halfway through a 40-year annual flood record; in order for the change to be discernible in a statistical test with 95 percent power, the change in the mean would have to be at least 45 percent. Discriminating such a modification is further complicated by the fact that watershed change, which may modify flow, may be gradual rather than abrupt.

Example time series of annual peak discharges for two rivers where man-induced change in hydrologic response over 70 to 90 years of record is highly unlikely are provided in Figures 1 and 2. They are the Merced River at Happy Isles Bridge near Yosemite, California, and the Saugeen River near Walkerton, Ontario. The Merced River drains a pristine 469-km^2 basin in the Sierra Nevada Range. It is among the streams in the U.S. Geological Survey's Hydrologic Benchmark network. The Saugeen River is a tributary of Lake Huron, and has a drainage area of 2,150 km^2. The basin is mainly pastoral, has several small towns, and has experienced no major changes in the level of development over the

Hirsch, R. M., Walker, J. F., Day, J. C., and Kallio, R., 1990, The influence of man on hydrologic systems, *in* Wolman, M. G., and Riggs, H. C., eds., Surface water hydrology: Boulder, Colorado, Geological Society of America, The Geology of North America, v. O-1.

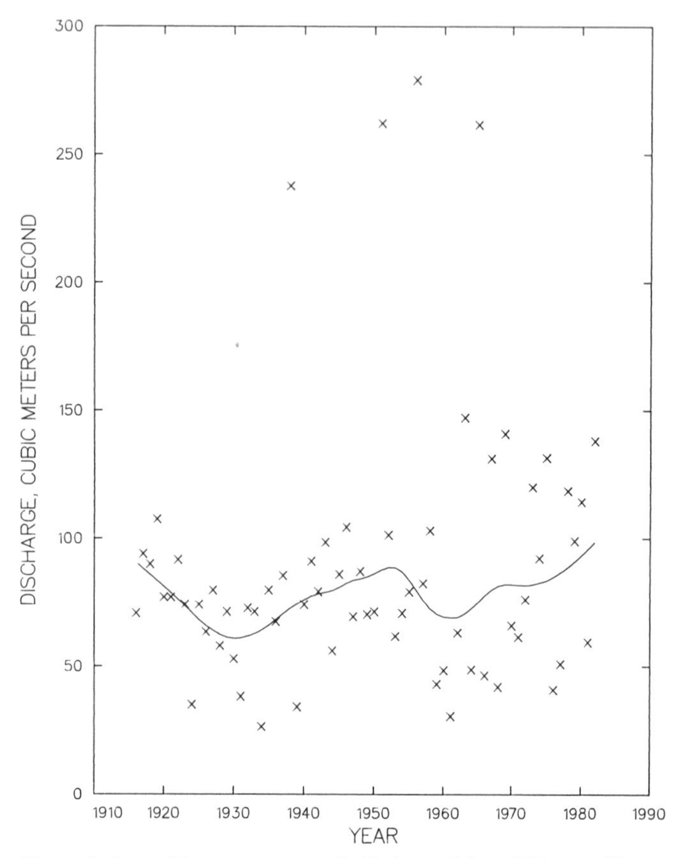

Figure 1. Annual instantaneous peak discharge, Merced River at Happy Isles Bridge near Yosemite, California.

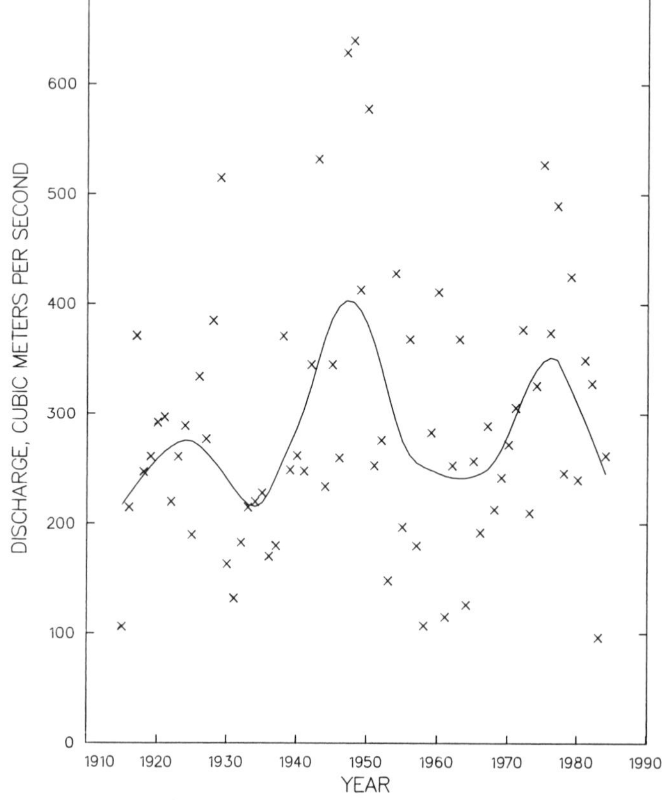

Figure 2. Annual maximum 1-day discharge, Saugeen River near Walkerton, Ontario.

course of history of the gage record. These examples are included as base lines to compare with flood records from other basins that have experienced profound changes related to alteration in land use, water diversion, and reservoir storage.

Figures 1 and 2 show the annual floods and a robustly fit moving average for the two time series. Because streamflow is an inherently statistical phenomenon, appropriate statistical techniques for exploring natural variability and changes in flow due to human influences are carefully sketched here. The weighted moving averages are computed from the logarithm of the flows and thus represent a good approximation of the 2-year flood. If the population skew of the logarithms were zero, then this would be precisely the 2-year flood, the flood exceeded by half the annual floods. In this robustly weighted moving average, described by Cleveland (1979), the weights are based on a 15-year window. That is, the level of the curve in each year is determined by the size of the floods in that year and the 14 closest years, the weights decreasing with the number of years of separation. A bisquare weight function is used (see Mosteller and Tukey, 1977). Robustness is achieved by a process of decreasing weights based on the degree to which the particular year's flood departs from the trend of floods in other years in the time window.

Despite the apparent scatter, these two time series reveal a strong pattern, in contrast to randomness, of certain periods of one or two decades showing groupings of particularly large floods or small floods, or large variance or small variance. These records convey an impression of nonstationarity, particularly when short parts of the record are considered. These apparent trends are probably best described as persistence.

One impediment to any analysis of hydrologic time series is the difficulty of discriminating persistence from long-term trend. Matalas and others (1982, p. 121) highlighted this concern. They stated:

> An initial look at a hydrologic record may suggest a behavior, e.g., upward trend or a point of discontinuity, not inconsistent with a prior conception of a nonstationary process. Thus, it is possible to postulate this process and develop analytical methods to describe the nature of the nonstationarity. However, it is also possible to accept the postulate of stationarity (albeit not necessarily as historically perceived in hydrology), maintaining that it is only the 'shortness' of the data record that gives it the appearance of nonstationarity. Neither position can be dismissed as false on the basis of the data alone...

Consideration of these examples is relevant in two ways. The first is the recognition of just how much natural variability there is in peak discharge, or any kind of flow record, which tends to obscure any anthropogenic signal that may exist. The second is the high probability of finding apparent trends in records that can be used as evidence that some particular human action has resulted in a change.

RESERVOIR CONSTRUCTION

Dams are built to store water for many purposes. They provide a reliable water supply for municipal, industrial, or agricultural pruposes, improve navigation, produce hydroelectric power, augment downstream flows to dilute wastes discharged to the river, control downstream flooding, and provide recreational opportunities. Many dams serve several of these purposes. Regardless of the purpose, all dams store water from one period of time for withdrawal or release at other times; in so doing they modify the distribution of river flows. Because of the shifting of flows from one season to another and a modified flow-duration curve, the downstream post-dam streamflow distribution commonly has decreased high flows and increased low flows.

Most dams have a certain portion of their total storage reservoir designated as a flood pool. This pool remains available virtually all the time to store water during flood events. Typically, water in the flood pool is released as rapidly as possible so long as its release does not cause downstream flooding. This decreases flood-peak magnitudes and increases flood lag times downstream. Williams and Wolman (1984) found that average annual floods below 29 dams in the central and western U.S. range from about 3 to about 90 percent of pre-dam values; the average is 40 percent.

Most dams also operate under minimum release requirements related to fish and wildlife habitat, waste dilution, recreation, or navigation. In most cases these minimum release requirements result in an increase in the magnitude of low flows occurring with a given frequency.

An additional aspect of flow regulation by dams is the practice of hydroelectric power peaking. Electric-power demands in most regions follow a diurnal cycle, typically peaking in the afternoon or evening on weekdays. If the electric power in a region is provided by some mix of coal, oil, nuclear, thermal, and hydroelectric sources, water power is always used to satisfy the peak demand. This is because the system can go from zero to full power in a few minutes with virtually no wasted energy, whereas thermal power requires long start-up and shut-down periods, and much energy is wasted. This peaking procedure can result in order-of-magnitude discharge changes in minutes, even on very large rivers. Furthermore, the high flows released during peak production times (e.g., 6 hours per day, 5 days per week) can be approximately as large as the mean annual flood that occurred in pre-dam conditions. Thus, even though a dam may reduce the size of large floods, smaller floods that may be largely responsible for channel formation may be increased.

Another important consequence of reservoirs is that they trap suspended sediment and prevent its movement downstream. Most reservoirs with large volumes (i.e., several months to years of inflow) trap nearly all of the sediment that enters them. Flow distribution changes and the trapping of sediment can result in considerable adjustments in the size and shape of downstream river channels (Williams and Wolman, 1984).

In this section the impact of reservoirs on the hydrologic and hydraulic characteristics of a stream will be examined in detail. By using a robust statistical technique the effect of reservoir construction on the seasonal variability of flow will be explored. Flood-peak and low-flow characteristics will also be considered. Next, detailed case studies for the Platte and Green rivers will be discussed. The section concludes with an examination of the spatial distribution of reservoir storage in North America.

Seasonal Variability

One way to analyze how flows have changed over the years due to reservoir regulation is with techniques known as "time series decomposition" (Cleveland and others, 1979). These techniques make it possible to decompose the overall variations in a time series of monthly (other time steps are possible) flows into three components. One is the seasonal component: a pattern that repeats itself from one year to the next, slowly evolving over a period of years. The next is the trend component: multi-month to multi-year variation, devoid of a regular seasonal pattern. The final component is the irregular component: that variation which is rapidly changing from month to month and not a part of a long-term trend or a seasonal cycle. The decomposition is based on judgmental factors relating to the choice of digital filters used to smooth out the trend and seasonal components. The decomposition of monthly streamflows is typically best accomplished in multiplicative form (except in cases where zero flows occur). Thus the mathematical model of flow becomes:

$$\text{monthly flow} = \text{trend} \cdot \text{seasonal} \cdot \text{irregular.}$$

A time-series flow decomposition for the Colorado River immediately below Hoover Dam in Arizona and Nevada reveals several important features (Fig. 3). During the period prior to 1935 there was virtually no regulation in the basin. In 1935, Lake Mead was completed and storage of water in Lake Mead began. Lake Mead has a capacity of 37 billion m^3, which is about 1.9 times the average annual inflow. In 1963, Glen Canyon Dam was completed 571 km upstream of Hoover Dam. Virtually all inflow into Hoover Dam is controlled by Glen Canyon Dam, and the combined storage of this two-reservoir system is 70 billion m^3. Thus, the ability to regulate flows increased greatly in 1963. The storage available is so large that it took from 1963 until the early 1980s to fill the combined system (Freeney, 1981). Water finally spilled from the filled system of reservoirs in 1983.

The trend component time series shows variation on the order of 1 or 2 years up to 1935, prior to regulation (Fig. 3). A protracted low period occurred from 1935 through 1943, the period of filling of Lake Mead. From 1943 to 1963, large variations existed, but they were of longer duration than those prior to 1935. Management of the reservoir does not react immediately to particularly high- or low-flow conditions; eventually, however, changed flow conditions must be accommodated if the reservoir is not to become either too full or too empty. In 1963, with the completion of Glen Canyon Dam, Lake Powell began to fill. The

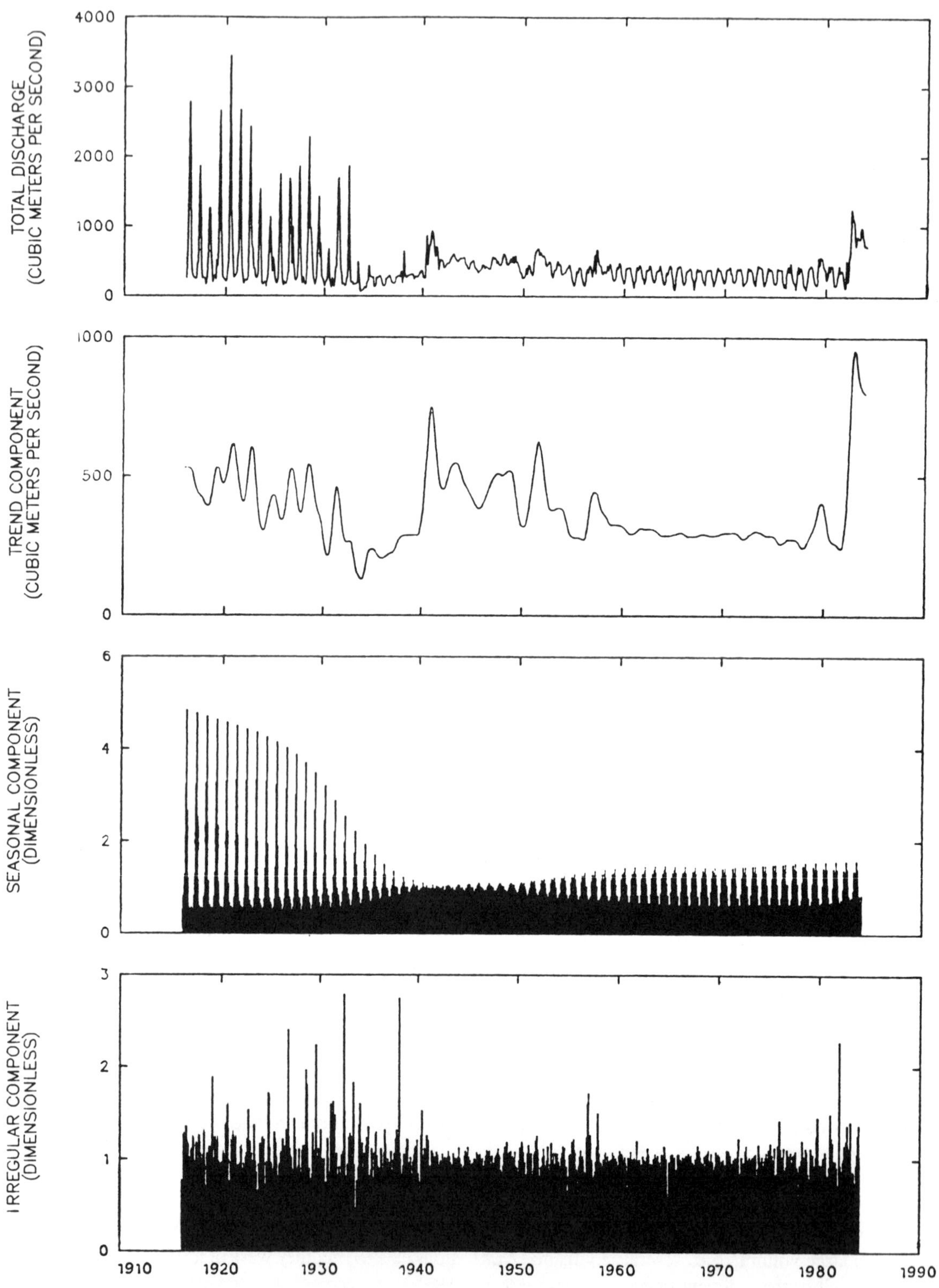

Figure 3. Time-series decomposition of monthly flows for the Colorado River at Topock, Arizona, below Hoover Dam (constructed in 1935) and Glen Canyon Dam (constructed in 1963).

trend shows a virtually unvarying record of low values as the water released from Hoover Dam was held to a level that would barely satisfy the demands of the downstream users. In the late 1970s, the trend began to rise with the recognition that the storage was filling and additional filling might result in a significant flood threat to those downstream of Hoover Dam, who had not experienced any uncontrolled flood flows in almost 50 years (Freeney, 1981). In 1983 and 1984, large runoff events occurred in the Colorado Basin, and large flows were spilled and released from both dams (see Matthai, this volume).

The seasonal component shows major changes over this record as well. This is highlighted in Figure 4, which is a rearrangement of the seasonal component shown in Figure 3. Prior to regulation, the annual hydrograph was dominated by May and June flows, the spring snowmelt flood. During the early years of regulation the seasonal cycle in the flows was almost nonexistent. In more recent years a new seasonal cycle, which is less pronounced, has come into existence; in this cycle, a broad peak occurs from April through September. This corresponds to the time of maximum irrigation and municipal demands. The irregular component shows a much higher degree of variability prior to the reservoir regulation (Fig. 3). The unusual spike in 1939 represents the first filling of Lake Mead and a deliberate test of the Hoover Dam spillways.

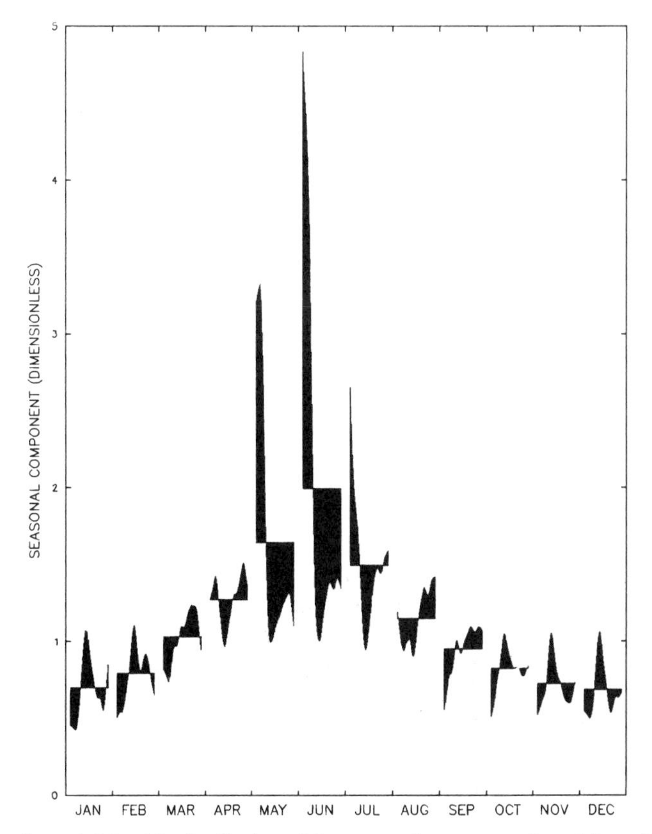

Figure 4. Monthly distribution of the seasonal component from Figure 3. Each month is plotted as an individual time series of 68-year duration. The horizontal line for each month is the 68-year average for that month's seasonal component.

Similar seasonal effects of artificial regulation on the monthly flow series for the Columbia River at Birchbank, British Columbia are shown in Figures 5 and 6. The river's source is in the icefields of the Rocky Mountains; it flows northwesterly in the Rocky Mountain Trench to the top of the "Big Bend," and then south into the state of Washington. Considerable reservoir development in the Columbia Basin has occurred in both the United States and Canada, primarily for hydroelectric and flood-control purposes. The first major dam was the Corra-Linn, finished in 1932 on the Kootenay River; it provided a storage of nearly 1 billion m^3. Three major reservoirs were constructed later: the Duncan, having a storage of 1.7 billion m^3; the Keenleyside, 8.76 billion m^3; and the Mica, 25 billion m^3. These projects were completed in 1967, 1968, and 1972, respectively. The Libby Dam on the Kootenay River began operation in 1973 and has a storage volume of 6.14 billion m^3. These developments total 42.6 billion m^3 or approximately 67 percent of the annual runoff.

In the mid 1960s, the variability of the seasonal component began to decrease, and reached a new, lower level by the early 1970s (Fig. 5). The seasonal component prior to regulation was dominated by high flows in May, June, and July, corresponding to snowmelt and meltwater from glaciers. Since artificial regulation has been in effect, the seasonal component has demonstrated a much broader and less-pronounced peak that extends from May through September.

Flood Characteristics

The effect of artificial regulation on instantaneous peak flows for the Gila River at Kelvin illustrates a feature common to many rivers with considerable reservoir storage (Fig. 7). The post-dam distribution of peak discharges has a lower mean and a much lower coefficient of variation than the pre-dam distribution. However, there are infrequent outliers; i.e., flows the dam is incapable of controlling. These outliers may be several standard deviations above the post-dam mean value. This system is highly controlled and yet in 1983 had a flood almost as large as the flood of record. This raises the possibility, but in no way confirms it, that although dams may decrease the discharge of the low-recurrence-interval floods, the right-hand tail of the distribution may be much less influenced. This possibility has a profound implication in terms of the benefits of flood-control impoundments.

Another example is the Peace River at Hudson Hope, Alberta (Fig. 8). The Peace River rises in the Rocky Mountains in British Columbia and flows across northeastern British Columbia and Alberta to join the outlet channels of Lake Athabasca and there forms the Slave River. The Peace River plays a role in the natural control of outflow from Lake Athabasca (Peace–Athabasca Delta Project Group, 1973). The discharge from the lake normally flows northward through the outlet channels toward the Peace River. The direction of flow in the outlet channels is reversed during high water levels on the Peace River, resulting in

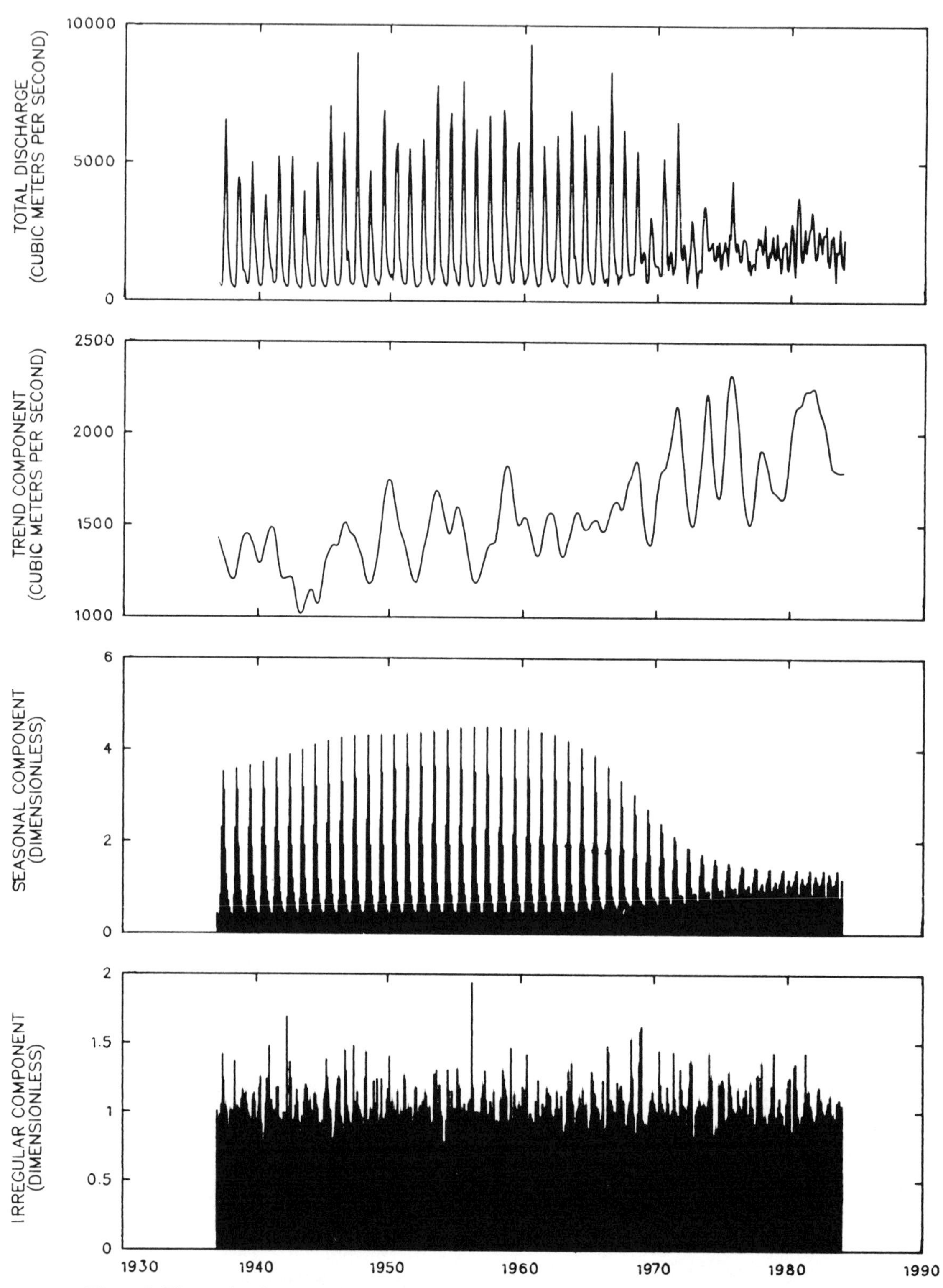

Figure 5. Time-series decomposition of monthly flows for the Columbia River at Birchbank, British Columbia.

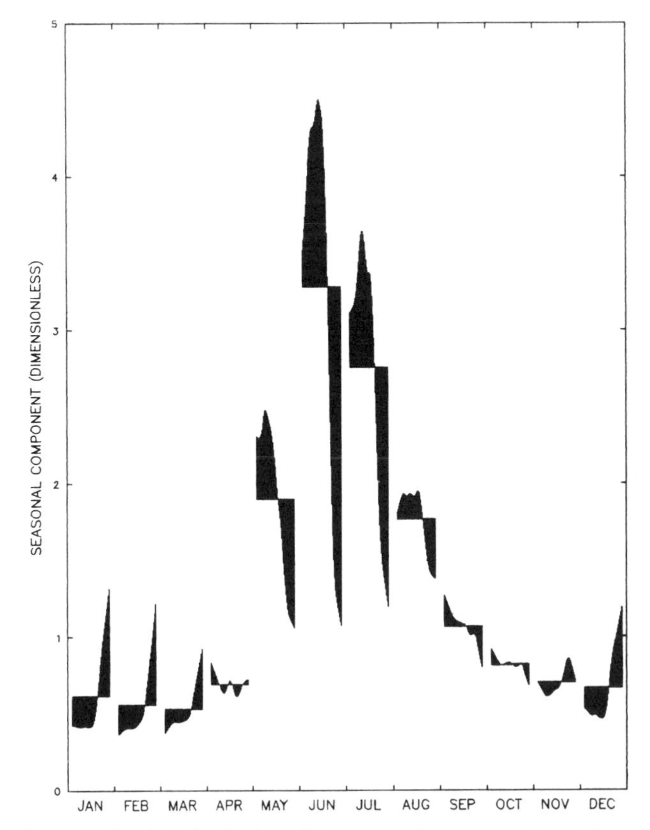

Figure 6. Monthly distribution of the seasonal component from Figure 5. Each month is plotted as an individual time series of 47-year duration. The horizontal line for each month is the 47-year average for that month's seasonal component.

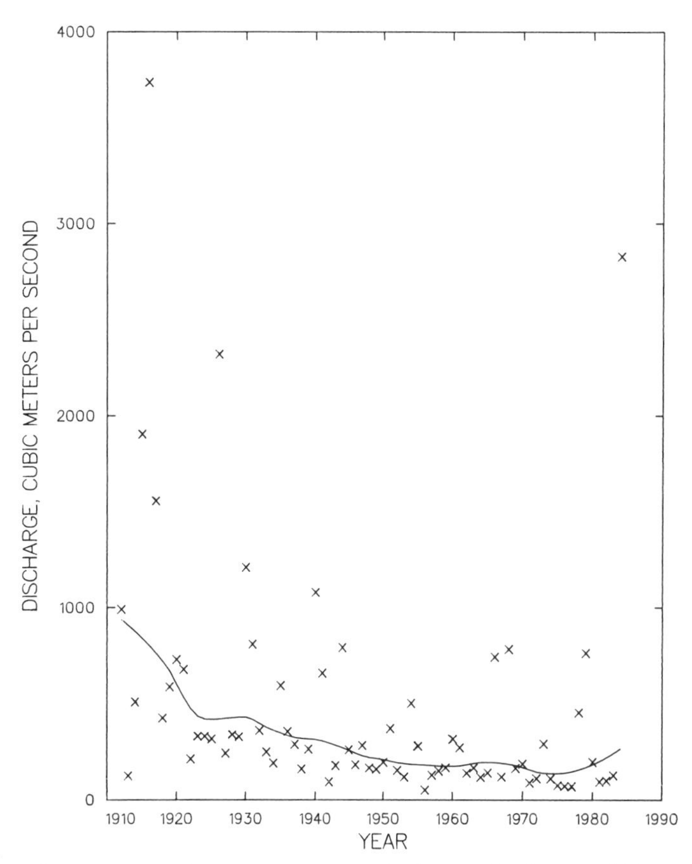

Figure 7. Annual instantaneous peak discharge, Gila River at Kelvin, Arizona.

storage in Lake Athabasca and the flooding of the Peace–Athabasca Delta. This is one of the largest freshwater deltas in the world.

The W.A.C. Bennett Dam was constructed and finished in 1968 on the Peace River in British Columbia, approximately 1,200 km upstream from the delta. The initial filling of Williston Lake (70 billion m^3 of storage) during 1968–1971 coincided with a few years of below-average runoff. Peace River levels were not high enough to significantly restrict flows in the Lake Athabasca outlet channels and, consequently, the lake and delta normal high levels were greatly reduced. With the generating plant in full operation, the regulation of spring floodwaters has resulted in changes to the natural level and outflow regime of the lake; however, the changes are not as drastic as during the initial filling of the reservoir.

Low-Flow Characteristics

Most dams operate under requirements for minimum releases. These minimum-release requirements are established for reasons of fish and wildlife habitat, waste dilution, recreation, or navigation. In most cases, these minimum-release requirements result in an increase in the magnitude of low flows occurring with a given recurrence interval. Significant increases in the annual 7-day low flows can be seen for the Tennessee and Columbia rivers (Figs. 9 and 10), two rivers with considerable amounts of reservoir storage.

Channel Characteristics

Cultural intervention in alluvial river systems to increase control for hydroelectricity, flood control, water abstraction, or navigation normally induces a complex set of changes both upstream and downstream from the point of flow-regime change (Galay, 1983). River-bed degradation progressing downstream may be triggered by many types of changes, such as: construction of a dam, stream-bed excavation, land-use changes, interbasin diversion, subsurface permafrost thawing, or rare floods. In these cases, downcutting is normally related to a decrease in the supply of bed material or to an increase in flood discharge. Conversely, upstream-progressing degradation may be set in motion by a lower base level related to a drop in lake level or the level of a main river. In other cases, a decrease in river length caused by naturally occurring cutoffs, channelization and regulation, or a horizontal shift of base level can produce the same result. Finally, natural erosion or the removal of a control point such as a dam can also induce upstream-progressing erosion. Downstream and upstream degradation can occur concurrently in the same river system.

Degradation below large dams can be extensive, due to a reduction in sediment supply. Reduced sediment supply in sand-bed rivers normally leads to degradation until either a stable, gravel-armored bed is created or until the slope is reduced until sediment is no longer removed (Kellerhals, 1982, p. 698) as a base-level control is reached. As a result, downcutting exceeding 7 m is reported below the Hoover Dam on the Colorado River, and downcutting is discernible up to 300 km downstream of the Sariyar Dam in Turkey (Galay, 1983, p. 1060).

River regulation normally decreases flood peaks. Regulation of a gravel-bed river can reduce or eliminate the ability of the river to move bed materials; thus channel degradation downstream of major storage dams might not occur. This phenomenon creates problems, however, when tributaries with significant gravel bedloads enter a regulated main channel; the stream is often no longer capable of removing the inflowing materials. For example, Kellerhals (1982, p. 685) reports the potential for massive long-term aggradation at the confluences of several major Canadian rivers, such as the Peace and Pine, downstream of the Bennett Dam.

Regulation can initiate a gradual reduction in channel size and capacity and, in Canada, normally implies dramatically increased winter flows. Gravel bars therefore become forested, and flood plains and side channels are blocked by debris and ultimately filled with fine sediments. The severity of ice jams during spring breakup also appears to increase, causing flooding of terraces that are normally too high to flood. At present, there are not theoretical models to estimate final channel morphology or size following regulation because the processes are too complex for quantification (Kellerhals, 1982, p. 695–699).

The predictability of downstream changes in channel stability and morphology is limited by the absence of theoretical, quantitative models for some processes involved and the enormous data requirements of existing theoretical models. The main difficulties relate to bed materials, sub-bed materials, and bed material transport rates. The prohibitive cost of sub-bed data collection, and the absence of an objective, practical sampling scheme for gravel-bed channels before and after regulation make accurate data collection extremely difficult (Kellerhals, 1982, p. 690–695).

If flooding of a river and its tributaries is out of phase, regulation can cause a tributary to flood when the main channel is far below its natural level. For example, the Pine River tributary to the Peace River, downstream of the W.A.C. Bennett Dam in British Columbia, experienced rapid bank erosion and channel shifting for a 2-km reach above the confluence.

Williams and Wolman (1984) documented channel changes

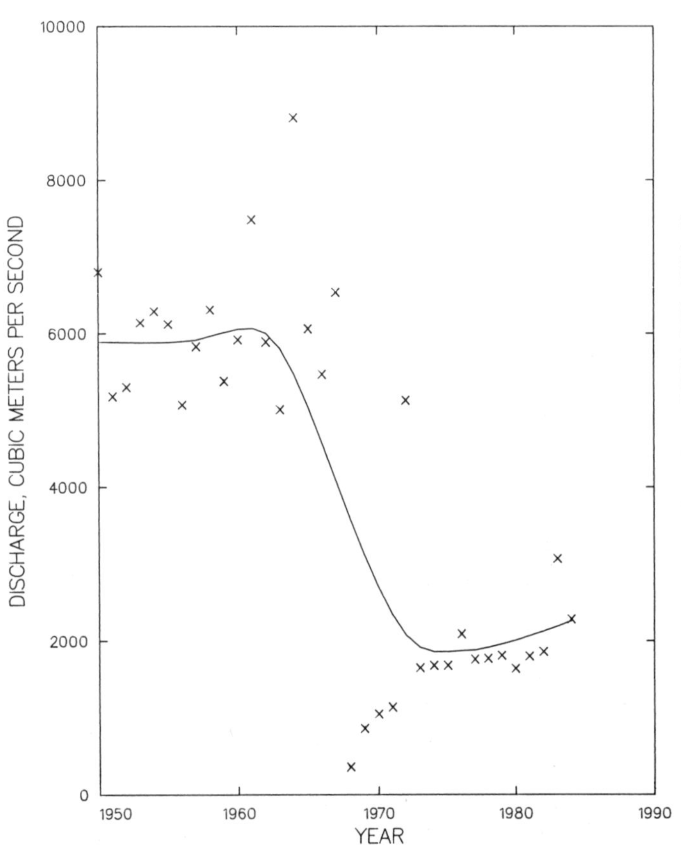

Figure 8. Annual maximum 1-day discharge, Peace River at Hudson Hope, Alberta.

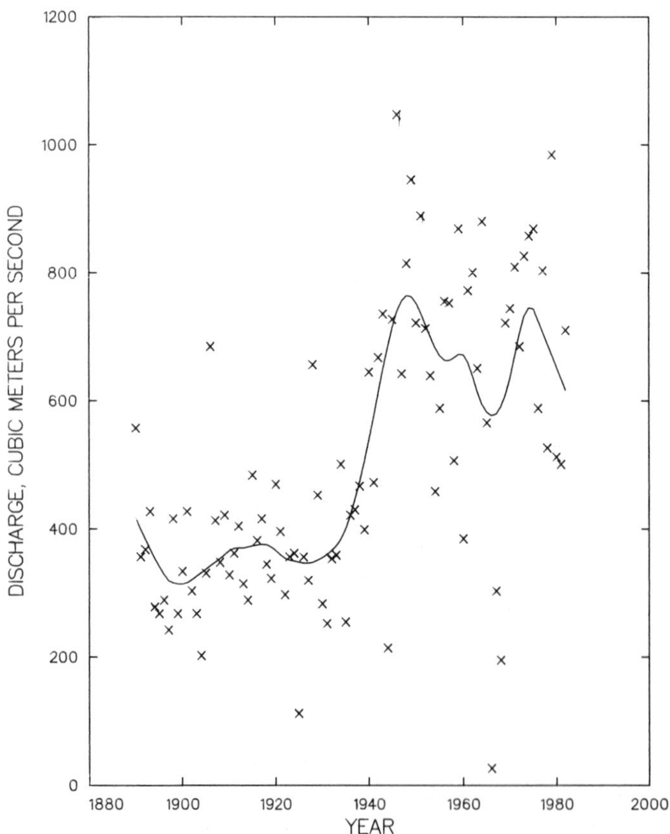

Figure 9. Annual 7-day low flows, Tennessee River near Paducah, Kentucky.

associated with 29 dams in the central and western United States. They note that the excess of sediment transport capacity in relation to sediment availability in the reach just below the dam often results in erosion of sediment from the channel bed and banks, bringing supply more nearly into equilibrium with capacity. Erosion causes degradation or lowering of the river bed. This was observed in 23 of the 29 dams in their study and attained a maximum of 7.6 m, averaging about 1.8 m. Degradation can result in a considerable decrease in channel gradient which, in turn, decreases the transport capacity of the river and leads eventually to a new equilibrium channel and transport rate. They describe the rate of bed degradation as a hyperbolic function of time

$$D = \frac{t}{C_1 + C_2 t} \tag{2}$$

where D is degradation in meters since the start of bed erosion, t is time in years, and C_1 and C_2 are empirical coefficients. Maximum bed degradation (equal to $1/C_2$) typically decreases with distance downstream from the dam, and the time from dam closure to the initiation of degradation increases with distance. By using the model described above, Williams and Wolman (1984) evaluated the maximum bed degradation for 111 cross sections. The modal value of expected maximum degradation was about 2 m, the modal time for degradation to attain 50 percent of maximum depth at a cross-section was about 7 years, and the modal time to attain 95 percent of maximum depth was about 150 years. Figure 11 illustrates the time series of bed degradation at six sites they examined. These are among the most regular of the sites considered.

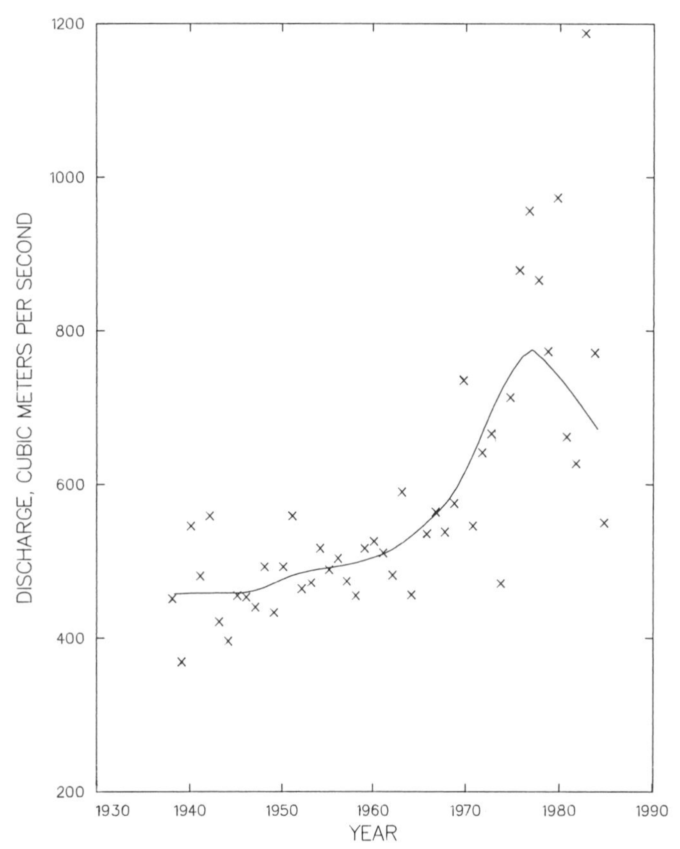

Figure 10. Annual 7-day low flows, Columbia River at Birchbank, British Columbia.

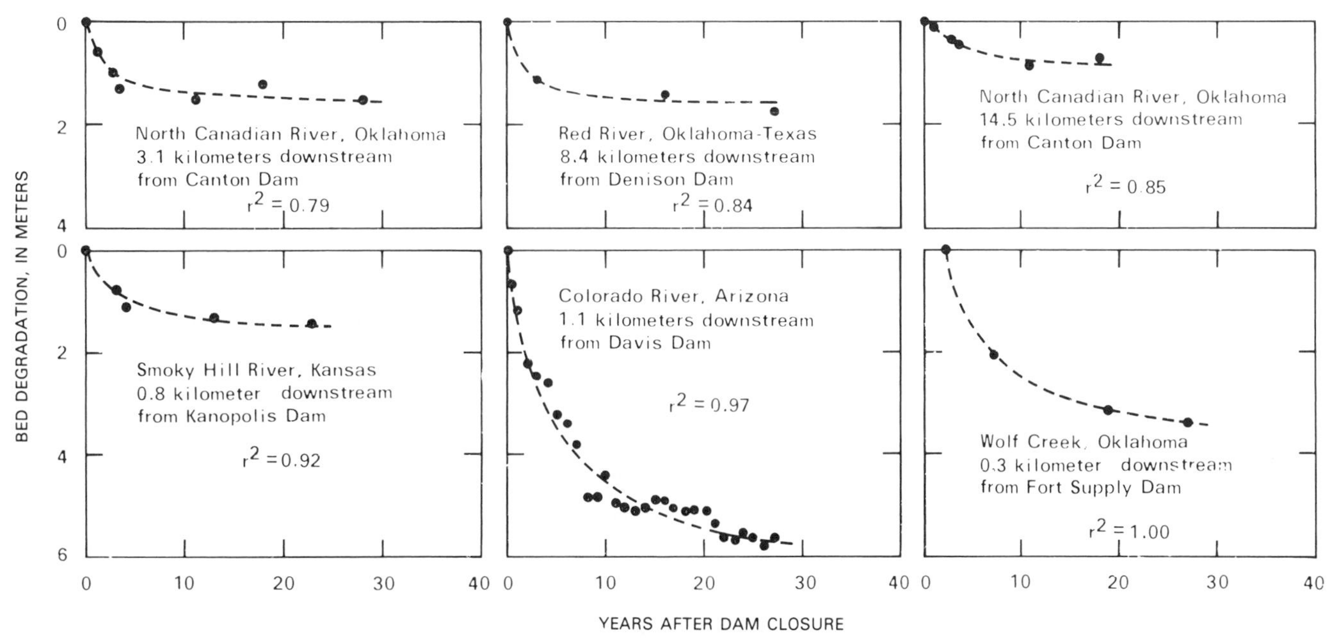

Figure 11. Representative regression curves (dashed lines) of bed degradation with time at selected sites (after Williams and Wolman, 1984).

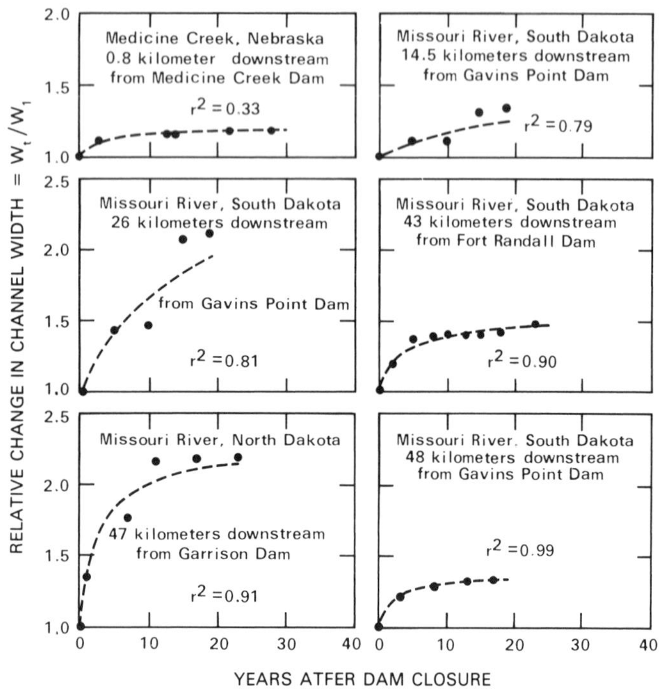

Figure 12. Examples of relative increase of channel width with time. The dashed lines represent fitted regression curves (after Williams and Wolman, 1984).

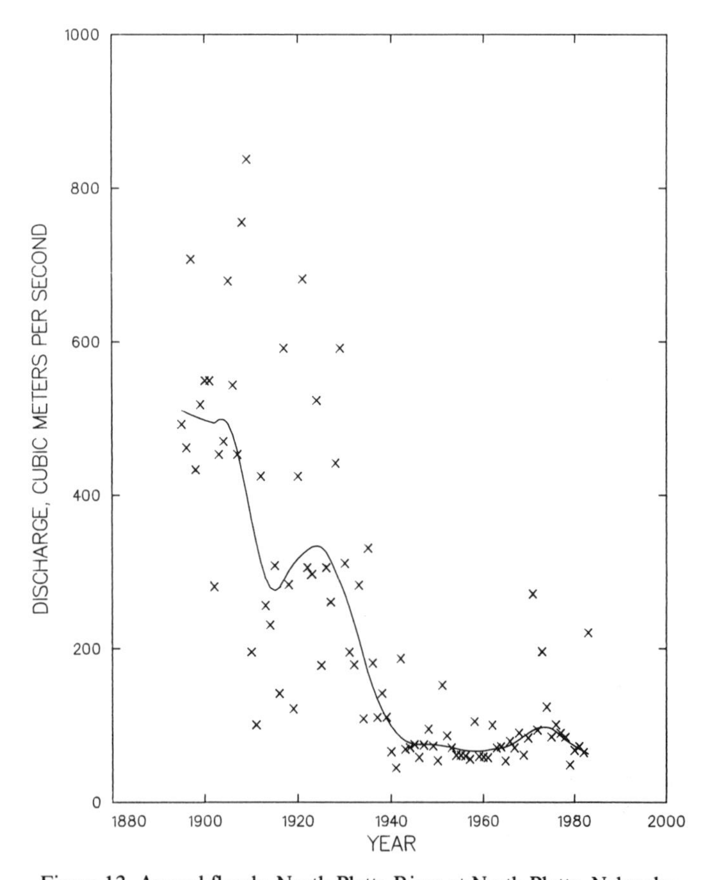

Figure 13. Annual floods, North Platte River at North Platte, Nebraska.

Channel-width changes below dams are less consistent. In cases where dams cause substantial decreases in flood flows, the channels typically narrow, and vegetation and sediment fill in parts of the old channel. However, where bedrock controls or armoring with coarse sediments prevents channel degradation, the new source of sediment to the river becomes the banks, and channel widening occurs. As the channel widens, flow depth decreases and the transport capacity of the river decreases; eventually a new equilibrium may be achieved. In cases of hydroelectric peaking, channel widening may occur from the increased frequency of wetting and drying of the bank and increased frequenty of flows capable of removing sloughed materials. Figure 12 shows examples of widening over time.

Platte River Example

One of the best-documented cases of changing floods and associated channel modification as related to dam construction, and to some extent consumptive use, is the Platte River of Wyoming, Colorado, and Nebraska. The North Platte River, which has its headwaters in the Rocky Mountains of Colorado and Wyoming and then flows through the arid high plains of Wyoming and Nebraska, is of particular interest. Annual precipitation in the lower reaches of the North Platte basin is about 460 mm/yr. Most runoff in the North Platte originates from spring snowmelt in the mountains of Colorado and Wyoming. The first significant storage to be built in the basin was Pathfinder Reservoir in Wyoming, finished in 1909. Other major reservoirs were added in 1938, 1939, 1941, and 1957. The total active storage capacity of these reservoirs is about 6.6 billion m^3, which is about 2.4 times the mean flow of the North Platte in pre-development conditions. This is among the highest levels of development of any North American river basin.

The ability of these reservoirs to store snowmelt waters from the Rocky Mountains and release them slowly during the irrigation season has caused the annual instantaneous peak discharge at North Platte to decrease from about 511 m^3/s before 1909 to about 72 m^3/s since 1957 (Williams, 1978; Fig. 13). This dramatic reduction in discharge has been accompanied by substantial channel changes (Fig. 14). The width of typical cross sections has decreased from an estimated average of 790 m in 1865 to an average of 90 m during 1965–1969 (Williams, 1978). Such decreases occur at each of 28 cross sections observed by Williams over a 576-km reach of the North Platte, and the Platte downstream of the confluence of the North Platte and South Platte. Eschner and others (1983) reported 1979 widths from 8 to 50 percent of the estimated 1860 width.

There have also been dramatic changes in vegetation. Prior to development, the river banks were relatively devoid of trees, although islands in the braided Platte River channel had some trees. As the channel has narrowed due to the lower discharge carried, dense vegetation succession—sedges, followed by grasses, then willows and other trees—has occupied a large part of the original channel. The absence of large floods has allowed en-

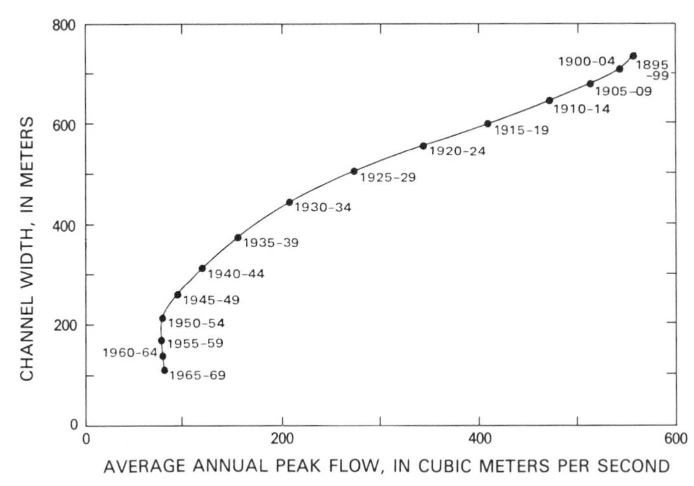

Figure 14. Relation of channel width to 5-year-average annual peak flows, North Platte River at North Platte (after Williams, 1978).

croachment of trees to take place unabated. In earlier times this process would have been reversed from time to time by large floods.

Green River Example

Another example of the effects of dams on the flow regime and the consequent effects on sediment transport and channel characteristics is the Green River, a major tributary of the Colorado River in Wyoming, Colorado, and Utah. The Flaming Gorge Reservoir on the Green River was completed in 1962 and controls part of the Green River basin. The basin supplied 38 percent of the water but only 21 percent of the sediment (Andrews, 1986). Downstream from the reservoir, the Green River basin is essentially free of reservoirs. Thus, although the supply of sediment from upstream of Flaming Gorge was essentially eliminated, the supply to the lower reaches has been reduced by only about 21 percent. Unlike the Platte River example, the mean discharge of the Green River has been virtually unchanged over time. The reservoir regulates the river flows, but no substantial changes in consumptive use have occurred since it was completed. Figure 15 shows the pre- and post-reservoir flow-duration curves at the Jensen gage, 169 km downstream of the dam. Similar pre- and post-reservoir flow-duration curves developed for gages farther downstream show similar absolute differences at various durations, although percentage differences diminish downstream.

Andrews (1986) evaluated the sediment transport-discharge relationship at three long-term daily sediment sampling sites downstream from Flaming Gorge Reservoir. There were no statistically significant differences in the sediment transport rate at a given discharge between pre-reservoir and post-reservoir periods. Transport follows an approximate power relationship to discharge, given as

$$L = aQ^b \tag{3}$$

where L = sediment load, Q is water discharge, and a and b are fitted regression constants (b is typically about 1.7 when all sediment-size fractions are considered). By integrating the sediment-transport rate over the probability distribution of flows (flow-duration curve), Andrews computed the pre- and post-reservoir mean annual transport rates. If mean discharge is essentially unchanged and its variance substantially decreases, then $\bar{L}$ (the mean annual load) will decrease,

$$\text{where } \bar{L} = \int_0^\infty aQ^b f(Q)dQ \tag{4}$$

and f(Q) is the probability density function of discharge. Figure 16 illustrates the pre- and post-reservoir transport at the Jensen gage located 169 km downstream from the reservoir. The total load at the Jensen gage has decreased by 54 percent since construction of the reservoir. Because the reservoir has trapped a portion of the sediment supply, the sediment-load equilibrium for the stream has been disrupted.

In the reach immediately below the dam, the capacity of the river to transport sediment, although reduced, is greater than the supply of sediment. The supply is greatly reduced because virtually all of the sediment is trapped by the reservoir above the dam. Thus this reach is in disequilibrium, and sediment is being removed from storage in the channel bed and banks. The channel has narrowed and the bed has been lowered in a 35-km reach.

Downstream from the Jensen gage (169 km downstream from the reservoir), the next reach has experienced a 54-percent reduction in sediment load transported, which is approximately equal to the reduction in sediment supply. Consequently, a 96-km reach is in quasi-equilibrium, but channel width is narrowing due

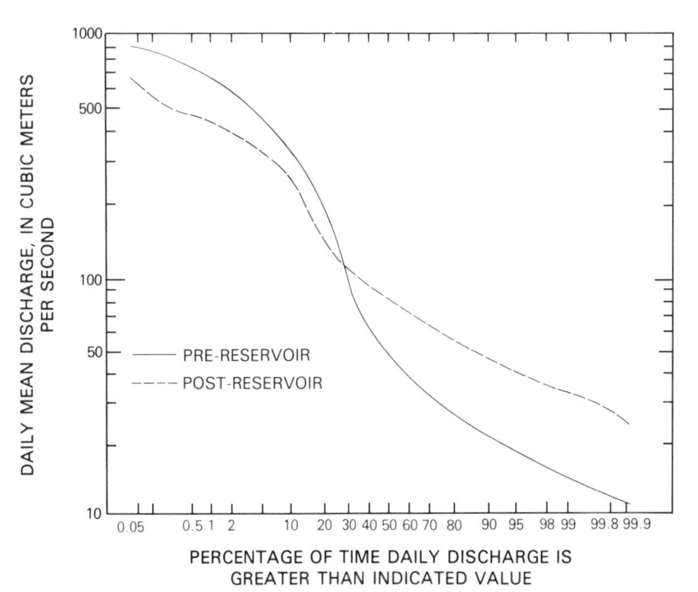

Figure 15. Comparison of the duration for mean daily discharge during the pre- and post-reservoir periods measured at the Green River near Jensen, Utah (after Andrews, 1986).

to reduced peak flow. In the lowest reach, located downstream from the Green River gage (464 km downstream from the reservoir), the transport capacity has decreased by a greater amount than the sediment supply. This is because sediment is supplied to the Green River primarily by the small tributaries in which the supply has not changed, but the transport capacity is determined by flows from Flaming Gorge Reservoir, which have changed a great deal. Thus, this 24-km reach is undergoing significant aggradation, and as a result, the channel is actively narrowing. The theory of alluvial channels (Mackin, 1948; Leopold and Maddock, 1953) indicates that adjustments must occur over time in this reach such that the rate of sediment transport for any given rate of discharge will increase, and sediment transport will eventually equal sediment supply. This will presumably occur through narrowing, deepening, and steepening of the channel slope. This example is undoubtedly not unique, but is perhaps the best-documented example of the effects of regulation by dams on the downstream flow regime, sediment transport, and channel disequilibrium and change.

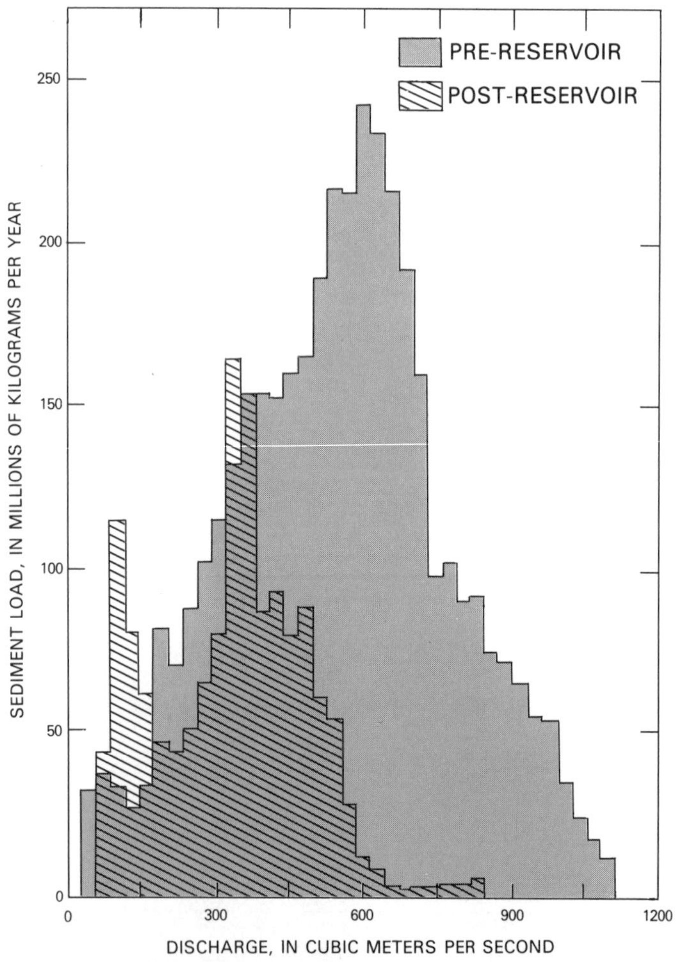

Figure 16. Comparison of the quantity of bed material transported by increments of discharge during the pre- and post-reservoir periods computed from the Green River in the vicinity of the Jensen, Utah, gage (after Andrews, 1986).

Areal Distribution of Reservoir Storage

The Colorado River illustrates a number of flow changes induced by dams. Modifications in this basin are more profound than in virtually any single large basin in North America due to the high level of regulation, measured by the ratio of storage to annual supply. Figure 17 shows this ratio for the 36 water-resources regions of the United States and Canada. Clearly, in both the United States and Canada, most of the area west of the 100th meridian has such a large concentration of reservoir storage that flows of virtually all large rivers are extensively regulated. This area of high regulation corresponds to those parts of the continent that are either arid, necessitating interseason or year-to-year storage, or are areas of high relief, providing great opportunities for hydroelectric development, or both.

The ability of reservoirs to provide and assure supply (safe yield) is subject to the "law of decreasing marginal returns." That is, greater reservoir capacity results in increased yields but, as capacity grows toward or exceeds 100 percent of annual flow, incremental yield decreases rapidly. Finally, as capacity reaches levels above about 150 percent of annual flow, yield gains are approximately offset by evaporation losses in regions of high evaporation; eventually, increased storage leads to decreasing yields (Hardison, 1972; Fig. 18). Langbein (1959) showed that the point of diminishing total returns would be exceeded in the Colorado basin with the completion of Glen Canyon Dam. Elsewhere there are opportunities to increase reliable yields by additional dam construction, although many of the best dam sites from hydrologic and engineering perspectives have already been used. As an indication of this, Langbein (1982) described the decline of reservoir capacity per unit volume of dam by decade for the United States; for Canada, the data was compiled by the Canadian National Committee (1984). This decline (Table 1) is indicative of a shift from the early gravity, arch, and buttress

TABLE 1. HISTORY OF RESERVOIR CAPACITY IN THE UNITED STATES AND CANADA, BY DECADE (after Langbein, 1982; Canadian National Committee, 1984)

	Reservoir Capacity per Unit Volume of Dam*	
Period	United States	Canada
Before 1930	16800	7960
1930-39	3390	699
1940-49	840	2900
1950-59	730	2280
1960-69	470†	1260
1970-79	—	1713

*Reservoir capacity and volume of dam are in m^3.
†Dams under construction in 1958.

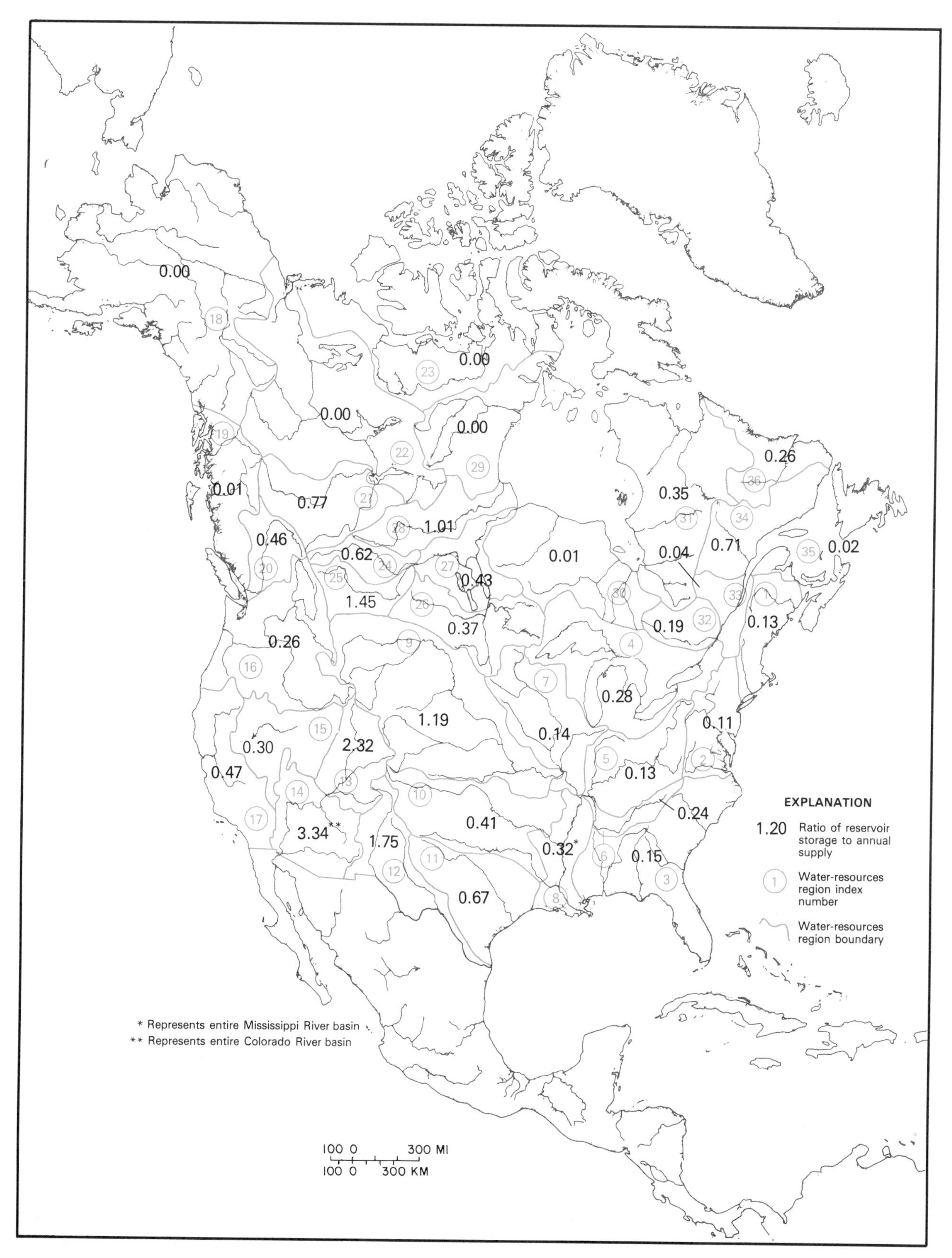

Figure 17. Ratio of reservoir storage to annual supply, by water-resources region.

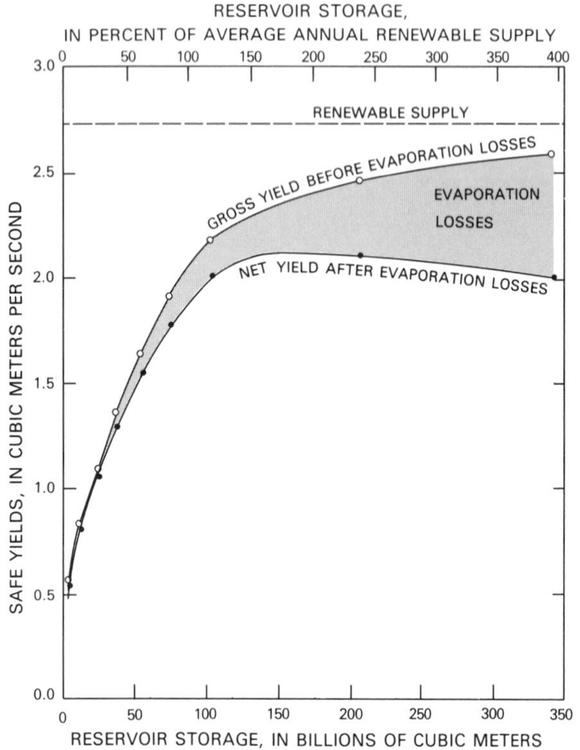

Figure 18. Relation of reservoir storage and regional safe yield for the upper Mississippi River region (after U.S. Geological Survey, 1984).

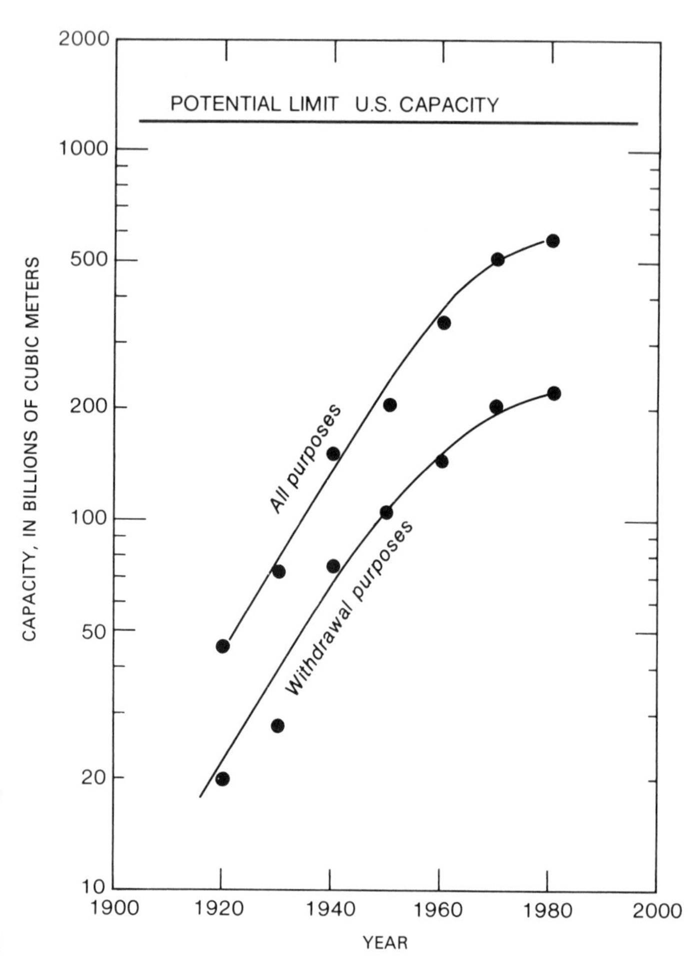

Figure 19. Trend in reservoir capacity in major reservoirs in the conterminous United States since 1920 (after Langbein, 1982).

masonry dams used in narrow valleys to broad-based, earth-fill dams.

Langbein (1982) reported on estimates of the upper limit on reservoir capacity in the United States and demonstrated that percentage increases in storage have slowed in recent years (Fig. 19); nor are storage increases keeping pace with withdrawals. In 1960, U.S. reservoir storage equaled 204 days of withdrawals; it grew to 216 days in 1970, but by 1980 it fell to 201 days. This was the first decline in six decades for which records are available. This suggests that surface water supplies are becoming less reliable, unless this decline is being compensated for by improved management efficiency (Toebes and Shepard, 1981).

URBANIZATION

Urbanization has a profound impact on the hydrologic response of a watershed. The process of construction results in temporary increases in sediment yield due to the increased erodibility of bare and disturbed soils. Following the construction phase, naturally pervious watershed surfaces are converted to impervious surfaces. Because of reduced infiltration and depression storage, these areas produce a substantially greater volume of runoff than they did before. Watershed surfaces and channels also become more efficient than the natural hillslopes and natural small channels. This causes an increased runoff velocity and decreases time from the center of mass of rainfall to the center of mass of runoff. The consequence of decreased lag time, even if total runoff is constant, is an increased peak discharge. Dunne and Leopold (1978) observed that the drainage density of a natural channel system—the length of all channels in the basin divided by the drainage area—typically decreases as a result of urbanization. This decrease results as small, ephemeral first- and second-order channels are typically graded over and paved, built upon, or replaced with buried storm sewers. However, if the entire drainage network is considered, including gutters, curbs, and storm sewers, the drainage density is typically greatly increased.

In this section, the effects of urbanization on stream characteristics are discussed. The best-understood impact of urbanization is on flood-peak characteristics, which is considered first. Next, various channel changes induced by urbanization are examined. Finally, changes in low-flow characteristics due to urbanization are briefly mentioned.

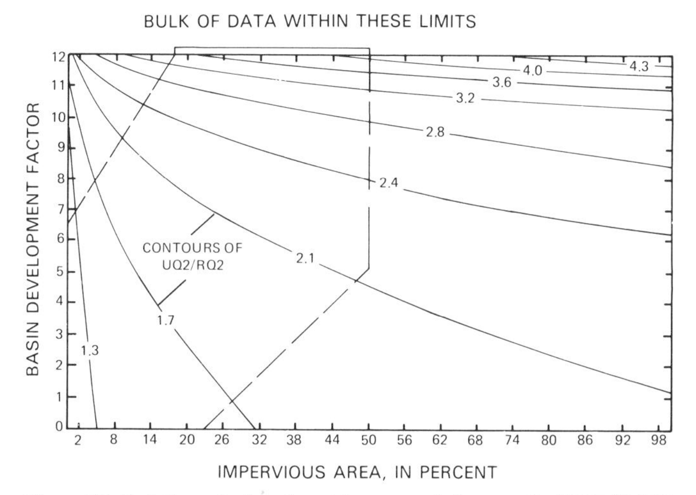

Figure 20. Relation of urban/rural 2-year peak-flows ratio (UQ2/RQ2) to basin development factor and impervious area (after Sauer and others, 1983).

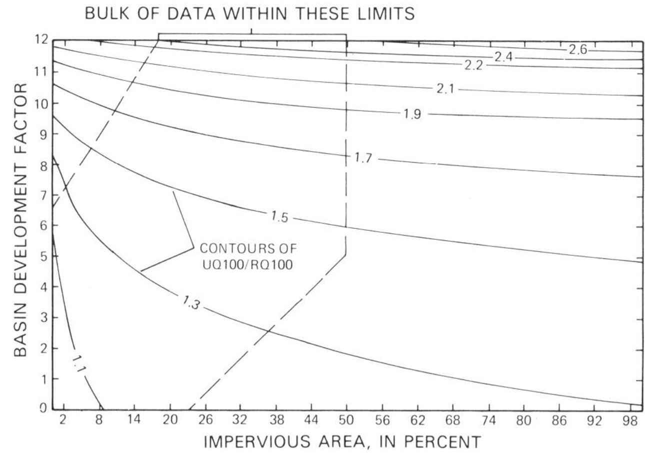

Figure 21. Relation of urban/rural 100-year peak-flow ratio (UQ100/RQ100) to basin-development factor and impervious area (after Sauer and others, 1983).

Flood Characteristics

Urbanization can induce important changes in the flood characteristics of affected basins. An early attempt to estimate the degree of increase in flood peak magnitudes for urban basins was made by Leopold (1968). He defined a ratio of urban to equivalent rural mean annual floods and related this ratio to the percentage of the basin served by storm sewers and the percentage of the basin covered by impervious surfaces. The most comprehensive attempt to relate urban-flood magnitudes is by Sauer and others (1983). This study used data from 269 sites throughout the United States, where at least 15 percent of the basin was in urban (commercial, industrial, or residential) uses, reliable flood frequency data existed (based on 10 or more years of flood records or synthesized from a calibrated rainfall-runoff mathematical model), and the degree of urbanization was relatively constant over the period of record. The sites covered 31 states and 56 metropolitan areas.

The study used two different measures of urbanization. The first is impervious area (IA), defined as the percentage of the drainage basin occupied by impervious surfaces. The other measure, which proved to be more significant than IA, was the basin development factor (BDF). The BDF is evaluated for separate thirds of the entire basin—upstream, middle, and lower—in terms of travel time. In each third, four factors are evaluated and given a zero or one score. The factors are: (1) Channel improvements. A score of one is given if channel straightening, enlarging, deepening, or clearing is prevalent on the main channel and principal tributaries. (2) Channel linings. A score of one is given if more than half of the main channel and principal tributaries have been lined. (3) Storm drains or storm sewers. A score of one is given if more than half of the secondary tributaries in the area are storm drains or sewers. (4) Curb and gutter streets. A score of one is given if more than half of the area is urbanized and streets of the area have curbs and gutters for at least half of their length. Given these definitions, the BDF has a possible range of zero for the least developed reaches to 12 for the most developed.

As more statistical and deterministic modeling of urban-flood frequency has occurred, there has been a growing recognition that total impervious area, which is typically easy to measure from air photos, is a poor predictor of urban effects on floods. Rather, the connectedness and efficiency of the entire drainage network are recognized to be of greater importance. Alley and Veenhuis (1983) confirm this assertion in their analysis of total impervious area and effective impervious area. The effective impervious area, which has a direct, continuous, impervious connection to the main channel system, is the most useful in terms of modeling basin response.

Sauer and others (1983) estimated several different kinds of regression equations that relate peak discharge to the IA and BDF and other variables that are not influenced by urbanization, such as drainage area, rainfall intensity, and natural storage. They reexpressed their basic estimating equations in a form that provides a general formula for predicting the ratio of urban to rural flood magnitudes of various recurrence intervals. There are:

$$R_2 = 2.27\,(13\text{-BDF})^{-0.32}\,(\text{IA})^{0.15}, \qquad (5)$$
$$R_{10} = 2.16\,(13\text{-BDF})^{-0.30}\,(\text{IA})^{0.09}, \qquad (6)$$
$$R_{100} = 2.05\,(13\text{-BDF})^{-0.28}\,(\text{IA})^{0.06}, \qquad (7)$$

where R_i is the ratio of urban to rural flood peaks with an i-year recurrence interval. Figures 20 and 21 show these relationships for the 2- and 100-year floods graphically, and also show the approximate limits of the data used for fitting the equations.

Two features of these relationships are worth highlighting. The first is the extent of change in flood magnitude that occurs with urbanization, and its variation with recurrence interval. For a highly developed condition (BDF = 12, IA = 50), the ratio of the 2-year flood to the original rural condition 2-year flood is 4.1. For the 10-year flood the ratio of urban to rural flood peaks is 3.1, and for the 100-year flood it is 2.6. Thus, expressed as a ratio, urbanization has its greatest effect on smaller floods. This is reasonable because, for the largest floods, the soil is typically saturated during the time of maximum runoff so infiltration losses can be nearly the same in the rural or urban condition. The peak discharges are different because of the greater efficiency, or shorter lag time, of the urban drainage system. Expressed as a difference (rather than a ratio), the increase in the 10-year flood may very well be greater than the increase in the 2-year flood.

Another way of looking at the same phenomenon is to consider the effect of a doubling of the amount of impervious area. A doubling of impervious area results in an 11-percent increase in the 2-year flood, a 6-percent increase in the 10-year flood, and only a 4-percent increase in the 100-year flood.

Sauer and others (1983) found that, for 12 percent of the sites studied, the urban 2-year flood is smaller than the estimated 2-year rural flood. Although some of these apparent discrepancies can be explained as model error or sampling error, there is reason to believe that such counter-intuitive changes are possible. One possible reason is different degrees of urbanization in different parts of the basin. If the lower reaches of the basin become urbanized and the upper reaches do not, it is possible that the runoff originating from that lower portion may, under these conditions, flow out of the basin before the arrival of the flow from upstream.

Another possibility is that many natural drainages are crossed by roadways as urbanization occurs. Culverts are installed under many of these roads and, either intentionally or unintentionally, the culvert openings are insufficient to pass peak discharges. Consequently, water is stored for short periods during flood events. Storage may also occur at storm-water inlets in parking lots, on streets, and also on roof tops and in basements. The design of these structures has the effect of imposing an approximate upper bound on flow from various parts of the basin and providing short-term storage, which attenuates the flood peak.

In the past 15 years there have been some selective widespread efforts to intentionally undersize inlets and build detention storage structures in urban development with the express intent of limiting flood peaks, either because of the flood damage they cause or because of the detrimental effects associated with channel widening. These detention structures are typically unmanned and have no moving parts. They are simply small earth-fill dam structures that have some combination of orifices and weirs to release the flood waters. In highly developed areas there are systems with gates or even pumps. The detention reservoir may or may not have a permanent pool of water.

The design requirements for urban drainage may require that the system, including the detention device, achieve a peak discharge from a specified storm event that is no greater than that which would have occurred prior to development from the same storm. McCuen (1974) and Hirsch (1980) noted that these regulations may not, in fact, produce the desired result in terms of the flood-frequency distribution in main channels of the watershed. This is because reductions in peak discharges from a variety of sub-basins are not additive in terms of reductions downstream, because of the different flood-peak arrival times from each sub-basin. McCuen (1974) identified cases where detention could increase downstream peaks. The principle involved here is the same as that brought out by Leopold and Maddock (1954) concerning basins that are orders of magnitude larger. The construction of a large amount of flood-control storage located throughout a basin on many tributaries is ineffective in controlling the size of flood peaks at downstream locations.

Channel Characteristics

Another motivation for urban runoff regulations is related to channel-geometry changes. The effect of sediment yield and flood-magnitude changes associated with urbanization has resulted in some cases in major channel-geometry changes. These have been documented in the middle Atlantic States by Wolman (1967), Hammer (1971), Leopold (1973), Robinson (1976), and Allen and Narramore (1985). Hammer (1971) developed a relationship between channel cross-sectional area and drainage area for rural streams in the southeastern Pennsylvania Piedmont and then compared urban channels with these. For equal drainage areas, he found that urban channels average at least twice as large; in cases where development is more than 4 years old and very dense, the ratio can be even greater. Hammer (1971) demonstrated that urban bankfull discharge on these enlarged adjusted channels has a recurrence interval of about 1.5 years (annual series), which is also the recurrence interval of bankfull discharge in rural catchments. Thus, if the increase in the 1.5-year flood is predictable by equations such as those produced by Sauer and others (1983) or by a mathematical rainfall-runoff model, then channel-dimension increases should be predictable. By using hydraulic geometry relationships described by Dunne and Leopold (1978) as $w \propto Q^{0.5}$ and $d \propto Q^{0.4}$, where w = bankfull width and d = bankfull mean depth, a doubling of the 1.5-year peak discharge should result in a 41-percent increase in width, a 32-percent increase in depth, and an 87-percent increase in cross-sectional area.

Urbanization can also dramatically increase flood risk due to channel morphology and sedimentation changes. Leopold's (1973) long-term detailed study of Watts Branch in the Maryland Piedmont showed an initial decrease in channel size as urbanization occurred. Wolman (1967) reported that an initial clogging of channels with sediments was common during and shortly after construction. He also estimated that sediment yields during construction can be as much as three orders of magnitude greater than from forested conditions and two orders of magnitude higher than for agricultural lands. A decrease in channel capacity, coupled with increased flood magnitudes once the construction, paving, and sewer installation have occurred, can result in vast increases in the frequency of overbank flooding. Leopold's (1973) data revealed an average of 1.9 overbank flow events per year for 8 years preceding the rapid phase of urbanization in Watts Branch versus an average of 6.5 events per year in the 4 years following urbanization.

The erosion rate decreases after urban development stabilizes, perhaps to levels lower than under any previous land use; thus the supply of sediment is markedly decreased, and sediment-transport capacity is increased. These conditions of frequent overbank flows and an imbalance of supply and capacity result in overbank deposition and channel-bank erosion, respectively. Together these processes increase channel capacity.

Robinson's (1976) study of streams draining 2.6-km^2 basins in the Maryland Piedmont shows a rather consistent pattern in channel size. He found that cross-sectional areas at bankfull stage were lowest during construction, about 1.1 m^2, largest after urbanization is established, around 5.6 m^2, and intermediate under mixed forested and farmed conditions, 1.9 to 3.7 m^2.

The channel-enlargement process has an additional effect on the hydrologic system. During the initial years of development, when channels have yet to enlarge or have even contracted, the propagation of most flood waves is significantly slowed and attenuated by overbank storage. This initially limits increases in flood peaks at downstream locations. However, as channel enlargement occurs over a period of several years, similar discharges are increasingly conveyed within the channel and are therefore much less attenuated, resulting in larger downstream flood peaks. This phenomenon is probably accounted for in the statistical summaries provided by Sauer and others (1983), but are not isolated in many modeling studies that base channel-flow parameters on existing channel characteristics, rather than on a reasonable expectation of future channel characteristics (Hirsch, 1977).

Low-Flow Characteristics

An additional aspect of the effect of urbanization on surface-water hydrology that has been much less studied than flooding is related to low-flow characteristics. In general, it is not intuitively apparent whether urbanization can be expected to increase or decrease low flows. Decreased infiltration caused by increased imperviousness should decrease base flow. However, the existence of septic tanks, the possibility of leaky water or sewer systems, and increased infiltration and dry-weather runoff from lawn watering all have the potential to increase streamflow during dry periods.

LAND DRAINAGE

Artificial land drainage is another significant human landscape modification that has attracted considerable attention as a potential cause of changing flow regimes. This includes the installation of tile drainage under agricultural fields, typically in areas of low land-surface gradients such as the prairie or plains regions of the central United States and Canada, in some low-lying coastal areas, and in areas of heavy, minimally porous soils. Land drainage also includes surface ditching to enhance wetland drainage, such as prairie potholes or coastal marshes, to allow them to be farmed or urbanized. Another form of land drainage involves major reshaping of the land surface, particularly in coastal areas, by transforming large areas that are perenially or intermittently subject to standing water into a mixture of high ground suitable for agricultural, residential, or industrial development, and canals that may or may not be navigable. Examples of this latter form of development are widespread on the Atlantic Coastal Plain and Gulf Coast, from Long Island to Texas; there is a particularly high concentration in Florida.

A recent assessment of wetlands in the United States (Tiner, 1984) indicates a net 54-percent loss of wetlands from the mid-1950s to mid-1970s. Significant gains in wetlands have occurred in the categories of unvegetated wetland flats (810 km^2) and small ponds (8,500 km^2). In fact, pond acreage nearly doubled from 9,300 km^2 to 17,800 km^2. This growth is primarily due to farm-pond construction. At the same time, about 44,500 km^2 were lost, of which about 4 percent was in estuarine areas and 96 percent was in freshwater areas. About 87 percent of these losses were caused by agricultural development; urbanization accounts for 8 percent, and other activities account for the remaining 5 percent. These freshwater wetland losses were 65 percent in forests, 30 percent in emergent marsh or meadow areas, and 5 percent in scrub-shrub wetlands. Iowa's loss is estimated at more than 95 percent. Such wetland conversion is an obvious source of concern because these lands are valuable fish and wildlife habitat and because of the effects of wetlands on the water budget and on flood flows.

The areal extent of Canadian wetlands is estimated to be approximately 1,270,000 km^2. No attempt has been made to monitor the rate of wetland conversion nationally. A recent study of land-use change between 1976 and 1981 in the vicinity of 22 urban centers across Canada (Kessel-Taylor, 1984) that embraced 37,000 km^2 revealed an increase in agricultural, urbanized, and recreational land at the expense of total wetland area (Table 2).

Determination of the effect that drainage will have on total runoff or on flood peaks is not easily deduced from first principles. Two opposing phenomena are at work when land is artificially drained. On the one hand, artificial drainage lowers the water table and reduces the amount of standing water. The increased thickness of the unsaturated zone means that greater volumes of rain must fall, or snow melt, before the available pore spaces are filled; only when this occurs will substantial runoff occur. In the pre-drainage condition, rain or snow must run off the surface. After drainage, more water infiltrates and contributes to very slow soil or ground-water drainage, or evapotranspiration, rather than rapid runoff. In fact, in the view of the variable source area theory of runoff (Dunne, 1982), the area initially saturated and the rapidity with which areas become saturated during rainfall or snowmelt events are directly related to the amount of runoff produced by a storm or snowmelt event. Con-

TABLE 2. LAND-USE CHANGE ON CANADIAN WETLANDS, 1976-1981*
(after Kessel-Taylor, 1984)

Land Use	1976	1981	Change
Agriculture	427,002	431,901	+4,899
Built-up	99,745	108,311	+8,566
Recreation	9,373	13,848	+4,475
Remaining wetland	173,031	155,275	- 17,756
Total wetland area	709,151	709,355	

*In hectares.

versely, artificial drainage increases the density and the efficiency of the drainage system. Increased drainage density typically leads to shorter basin lag times. If flood volume were constant and lag time decreased, the result would be higher peak discharge. Furthermore, this improvement in drainage efficiency may have the result that some portions of a basin that previously did not contribute any runoff will contribute once the artificial drainage has been implemented. Thus, even if runoff per unit contributing area were not increased, the total basin runoff may be increased.

The consensus stated by a Canadian study of this question (National Hydrology Research Institute, 1982) is that surface or ditch drainage increases peak discharges; the effect of subsurface tile drainage is much less clear. Some factors that complicate attempts to analyze induced changes are that many drainage systems decline in efficiency rapidly, so that they may have a significant hydrological effect for only two or three years before the drains fill with sediment and ditches fill with sediment or vegetation. Even if an entire basin were drained simultaneously, the two or three years of substantial effects may be insufficient to obtain a hydrologic record to reveal statistically significant change. In practice, drainage occurs gradually in various parts of a river basin and, given declines in drainage efficiency over time, the aggregate effect may be difficult to discern. Hill (1976, p. 261) stated that "small drainage schemes are not likely to produce major changes in flow at downstream sites. The location of drained land within the river basin is also a significant factor. The drainage of considerable areas in the lower portion of a river basin may actually reduce flood peaks at the basin outlet because water from this area will be removed before arrival of flood peaks from the headwater areas (Wisler and Brater, 1965)."

Red River of the North

A major agricultural drainage region in North America is the Lake Agassiz basin in North Dakota, Minnesota, and Manitoba. This flat area, dominated by glacial drift and lacustrine clays, is drained by the Red River of the North. The record of the past 100 years of annual peak flows for the Red River at Grand Forks, North Dakota, is shown in Figure 22. Miller and Frink (1984) characterized the history of drainage activity as a gradual increase from about 1897 through 1926, relatively static conditions to 1942, followed by a precipitously rapid rise to the present. By 1944 there were 1,120 km of major drainage; by 1955 the total was 5,110 km, and since then great increases in farm drainage have occurred. A total of 18.2 billion km^2 or 18 percent of the United States basin, has been drained.

The abrupt increase in average flood magnitudes in the mid-1940s in comparison to the most recent two decades suggests the possibility of a relationship between land drainage and flood peaks. Except for a few years in the late 1950s, the moving average of annual instantaneous peak discharges has been high for the past four decades. This has resulted in a strong public perception of a drainage-flood relationship (Miller and Frink, 1984). However, records of four extraordinarily large floods between 1880 and 1900 demonstrate that the land drainage is not a necessary condition for the occurrence of these large discharges.

The year-to-year variation in maximum flows is so great that discerning a long-term trend from available data is difficult (Fig. 22). Miller and Frink (1984) used multiple regression methods to evaluate the alleged trend. Regression equations for the

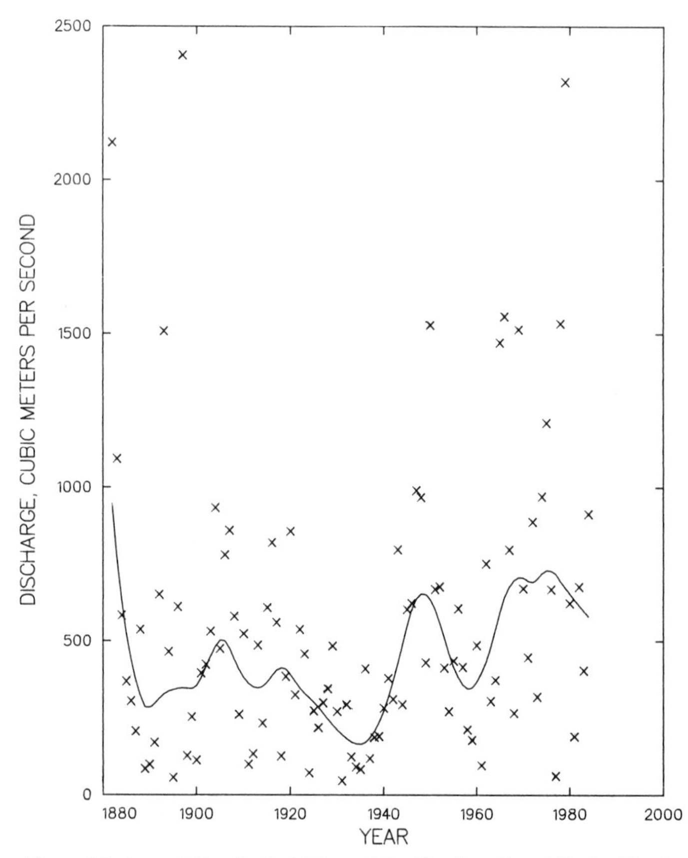

Figure 22. Annual floods, Red River of the North at Grand Forks, North Dakota.

annual peak discharge at Grand Forks, and another similar one for Fargo, North Dakota, include variables related to winter precipitation for the basin, a winter temperature index, an index of the snowmelt available in the spring, peak flows of the previous year, and an index of land drainage as explanatory variables. The index of land drainage was statistically significant ($\alpha = 0.05$) but only explained 5 percent of the variance in the annual peak flows when used in the equation along with the climatic variables. Miller and Frink concluded that there has probably been some change in the maximum-discharge response of the basin over the 1897 to 1980 period, although between 1950 and 1980, drainage changes did not have a discernible effect on peak flows.

This case is characteristic of many in which several extraordinarily large maximum discharges occur in a short period after a long interval devoid of large peak flows. This typically leads to suggestions by the public and public officials that human actions contribute to the large flows. This, in turn, can lead to investigations of possible trends. The very act of conducting such a study is dependent on the occurrence of large recent peak flows introducing a bias in favor of finding statistically significant trends, even when such relationships do not exist. The potential for such an error (called a type I error by statisticians) is exacerbated if there is year-to-year persistence in maximum-discharge magnitudes. For example, a record such as that of the Red River at Grand Forks suggests that considerable persistance may exist. Such a basin, having low gradients and thick unconsolidated deposits, may be particularly conducive to such persistent hydrologic conditions. As noted previously (Matalas and others, 1982), there is considerable difficulty in discriminating between trend and persistence.

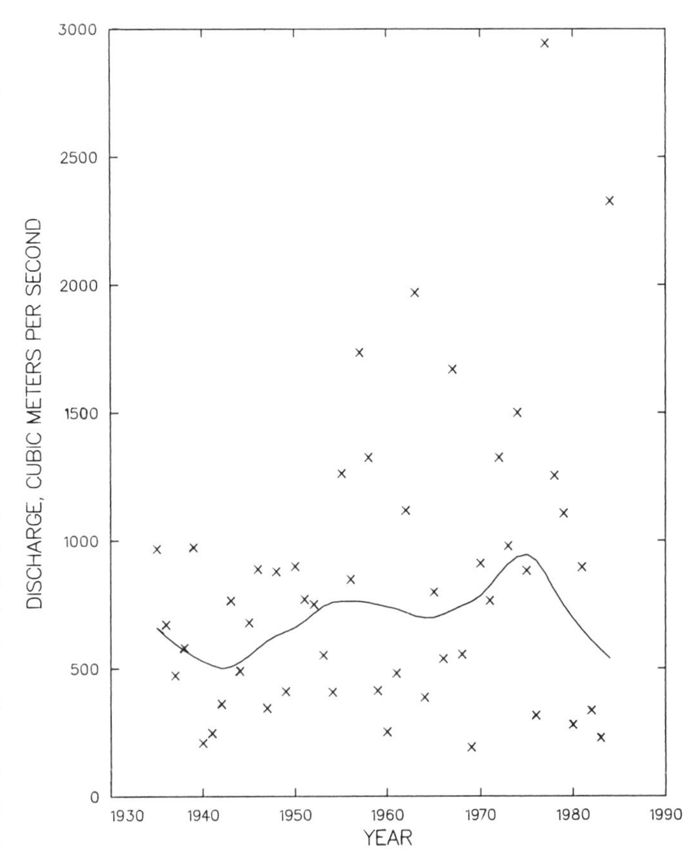

Figure 23. Annual floods, Tug Fork near Kermit, West Virginia.

SURFACE MINING

Surface mining, typically for coal, is also thought to produce changes in the magnitude of peak discharge. As in the case of land drainage, it is not immediately apparent what sort of change can be expected to result from mining. Soil materials are removed and are replaced by bare bedrock surfaces. Vegetation is also changed, at least in the short run, which modifies its interception characteristics. In areas of shallow soils, landscape disturbance can increase the porosity and permeability of surficial materials. Shallow soils over bedrock are replaced by coarse deposits of blasted rock, either dumped over slopes or carefully graded and arranged at the conclusion of mining activities.

Channel sedimentation is also common in many mining areas. This may be widespread or, if mining practices are particularly careless, may be concentrated in specific slide areas that overfill old channels, causing backwater flooding or catastrophic water and sediment inundations of the flood plain.

Tug Fork

In the very steep terrain of the Allegheny Plateau, where surface-coal mining is extensive, there is a widespread perception that peak discharges have increased over the years. Bryan and Hewlett (1981), and Hirsch and others (1982) are two examples of studies that address this question. The latter considered the Tug Fork basin of Virginia, West Virginia, and Kentucky. The annual peak flow series for Tug Fork near Kermit, West Virginia (Fig. 23) indicates a large peak discharge in 1977, together with a clear increase in average peak flow magnitudes from the mid-1950s to 1977 when compared to the mid-1930s to mid-1950s. The annual surface-mining coal production increased from zero prior to 1942 to a peak in the mid-1950s and a higher peak in the 1970s. This strong association lends credence to the hypothesis that mining leads to increased peak-discharge magnitudes. However, only about 3.5 percent of the basin had been mined or disturbed by mining as of the mid-1970s; over 90 percent remained in deciduous forest cover.

A nonparametric trend test applied to 45 years of annual maximum discharge data for the Kermit gage indicated a significant upward trend ($\alpha = 0.025$). The data were further analyzed in a way that attempted to remove much of the randomness of annual peak flows and focus on the possibility of change in the response of the basin to storm rainfall. By using discharge records from a nearby control stream gage at the mouth of an unmined basin, it was determined that there has been an upward trend in flood response. That is, for a given size flood at the control gage and given storm rainfall, the Kermit gage experienced larger floods in the later years of the record than it did in the earlier years.

The use of statistical controls, either precipitation inputs or concurrent discharges at nearby unaffected gages, is essential in analyzing changes in basin hydrologic response and discerning those changes attributable to the activities of man. A demonstration of changing response exemplified by different discharges at different times, given the same hydrologic inputs, is tantamount to a demonstration of trend in the hydrologic variable of interest, such as annual peak discharge, only if the inputs can be considered to be stationary over the period of record. Thus, the procedure of adjusting for or removing variation due to variations in the inputs can be useful in terms of increasing statistical power to detect trends. It can also protect against cases where the trend is an artifact of trend in the inputs.

VEGETATION

The influence of vegetation on hydrologic systems can be substantial. Considerable attention has been paid to the effect of vegetative change brought about by human activity on the water yield of a river basin. Ponce and Meiman (1983), summarizing the impact of managing vegetation, conclude that increased water yields can be achieved in the first few years after timber harvesting, although sustaining these increases for several years can be difficult. In addition to changes associated with decreased evapotranspiration from soils throughout the basin, there are potential hydrologic changes associated with flood-plain vegetation change. Such modification can affect not only evapotranspiration, but the movement of flood waves by altering the flood-plain roughness characteristics. Phreatophytes, plants that rely on ground water for growth, are a striking example. By extending their roots into either the capillary fringe or saturated zone, flood-plain phreatophytes use water that would otherwise contribute to baseflow and ultimately move downstream. The Gila River example discussed below is a good demonstration of the relationship between vegetation and hydrologic systems.

Gila River

In the upper Gila River basin, there has been a steady increase in phreatophytes, primarily saltcedar and mesquite, since the late 1930s. Saltcedar dominated the flood plain by the mid-1960s. As a result, the U.S. Geological Survey initiated a study to evaluate the effects of complete saltcedar removal along a 24-km reach east of the Gila River and San Carlos River junction.

Using vegetation maps made during 1914, 1937, 1944, and 1964, Turner (1974) established the vegetation succession in the study area. Coolidge Dam reservoir backwater appears to have contributed to the establishment of saltcedar in the lower two-thirds of the study reach. Turner argued that saltcedar, introduced into the region in the mid-19th century, successfully dominated the native plant species.

Culler and others (1982) determined evapotranspiration (ET) as the residual of a water budget for four reaches within the study area. Evapotranspiration was determined for periods before and after complete saltcedar removal. Before vegetation removal, the average annual ET was 1090 mm and the ET range was from 630 to 1,420 mm, corresponding to areas of 0 to 100 percent vegetation coverage. After clearing, the average annual decrease in ET was 480 mm, ranging from 360 to 660 mm. These water-loss reductions corresponded to cleared conditions; establishment of a replacement crop must reduce the potential savings from phreatophyte removal.

On the basis of the Blaney-Criddle equation (Blaney and others, 1942), ET estimates for replacement crops are as follows: alfalfa, 1,750 mm; blue-panic grass, 1,240 mm; and bermuda grass, 1,070 mm. Since the pre-clearing maximum value of ET was 1,420 mm, use of alfalfa would result in an increased water loss. Selective introduction of either blue-panic or bermuda grass would result in a slight reduction of the water lost, thus constituting a savings of water. The estimated figures are for optimum production; because the depth to ground water exceeds 2.4 m, it is likely that consumptive use would be less, slightly increasing the amount of water saved.

Using discharge data for stations above and below the study reach, Burkham (1976b) examined flood-wave travel times for five periods: 1914–1927, 1930–1940, 1941–1950, 1951–1960, and 1961–1970. For each period the travel time for the center of mass of flood waves was determined for runoff events, and a general trend line drawn. Differences between the trend lines revealed that, as the flood-plain vegetation increases, the lag time for overbank discharges increases drastically, as expected. Consequently, the velocity of the center of mass of the flood wave decreases; in 1964, for discharges between 283 and 566 m^3/sec, the velocity of the center of mass was one-third smaller than the 1914–1918 period.

The flood-plain vegetation also affects the flood hydrograph shape. During the period 1914–1927, the flood hydrographs retained their shapes as they moved through the study reach. Pronounced peak attenuation was observed during later periods; thus for large discharges the study reach began to behave as a reservoir.

Burkham (1976a) assessed the impact of vegetation on the hydraulic characteristics of the study reach. Three floods, in 1965, 1967, and 1972 were considered. The floods had peak discharges of 1,100, 1,130, and 2,270 m^3/sec. The study area was divided into two reaches, and hydraulic properties were evaluated at several cross sections within each reach. Saltcedar was removed from the first reach between 1965 and 1967. Comparison of the 1965 and 1967 floods revealed a 0.34 m average decrease in stage, a 0.24 m/sec increase in average velocity, and a 0.026 average decrease in Manning's n, representing a 33 percent decrease from pre-clearing conditions. Saltcedar was removed from the second reach between 1967 and 1972. Comparison of the 1967 and 1972 floods revealed a 0.3 m average decrease in stage, a 0.15 m/sec average increase in velocity, and a 0.021 average decrease in Manning's n, corresponding to a 26 percent decrease in roughness compared to conditions in 1965. Thus, even with a doubling of the discharge, the cleared reach exhibited lower depth, increased velocity, and reduced roughness.

CHANNEL ALTERATION

Stream channels are complex systems that are shaped by the flow of water and sediment. Stream-channel characteristics, such as width, depth, bed material, and sinuosity, are continually reacting to various environmental stresses in an attempt to reach a quasi-equilibrium state. The balance of some of the various stresses was described by Lane (1955) in terms of the stream-power equation:

$$QS \propto Q_s d_{50} \quad (8)$$

where Q = water discharge, in m^3/sec; S = channel gradient, in m/m; Q_s = bed-material discharge, in kg/m^3; and d_{50} = median grain size of bed-material load, in millimeters.

Man directly alters the quasi-equilibrium of stream channels by two main activities: channelization and gravel mining. Channelization typically involves widening, deepening, and straightening a stream channel. The result is a local increase in the stream gradient, which increases sediment-carrying capacity. The increased sediment-carrying capacity causes progressive degradation upstream of the disturbance; downstream of the disturbance, the sediment-carrying capacity is less than the sediment load, and thus aggradation occurs. Large changes in bed elevation and stage-discharge relationships caused by channelization have been documented on the Mississippi River near St. Louis (Belt, 1975), and at several locations on the Missouri River (Nordin and Meade, 1974; Water and Environment Consultants, Inc., 1976a, b).

The effects of gravel mining are somewhat similar to those resulting from channelization. If mining occurs in an active channel, the increased local channel gradient results in progressive degradation upstream of the mining operation. If the mined area is large enough, most of the sediment load can be trapped in the pit, effectively reducing the sediment load downstream of the disturbance. Thus degradation will also occur in the downstream direction. The impacts from mining operations in inactive channels are generally minor. However, if the pit is breached and assumes the flow of the river, the results can be catastrophic.

In this section, examples will be presented illustrating the impact of channelization and gravel mining on stream-channel behavior. A detailed description of channelization work on the South Fork of the Forked Deer River in Tennessee will be followed by a general study of channelization work at several western Tennessee sites. Finally, the consequences of gravel mining along a portion of Tujunga Wash in California will be discussed.

Channelization

A good example of channelization effects is found on the South Fork of the Forked Deer River in western Tennessee. Due to problems associated with flood plain drainage, there has been periodic channelization of the South Fork since 1900. Typical responses included upstream degradation, downstream aggradation, and bank failure. The U.S. Geological Survey recently monitored responses from channelization performed in 1968 and 1969 (A. Simon, 1986, personal communication).

There has been a long history of river alterations. Between 1914 and 1920, 129 km of the river was straightened, reducing its sinuosity from 2.1 to 1.3. From 1964 to 1966, the main stem from the confluence of the North and South Forks to the mouth was enlarged and straightened. Starting in 1968, 11.3 km of the South Fork was deepened by 3 m, widened on average by 15 m, and shortened by 3.7 km. In all, seven major cutoffs were employed. In 1969, two more cutoffs further decreased the river length by 0.6 km.

To quantify longitudinal adjustment of the stream bed, specific gage—defined as the water-surface elevation corresponding to a given discharge—was determined at three U.S. Army Corps of Engineers gaging stations. Annual specific-gage values for the South Fork of the Forked Deer River and a discharge of 28.3 m^3/sec were determined for 1950–1981. The gaging station farthest upstream is located at Gates, followed downstream by Halls and Yellow Bluff.

The resulting history of specific-gage values is depicted in Figure 24. In 1969, the bed at Yellow Bluff was lowered nearly 3.7 m. By 1970, degradation had progressed to Halls (7.4 km upstream). Lowering of the stream bed in the South Fork also resulted in headward erosion in the tributary streams. As the tributaries eroded, channel widening and bank failures were noted. Increased sediment load in the tributaries led to aggradation at Yellow Bluff.

The net result of the downstream aggradation was a decrease in the channel slope in the reach between Yellow Bluff and Halls (Fig. 25). The slope in the Yellow Bluff–Halls reach tripled after the 1968 disturbance (Fig. 25), then began to decrease. The slope in the Yellow Bluff–Halls reach eventually decreased to a level below the pre-disturbed conditions. This is not surprising, because a straightened channel requires less stream gradient for a given sediment load. The net result has been a wider, slower moving river with a flatter slope.

Simon and Hupp (1986) summarize work carried out at 40 channelization sites in western Tennessee. Using bed-elevation data and specific-gage values where available, gradation rates are developed using simple power relationships:

$$E = at^b \quad (9)$$

where E = elevation of the bed or specific gage for a given year, in meters; a = pre-modified elevation of the bed or specific gage, in meters; t = time since channel work, in years; and b = exponent representing gradation rate.

The variation of the aggradation rate over time and space can be used to predict degradation and aggradation. For the Obion River, the exponent representing gradation rate (b) exhibits a consistent relation with distance above the mouth (Fig. 26). As expected, the maximum degradation rate occurs at the area of maximum disturbance (point A, Fig. 26), and diminishes in the upstream direction (line C). Downstream from the disturbance, the stream is aggrading (line B). Overadjustment of the stream just upstream from the disturbance results in aggradation rates (line D) consistent with downstream (line B) and premodified (line E) rates. Similar relationships have been observed in other western Tennessee streams (A. Simon, 1986, personal communication).

Gravel Mining

Gravel mining typically occurs in either active or inactive channels. Active channels carry water for a substantial portion of the year; conversely, inactive channels rarely contain water. Although the hydrologic changes associated with mining in inactive channels are usually minor, a reactivation of the channel can have a pronounced effect. An example is Tujunga Wash, as described by Bull and Scott (1974).

Tujunga Wash is the major drainage from the western San Gabriel Mountains to the San Fernando Valley in California. Two channels, north and south, are distributary from the apex of an alluvial fan. Most flow prior to 1969 was confined to the north channel. In the south channel a gravel pit 300 by 460 m and 15–23 m deep had been mined sporadically since 1925.

In 1969, two severe storms in rapid succession produced a peak discharge in excess of what was previously estimated to be the 50-year flood, reactivating the south channel. Flow in the south channel broke through small levees designed to contain local runoff, and entered the gravel pit. The remaining north channel flow also broke through levees and entered the pit. Scour immediately proceeded upstream, causing failure of three major highway bridges.

Downstream of the pit, the combined flow of the north and south channels followed the south channel course. Lateral scour destroyed seven homes along the heavily urbanized south bank. A long section of a four-lane highway was destroyed farther downstream, and net fill of up to 3.7 m was observed in the south channel.

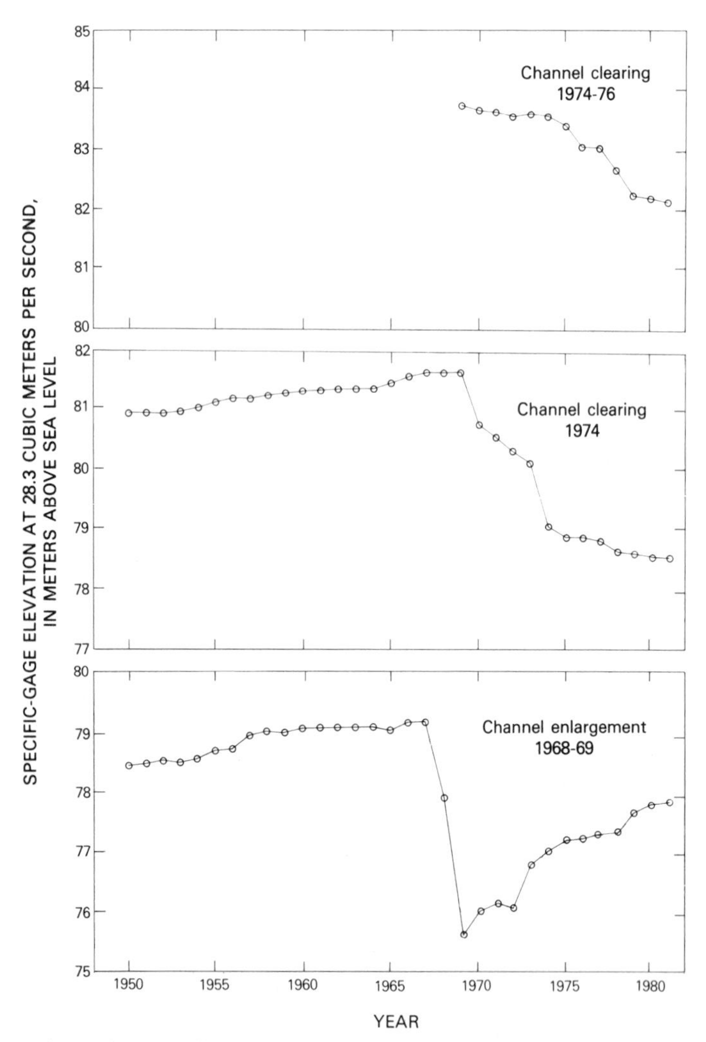

Figure 24. Specific-gage elevations for stations on the South Fork Forked Deer River, Tennessee (modified from Robbins and Simon, 1983).

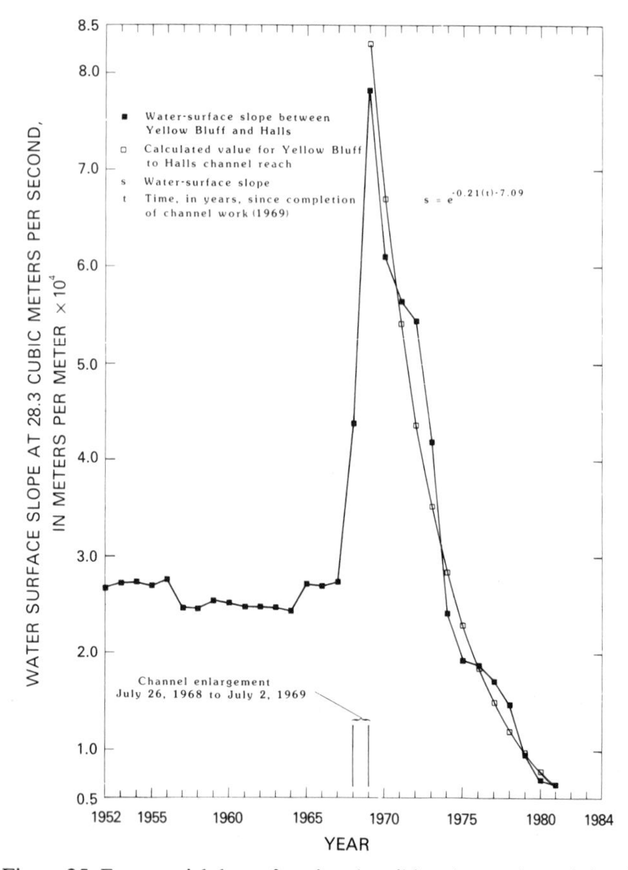

Figure 25. Exponential decay function describing the yearly variation in water-surface slope fluctuation associated with artificial channel enlargement for the South Fork Forked Deer River, Tennessee (modified from Robbins and Simon, 1983).

INTERBASIN TRANSFER

One situation in which man has a direct impact on the distribution of water is through diversions from one location to another. In this section, interbasin transfers in the United States and Canada are discussed in general. After a brief consideration of the various ramifications of interbasin transfers, two Canadian examples are explored.

Approximately 90 percent of the Canadian population currently lives within 240 km of the southern Canadian border, and 60 percent of the total river discharge flows toward the north. Thus, there are places where, and times when, water shortage occurs in intensively settled areas or when water flow can be beneficially transferred from one river to another. Interbasin or intrabasin water transfer combined with storage have been frequently used to increase the discharge of a receiving river (at the expense of one or more contributing rivers) to generate hydroelectricity, to enhance the capacity for navigation, or to support economic development based on irrigation or industrial water use. To date, existing transfers have been confined within provincial boundaries, so that serious jurisdictional conflicts have not yet developed (Quinn, 1981, p. 65; Quinn, 1983, p. 134).

Water diversions are relatively easy to construct throughout Canada, except through the high western mountains. A legacy of the Pleistocene glacial periods is low divides separating most drainages, and abandoned meltwater channels that often provide economical sites for constructing diversion channels. Until 1983, 54 major water transfers were identified in nine provinces, but none as yet in the northern territories (Table 3), based on the criteria that transferred water does not return to the donor stream within 20 km from the point of abstraction, and mean annual transferred flow is at least 0.5 m^3/s or has an annual volume of 16 million m^3 (Day, 1985).

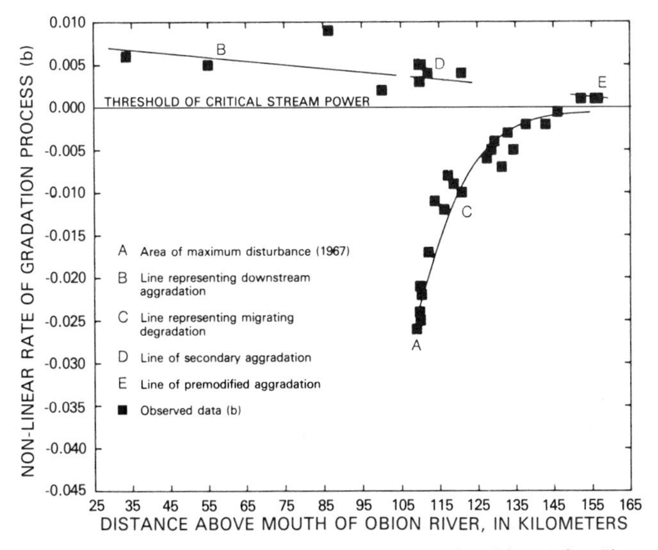

Figure 26. Model of bed-level response for the Obion River (after Simon and Hupp, 1986).

The principal reason for Canadian water transfer has been to generate hydroelectric energy. Quinn explained (1983, p. 129):

> . . . hydropower dominates overwhelmingly in number and scale of transfers with irrigation, flood control, and municipal uses assuming some importance regionally. With almost 70 percent of our electrical energy generated by falling water and 96 percent of our total water transfer attributable to hydroelectric projects, Canada is still very much hydro country.
>
> The total flow involved in water transfer—4400 m^3/s—is significant. If all of this flow were concentrated into a mythical "New River," it would be Canada's third largest, behind only the St. Lawrence and Mackenzie rivers!
>
> . . . Canadian water transfer is considerably greater than that of the next two countries, the United States and the Soviet Union, combined . . .

Three enormous hydroelectric projects completed in the 1970s incorporate five diversions and cumulatively compose two-thirds of all the volume of Canadian water transferred to 1986. The three projects are the Churchill Falls project in Newfoundland, the Lake Winnipeg, Churchill, and Nelson regulation and diversion project in Manitoba, and phase 1 of the La Grande project in Quebec. Water transfers have become a cornerstone of Canadian hydroelectric development policy over the past two decades.

Seventeen major Canadian diversions each exceeded 25 m^3/s in 1980 (Fig. 27, Table 4). These megascale projects are located in the provinces of British Columbia, Saskatchewan,

TABLE 3. CANADIAN WATER TRANSFERS EXISTING OR UNDER CONSTRUCTION: 1986
(adapted from Quinn, 1983)

Province	Number of Transfers	Average Annual Flow* (m3/s)	Major Use
Newfoundland	5	725	Hydro
Nova Scotia	4	18	Hydro
New Brunswick	2	2	Municipal
Quebec	6	1854†	Hydro
Ontario	9	564	Hydro
Manitoba	5	775§	Hydro
Saskatchewan	5	30	Hydro
Alberta	9	67	Irrigation
British Columbia	9	367	Hydro
Canada	54	4400	Hydro

*Estimates subject to revision.
†Excludes Beauharnois Canal flows from St. Lawrence River.
§Excludes floodway flows (Portage Diversion, Winnipeg Floodway, Seine Diversion) of short duration.

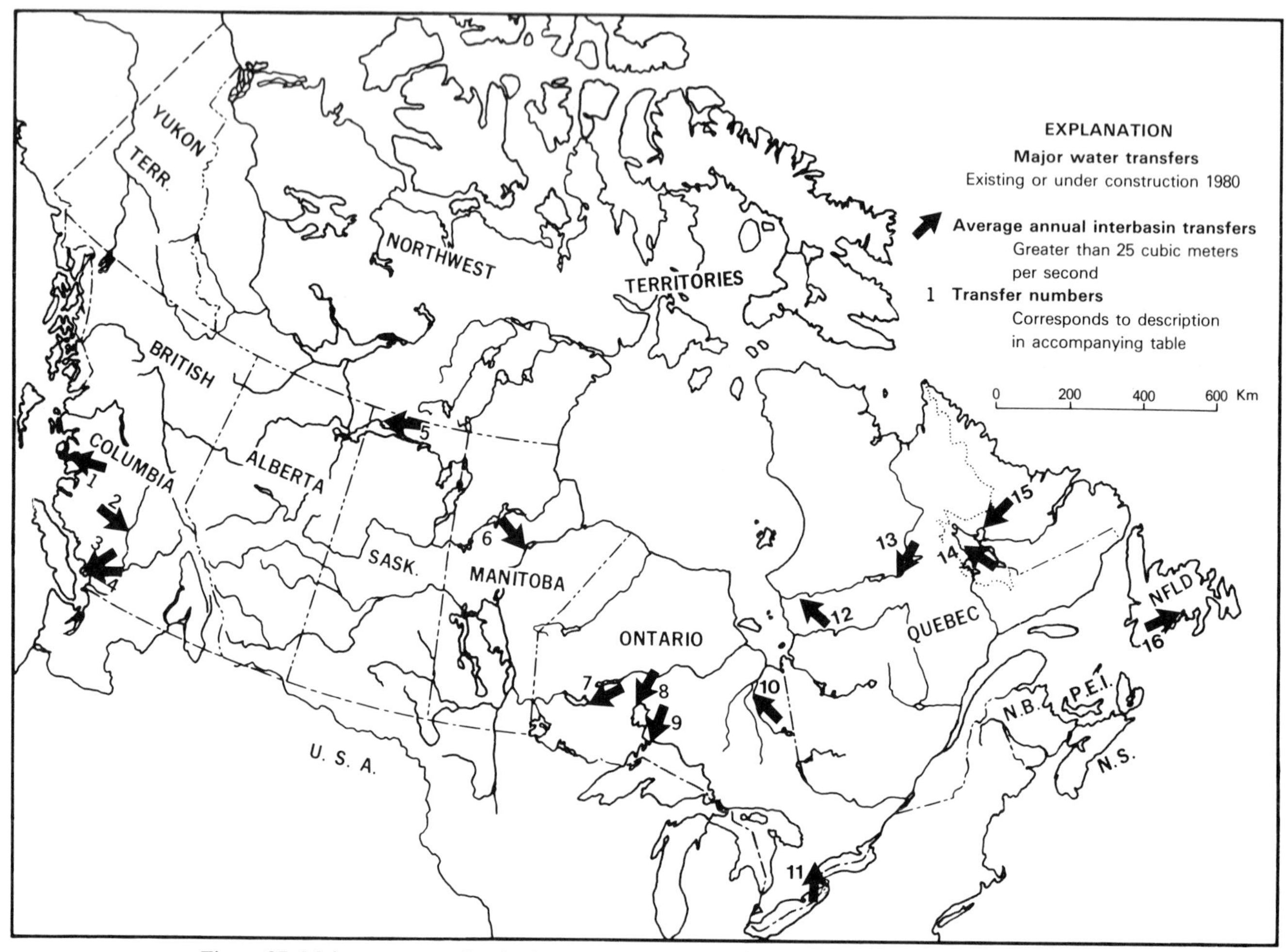

Figure 27. Major water transfers in Canada existing or under construction in 1980 (see Table 4 for diversion descriptions).

Manitoba, Ontario, Quebec, and Newfoundland. There was very little independent critical analysis of the consequences of undertaking the work prior to construction. Diversion-related projects will undoubtedly be undertaken in coming decades and they will reinforce the established trend across Canada to generate hydroelectricity based partly on diverted waters. But the prairie situation, and particularly that of Alberta, differs significantly.

Interbasin transfers in the United States were inventoried by Petsch (1985) for 19 western states, and by Mooty and Jeffcoat (1986) for 31 eastern states. Prior to 1982, 56 major water transfers having mean annual transferred flows of at least 0.5 m^3/s were identified (Table 5). The total average annual interbasin transfer of water in the United States is 694 m^3/sec, nearly an order of magnitude lower than the 4,400 m^3/sec from the Canadian diversions. Contrary to the Canadian cases, interbasin transfers in the United States are for irrigation and municipal use.

Nine transfers in the United States each averaged at least 25 m^3/sec in the 10 years prior to 1982 (Table 6). The major United States diversions are rather small in comparison with the Canadian efforts.

Downstream Effects of Increased Flows

Kellerhals (1982) studied two diversions into gravel-bed channels in British Columbia that reveal the complexity of geomorphic responses that may be induced. On the lower Kemano, the mean flow of this highly unstable braided channel was increased threefold by the Kemano hydroelectric project without appreciably changing the largest floods or increasing sediment flow from upstream reaches. In response, the channel zone widened and the course straightened. Dramatic increases in the average pink- and chum-salmon runs may be a result of improved spawning and incubation environment (Jackson, 1984, p. 41–46).

As part of the same project, the Cheslatta River discharge was increased by two orders of magnitude and caused the formation of a new gravel-bed river by degradation into glaciofluvial gravels. The new channel is controlled by several bedrock sills; the intervening reaches are gravel-paved and relatively straight uniform channels that are incised 5–10 m below the original valley.

TABLE 4. MAJOR INTERBASIN TRANSFERS IN CANADA EXISTING OR UNDER CONSTRUCTION: 1980
(adapted from Quinn, 1983)

Name	Province	Contributing Basin(s)	Receiving Basin	Average Annual Transfer (m3/s)*	Operational Uses	Operational Date	Owner
Kemano	B.C.	Nechako (Fraser)	Kemano	102	Hydro	1952	Alcan Ltd.
	B.C.	Bridge	Seton Lake	92	Hydro	(1934) 1959	B.C. Hydro
	B.C.	Cheakamus	Squamish	37	Hydro	1957	B.C. Hydro
	B.C.	Coquitlam Lake	Buntzen Lake	28	Hydro	(1902) 1912	B.C. Hydro
	Sask.	Tazin Lake	Charlot (L. Athabasca)	25	Hydro	1958	Eldorado Nuclear
Churchill Diversion	Man.	Churchill (Southern Indian Lake)	Rat-Burntwood (Nelson)	752	Hydro	1976	Manitoba Hydro
	Ont.	L.St. Joseph (Albany)	Root (Winnepeg)	86	Hydro	1957	Ontario Hydro
	Ont.	Ogoki (Albany)	Lake Nipigon (Superior)	120	Hydro	1943	Ontario Hydro
	Ont.	Long Lake (Albany)	Lake Superior	38	Hydro/ Logging	1939	Ontario Hydro
	Ont.	Little Abitibi (Moose)	Abitibi (Moose)	40	Hydro	1963	Ontario Hydro
Welland Canal	Ont.	Lake Erie	Lake Ontario	250	Hydro/ Navig	1829ff	Govt. of Canada
James Bay	Que.	Eastmain-Opinaca	La Grande	798	Hydro	1980	James Bay Energy Corp.
James Bay	Que.	Caniapiscau	La Grande	771	Hydro	1983	James Bay Energy Corp.
James Bay	Que.	Frigate	La Grande	31	Hydro	1982	Nfld. & Lab.
Churchill Falls	Nfld.	Julian-Unknown	Ashuanipi-Smallwood Res.	196	Hydro	1971	Nfld. & Lab.
Churchill Falls	Nfld.	Naskaupi, Kanairiktok	Churchill	330	Hydro	1971	Nfld. & Lab.
Bay d'Espoir	Nfld.	Victoria, White Bear, Grey, and Salmon	Northwest Brook (Bay d'Espoir)	185	Hydro	1969	Nfld. & Lab.

*Estimates subject to revision.

Prairie and Northern Waters

Most Saskatchewan-Nelson basin rivers are controlled for hydroelectricity, agricultural, municipal, industrial, and recreational uses. Some of these rivers cannot always meet the demand for water, particularly in the semiarid regions of southern Alberta and Saskatchewan. This problem relates to the pattern of discharge; 60 percent of the natural flow passes through the region in a 3-month period in spring. This necessitates storage for later use.

Consumptive use of 17 percent of the annual discharge of the south Saskatchewan basin in Alberta has supported settlement and economic developments in the southern third of that province. However, water consumption for irrigation and increasing waste loads contribute to inadequate streamflow and poor water quality.

The Churchill-Nelson Diversion

The Missi Control Dam was completed at the outlet of Southern Indian Lake into the Churchill River in 1976 (Fig. 28). Southern Indian Lake (SIL) was raised 3 m, and 775 m^3/s was transferred south to the Nelson River via the Rat and Burntwood rivers. Although the effects of the diversion on the river were not

TABLE 5. UNITED STATES WATER TRANSFERS PRIOR TO 1982
(adapted from Petsch, 1985; and Mooty and Jeffcoat, 1986)

Region of Origin	Number of Transfers	Average Total Annual Flow (m^3/s)
Mid-Atlantic	10	129
South Atlantic-Gulf	5	7
Great Lakes	1	88
Ohio	1	1
Lower Mississippi	2	2
Missouri	6	81
Arkansas-White-Red	2	3
Texas-Gulf	2	3
Upper Colorado	12	31
Lower Colorado	9	212
California	6	137
Total	54	694

generally documented, 10 years of measurement were undertaken on SIL.

The induced biophysical change was significant. Modifications occurred in Southern Indian Lake, such as increased shoreline erosion, littoral sedimentation, turbidity, and phosphorus availability, as well as decreased light penetration, visibility, and light limitation of primary production. There was also a decrease in the lake-water temperature and socially significant changes related to increased mercury content in fish flesh, rapid declines in the quantity and quality of whitefish taken in the commercial fishery, and the need for compensation programs to keep the commercial fishery economically viable (Hecky and others, 1984).

As a result of impoundment, the area of Southern Indian Lake increased from 1977 to 2391 km^2, the total volume from 16.84 to 23.38 billion m^3, and the total shoreline length from 3,665 to 3,788 km. Prior to flooding, 88 percent of the shoreline was bedrock controlled and only 5 percent was actively eroding. Immediately following inundation, bedrock occurred on only 15 percent of the shoreline because the post-impoundment water surface intersected permafrost-affected glacial and organic deposits on 85 percent of the new shorelines. Onshore waves initiated substantial erosion on all shores exposed to more than 1 km of offshore fetch. This erosion has caused retreats of up to 10 m/yr and has annually removed up to 25 m^3 of material per meter of shoreline (Newbury and McCullough, 1984). It is not yet clear how long it will take to reestablish a stable shoreline around most of Southern Indian Lake. As much as 80 percent of this eroded material was initially deposited near shore; the remainder went into suspension and significantly increased offshore sediment concentrations by 2–5 times. This could dminish fish reproduction.

THE WATER BUDGET

A water-budget analysis is an attempt to compute average rates of input and output of water to and from a region such as a river basin or collection of river basins. The inputs of water to the region are precipitation from the atmosphere and imports from other regions. The outputs are natural evapotranspiration, exports to other regions, consumptive use, and the outflow from the region to another region or the ocean. Consumptive use is the difference between the actual amount of water lost to the atmosphere by evapotranspiration, evaporation from cooling systems, and reservoir evaporation, minus the natural evapotranspiration, or that which would occur in the absence of development. There is one additional component of the budget, change in storage. The only aspect of storage change of any practical significance is decrease in ground-water storage (ground-water mining).

There are several reasonable ways to define "water supply" in a water-budget analysis. One of these might be called "natural supply," given as precipitation minus natural evapotranspiration. However, evapotranspiration is usually estimated as a residual,

TABLE 6. MAJOR INTERBASIN TRANSFERS IN THE UNITED STATES PRIOR TO 1982
(adapted from Petsch, 1985; and Mooty and Jeffcoat, 1986)

Name	State of Origin	Contributing Basin	Receiving Basin	Average Annual Flow (m3/s)	Operation Date	Owner
Delaware Aqueduct	New York	Upper Hudson	Lower Hudson	36	1951	City of New York
Diversion Canal	New York	Susquehanna	SE Lake Ontario	25	1931	New York State Electric and Gas Corp.
Chicago Sanitary and Ship Canal	Illinois	NE Lake Michigan	Upper Illinois	88	1900	Metropolitan Sanitary District
Sutherland Canal	Nebraska	North Platte	South Platte	25	1935	Nebraska Public Power District
Loup River Power Canal	Nebraska	Loup	Platte	47	1936	Loup River Public Power District
Colorado River Aqueduct	California	Lower Colorado	Santa Ana	36	1939	Metropolitan Water District of Southern California
All American Canal	California	Lower Colorado	S. Mohave-Salton Sea	131	1940	Imperial Irrigation District
Lewiston-Whiskeytown Tunnel	California	Klamath-North California coastal	Sacramento	39	1963	U.S. Bureau of Reclamation
Friant-Kern Tunnel	California	San Joaquin	Tulare-Buena Vista Lakes	47	1949	U.S. Bureau of Reclamation

*Averaged over the period 1973-1982.

the difference between rainfall and outflow, if no imports, exports, ground-water depletion, or consumptive use are assumed. In practice, natural supply is calculated as current outflow of a basin plus exports and consumptive use, minus imports and ground-water depletions. The current outflow is largely measurable, except for the ground-water components of outflow. However, the period over which outflow should be averaged is not clearly defined. A short period, for example 5 years, may be a poor estimate of the mean, but a long period, say 20 years, may be subject to considerable man-induced trend. Thus, these numbers are estimated with some subjectivity even when the original measurements are highly accurate.

Imports and exports are typically well known. Consumption estimates are subject to substantial error because the agricultural component, which dominates consumption, is based on a difference between two unmeasured evapotranspiration values (with and without the land-use change). The consumption by irrigation is typically estimated as a product of irrigated acreage and ideal water-consumption rates for the given crop. The fact that the irrigation may be suboptimal (too little or too much) is largely ignored. Finally, ground-water mining rates are determined by regional ground-water studies, using declines in water table, aquifer-storage coefficients, and, where possible, an estimate of the amount of runoff capture occurring as a result of the water-level declines (see Bredehoeft and others, 1982). These rates of ground-water mining are typically uncertain (±25%).

Another measure of the resource is the "renewable supply," which is the current outflow plus consumption minus ground-water mining. It is the supply that is potentially available on a sustained basis. It assumes that the present pattern of imports and exports will persist.

Compilations of measures of supply and consumption for the United States are published in the U.S. Geological Survey National Water Summary (U.S. Geological Survey, 1984, p. 23–30) and for Canada in Currents of Change (Pearse and others, 1985, p. 23–47). These compilations reveal that, for most of these regions, current outflow is between 90 and 100 percent of natural outflow, as illustrated in Figure 29 and Table 7. There are 11 regions where the current outflow is less than 90 percent of natural. Of these, the most extreme are the South Saskatchewan (78%), California (71%), Missouri (70%), Great Basin (69%), upper Colorado (68%), Churchill (48%), Rio Grande (41%), and the lower Colorado (8%). Only the lower Saskatchewan–Nelson region had a substantial increase in outflow (119%).

On the basis of average flows over broad regions (not flood or low flows), the substantial effects of man in North America are

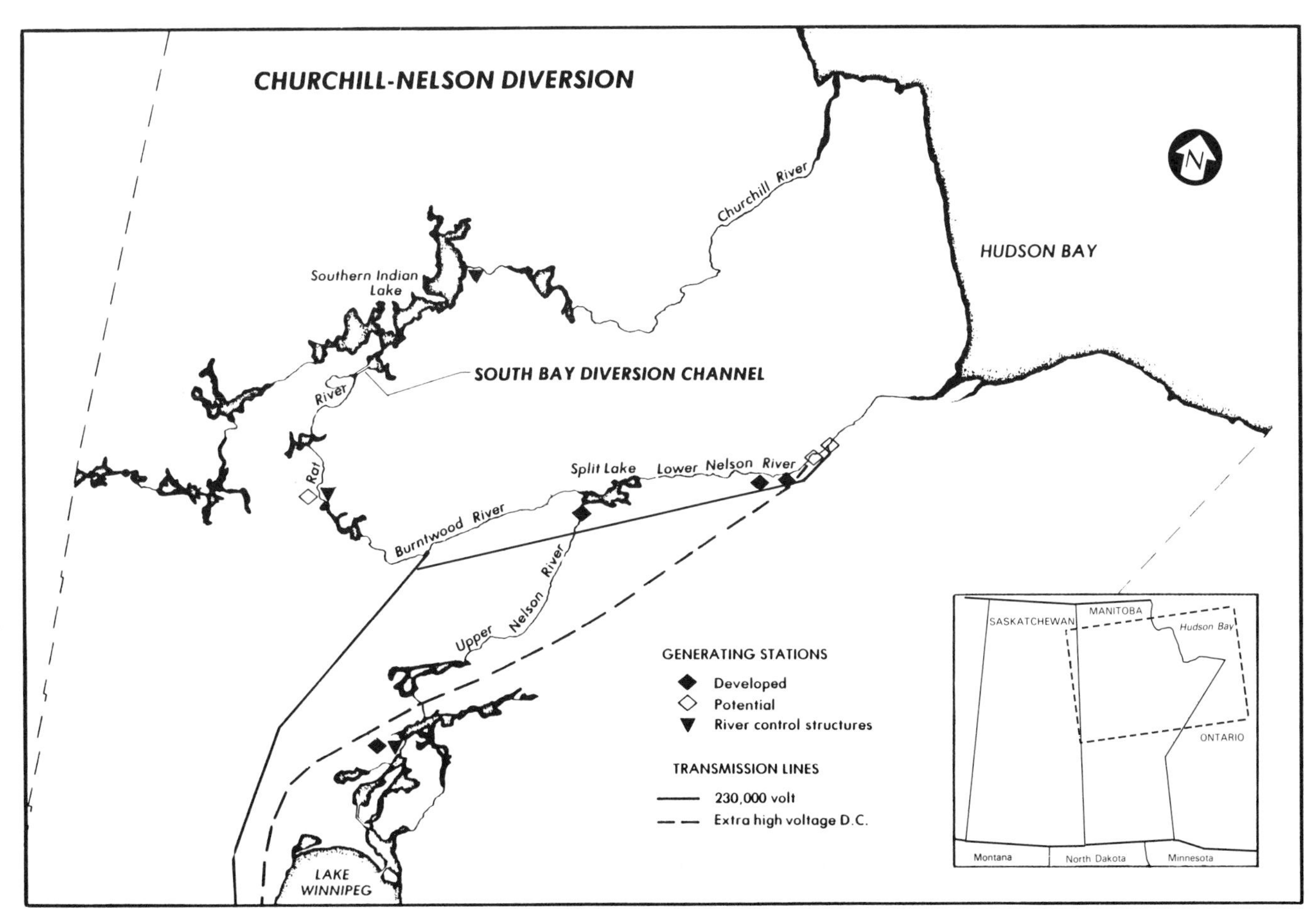

Figure 28. The Churchill-Nelson Diversion.

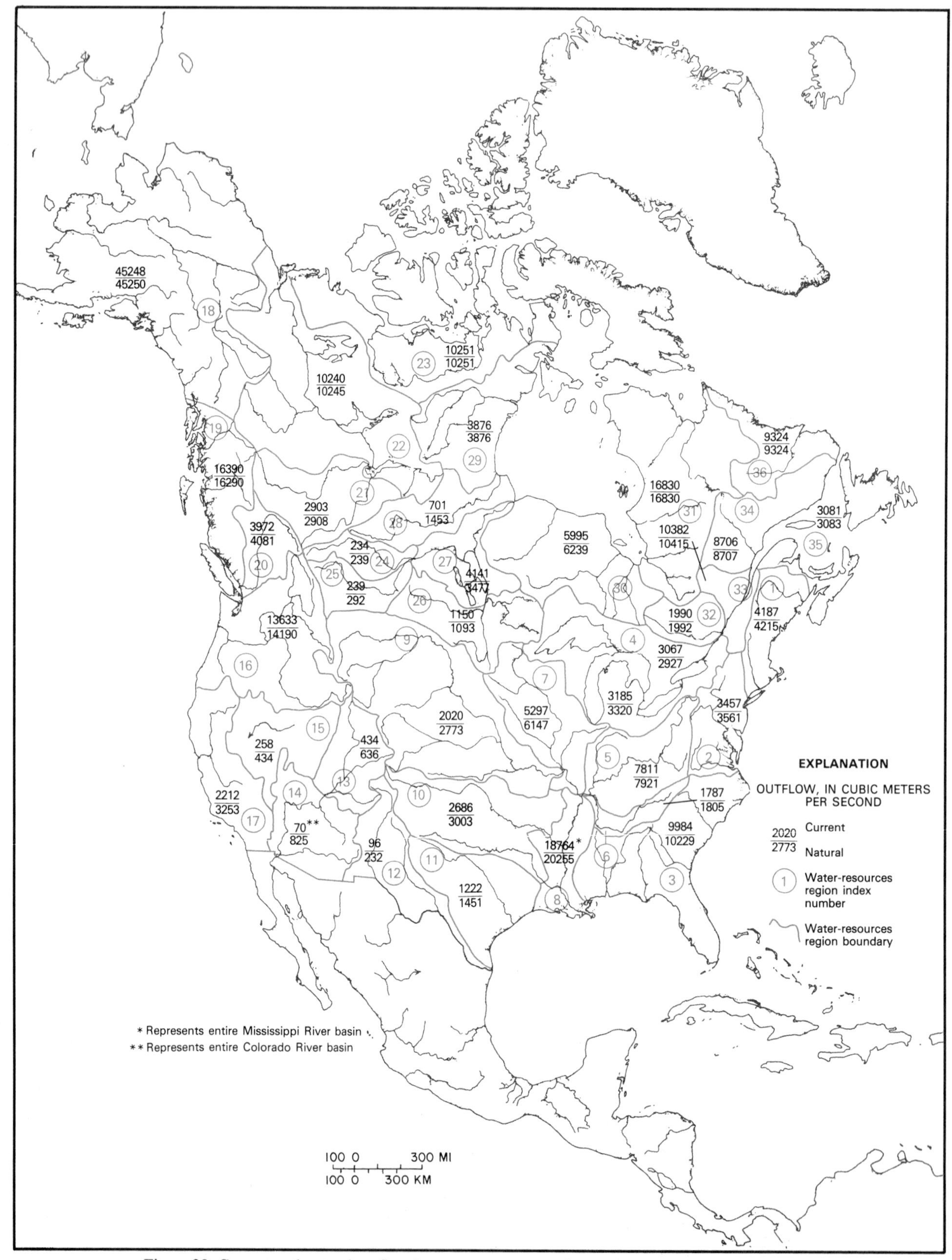

Figure 29. Current and natural outflow by water resources region. Index numbers refer to Table 7.

TABLE 7. CHANGE IN WATER BUDGET, UNITED STATES AND CANADA

Index	Region	Flows (m /s)						Current Natural (%)
		Current Outflow	Consumed	Import	Export	GW Mining	Natural Outflow	
1	N. Atl. U.S.+ St. John-St. Croix, Can.	4,187	28	0	0	0	4,215	99
2	M. Atlantic U.S.	3,457	79	0	25	0	3,561	97
3	S. Atl.-Gulf Coast, U.S.	9,984	245	0	0	0	10,229	98
4	Great Lakes, U.S. (inflows)	3,185	70	25	90	0	3,320	96
4	Great Lakes, Canada (inflows)	3,067	18	158	0	0	2,927	105
5	Ohio U.S.	7,811	110	0	0	0	7,921	99
6	Tennessee U.S.	1,787	18	0	0	0	1,805	99
7	Upper Mississippi U.S.	5,297	938	90	2	0	6,147	86
8	Mississippi U.S.	18,764	1,854	112	3	254	20,255	93
9	Missouri U.S. + Missouri Canada	2,020	847	0	2	96	2,773	73
10	Arkansas-White-Red U.S.	2,686	482	7	0	158	3,003	89
11	Texas Gulf U.S.	1,222	364	0	1	136	1,451	84
12	Rio Grande U.S.	96	140	4	0	0	232	41
13	Upper Colorado U.S.	434	175	1	28	0	636	68
14	Colorado U.S.	70	473	1	375	92	825	8
15	Great Basin U.S.	258	180	4	0	0	434	59
16	Pac. NW U.S. + Okan. + Colum.Can.	13,633	558	1	0	0	14,190	96
17	California U.S.	2,212	1,117	16	1	61	3,253	68
18	Alaska, U.S. + Yukon, Canada	45,248	2	0	0	0	45,250	100
19	Pacific Coastal, Canada	16,390	2	102	0	0	16,290	101
20	Fraser Lwr. Mainland, Canada	3,972	7	0	102	0	4,081	97
21	Peace-Athabasca, Canada	2,903	5	0	0	0	2,908	100
22	Mackenzie, Canada	10,240	5	0	0	0	10,245	100
23	Arctic Coast Isl., Canada	10,251	0	0	0	0	10,251	100
24	N. Saskatchewan, Canada	234	5	0	0	0	239	98
25	S. Saskatchewan, Canada	239	53	0	0	0	292	82
26	Assinboine-Red-Winnipeg, Canada Souris, Rainy Red, U.S.	1,150	29	86	0	0	1,093	105
27	Lower Saskatchewan-Nelson, Canada	4,141	88	752	0	0	3,477	119
28	Churchill, Canada	701	0	0	752	0	1,453	48
29	Keewatin, Canada	3,876	0	0	0	0	3,876	100
30	Northern Ontario, Canada	5,995	0	0	0	244	6,239	96
31	Northern Quebec, Canada	16,830	0	0	0	0	16,830	100
32	Ottawa, Canada	1,990	2	0	0	0	1,992	100
33	St. Lawrence, Canada	10,382	101	158	90	0	10,415	100
34	North Shore-Gaspe', Canada	8,706	1	0	0	0	8,707	100
35	Maritime Coastal, Canada	3,081	2	0	0	0	3,083	100
36	Newfoundland-Labrador, Canada	9,324	0	0	0	0	9,324	94
	Puerto Rico, U.S.	210	13	0	0	0	233	94

concentrated in the arid to semiarid plains and intermountain basins of the West. One can find substantial effects of man on the water budget in areas of intensive water use throughout the United States and Canada, but these shortages are typically alleviated by interbasin transfers. In the arid to semiarid regions, development of water is beginning to, or already has, reached levels where additional increments of supply would require very costly long-distance import schemes.

In summary, progress in understanding the full array of impacts induced by human intervention into hydrologic systems is uneven. There has been major progress made in documenting the many effects of dams, urbanization, and consumption for agriculture and industry. More limited progress has been made in interpreting the effects of interbasin diversions, land drainage, surface mining, and land use, and anthropogenic changes to water chemistry, aquatic life, and vegetation. The broader question of the comprehensive biophysical and socioeconomic effects of major projects has seldom been studied and never analyzed from the perspective of cumulative long-term consequences of all water-related management effects within major hydrologic systems. The stability of major river-basin systems over the coming decades in terms of sediment movement, water yield, flood-control capacity, climate change, and the risk of human occupancy is worthy of detailed research.

REFERENCES CITED

Allen, P. M., and Narramore, R., 1985, Bedrock controls on stream channel enlargement with urbanization, north-central Texas: Water Resources Bulletin, v. 21, p. 1037–1048.

Alley, W. M., and Veenhuis, J. E., 1983, Effective impervious area in runoff modeling: American Society of Civil Engineers Journal of Hydraulic Engineering, v. 109, p. 313–319.

Andrews, E. D., 1986, Downstream effects of Flaming Gorge Reservoir on the Green River: Geological Society of America Bulletin, v. 97, p. 1012–1023.

Belt, C. B., Jr., 1975, The 1973 flood and man's constriction of the Mississippi River: Science, v. 189, no. 4204, p. 681–684.

Blaney, H. F., Ewing, P. A., Morin, K. V., and Criddle, W. P., 1942, Consumptive water use and requirements: National Resources Planning Board, Pecos River Joint Investigation Reports of Participating Agencies, p. 170–200.

Bredehoeft, J. D., Papadopulos, S. S., and Cooper, H. H., Jr., 1982, Groundwater; The water-budget myth, *in* Scientific basis of water-resource management: Washington, D. C., National Academy Press, p. 51–57.

Bryan, B. A., and Hewlett, J. D., 1981, Effects of surface mining on storm flow and peak flow from six small basins in eastern Kentucky: Water Resources Bulletin, V. 17, p. 290–299.

Bull, W. B., and Scott, K. M., 1974, Impact of mining gravel from urban stream beds in the southwestern United States: Geology, v. 2, no. 4, p. 171–174.

Burkham, D. E., 1976a, Hydraulic effects of changes in bottom-land vegetation on three major floods, Gila River in southeastern Arizona: U.S. Geological Survey Professional Paper 655-J, 14 p.

—— , 1976b, Effects of changes in an alluvial channel on the timing, magnitude, and transformation of flood waves, southeastern Arizona: U.S. Geological Survey Professional Paper 655-K, 25 p.

Canadian National Committee, 1984, Register of dams in Canada: Published under the auspices of the Canadian National Committe, International Commission on Large dams, 104 p.

Cleveland, W. S., 1979, Robustly weighted regression and smoothing scatter plots: Journal of American Statistical Association, no. 74, p. 829–836.

Cleveland, W. S., Dunn, D. M., and Terpenning, I. J., 1979, SABL; A resistant seasonal adjustment procedure with graphical methods for interpretation and diagnosis, *in* Zellner, A., ed., Seasonal analysis of economic time series: Washington, D.C., U.S. Government Printing Office, 485 p.

Culler, R. C., Hanson, R. L., Myride, R. M., Turner, R. M., and Kepple, F. P., 1982, Evapotranspiration before and after clearing phreatophytes, Gilla River flood plain, Graham County, Arizona: U.S. Geological Survey Professional Paper 655-p, 67 p.

Day, J. C., 1985, Canadian interbasin diversions: Ontario, Environment Canada, Inquiry on Federal Water Policy Research Paper no. 6, 111 p.

Dunne, T., 1982, Models of runoff processes and their significance, *in* Scientific basis of water-resource management: Washington, D.C., National Academy Press, p. 17–30.

Dunne, T., and Leopold, L. B., 1978, Water in environmental planning: San Francisco, California, W. H. Freeman and Company, 818 p.

Eschner, T. R., Hadley, R. F., and Crawley, K. D., 1983, Hydrologic and morphologic changes in channels of the Platte River basin in Colorado, Wyoming, and Nebraska; A historical perspective: U.S. Geological Survey Professional Paper 1277, p. A1–A39.

Freeney, G. B., 1981, Managing conflicts on the lower Colorado River system, *in* Joebes, G. H., and Shepard, A. A., eds., Proceedings of the National Workshop on reservoir Systems Operations: American Society of Civil Engineers, p. 450–485.

Galay, V. J., 1983, Causes of river bed degradation: Water Resources Research, v.19, no. 5, p. 1057–1090.

Hammer, T. R., 1971, Urbanization and stream channel enlargement: Philadelphia, University of Pennsylvania Regional Science Research Institute, 20 p.

Hardison, C. H., 1972, Potential United States water-supply development: Proceedings of the American Society of Civil Engineers, Journal of Irrigation and Drainage Division, paper 9214, p. 479–492.

Hecky, R. E., Newbury, R. W., Bodaly, R. A., Patalas, K., and Rosenberg, D. M., 1984, Environmental impact prediction and assessment; The Southern Indian Lake experience: Canadian Journal of Fisheries and Aquatic Sciences, v. 41, no. 4, p. 720–732.

Hill, A. R., 1976, The environmental impacts of agricultural land drainage: Journal of Environmental Management, v. 4, p. 251–274.

Hirsch, R. M., 1977, The interaction of channel size and flood discharge for basins undergoing urbanization, *in* Proceedings of a Symposium on the Effects and Organization and Indstrialization on the Hydrologic Regime and Water Quality: Amsterdam, International Association for Hydrologic Sciences Publication 123, p. 83–92.

—— , 1980, Pond design alternatives and stormwater management policy, *in* Proceedings, Stormwater Hydrology Symposium, March, 1980: University Park, Pennsylvania State University, 30 p.

Hirsch, R. M., Scott, A. G., and Wyant, T., 1982, Investigation of trends in flooding in the Tug Fork basin of Kentucky, Virginia, and West Virginia: U.S. Geological Survey Water-Supply Paper 2203, 37 p.

Jackson, K., ed., 1984, Toward a fish habitat decision on the Kemano Completion Project; A discussion paper: Vancouver, British Columbia, Department of Fisheries and Oceans, 79 p.

Kellerhals, R., 1982, Effects of river regulation on channel stability in gravel bed rivers, *in* Hay, R. D., Bathurst, J. C., and Thorne, C. R., eds., Gravel bed rivers: New York, John Wiley and Sons, Ltd., p. 685–715.

Kessel-Taylor, I., 1984, The application of the Canada Land Data System for quantitative analysis of land use dynamics on wetlands for twenty-three urban-centered regions in Canada: Ottawa, Environment Canada, Lands Directorate, Canada Land Data System Report R003200, 143 p.

Lane, E. W., 1955, The importance of fluvial morphology in hydraulic engineering: American Society of Civil Engineers Proceedings, v. 81, no. 795, 17 p.

Langbein, W. B., 1959, Water yield and reservoir storage in the United States: U.S. Geological Survey Circular 409, 5 p.

—— , 1982, Dams, reservoirs, and withdrawals for water supply: Historic trends: U.S. Geological Survey Open-File Report 82–256, 9 p.

Leopold, L. B., 1968, Hydrology for urban land planning; A guide-book on the hydrologic effects of urban land use: U.S. Geological Survey Circular 554, 18 p.

—— , 1973, River channel change with time; An example: Geological Society of America Bulletin, v. 84, p. 1845–1860.

Leopold, L. B., and Maddock, T., Jr., 1953, The hydraulic geometry of stream channels and some physiographic implications: U.S. Geological Survey Professional Paper 252, 57 p.

—— , 1954, The flood control controversy: New York, Ronald Press, 275 p.

Mackin, J. H., 1948, Concept of the graded river: Geological Society of America Bulletin, v. 59, p. 463–512.

Matalas, N. C., Landwehr, J. M., and Wolman, M. G., 1982, Prediction in water management, *in* Scientific basis of water-resource management: Washington, D.C., National Academy Press, p. 118–127.

McCuen, R. H., 1974, A regional approach to urban storm water detention: Geophysical Research Letters, v. 1, p. 321–322.

Miller, J. E., and Frink, D. L., 1984, Changes in flood response of the Red River of the North basin, North Dakota–Minnesota: U.S. Geological Survey Water Supply Paper 2243, 103 p.

Mooty, W. S., and Jeffcoat, H. H., 1986, Inventory of interbasin transfers of water in the eastern United States: U.S. Geological Survey Open-File Report 86–148, 47 p.

Mosteller, G. H., and Tukey, J. W., 1977, Data analysis and regression: Reading, Massachusetts, Addison-Wesley, 588 p.

National Hydrology Research Institute, 1982, NHRI research project; Agricultural land drainage: Ottawa, Environment Canada, 38 p.

Newbury, R. W., and McCullough, G. K., 1984, Shoreline erosion and restabilization in the Southern Indian Lake Reservoir: Canadian Journal of Fishery and Aquatic Science, v. 41, no. 4, p. 558–565.

Nordin, C. F., and Meade, R. H., 1974, Missouri River stage trends [abs.]: Paper presented at American Society of Civil Engineers Hydraulics Division meeting, August 1974, Knoxville, Tennessee.

Peace-Athabasca Delta Project group, 1973, The Peace-Athabasca delta project technical report: Edmonton, Environmental Ministers of Alberta and Saskatchewan, 176 p.

Pearse, P. H., Bertrand, F., and MacLaren, J. W., 1985, Currents of change; final report: Ottawa, Inquiry on Federal Water Policy, 222 p.

Petsch, H. E., Jr., 1985, Inventory of interbasin transfers of water in the western United States: U.S. Geological Survey Open-File Report 85–166, 47 p.

Ponce, S. L., and Meiman, J. R., 1983, Water yield augmentation through forest and range management; Issues for the future: Water Resources Bulletin, v.19, p. 415–419.

Quinn, F., 1981, Water transfers—Canadian style: Canadian Water Resources Journal, v. 6, no. 1, p. 64–76.

——, 1983, Changing patterns of Canadian water transfer, *in* Gentilcore, R. L., ed., China in Canada; A dialogue on resources and development: Hamilton, Canada, McMaster University, p. 126–135.

Robbins, C. H., and Simon, A., 1983, Man-induced channel adjustment in Tennessee streams: U.S. Geological Survey Water-Resources Investigation Report 82–4098, 129 p.

Robinson, A. M., 1976, Effects of urbanization on stream channel morphology, *in* Barfield, B. J., National symposium on urban hydrology, hydraulics, and sediment control: Lexington, University of Kentucky, College of Engineering Publication III, p. 115–127.

Sauer, V. B., Thomas, W. O., Jr., Strickler, V. A., and Wilson, K. V., 1983, Flood characteristics of urban watersheds in the United States: U.S. Geological Survey Water Supply Paper 2207, 63 p.

Simon, A., and Hupp, C. R., 1986, Channel evolution in modified Tennessee channels, *in* Proceedings of the Fourth Interagency Sedimentation Conference: Subcommittee on Sedimentation of the Interagency Advisory Committee on Water Data, v. 2, p. 5-71–5-82.

Tiner, R. W., Jr., 1984, Wetlands of the United States; Current status and recent trends: U.S. Department of the Interior, Fish and Wildlife Service, 59 p.

Toebes, G. H., and Shepard, A. A., eds., 1981, Proceedings of the National Workshop on Reservoir Systems Operation: New York, American Society of Civil Engineers, 595 p.

Turner, R. M., 1974, Quantitative and historical evidence of vegetation along the upper Gila River, Arizona: U.S. Geological Survey Professional Paper 655-H, 20 p.

U.S. Geological Survey, 1984, National water summary 1983; Hydrologic events and issues: U.S. Geological Survey Water-Supply Paper 2250, 243 p.

Water and Environment Consultants, Inc., 1976a, Potamology investigation; A study of the shift in the stage-discharge relationship of the Missouri River at Sioux City, Iowa: Fort Collins, Colorado, Water and Environment Consultants, Inc., 42 p.

——, 1976b, Missouri River–Gavins Point to Omaha, Nebraska; Historical records research: Fort Collins, Colorado, Water and Environment Consultants, Inc., 52 p.

Williams, G. P., 1978, The case of the shrinking channels; The North Platte and Platte rivers in Nebraska: U.S. Geological Survey Circular 781, 48 p.

Williams, G. P., and Wolman, M. G., 1984, Effects of dams and reservoirs on surface-water hydrology; Changes in rivers downstream from dams: U.S. Geological Survey Professional Paper 1286, 83 p.

Wisler, C. O., and Brater, E. F., 1965, Hydrology: New York, John Wiley and Sons, p. 55–56.

Wolman, M. G., 1967, A cycle of sedimentation and erosion in urban river channels: Geografiska Annaler, v. 49A, p. 385–395.

Manuscript Accepted by the Society May 14, 1987

Printed in U.S.A.

Index

[Italic page numbers indicate major references]

Absaroka Mountains, Wyoming, 197
Acari, 241
Achnanthes, 235
acid rain, 204
Acipenser oxyrhynchus, 246
acrisols, *69*
Acrocheilus alutaceus, 245
acrotelm, 161, 162, 176
Adirondack Mountains, 170
adsorption, 203, 207
aerosols, 40, 201
Aeshna, 249
Aeshnidae, 249
Africa, northern, 47
Agardhiella, 239
Agrion spp., 249
Ainsworth Irrigation Unit, 317
air masses, 23
airstreams, 43, *44*
Alabama River, 223
Alaska Range, *145*, 262
Alaska, 25, 58, 69, 70, 86, 87, 88, 104, *105*, *106*, 141, 144, 146, 152, *153*, 156, 166, 167, 244, 245, 249, 256, 258, 274
  central, 162
  coast, 144
  Highland and Basin region, *58*
  northern, 161
  southeastern, 146
Albany River, 243
albedo, *149*
Alberta, 5, *24*, 27, 88, *92*, 94, 104, 241, 243, 244, 245, 284, *333*, 336, 353
Aleutian low-pressure area, 144
algae, 198, 200, 233, *234*
Allegheny Plateau, 347
Allocapnia, 249
alluvial channels, 101, 111, *117*, 340
alluvial fans, 102, 111, 284, 286
alluvial rivers, 281, 282
alluvial sediments, 58
alluvial streams, 205
alluvium, 102, 309, 317, 321
Alsek River, 115
Alsek Valley, 115
Altamaha River basin, Georgia, *271*
alteration, *195*
Amazon River, 208, 223
Ambrosia pollen, 274
Amphipoda, *251*
Analatris eximia, 248
Anguillidae, 246
anions, 185
Anisoptera, 249
anisotropy, 163
Anodonta
  cataracta, 251
  nuttaliana, 250
anomalies
  pressure, 36
  rainfall, *31*
  streamflow, 39
Anquilla rostrata, 246
Antarctic Ice Sheet, 131, 132
anticyclonic circulation, 23, 24
Api, 133
Appalachian Highlands region, *56*
Appalachian Mountains, 69, 74, 175, 226, 249, 299
  northern, 160
Appalachian system, 42
Appalachicola, 223
aquatic ecosystem, 183
aquatic environmental systems, *205*, 207
aquifers, 124
  Floridan, 123, *183*
  limestone, 90, 123
  sand, 183
  storage, 89
Arctic, *134*, *163*
  airstreams, 44
  drainage areas, 250
  lakes, *165*
Arctic Archipelago, 244
Arctic Lowland and Coastal Plain region, *56*
Arctic Ocean drainage basin, *241*
aridisols, 59
Arizona, 74, 77, 84, 103, 104, 108, 115, 140, 227, 267, *304*, 307, 333
Arkansas, 160, 180
Arkansas River, Colorado, 109
Arkansas River basin, 226
Arrow Lakes, British Columbia, 160
arroyo formation, 309
arthropods, 239
Ascophyllum, 239
Assiniboine River, 243
  diversion, 110
Astacidae, 251
Atchafalaya River, 109
  diversion, 225, 324
Athabasca Lake, 41
Athabasca River, 148
Atlantic coast, 27, 58, 70, 116, 239, 274
Atlantic–Gulf of Mexico Coastal Plains region, 42, *58*
Atlantic Ocean
  drainage basin, *243*, 245
  western, 27
Atlantic provinces, 163
atmosphere, *11*, 40, 46, *193*
  circulation, *17*, 18, *19*, *34*, *39*
Attawapiskat River, 243
aufeis, 86, *104*
Aulodrilus americanus, 250
Autocha, 250
avalanches
  debris, 112, 160, 312
  rock, 160
  snow, 134
Axel Heiberg Island, 148
Azolla, 233

backscattering, 40
badlands, 264
Baetis, 248
Baetisca, 248
Baffin Bay, 148
Baffin Island, 41, 135, 244, 249, 250, 255, 263
Baker Lake, 243
Baltimore, Maryland, *266*
Balzac gaging station, 107
Banks Island, 244
Barbara Creek, 88
barnacles, 239
barrier effect, *92*
barrier islands, 239
Barrow, Alaska, 166
bars
  linguoid, *317*
  riverine, *160*
basalts, 91, 102
base-flow recession, 89
Basiaeschna, 249
Basin and Range province, *58*, 160, 180, 199
basins, 77
  closed, 92, 93, *116*
  drainage, 76, 92, *169*, 170, 180, 197, *241*, *307*, *321*
  shape, *77*, *103*
  slope, 77, 92
  urban, 343
  water loss, 4
Bassano Reservoir, 267
Batracospermum, 237
Battle River, 94
Bay of Fundy, 243
beaches, raised, 167
Bear Creek, Colorado, 92
Bear River, Utah, 219, 222
Beaufort Sea, 241, 292
Beauharnois dam, 300
bed load, *255*
  discharge, *260*
  measurement, *259*
  sampling, 260
  sediments, *255*
  transport, *259*
  traps, 260
bedrock, 167, 170, 174, *175*, 178, 243, 281, 282, 284, 286, 311, 314, 338, 347, 354
  crystalline, *169*
beetles, *250*
Belize, 91
Bella Coola River, British Columbia, 270
Bennett Dam, 336
Bering Glacier, 145
Bering River, 105

Bering Sea, 25
Beringia, *244*, 249
Beringian Refugium, 250
bicarbonate, 185, *189*, 223, 224
Big Otter Creek, Ontario, 90
Big Slide area, Quesnel, 286
Big Spring, Missouri, 102
Bijou Creek basin, Colorado, 107
biota, 6, 199, 203, *205*, 226, *233*, *244*
  aquatic, 200, *233*
  Canada, *240*
  United States, *233*
Birch Dam, Montana, 106
Birchbank, British Columbia, 333
Black Butte River, 313
Black River, South Carolina, 216, 241
Blaney-Criddle equation, 348
Blephariceridae, 248, 250
Bloomington Creek, Idaho, 91
Blue Hole, Belize, 91
bogs, 69, 163, 167, *176*, 206, 243, 288
Boise, Idaho, 115
Boise River, Idaho, 87
Bonnet Carre spillway, 110, 324
Bonneville Flood, *115*
Boreal Forest and Tall-Grass Prairie region, *71*
Boreal Forest region, *71*
boreal forests, 69, *71*, 242, 248
bottomlands, wet, 167
Bowen ratios, 13
Bowron Lake, British Columbia, *172*
Bowron Lake basin, 172
Box Butte Reservoir, 316, 317
Boyeria, 249
Brazos River, Texas, *82*, 274
British Columbia, 71, 84, 105, 106, 133, 144, 146, 156, 159, 160, *172*, *174*, 243, 244, 245, 256, 257, 263, 270, 281, *282*, 333, 352
  central, *282*
  coast, 24
Broad River, Georgia, 263
Broadleaf Handwood Forest and Tall-Grass Prairie reigon, *74*
Brooks Range, 146
bryozoans, 239
budgets, soil-water, *63*
buffering, *203*
Burntwood River, 353
Bylot Island, 148

caddis flies, *238*, *249*
calcium, 185, 189
calderas, 160
California, 74, 102, *106*, 109, 111, *116*, *129*, 160, 200, 216, 256, 258, 266, *270*, *272*, *281*, *311*, 312, 329
  southern, *111*
California Coast Range, 256
California Steppe region, *74*
calving, iceberg, 152, *153*
Cambarellus, 251
Cambarinae, 251
Cambaroidinae, 251
Cambarus sp., 251
cambisols, *69*
Canada, 25, 87, 105, 148, 160, 161, *163*, *175*, 176, *233*, *240*, 245, *255*, *340*
  arctic, 41
  biota, *240*
  drainage basin, *241*
  eastern, 135
  northern, 69, 104
  northwestern, 69, 70
  southeastern, *134*
  southern, 121
  southwestern, 101
  western, 58
  wetlands, *345*
  See also specific provinces
Canada-U.S. border, 24, 26, 69, *111*
Canadian archipelago, 56
Canadian Arctic, *148*, 240, 249
Canadian Cordillera, 174
Canadian Experimental Lakes Area, 169
Canadian plains, 69
Canadian Prairies, *133*, 249
Canadian Rockies, 148, 282
Canadian Shield, *41*, 44, *56*, 92, 93, 159, *163*, *169*, 223, 243, 274, 281, 298, 299
Cane Branch, Kentucky, *266*
Canoochee River, Georgia, 84
Cape Bathurst, 161
capillary action, 64
Capnia neartica, 249
carbon, 189
carbon dioxide, *46*, 48, 132, 189, 193, *198*, *204*, 223
carbonates, 160, 196, 221
Caribbean Islands, 121
Cariboo River, 282
Caribou Hills, 292
Cascades, 58, 102, 106, *133*, 195, 197
Catastomidae, 246
cations, 185, 203, 207
  solute, 196
Catostomidae, *245*
Catostomus, spp., 245
catotelm, 161
caves, 90
cell dialysis, 205, 206
cenotes, 183
Central America, 121
Central Lowlands region, *58*
Central North Pacific (CNP) pressure center, *36*
Centrarchidae, 245
centrifugation, 206
Ceramium, 239
Ceratopogonidae, 241, 243
Chaco River, 306
Champlain Sea, 299
channelization, *349*
channels, 41, *107*, 200, 216, *237*, 239, 255, 272, 292, 311, *313*, *318*, 322, *335*, 342, *344*, *350*
  active, 350
  alluvial, 101, 111, *117*, 340
  arroyo, 310
  braided, 237, *317*
  changes, *307*, 336, 338, *349*
  critical depth, 111
  cutoff, 323
  degradation, 336
  drainage, 102
  englacial drainage, 156
  gravel-bed, 352
  highwater, 323
  inactive, 350
  network, 292
  patterns, 318
  realignment, 115
  recovery, *313*
  river, 1, 216, 233, *234*, 255, 271, 281, 287, 290, 291, 307, *316*
  root, 70
  scour, 309
  sedimentation, 347
  short-surface, 91
  slope, 112
  stability, 324
  storage, *107*
  stream, 4, 58, 66, 86, 274, 271, 311, *349*
  urban, *344*
Charlesvoix region, 301
Chase Lake, North Dakota, 178
chemical processes, *202*
chernozems, *69*
Cherry Creek, Colorado, 92
Chesapeake Bay, 239
Cheslatta River discharge, 352
Chihuahua, 92, 247
Chilcotin River, 284
Chinle Formation, 306
Chinle River, 306
Chironomidae, 236, 239, 241, 250
chironomids, 235, 238, 242, 250
Chironominae, 241, 250
chloride, 189, 193, *223*, *225*, *226*
chlorofluorocarbons, 48
Chromagrion, 249
Chrosomus eos, 245
Chugach Mountains, 145
Churchill Falls project, 351
Churchill River, Labrador, 223, 243, 245, 289, 353
  diversion, 291
Churchill-Nelson Diversion, *353*
Chyranda centralis, 249
Cinygma, 248
circulation
  anticyclonic, 23, 24
  atmospheric, *17*, 18, *19*, *34*, *39*
  cyclonic, 23
  mid-tropospheric, 19
  patterns, *17*, *34*
Cisco, Utah, 228
Cladocera, 238
Cladophora glomerata, 238
clams, 238
clay, 62, 196, 237
Clear Lake, California, 116
Clearwater River, Idaho, 94, 256
climate, *4*, *17*, *22*, *31*, *45*, *84*, *131*, 163, 170, *177*, *194*, 243, *261*, 282, *311*

change, *45*, 47, 317
climatology, storms, *22*
clouds, 40
cumuliform, 43
seeding, 134
stratocumulus, 43
CNP. *See* Central North Pacific pressure center
Coahuila, 92
Coast Range, *146*, 284, 311, 313
Coastal Plain, 103, 237, 258
coastal ranges, 58
coastal zones, river sediment, *274*
Cocconesi, 235
cofferdams, 290
Coleoptera, 236, 241, *250*
colonization rates, 252
Colorado, *24*, 92, 102, 107, 109, 111, 148, 174, 223, 227, 338
Colorado Plateau, 58, 309
Colorado River, 223, 228, *267*, 269, 274, 281, *282*, *304*, 336
flow, *307*, *331*
peak flows, *306*
reservoir storage, *307*, *340*
sediment, *304*
water discharge, *304*
Colorado River basin, *115*
upper, 140
Colorado River channel, 309
Colorado River compact, 304, 306
Colorado River delta, 274
Columbia Forest region, *71*
Columbia Glacier, Alaska, *153*
Columbia Ice Fields, 93, 127
Columbia Plateau, 58
Columbia River, 105, 115, 127, 148, 152, 212, *227*, 243, 269, 335
monthly flow series, 333
Columbia River basin, 101, 129, 263, 265
combustion, fossil-fuel, 132
communities
aquatic, *233*, *235*
types, *234*
Compsopogen, 237
concentrations
sediment, *216*
suspended-sediment, *256*, *276*, 286, 304, *306*, *310*
conglomerates, 282
coniferous forests, 69, 169
contaminants, 255, 302
convection, 13, *43*
Coolidge Dam reservoir, 348
Coon Creek, Wisconsin, *272*
copepods, 238
copper, toxicity, 207
Copper Lake, 174
Copper River, *262*
Cordillera, *44*, *58*, 144
Cordulegasteridae, 248
Corduliidae, 249
Coregoninae, 245
Coregonus spp., 245
Coriolis effect, 27
Corixidae, 249
Cornwall, Ontario, 172
Cornwallis Island, 165
Corra-Linn dam, 333
Corydalidae, 249
Coshocton, Ohio, 77
Cottidae, *245*
Cottonwood Lake area, North Dakota, *185*
Cottus spp., 245, 246
Couesius plumbeus, 245
Cowlitz River channel, 105
crabs, 235, 238, 239
Crangonyx, 251
Crater Lake, Oregon, 160
Crescent Lake National Wildlife Refuge, 178
Crooked River, 284
cropland, *76*, *264*
sediment yield, *265*
crops, replacement, 348
crustaceans, 234, 235
cryptogams, 233
Cryptolabis, 250
Crystal Springs area, *177*
crystals, 291
Cub River, Idaho, 91
Culicidae, 241
curves, base-flow recession, 123
cumulative frequency, 121
duration, *121*, 125
flood-frequency, 99
flow-duration, 329, 339
frequency, 82, 123, 124
low-flow frequency, 122
rating, 8
sediment rating, 257
sediment-transport, 257
cyclogenesis, *24*, *27*, 45, 133
cyclones
circulation, 23
development, 23, 24
extratropical, *23*
frequency, oceanic, 27
movement, patterns, 24
surface, 24
tropical, *27*
waves, *23*
Cyclopoida, 241
Cyprinidae, *245*, 246
Cyprinodon, 247

dams, *238*, *287*, *311*, 335, *337*, *339*
hanging ice, *291*
landslide, *160*
low-flow, *335*
sediment loads, *266*
Davis Strait, 25
debris
dead organic, 199
fan, 309
hydraulic-mining, *271*
lobes, 112
torrents, *111*, *112*
Decapoda, *251*
deciduous forests, 69, 169, 240, 243
deglaciation, 167
degradation
downstream, 335, 349
river-bed, *325*, 349, 352
upstream, 335, 349
Delaware Bay, 239
Delaware River, *128*
deltas, *274*
arctic, *292*
estuary, *292*
plain, 292
submergence, 292
Denali fault zone, 145
denudation, chemical, 197
deposits, 108, *159*
acid, 134
debris-torrent, *112*
flood plain, *217*
glacial, 169, *174*
slack-water, *115*, 309
depressions
fault-block, 180
glacial deposits, 169
topographic, I159
Dero digitata, 250
Des Cedres dam, 300
Desert Shrubs and Grasslands region, *74*
desmids, 237
detritus, 233, *234*, 239
Deuterophlebiidae, 248, 250
Devon Island, 148, 166
Dezadeash Valley, 115
Diamesinae, 242, 250
diatoms, 226, 227, 233, 234, 236, 239
Didymops, 249
Diptera, 234, 238, 239, 241, *250*
Dirty Devil River, 306
discharge
bedload, *260*
flood, 103, *106*, 112, *117*, 156, 313
ground-water, *66*, *122*, *160*, *165*, 167, 177
high, 324
long-term sediment, *257*, 285
mean annual, 317
peak, 99, 100, 101, 102, 103, 106, 107, *112*, *117*, 242, 317, *329*, 333, 338, 343, 344, 347, 348, 350
river, 200
river-sediment, *255*, *267*, 276
sediment, 8, *255*, *256*, *257*, *258*, *267*, *274*, 276, *311*, 313, 317, 339
stream, 5, 9, 97, *216*, 243
suspended-sediment, 256, *257*, *258*, *311*, 313
water, *66*, *122*, *160*, 167, *177*, 244, 257, 258, 269, *270*, *276*, *304*, *306*, 307, 310
wetland, *166*, 167, 174
discontinuities, slope, *160*
Dismal Swamp, 160
Diura Bicaudata, 249
dolomite, 102, 264
downcutting, 325
dragonflies, *249*
drainage, *100*, *160*, 200
basin, 76, 92, *169*, 170, 172, 180, 197, *241*, *243*, 245, 263, 265, *307*, 317

  channels, 102
  density, *79*
  grain-parallel, 292
  impeded, 161
  internal, 1
  land, *345*
  natural, 344
  network, 342
  ocean, 1
  relationship to floods, *346*
  reversal, Frazer River, *284*
  urban, *342*, 344
  wetlands, 345
dredging, 110, 324
drift, *234*
  glacial, 58, 170, 174, *175*, 263
  wind, 134
droughts, 32, *121*, 317, *318*
  agricultural, 31, *126*
  atmospheric, *126*
  climatological, *31*, *126*
  Dust Bowl, 127
  history, *127*
  hydrologic, 31, *121*, *126*
  meteorological, 31, *126*
  multiyear, *127*
  1976/1977, *128*
  Northeast (1960s), *127*
  one-year, *127*
  patterns, *33*
  periodicities, 33
  variability, *33*
  water-management, *126*
dryness ratios, 13
Dubawnt Lake, *165*
Dubawnt River drainage system, 165
Dubiraphia, 250
Duncan reservoir, 333
dune fields, 160, 178
Durango, Mexico, 247
Dytiscidae, 250

earthflows, 285
earthquakes, 117, 160
East Channel, 294
East Coast, *24*, 133
East Fork River, Wyoming, 260
East Fork San Gabriel River, 188
East Pacific (EP) pattern, *22*
East Twin Lake, Ohio, 175
Eastern Hardwood-Conifer Forest region, *74*
Echimamish, 288
ecosystem, aquatic, 183
Eel River, California, 258, *281*, *311*
  basin, *311*, 312, 313, 314
  hydrology, *311*
  sediment yields, *311*
Ekwan River, Ontario, 169
El Chichon, Mexico, 201
El Dorado Canyon, Nevada, 108
El Niño–Southern Oscillation (ENSO) events, 36, 46
Elbert, Colorado, 102
elevation, snow accumulation, *149*
Ellesmere Island, 148
Elliptera, 250
Elmidae, 241, 248, 250
Enallagma, 249
energy, *11*, 18, 131, 162, 169, 234, 235
  balance, 12
  heat, 141
  hydroelectric, 351
  solar, 11
ENSO. *See* El Niño–Southern Oscillation events
EP. *See* East Pacific pattern
Ephemeroptera, 234, 241, *248*
Ephydridae, 241
equisetums, 233
Equisteals, 233
erosion, 6, 56, 102, 109, 116, *159*, 270, 337, 345, 354
  natural, 335
  solute, 195
  upland, *269*
  wave, 161
eruptions, 103, *105*, 115, 201
  Mount St. Helens, *105*, 201
Escalante River, 306, *309*
Esocidae, 246
estuaries, *238*, *274*
  barrier-island, *239*
  fjord-type, *239*
  open-mouth, *239*
Etheostoma spp., 245, 247
Eucrenon, *240*
evaporation, 4, 5, *13*, 42, 63, 70, 116, 125, 134, *162*, 165, 166, 169, *171*, 174, 175, 177, *182*, 340, 354
  mean annual, 15
evapotranspiration, *4*, *13*, 42, 63, 65, 69, 85, *88*, 92, 121, 124, 166, 174, *176*, 180, 227, *348*, 354
  mean annual, 15
  potential, 13
exchange
  cation, 203
  ion, 203
extraction techniques, 208

fans, alluvial, 102, 111, 284, 286
Fargo, North Dakota, 347
farming, crop-land, *264*
farmland, 69
faunas, interstitial, 240
feldspars, 196
fens, 163, 167, *176*, 288
Ferrissia, 236, 237
fertilizers, nitrogen, 226
field capacity, 17
filtration, 205, *206*, 224
fish, 237, *244*, *246*
Flaming Gorge Reservoir, *228*, *339*
Flathead River, British Columbia, 245
Flatrock Creek, Illinois, 122, 123
Flood Lake, British Columbia, 106
flooding
  lakes, *116*
  oceans, *116*
  rivers, 336
floods, 32, *97*, *107*, *180*, 200, *262*, 284, *302*, 306, *309*, *313*, 318, *324*, *329*, *333*, 338, *343*, 346, 348
  annual, *100*
  control systems, *109*
  dam-failure, *106*
  damage, 110
  discharges, 103, *106*, 112, *117*, 156, 313
  flows, 77, 79, 97, 101, 102. 106, *111*, 311, 329
  frequency, *99*, *309*, 343
  glacier outburst, *105*, 106, 115, *156*
  high annual, 99
  historic, *114*
  100-year, 344
  hurricane, 101
  ice-jam, *104*
  magnitudes, 99, *100*
  peak, *109*, 110, 117, 331, 333, 336, *343*
  plain, 103, 105, *178*, 200, 269, 270, 271, 272, 309, 317, 336, *348*
  pool, 331
  prehistoric, *114*
  rural, *344*
  series, *99*
  snowmelt, *101*, 133
  spring, 307
  stages, *117*, *113*
  10-year, 344
  travel times, *348*
  tropical cyclones, 28
  2-year, 344
  urban, 110, *343*, *344*
  variability, *100*
  waves, *348*
flora, vascular, 167
Florida, 58, 91, 121, 123, 160, 163, *183*
  southern, *239*
Florida Marsh and Swamp region, *75*
Floridan aquifer, 123, *183*
flows
  annual low, 121, 122
  annual maximum, 81
  annual mean, *81*, 85, 90
  changes, *329*
  classification, *84*
  continental variation, 83
  control, 109, *227*
  daily mean, 81
  debris, *111*, *156*
  decomposition, 331
  flood, 77, 79, 97, 101, 102, 106, *111*, 311, 329
  glacial, 149
  ground-water, 86, *163*, *177*
  high, 317
  human influence, *329*
  hydraulics, 65
  ice, 153, 160, 284
  increased, 352
  lava, 58
  low, 77, *121*, *125*, *126*, 176, 324, 331, *345*

mean, 1, 81
minimum, 89
monthly mean, 81, 83, 84, *93*
overland, 65
patterns, *234*
peak, 284, *306*, *343*
regimes, *83*, 121, 339
regulations, *307*, *331*
river, *200*
river discharge, 351
seasonal variability, *331*
sediment, 349
series, Columbia River, 333
sheet, *111*
spring, 169
subcritical, 111
subsurface, 163, 169
summer, 85, 90, 152
supercritical, *111*
surface water, 65, 67, 90
systems, ground water, *163*
tidal, 301
tributary, 103
variability, *83*, *331*
water, *143*, *240*, *255*, *267*, 349, 351
winter, 86, 336
zero, 89
fluctuations
climatic, 317
diurnal, 90
lake discharge, 167
water-level, *165*, 167, *172*, 178
water table, 176
fluoride, 189
fog drip, 84
food web, 233, *234*
forage, 69
forests
boreal, 69, 242, 248
coniferous, 69, 169
deciduous, 69, 169, 240, 243
mangrove, 239
rain, 71
forestland, *76*
Forked Deer River, South Fork, *349*
Formicidae, 241
Fort Randall Dam, 267
Fort Seward station, 313
Foxe Basin, *41*
Frazer River, British Columbia, 243, 256, 257, 281, *282*
hydrography, *282*
hydrology, *284*
northern drainage reversal, *284*
sediment transport, *284*
suspended-sediment concentration, 286
Frazer River basin, 284
Frazer River delta, 274
Frazer Valley, 243
Fremont River, 309
fronts, 23
polar, 23
frost, 133, 169
damage, 133
impermeable, 167
winter, 285
Fucus, 239
gages, precipitaion, 140
gaging stations, 107, 304, 306, 309, 310, 313, 317, 322, 339, 340, 347, 349
Gammarus spp., 251
Gandil River, 105
gases, greenhouse, 132
Gastropoda, 234
Gavins Point Dam, 267
GEMS/WATER. *See* Global Environmental Monitoring System–Water
geochemical processes, *202*
Georgia, 84, *106*, *263*, *271*
Gerridae, 236
Gila River, 103, *227*, *333*, *348*
Gila River basin, upper, 348
glacial drift, 58, 170, 174, *175*, 263, 346
glacial erosion, 270, 274
glacial flow, 149
glacial lakes, 298
glacial melt, 88, *132*, 172, 174, 243
glacial scour, 115, *159*, 169, 350
glacial terrain, 159, *174*, *176*, 178
glacial till, 66, 175, *176*, 287
glaciation, 58, *244*, *284*, 286, *298*
mountain, 56
glacier-fed streams, 127, 255
glaciers, 3, 41, 87, *88*, *105*, *132*, *144*, *148*, 159, *160*, 176, 195, *235*, *262*, 284, 298
advance, 152, *153*
alpine, 195, 270
annual runoff, 149
cirque, 146, *148*
distribution, i144
floods, *105*, 106, 115, *156*
headwater, 127
ice, *131*, *144*, 148, 151, 159, 255, 258, 284
icefield, 145
instabilities, *152*
mountain, 131, 146
polar, 148
recession, 152, *153*
runoff, *148*, *151*
subpolar, 146, 149
surge-type, 152, 153
temperate, 148
thermal regime, *148*
tidal, 146
tidewater, *153*
valley, 56, 145, 146, *148*, 174
vertical, 148
wastage, 132, 152
water resources, *149*
Glen Canyon Dam, *307*, *331*, 340
Glenn Creek, Alaska, 167
gleysols, *70*
Global Envirnomental Monitoring System–Water (GEMS/WATER), *210*, 222
global ice cover, 132
global temperatures, 45, *46*
global warming, *48*
global wind system, 17
gold, placer, 282
Gomphidae, 237, 248, 249
Gomphus, 249
Gonidae angulata, 250
Goniobasis Virginica, 237
Goose Creek, North Dakota, 85, 87
grab samples, *213*, 222
grains, 69
Grand Canyon, 227, 282
gaging station, 310
Grand Forks, 347
Grande Riviere de la Baleine, 243
grasses, 69
grasslands, 71
gravel-bed river, 336
gravity, 64
Great Basin, Alberta, 5, *24*, *92*
Great Bear Lake, 41, 159, *167*, 241
Great Lakes, *41*, *116*, 134, 135, *159*, *170*, 224, 243, 245, *274*, 281, *299*, *303*
drainage basin, 170, 172, 265, 300
drainage system, 159
evaporation, *171*
precipitation, *170*
water level, *172*
wetlands, 172
Great Lakes–St. Lawrence River system, composition data, *224*, 228
Great Plains, *31*, 44, *133*, 263, 282
northern, *39*
Great Plains province, *314*
Great Salt Lake, Utah, 92, *116*, 160, *182*
lake level, *182*
Great Slave Lake, 41, 104, 159, *167*, 241
Green River, 228, 306, *339*
gage, 340
sediment discharge, 339
sediment transport, 339
Green River basin, 339
greenhouse effect, *45*, *132*
Greenland, 19, 140, *148*
Greenland ice cap, 195
Greenland Ice Sheet, 131, 132, *148*
Greer Spring, Missouri, 91
Grimsvotn, Iceland, 156
Gros Ventre River, Wyoming, 105
ground water, *65*, *67*, *90*, *122*, *160*, *162*, 165, *169*, *175*, *177*, 180, 183, *200*, 240, 317, 322
discharge, *66*, *122*, *160*, *165*, 167, *177*, 224
flow systems, *163*, *177*
inflow, 86, *165*, 170, 181, 183, 201
outflow, *165*, 170
recharge, *65*, *123*, 160, *177*
seepage, *125*, 160, 162
storage, 67
groundthaw, 161
Guadalupe River, Texas, 235, *237*, 238
Gulf airstreams, 43
Gulf and Atlantic Coastal Plains region, I58
Gulf Coast, 239
Gulf Coast Prairie region, *75*
Gulf of Alaska, 24, 132, 276

Gulf of Mexico, *43*
  coast, 58, 70, 116
  drainage basin, *243*
  northern, 27
  northwestern, *24*
  western, 27
gullying, 274, 311
Gyrinidae, 250
Gyrinus sp., 236

habitats
  invertebrates, 234
  runing-water, *240*
Hagenius, 249
Haliplidae, 250
Halls, Tennessee, 349
Happy Isles Bridge, 329
Harney Lake, Oregon, 116
Harp Lake, Ontario, 169
Harpactcoida, 241
Harrison Lake, British Columbia, *174*
Hawthorne Formation, 183
Hay River, Northwest Territories, 104, 243
Hayes River, 243, *288*
Hayes-Echimamish route, 289
Hazard Lakes, Yukon Territory, 106, 156
headwaters, 124, 127, 148, 180, 234, *235*, 304
heat
  energy, 141
  fluxes, 142
  latent, 27, 144
  net, 12
  sensible, 11
  storage, 171
  transfer, 141
  transport, meridional, 23
heating, flood, 156
heavy metals, 255
Hebgen Lake, Montana, 160
Helley-Smith sampler, 260
Helton Branch, Kentucky, 266
Hemiptera, 241, 249
Hesperoconopa, 250
Heteroceridae, 241
Heteroptera, 249
Hexagernia, 237, 238
hillslopes, 269, 282, 311, *313*, 342
Hiodon alosoides, 246
histosols, *69*
hoar, depth, 134
Hochelaga archipelago, 300
holes, kettle, 160, 174
Hoover Dam, 227, 267, *331*, 336
Horse Lake, British Columbia, *174*
Hubbard Glacier, 146
Hudson Bay, 25, *41*, 56, *169*, *243*, 245, 250
  drainage basin, *243*, 245, 251
Hudson Bay Lowlands, *167*, 243
Hudson Hope, Alberta, 333
Hudson Strait, *41*, 148
humates, 237
Humboldt-Carson sink, Nevada, 116
humic substances, aquatic, *206*
humidity, air, 70
Hunting Creek, Maryland, 235
Hurricane Agnes, 101, 258
Hurricane Camille, 101
Hurricane Hazel, 101
hurricanes, 101, 258
Hyalella, 251
Hybognathus hankinsoni, 247
Hydrachnellae, 241
Hydraenidae, 250
hydrocarbons, 201
hydroclimatology, *34*
hydroelectric projects, *351*
hydrogeochemistry, *189*, *198*
hydrologic cycle, *4*, 11, *39*, 58, 76, 81, 85, 88, *131*, *191*
hydrologic measurements, *6*
hydrologic systems, *329*
hydrology, *159*
  processes, *162*
  wetlands, *159*
Hydrophilidae, *236*
hydrosphere, 34
Hymenoptera, 241
Hypocrenon, *240*
hyporheos, *240*

ice, *85*, *118*, *131*, *148*, *171*, 192, 195, 235, 242, *244*, *290*, 292, *298*, *302*, *319*
  anchor, 86
  caps, 131, 148, 174
  cores, 148, 195
  cover, 104, 105, 132, 133, 148, *165*, 167, *291*, 302, 319
  dams, *291*
  drift, 303
  flow, 153, 160, 284
  frazil, 291, 303
  glacier, *131*, *144*, 148, 151, 159, 255, 258, 284
  jams, *103*, 290, 303
  lake, 104
  mass, 131, 148
  pans, 291
  plateau, 284
  river, 180
  sheets, 41, 131, *148*, 195, *272*, 299
  shelves, 148
  shore, 303
  slush, 117, *219*
  streams, 148
  subpolar, 146
  surface, 86
  wastage, 131
icebergs, 153
  calving, 152, *153*
icebreakers, 303
icefields, 56, *145*, *148*, 195
Iceland, 156
icing, 169
Ictaluridae, 245
Icy Bay, 145
Idaho, 71, 86, 87, 91, 94, 102, 104, *106*, 115, 148, 222, 256, *317*
igneous rocks, 56, 243, 298
Ile d'Orleans, 301
Iles de Contrecoeur, 300
Iles de Vercheras, 300
Illinois, 122, 123
Illinois River, 226
Ilyodrilus mastix, 250
Imperial Dam 228
Indiana, 226
Indonesia, 201
industrialization, 201, 223
infiltration, 4, 55, 58, 60, *63*, 69, 77, 89, 92, 100, *101*, 163, 345
  defined, 63
  rate, 64, 70, 90
inflow
  ground-water, 86, *165*, 170, 181, 183, 201
  stream, 169, 177, 182
  surface-water, 183
Innuitian region, *56*
insects, *248*
  See also specific insects
interbasin transfers, 329, *351*
Interior Highlands region, *58*
Interior Montane Forest, *17*
Interior Plains region, *58*, 93, 94
Interior Plateau, *282*
Intermontane Plateaus region, *58*
interstices, 60
invertebrates, 233, 239, I248
  habitat 234
ions, *203*, *207*, 228
  exchange, 203
Iowa, 176, 226
Iowa River, Iowa, 226
irradiance, solar, 40
irrigation, 69
Isle Verte, 298
Isoperla longiseta, 249

Jakobsharn Ice Stream, 148
James Bay, 25, 243, 274
James River, 4
jams, ice, *104*
Jensen gage, *339*
jet stream, 38
  mean, 20
John Martin Dam, 109
Johnstown flood, Pennsylvania, 106
jökulhlaup, 105, 156, 160
Juneau, Alaska, 141
Juneau ice field, 106
Juniata River, 87

Kakiska Lake, *167*
Kamchatka anomaly band, 21
Kanab Creek, 309
Kansas, 84
Kansas River, 238
kaolinite, 196
karst, 91, 102, 183
kastanozems, *69*
Kazan River, 243
Keenleyside reservoir, 333
Keewatin, Northwest Territories, 165, 166
Kelly Barnes Dam, Georgia, *106*
Kelvin, Arizona, 333
Kemano hydroelectric project, 352
Kemano River, 352
Kenai Mountains, 145
Kenai Peninsula, Alaska, 88
Kentucky, *266*

Kermit, West Virginia, 347
gage, 347
Kettle Rapids, 287, 289, *290*
Kettle Rapids cofferdams, *290*
Kettle Rapids dam, 291
Kidder County, North Dakota, 177
Kigluaik Mountains, 148
Kiowa Creek, Colorado, 102
Klamath River, California, 216
Knik Glacier, 105, 156
Knik River valley, 105
Knob Lake basin, 167
Kootenay River, 243, 333
Kugmallit Bay, *292*
Kuparuk River, Alaska, 86, 88
Kuskokwim River, 262

La Grande project, 351
La Grande Riviere, 243
laboratory methodology, *212*
Labrador, 148, 223, 243, 245, 289, 353
Labrador Coast, 69
Labrador Sea, 243
Lac Saint-François, 300
Lac Saint-Louis, 300
Lac Saint-Pierre, *300*
Lachine rapids, 300
lagoons, *160*
Lake Agassiz, Manitoba, 115, 287, 346
Lake Agassiz plain, glacial, 174
Lake Alsek, neoglacial, 115
Lake Athabasca, 241, 333
Lake Berg, *103*
Lake Bonneville, Utah, 115, *182*
Lake Chicot, Arkansas, 160, 180
Lake Diefenbaker, 267
Lake Donjek, Yukon Territory, 106
Lake Elsinor, California, *116*
flood, 116
Lake Erie, 116, *171*, 226
sediment, 209
Lake George, Alaska, *105*, 156
Lake Hefner study, 15
Lake Hind, Manitoba, 115
Lake Huron, *170*, 226, 329
Lake Kootenay, British Columbia, 160
Lake Manitoba, 41
Lake Mead, 227, 267, 307, *331*
Lake Michigan, *171*, 226
Lake Missoula, glacial, 115
flood, *115*
Lake of the Woods, 243
Lake Okanagan, British Columbia, 160
Lake Ontario, *170*, 224, *226*
Lake Pepin, Minnesota, 160
Lake Pepin, Wisconsin, 160
Lake Pontchartrain, 110
Lake Powell, 267, *307*, 331
Lake Regina, Saskatchewan, 115
Lake Sallie, Minnesota, 175
Lake Shuswap, British Columbia, 160
Lake Souris, North Dakota, 115
Lake Superior, 116, *170*
Lake Tahoe, California, 116, 160
Lake Warren, New York, *105*
Lake Winnipeg, Manitoba, 159, 243, 287, 289
Lake Winnipeg, Churchill, and Nelson project, 351
lakes, 4, *41*, 58, *105*, *115*, *159*, *163*, *189*
artificial, 213
basins, geologic settings, *159*
chemical composition, 201, *224*
chemical reactions, 202
chemistry, *183*, 199
cirque, 159
clay, 175
closed-basin, 181, 199, 213
composition trends, *224*
covered drainage basins, 169
deep, 299
desert, *180*, *183*
effects, 170
eutrophic, 226
evaporation, *13*, *162*
fiord, 160
flooding, *116*
glacial, 298
glacial terrain, *174*
glacier-dammed, 103, 105, 106, 115, 160
hydrogeochemistry, *189*, *198*
hydrology, *159*
inflow, *165*
kettle hole, 160
landslide-dammed, 160
natural organic matter, *206*
oligotrophic, 200
outflow, *165*, 169
outlet streams, 299
oxbow, 160, 180
paternoster, 159
plains, *177*
pollutants, *209*
prairie, 162, 175, I176, 185
riverine, 162, *178*
saline, 178, 200, 213
sediments, 203, *207*
seepage, *178*, 181
sinkhole, *183*
solution, 160
storage, 162, *167*, 292
subarctic, *167*
terminal, 5
thermokarst, 161
water balance, *162*, *165*, *169*, 170, 172, 174, *175*, *177*, 178, 181, 183
water quality, 201, *205*, *212*, *221*, *224*
See also specific lakes
Lamoille Creek, Nevada, 87
Lampsilis ochracea, 251
land drainage, *345*
artificial, 345
land-use regions, *75*
landscape, 286, *313*
landslides, 103, *105*, *160*
Langbein-Schumm rule, *261*
Lanthus, 249
larvae, insect, 236, 240
Laurentian drainage system 299
Laurentide Ice Sheet, 148, 244, 249, 295
Laurentides, Canadian Shield, 299
lavas, 282
lead, 193, 209
toxicity, 207
Lees Ferry, Arizona, 140, *304*, 307
channel changes, *307*
gaging station, 309
Lepidostoma, 249
Lepisosteidae, 247
levees, *110*, *322*
Levis, Quebec, 226
Liard River, 241
Libby Dam, 333
Libellulidae, 249
lichens, 166
Limestone Rapids, 291
limestone, 58, *90*, 183, 221, 264
aquifers, 123
spring, 91
Limnaea, 237
Limnephilidae, 249
Limnodrilus hoffmeisteri, 250
Limnohalacarinae, 241
Limonia, 250
Limoniinae, 250
lithosols, *69*
Little Beaver Kill, New York, 89
Little Colorado River, 307, 309
Little Ice Age, *45*, 48
loess, 58, 66
Logan fault, 298
logging, clearcut, 106, 112
Long Spruce Rapids, 287
Lost River, West Virginia, 91
Louisiana, 201, *225*
Lowell Glacier, *115*
Lumbriculidae, 250
Lumbrilicus variegatus, 250
Lusk, Wyoming, 316
luvisols, *69*

MacKenzie Bay, 241
MacKenzie Delta, 281, *292*
channels, *292*, *294*
MacKenzie Mountains, 148
MacKenzie River, Northwest Territories, 241, 244, *292*
annual discharge, *292*
sediment load, *292*
Macromia, 249
Macromiidae, 249
Mad River, Ohio, 90
Magdalena, 233
magnesium, 185, 189
Malheur Lake, Oregon, 116
Malheur National Wildlife Refuge, 116
Mancos Formation, 306
mangrove forests, 239
Manitoba, 92, 93, 110, 115, 130, 159, 167, 226, 243, 245, 258, 281, *287*, 289, 346
Marble Canyon, 304
marcasite, 226
Margaretifera margaretifera, 251
Maritime Provinces, 176, 245
maritime region, 243
Marjorie Lake, *165*

marshes, 163
  coastal, 167
  grass, *239*
Maryland, 235, *237*, 265, *266*
Maryland Piedmont, 344
Masseau, New York, 276
Matamek River, Quebec, 240
material flux, *220*
Mattole River, California, 200
mayflies, 238, *242*
McClintock Channel, 241
meadows, alpine, 71
Megaloptera, 236, 237, 248, 249
Melosira, 236
melt, 88, 141, 115, 298
  glacial, 88, *132*, 172, 174, 243
  rate, *141*
  spring, 69, 104, 167
  See also snowmelt
melting, 87, *101*, 160, 161, 165, *272*
meltwater, *149*, 175, 195, 235
  glacial, 263
  runoff, 151
Melville Island, 148
Memphis reach, *326*
Merced River, California, 329
mercury, *209*
Merritt Reservoir, 316, 317
meters
  current, 117
  neutron, *61*
  soil-water, *61*
methane, 48
Mexico, 85, 90, 101, 121, 146, 183, 201, 247
  eastern, 27
mélange terrain, *312*
Mica reservoir, 333
Michigan, 116, *125*
microflora, 234
mid-Atlantic coast, 25
Middle Channel, *292*
midges, 242
Midwest, 44
mining, *266*
  gravel, 324, *350*
  strip, 266
  surface, *347*
Minnesota, 160, 175, *176*, 185, 226, 346
Minnesota River, Minnesota, 226
Mirage Flats Irrigation Project, 317
Mirror Lake, New Hampshire, *170*
Mirror Lake, Wisconsin, *175*
Missi Control Dam, 353
Mississippi, *82*, 265
Mississippi basin, 42
Mississippi drainage, 263
Mississippi–Missouri River system, 245
Mississippi River, *180*, 201, *225*, 251, *256*, 257, *267*, *269*, 349
  bank width, 325
  flood plain, 180
  length variation, *326*
  lower 109, 228, *282*, *321*
  meander belt, 321, 325
Mississippi River basin, upper, 265
Mississippi River Commission, 322
Mississippi River delta, *274*
Mississippi River Valley, 7, 43, 160, 272
  central, 35
  lower, 321, *322*
Missouri, 91, 102, 256
Missouri Basin, 129
  upper, 267
Missouri Coteau, 177
Missouri River, 223, 227, 238, 243, 251, 256, *267*, 269, 349
mites, oribatid 241
models
  general circulation, I48
  hydrologic, *51*
  Soil-Plant-Air-Water (SPAW), 63
  water quality, 202
Moenkopi Formation, 306
Moira River, Ontario, 104
Moisie River, 243
moisture, *11*
  soil, 13, *15*, 42, 126, 133, 141, 198
  storage, 167
mollisols, 59
Mollusca, *250*
molluscs, 239
monitoring
  data retrieval, *215*
  data storage, *215*
  evaluation parameters, *211*
  frequency sampling, *211*
  programs, *209*, *210*, *221*
  sampling stations, *211*
  water quality, *209*, *221*
Monroe, Michigan, 116
Montana, 25, *106*, 148, 160, 223, 226
Montreal, 300, 303
Moose River, Ontario, 87, 243
moraines, end, *176*
Morganza Control Structure, 109
Morrison Formation, 306
Moses Saunders dam, 300
mosses, 166, 167, 233
Mougeotia, 237
Mount Baker, 106
Mount Hood, 106
Mount Lassen, California, 102
Mount McKinley, 145
Mount Palmer, Alaska, 105
Mount Rainer, Washington, 105, 156
Mount Robson, 282
Mount St. Helens, Washington, *105*, 156, 201
Mount Tambora, Indonesia, 201
mountains
  non-Arctic, *172*
  volcanic, 106
  See also specific mountains
Moxostoma, 245
mudflows, *111*, 156, 160, 286
muds, 276
  surface, 239
mudstone, 306
Muskeg River, British Columbia, 174
Muskegs, 159, 163, 243
mussels, 239
Myoxocephalus quadricornis, 245

Naididae, 250
Nais spp., 250
Nanaimo River, British Columbia, 84
NAO. *See* North Atlantic Oscillation
Napoiak Channel, 294
NASQAN. *See* National Stream Quality Accounting Network
Nass River, 243
Natchez gage, 322
National Stream Quality Accounting Network (NASQAN), *210*
Nebraska, 89, 101, 107, 123, *178*, *314*
Nebraska sandhills, 160
Neches River, 4
nekton, 235, 239
Nelson basin, 288
Nelson estuary, 288, 289
Nelson River, Manitoba, 92, 243, 281, *287*
Nelson River basin, Manitoba, 130, *287*
Nelson River valley, 287
Nemoura arctica, 249
Nenana River, Alaska, 258
Neoephemera, 248
Nevada, 87, 92, 108, 111, 116, 148, 160
New Brunswick, 101, 246
New Hampshire, *170*, 175
New Jersey, 128, 237
New Madrid earthquake, 322, *325*
New Mexico, 74
New Orleans, *225*
New Orleans reach, 326
New York, 89, *105*, 108, *128*, 170
New York City reservoir system, 128
Newfoundland, 101, 135, 176, 243, 244
Niobrara drainage basin, 317
Niobrara River, 281, 282, *314*
Nisqually Glacier, 105
Nisqually River, Washington, 156
Nitocris carinatus, 237
nitrate, 189, 216, 223, 233
nitrogen, 185, 189, *226*, 239
Nocomis spp., 245
nodal tide, lunar, *33*
non-stationarity, 257
Norden, Nebraska, 314
  gage, *317*, 318
North American Cordillera, 144
North Anna River, Virginia, 124
North Atlantic Oscillation (NAO), *19*
North Carolina, 258
North Cascade Range, 146, 174
North Dakota, 85, 87, 92, 115, 176, *177*, 178, *185*, 346, 347
North Pacific (NP) pattern, *22*
North Pacific high-pressure system, 144
North Pacific Ocean, 36, 144
North Platte River, 338
North Saskatchewan River, Canadian prairies, 87, 94, 243

Northwest Territories, *24*, 104, 159, 165, 166, 241, 243, 244, *292*
Nostocaceae, 233
Notropis spp., 245
Nova Scotia, 101, 243, 245
NP. *See* North Pacific pattern
nunataks, 249
nutrients, 234, 235

Obion River, 349
oceans, flooding, *116*
Octogomphus, 249
Odonata, 236, 241, 249
Ogeechee River, Georgia, 263
Ohio River, 226, 251, 267
Ohio River basin, 265
Okefenokee Swamp, 123
Old River, 109
Oligochaeta, 236, 238, 243, *250*
Olympic Mountains, 146
Oncorhynchus spp., 244, 282
Ontario, 41, 87, 90, 104, 116, *169*, 172, 240, 243, 245, 250, 251, 329
  northwestern, 167
  southern, 101, 104, 176, 245
Ophiogomphus, 249
Optioservus, 250
Orconectes spp., 251
Oregon, 71, 101, 116, 123, 133, 148, 160, 201, 222
  coast, 24
organic matter, 206
organics, suspended, 205
Orimarga, 250
Orinoco, 223
Orthocladiinae, 241, 242, 250
Ostracoda, 241
Ottawa River, Ohio, 234, 243
Ottawa River, Quebec, 226, 300
Outardes River, 243
outflow
  current 355
  ground-water, *165*, 170
  lake, 169
  natural, 355
  peak, *165*, 167
  stream, 177
oxbows, 103, 238
oxygen, 189, 196, *203*, 228, 233
oysters, 239
Ozarks, 102

Pacifasticus, 251
Pacific airstreams, 44
Pacific coast, 69, 144, 239
Pacific Mountain system, *144*
Pacific Northwest, *35*, 160
Pacific Ocean drainage basin *243*
Pacific track, 27
Pacific Transition (PT) pattern, *22*
Pacific/North America, (PNA) pattern *31*, *36*
paleofloods, *114*
palsas, 288
palynomorphs, 282
Panama, 84
Parapsyche, 249
Parastenocaris starretti, 241
Paria River, *306*, 307, *309*
Paria River basin, *306*, 309
Paria River channel, 309
Pascagoula River, Mississippi, *82*
Pathfinder Reservoir, 338
patterns
  circulation, *17*, *34*
  drought, *33*
  stream, 102, 234
  streamflow variability, *35*
Peace-Athabasca Delta, 335
Peace River, Alberta, 104, 241, 284, *333*, 336
peat, *160*, 162, *167*, 176, 288
Pecos River, Texas, 99, 113, *115*, 194, 222, 223, 226
Pediciinae, 250
pedon, *59*
Peel River, 241
  drainage, 292
peneplain, 56
Pennsylvania, 106, 108, 236, *237*
Pentagenia sp., 248
Perca flavescens, 245
Perch Lake, Ontario, *169*
Percidae, *245*
Percina copelandi, 245
percolation, *65*, 144, 149
periodicities
  drought, 33
  rainfall, 32
  streamflow, 32
permafrost, *3*, 69, 70, *87*, *161*, 163, 165, 167, 194, 201, 291
  continuous, *161*, *169*
  discontinuous, *161*, *166*, 292
pesticides, *209*, 255
petroleum, 292
Peyto Creek, 88
Peyto Glacier, 88
phaeozems, *69*
phosphorus, 185, 189, *216*, 239
photosynthesis, 198
phreatophytes, *71*, 125, *348*
Physa, 237
  columbiana, 250
physiochemical systems, *205*
physiographic regions, *56*
physiography, *56*
phytoplankton, 239
Pickerel Lake, Wisconsin, 175
Piedmont Province, Virginia, 77
Piedmont region, 257, 270
Pierre Shale, 317
pillows, snow, 140
Pimephales notatus, 245
Pine Barrens, New Jersey, 237
Pine River tributary, 336
Pinus banksiana, 288
Pinyon-Juniper Woodland region, *74*
Pisidium, 250
Planaria, 236
plankton, 239
plants, vascular, 166, *233*, 234
plateaus, intermountain, 174
Platte River, *338*
  channel, 228]
Platte River Valley, 180
Platygobio gracilis, 245, 246
playas, 5, 92
Plecoptera, 234, 241, *249*
PNA. *See* Pacific/North American pattern
podzols, *69*
Point Barrow, Alaska, 249
Point Separation, 292
Pointe des Monts, 301
poljes, 160
pollutants
  organic, 203, *209*
  soluble, 134
polychaetes, 239
Polysiphonia, 239
Populus tremuloides, 288
Porcupine River, Yukon Teritory, 86, 243
pores, *60*, 64
Port Nelson, 289
Potamanthus, 248
Potamon, *240*
potassium, 189
potholes, prairie, 103, 107
Potomac River, Maryland, *237*, 265, 266
Powder River basin, Wyoming, 264
power, hydroelectric, 331
prairie provinces, 160
Precambrain Shield, 287
precipitation, 4, 5, 22, *31*, 55, 63, 79, *84*, 93, *100*, 126, 135, 144, *148*, 151, *162*, 169, *170*, 175, 177, 183, 197, 235, 243, *261*, 263, 311, 316, 317, 338
  annual, 84, *221*
  data analysis, 47
  infiltration, 15
  mean annual, 41, 84, 261
  U.S. annual average, 47
  variability, *31*, 33, 82
  winter, 144
pressure
  atmospheric, *18*
  ridge, *104*
Price River, 306
Pristina idrensis, 250
Procambarus, 251
Prosimulium spp., 242, 250
Prosopium williamsoni, 245
proto-Frazer River, 284
protozoans, 236
Psephenidae, 248, 250
Psychodidae, 241
Psychoglypha, 249
PT. *See* Pacific Transition pattern
Ptilostomis ocellifera, 249
Ptychocheilus oregonenss, 245
Puget Sound, 239, 274
Pyramid Lake, Nevada, 92, 160
pyrite, 226

quality assurance processes, *214*
quartz, 196
Quebec, 41, 167, 226, 240, 246, 300
  eastern, 135
  northern, 69
  southern, 38, 101
Queen Elizabeth Islands, 148

Quesnel, 286
Quesnel River, 282

radiation, 11, 86, 133, 151
  balance, 11
  longwave, 141
  mean annual net, 40
  net, 12, 13, 14, *39*, *141*
  solar, 18, *39*, 70, 127, 131, 133, 141, 149, 165, 174
  terrestrial, *39*
radionuclides, 255
rain, 193
  acid, 204
  composition, *193*
  forests, 71
  winter, 257
rainfall, 71, 100, 111, 115, 116, 122, 123, 126, 135, 160, 162, 165, 170, 172, *176*, *193*, 261, 282, 311, 313, 314
  periodicities, 32
rainstorms, 126, 167, 169, *258*, *284*
rangeland, *76*
Rat River, 353
ravel, dry, 314
Rawson Lake, Ontario, *169*
recession, base-flow, 89
recharge, ground-water, *65*, *123*, 160, *177*
reclamation, wetland, 108
Red Bluff, Texas, 99, 113, *115*, 194, 222, 223, 226
Red Deer River, Alberta, 94
Red Deer River basin, 264
Red Lake peatland, Minnesota, 176
Red River, 243, 250, 287, *346*
Red River basin, Manitoba, 110, 226
Red River confluence, 324
Red River Floodway, *110*
Red Rock Pass, Utah, *115*
redox processes, 196, *203*
Redwood Creek, California, *272*, 312
Redwood Creek basin, *272*
Redwood Forest region, *71*
Reelfoot Lake, Tennessee, 160
regosols, *69*
remote sensing, *140*
reservoirs, 124, *227*, 262, *299*, *307*, 324, *331*, *333*, *338*, *340*
  artificial, 269
  capacity, United States, *342*
  carryover storage, 129
  construction, *331*
  hydrogeochemistry, *198*
  sediment loads, *266*
  storage, 227, *307*, *331*, *333*, *338*, *340*
Resolute, Cornwallis Island, 165
Rhafionidae, 250
Rhine River, 223
Rhynchelmis elrodi, 250
Richardson Mountains, 252
Richardsonius balteatus, 245
Richelieu rapids, 300
Richelieu River, 300
riffles, *235*
rille, 169
Rio Cocle del Norte, Panama, 84
Rio Conches basin, 247
Rio Grande, *227*, 257, 269
Rio Grande basin, 39
Rio Santiago, Mexico, 85
rithron, 240
River Bottom Forest region, *74*
rivers, 1, 178, *189*, 212, 216, *220*, *240*, 241, *255*
  alluvial, 281, 282
  alterations, 349
  aquatic communities in, *233*
  beds, 237, 238, *325*, 349, 352
  Canada, *233*, *240*
  channels, 1, 216, 233, *234*, 255, 271, 281, 287, 290, 291, 307, *316*
  chemical composition, 201, *224*
  chemical reactions, *202*
  coastal zones, *274*, 276
  coastal-plain, *238*
  composition trends, *224*
  discharge, 200
  flooding, 336
  flow, *200*
  food web, 233, *234*
  geochemical character, *221*
  gravel-bed, 336
  hydrogeochemistry, *189*
  ice, 180
  mainstem, *237*
  natural organic matter, *206*
  pollutants, *209*
  sediment, *207*, *274*, 276
  sediment discharge, *274*, *255*, *267*, 276
  sediment movement, *255*, *269*
  sediment storage, *255*, *269*
  United States, *233*
  water quality, 201, *205*, *212*, *221*, *224*
riverscape, *281*
Riviere Aranud, 243
Riviere aux Feuilles, 243
Riviere à la Baleinè, 243
Riviere Conaipiscan, 243
Riviere de Rupert, 243
Riviere des Milles, 300
Riviere des Prairies, 300
Riviere du Peut-Mecatina, 243
Riviere Eastmain, 243
Riviere George, 243
Riviere Harricana, 243
Riviere Natashquan, 243
Riviere Nottaway, 243
Riviere Saint Maurice, 300
rivulets, 169
Rock River, Iowa, 89
rocks
  carbonate, 160, 221
  crystalline, *169*, 170, 263, 281
  dolomitic, 102
  igneous, 56, 243, 298
  impervious, 162
  metamorphic, 287, 298, 311
  permeable, 161, 162
  sedimentary, 56, *58*, 223, 243, 264, 281
  silicic, 91
  slumps, 160
  weathering, *195*, 201, *220*, 263
Rocky Mountain Coniferous Forest region, *71*
Rocky Mountain Trench, 333
Rocky Mountains, 24, 58, 69, 71, 88, 92, 93, 101, 111, *134*, 144, *146*, 148, 160, 249, *292*, 304, 306, 333, 338
Rogue River, Ontario, 250
Rogue River, Oregon, 201, 222
rotation, of Earth, 18
Roystown Branch, Juanita River, 87
runoff, *1*, *15*, 31, *63*, 77, *85*, 128, 131, 167, *171*, 174, *176*, 216, *310*, 311, 338
  amount, 3
  annual, 1, 3, 5, *81*, *82*, 144
  composition, *194*
  forecasting, *140*
  glacier, *148*, *151*
  mean annual, 15, 77, *81*, 92, 307
  meltwater, 151
  nonglacier, 151
  rate, 79
  ratios, 13
  regimes, *83*
  snowmelt, 69, *87*, 93, 94, 100, *101*, 116, 126, *133*, *140*, 160, 162, 165, 169, 172, *174*, 176, 194, 284
  spring, 180, 284
  summer, 126
  surface, 64, 65, 88
  variability, 3, *81*, 82

Sacramento River, California, 109, 129, 266, 271
Sacramento River Delta, 129
Sacramento River valley, California, 109, 266, *270*
Safford Valley, Arizona, 103
Saguenay fjord, 298
Saguenay River, 243
St. Clair Lake, 116
St. Clair River, 172
St. Elias Mountains, 115, 145
St. Francis Dam, California, *106*
St. Francisville, Louisiana, 201, *225*
St. John River, 243
St. Lawrence basin, 246
St. Lawrence estuary, *300*
St. Lawrence River, *92*, *107*, 221, 222, 224, 226, 243, 245, 276, *295*, *300*
  floods, *302*
  flow, *301*
  gulf, *301*
  ice, *302*
  lower, 281
  sedimentation, *301*
St. Lawrence Seaway, 300, *303*
  locks, *304*
St. Lawrence Valley, 298
St. Lawrence–Great Lakes region, 250
St. Louis, Missouri, 256
St. Maurice River, 243
Sainte-Anne channel, 300
Sainte-Anne de-Bellevue, Quebec, 226
Saint-Laurent, *298*

Salamonie River, Illinois, 123
Saldidae, 249
Saline Branch Watershed Study, 209
salinity, *183*, 199, 227, 301
Salix spp., 288
Salmo, 244
  clarki, 244
  gairdneri, 244
  salar, 245, 246, 301
  trutta, 245
Salmon Glacier, 156
Salmon River, Idaho, 86, 148
Salmonidae, *244*
Salt River, Arizona, 115
saltcedar, 309, *348*
Salvelinus, 245
  alpinus, 245, 246
  fontinalis, 245, 247
  malma, 245
  namaycush, 245
Sam Rayburn Reservoir, 228
sampling
  composite, *214*
  depth-integrated, *257*
  point, *257*
San Fernando Valley, 350
San Francisco Bay, 274
San Gabriel Mountains, 350
San Jacinto River, 116
San Joaquin River, 129
San Juan River, 306
San Pedro River, Arizona, 84
San Rafael River, 306
Sand Hills, Nebraska, 89, 123, *178*
sand plains, 62
sandhills, 316, 317
sandstone, 58, 70, 176, 178, 221, 256, 312
Santa Clara River, California, 258
Santa Cruz River, Arizona, 108
Saskatchewan, 92, 115, 167, 243, 245, 353
Saskatchewan River, *93*, 127, 148, 248, 287
Saskatchewan River basin, *93*
Saskatchewan-Nelson basin, 41
  rivers, 353
Saskatoon, 267
Saugeen River, Ontario, 329
Savannah River, 223, 238
Scarabaeidae, 241
Schefferville, Quebec, 167
scour
  glacial, 115, *159*, 169, 350
  holes, 295
sea level, 131, *132*
sedges, 166, 167
sedimentary rocks, 56, *58*, 223, 243, 264, 281
sedimentation
  channel, 347
  lake, 203, 208
  river, *207*
  valley, 255
sediments, 6, *8*, 167, 209, 298, *307*, 336, 339, 344
  alluvial, 58
  bed, 203, 208, *317*
  bed load, *255*
  bottom, 207, 208
  concentrations, *216*, 257
  dams, *266*
  discharge, 8, *255*, 256, *257*, *258*, *267*, *274*, 276, *311*, 313, 317, 339
  estuarine entrapment, *274*
  flow, 349
  fluvial, 9
  loads, 257, *264*, *266*, 304, 307, 310, 349
  movement, *255*
  paraglacial, 286
  routing, *269*, *270*, *273*
  seasonal storage, *269*
  storage, *255*
  streams, 203, *207*, 208
  suspended load, *255*
  suspended, 203, 207, 224, 307
  toxic substances, *208*
  transport, 310, *313*, 337, 339
  yield, *260*, *264*, *265*, 286, *311*, 312, 314, 342, 344
seepage, 185
  ground-water, *125*, 160, 162
  lake, *178*, 181
Selwyn Mountains, 148
Sepsidae, 241
Severn River, 243
Seward-Malaspinia Glacier, 146
Shadow Lake, Wisconsin, *175*
shale, 264, 306, 175, 176
Shallow Bay, *292*
sheet flow, *111*
shelf, continental, 276, 301
Shellmouth Reservoir, 110
shorelines, 159
Short-Grass Prairie region, *75*
shrimp, 239
Sialidae, 249
Sialis velata, 249
Sierra Madre, California, 111
Sierra Nevada, 58, 101, 129, 146, 266, 329
silica, 185, 189, 195, *226*, *227*, 228
silicic rocks, 91
silt-loam soils, 63
siltstone, 176, 264
Silver Springs, Florida, 91, 123
Simuliidae, 241, 235, 248, 250
sinkholes, 58, 90, 102, 160
sinks, 5, 92, *274*, 276
Siqoq, 133
Skeens River, 243
Slave River, 241, 333
slides, streamside, 312
slopes
  basin, 77
  channel, 112
  discontinuities, *160*
  land, *103*
  windward, 162
sloughs, 163, 238
slumps, 312
Small Lake, *165*
Smoky River, 241
snails, 236
Snake River, Idaho, 102, 104, 222, 256, *317*
Snake River basin, 115
Snake River Plain, 91, 102
snow course, 140
snow cover, 131, *135*, 151, *170*
  runoff log, 143
  seasonal, 3, *133*
snow fences, 134
snow pillows, 140
snow, 43, *85*, 105, 122, *131*, 162, 194, 243, 284
  albedo, *149*
  composition, *193*
  crystals, 133
  density, 133
  elevation, *149*
  facies, *148*
  remote sensing, *140*
  types, *133*
  volume forecasts, *140*
  water flow, *143*
snowbanks, 166
snowfall, 115, 162, *169*, 170, 175
  average annial, 135
  summer, 144
snowfields, 3, 87, 88, 127
snowmelt, 69, *87*, 93, 94, 100, *101*, 126, *133*, *140*, 151, 160, 162, 165, 169, 172, *174*, 176, 194, 242, 257, 284, 333, 338, 347
  process, *141*
  runoff, 69, *87*, 93, 94, 100, *101*, 116, 126, *133*, *140*, 160, 162, 165, 169, 172, *174*, 176, 194, 284
  spring, 180, 306
snowpack, 69, 87, 101, 126, *133*
  measurement, *140*
sodium, 189
soil, 6, *58*, *198*, 235, 238, 312, 348
soil creep, 314
soil-plant-air system, *59*
Soil-Plant-Air-Water (SPAW) model, 63
soils
  classification, *59*
  clay, 62
  defined, *60*
  erosion, 255, 270
  formation, *59*
  gleyic, 70
  horizons, 59, 60
  mantle, 55
  moisture, 13, *15*, 42, 126, 133, 141, 198
  organic, 243
  piedmont, 263
  pores, *6*, 64
  profiles, 59, 67
  regions, *67*
  silt-loam, 63
  surface, 55
  texture, *61*
  thin, 243
  water content, *59*, *60*
  water-holding capacities, 62, 67
  wetness, 64
  See also specific soils
soil-water budgets, *63*

solar energy, 11
solar radiation, 18, *39*, 70, 127, 131, 133, 141, 149, 165, 174
solid-phase components analysis, *207*
solutes, 194, 196, 200, *205*, *207*, 221, 222
  flux, *220*
Sorel Islands, 300
South Branch Raritan River, New Jersey, 128
South Carolina, 216, 241
South Dakota, 176, 267
South Fork Little Conemaugh River, 106
South Platte River, Colorado, 92, 223, 227, 338
South Platte River basin, Colorado, 102
South Saskatchewan River, 243, 245, 267
Southeast Asia, 47
Southeastern Broadleaf Hardwood Forest region, *74*
Southeastern Mixed Forest region, *75*
Southern Indian Lake, 291, *353*
Southhampton Island, 249
Southwestern Desert Shrubs region, *75*
Soviet Union, 47
Spartina, 239
SPAW. *See* Soil-Plant-Air-Water model
Specaria fraseri, 250
species interaction, *234*
Speed River, Ontario, 240
Spencer Dam, 316, *317*
Sphaeridae, 236
Sphaerium nitidum, 250
Sphaeroceridae, 241
Spit Lake, 287
sponges, 237, 239
spring runoff, 180, 284
spring snowmelt, 180
springs, *90*, 123, 178, 240
stage-discharge relation, *117*
Stagnicola actica, 250
Staphylinidae, 241
stationarity, 257
Steele Glacier, Yukon Territory, 156
Stenelmis, 250
Stigeoclonium lubricum, 238
Stikine River, 243
Stizostedion spp., 245
stomatal resistance, 13
Stomatochlora, 249
stoneflies, 235, 238, *242*
storage
  aquifer, 89
  bank, 180
  capacity, 227
  carryover, 129
  change, 354
  channel, *107*
  drainage basin, 167
  flood-control, 344
  ground-water, 354
  heat, 171
  lake, *167*, 292
  lake water, 162
  moisture, 167
  natural basin, *103*
  off-stream, 107
  overbank, 345
  reservoir, 124, 227, 262, *299*, *307*, 324, *331*, *333*, *338*, *340*
  seasonal sediment, *269*
  sediment, *269*, *273*
  surface, *92*
  urban, *344*
  water, 351
storms, 162
  climatology of, *22*
Straits of Georgia, 243
stratigraphy, subsurface, *160*
stream channels, 4, 58, 66, 86, 274, 271, 311, *349*
streambanks, 167, 169
streamflow, 1, 4, *35*, *55*, *59*, 65, 67, *70*, 79, *81*, *84*, *126*, *168*, 174, *181*, *216*, 306, 317, *329*
  natural, *329*
  periodicities, 32
  physiographic impact, *56*
  spatial variability, *81*
  summer, 88
  temporal variability, *81*
  variability, *35*, *81*, *93*, *329*
  watershed topography, 55
  See also flows
streams, 125, 126, *162*, 166, 174, 198, *220*, 240, 345
  alluvial, 205
  annual low flows, 125
  annual mean flows, 5
  aquatic communities, *235*
  Arctic, 222
  Canada, *222*
  channels, 4, 58, 66, 86, 271, 274, 311, *349*
  composition, 201
  discharge, 5, 9, 97, *216*, 243
  ephemeral, 194
  food web, 233, *234*
  glacier-fed, 127, 255
  headwater, *235*, *237*
  high flow stages, 200
  ice, 148
  inflow, 169, 177, 182
  lowland, 242
  mainstem, *237*
  mountain, *242*
  outflow, 177
  patterns, *102*, 234
  perennial, *200*
  permanent, 241
  sections, *235*
  sediments, 203
  stage, 87
  temporary, 241
  water quality, 201
Stutsman County, North Dakota, 177
Stygobromus, 251
Stylodrilus heringianus, 250
Subalpine Forest, *71*
subarctic lakes, *166*
subarctic wetlands, *166*
subsidence, *160*
substrates, 236, 237, 238
Subtropical Zonal (SZ) pattern, *22*
sulfate, 185, 189, 223, *226*, 228
  concentrations, *225*
sulfur, 201, 223
summer runoff, 126
Summit Lake, British Columbia, 105, 156
surface chemical processes, *202*
surface soils, 55
surface storage, *92*
surface water, *1*, 6, *9*, *55*, 67, *89*, *162*, *170*, *189*, *200*, 322, *329*
  generation, 65
  inflow, 183
  quality, *205*
  sample types, *213*
  trace element analysis, *207*
surge, glacier, 152, 153
Susitna River, Alaska, 256
Susitna River basin, Alaska, 152
suspended load, *255*, *256*, *257*
suspended sediment
  discharge, 256, *257*, *258*, *311*, 313
  transport, *258*, 306
Susquehanna River, Pennsylvania, *108*, *237*
swamps, 163, 206, 323
Swampy Cree people, 288
Swartz Creek basin, Michigan, *125*
Swift Creek, 103
Swift Dam, Montana, *106*
Syrphidae, 241
SZ. *See* Subtropical Zonal pattern

taliks, 292
Tall-Grass Prairie region, *75*
talus, 309
Tanana River, Alaska, 256
Tanypodinae, 241, 250
Tanytarsini, 250
Tar River, North Carolina, 258
Tar River drainage, 103
tectonism, *160*, 264
teleconnections, *34*
Temescal Wash, 116
temperatures, *85*, 177
  air, 140, *141*
  annual, 46
  global, *45*, *46*
  gradients, meridional, 23
  mean annual, 4
  polar, 23
  surface-air, 46
  tropical, 23
  variability, 46, *48*
Tennessee, 160, *349*
Tennessee River, 247, 335
terrain
  glacial, 159, *174*, *176*, 178
  mélange, *312*
Teton Dam, Idaho, *106*
Texas, *82*, 99, 113, *115*, 194, 222, 223, 226, 235, *237*, 238, 274
thalweg, *318*, 321
Thames River, Ontario, 104
thaw, 285

The Dalles, Oregon, 101
The Pas, Manitoba, *93*
Thelon River, 243
thermal regime, 148
Thompson River, 243
Thousand Islands, 300
Thousand Springs, Idaho, 102
thunderstorm systems, 27
Thymallinae, 245
Thymallus arcticus, 245, 246
tides, 239
  lunar nodal, *33*
till, glacial, 66
timber harvesting, *264*
time scales, *270, 271, 272*
Tipulidae, 241, 250
TNH. *See* Tropical/Northern Hemisphere pattern
Tombstone, Arizona, 77
Tongue River, Montana, 223, 226
topography, *76, 92, 123, 159*, 282
Tortopus primus, 248
Toutle River basin, Wahsington, 105
Toutle River channel, Washington, 105
toxic substances, *208*
  copper, 207
  lead, 207
  zinc, 207
trace element analysis, surface water, *207*
tracks, extratropical cyclone, *24*
transpiration, 4, 13, 14, 43, 70, 126, 151, 162, 176, 180
  vegetative, *70*
transport
  rates, 221
  suspended-sediment, *258*, 306
traps, bed load, 260
tree rings, 115
treeline, 166, 242
tributaries, 102, 103
Trichoptera, 241, 249
trimlines, 291
Trinity River, California, 256
Tropical/Northern Hemisphere (TNH) pattern, *22*
troposphere, 23
Truelove Lowland, Devon Island, 166
tsunamis, 117
Tubifex spp., 250
Tubificidae, 250
Tucson basin, Arizona, 102
Tug Fork, *347*
Tug Fork basin, 347
Tujunga Wash, *350*
Tuktoyatuk Peninsula, 161
Tulare Lake, California, 116
Tulsequah Glacier, 106, 156
Tulsequah Lake, British Columbia, *106*, 156
Tundra and Alpine Meadow region, *74*
Tundra and Boreal Forest region, *74*
tundra, 69, 70, *74*, 206, 242, 248
  vegetation, 163
turbidity, 237, 302
Tusket River system, Nova Scotia, 245
Tyrrell Sea, 167
Ungava Bay, 243
Unionidae, 234, 238
United States
  eastern, *24*, 133, *175*
  north-central, 163
  northeastern, *38, 127, 134*
  northern, 121
  northwestern, 74, 101
  southeastern, 268
  southern, 70
  southwestern, 70
  western, *35*
  See also specific states
uplands, *269*, 274
uplifts, 58, 167, 284, 311, 314
Upper Bradley River, 88
Upper Colorado Basin, 129
Upper Limestone Rapids, 287
Upsik, 133
urban basins, 343
urban channels, *344*
urban drainage, *342*, 344
urban storage, *344*
urbanization, 201, 223, *266*, 329, *342*
Utah, 74, *92*, 111, *115, 116*, 148, 160, *182*, 219, 222, 228

valley sedimentation, 255
Van Duzen basin, 312
Van Duzen River, *312*
Van Duzen station, 313
Vancouver, 282
Vancouver Island, 243
Variegated Glacier, Alaska, *153*
Vaudreuil channel, 300
vegetation, *55, 70, 103*, 175, *198*, 263, *288*, 309, 338, 347, *348*
  flood-plain, 348
  regions, *71*
  riparian, 314
  tundra, 163
Verde River, Arizona, 115
vertisols, *70*
Vicksburg reach, *326*
Victoria Island, 249, 250
Virginia, 77, 124
volcanics, 90, 102, 123
volcanism, 201, 282, 286
volcanoes, 146, 156, 201

W.A.C. Bennett Dam, 335
Wabash River, Indiana, 226
Wakarusa River, Kansas, 84
Walker Lake, Nevada, 92, 116
Walkerton, Ontario, 329
Wallowa Mountains, 148
warming, global, 46, *48*, 132, 284
Wasatch Formation, 306
Wasatch Mountains, 148
wash load, 256
Washington, 105, 115, 133, 135, 144, 146, 156, 201
  coast, 24
Washington, D.C., *266*
Washington-Baltimore area, 266
waste, untreated organic, 201
water, *191*, 194, 350, *353*
  acidic, 143
  alkaline, 243
  analysis methods, *214*
  atmospheric, *162*, 175, 194
  availability, 17
  chemical properties, *191, 224*
  consumption, 329
  discharge, *66, 122, 160*, 167, *177, 224*, 257, 258, 269, *270*, 276, *304, 306*, 307, 310
  diversions, 351
  flows, *143, 240, 255, 267*, 349, 351
  geochemistry, *189, 198*, 228
  ground. *See* ground water
  interbasin transfers, 329
  isotopic composition, *192, 216*
  levels, 165, 170
  loss, 4, 162, 165, 180 *See also* evapotranspiration
  molecules, 40, 64, *192*
  monitoring, *209*. *See also* monitoring
  natural, 195, 196, *203*, 223
  oxygen deposits, 203
  physical properties, *191*
  properties, *191*
  resources, *149*, 355
  samples, *213, 214*
  sanitary analyses, *205*
  solvent power, 196
  stagnant, 169
  storage, 162, *167, 331*
  stream discharge, *216*
  supply, defined, 354
  surface. *See* surface water
  toxic substances, 208
  transfers, *351*
  urbanization, 329
  velocity, *256*
water balance, 11, *162*, 182
  lakes, *162, 165, 169*, 170, 172, 174, *175, 177*, 178, 181, 183
  wetlands, *162*, 163, *175, 177*, 178
water bodies, temporary, *241*
water budget, *354, 355*
water quality, *202, 227*
  chemical parameters, *211*
  data, *204, 215*
  data collection methods, *204*
  evalution, *211*
  field methodology, *212*
  human activities, *201, 223*
  interpreting, *215*
  laboratory methodology, *212*
  lakes, 201
  modeling, 202
  monitoring programs, *209, 221*
  physical parameters, *211*
  reporting, *215*
  rivers, *201*
  streams, 201
  trend analysis, *218*
water table, 160, 162, 163, 167, 169, *176*, 177, 183
water vapor, 46, 201
watershed, 56, *65*, 67, *70*, 87, *257*, 262, 265, 282, 342
  drainage patterns, 92
  topography, 55, *92*

waves
  cyclone, *23*
  erosion, 161
  long, 24
  short, 24
  standing, 111
  tidal, 117
weathering, 64
  geochemistry, *220*
  rates, chemical, *197*
  rocks, *195*, 201, *220*, 263
weirs, gated, 109
Welland Canal, 246
West Channel, 294
West Pacific Oscillation (WPO) pattern, *21*
West Virginia, 91, 347
westerlies, 44
Western Coastal Forest region, *71*
Western Cordillera, 44, 93, 172
Western Sagebrush Steppe region, *74*
wet periods, *31*
wetlands, *4*, *41*, *107*, *159*, *163*, 172, 175, *176*, *178*
  Canada, *345*
  chemistry, *183*
  covered drainage basins, *169*
  discharges, *166*, 167, 174
  drainage, 345
  formation, *160*, 162, 167
  hydrology, *159*
  minerotrophic, 163, 176
  ombrotrophic, 163, 176
  prairie, 185
  reclamation, 108
  runoff, *176*
  subsurface stratigraphy, *160*
  United States, *345*
  water balance, 183
Wheeler Peak, 148
White Clay Creek, Pennsylvania, 236
White Mountains, 170, 175
Williams Lakes, Minnesota, 175, 185
Williston Lake, 335
Wilson Creek, Manitoba, 258
Wind River basin, Wyoming, 264
Wind River Mountains, 148
Wind River Range, Wyoming, 256
wind shear, 23
winds, 162
  global, 17
  mid-tropospheric, 24
Winisk River, 243
Winnipeg, 110
Winnipeg River, 287
Wisconsin, 160, *175*, 272
Wood River Mountains, 148
worms
  oligochaete, 242
  tubificid, 234
WPO. *See* West Pacific Oscillation pattern
Wrangell Mountains, 145
Wrightwood, California, 111
Wyoming, 92, 148, 165, 197, 256, 260, 264, 284, 316

xerosols, *69*

Yankton, South Dakota, 267
Yellow Bluff, 349
yermosols, *70*
Ylodes kaszabi, 249
York Factory, *288*
Yosemite, California, 329
Youghiogheny River, *226*
Yuba River, 271
Yucatan Peninsula, 90, 183
Yukon, 244, 250
Yukon River, 206, 243
  drainage basin, 58, 249
Yukon River valley, 249
Yukon Territory, 86, 106, 115, 156, 162, 243
Yuma, Arizona, 227, 267

zinc toxicity, 207
zooplankton, 239
Zygnemia, 237
Zygoptera, 249

Typeset by WESType Publishing Services, Inc., Boulder, Colorado
Printed in U.S.A. by Malloy Lithographing, Inc., Ann Arbor, Michigan